T0205252

Lecture Notes in Computer Science　　12535

More information about this subseries at http://www.springer.com/series/7412

Adrien Bartoli · Andrea Fusiello (Eds.)

Computer Vision – ECCV 2020 Workshops

Glasgow, UK, August 23–28, 2020
Proceedings, Part I

 Springer

Editors
Adrien Bartoli
University of Clermont Auvergne
Clermont Ferrand, France

Andrea Fusiello
Università degli Studi di Udine
Udine, Italy

ISSN 0302-9743   ISSN 1611-3349 (electronic)
Lecture Notes in Computer Science
ISBN 978-3-030-66414-5   ISBN 978-3-030-66415-2 (eBook)
https://doi.org/10.1007/978-3-030-66415-2

LNCS Sublibrary: SL6 – Image Processing, Computer Vision, Pattern Recognition, and Graphics

This Springer imprint is published by the registered company Springer Nature Switzerland AG
The registered company address is: Gewerbestrasse 11, 6330 Cham, Switzerland

Foreword

Hosting the 2020 European Conference on Computer Vision was certainly an exciting journey. From the 2016 plan to hold it at the Edinburgh International Conference Centre (hosting 1,800 delegates) to the 2018 plan to hold it at Glasgow's Scottish Exhibition Centre (up to 6,000 delegates), we finally ended with moving online because of the COVID-19 outbreak. While possibly having fewer delegates than expected because of the online format, ECCV 2020 still had over 3,100 registered participants.

Although online, the conference delivered most of the activities expected at a face-to-face conference: peer-reviewed papers, industrial exhibitors, demonstrations, and messaging between delegates. As well as the main technical sessions, the conference included a strong program of satellite events, including 16 tutorials and 44 workshops.

On the other hand, the online conference format enabled new conference features. Every paper had an associated teaser video and a longer full presentation video. Along with the papers and slides from the videos, all these materials were available the week before the conference. This allowed delegates to become familiar with the paper content and be ready for the live interaction with the authors during the conference week. The 'live' event consisted of brief presentations by the 'oral' and 'spotlight' authors and industrial sponsors. Question and Answer sessions for all papers were timed to occur twice so delegates from around the world had convenient access to the authors.

As with the 2018 ECCV, authors' draft versions of the papers appeared online with open access, now on both the Computer Vision Foundation (CVF) and the European Computer Vision Association (ECVA) websites. An archival publication arrangement was put in place with the cooperation of Springer. SpringerLink hosts the final version of the papers with further improvements, such as activating reference links and supplementary materials. These two approaches benefit all potential readers: a version available freely for all researchers, and an authoritative and citable version with additional benefits for SpringerLink subscribers. We thank Alfred Hofmann and Aliaksandr Birukou from Springer for helping to negotiate this agreement, which we expect will continue for future versions of ECCV.

August 2020

Vittorio Ferrari
Bob Fisher
Cordelia Schmid
Emanuele Trucco

Preface

Welcome to the workshops proceedings of the 16th European Conference on Computer Vision (ECCV 2020), the first edition held online. We are delighted that the main ECCV 2020 was accompanied by 45 workshops, scheduled on August 23, 2020, and August 28, 2020.

We received 101 valid workshop proposals on diverse computer vision topics and had space for 32 full-day slots, so we had to decline many valuable proposals (the workshops were supposed to be either full-day or half-day long, but the distinction faded away when the full ECCV conference went online). We endeavored to balance among topics, established series, and newcomers. Not all the workshops published their proceedings, or had proceedings at all. These volumes collect the edited papers from 28 out of 45 workshops.

We sincerely thank the ECCV general chairs for trusting us with the responsibility for the workshops, the workshop organizers for their involvement in this event of primary importance in our field, and the workshop presenters and authors.

August 2020

Adrien Bartoli
Andrea Fusiello

Organization

General Chairs

Vittorio Ferrari	Google Research, Switzerland
Bob Fisher	The University of Edinburgh, UK
Cordelia Schmid	Google and Inria, France
Emanuele Trucco	The University of Dundee, UK

Program Chairs

Andrea Vedaldi	University of Oxford, UK
Horst Bischof	Graz University of Technology, Austria
Thomas Brox	University of Freiburg, Germany
Jan-Michael Frahm	The University of North Carolina at Chapel Hill, USA

Industrial Liaison Chairs

Jim Ashe	The University of Edinburgh, UK
Helmut Grabner	Zurich University of Applied Sciences, Switzerland
Diane Larlus	NAVER LABS Europe, France
Cristian Novotny	The University of Edinburgh, UK

Local Arrangement Chairs

Yvan Petillot	Heriot-Watt University, UK
Paul Siebert	The University of Glasgow, UK

Academic Demonstration Chair

Thomas Mensink	Google Research and University of Amsterdam, The Netherlands

Poster Chair

Stephen Mckenna	The University of Dundee, UK

Technology Chair

Gerardo Aragon Camarasa	The University of Glasgow, UK

Tutorial Chairs

Carlo Colombo   University of Florence, Italy
Sotirios Tsaftaris   The University of Edinburgh, UK

Publication Chairs

Albert Ali Salah   Utrecht University, The Netherlands
Hamdi Dibeklioglu   Bilkent University, Turkey
Metehan Doyran   Utrecht University, The Netherlands
Henry Howard-Jenkins   University of Oxford, UK
Victor Adrian Prisacariu   University of Oxford, UK
Siyu Tang   ETH Zurich, Switzerland
Gul Varol   University of Oxford, UK

Website Chair

Giovanni Maria Farinella   University of Catania, Italy

Workshops Chairs

Adrien Bartoli   University Clermont Auvergne, France
Andrea Fusiello   University of Udine, Italy

Workshops Organizers

W01 - Adversarial Robustness in the Real World

Adam Kortylewski   Johns Hopkins University, USA
Cihang Xie   Johns Hopkins University, USA
Song Bai   University of Oxford, UK
Zhaowei Cai   UC San Diego, USA
Yingwei Li   Johns Hopkins University, USA
Andrei Barbu   MIT, USA
Wieland Brendel   University of Tübingen, Germany
Nuno Vasconcelos   UC San Diego, USA
Andrea Vedaldi   University of Oxford, UK
Philip H. S. Torr   University of Oxford, UK
Rama Chellappa   University of Maryland, USA
Alan Yuille   Johns Hopkins University, USA

W02 - BioImage Computation

Jan Funke   HHMI Janelia Research Campus, Germany
Dagmar Kainmueller   BIH and MDC Berlin, Germany
Florian Jug   CSBD and MPI-CBG, Germany
Anna Kreshuk   EMBL Heidelberg, Germany

Peter Bajcsy	NIST, USA
Martin Weigert	EPFL, Switzerland
Patrick Bouthemy	Inria, France
Erik Meijering	University New South Wales, Australia

W03 - Egocentric Perception, Interaction and Computing

Michael Wray	University of Bristol, UK
Dima Damen	University of Bristol, UK
Hazel Doughty	University of Bristol, UK
Walterio Mayol-Cuevas	University of Bristol, UK
David Crandall	Indiana University, USA
Kristen Grauman	UT Austin, USA
Giovanni Maria Farinella	University of Catania, Italy
Antonino Furnari	University of Catania, Italy

W04 - Embodied Vision, Actions and Language

Yonatan Bisk	Carnegie Mellon University, USA
Jesse Thomason	University of Washington, USA
Mohit Shridhar	University of Washington, USA
Chris Paxton	NVIDIA, USA
Peter Anderson	Georgia Tech, USA
Roozbeh Mottaghi	Allen Institute for AI, USA
Eric Kolve	Allen Institute for AI, USA

W05 - Eye Gaze in VR, AR, and in the Wild

Hyung Jin Chang	University of Birmingham, UK
Seonwook Park	ETH Zurich, Switzerland
Xucong Zhang	ETH Zurich, Switzerland
Otmar Hilliges	ETH Zurich, Switzerland
Aleš Leonardis	University of Birmingham, UK
Robert Cavin	Facebook Reality Labs, USA
Cristina Palmero	University of Barcelona, Spain
Jixu Chen	Facebook, USA
Alexander Fix	Facebook Reality Labs, USA
Elias Guestrin	Facebook Reality Labs, USA
Oleg Komogortsev	Texas State University, USA
Kapil Krishnakumar	Facebook, USA
Abhishek Sharma	Facebook Reality Labs, USA
Yiru Shen	Facebook Reality Labs, USA
Tarek Hefny	Facebook Reality Labs, USA
Karsten Behrendt	Facebook, USA
Sachin S. Talathi	Facebook Reality Labs, USA

W06 - Holistic Scene Structures for 3D Vision

Zihan Zhou	Penn State University, USA
Yasutaka Furukawa	Simon Fraser University, Canada
Yi Ma	UC Berkeley, USA
Shenghua Gao	ShanghaiTech University, China
Chen Liu	Facebook Reality Labs, USA
Yichao Zhou	UC Berkeley, USA
Linjie Luo	Bytedance Inc., China
Jia Zheng	ShanghaiTech University, China
Junfei Zhang	Kujiale.com, China
Rui Tang	Kujiale.com, China

W07 - Joint COCO and LVIS Recognition Challenge

Alexander Kirillov	Facebook AI Research, USA
Tsung-Yi Lin	Google Research, USA
Yin Cui	Google Research, USA
Matteo Ruggero Ronchi	California Institute of Technology, USA
Agrim Gupta	Stanford University, USA
Ross Girshick	Facebook AI Research, USA
Piotr Dollar	Facebook AI Research, USA

W08 - Object Tracking and Its Many Guises

Achal D. Dave	Carnegie Mellon University, USA
Tarasha Khurana	Carnegie Mellon University, USA
Jonathon Luiten	RWTH Aachen University, Germany
Aljosa Osep	Technical University of Munich, Germany
Pavel Tokmakov	Carnegie Mellon University, USA

W09 - Perception for Autonomous Driving

Li Erran Li	Alexa AI, Amazon, USA
Adrien Gaidon	Toyota Research Institute, USA
Wei-Lun Chao	The Ohio State University, USA
Peter Ondruska	Lyft, UK
Rowan McAllister	UC Berkeley, USA
Larry Jackel	North-C Technologies, USA
Jose M. Alvarez	NVIDIA, USA

W10 - TASK-CV Workshop and VisDA Challenge

Tatiana Tommasi	Politecnico di Torino, Italy
Antonio M. Lopez	CVC and UAB, Spain
David Vazquez	Element AI, Canada
Gabriela Csurka	NAVER LABS Europe, France
Kate Saenko	Boston University, USA
Liang Zheng	The Australian National University, Australia

Xingchao Peng   Boston University, USA
Weijian Deng   The Australian National University, Australia

W11 - Bodily Expressed Emotion Understanding

James Z. Wang   Penn State University, USA
Reginald B. Adams, Jr.   Penn State University, USA
Yelin Kim   Amazon Lab126, USA

W12 - Commands 4 Autonomous Vehicles

Thierry Deruyttere   KU Leuven, Belgium
Simon Vandenhende   KU Leuven, Belgium
Luc Van Gool   KU Leuven, Belgium, and ETH Zurich, Switzerland
Matthew Blaschko   KU Leuven, Belgium
Tinne Tuytelaars   KU Leuven, Belgium
Marie-Francine Moens   KU Leuven, Belgium
Yu Liu   KU Leuven, Belgium
Dusan Grujicic   KU Leuven, Belgium

W13 - Computer VISion for ART Analysis

Alessio Del Bue   Istituto Italiano di Tecnologia, Italy
Sebastiano Vascon   Ca' Foscari University and European Centre for Living
                              Technology, Italy
Peter Bell   Friedrich-Alexander University Erlangen-Nürnberg,
                              Germany
Leonardo L. Impett   EPFL, Switzerland
Stuart James   Istituto Italiano di Tecnologia, Italy

W14 - International Challenge on Compositional and Multimodal Perception

Alec Hodgkinson   Panasonic Corporation, Japan
Yusuke Urakami   Panasonic Corporation, Japan
Kazuki Kozuka   Panasonic Corporation, Japan
Ranjay Krishna   Stanford University, USA
Olga Russakovsky   Princeton University, USA
Juan Carlos Niebles   Stanford University, USA
Jingwei Ji   Stanford University, USA
Li Fei-Fei   Stanford University, USA

W15 - Sign Language Recognition, Translation and Production

Necati Cihan Camgoz   University of Surrey, UK
Richard Bowden   University of Surrey, UK
Andrew Zisserman   University of Oxford, UK
Gul Varol   University of Oxford, UK
Samuel Albanie   University of Oxford, UK

Kearsy Cormier   University College London, UK
Neil Fox   University College London, UK

W16 - Visual Inductive Priors for Data-Efficient Deep Learning

Jan van Gemert   Delft University of Technology, The Netherlands
Robert-Jan Bruintjes   Delft University of Technology, The Netherlands
Attila Lengyel   Delft University of Technology, The Netherlands
Osman Semih Kayhan   Delft University of Technology, The Netherlands
Marcos Baptista-Ríos   Alcalá University, Spain
Anton van den Hengel   The University of Adelaide, Australia

W17 - Women in Computer Vision

Hilde Kuehne   IBM, USA
Amaia Salvador   Amazon, USA
Ananya Gupta   The University of Manchester, UK
Yana Hasson   Inria, France
Anna Kukleva   Max Planck Institute, Germany
Elizabeth Vargas   Heriot-Watt University, UK
Xin Wang   UC Berkeley, USA
Irene Amerini   Sapienza University of Rome, Italy

W18 - 3D Poses in the Wild Challenge

Gerard Pons-Moll   Max Planck Institute for Informatics, Germany
Angjoo Kanazawa   UC Berkeley, USA
Michael Black   Max Planck Institute for Intelligent Systems, Germany
Aymen Mir   Max Planck Institute for Informatics, Germany

W19 - 4D Vision

Anelia Angelova   Google, USA
Vincent Casser   Waymo, USA
Jürgen Sturm   X, USA
Noah Snavely   Google, USA
Rahul Sukthankar   Google, USA

W20 - Map-Based Localization for Autonomous Driving

Patrick Wenzel   Technical University of Munich, Germany
Niclas Zeller   Artisense, Germany
Nan Yang   Technical University of Munich, Germany
Rui Wang   Technical University of Munich, Germany
Daniel Cremers   Technical University of Munich, Germany

W21 - Multimodal Video Analysis Workshop and Moments in Time Challenge

Dhiraj Joshi	IBM Research AI, USA
Rameswar Panda	IBM Research, USA
Kandan Ramakrishnan	IBM, USA
Rogerio Feris	IBM Research AI, MIT-IBM Watson AI Lab, USA
Rami Ben-Ari	IBM-Research, USA
Danny Gutfreund	IBM, USA
Mathew Monfort	MIT, USA
Hang Zhao	MIT, USA
David Harwath	MIT, USA
Aude Oliva	MIT, USA
Zhicheng Yan	Facebook AI, USA

W22 - Recovering 6D Object Pose

Tomas Hodan	Czech Technical University in Prague, Czech Republic
Martin Sundermeyer	German Aerospace Center, Germany
Rigas Kouskouridas	Scape Technologies, UK
Tae-Kyun Kim	Imperial College London, UK
Jiri Matas	Czech Technical University in Prague, Czech Republic
Carsten Rother	Heidelberg University, Germany
Vincent Lepetit	ENPC ParisTech, France
Ales Leonardis	University of Birmingham, UK
Krzysztof Walas	Poznan University of Technology, Poland
Carsten Steger	Technical University of Munich and MVTec Software GmbH, Germany
Eric Brachmann	Heidelberg University, Germany
Bertram Drost	MVTec Software GmbH, Germany
Juil Sock	Imperial College London, UK

W23 - SHApe Recovery from Partial Textured 3D Scans

Djamila Aouada	University of Luxembourg, Luxembourg
Kseniya Cherenkova	Artec3D and University of Luxembourg, Luxembourg
Alexandre Saint	University of Luxembourg, Luxembourg
David Fofi	University Bourgogne Franche-Comté, France
Gleb Gusev	Artec3D, Luxembourg
Bjorn Ottersten	University of Luxembourg, Luxembourg

W24 - Advances in Image Manipulation Workshop and Challenges

Radu Timofte	ETH Zurich, Switzerland
Andrey Ignatov	ETH Zurich, Switzerland
Kai Zhang	ETH Zurich, Switzerland
Dario Fuoli	ETH Zurich, Switzerland
Martin Danelljan	ETH Zurich, Switzerland
Zhiwu Huang	ETH Zurich, Switzerland

Hannan Lu	Harbin Institute of Technology, China
Wangmeng Zuo	Harbin Institute of Technology, China
Shuhang Gu	The University of Sydney, Australia
Ming-Hsuan Yang	UC Merced and Google, USA
Majed El Helou	EPFL, Switzerland
Ruofan Zhou	EPFL, Switzerland
Sabine Süsstrunk	EPFL, Switzerland
Sanghyun Son	Seoul National University, South Korea
Jaerin Lee	Seoul National University, South Korea
Seungjun Nah	Seoul National University, South Korea
Kyoung Mu Lee	Seoul National University, South Korea
Eli Shechtman	Adobe, USA
Evangelos Ntavelis	ETH Zurich and CSEM, Switzerland
Andres Romero	ETH Zurich, Switzerland
Yawei Li	ETH Zurich, Switzerland
Siavash Bigdeli	CSEM, Switzerland
Pengxu Wei	Sun Yat-sen University, China
Liang Lin	Sun Yat-sen University, China
Ming-Yu Liu	NVIDIA, USA
Roey Mechrez	BeyondMinds and Technion, Israel
Luc Van Gool	KU Leuven, Belgium, and ETH Zurich, Switzerland

W25 - Assistive Computer Vision and Robotics

Marco Leo	National Research Council of Italy, Italy
Giovanni Maria Farinella	University of Catania, Italy
Antonino Furnari	University of Catania, Italy
Gerard Medioni	University of Southern California, USA
Trivedi Mohan	UC San Diego, USA

W26 - Computer Vision for UAVs Workshop and Challenge

Dawei Du	Kitware Inc., USA
Heng Fan	Stony Brook University, USA
Toon Goedemé	KU Leuven, Belgium
Qinghua Hu	Tianjin University, China
Haibin Ling	Stony Brook University, USA
Davide Scaramuzza	University of Zurich, Switzerland
Mubarak Shah	University of Central Florida, USA
Tinne Tuytelaars	KU Leuven, Belgium
Kristof Van Beeck	KU Leuven, Belgium
Longyin Wen	JD Digits, USA
Pengfei Zhu	Tianjin University, China

W27 - Embedded Vision

| Tse-Wei Chen | Canon Inc., Japan |
| Nabil Belbachir | NORCE Norwegian Research Centre AS, Norway |

Stephan Weiss   University of Klagenfurt, Austria
Marius Leordeanu   Politehnica University of Bucharest, Romania

W28 - Learning 3D Representations for Shape and Appearance

Leonidas Guibas   Stanford University, USA
Or Litany   Stanford University, USA
Tanner Schmidt   Facebook Reality Labs, USA
Vincent Sitzmann   Stanford University, USA
Srinath Sridhar   Stanford University, USA
Shubham Tulsiani   Facebook AI Research, USA
Gordon Wetzstein   Stanford University, USA

W29 - Real-World Computer Vision from inputs with Limited Quality and Tiny Object Detection Challenge

Yuqian Zhou   University of Illinois, USA
Zhenjun Han   University of the Chinese Academy of Sciences, China
Yifan Jiang   The University of Texas at Austin, USA
Yunchao Wei   University of Technology Sydney, Australia
Jian Zhao   Institute of North Electronic Equipment, Singapore
Zhangyang Wang   The University of Texas at Austin, USA
Qixiang Ye   University of the Chinese Academy of Sciences, China
Jiaying Liu   Peking University, China
Xuehui Yu   University of the Chinese Academy of Sciences, China
Ding Liu   Bytedance, China
Jie Chen   Peking University, China
Humphrey Shi   University of Oregon, USA

W30 - Robust Vision Challenge 2020

Oliver Zendel   Austrian Institute of Technology, Austria
Hassan Abu Alhaija   Interdisciplinary Center for Scientific Computing
                              Heidelberg, Germany
Rodrigo Benenson   Google Research, Switzerland
Marius Cordts   Daimler AG, Germany
Angela Dai   Technical University of Munich, Germany
Andreas Geiger   Max Planck Institute for Intelligent Systems
                              and University of Tübingen, Germany
Niklas Hanselmann   Daimler AG, Germany
Nicolas Jourdan   Daimler AG, Germany
Vladlen Koltun   Intel Labs, USA
Peter Kontschieder   Mapillary Research, Austria
Yubin Kuang   Mapillary AB, Sweden
Alina Kuznetsova   Google Research, Switzerland
Tsung-Yi Lin   Google Brain, USA
Claudio Michaelis   University of Tübingen, Germany
Gerhard Neuhold   Mapillary Research, Austria

Matthias Niessner	Technical University of Munich, Germany
Marc Pollefeys	ETH Zurich and Microsoft, Switzerland
Francesc X. Puig Fernandez	MIT, USA
Rene Ranftl	Intel Labs, USA
Stephan R. Richter	Intel Labs, USA
Carsten Rother	Heidelberg University, Germany
Torsten Sattler	Chalmers University of Technology, Sweden and Czech Technical University in Prague, Czech Republic
Daniel Scharstein	Middlebury College, USA
Hendrik Schilling	rabbitAI, Germany
Nick Schneider	Daimler AG, Germany
Jonas Uhrig	Daimler AG, Germany
Jonas Wulff	Max Planck Institute for Intelligent Systems, Germany
Bolei Zhou	The Chinese University of Hong Kong, China

W31 - The Bright and Dark Sides of Computer Vision: Challenges and Opportunities for Privacy and Security

Mario Fritz	CISPA Helmholtz Center for Information Security, Germany
Apu Kapadia	Indiana University, USA
Jan-Michael Frahm	The University of North Carolina at Chapel Hill, USA
David Crandall	Indiana University, USA
Vitaly Shmatikov	Cornell University, USA

W32 - The Visual Object Tracking Challenge

Matej Kristan	University of Ljubljana, Slovenia
Jiri Matas	Czech Technical University in Prague, Czech Republic
Ales Leonardis	University of Birmingham, UK
Michael Felsberg	Linköping University, Sweden
Roman Pflugfelder	Austrian Institute of Technology, Austria
Joni-Kristian Kamarainen	Tampere University, Finland
Martin Danelljan	ETH Zurich, Switzerland

W33 - Video Turing Test: Toward Human-Level Video Story Understanding

Yu-Jung Heo	Seoul National University, South Korea
Seongho Choi	Seoul National University, South Korea
Kyoung-Woon On	Seoul National University, South Korea
Minsu Lee	Seoul National University, South Korea
Vicente Ordonez	University of Virginia, USA
Leonid Sigal	University of British Columbia, Canada
Chang D. Yoo	KAIST, South Korea
Gunhee Kim	Seoul National University, South Korea
Marcello Pelillo	University of Venice, Italy
Byoung-Tak Zhang	Seoul National University, South Korea

W34 - "Deep Internal Learning": Training with no prior examples

Michal Irani	Weizmann Institute of Science, Israel
Tomer Michaeli	Technion, Israel
Tali Dekel	Google, Israel
Assaf Shocher	Weizmann Institute of Science, Israel
Tamar Rott Shaham	Technion, Israel

W35 - Benchmarking Trajectory Forecasting Models

Alexandre Alahi	EPFL, Switzerland
Lamberto Ballan	University of Padova, Italy
Luigi Palmieri	Bosch, Germany
Andrey Rudenko	Örebro University, Sweden
Pasquale Coscia	University of Padova, Italy

W36 - Beyond mAP: Reassessing the Evaluation of Object Detection

David Hall	Queensland University of Technology, Australia
Niko Suenderhauf	Queensland University of Technology, Australia
Feras Dayoub	Queensland University of Technology, Australia
Gustavo Carneiro	The University of Adelaide, Australia
Chunhua Shen	The University of Adelaide, Australia

W37 - Imbalance Problems in Computer Vision

Sinan Kalkan	Middle East Technical University, Turkey
Emre Akbas	Middle East Technical University, Turkey
Nuno Vasconcelos	UC San Diego, USA
Kemal Oksuz	Middle East Technical University, Turkey
Baris Can Cam	Middle East Technical University, Turkey

W38 - Long-Term Visual Localization under Changing Conditions

Torsten Sattler	Chalmers University of Technology, Sweden, and Czech Technical University in Prague, Czech Republic
Vassileios Balntas	Facebook Reality Labs, USA
Fredrik Kahl	Chalmers University of Technology, Sweden
Krystian Mikolajczyk	Imperial College London, UK
Tomas Pajdla	Czech Technical University in Prague, Czech Republic
Marc Pollefeys	ETH Zurich and Microsoft, Switzerland
Josef Sivic	Inria, France, and Czech Technical University in Prague, Czech Republic
Akihiko Torii	Tokyo Institute of Technology, Japan
Lars Hammarstrand	Chalmers University of Technology, Sweden
Huub Heijnen	Facebook, UK
Maddern Will	Nuro, USA
Johannes L. Schönberger	Microsoft, Switzerland

| Pablo Speciale | ETH Zurich, Switzerland |
| Carl Toft | Chalmers University of Technology, Sweden |

W39 - Sensing, Understanding, and Synthesizing Humans

Ziwei Liu	The Chinese University of Hong Kong, China
Sifei Liu	NVIDIA, USA
Xiaolong Wang	UC San Diego, USA
Hang Zhou	The Chinese University of Hong Kong, China
Wayne Wu	SenseTime, China
Chen Change Loy	Nanyang Technological University, Singapore

W40 - Computer Vision Problems in Plant Phenotyping

Hanno Scharr	Forschungszentrum Jülich, Germany
Tony Pridmore	University of Nottingham, UK
Sotirios Tsaftaris	The University of Edinburgh, UK

W41 - Fair Face Recognition and Analysis

Sergio Escalera	CVC and University of Barcelona, Spain
Rama Chellappa	University of Maryland, USA
Eduard Vazquez	Anyvision, UK
Neil Robertson	Queen's University Belfast, UK
Pau Buch-Cardona	CVC, Spain
Tomas Sixta	Anyvision, UK
Julio C. S. Jacques Junior	Universitat Oberta de Catalunya and CVC, Spain

W42 - GigaVision: When Gigapixel Videography Meets Computer Vision

Lu Fang	Tsinghua University, China
Shengjin Wang	Tsinghua University, China
David J. Brady	Duke University, USA
Feng Yang	Google Research, USA

W43 - Instance-Level Recognition

Andre Araujo	Google, USA
Bingyi Cao	Google, USA
Ondrej Chum	Czech Technical University in Prague, Czech Republic
Bohyung Han	Seoul National University, South Korea
Torsten Sattler	Chalmers University of Technology, Sweden and Czech Technical University in Prague, Czech Republic
Jack Sim	Google, USA
Giorgos Tolias	Czech Technical University in Prague, Czech Republic
Tobias Weyand	Google, USA

Xu Zhang	Columbia University, USA
Cam Askew	Google, USA
Guangxing Han	Columbia University, USA

W44 - Perception Through Structured Generative Models

Adam W. Harley	Carnegie Mellon University, USA
Katerina Fragkiadaki	Carnegie Mellon University, USA
Shubham Tulsiani	Facebook AI Research, USA

W45 - Self Supervised Learning – What is Next?

Christian Rupprecht	University of Oxford, UK
Yuki M. Asano	University of Oxford, UK
Armand Joulin	Facebook AI Research, USA
Andrea Vedaldi	University of Oxford, UK

Contents – Part I

W02 - BioImage Computation

W03 - Egocentric Perception, Interaction, and Computing

W05 - Eye Gaze in VR, AR, and in the Wild

W10 - TASK-CV Workshop and VisDA Challenge

W11 - Bodily Expressed Emotion Understanding

W01 - Adversarial Robustness in the Real World

W01 - Adversarial Robustness in the Real World

Computer vision systems nowadays have advanced performance, but research in adversarial machine learning also shows that they are not as robust as the human vision system. Recent work has shown that real-world adversarial examples exist when objects are partially occluded or viewed in previously unseen poses and environments (such as different weather conditions). Discovering and harnessing those adversarial examples provides opportunities for understanding and improving computer vision systems in real-world environments. In particular, deep models with structured internal representations seem to be a promising approach to enhance robustness in the real world, while also being able to explain their predictions.

In this workshop, we aim to bring together researchers from the fields of adversarial machine learning, robust vision, and explainable AI to discuss recent research and future directions for adversarial robustness and explainability, with a particular focus on real-world scenarios.

For this first edition of Adversarial Robustness in the Real World (AROW 2020), held on August 23, 2020, we had nine great invited speakers: Prof. Andreas Geiger from the University of Tübingen, Germany, Dr. Wieland Brendel from the University of Tübingen, Prof. Alan Yuille from the Johns Hopkins University, USA, Prof. Raquel Urtasun from the University of Toronto, Canada, Alex Robey from the University of Pennsylvania, USA, Prof. Judy Hoffman from Georgia Institute of Technology, USA, Prof. Honglak Lee from the University of Michigan, USA, Prof. Bo Li from the University of Illinois at Urbana-Champaign, USA, and Prof. Daniel Fremont from the University of California Santa Cruz, USA. With their diverse backgrounds and affiliations, they covered a wide range of themes within the general topic of the workshop.

When it comes to paper contributions, we initially received 31 submissions (long papers and extended abstracts). Each submission was reviewed by two independent referees, drawn from a pool of 35 Program Committee members and assigned to the papers by the workshop organizers. The reviews were then moderated by the workshop organizers, and we ended up with 24 accepted submissions (14 long papers). These papers were presented during two 2 hour poster sessions at the workshop, the goal of favoring posters over oral presentations being to facilitate discussions in this online workshop.

Altogether, we believe that this edition of AROW was a success, with high-quality talks and contributed papers. We are glad to see that this topic attracted much interest in the community and hope that this trend will continue in the future.

August 2020

Adam Kortylewski   Wieland Brendel
Cihang Xie   Nuno Vasconcelos
Song Bai   Andrea Vedaldi
Zhaowei Cai   Philip H. S. Torr
Yingwei Li   Rama Chellappa
Andrei Barbu   Alan Yuille

A Deep Dive into Adversarial Robustness in Zero-Shot Learning

Mehmet Kerim Yucel[1]([⊠]), Ramazan Gokberk Cinbis[2], and Pinar Duygulu[1]

[1] Department of Computer Engineering, Hacettepe University, Ankara, Turkey
mkerimyucel@gmail.com
[2] Department of Computer Engineering, Middle East Technical University (METU),
Ankara, Turkey

Abstract. Machine learning (ML) systems have introduced significant advances in various fields, due to the introduction of highly complex models. Despite their success, it has been shown multiple times that machine learning models are prone to imperceptible perturbations that can severely degrade their accuracy. So far, existing studies have primarily focused on models where supervision across all classes were available. In contrast, Zero-shot Learning (ZSL) and Generalized Zero-shot Learning (GZSL) tasks inherently lack supervision across all classes. In this paper, we present a study aimed on evaluating the adversarial robustness of ZSL and GZSL models. We leverage the well-established label embedding model and subject it to a set of established adversarial attacks and defenses across multiple datasets. In addition to creating possibly the first benchmark on adversarial robustness of ZSL models, we also present analyses on important points that require attention for better interpretation of ZSL robustness results. We hope these points, along with the benchmark, will help researchers establish a better understanding what challenges lie ahead and help guide their work.

1 Introduction

The meteoric rise of complex machine learning models in the last decade sparked a whole new wave of state-of-the-art (SOTA) results in numerous fields, such as computer vision, natural language processing and speech recognition. Due to the increase in available data, compute power and architectural improvements, these fields are still seeing rapid improvements with no signs of slowing down.

However, it has been shown [8] that ML models are prone to adversarial examples, which are perturbations aimed to guide models into inaccurate results. Such perturbations can successfully misguide models while introducing imperceptible perturbations to a query data. Starting with computer vision, such attacks have been extended to speech recognition [12], natural language processing [16] and various other tasks/modalities [10]. Naturally, equal attention has been given to defend the models against these attacks, either by designing robust models or introducing mechanisms to detect and invalidate adversarial examples [42]. Adversarial machine learning initially focused on small datasets

© Springer Nature Switzerland AG 2020
A. Bartoli and A. Fusiello (Eds.): ECCV 2020 Workshops, LNCS 12535, pp. 3–21, 2020.
https://doi.org/10.1007/978-3-030-66415-2_1

in computer vision, such as MNIST and CIFAR-10, but it has been extended to large datasets such as ImageNet and even commercial products [40,66].

The majority of the adversarial ML literature has so far focused on supervised models and aimed to improve their robustness using various approaches. In Zero-shot learning (ZSL) and Generalized Zero-shot learning (GZSL) settings, however, the task differs from a generic supervised approach; the aim is to learn from a set of classes such that we can optimize the knowledge transfer from these classes to a set of previously unseen classes, which we use during the evaluation phase. As a result, the notorious problem of ZSL is far from being solved despite significant advances. The introduction of adversarial examples to ZSL models would theoretically further exacerbate the problem.

In this paper, we present an exploratory adversarial robustness analysis on ZSL techniques[1]. Unlike the most recent approaches where ZSL problem is effectively reduced to a supervised problem [9,47], we take a step back and focus on label embedding model [1,56] and analyze its robustness against several prominent adversarial attacks and defenses. Through rigorous evaluation on most widely used ZSL datasets, we establish a framework where we analyse not only the algorithm itself, but also the effect of each dataset, effect of per-class sample count, the trends in boundary transitions as well as how the existing knowledge transfer between seen and unseen classes are effected by adversarial intrusions. We hope the presented framework will focus the community's attention on the robustness of ZSL models, which has largely been neglected. Moreover, this study will serve as a benchmark for future studies and shed light on important trends to look out for when analysing ZSL models for adversarial robustness.

2   Related Work

Adversarial Attacks. Adversarial ML has been an integral part of ML research in the last few years as it exposed significant robustness-related problems with existing models. It has first been shown in [8] that a perturbation can be crafted by optimizing for misclassification of an image; under certain ℓ-norm constraints this perturbation can even be imperceptible to humans. A fast, one-step attack that exploits the gradients of the cost function w.r.t the model parameters to craft a perturbation was shown in [20]. An iterative algorithm that approximates the minimum perturbation required for misclassification in ML models has been shown in [38]. Carlini and Wagner [11] showed that a much improved version of [8] can be tailored into three different attacks, each using a different ℓ-norm, that can successfully scale to ImageNet and invalidate the distillation defense [42].

Transformation attacks [59], universal attacks for a dataset [37], one-pixel attacks [52], attacks focusing on finding the pixels that will alter the output the most [41], unrestricted attacks [6], black-box attacks where attacker has no information about the model [40] and attacks transferred to the physical world [29] are some highlights of adversarial ML. Moreover, these attacks have

---

[1] Code is available at https://github.com/MKYucel/adversarial_robustness_zsl

been extended to various other ML models with various modalities, such as recurrent models [24], reinforcement learning [5], object detection [60], tracking [15,25], natural language processing [16], semantic segmentation [4], graph neural networks [71], networks trained on LIDAR data [10], speech recognition [12], visual question answering [50] and even commercial systems [33].

Adversarial Defenses. In order to address the robustness concerns raised by adversarial attacks, various defense mechanisms have been proposed, resulting into an arms-race in the literature between attacks and defenses. Several prominent defense techniques include network distillation [42], adversarial training [8], label smoothing [22], input-gradient regularization [45], analysis of ReLu activation patterns [34], feature regeneration [7], generalizable defenses [39], exploitation of GANs [46] and auto-encoders [36] for alleviation of adversarial changes and feature space regularization via metric learning [35]. These defense methods rely on either re-training of the network or preparing an additional module that would either detect or alleviate the adversarial perturbations. Another segment of adversarial defenses have borrowed several existing techniques and used them towards adversarial robustness; JPEG compression [49], bit-depth reduction, spatial smoothing [62] and total variance minimization [21] are some examples of such techniques. Adversarial ML field is quite vast; readers are referred to [3,61] for a more in-depth discussion.

Zero-Shot Learning. In majority of the ML fields, SOTA results are generally held by supervised models, where all classes have a form of strong supervisory signal (i.e. ground-truth labels) that guides the learning. However, the collection of a supervised training set quickly turns into a bottleneck in semantically scaling up a recognition model. The problem of strong supervision through ground-truth labels can be somewhat alleviated by the (transductive) self-supervised learning [27], but the unlabeled data or the auxiliary supervision may not be available for every task. Zero-shot learning aims to address this issue by bridging the gap between *seen* (i.e. classes available during training) and *unseen* (i.e. classes unavailable during training) classes by transferring the knowledge learned during training. Generalized Zero-Shot learning, on the other hand, aims to facilitate this knowledge transfer while keeping the accuracy levels on the *seen* classes as high as possible. The auxiliary information present for both seen and unseen classes (i.e. class embeddings) is exploited to bridge the gap.

In earlier years, ZSL methods consisted of a two-stage mechanism, where the attributes of an image were predicted and these attributes were used to find the class with the most similar attributes [30,31]. By directly learning a linear [1,2,18,32] or a non-linear compability [14,51,58,65] function to map from visual to semantic space, later models transitioned to a single-stage format. The reverse mapping, from semantic to the visual space [55,68], has also been explored for ZSL problem. Embedding both visual and semantic embeddings into a common latent space for ZSL have proven to be successful [13,70]. Transductive approaches using visual or semantic information on unlabeled unseen classes [63,64] are also considered. In recent years, in addition to discriminative approaches [26,67], generative approaches [9,17,28,47] which model the mapping between visual and semantic spaces are increasingly being used to generate samples for unseen classes, slowly reducing ZSL to a supervised problem. For further information on ZSL, readers are referred to [54,57].

Only a recent unpublished study [69] proposed a ZSL model that is robust to several adversarial attacks by formulating an adversarial training regime. Our study, on the other hand, concentrates on setting up a framework and creates a benchmark to guide researchers' efforts towards adversarially robust ZSL/GZSL models, by presenting a detailed analysis of existing datasets and the effects of several well-established attacks and defenses. To the best of our knowledge, our study is the first to establish such a benchmark with a detailed analysis.

3   Methodology

3.1   Model Selection

For the model selection, we take a step back and focus on the models that aim to transfer the knowledge learned from seen classes to unseen classes, unlike generative approaches aiming to reduce ZSL to a supervised problem. We hypothesize that concentrating on the latter would mean evaluating the sample generation part for adversarial robustness, rather than evaluating the robustness of the model that aims directly to facilitate seen/unseen class knowledge transfer.

As presented in Sect. 2, there are numerous suitable candidates for our goal. Towards this end, we select the label-embedding model [1], which has been shown to be a stable and competitive model in modern benchmarks [57]. Attribute-label embedding (ALE) model is formulated as

$$F(x, y; W) = \theta(x)W^T \phi(y) \tag{1}$$

where $\theta(x)$ is the visual and $\phi(y)$ is the class embeddings. These two modalities are associated through the compatibility function $F()$, which is parametrized by learnable weights W.

The reason for selecting ALE is that it is one of the earlier studies that showed direct mapping by exploiting data and auxiliary information is more effective than intermediate attribute prediction stages. Although there are several studies which build on what ALE does[2], we believe results of ALE will be representative of the adversarial robustness of this family of ZSL approaches. Individual analyses of more approaches are certainly welcome, but is not in our scope. It must be noted that we focus on an inductive setting for ZSL.

3.2   Attacks and Defenses

Threat Model. Our evaluation makes several assumptions on the threat model. We choose three white-box, per-image attacks (i.e. non-universal) where the

---

[2] As noted in [57], models focusing on linear compability functions have the same formulation, but different optimization objectives.

attacker has access to model architecture and its parameters. The attack model operates under a setting where *all* images are attacked, regardless of their original predictions (i.e. whether they were classified correctly by the model or not). We choose a training-time defense (i.e. robustifying the model by re-training) and two data-independent, pre-processing defenses, where input images are processed before being fed to the network. The defense model operates under a *blind* setting, where none of the defenders have access to attack details or the attack frequency (i.e. defenses are applied to all images; regardless of the fact that attacks introduced misclassifications or not). In the next sections, we briefly present attacks and defenses considered in this work.

Attacks. *The first attack* we select is the widely-used Fast Gradient Sign method (FGSM) attack [20] that is based on the *linearity hypothesis*. By taking the gradient of the loss function with respect to the input, change of the output with respect to each input component is effectively estimated. This information is then used to craft adversarial perturbations that will guide the image towards these directions, which means *maximizing* the loss with respect to input components. We select FGSM due to its one-shot nature (i.e. no optimization), its low computational complexity and the fact that it is inherently *not* optimized for the minimum possible perturbation.

The second attack is the *DeepFool* [38] attack. DeepFool concentrates on the distance of an image to the closest decision boundary. Essentially, DeepFool calculates the distance to select number of decision boundaries, finds the closest one and takes the step towards this boundary. For non-linear classifiers, this step is approximated by an iterative regime that repeatedly tries to cross the boundary, until an iteration limit is reached or the boundary is crossed. We select DeepFool due to several reasons; i) it is an optimization based attack, ii) it directly aims for the minimum perturbation, iii) it operates under the linearity hypothesis assumption and iv) it is inherently indicative of the decision boundary characteristics. The version of DeepFool we experiment with is the original *untargeted* version that controls the perturbation with the ℓ_2 norm.

The last attack is the *Carlini-Wagner* [11] attack. In their paper, authors essentially refine the objective function proposed in [8] via several improvements and propose three different attacks where each attack uses a different ℓ-norm constraint to control the perturbation. We select Carlini-Wagner attacks due to several reasons; i) it is one of the first attacks that is shown to beat an adversarial defense, ii) one of the first to scale to ImageNet and iii) it is still one of the high-performing attacks in the literature. We use the *untargeted*, ℓ_2-norm version to have a better comparison with DeepFool.

Defenses. *The first defense* analyzed is the well-known *label smoothing*. It is a well-known regularization technique that prevents over-confident predictions. Label smoothing has been shown to be a good defense mechanism [19,22] and its success is tied to the prevention of confident classifier predictions when faced with an out-of-distribution sample. We select label smoothing as it is i) a training-time defense, ii) conceptually easy and iii) it is a good use case of ZSL models.

The second defense is *local spatial smoothing*. It has been reported that feature squeezing techniques [62] can provide robustness against adversarial examples as they effectively shrink the feature space where adversarial examples can reside. Similar to the original paper, we use median-filter with reflect-padding to pre-process images before they are fed to the model. We select spatial smoothing due to i) its data and attack-independent and ii) its inexpensive nature. Moreover, testing this against non-l_0 attacks is a good use case for its efficiency[3]. We do not use the detection mechanism in the original paper, but just the spatial smoothing operation.

The last defense is the *total variance minimization* defense. It has been proposed [21] as an input transformation defense, where the aim is to remove perturbations by image reconstruction. Initially, several pixels are selected with a Bernoulli random variable from the perturbed image. Using the selected pixels, the image is reconstructed by taking into account the total variation measure. Total-variance minimization is shown to be an efficient defense as it encourages the removal of small and localized perturbations. We select this defense due to its simple and data/attack independent nature. It is also a good candidate to evaluate different attacks due to its localized perturbation removal ability.

4 Experimental Results

4.1 Dataset and Evaluation Metrics

We perform our evaluation on three widely used ZSL/GZSL datasets; Caltech-UCSD-Birds 200–2011 (CUB) [53], Animals with Attributes 2 (AWA2) [57] and SUN [44]. CUB is a medium-sized fine-grained dataset with 312 attributes, where a total number of 200 classes are presented with a total of 11788 images. CUB is a challenging case as intra-class variance is quite hard to model due to similar appearances and low number of samples. SUN is another medium-sized fine-grained dataset with 102 attributes. SUN, similar to CUB, is a challenging case as it consists of 14340 images of 717 classes, resulting into even fewer images per class compared to CUB. AWA2 is a larger-scale dataset with 85 attributes, where a total of 50 classes are presented with 37322 images. AWA2, although it has a higher amount of images with fewer classes, inherently makes generalization to unseen classes harder. Throughout the experiments, we use the splits proposed in [57] for both ZSL and GZSL experiments. We use the standard per-class top-1 accuracy for ZSL evaluation. For GZSL, per-class top-1 accuracy values for seen and unseen classes are used to compute harmonic-scores.

4.2 Implementation Details

In order to make the computational graph end-to-end differentiable, we merge the ResNet-101 [23] (used to produce AWA2 [57] dataset embeddings) feature

---

[3] It has been noted in [62] that this defense is inherently more effective against l_0-norm attacks.

Table 1. Results when *all* images are attacked. *C*, *S* and *A* stand for CUB, SUN and AWA2 datasets, respectively. Parameters: $[FGSM_{1-3}$ ϵ: 0.001, 0.01, 0.1] $[DeepFool_{1-3}$ max_iter, ϵ: (3,1e−6), (3,1e−5), (10,1e−6)] $[C\&W_{1-3}$ max_iter: 3,6,10]. *Top-1* is the top-1 accuracy, where *u*, *s* and *h* are unseen, seen and harmonic accuracy values, respectively.

	Zero shot			Generalized zero shot								
	C	S	A	C			S			A		
Attack	Top-1			u	s	h	u	s	h	u	s	h
Original	54.5	57.4	62.0	25.6	64.6	36.7	20.5	32.3	25.1	15.3	78.8	25.7
$FGSM_1$	40.3	47.7	42.5	18.5	45.4	26.3	17.7	25.9	21.0	10.7	58.9	18.1
$FGSM_2$	18.5	16.3	14.8	10.8	11.7	11.2	8.1	9.8	8.9	3.4	10.0	5.1
$FGSM_3$	15.2	11.8	16.4	9.0	10.2	9.6	4.3	5.5	4.9	2.2	11.2	3.7
$DEFO_1$	30.9	25.6	50.6	9.1	19.1	12.3	6.4	7.2	6.8	13.3	41.2	20.1
$DEFO_2$	30.8	25.5	50.5	9.1	18.9	12.3	6.4	7.2	6.8	13.4	41.2	20.2
$DEFO_3$	22.4	17.8	41.4	7.6	11.5	9.2	6.3	6.3	6.3	13.0	30.2	18.2
$CaWa_1$	28.9	43.1	43.2	17.0	29.0	21.4	17.7	24.9	20.7	15.2	56.3	24.0
$CaWa_2$	25.9	40.9	36.9	16.4	24.4	19.6	17.7	23.9	20.3	15.2	46.6	22.9
$CaWa_3$	24.6	39.8	34.7	15.9	23.1	18.9	17.5	23.4	20.0	15.2	43.6	22.5

extractor with ALE model. To reproduce the results of ALE reported in [57], we freeze the feature extractor and only train ALE for each dataset. In our tables, the reproduced values of ALE are denoted as *original*, although there are slight variations compared to the original results reported by the authors in [57]. We use PyTorch [43] for our experiments.

For FGSM, we sweep with a large range of ϵ values where we end up with visible perturbations. We primarily sweep with *maximum iteration* and ϵ (added value to cross the boundary) parameter for DeepFool (DEFO) and Carlini-Wagner (CaWa, C&W) attacks, as we observe diminishing returns (i.e. not producing strong attacks despite reaching intractable compute time) for other parameters. We assign *0.9* to the ground-truth class in label smoothing defense. For spatial smoothing and total-variance minimization, we use 3×3 windows and maximum iteration of 3, respectively. We apply the same attack and defense parameters for every dataset to facilitate a better comparison of dataset characteristics.

4.3   Results

Attacks. First, we present the effect of each attack setting on ZSL/GZSL performance metrics. Results are shown in Table 1.

In *ZSL* setting, we see every attack has managed to introduce a visible detrimental effect on accuracy values across all datasets. As expected, stronger attacks introduce more pronounced attacks, FGSM being the most effective across all

Table 2. Results where all images are defended (without any attacks). SpS, LbS and TVM are spatial smoothing, label smoothing and total-variance minimization, respectively.

	Zero shot			Generalized zero shot								
	C	S	A	C			S			A		
Attack	Top-1			u	s	h	u	s	h	u	s	h
Original	54.5	57.4	62.0	25.6	64.6	36.7	20.5	32.3	25.1	15.3	78.8	25.7
SpS	49.3	53.2	59.3	21.5	56.5	31.1	20.1	28.0	23.4	14.3	75.5	24.1
LbS	52.2	55.2	60.6	22.7	56.2	32.4	18.4	31.6	23.3	16.3	74.2	26.8
TVM	51.4	54.0	60.3	24.4	60.7	34.8	19.9	29.5	23.8	12.9	76.4	22.1

datasets. This is an expected behaviour as we effectively introduce visible and quite strong attacks in the last FGSM setting. In CUB, we see C&W attack leading in low maximum iterations, but it starts losing out to DeepFool in higher maximum iterations. In SUN, although introducing some effects, C&W fails to impress and scale with the increasing maximum iteration values, where Deep-Fool manages to do a better job. In AWA2, C&W actually does a better job than DeepFool across all parameter settings. FGSM introduces an upward accuracy spike in AWA2, despite its increasing strength. This is primarily caused by actually changing originally incorrectly predicted labels to their correct labels, thereby increasing the accuracy. Lastly, DeepFool produces diminishing returns except the highest maximum iteration setting, across all datasets.

In *GZSL* setting, we again see an across the board reduction of accuracy values in all datasets. In CUB, DeepFool is the best performing attack, despite FGSM producing significantly more visible perturbations. In SUN, DeepFool loses out to FGSM slightly, though the produced perturbation is still significantly less visible. For CUB and SUN, DeepFool actually takes about the same time to produce the attack regardless of the maximum iteration value, indicating that it manages to cross the boundary in really few iterations, basically making 10 maximum iterations unnecessary. This means the class boundaries are close to each other and easy to cross, which makes sense as SUN and CUB has significantly more classes compared to AWA2. However, we do not see that effect for C&W, meaning it still needs more iterations to successfully cross the boundary despite needing the highest compute time. In AWA2, FGSM has a significant lead; DeepFool is somewhat effective but fails to impress. C&W, on the other hand, basically fails to introduce any meaningful degradation in accuracy, especially in unseen accuracy values. As can be seen from the Table 1, this is actually a wider phenomenon; unseen accuracies are less effected compared to their seen counterparts. We investigate this issue in depth in the following sections.

Defenses. Before going through the recovery rates of each defense, we first apply the defenses *without* any attacks to see what the effects of defenses are; a defense that is actually degrading the results are naturally not suitable for

Table 3. Results when *all* images are attacked and then defended with *spatial smoothing*. Parameter sets of the attacks are same as Table 1.

Attack	Zero shot			Generalized zero shot								
	C	S	A	C			S			A		
	Top-1			u	s	h	u	s	h	u	s	h
Original	54.5	57.4	62.0	25.6	64.6	36.7	20.5	32.3	25.1	15.3	78.8	25.7
$FGSM_1$	47.9	51.1	54.5	20.3	53.5	29.4	19.8	26.0	22.5	12.7	70.0	21.5
$FGSM_2$	31.9	36.0	24.6	14.5	30.5	19.7	14.5	16.6	15.5	6.2	25.3	10.0
$DEFO_1$	46.4	49.0	58.0	18.8	50.1	27.3	15.9	21.0	18.1	13.5	69.3	22.6
$DEFO_3$	46.2	48.8	58.0	18.7	50.0	27.2	15.9	21.0	18.1	13.1	68.8	22.1
$CaWa_1$	48.3	52.7	58.2	21.0	55.0	30.5	20.2	27.3	23.2	14.2	73.6	23.9
$CaWa_3$	48.4	52.3	58.2	21.0	54.9	30.4	20.0	27.2	23.1	14.2	73.3	23.8

use. Results are shown in Table 2. We see modest detrimental effects of defenses across the board, which we believe to be acceptable given the improvements they bring. We also see that in AWA2, label smoothing actually improves the GZSL performance compared to its original value. There is no winner in this regard, although label smoothing and total-variance minimization tend to do a better job than spatial smoothing. We now analyze the effects of each defense under several attack settings; we note that we omit one setting per each attack algorithm from our defense analysis; they either introduce extreme perturbations ($FGSM_3$) or negligible effects compared to their weaker counterpart ($DeepFool_2$ and $C\&W_2$).

Spatial smoothing results are shown in Table 3[4]. In *ZSL* setting, we see quite good recoveries across all datasets. The recovered accuracy values are naturally better for weaker attacks. We see quite similar recovered accuracy values for each DeepFool and C&W settings ($DeepFool_1$ vs $DeepFool_3$, $C\&W_1$ vs $C\&W_3$), unlike what we see for FGSM. This is potentially due to the nature of the attacks; FGSM strength scales proportionally with the coefficient ϵ, whereas maximum iteration for C&W and DeepFool acts like a binary switch indicating whether the attacks will function or not. In *GZSL* setting, results generally show the same trends with ZSL. However, we see negligible recoveries for C&W and DeepFool in AWA2, especially in unseen accuracy values. Surprisingly, spatial smoothing *degrades* the unseen and harmonic scores of $C\&W_1$ compared to its original (unattacked) values. This will be investigated in the following sections.

Label smoothing results are shown in Table 4. Although we can not directly compare recovered accuracies to ones reported in Table 1 as the original accuracy is different, we compare the trends to gain insights for label smoothing. In *ZSL* setting, we do not see gains for FGSM and even reductions in accuracy for some cases. DeepFool results are improved for some cases and C&W sees the highest improvements among all attacks. In *GZSL*, similar trends with ZSL is observed;

---

[4] Tables 3, 4 and 5 should be compared to Table 1.

FGSM seems unaffected, DeepFool is slightly recovered but C&W is the most recovered attack.

Total-variance minimization results are shown in Table 5. In *ZSL* setting, we observe recoveries for every attack setting. Similar to what we observed in *spatial smoothing*, recovered values for DeepFool and C&W are similar in values. Compared to other defenses, TVM does the best job in ZSL accuracy recovery. In *GZSL* setting, we see similar trends with ZSL. However, we observe in AWA2 that unseen accuracies actually go down when TVM is applied, especially for DeepFool and C&W. For C&W, this effect is also present for harmonic scores. This is similar to what we observed in spatial smoothing, however the effect is more pronounced. This will be investigated later in the paper.

Summary. In *attacks*, an unbounded, high epsilon FGSM attack is the strongest and the fastest one, as expected. However, when minimum perturbation is considered, FGSM loses out to DeepFool and C&W significantly. Across all datasets, DeepFool seems to be the best trade-off between perturbation magnitude and attack success. In *defenses*, we see varying degrees of success for each dataset. In CUB, we see spatial smoothing to the best for FGSM attacks, whereas TVM is the best for the rest. In AWA2, spatial smoothing is the best all-around defense for every attack setting. For SUN, spatial smoothing is still the best for FGSM, however TVM has a lead in C&W and DeepFool. Label smoothing is the worst defense all around and TVM is the most compute-heavy one, as expected. We present qualitative samples in Fig. 1.

Table 4. Results when *all* images are attacked and then defended with *label smoothing*. Parameter sets of the attacks are same as Table 1. *Original* results are results obtained by training ALE with label-smoothing.

| | Zero shot | | | Generalized zero shot | | | | | | | | |
| | C | S | A | C | | | S | | | A | | |
Attack	Top-1			u	s	h	u	s	h	u	s	h
Original	52.2	55.2	60.6	22.7	56.2	32.4	18.4	31.6	23.3	16.3	74.2	26.8
$FGSM_1$	39.8	46.8	41.1	17.4	43.7	24.9	15.7	25.2	19.4	12.1	59.8	20.1
$FGSM_2$	11.7	15.2	14.1	6.7	9.8	0.80	5.5	7.7	6.4	2.7	10.3	4.4
$DEFO_1$	29.8	30.4	49.6	10.0	20.3	13.4	6.2	8.8	7.3	14.01	42.4	21.1
$DEFO_3$	19.8	19.2	41.7	8.2	11.9	9.7	5.2	7.1	6.0	13.1	25.5	17.3
$CaWa_1$	38.8	45.6	46.6	19.4	40.4	26.3	16.2	26.6	20.1	16.4	61.0	25.9
$CaWa_3$	34.1	42.7	40.6	18.8	34.8	24.5	16.2	25.1	19.7	16.2	52.0	24.7

4.4   Analysis

It is clear that adversarial examples can be considered as out-of-distribution samples which we fail to recognize properly. As they do not have their own

Table 5. Results when *all* images are attacked and then defended with *total-variance minimization*. Parameter sets of the attacks are same as in Table 1.

Attack	Zero shot			Generalized zero shot								
	C	S	A	C			S			A		
	Top-1			u	s	h	u	s	h	u	s	h
Original	54.5	57.4	62.0	25.6	64.6	36.7	20.5	32.3	25.1	15.3	78.8	25.7
$FGSM_1$	49.1	53.2	53.8	23.0	57.5	32.8	18.9	28.3	22.7	11.7	71.8	20.1
$FGSM_2$	25.3	32.8	21.1	12.6	21.9	16.0	12.6	15.3	13.8	5.0	22.5	8.2
$DEFO_1$	48.4	50.3	59.0	19.7	52.3	28.6	15.2	20.8	17.5	12.5	70.9	21.4
$DEFO_3$	48.3	50.3	59.0	19.5	52.3	28.4	15.1	20.8	17.5	12.5	70.6	21.3
$CaWa_1$	50.9	53.3	58.8	24.0	60.3	34.3	20.0	29.2	23.8	12.7	75.6	21.7
$CaWa_3$	51.2	53.4	58.9	24.2	60.2	34.4	19.9	29.1	23.6	12.6	75.6	21.6

class prototypes, the learned ranking system incorrectly assigns them to a class. Effectively, we require a mechanism to transfer knowledge from clean to adversarial images, on top of the seen-to-unseen transfer we need to tackle already. Moreover, possibly from a simpler perspective, ZSL models can be considered as immature compared to supervised models; accuracy levels are not on the same level. The second perspective harbors interesting facts. Assuming a model with the perfect accuracy, we know attacks can only degrade the results, assuming they are effective. Defenses can still degrade the results without any attacks, but we know they alleviate the issues to a certain degree, assuming they are effective. What happens when the model is far from perfect is what we focus on now.

Class-Transitions: False/Correct. It is observed throughout the attacks that in *GZSL* setting, unseen accuracies are less effected compared to seen accuracies. We further investigate this by looking at the class-transitions during each attack setting. For each class, we calculate the ratio of class transitions; out of all (originally) correctly predicted samples, what percentage have transitioned to false? Out of all (originally) falsely predicted samples, what percentage have transitioned to correct or *other* false classes? Our results are shown in Table 6.

Stronger the attack, higher *correct-to-false* (CF) percentages we observe. Moreover, stronger attacks also introduce higher *false-to-other-false* (FF) ratios. This means regardless of the success of original predictions, stronger attacks induce more class transitions. Statistically, there is the possibility of an attack *correcting* an originally incorrect prediction. We observe the highest FC ratios in C&W attacks and the lowest in DeepFool attacks. Coupled with the lowest CF ratios, this can explain why C&W performed the worst in our attack scenarios.

When we compare seen and unseen classes, we see higher FC ratios for seen classes. Moreover, seen classes have smaller CF ratios which means seen classes are less effected detrimentally and more effected positively. This contradicts with our starting point; unseen classes being less effected by an attack than seen classes. However, unseen classes have a lot of initially zero accuracy classes and

Table 6. Categorization of prediction changes induced by each attack. U and S columns are results for unseen and seen classes, respectively. CF, FC and FF are *correct-to-false* (as the percentage of all originally correct predictions), *false-to-correct* and *false-to-other-false* (as the percentage of all originally incorrect predictions) changes in %, represented as per-class normalized ratio averages. Classes having no originally correct or incorrect predictions have not been included in the calculation.

	Generalized zero shot																	
	C						S						A					
Class type	U			S			U			S			U			S		
Transitions	CF	FC	FF	CF	FC	FF	CF	FC	FF	CF	FC	FF	CF	FC	FF	CF	FC	FF
$FGSM_1$	81	20	69	55	40	47	65	11	65	54	16	59	89	12	69	39	49	43
$FGSM_2$	99	18	82	99	36	64	99	12	87	98	16	84	100	5	93	94	33	67
$DEFO_1$	91	10	82	83	21	70	93	8	89	95	10	87	53	7	61	55	26	44
$DEFO_3$	94	11	84	93	22	72	94	8	90	98	10	88	60	7	72	74	28	49
$CaWa_1$	92	24	69	78	39	51	63	12	61	55	15	57	81	16	58	43	53	36
$CaWa_3$	95	24	70	86	40	53	67	13	66	61	10	90	89	18	64	60	57	38

they are not taken into account in our calculation. This leads to fewer number of classes with higher than 0 accuracy and fewer correctly predicted samples for each (as unseen accuracies are low all around). Once these samples are effected, we observe higher FC rates. However, high FC rates in seen classes tells us that in an event of misclassification, the algorithm predicts the correct class with a high probability, but not high enough to be the highest prediction. This is an interesting effect of the adversarial attacks, which means softmax probabilities are quite close to each other, and class boundary transition is easier. Conversely, one can expect high CF ratios for seen classes, but this is not the case. This means that the model, for seen classes, is robust against attacks when it comes to correct predictions, but its false predictions are not really confident.

Class-Transitions: Seen/Unseen. We also analyze the effect of attacks from a seen/unseen class perspective. For each class, we calculate the following for each class and average it for seen and unseen classes; out of *all* changed samples, what percent went to a seen or an unseen class? Our results are shown in Table 7. Results show that except FGSM, attack behaviour in terms of seen/unseen class transition seems to be stable. For FGSM, however, we see increase in unseen-to-seen transitions, which is in line with the further decrease of accuracy values (i.e. unseen-to-unseen can have false to correct transitions for unseen). This behaviour bodes well with the attack settings; FGSM scales its attack with the ϵ coefficient whereas DeepFool and C&W simply have more time to *solve* for the minimum perturbation with higher iterations. Regardless of the dataset and the attack setting, the majority of the transitions happen towards seen classes. This is likely due to the fact that the model trains exclusively on seen classes and naturally is more confident about its predictions, and this can cause a bias towards seen classes in the class boundary transitions.

Table 7. Attack-induced, per-class normalized class transition averages (in %) for different attack settings. UU, US, SU and SS are unseen-to-unseen, unseen-to-seen, seen-to-unseen and seen-to-seen transitions, respectively.

	Generalized zero shot											
	C				S				A			
Class transition	UU	US	SU	SS	UU	US	SU	SS	UU	US	SU	SS
$FGSM_1$	30	70	16	84	22	78	10	90	17	83	7	93
$FGSM_2$	28	72	18	82	17	83	10	90	12	88	7	93
$DEFO_1$	24	76	20	80	16	84	10	90	13	87	7	93
$DEFO_3$	24	76	20	80	16	84	10	90	14	86	7	93
$CaWa_1$	31	69	17	83	22	78	10	90	24	76	8	92
$CaWa_3$	31	69	17	83	22	78	10	90	25	75	9	91

Adverse Effects of Defenses. As observed in Sect. 4.3, there have been cases where defenses actually reduced the accuracy after the attacks rather than recovering it. Following the work we've presented in Table 6, we observe the effect of defenses (i.e. we add another layer to CF, FC, FF transitions, such as CFC, FCF, FFC, etc.)[5]. Logically, we can analyse the effect of defenses in four main categories; it corrects a mistake (CFC, FFC), it preserves the results (CCC, FFF) and it has detrimental effects (CCF, FCF) and it fails to recover (CFF, FCC). It must be noted that recovery here means recovering the *original* label, not necessarily the correct one. Across all experiments we observe every category of effect up to some degree, with correct-recoveries (CFC) spearheading the overall recovery of accuracy values. However, we observe that the defense-induced reduction of accuracies strongly correlate with high FCF ratios. This means alleviating the *positive* effects of attacks. Although the defense does its job by recovering the original predictions, the reductions occur nonetheless.

Attacking Only Correct Predictions. We investigate attacking only the originally correct predictions and only defending them. This is not under the threat model we assumed in the beginning, but it is valuable to decouple weak model effects (i.e. low accuracy) with potential ZSL-specific effects in our results.

We observe that the *unintuitive* effects such as attacks *improving* or defenses *degrading* the results eliminated. Across all attacks, we see more dramatic accuracy reductions and we see improvements across all defenses. The overall *rankings* for best attack and defense follow our previous *all images* attack settings. In this setting, results are more reminiscent of a supervised model, however due to the extreme bias between seen and unseen classes, GZSL-specific effects still remain (i.e. unseen and seen classes being effected differently, class transition trends).

Dataset Characteristics. The datasets considered are inherently different; SUN and CUB have fewer samples per class and consist of high number of

---

[5] We do not include a table for this analysis due to space restrictions.

classes, whereas AWA2 has high number of samples per class but consist of fewer classes. In AWA2, we see attacks failing to effect in their weakest setting; especially DeepFool and C&W performing their worst among other datasets. This is likely related to the sample count of AWA2; a larger distribution per class helps robustness, as suggested in [48]. We see FC transitions happening more frequently compared to other datasets; this is likely an effect of multiple confident predictions as this effect is more pronounced in seen classes. Upward accuracy spikes that occur in FGSM attacks (this analysis is performed with a wide range of parameters for FGSM and not included in detail due to page limitations) are more frequent here as well (especially in ZSL setting); this is likely an effect of having fewer number of classes as misclassifications are statistically more likely to fall into the originally correct classes. We see similar trends for SUN and CUB in general; SUN has the fewest transitions to unseen classes. This correlates strongly with the really high number of classes in SUN. In overall, we see SUN and CUB get better returns from all defenses, compared to AWA2.

Fig. 1. Example images from AWA2 dataset. The first and second rows show (in pairs) original and perturbed images, where attacks have induced misclassification and converted mispredictions into correct predictions, respectively. Third and fourth rows show (in triples) original, attacked and defended images where defenses have corrected and induced misclassifications, respectively. We use a powerful attack to show a more visible perturbation. I and C indicate incorrect and correct predictions.

5   Conclusion and Future Work

Despite their stunning success, it is shown that machine learning models can be fooled with carefully crafted perturbations. Adversarial robustness have generally been studied from a fully supervised perspective. ZSL and GZSL algorithms that lack supervision for a set of classes have not received attention for their adversarial robustness. In this paper, we introduce a study aiming to fill this gap by assessing a well-known ZSL model for its adversarial robustness, both ZSL and GZSL evaluation set-ups. We subject the model to several attacks and defenses across widely-used ZSL datasets. Our results indicate that adversarial robustness for ZSL has its own challenges, such as the extreme data bias and the comparably immature state of the field (compared to supervised learning). We highlight and analyse several points, especially in GZSL settings, to guide future researchers as to what needs attention in making ZSL models robust and also what points could be important for interpreting the results.

References

1. Akata, Z., Perronnin, F., Harchaoui, Z., Schmid, C.: Label-embedding for attribute-based classification. In: Proceedings of the IEEE Conference on Computer Vision and Pattern Recognition, pp. 819–826 (2013)
2. Akata, Z., Reed, S., Walter, D., Lee, H., Schiele, B.: Evaluation of output embeddings for fine-grained image classification. In: The IEEE Conference on Computer Vision and Pattern Recognition (CVPR), June 2015
3. Akhtar, N., Mian, A.: Threat of adversarial attacks on deep learning in computer vision: a survey. IEEE Access **6**, 14410–14430 (2018)
4. Arnab, A., Miksik, O., Torr, P.H.: On the robustness of semantic segmentation models to adversarial attacks. In: Proceedings of the IEEE Conference on Computer Vision and Pattern Recognition, pp. 888–897 (2018)
5. Behzadan, V., Munir, A.: Vulnerability of deep reinforcement learning to policy induction attacks. In: Perner, P. (ed.) MLDM 2017. LNCS (LNAI), vol. 10358, pp. 262–275. Springer, Cham (2017). https://doi.org/10.1007/978-3-319-62416-7_19
6. Bhattad, A., Chong, M.J., Liang, K., Li, B., Forsyth, D.: Unrestricted adversarial examples via semantic manipulation. In: International Conference on Learning Representations (2020)
7. Borkar, T., Heide, F., Karam, L.: Defending against universal attacks through selective feature regeneration. In: Proceedings of the IEEE/CVF Conference on Computer Vision and Pattern Recognition (CVPR), June 2020
8. Bruna, J., et al.: Intriguing properties of neural networks (2013)
9. Bucher, M., Herbin, S., Jurie, F.: Generating visual representations for zero-shot classification. In: Proceedings of the IEEE International Conference on Computer Vision Workshops, pp. 2666–2673 (2017)
10. Cao, Y., et al.: Adversarial objects against lidar-based autonomous driving systems. arXiv preprint. arXiv:1907.05418 (2019)
11. Carlini, N., Wagner, D.: Towards evaluating the robustness of neural networks. In: 2017 IEEE Symposium on Security and Privacy (SP), pp. 39–57. IEEE (2017)
12. Carlini, N., Wagner, D.: Audio adversarial examples: targeted attacks on speech-to-text. In: 2018 IEEE Security and Privacy Workshops (SPW), pp. 1–7. IEEE (2018)
13. Changpinyo, S., Chao, W.L., Gong, B., Sha, F.: Synthesized classifiers for zero-shot learning. In: The IEEE Conference on Computer Vision and Pattern Recognition (CVPR), June 2016
14. Chen, L., Zhang, H., Xiao, J., Liu, W., Chang, S.F.: Zero-shot visual recognition using semantics-preserving adversarial embedding networks. In: The IEEE Conference on Computer Vision and Pattern Recognition (CVPR), June 2018

15. Chen, X., et al.: One-shot adversarial attacks on visual tracking with dual attention. In: Proceedings of the IEEE/CVF Conference on Computer Vision and Pattern Recognition, pp. 10176–10185 (2020)
16. Ebrahimi, J., Rao, A., Lowd, D., Dou, D.: HotFlip: white-box adversarial examples for text classification. In: Proceedings of the 56th Annual Meeting of the Association for Computational Linguistics (Volume 2: Short Papers), Melbourne, Australia, pp. 31–36. Association for Computational Linguistics, July 2018. https://doi.org/10.18653/v1/P18-2006, https://www.aclweb.org/anthology/P18-2006
17. Felix, R., Kumar, V.B., Reid, I., Carneiro, G.: Multi-modal cycle-consistent generalized zero-shot learning. In: Proceedings of the European Conference on Computer Vision (ECCV), pp. 21–37 (2018)
18. Frome, A., et al.: Devise: a deep visual-semantic embedding model. In: Burges, C.J.C., Bottou, L., Welling, M., Ghahramani, Z., Weinberger, K.Q. (eds.) Advances in Neural Information Processing Systems, vol. 26, pp. 2121–2129. Curran Associates, Inc. (2013). http://papers.nips.cc/paper/5204-devise-a-deep-visual-semantic-embedding-model.pdf
19. Goibert, M., Dohmatob, E.: Adversarial Robustness via Label-Smoothing, January 2020. https://hal.archives-ouvertes.fr/hal-02437752. Working paper or preprint
20. Goodfellow, I.J., Shlens, J., Szegedy, C.: Explaining and harnessing adversarial examples. arXiv preprint arXiv:1412.6572 (2014)
21. Guo, C., Rana, M., Cisse, M., van der Maaten, L.: Countering adversarial images using input transformations. In: International Conference on Learning Representations (2018). https://openreview.net/forum?id=SyJ7ClWCb
22. Hazan, T., Papandreou, G., Tarlow, D.: Perturbations, Optimization, and Statistics. MIT Press, Cambridge (2016)
23. He, K., Zhang, X., Ren, S., Sun, J.: Deep residual learning for image recognition. In: The IEEE Conference on Computer Vision and Pattern Recognition (CVPR), June 2016
24. Hu, W., Tan, Y.: Black-box attacks against RNN based malware detection algorithms. In: Workshops at the Thirty-Second AAAI Conference on Artificial Intelligence (2018)
25. Jia, Y., et al.: Fooling detection alone is not enough: adversarial attack against multiple object tracking. In: International Conference on Learning Representations (2019)
26. Jiang, H., Wang, R., Shan, S., Chen, X.: Transferable contrastive network for generalized zero-shot learning. In: Proceedings of the IEEE International Conference on Computer Vision, pp. 9765–9774 (2019)
27. Jing, L., Tian, Y.: Self-supervised visual feature learning with deep neural networks: a survey. IEEE Trans. Pattern Anal. Mach. Intell. 1 (2020)
28. Kumar Verma, V., Arora, G., Mishra, A., Rai, P.: Generalized zero-shot learning via synthesized examples. In: Proceedings of the IEEE Conference on Computer Vision and Pattern Recognition, pp. 4281–4289 (2018)
29. Kurakin, A., Goodfellow, I., Bengio, S.: Adversarial examples in the physical world. arXiv preprint. arXiv:1607.02533 (2016)
30. Lampert, C.H., Nickisch, H., Harmeling, S.: Learning to detect unseen object classes by between-class attribute transfer. In: 2009 IEEE Conference on Computer Vision and Pattern Recognition, pp. 951–958 (2009)
31. Lampert, C.H., Nickisch, H., Harmeling, S.: Attribute-based classification for zero-shot visual object categorization. IEEE Trans. Pattern Anal. Mach. Intell. **36**(3), 453–465 (2014)

32. Li, Y., Zhang, J., Zhang, J., Huang, K.: Discriminative learning of latent features for zero-shot recognition. In: The IEEE Conference on Computer Vision and Pattern Recognition (CVPR), June 2018
33. Liu, Y., Chen, X., Liu, C., Song, D.: Delving into transferable adversarial examples and black-box attacks. CoRR abs/1611.02770 (2016). http://arxiv.org/abs/1611.02770
34. Lu, J., Issaranon, T., Forsyth, D.: SafetyNet: detecting and rejecting adversarial examples robustly. In: Proceedings of the IEEE International Conference on Computer Vision, pp. 446–454 (2017)
35. Mao, C., Zhong, Z., Yang, J., Vondrick, C., Ray, B.: Metric learning for adversarial robustness. In: Advances in Neural Information Processing Systems, pp. 478–489 (2019)
36. Meng, D., Chen, H.: Magnet: a two-pronged defense against adversarial examples. In: Proceedings of the 2017 ACM SIGSAC Conference on Computer and Communications Security, pp. 135–147 (2017)
37. Moosavi-Dezfooli, S.M., Fawzi, A., Fawzi, O., Frossard, P.: Universal adversarial perturbations. In: Proceedings of the IEEE Conference on Computer Vision and Pattern Recognition, pp. 1765–1773 (2017)
38. Moosavi-Dezfooli, S.M., Fawzi, A., Frossard, P.: DeepFool: a simple and accurate method to fool deep neural networks. In: Proceedings of the IEEE Conference on Computer Vision and Pattern Recognition, pp. 2574–2582 (2016)
39. Naseer, M., Khan, S., Hayat, M., Khan, F.S., Porikli, F.: A self-supervised approach for adversarial robustness. In: Proceedings of the IEEE/CVF Conference on Computer Vision and Pattern Recognition, pp. 262–271 (2020)
40. Papernot, N., McDaniel, P., Goodfellow, I., Jha, S., Celik, Z.B., Swami, A.: Practical black-box attacks against machine learning. In: Proceedings of the 2017 ACM on Asia Conference on Computer and Communications Security, pp. 506–519 (2017)
41. Papernot, N., McDaniel, P., Jha, S., Fredrikson, M., Celik, Z.B., Swami, A.: The limitations of deep learning in adversarial settings. In: 2016 IEEE European Symposium on Security and Privacy (EuroS&P). pp. 372–387. IEEE (2016)
42. Papernot, N., McDaniel, P., Wu, X., Jha, S., Swami, A.: Distillation as a defense to adversarial perturbations against deep neural networks. In: 2016 IEEE Symposium on Security and Privacy (SP), pp. 582–597. IEEE (2016)
43. Paszke, A., et al.: PyTorch: an imperative style, high-performance deep learning library. In: Advances in Neural Information Processing Systems, pp. 8026–8037 (2019)
44. Patterson, G., Xu, C., Su, H., Hays, J.: The SUN attribute database: beyond categories for deeper scene understanding. Int. J. Comput. Vis. **108**(1), 59–81 (2014). https://doi.org/10.1007/s11263-013-0695-z
45. Ross, A.S., Doshi-Velez, F.: Improving the adversarial robustness and interpretability of deep neural networks by regularizing their input gradients. In: Thirty-Second AAAI Conference on Artificial Intelligence (2018)
46. Samangouei, P., Kabkab, M., Chellappa, R.: Defense-GAN: protecting classifiers against adversarial attacks using generative models. In: International Conference on Learning Representations (2018). https://openreview.net/forum?id=BkJ3ibb0-
47. Sariyildiz, M.B., Cinbis, R.G.: Gradient matching generative networks for zero-shot learning. In: Proceedings of the IEEE Conference on Computer Vision and Pattern Recognition, pp. 2168–2178 (2019)
48. Schmidt, L., Santurkar, S., Tsipras, D., Talwar, K., Madry, A.: Adversarially robust generalization requires more data. In: Advances in Neural Information Processing Systems, pp. 5014–5026 (2018)

49. Shaham, U., et al.: Defending against adversarial images using basis functions transformations. arXiv preprint. arXiv:1803.10840 (2018)
50. Sharma, V., Vaibhav, A., Chaudhary, S., Patel, L., Morency, L.: Attend and attack: attention guided adversarial attacks on visual question answering models (2018)
51. Socher, R., Ganjoo, M., Manning, C.D., Ng, A.: Zero-shot learning through cross-modal transfer. In: Burges, C.J.C., Bottou, L., Welling, M., Ghahramani, Z., Weinberger, K.Q. (eds.) Advances in Neural Information Processing Systems, vol. 26, pp. 935–943. Curran Associates, Inc. (2013). http://papers.nips.cc/paper/5027-zero-shot-learning-through-cross-modal-transfer.pdf
52. Su, J., Vargas, D.V., Sakurai, K.: One pixel attack for fooling deep neural networks. IEEE Trans. Evol. Comput. **23**(5), 828–841 (2019)
53. Wah, C., Branson, S., Welinder, P., Perona, P., Belongie, S.: The Caltech-UCSD birds-200-2011 dataset. Technical report (2011)
54. Wang, W., Zheng, V.W., Yu, H., Miao, C.: A survey of zero-shot learning: Settings, methods, and applications. ACM Trans. Intell. Syst. Technol. **10**(2) (2019). https://doi.org/10.1145/3293318
55. Wang, X., Ye, Y., Gupta, A.: Zero-shot recognition via semantic embeddings and knowledge graphs. In: The IEEE Conference on Computer Vision and Pattern Recognition (CVPR), June 2018
56. Weston, J., Bengio, S., Usunier, N.: Large scale image annotation: learning to rank with joint word-image embeddings. Mach. Learn. **81**(1), 21–35 (2010). https://doi.org/10.1007/s10994-010-5198-3
57. Xian, Y., Lampert, C.H., Schiele, B., Akata, Z.: Zero-shot learning–a comprehensive evaluation of the good, the bad and the ugly. IEEE Trans. Pattern Anal. Mach. Intell. **41**(9), 2251–2265 (2019)
58. Xian, Y., Akata, Z., Sharma, G., Nguyen, Q., Hein, M., Schiele, B.: Latent embeddings for zero-shot classification. In: The IEEE Conference on Computer Vision and Pattern Recognition (CVPR), June 2016
59. Xiao, C., Zhu, J.Y., Li, B., He, W., Liu, M., Song, D.: Spatially transformed adversarial examples. In: International Conference on Learning Representations (2018). https://openreview.net/forum?id=HyydRMZC-
60. Xie, C., Wang, J., Zhang, Z., Zhou, Y., Xie, L., Yuille, A.: Adversarial examples for semantic segmentation and object detection. In: Proceedings of the IEEE International Conference on Computer Vision, pp. 1369–1378 (2017)
61. Xu, H., et al.: Adversarial attacks and defenses in images, graphs and text: a review. ArXiv abs/1909.08072 (2019)
62. Xu, W., Evans, D., Qi, Y.: Feature squeezing: detecting adversarial examples in deep neural networks. CoRR abs/1704.01155 (2017). http://arxiv.org/abs/1704.01155
63. Xu, X., Shen, F., Yang, Y., Shao, J., Huang, Z.: Transductive visual-semantic embedding for zero-shot learning. In: Proceedings of the 2017 ACM on International Conference on Multimedia Retrieval, pp. 41–49. ICMR 2017. Association for Computing Machinery, New York (2017). https://doi.org/10.1145/3078971.3078977
64. Ye, M., Guo, Y.: Zero-shot classification with discriminative semantic representation learning. In: The IEEE Conference on Computer Vision and Pattern Recognition (CVPR), July 2017
65. Yu, Y., Ji, Z., Fu, Y., Guo, J., Pang, Y., Zhang, Z.M.: Stacked semantics-guided attention model for fine-grained zero-shot learning. In: Bengio, S., Wallach, H., Larochelle, H., Grauman, K., Cesa-Bianchi, N., Garnett, R. (eds.) Advances in Neural Information Processing Systems, vol. 31, pp. 5995–6004 (2018)

66. Yuan, X., He, P., Zhu, Q., Li, X.: Adversarial examples: attacks and defenses for deep learning. IEEE Trans. Neural Netw. Learn. Syst. **30**(9), 2805–2824 (2019)
67. Zhang, H., Long, Y., Guan, Y., Shao, L.: Triple verification network for generalized zero-shot learning. IEEE Trans. Image Process. **28**(1), 506–517 (2018)
68. Zhang, L., Xiang, T., Gong, S.: Learning a deep embedding model for zero-shot learning. In: The IEEE Conference on Computer Vision and Pattern Recognition (CVPR), July 2017
69. Zhang, X., Gui, S., Zhu, Z., Zhao, Y., Liu, J.: ATZSL: defensive zero-shot recognition in the presence of adversaries. ArXiv abs/1910.10994 (2019)
70. Zhang, Z., Saligrama, V.: Zero-shot learning via semantic similarity embedding. In: The IEEE International Conference on Computer Vision (ICCV), December 2015
71. Zügner, D., Günnemann, S.: Adversarial attacks on graph neural networks via meta learning. arXiv preprint arXiv:1902.08412 (2019)

Towards Analyzing Semantic Robustness of Deep Neural Networks

Abdullah Hamdi$^{(\boxtimes)}$ and Bernard Ghanem

King Abdullah University of Science and Technology (KAUST),
Thuwal, Saudi Arabia
{abdullah.hamdi,bernard.ghanem}@kaust.edu.sa

Abstract. Despite the impressive performance of Deep Neural Networks (DNNs) on various vision tasks, they still exhibit erroneous high sensitivity toward semantic primitives (*e.g.* object pose). We propose a theoretically grounded analysis for DNN robustness in the semantic space. We qualitatively analyze different DNNs' semantic robustness by visualizing the DNN global behavior as semantic maps and observe interesting behavior of some DNNs. Since generating these semantic maps does not scale well with the dimensionality of the semantic space, we develop a bottom-up approach to detect robust regions of DNNs. To achieve this, we formalize the problem of finding robust semantic regions of the network as optimizing integral bounds and we develop expressions for update directions of the region bounds. We use our developed formulations to quantitatively evaluate the semantic robustness of different popular network architectures. We show through extensive experimentation that several networks, while trained on the same dataset and enjoying comparable accuracy, do not necessarily perform similarly in semantic robustness. For example, InceptionV3 is more accurate despite being less semantically robust than ResNet50. We hope that this tool will serve as a milestone towards understanding the semantic robustness of DNNs.

1 Introduction

As a result of recent advances in machine learning and computer vision, deep neural networks (DNNS) have become an essential part of our lives. DNNs are used to suggest articles to read, detect people in surveillance cameras, automate big machines in factories, and even diagnose X-rays for patients in hospitals. So, what is the catch here? These DNNs struggle with a detrimental weakness on specific naive scenarios, despite having strong on-average performance. Figure 1 shows how a small perturbation in the view angle of the teapot object results in a drop in InceptionV3 [36] confidence score from 100% to almost 0%. The *softmax* confidence scores are plotted against one semantic parameter (*i.e.* the azimuth

---

The code is available at https://github.com/ajhamdi/semantic-robustness.

Electronic supplementary material The online version of this chapter (https://doi.org/10.1007/978-3-030-66415-2_2) contains supplementary material, which is available to authorized users.

© Springer Nature Switzerland AG 2020
A. Bartoli and A. Fusiello (Eds.): ECCV 2020 Workshops, LNCS 12535, pp. 22–38, 2020.
https://doi.org/10.1007/978-3-030-66415-2_2

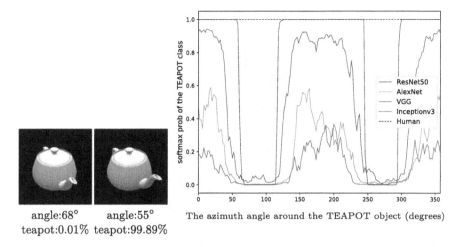

angle:68°   angle:55°
teapot:0.01% teapot:99.89%

Fig. 1. Semantic robustness of deep networks. Trained neural networks can perform poorly when subject to small perturbations in the semantics of the image. (*left*): We show how perturbing the azimuth viewing angle of a simple *teapot* object can dramatically affect the score of a pretrained InceptionV3 [36] for the *teapot* class. (*right*): We plot the softmax confidence scores of different DNNs on the same *teapot* object viewed from 360° around the object. For comparison, lab researchers identified the object from all angles.

angle around the teapot) and it fails in such a simple task. Similar behaviors are consistently observed across different DNNs (trained on ImageNet [31]) as noted by other concurrent works [1].

Furthermore, because DNNs are not easily interpretable, they work well without a complete understanding of *why* they behave in such a manner. A whole research direction is dedicated to studying and analyzing DNNs. Examples of such analysis include activation visualization [8,25,40], noise injection [3,11,26], and studying the effect of image manipulation on DNNs [12,13,13]. By leveraging a differentiable renderer **R** and evaluating rendered images for different semantic parameters **u**, we provide a new lens of semantic robustness analysis for such DNNs as illustrated in Fig. 2. These Network Semantic Maps (NSM) demonstrate unexpected behavior of some DNNs, in which adversarial regions lie inside a very confident region of the semantic space. This constitutes a "trap" that is hard to detect without such analysis and can lead to catastrophic failure for the DNN.

Recent work in adversarial network attacks explores DNN sensitivity and performs gradient updates to derive targeted perturbations [5,14,23,38]. In practice, such attacks are less likely to naturally occur than semantic attacks, such as changes in camera viewpoint and lighting conditions. The literature on semantic attacks is sparser, since they are more subtle and challenging to analyze [1,16,24,41]. This is due to the fact that we are unable to distinguish between failure cases that result from the network structure, and learning, or from the data bias [39]. Current methods for adversarial semantic attacks either work on individual examples [1], or try to find distributions but rely on sampling methods, which do not scale with dimensionality [16]. We present a novel approach to find

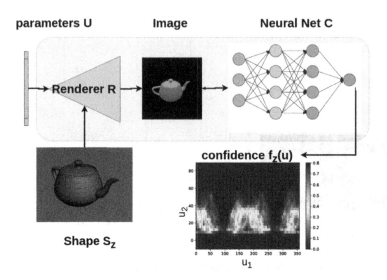

Fig. 2. Analysis pipeline: we leverage neural mesh renderer **R** [20] to render shape S_z of class z according to semantic scene parameters **u**. The resulting image is passed to trained network **C** that is able to identify the class z. The behaviour of the softmax score at label z (dubbed $f_z(\mathbf{u})$) is analyzed for different parameters **u** and for the specific shape S_z. In our experiments, we pick u_1 and u_2 to be the azimuth and elevation angles, respectively.

robust/adversarial regions in the n-dimensional semantic space. The proposed method scales better than sampling-based methods [16]. We use this method to quantify semantic robustness of popular DNNs on a collected dataset.

Contributions. (1) We analyze popular deep networks from a semantic lens showing unexpected behavior in the 1D and 2D semantic space. **(2)** We develop a novel bottom-up approach to detect robust/adversarial regions in the semantic space of a DNN, which scales well with increasing dimensionality. The method specifically optimizes for the region's bounds in semantic space (around a point of interest), such that the continuous region offers semantic parameters that confuse the network. **(3)** We develop a new metric to quantify semantic robustness of DNNs that we dub Semantic Robustness Volume Ratio (SRVR), and we use it to benchmark popular DNNs on a collected dataset.

2   Related Work

2.1   Understanding Deep Neural Networks

There are different lenses to analyze DNNs depending on the purpose of analysis. A popular line of work tries to visualize the network hidden layers by inverting the activations to get a visual image that represents a specific activation [8,25,40]. Others observe the behavior of these networks under injected noise [2–4,11,15,37]. Geirhos *et al.* show that changing the texture of the object while keeping the borders can hugely deteriorate the recognizability of the object by the DNN [13]. More closely related to our work is the work of Fawzi et al., which shows that geometric changes in the image greatly affect the performance of the classifier [12]. The work of Engstrom et al. [10] studies the robustness of a network under natural 2D transformations (i.e. translation and planar rotation).

2.2 Adversarial Attacks on Deep Neural Networks

Pixel-Based Adversarial Attacks. The way that DNNs fail for some noise added to the image motivated the adversarial attacks literature. Several works formulate attacking neural networks as an optimization on the input pixels [5,14, 23,27,38]. However, all these methods are limited to pixel perturbations and only fool classifiers, while we consider more general cases of attacks, e.g. changes in camera viewpoint to fool a DNN by finding adversarial regions. Most attacks are white-box attacks, in which the attack algorithm has access to network gradients. Another direction of adversarial attacks treats the classifier as a black-box, where the adversary can only probe the network and get a score from the classifier without backpropagating through the DNN [7,28]. We formulate the problem of finding the robust/adversarial region as an optimization of the corners of a hyper-rectangle in the semantic space for both black-box and white-box attacks.

Semantic Adversarial Attacks. Other works tried to move away from pixel perturbation to semantic 3D scene parameters and 3D attacks [1,16,17,24,41]. Zeng et al. [41] generate attacks on deep classifiers by perturbing scene parameters like lighting and surface normals. Hamdi et al. propose generic adversarial attacks that incorporate semantic and pixel attacks, in which an adversary is sampled from some latent distribution that is produced from a GAN trained on example semantic adversaries [16]. However, their work used a sampling-based approach to learn these adversarial regions, which does not scale with the dimensionality of the problem. Another recent work by Alcorn et al. [1] tries to fool trained DNNs by changing the pose of the object. They used the Neural Mesh renderer (NMR) by Kato et al. [21] to allow for a fully differentiable pipeline that performs adversarial attacks based on the gradients to the parameters. Our work differs in that we use NMR to obtain gradients to the parameters $\mathbf{u}$ not to attack the model, but to detect and quantify the robustness of different networks as shown in Sect. 4.4. Furthermore, Dreossi et al. [9] used adversarial training in the semantic space for self-driving, whereas Liu et al. [24] proposed a differentiable renderer to perform parametric attacks and the *parametric-ball* as an evaluation metric for physical attacks. The work by Shu et al. [33] used an RL agent and a Bayesian optimizer to asses the DNNs behaviour under good/bad physical parameters for the network. While we share similar insights as [33], we try to study the global behaviour of DNNs as collections of regions, whereas [33] tries to find individual points that pose difficulty for the DNN.

Network Confidence under 1 Parameter   Network Confidence under 2 Parameters

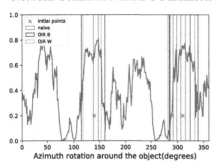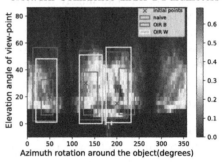

Fig. 3. Semantic robust region finding: we find robust regions of semantic parameters for Rsnet50 [18] and for a *bathtub* object by the three bottom-up formulations (naive , OIR_W , and OIR_B). (*left*): Semantic space is 1D (azimuth angle of camera) with three initial points. (*right*): Semantic space is 2D (azimuth angle and elevation angle of camera) with four initial points. We note that the naive approach usually predicts smaller regions, while the OIR formulations find more comprehensive regions.

2.3 Optimizing Integral Bounds

Naive Approach. To develop an algorithm for robust region finding, we adopt an idea from weakly supervised activity detection in videos by Shou *et al.* [32]. The idea is to find bounds that maximize the inner average of a continuous function while minimizing the outer average in a region. This is achieved because optimizing the bounds to exclusively maximize the area can lead to diverging bounds of $\{-\infty, \infty\}$. To solve the issue of diverging bounds, the following naive formulation is to simply regularize the loss by adding a penalty on the region size. The expressions for the loss of $n = 1$ dimension is: $L = -\text{Area}_{\text{in}} + \frac{\lambda}{2} |b - a|_2^2 = \int_a^b f(u)du + \frac{\lambda}{2} |b - a|_2^2$, where $f : \mathbb{R}^1 \rightarrow (0, 1)$ is the function of interest and (a, b) are the left and right bounds respectively and λ is a hyperparameter. The update directions to minimize the loss are: $\frac{\partial L}{\partial a} = f(a) - \lambda(b - a)$; $\frac{\partial L}{\partial b} = -f(b) + \lambda(b - a)$. The regularizer will prevent the region from growing to ∞ and the best bounds will be found if the loss is minimized with gradient descent or any similar approach.

Trapezoidal Approximation. To extend the naive approach to n-dimensions, we face more integrals in the update directions (hard to compute). Therefore, we deploy the following first-order trapezoid approximation of definite integrals. The Newton-Cortes formula for numerical integration [35] states that: $\int_a^b f(u)du \approx (b - a)\frac{f(a)+f(b)}{2}$. An asymptotic error estimate is given by $-\frac{(b-a)^2}{48} [f'(b) - f'(a)] + \mathcal{O}\left(\frac{1}{8}\right)$. So, as long as the derivatives are bounded by some Lipschitz constant $\mathbb{L}$, then the error becomes bounded such that $|\text{error}| \le \mathbb{L}(b - a)^2$.

| AlexNet[22] | VGG[34] | Resnet50[18] | InceptionV3[36] |

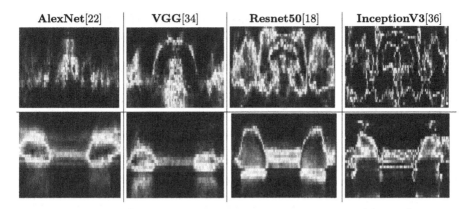

Fig. 4. Network semantic maps: We plot the 2D semantic maps (as in Fig. 3 *right*) of four different networks on two shapes of a *chair* class (*top*) and *cup* class (*bottom*). InceptionV3 is very confident about its decision, but at the cost of creating semantic "traps", where sharp performance degradation happens in the middle of a robust region. This behaviour is more apparent for complex shapes (*e.g.* the chair in *top* row).

3 Methodology

Typical adversarial pixel attacks involve a neural network $\mathbf{C}$ (*e.g.* classifier or detector) that takes an image $\mathbf{x} \in [0,1]^d$ as input and outputs a multinoulli distribution over K class labels with softmax values $[l_1, l_2, ..., l_K]$, where l_j is the softmax value for class j. The adversary (attacker) tries to produce a perturbed image $\mathbf{x}' \in [0,1]^d$ that is as close as possible to $\mathbf{x}$, such that $\mathbf{x}$ to $\mathbf{x}'$ have different class predictions through $\mathbf{C}$. In this work, we consider a more general case where we are interested in parameters $\mathbf{u} \in \Omega \subset \mathbb{R}^n$, a latent parameter that generates the image via a scene generator (*e.g.* a renderer function $\mathbf{R}$). This generator/renderer takes the parameter $\mathbf{u}$ and an object shape $\mathbf{S}$ of a class that is identified by $\mathbf{C}$. Ω is the continuous semantic space for the parameters that we intend to study. The renderer creates the image $\mathbf{x} \in \mathbb{R}^d$, and then we study the behavior of a classifier $\mathbf{C}$ of that image across multiple shapes and multiple popular DNN architectures. Now, this function of interest is defined as follows:

$$f(\mathbf{u}) = \mathbf{C}_z(\mathbf{R}(\mathbf{S}_z, \mathbf{u})), \quad 0 \le f(\mathbf{u}) \le 1 \tag{1}$$

where z is a class label of interest to study and we observe the network score for that class by rendering a shape $\mathbf{S}_z$ of the same class. The shape and class labels are constants and only the parameters $\mathbf{u}$ vary for f during analysis.

3.1 Region Finding as an Operator

We can visualize the function in Eq. (1) for any shape $\mathbf{S}_z$ as long as the DNN can identify the shape at some region in the semantic space Ω of interest, as we show in Fig. 1. However, plotting these figures is expensive and the complexity of plotting them increases exponentially with a big base. The complexity of plotting these plots of semantic maps, which we call Network Semantic Maps (NSM), is N for $n = 1$, where N is the number of samples needed for that dimension to be fully characterized. The complexity is N^2 for $n = 2$, and we can see that for a general dimension n, the complexity of plotting the NMS to adequately fill the semantic space Ω is N^n. This number is intractable even if we have only moderate dimensionality. To tackle this issue, we use a bottom-up approach to detect regions around some initial parameters $\mathbf{u}_0$, instead of sampling in the entire space of parameters Ω. Explicitly, we define region finding as an operator Φ that takes the function of interest in Eq. (1), initial point in the semantic space $\mathbf{u}_0 \in \Omega$, and a shape $\mathbf{S}_z$ of some class z. The operator will return the hyper-rectangle $\mathbb{D} \subset \Omega$, where the DNN is robust in the region and does not sharply drop the score of the intended class. It also keeps identifying the shape with label z as illustrated in Fig. 4. The robust-region-finding operator is then defined as follows:

$$\Phi_{\text{robust}}(f(\mathbf{u}), \mathbf{S}_z, \mathbf{u}_0) = \mathbb{D} = \{\mathbf{u} : \mathbf{a} \leq \mathbf{u} \leq \mathbf{b}\}$$
$$\text{s.t. } \mathbb{E}_{\mathbf{u} \sim \mathbb{D}}[f(\mathbf{u})] \geq 1 - \epsilon_m , \quad \mathbf{u}_0 \in \mathbb{D} , \quad \text{VAR}[f(\mathbf{u})] \leq \epsilon_v \tag{2}$$

where the left and right bounds of $\mathbb{D}$ are $\mathbf{a} = [a_1, a_2, ..., a_n]$ and $\mathbf{b} = [b_1, b_2, ..., b_n]$, respectively. The two small thresholds (ϵ_m, ϵ_v) are needed to ensure high performance and low variance of the DNN in that robust region. We can define the complementary operator, which finds adversarial regions as:

$$\Phi_{\text{adv}}(f(\mathbf{u}), \mathbf{S}_z, \mathbf{u}_0) = \mathbb{D} = \{\mathbf{u} : \mathbf{a} \leq \mathbf{u} \leq \mathbf{b}\}$$
$$\text{s.t. } \mathbb{E}_{\mathbf{u} \sim \mathbb{D}}[f(\mathbf{u})] \leq \epsilon_m , \quad \mathbf{u}_0 \in \mathbb{D} , \quad \text{VAR}[f(\mathbf{u})] \geq \epsilon_v \tag{3}$$

We can clearly show that Φ_{adv} and Φ_{robust} are related:

$$\Phi_{\text{adv}}(f(\mathbf{u}), \mathbf{S}_z, \mathbf{u}_0) = \Phi_{\text{robust}}(1 - f(\mathbf{u}), \mathbf{S}_z, \mathbf{u}_0) \tag{4}$$

So, we can just focus our attention on Φ_{robust} to find robust regions, and the adversarial regions follow directly from Eq. (4). We need to ensure that $\mathbb{D}$ has a positive size: $\mathbf{r} = \mathbf{b} - \mathbf{a} > 0$. The volume of $\mathbb{D}$ normalized by the exponent of dimension n is expressed as follows:

$$\text{volume}(\mathbb{D}) = \triangle = \frac{1}{2^n} \prod_{i=1}^{n} \mathbf{r}_i \tag{5}$$

The region $\mathbb{D}$ can also be defined in terms of the matrix $\mathbf{D}$ of all the corner points $\{\mathbf{d}^i\}_{i=1}^{2^n}$ as follows:

$$\text{corners}(\mathbb{D}) = \mathbf{D}_{n \times 2^n} = \left[\mathbf{d}^1 | \mathbf{d}^2 | .. | \mathbf{d}^{2^n}\right] = \mathbf{1}^T \mathbf{a} + \mathbf{M}^T \odot (\mathbf{1}^T \mathbf{r})$$
$$\mathbf{M}_{n \times 2^n} = \left[\mathbf{m}^0 | \mathbf{m}^1 | .. | \mathbf{m}^{2^n - 1}\right], \text{ where } \mathbf{m}^i = \text{binary}_n(i) \tag{6}$$

and $\mathbf{1}$ is the all-ones vector of size 2^n, $\odot$ is the Hadamard (element-wise) product of matrices, and $\mathbf{M}$ is a constant masking matrix defined as the permutation matrix of binary numbers of n bits that range from 0 to $2n - 1$.

3.2   Deriving Update Directions

Extending Naive to n-dimensions. We start by defining the function vector $\mathbf{f}_{\mathbb{D}}$ of all function evaluations at all corner points of $\mathbb{D}$.

$$\mathbf{f}_{\mathbb{D}} = \left[f(\mathbf{d}^1), f(\mathbf{d}^2), ..., f(\mathbf{d}^{2^n}) \right]^T , \quad \mathbf{d}^i = \mathbf{D}_{:,i} \tag{7}$$

Then, using Trapezoid approximation and Leibniz rule of calculus, the loss expression and the update directions become as follows:

$$L(\mathbf{a}, \mathbf{b}) = - \int \cdots \int_{\mathbb{D}} f(u_1, \ldots, u_n)\, du_1 \ldots du_n + \frac{\lambda}{2} |\mathbf{r}|^2 \approx - \triangle \mathbf{1}^T \mathbf{f}_{\mathbb{D}} + \frac{\lambda}{2} |\mathbf{r}|^2$$

$$\nabla_{\mathbf{a}} L \approx 2 \triangle \mathrm{diag}^{-1}(\mathbf{r}) \overline{\mathbf{M}} \mathbf{f}_{\mathbb{D}} + \lambda \mathbf{r} \quad ; \quad \nabla_{\mathbf{b}} L \approx - 2 \triangle \mathrm{diag}^{-1}(\mathbf{r}) \mathbf{M} \mathbf{f}_{\mathbb{D}} - \lambda \mathbf{r} \tag{8}$$

We show all the derivations for $n \in \{1, 2\}$ and for general n in the **supplement**.

Outer-Inner Ratio Loss (OIR). We introduce an outer region (A, B) with that contains the small region (a, b). We follow the following assumption to ensure that the outer area is always positive: $A = a - \alpha \frac{b-a}{2}; B = b + \alpha \frac{b-a}{2}$. Here, α is the small boundary factor of the outer area to the inner. We formulate the problem as a ratio of outer over inner areas and we try to make this ratio $(L = \frac{\mathrm{Area_{out}}}{\mathrm{Area_{in}}})$ as close as possible to 0. We utilize the Dinkelbach technique for solving non-linear fractional programming problems [30] to transform L as follows.

$$L = \frac{\mathrm{Area_{out}}}{\mathrm{Area_{in}}} = \mathrm{Area_{out}} - \lambda\, \mathrm{Area_{in}}$$

$$= \int_A^B f(a)du - \int_a^b f(a)du - \lambda \int_a^b f(a)du \tag{9}$$

where $\lambda^* = \frac{\mathrm{Area_{out}^*}}{\mathrm{Area_{in}^*}}$ is the Dinkelbach factor that is the best objective ratio.

Black-Box (OIR_B). Here we set $\lambda = 1$ to simplify the problem. This yields the following expression of the loss $L = \mathrm{Area_{out}} - \mathrm{Area_{in}} = \int_A^B f(u)du - 2 \int_a^b f(u)du$, which is similar to the area contrastive loss in [32]. The update rules would be $\frac{\partial L}{\partial a} = -(1 + \frac{\alpha}{2})f(A) - \frac{\alpha}{2} f(B) + 2f(a); \frac{\partial L}{\partial b} = (1 + \frac{\alpha}{2})f(B) + \frac{\alpha}{2} f(A) - 2f(b)$. To extend to n-dimensions, we define an outer region $\mathbb{Q}$ that includes the smaller region $\mathbb{D}$ and defined as: $\mathbb{Q} = \{\mathbf{u} : \mathbf{a} - \frac{\alpha}{2}\mathbf{r} \leq \mathbf{u} \leq \mathbf{b} + \frac{\alpha}{2}\mathbf{r}\}$, where $(\mathbf{a}, \mathbf{b}, \mathbf{r})$ are defined as before, while α is defined as the boundary factor of the outer region for all the dimensions. The inner region $\mathbb{D}$ is defined as in Eq. (6), while the outer region can be defined in terms of the corner points as follows:

$$\mathrm{corners}(\mathbb{Q}) = \mathbf{Q}_{n \times 2^n} = \left[\mathbf{q}^1 | \mathbf{q}^2 | .. | \mathbf{q}^{2^n} \right]$$

$$\mathbf{Q} = \mathbf{1}^T (\mathbf{a} - \frac{\alpha}{2}\mathbf{r}) + (1 + \alpha)\mathbf{M}^T \odot (\mathbf{1}^T \mathbf{r}) \tag{10}$$

Let $\mathbf{f}_{\mathbb{D}}$ be a function vector as in Eq. (7) and $\mathbf{f}_{\mathbb{Q}}$ be another function vector evaluated at all possible outer corner points: $\mathbf{f}_{\mathbb{Q}} = \left[f(\mathbf{q}^1), f(\mathbf{q}^2), ..., f(\mathbf{q}^{2^n}) \right]^T , \quad \mathbf{q}^i = \mathbf{Q}_{:,i}$.

Now, the loss and update directions for the n-dimensional case becomes:

$$L(\mathbf{a}, \mathbf{b}) = \int \cdots \int_{\mathbb{Q}} f(u_1, \ldots, u_n)\, du_1 \ldots du_n - 2 \int \cdots \int_{\mathbb{D}} f(u_1, \ldots, u_n)\, du_1 \ldots du_n$$

$$\approx \triangle \left((1+\alpha)^n \mathbf{1}^T \mathbf{f}_{\mathbb{Q}} - 2\,\mathbf{1}^T \mathbf{f}_{\mathbb{D}} \right) \tag{11}$$

$$\nabla_{\mathbf{a}} L \approx 2\triangle \mathrm{diag}^{-1}(\mathbf{r}) \left(2\overline{\mathbf{M}}\mathbf{f}_{\mathbb{D}} - \overline{\mathbf{M}}_{\mathbb{Q}}\mathbf{f}_{\mathbb{Q}} \right); \quad \nabla_{\mathbf{b}} L \approx 2\triangle \mathrm{diag}^{-1}(\mathbf{r}) \left(-2\mathbf{M}\mathbf{f}_{\mathbb{D}} + \mathbf{M}_{\mathbb{Q}}\mathbf{f}_{\mathbb{Q}} \right),$$

where diag(.) is the diagonal matrix of the vector argument or the diagonal vector of the matrix argument. $\overline{\mathbf{M}}_{\mathbb{Q}}$ is the outer region scaled mask defined as follows:

$$\overline{\mathbf{M}}_{\mathbb{Q}} = (1+\alpha)^{n-1}\left((1+\tfrac{\alpha}{2})\overline{\mathbf{M}} + \tfrac{\alpha}{2}\mathbf{M} \right) \; ; \; \mathbf{M}_{\mathbb{Q}} = (1+\alpha)^{n-1}\left((1+\tfrac{\alpha}{2})\mathbf{M} + \tfrac{\alpha}{2}\overline{\mathbf{M}} \right) \tag{12}$$

White-Box OIR (OIR_W). Here, we present the white-box formulation of Outer-Inner-Ratio. This requires access to the gradient of the function f in order to update the current estimates of the bound. As we show in Sect. 4, access to gradients enhance the quality of the detected regions. To derive the formulation, We set $\lambda = \tfrac{\alpha}{\beta}$ in Eq. (9), where α is the small boundary factor of the outer area and β is the gradient emphasis factor. Hence, the objective in Eq. (9) becomes:

$$\arg\min_{a,b} L = \arg\min_{a,b} \; \mathrm{Area}_{\mathrm{out}} - \lambda\,\mathrm{Area}_{\mathrm{in}}$$

$$= \arg\min_{a,b} \; \int_A^a f(u)du + \int_b^B f(u)du - \frac{\alpha}{\beta}\int_a^b f(u)du$$

$$= \arg\min_{a,b} \; \frac{\beta}{\alpha} \int_{a-\alpha\frac{b-a}{2}}^{b+\alpha\frac{b-a}{2}} f(u)du - (1+\frac{\beta}{\alpha})\int_a^b f(u)du \tag{13}$$

$$\frac{\partial L}{\partial a} = \frac{\beta}{\alpha}\left(f(a) - f\left(a - \alpha\frac{b-a}{2} \right) \right)$$

$$- \frac{\beta}{2} f\left(b + \alpha\frac{b-a}{2} \right) - \frac{\beta}{2} f\left(a - \alpha\frac{b-a}{2} \right) + f(a)$$

Now, since λ^* should be small for the optimal objective as $\lambda \to 0$, $\alpha \to 0$ and hence the derivative in Eq. (13) becomes the following:

$$\lim_{\alpha \to 0} \frac{\partial L}{\partial a} = \frac{\beta}{2}\left((b-a)f'(a) + f(b) \right) + (1 - \frac{\beta}{2})f(a)$$

$$\lim_{\alpha \to 0} \frac{\partial L}{\partial b} = \frac{\beta}{2}\left((b-a)f'(b) + f(a) \right) - (1 - \frac{\beta}{2})f(b) \tag{14}$$

We can see that the update rule for a and b depends on the function value **and** the derivative of f at the boundaries a and b respectively, with β controlling the dependence. If $\beta \to 0$, the update directions in Eq. (14) collapse to the unregularized naive update. To extend to n-dimensions, we have to define a term that involves the gradient of the function, *i.e.* the all-corners gradient matrix $\mathbf{G}_{\mathbb{D}}$.

$$\mathbf{G}_{\mathbb{D}} = \left[\nabla f(\mathbf{d}^1) \mid \nabla f(\mathbf{d}^2) \mid \ldots \mid \nabla f(\mathbf{d}^{2^n}) \right]^T \tag{15}$$

Algorithm 1: Robust n-dimensional Region Finding for Black-Box DNNs by Outer-Inner Ratios

---

Requires: Semantic Function of a DNN $f(\mathbf{u})$ in Eq (1), initial semantic parameter $\mathbf{u}_0$, number of iterations T , learning rate η , object shape $\mathbf{S}_z$ of class label z, boundary factor α, Small ϵ

Form constant binary matrices $\mathbf{M}, \overline{\mathbf{M}}, \mathbf{M}_\mathbb{Q}, \overline{\mathbf{M}_\mathbb{Q}}, \mathbf{M}_\mathbb{D}, \overline{\mathbf{M}_\mathbb{D}}$

Initialize bounds $\mathbf{a}_0 \leftarrow \mathbf{u}_0 - \epsilon\mathbf{1}$, $\mathbf{b}_0 \leftarrow \mathbf{u}_0 + -\epsilon\mathbf{1}$

$\mathbf{r}_0 \leftarrow \mathbf{a}_0 - \mathbf{b}_0$, update region volume $\triangle_0$ as in Eq (5)

for $t \leftarrow 1$ **to** T **do**

    form the all-corners function vectors $f_\mathbb{D}, f_\mathbb{Q}$ as in Eq (7,10)

    $\nabla_\mathbf{a}L \leftarrow 2\triangle_{t-1}\text{diag}^{-1}(\mathbf{r}_{t-1})\left(2\overline{\mathbf{M}}f_\mathbb{D} - \overline{\mathbf{M}_\mathbb{Q}}f_\mathbb{Q}\right)$

    $\nabla_\mathbf{b}L \leftarrow 2\triangle_{t-1}\text{diag}^{-1}(\mathbf{r}_{t-1})\left(-2\mathbf{M}f_\mathbb{D} + \mathbf{M}_\mathbb{Q}f_\mathbb{Q}\right)$

    update bounds: $\mathbf{a}_t \leftarrow \mathbf{a}_{t-1} - \eta\nabla_\mathbf{a}L$, $\mathbf{b}_t \leftarrow \mathbf{b}_{t-1} - \eta\nabla_\mathbf{b}L$

    $\mathbf{r}_t \leftarrow \mathbf{a}_t - \mathbf{b}_t$, update region volume $\triangle_t$ as in Eq (5)

end

Returns: robust region bounds: $\mathbf{a}_T, \mathbf{b}_T$.

Now, the loss and update directions are given as follows.

$$L(\mathbf{a}, \mathbf{b}) \approx \frac{(1+\alpha)^n \mathbf{1}^T f_\mathbb{Q}}{\mathbf{1}^T f_\mathbb{D}} - 1$$

$$\nabla_\mathbf{a}L \approx \triangle\left(\text{diag}^{-1}(\mathbf{r})\overline{\mathbf{M}}_\mathbb{D}f_\mathbb{D} + \beta\text{diag}(\overline{\mathbf{M}}\mathbf{G}_\mathbb{D}) + \beta\overline{\mathbf{s}}\right) \tag{16}$$

$$\nabla_\mathbf{b}L \approx \triangle\left(-\text{diag}^{-1}(\mathbf{r})\mathbf{M}_\mathbb{D}f_\mathbb{D} + \beta\text{diag}(\mathbf{M}\mathbf{G}_\mathbb{D}) + \beta\mathbf{s}\right)$$

where the mask is the special mask

$$\overline{\mathbf{M}}_\mathbb{D} = \left(\gamma_n\overline{\mathbf{M}} - \beta\mathbf{M}\right); \ \mathbf{M}_\mathbb{D} = \left(\gamma_n\mathbf{M} - \beta\overline{\mathbf{M}}\right); \ \gamma_n = 2 - \beta(2n-1) \tag{17}$$

$\mathbf{s}$ is a weighted sum of the gradient from other dimensions $(i \neq k)$ contributing to the update direction of dimension k, where $k \in \{1, 2, ..., n\}$.

$$\mathbf{s}_k = \frac{1}{\mathbf{r}_k}\sum_{i=1, i\neq k}^{n} \mathbf{r}_i((\overline{\mathbf{M}}_{i,:} - \mathbf{M}_{i,:}) \odot \overline{\mathbf{M}}_{k,:})\mathbf{G}_{:,i}$$

$$\tag{18}$$

$$\overline{\mathbf{s}}_k = \frac{1}{\mathbf{r}_k}\sum_{i=1, i\neq k}^{n} \mathbf{r}_i((\mathbf{M}_{i,:} - \overline{\mathbf{M}}_{i,:}) \odot \mathbf{M}_{k,:})\mathbf{G}_{:,i}$$

Algorithms 1, and 2 summarize the techniques explained above, which we implement in Sect. 4. The derivation of the 2-dimensional case and n-dimensional case of the OIR formulation, as well as other unsuccessful formulations are all included in the **supplement**.

Algorithm 2: Robust n-dimensional Region Finding for White-Box DNNs by Outer-Inner Ratios

---

Requires: Semantic Function of a DNN $f(\mathbf{u})$ in Eq (1), initial semantic
    parameter $\mathbf{u}_0$, , learning rate η , object shape $\mathbf{S}_z$ of class label z, emphasis
    factor β, Small ϵ
Form constant binary matrices $\mathbf{M}, \overline{\mathbf{M}}, \mathbf{M}_{\mathbb{D}}, \overline{\mathbf{M}_{\mathbb{D}}}$
Initialize bounds $\mathbf{a}_0 \leftarrow \mathbf{u}_0 - \epsilon\mathbf{1}$, $\mathbf{b}_0 \leftarrow \mathbf{u}_0 + -\epsilon\mathbf{1}$
$\mathbf{r}_0 \leftarrow \mathbf{a}_0 - \mathbf{b}_0$, update region volume $\triangle_0$ as in as in Eq (5)
for $t \leftarrow 1$ **to** T **do**
| form the all-corners function vector $f_{\mathbb{D}}$ as in Eq (7)
| form the all-corners gradients matrix $\mathbf{G}_{\mathbb{D}}$ as in Eq (15)
| form the gradient selection vectors $\mathbf{s}, \bar{\mathbf{s}}$ as in Eq (18)
| $\nabla_{\mathbf{a}}L \leftarrow \triangle_{t-1} \left(\text{diag}^{-1}(\mathbf{r}_{t-1})\overline{\mathbf{M}}_{\mathbb{D}}f_{\mathbb{D}} + \beta\text{diag}(\overline{\mathbf{M}\mathbf{G}}_{\mathbb{D}} + \beta\bar{\mathbf{s}}\right)$
| $\nabla_{\mathbf{b}}L \leftarrow \triangle_{t-1} \left(-\text{diag}^{-1}(\mathbf{r}_{t-1})\mathbf{M}_{\mathbb{D}}f_{\mathbb{D}} + \beta\text{diag}(\mathbf{M}\mathbf{G}_{\mathbb{D}}) + \beta\mathbf{s}\right)$
| update bounds: $\mathbf{a}_t \leftarrow \mathbf{a}_{t-1} - \eta\nabla_{\mathbf{a}}L$, $\mathbf{b}_t \leftarrow \mathbf{b}_{t-1} - \eta\nabla_{\mathbf{b}}L$
| $\mathbf{r}_t \leftarrow \mathbf{a}_t - \mathbf{b}_t$, update region volume $\triangle_t$ as in Eq (5)
end
Returns: robust region bounds: $\mathbf{a}_T, \mathbf{b}_T$.

---

4 Experiments

4.1 Setup and Data

In this paper, we chose the semantic parameters $\mathbf{u}$ to be the azimuth rotations of the viewpoint and the elevation angle from the horizontal plane, where the object is always at the center of the rendering. This is common practise in the literature [16,19]. We use 100 shapes from 10 different classes from ShapeNet [6], the largest dataset for 3D models that are normalized from the semantic lens. We pick these 100 shapes specifically such that: (1) the class label is available in ImageNet [31] and that ImageNet classifiers can identify the exact class, and (2) the selected shapes are identified by the classifiers at some part of the semantic space. To do this, we measured the average score in the space and accepted the shape only if its average Resnet softmax score is 0.1. To render the images, we use a differentiable renderer NMR [21], which allows obtaining the gradient to the semantic input parameters. The networks of interest are Resnet50 [18], VGG [34], AlexNet [22], and InceptionV3 [36]. We use the official PyTorch implementation for each network [29].

4.2 Mapping the Networks

Similar to Fig. 1, we map the networks for all 100 shapes on the first semantic parameter (the azimuth rotation), as well as the joint (azimuth and elevation). We show these results in Fig. 4. The ranges for the two parameters were $[0°,360°]$, $[-10°,90°]$, with a 3×3 grid. The total number of network evaluations is 4K forward passes from each network for every shape (total of 1.6M forward passes). We show all of the remaining results in the **supplement**.

Table 1. Benchmarking popular DNNs in semantic robustness vs error rate.
We develop the Semantic Robustness Volume Ratio (SRVR) metric to quantify and
compare the semantic robustness of well-known DNNs in Sect. 4.4. We see that semantic robustness does not necessarily depend on the accuracy of the DNN. This motivates studying it as an independent metric from the classification accuracy. Results are
reported based on the official PyTorch implementations of these networks [29].

Deep networks	SRVR	Top-1 error	Top-5 error
AlexNet [22]	8.87%	43.45	20.91
VGG-11 [34]	9.72%	30.98	11.37
ResNet50 [18]	**16.79%**	23.85	7.13
Inceptionv3 [36]	7.92%	**22.55**	**6.44**

Table 2. Semantic analysis techniques: we compare different approaches to analyse
the semantic robustness of DNNs.

Analysis approach	Paradigm	Total sampling complexity	Black-box functions	Forward pass/step	Backward pass/step	Identification capability
Grid sampling	Top-down	$\mathcal{O}(N^n), N \gg 2$	✓	–	–	Fully identifies the semantic map of DNN
Naive	Bottom-up	$\mathcal{O}(2^n)$	✓	2^n	0	Finds strong robust regions only around $\mathbf{u}_0$
OIR_B	Bottom-up	$\mathcal{O}(2^{n+1})$	✓	2^{n+1}	0	Finds strong and week robust regions around $\mathbf{u}_0$
OIR_W	Bottom-up	$\mathcal{O}(2^n)$	✗	2^n	2^n	Finds strong and week robust regions around $\mathbf{u}_0$

4.3 Growing Semantic Robust Regions

We implement the three bottom-up approaches in Table 2 and Algorithms 1 and
2. The hyper-parameters were set to $\eta = 0.1, \alpha = 0.05, \beta = 0.0009, \lambda = 0.1, T =$
800. We can observe in Fig. 3 that multiple initial points inside the same robust
region converge to the same boundary. One key difference to be noted between
the naive approach in Eq. (8) and the OIR formulations in Eq. (11 and 16) is
that the naive approach fails to capture robust regions in some scenarios and
fall for trivial regions (see Fig. 3).

4.4 Applications

Quantifying Semantic Robustness

Looking at these NSM can lead to insights about the network, but we would like
to develop a systemic approach to quantify the robustness of these DNNs. To
do this, we develop the Semantic Robustness Volume Ratio (SRVR) metric. The
SRVR of a network is the ratio between the expected size of the robust region

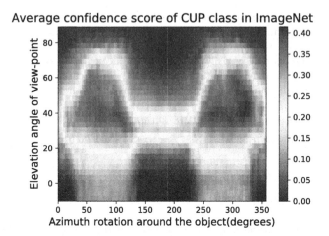

Fig. 5. Semantic bias in ImageNet. By taking the average semantic maps over 10 shapes of *cup* class and over different networks, we can visualize the bias of the training data. The angles of low score are probably not well-represented in ImageNet [31].

obtained by Algorithms 1 and 2 over the nominal total volume of the semantic map of interest. Explicitly, the SRVR of network $\mathbf{C}$ for class label z is defined as follows:

$$\text{SRVR}_z = \frac{\mathbb{E}[\text{Vol}(\mathbb{D})]}{\text{Vol}(\Omega)} = \frac{\mathbb{E}_{\mathbf{u}_0 \sim \Omega, \mathbf{S}_z \sim \mathbb{S}_z}[\text{Vol}(\mathbf{\Phi}(f, \mathbf{S}_z, \mathbf{u}_0))]}{\text{Vol}(\Omega)}, \tag{19}$$

where f, Φ are defined in Eq. (1 and 2) respectively. We take the average volume of all the adversarial regions found for multiple initializations and multiple shapes of the same class z. Then, we divide by the volume of the entire space. This provides a percentage of how close the DNN is from the ideal behaviour of identifying the object robustly in the entire space. The SRVR metric is not strict in its value, since the analyzer defines the semantic space of interest and the shapes used. However, comparing SRVR scores among DNNs is of extreme importance, as this relative analysis conveys insights about a network that might not be evident by only observing the accuracy of this network. For example, we can see in Table 1 that while InceptionV3 [36] is the best in terms of accuracy, it lags behind Resnet50 [18] in terms of semantic robustness. This observation is also consistent with the qualitative NSMs in Fig. 4, in which we can see that while Inception is very confident, it can fail completely inside these confident regions. Note that the reported SRVR results are averaged over all 10 classes and over all 100 shapes. We use 4 constant initial points for all experiments and the semantic parameters are the azimuth and elevation as in Fig. 3 and 4. As can be seen in Fig. 3, different methods predict different regions, so we take the average size of the the three methods used (naive, OIR_W, and OIR_B) to give an overall estimate of the volume used in the SRVR results reported in Table 1.

Finding Semantic Bias in the Data

While observing the above figures uncovers interesting insights about the DNNs and the training data of ImageNet [31], it does not allow to make a conclusion about the network nor about the data. Therefore, we can average these semantic maps of these networks to factor out the effect of the network structure and training and maintain only the effect of training data. We show two such maps called Data Semantic Maps (DSMs). We observe that networks have holes in their semantic maps and these holes are shared among DNNs, indicating bias in the data. Identifying this gap in the data can help train a more semantically robust network. This can be done by leveraging adversarial training on these data-based adversarial regions as performed in the adversarial attack literature [14]. Figure 5 shows an example of this semantic map, which points to the possibility that ImageNet [31] may not contain angles of the easy *cup* class.

5   Analysis

First, we observe in Fig. 4 that some red areas (adversarial regions the DNN can not identify the object within) are surrounded by blue areas (robust regions the DNN can identify the object within). These "semantic traps" are dangerous in ML, since they are hard to identify (without NSM) and they can cause failure cases for some models. These traps can be attributed to either the model architecture, training, and loss, or bias in the dataset, on which the model was trained (*i.e.* ImageNet [31].

Note that plotting NMS is extremely expensive even for a moderate dimensionality *e.g.* $n = 8$. For example, for the plot in Fig. 1, we use $N = 180$ points in the range of $360°$. If all the other dimensions require the same number of samples for their individual range, the total joint space requires $180^n = 180^8 = 1.1 \times 10^{18}$ samples, which is enormous. Evaluating the DNN for that many forward passes is intractable. Thus, we follow a bottom-up approach instead, where we start from one point in the semantic space $\mathbf{u}_0$ and we grow an n-dimensional hyperrectangle around that point to find the robust "neighborhood" of that point for this specific DNN. Table 2 compares different analysis approaches for semantic robustness of DNNs.

6   Conclusion

We analyse DNN robustness with a semantic lens and show how more confident networks tend to create adversarial semantic regions inside highly confident regions. We developed a bottom-up approach to semantically analyse networks by growing adversarial regions. This approach scales well with dimensionality, and we use it to benchmark the semantic robustness of several well-known and popular DNNs.

Acknowledgments. This work was supported by the King Abdullah University of Science and Technology (KAUST) Office of Sponsored Research under Award No. OSR-CRG2018-3730

References

1. Alcorn, M.A., et al.: Strike (with) a pose: neural networks are easily fooled by strange poses of familiar objects. In: The IEEE Conference on Computer Vision and Pattern Recognition (CVPR), June 2019
2. An, G.: The effects of adding noise during backpropagation training on a generalization performance. Neural Comput. **8**(3), 643–674 (1996)
3. Bibi, A., Alfadly, M., Ghanem, B.: Analytic expressions for probabilistic moments of PL-DNN with Gaussian input. In: The IEEE Conference on Computer Vision and Pattern Recognition (CVPR), June 2018
4. Bishop, C.M.: Training with noise is equivalent to Tikhonov regularization. Neural Comput. **7**(1), 108–116 (2008)
5. Carlini, N., Wagner, D.: Towards evaluating the robustness of neural networks. In: IEEE Symposium on Security and Privacy (SP) (2017)
6. Chang, A.X., et al.: ShapeNet: an information-rich 3D model repository. Technical Report. arXiv:1512.03012 [cs.GR]. Stanford University – Princeton University – Toyota Technological Institute at Chicago (2015)
7. Chen, P.Y., Zhang, H., Sharma, Y., Yi, J., Hsieh, C.J.: Zoo: zeroth order optimization based black-box attacks to deep neural networks without training substitute models. In: Proceedings of the 10th ACM Workshop on Artificial Intelligence and Security, AISec 2017, pp. 15–26. ACM, New York (2017)
8. Dosovitskiy, A., Brox, T.: Inverting visual representations with convolutional networks. In: Proceedings of the IEEE Conference on Computer Vision and Pattern Recognition, pp. 4829–4837 (2016)
9. Dreossi, T., Jha, S., Seshia, S.A.: Semantic adversarial deep learning. In: Chockler, H., Weissenbacher, G. (eds.) CAV 2018. LNCS, vol. 10981, pp. 3–26. Springer, Cham (2018). https://doi.org/10.1007/978-3-319-96145-3_1
10. Engstrom, L., Tran, B., Tsipras, D., Schmidt, L., Madry, A.: Exploring the landscape of spatial robustness. In: Proceedings of the 36th International Conference on Machine Learning (ICML) (2019)
11. Fawzi, A., Moosavi-Dezfooli, S.M., Frossard, P.: Robustness of classifiers: from adversarial to random noise. In: Advances in Neural Information Processing Systems (2016)
12. Fawzi, A., Moosavi Dezfooli, S.M., Frossard, P.: The robustness of deep networks - a geometric perspective. IEEE Sig. Process. Mag. **34**, 50–62 (2017)
13. Geirhos, R., Rubisch, P., Michaelis, C., Bethge, M., Wichmann, F.A., Brendel, W.: ImageNet-trained CNNs are biased towards texture; increasing shape bias improves accuracy and robustness. arXiv preprint arXiv:1811.12231 (2018)
14. Goodfellow, I., Shlens, J., Szegedy, C.: Explaining and harnessing adversarial examples. In: International Conference on Learning Representations (ICLR) (2015)
15. Grandvalet, Y., Canu, S., Boucheron, S.: Noise injection: theoretical prospects. Neural Comput. **9**(5), 1093–1108 (1997)
16. Hamdi, A., Muller, M., Ghanem, B.: SADA: semantic adversarial diagnostic attacks for autonomous applications. In: AAAI Conference on Artificial Intelligence (2020)
17. Hamdi, A., Rojas, S., Thabet, A., Ghanem, B.: AdvPC: transferable adversarial perturbations on 3D point clouds (2019)
18. He, K., Zhang, X., Ren, S., Sun, J.: Deep residual learning for image recognition. CoRR abs/1512.03385 (2015)
19. Hosseini, H., Poovendran, R.: Semantic adversarial examples. CoRR abs/1804.00499 (2018)

20. Kato, H., Ushiku, Y., Harada, T.: Neural 3D mesh renderer. In: The IEEE Conference on Computer Vision and Pattern Recognition (CVPR) (June 2018)
21. Kato, H., Ushiku, Y., Harada, T.: Neural 3D mesh renderer. In: Proceedings of the IEEE Conference on Computer Vision and Pattern Recognition, pp. 3907–3916 (2018)
22. Krizhevsky, A., Sutskever, I., Hinton, G.E.: ImageNet classification with deep convolutional neural networks. In: Pereira, F., Burges, C.J.C., Bottou, L., Weinberger, K.Q. (eds.) Advances in Neural Information Processing Systems, vol. 25, pp. 1097–1105. Curran Associates, Inc. (2012)
23. Kurakin, A., Goodfellow, I.J., Bengio, S.: Adversarial machine learning at scale. CoRR abs/1611.01236 (2016)
24. Liu, H.T.D., Tao, M., Li, C.L., Nowrouzezahrai, D., Jacobson, A.: Beyond pixel norm-balls: parametric adversaries using an analytically differentiable renderer. In: International Conference on Learning Representations (2019)
25. Mahendran, A., Vedaldi, A.: Understanding deep image representations by inverting them. CoRR abs/1412.0035 (2014)
26. Moosavi-Dezfooli, S.M., Fawzi, A., Fawzi, O., Frossard, P.: Universal adversarial perturbations. In: The IEEE Conference on Computer Vision and Pattern Recognition (CVPR) (2017)
27. Moosavi-Dezfooli, S.M., Fawzi, A., Frossard, P.: DeepFool: a simple and accurate method to fool deep neural networks. In: The IEEE Conference on Computer Vision and Pattern Recognition (CVPR), June 2016
28. Bhagoji, A.N., He, W., Li, B., Song, D.: Practical black-box attacks on deep neural networks using efficient query mechanisms. In: Ferrari, V., Hebert, M., Sminchisescu, C., Weiss, Y. (eds.) ECCV 2018. LNCS, vol. 11216, pp. 158–174. Springer, Cham (2018). https://doi.org/10.1007/978-3-030-01258-8_10
29. Paszke, A., et al.: Automatic differentiation in PyTorch. In: NIPS-W (2017)
30. Ródenas, R.G., López, M.L., Verastegui, D.: Extensions of Dinkelbach's algorithm for solving non-linear fractional programming problems. Top 7(1), 33–70 (1999). https://doi.org/10.1007/BF02564711
31. Russakovsky, O., et al.: ImageNet large scale visual recognition challenge. CoRR abs/1409.0575 (2014)
32. Shou, Z., Gao, H., Zhang, L., Miyazawa, K., Chang, S.F.: AutoLoc: weakly-supervised temporal action localization in untrimmed videos. In: Proceedings of the European Conference on Computer Vision (ECCV), pp. 154–171 (2018)
33. Shu, M., Liu, C., Qiu, W., Yuille, A.: Identifying model weakness with adversarial examiner. In: Proceedings of the AAAI Conference on Artificial Intelligence, vol. 34, no. 07, pp. 11998–12006 (2020)
34. Simonyan, K., Zisserman, A.: Very deep convolutional networks for large-scale image recognition. arXiv preprint arXiv:1409.1556 (2014)
35. Stroud, A.H.: Methods of numerical integration (Philip J. Davis and Philip Rabinowitz). SIAM Rev. 18(3), 528–529 (1976)
36. Szegedy, C., Vanhoucke, V., Ioffe, S., Shlens, J., Wojna, Z.: Rethinking the inception architecture for computer vision. In: Proceedings of the IEEE Conference on Computer Vision and Pattern Recognition, pp. 2818–2826 (2016)
37. Szegedy, C., et al.: Intriguing properties of neural networks. ICLR (2013)
38. Szegedy, C., et al.: Intriguing properties of neural networks. CoRR abs/1312.6199 (2013)
39. Torralba, A., Efros, A.A.: Unbiased look at dataset bias. In: CVPR (2011)

40. Vondrick, C., Khosla, A., Malisiewicz, T., Torralba, A.: HOGgles: visualizing object detection features. In: Proceedings of the IEEE International Conference on Computer Vision, pp. 1–8 (2013)
41. Zeng, X., et al.: Adversarial attacks beyond the image space. In: The IEEE Conference on Computer Vision and Pattern Recognition (CVPR), June 2019

Likelihood Landscapes: A Unifying Principle Behind Many Adversarial Defenses

Fu Lin[✉], Rohit Mittapalli, Prithvijit Chattopadhyay, Daniel Bolya, and Judy Hoffman

Georgia Institute of Technology, Atlanta, USA
{flin68,rmittapalli3,prithvijit3,dbolya,judy}@gatech.edu

Abstract. Convolutional Neural Networks have been shown to be vulnerable to adversarial examples, which are known to locate in subspaces close to where normal data lies but are not naturally occurring and of low probability. In this work, we investigate the potential effect defense techniques have on the geometry of the likelihood landscape - likelihood of the input images under the trained model. We first propose a way to visualize the likelihood landscape leveraging an energy-based model interpretation of discriminative classifiers. Then we introduce a measure to quantify the flatness of the likelihood landscape. We observe that a subset of adversarial defense techniques results in a similar effect of flattening the likelihood landscape. We further explore directly regularizing towards a flat landscape for adversarial robustness.

Keywords: Adversarial robustness · Understanding robustness · Deep learning

1 Introduction

Although Convolutional Neural Networks (CNNs) have consistently pushed benchmarks on several computer vision tasks, ranging from image classification [15], object detection [8] to recent multimodal tasks such as visual question answering [1] and dialog [3], they are not robust to small adversarial input perturbations. Prior work has extensively demonstrated the vulnerability of CNNs to adversarial attacks [2,10,24,29,39] and has therefore, exposed how intrinsically unstable these systems are. Countering the susceptibility of CNNs to such attacks has motivated a number of defenses in the computer vision literature [16,18,24,27,33,35,43,49].

In this work we explore the questions, why are neural networks vulnerable to adversarial attacks in the first place, and how do adversarial defenses protect

Electronic supplementary material The online version of this chapter (https:// doi.org/10.1007/978-3-030-66415-2_3) contains supplementary material, which is available to authorized users.

A. Bartoli and A. Fusiello (Eds.): ECCV 2020 Workshops, LNCS 12535, pp. 39–54, 2020.
https://doi.org/10.1007/978-3-030-66415-2_3

against them? Are there some inherent deficiencies in vanilla neural network training that these attacks exploit and that these defenses counter? To start answering these questions, we explore how adversarial and clean samples fit into the marginal input distributions of trained models. In doing so, we find that although CNN classifiers implicitly model the distribution of the clean data (both training and test), standard training induces a specific structure in this marginal distribution that adversarial attacks can exploit. Moreover, we find that a subset of adversarial defense techniques, despite being wildly different in motivation and implementation, all tend to modify this structure, leading to more robust models.

A standard discriminative CNN-based classifier only explicitly models the conditional distribution of the output classes with respect to the input image (namely, $p_\theta(y|x)$). This is in contrast to generative models (such as GANs [9]), which go a step further and model the marginal (or joint) distribution of the data directly (namely, $p_\theta(x)$ or $p_\theta(x, y)$), so that they can draw new samples from that distribution or explain existing samples under the learned model. Recently, however, Grathwohl et al. [12] have shown that it is possible to interpret a CNN classifier as an energy-based model (EBM), allowing us to infer the conditional ($p_\theta(y|x)$) as well as the marginal distribution ($p_\theta(x)$). While they use this trick to encode generative capabilities into a discriminative classifier, we are interested in exploring the marginal distribution of the input, $p_\theta(x)$ for models trained with and without adversarial defense mechanisms.

To do this, we study the "relative likelihood landscape" ($\Delta \log p_\theta(x)$ in a neighborhood around a test example), which lets us freely analyze this marginal distribution. In doing so, we notice a worrying trend – for most training or test examples, a small random perturbation in pixel space can cause the modeled $\log p_\theta(x)$ to drop significantly (see Fig. 1a). Despite training and test samples (predicted correctly by the model) being highly likely under the marginal data distribution of the model (high $\log p_\theta(x)$), slight deviations from these examples significantly moves the perturbed samples out of the high-likelihood region. Whether this is because of dataset biases such as chromatic aberration [4] and JPEG artifacts [38] or some other factor [6,17,40] isn't clear. Yet, what is clear is that adversarial examples exploit this property (see Fig. 3a). Moreover, we observe that adversarial defenses intrinsically address this issue.

Specifically, we study two key adversarial defense techniques – namely adversarial training [10,24,49] and Jacobian regularization [16,18]. Although these defense mechanisms are motivated by different objectives, both of them result in elevating the value of $\log p_\theta(x)$ in the region surrounding training and test examples. Thus, they all intrinsically tend to "flatten" the likelihood landscape, thereby patching the adversarially-exploitable structures in the network's marginal data distribution (see Fig. 1). Moreover, we find that the stronger the adversarial defense, the more pronounced this effect becomes (see Fig. 2).

To quantify this perceived "flatness" of the likelihood landscapes, we build on top of [12] and devise a new metric, Φ-flatness, that captures how rapidly the marginal likelihood of clean samples change in their immediate neighborhoods.

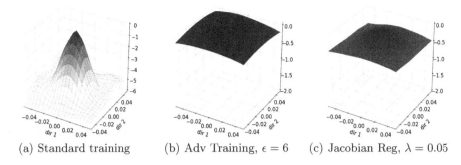

(a) Standard training   (b) Adv Training, $\epsilon = 6$ (c) Jacobian Reg, $\lambda = 0.05$

Fig. 1. Relative likelihood landscapes ($\Delta \log p_\theta(x)$) plotted over two random pertur- bation directions for three DDNet models trained on CIFAR10 with the clean sample in the center. (a) follows standard training without additional defense. (b) uses PGD based adversarial training for 10 iterations with $\epsilon = 6/255$. (c) uses Jacobian regu- larization with a strength of $\lambda = 0.05$. Models trained to defend against adversarial attacks tend to have much fatter landscapes.

As predicted by our qualitative observations, we find that stronger adversarial defenses correlate well with higher Φ-flatness and a flatter likelihood landscape (see Fig. 4). This supports the idea that deviations in the $\log p_\theta(x)$ landscape are connected to what give rise to adversarial examples.

In order to fully test that hypothesis, we make an attempt to regularize $\log p_\theta(x)$ directly, thereby explicitly enforcing this "flatness". Our derivations indicate that regularizing $\log p_\theta(x)$ directly under this model is very similar to Jacobian regularization. We call the resulting regularization scheme AMSReg (based on the Approximate Mass Score proposed in [12]) and find that it results in adversarial performance similar to Jacobian regularization. While this defense ends up being less robust than adversarial training, we show that this may simply be because adversarial training prioritizes smoothing out the likelihood landscape in the directions chosen by adversarial attacks (see Fig. 5b) as opposed to random chosen perturbations. The regularization methods, on the other hand, tend to smooth out the likelihood in random directions very well, but are not as successful in the adversarial directions (see Fig. 5). Concretely, we make the following contributions:

- We propose a way to visualize the relative marginal likelihoods of input sam- ples (clean as well as perturbed) under the trained discriminative classifier by leveraging an interpretation of a CNN-based classifier as an energy-based model.
- We show that the marginal likelihood of a sample drops significantly with small pixel-level perturbations (Fig. 1a), and that adversarial examples lever- age this property to attack the model (Fig. 3a).
- We empirically identify that a subset of standard defense techniques, includ- ing adversarial training and Jacobian regularization, implicitly work to "flat- ten" this marginal distribution, which addresses this issue (Fig. 1). The stronger the defense, "flatter" the likelihood landscape (Fig. 2).

– We devise a method for regularizing the "flatness" of this likelihood land-scape directly (called AMSReg) and arrive at a defense similar to Jacobian regularization that works on par with existing defense methods.

2   Related Work

Adversarial Examples and Attacks. Adversarial examples for CNNs were originally introduced in Szegedy et al. [39], which shows that deep neural net-works can be fooled by carefully-crafted imperceptible perturbations. These adversarial examples locate in a subspace that is very close to the naturally occuring data, but they have low probability in the original data distribution [23]. Many works have then been proposed to explore ways finding adversarial exam-ples and attacking CNNs. Common attacks include FGSM [10], JSMA [29], C&W [2], and PGD [24] which exploit the gradient in respect to the input to deceitfully perturb images.

Adversarial Defenses. Several techniques have also been proposed to defend against adversarial attacks. Some aim at defending models from attack at inference time [13,14,28,33,35,46], some aim at detecting adversarial data input [7,21,26,32,50], and others focus on directly training a model that is robust to a perturbed input [16,18,24,27,43,49].

Among those that defend at training time, adversarial training [10,19,24] has been the most prevalent direction. The original implementation of adver-sarial training [10] generated adversarial examples using an FGSM attack and incorporated those perturbed images into the training samples. This defense was later enhanced by Madry et al. [24] using the stronger projected gradient descent (PGD) attack. However, running strong PGD adversarial training is computa-tionally expensive and several recent works have focused on accelerating the process by reducing the number of attack iterations [36,44,48].

Gradient regularization, on the other hand, adds an additional loss in attempt to produce a more robust model [5,16,18,34,37,42]. Of these methods, we con-sider Jacobian regularization, which tries to minimize the change in output with respect to the change in input (i.e., the Jacobian). In Varga et al. [42], the authors introduce an efficient algorithm to approximate the Frobenius norm of the Jacobian. Hoffman et al. [16] later conducts comprehensive analysis on the efficient algorithm and promotes Jacobian Regularization as a generic scheme for increasing robustness.

Interpreting Robustness via Geometry of Loss Landscapes. In a similar vein to our work, several works [25,31,47] have looked at the relation between adversarial robustness and the geometry of *loss landscapes*. Moosavi-Dezfooli et al. [25] shows that small curvature of loss landscape has strong relation to large adversarial robustness. Yu et al. [47] qualitatively interpret neural network models' adversarial robustness through loss surface visualization. In this work, instead of looking at *loss landscapes*, we investigate the relation between adver-sarial robustness and the geometry of *likelihood landscapes*. One key difference between the two lies in that loss is usually with respect to a specific target class while the marginal likelihood is not specific to a class. The likelihood landscape usually can't be computed for discrimitive models, but given recent work into interpreting neural network classifiers as energy-based models (EBMs) [12], we can now study the likelihood of data samples under trained classifiers.

3 Approach and Experiments

We begin our analysis of adversarial examples by exploring the marginal log-likelihood distribution of a clean sample relative to perturbed samples in a surrounding neighborhood (namely, the "relative likelihood landscape"). Next, we describe how to quantify the "flatness" of the landscape and then describe a regularization term in order to optimize for this "flatness". Finally, we present our experiment results and observations.

3.1 Computing the Marginal Log-Likelihood ($\log p_\theta(x)$)

As mentioned before, CNN-based classifiers are traditionally trained to model the conditional, $\log p_\theta(y|x)$, and not the marginal likelihood of samples under the model, $\log p_\theta(x)$ (where θ are the parameters of the model). To get around this restriction, we leverage the interpretation presented by Grathwohl et al. [12], which allows us to compute the marginal likelihood based on an energy-based graphical model interpretation of discriminative classifiers. More specifically, let $f_\theta : \mathbb{R}^D \to \mathbb{R}^K$ denote a classifier that maps input images to output pre-softmax logits. Grathwohl et al. model the joint distribution of the input images x and output labels y as:

$$p_\theta(x, y) = \frac{\exp(-E_\theta(x, y))}{Z(\theta)} = \frac{\exp(f_{\theta,y}(x))}{Z(\theta)}, \tag{1}$$

where $E_\theta(x, y)$ is known as the energy function, $f_{\theta,y}(x)$ denotes the yth index of the logits, $f_\theta(x)$, and $Z(\theta) = \int_x \int_y \exp(-E_\theta(x, y))$ is an unknown normalizing constant (that depends only on the parameters, and not the input). The marginal likelihood of an input sample x can therefore be obtained as,

$$\log p_\theta(x) = \log \sum_y p_\theta(x, y) = \log \left(\sum_y \exp\left(f_{\theta,y}(x)\right) \right) - \log Z(\theta). \tag{2}$$

However, since $Z(\theta)$ involves integrating over the space of all images, it is intractable to compute in practice. Thus, we focus on *relative* likelihoods instead, where we can cancel out the $\log Z(\theta)$ term. Specifically, given a perturbed sample x' and clean sample x, we define the relative likelihood of x' w.r.t. x as

$$\Delta \log p_\theta(x') = \log p_\theta(x') - \log p_\theta(x) = \sum_y \exp\left(f_{\theta,y}(x')\right) - \sum_y \exp\left(f_{\theta,y}(x)\right) \tag{3}$$

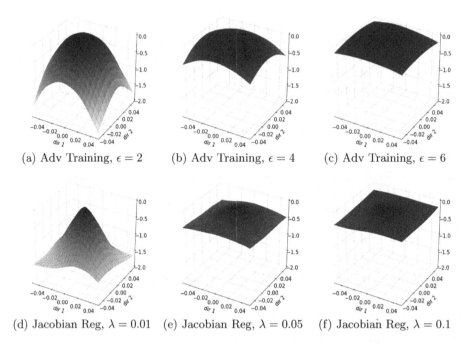

(a) Adv Training, $\epsilon = 2$ (b) Adv Training, $\epsilon = 4$ (c) Adv Training, $\epsilon = 6$

(d) Jacobian Reg, $\lambda = 0.01$ (e) Jacobian Reg, $\lambda = 0.05$ (f) Jacobian Reg, $\lambda = 0.1$

Fig. 2. Relative likelihood landscapes for Adversarial Training and Jacobian regularization with increasing strength. All models used DDNet trained on CIFAR10 and adversarial training uses PGD for 10 iterations. We see that stronger defenses tend to have have flatter landscapes.

3.2 Relative Likelihood Landscape Visualization

We now describe how to utilize Eq. 3 to visualize relative likelihood landscapes, building on top of prior work focusing on visualizing *loss-landscapes* [11,22, 47]. To make these plots, we visualize the relative likelihood with respect to a clean sample in a surrounding neighborhood. Namely, for each point x' in the neighborhood $N(x)$ surrounding a clean sample x, we plot $\Delta \log p_\theta(x')$.

Consistent with prior work, we define the neighborhood $N(x)$ around a clean sample as

$$N(x) = \{x'|x' = x + \epsilon_1 d + \epsilon_2 d^\perp\} \tag{4}$$

where d and $d^\perp$ are the two randomly pre-selected orthogonal signed vectors, and ϵ_1 and ϵ_2 represent the perturbation intensity along the axes d and $d^\perp$. Visualizing $\Delta \log p(x')$ for samples x' surrounding the clean sample x allows us to understand the degree to which structured perturbations affect the likelihood of the sample under the model.

Figure 1 shows such visualizations on a typical test-time example for standard training (no defense), adversarial training, and training with Jacobian regularization. Note that while we show just one example in these figures, we observe the same general trends over many samples. More examples of these plots can be found in the Appendix. Then as shown in Fig. 1a, for standard training small perturbations from the center clean image drops the likelihood of that sample considerably. While likelihood doesn't necessarily correlate one to one with accuracy, it's incredibly worrying that standard training doesn't model data samples that are slightly perturbed from test examples. Furthermore, these images are *test* examples that the model hasn't seen during training. This points to an underlying dataset bias, since the model has high likelihood for exactly the test example, but not for samples a small pixel perturbation away. This supports the recent idea that adversarial examples might be more of a property of datasets rather than the model themselves [17].

The rest of Fig. 1 shows that adversarial defenses like adversarial training and Jacobian regularization tend to have a much more uniform likelihood distribution, where perturbed points have a similar likelihood to the clean sample (i.e., the likelihood landscape is "flat"). Moreover, we also observe in Fig. 2 that the stronger the defense (for both adversarial training and Jacobian regularization), the lower the variation in the resulting relative likelihood landscape. Specifically, if we increase the adversarial training attack strength (ϵ) or increase the Jacobian regularization strength (λ), we observe a "flatter" likelihood landscape. We measure this correlation quantitatively in Sect. 3.3.

To get an understanding of the impact different defenses have on the likelihood of overall data distribution, we further plot the histogram of $\log p_\theta(x)$ for all clean test samples and randomly perturbed samples near them (Fig. 3). Adversarial training and Jacobian regularization share several common effects on the log-likelihood distribution: (1) both techniques induce more aligned distribution between clean samples and randomly perturbed samples (likely as a consequence of likelihood landscape flattening). (2) both techniques tend to decrease the log-likelihood of clean samples relatively to randomly perturbed samples. (3) The stronger the regularization, the more left-skewed the log-likelihood distribution of clean samples is.

3.3   Quantifying Flatness

To quantify the degree of variation in the likelihood landscape surrounding a point, we use the *Approximate Mass Score* (AMS), which was originally introduced in [12] as an out of distribution detection score. AMS depends on how the marginal likelihood changes in the immediate neighborhood of a point in the input space. The Approximate Mass Score can be expressed as,

$$s_\theta(x) = -\left\|\frac{\partial \log p_\theta(x)}{\partial x}\right\|_F \tag{5}$$

We use the gradient for all $x' \in \mathbb{N}(x)$ as an indicator of the "flatness" of the likelihood landscape surrounding the clean sample. While $\log p_\theta(x)$ is intractable to compute in practice, we can compute $\frac{\partial \log p_\theta(x)}{\partial x}$ exactly,

$$\frac{\partial \log p_\theta(x)}{\partial x} = \frac{\partial}{\partial x} \log\left(\frac{\sum_y \exp\left(f_{\theta,y}(x)\right)}{Z(\theta)}\right) = \sum_y p_\theta(y|x)\frac{\partial f_{\theta,y}(x)}{\partial x} \tag{6}$$

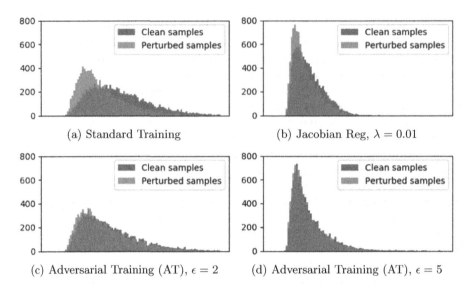

(a) Standard Training

(b) Jacobian Reg, $\lambda = 0.01$

(c) Adversarial Training (AT), $\epsilon = 2$

(d) Adversarial Training (AT), $\epsilon = 5$

Fig. 3. Histograms of log-likelihood $\log p_\theta(x)$ for clean samples and randomly perturbed samples. Blue corresponds to the log-likelihood on clean samples in CIFAR10 while orange corresponds to the log-likelihood on the perturbed samples. Note that the x-axis for plots might be differently shifted due to the different unknown normalizing constant $Z(\theta)$ for each model, but the scale of x-axis are the same across plots. (a) visualizes DDNet trained without additional defense, (b) uses a Jacobian regularized DDNet and (c)(d) use DDNet trained using PGD based adversarial training for 10 iterations. We see in adversarially robust approaches, clean and perturbed samples have more aligned distributions. We also note that with stronger regularization the distributions become increasingly left-skewed.

To quantify the flatness in the surrounding area of x, we sample this quantity at each point in the neighborhood of x (see Eq. 4) and average over n random choices for d and $d^\perp$ to obtain:

$$\phi(x) = \sum_{j=1}^{n} \sum_{x' \in \mathbb{N}_j(x)} \frac{s_\theta(x)}{n} = \sum_{j=1}^{n} \sum_{x' \in \mathbb{N}_j(x)} \frac{-1}{n} \left\| \frac{\partial \log p_\theta(x')}{\partial x} \right\|_F \tag{7}$$

The flatness of likelihood landscape for a model is then calculated by averaging $\phi(x)$ for all testing samples.

$$\Phi = \sum_{i=1}^{N} \frac{\phi(x_i)}{N} \tag{8}$$

Figure 4 shows the relationship between flatness score Φ and defense strength. For both adversarial training (Fig. 4a) and Jacobian regularization (Fig. 4b), the stronger the defense, the lower the score and the higher the adversarial accuracy. This observation motivates us to explore directly regularizing this AMS term (AMSReg) to induce a flat likelihood landscape.

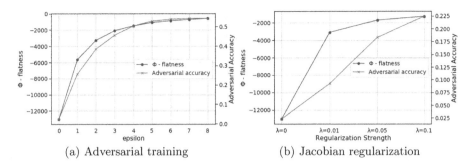

(a) Adversarial training   (b) Jacobian regularization

Fig. 4. Visualizes the relationship between the likelihood landscape flatness (as defined by Eq. 8) and adversarial accuracy. (a) showcases DDNet trained on CIFAR10 using PGD based adversarial training for 10 iterations with increasing epsilon strength. (b) showcases DDNet trained on CIFAR10 using Jacobian regularization of increasing strength. We see that adversarial accuracy and flatness of the relative likelihood landscape are correlated in the case of both defenses.

3.4 Approximate Mass Score Regularization (AMSReg)

In our experiments, we observe that high adversarial robustness corresponds with a low Approximate Mass Score in the case of both adversarial training and Jacobian regularization. This begs the question, is it possible to improve adversarial robustness by directly regularizing this term? That is, can we regularize $\left\lVert \frac{\partial \log p_\theta(x)}{\partial x} \right\rVert_F^2$ directly in addition to the cross-entropy loss to encourage robust predictive performance? Note however that naively computing this term during training would require double backpropagation, which would be inefficient.

Instead we make the observation that the right hand side of Eq. 6 (used to compute $\frac{\partial \log p_\theta(x)}{\partial x}$) is reminiscent of gradient regularization techniques – specifically, Jacobian Regularization [16].

Connection with Jacobian Regularization [16]. As discussed in [16], Jacobian regularization traditionally emerges as a consequence of stability analysis of model predictions against input perturbations – the high-level idea being that small input perturbations should minimally affect the predictions made by a network up to a first order Taylor expansion. Hoffman et al. [16] characterize this by regularizing the Jacobian of the output logits with respect to the input images. The difference between each element being summed over in the right hand side of Eq. 6 and the input-output Jacobian term used in [16] is that now the Jacobian term is weighted by the predicted probability of each class. Namely, optimizing $\left\lVert \frac{\partial \log p_\theta(x)}{\partial x} \right\rVert_F^2$ results in regularizing $\lVert \sum_c p_\theta(c \mid x) \frac{\partial f_{\theta,c}(x)}{\partial x} \rVert_F^2$, while Jacobian regularization regularizes just $\lVert \frac{\partial f_{\theta,c}(x)}{\partial x} \rVert_F^2$. The following proposition provides an upper bound of the squared norm that can be efficiently calculated in batch when using the efficient algorithm proposed for Jacobian regularization in [16].

Proposition 3.1. *Let C denote the number of classes, and $J^w(x) = p_\theta(c \mid x) \frac{\partial f_{\theta,c}(x)}{\partial x_i}$ be a weighted variant of the Jacobian of the class logits with respect to the input. Then, the frobenius norm of $J^w(x)$ multiplied by the total number of classes upper bounds the AMS score:*

$$\left\|\frac{\partial \log p_\theta(x)}{\partial x}\right\|_F^2 \leq C \, \|J^w(x)\|_F^2 \tag{9}$$

Proof. Recall that as per the EBM [12] interpretation of a discriminative classifier, $\frac{\partial \log p_\theta(x)}{\partial x}$ can be expressed as,

$$\frac{\partial \log p_\theta(x)}{\partial x} = \frac{\partial}{\partial x} \log \left(\frac{\sum_y \exp\left(f_{\theta,y}(x)\right)}{Z(\theta)} \right) \tag{10}$$

$$= \frac{\partial}{\partial x} \log \sum_y \exp(f_{\theta,y}(x)) \tag{11}$$

$$= \sum_y \frac{\exp(f_{\theta,y}(x))}{\sum_y \exp(f_{\theta,y}(x))} \frac{\partial f_{\theta,y}(x)}{\partial x} \tag{12}$$

Or,

$$\frac{\partial \log p_\theta(x)}{\partial x} = \sum_y p_\theta(y \mid x) \frac{\partial f_{\theta,y}(x)}{\partial x} \tag{13}$$

We utilize the Cauchy-Schwarz inequality, as stated below to obtain an upper bound on $\frac{\partial \log p_\theta(x)}{\partial x}$

$$\left(\sum_{k=1}^N u_k v_k\right)^2 \leq \left(\sum_{k=1}^N u_k^2\right)\left(\sum_{k=1}^N v_k^2\right) \tag{14}$$

where $u_k, v_k \in \mathbb{R}$ for all $k \in \{1, 2, ..., N\}$. If we set $u_k = 1$ for all k, the inequality reduces to,

$$\left(\sum_{k=1}^N v_k\right)^2 \leq N\left(\sum_{k=1}^N v_k^2\right) \tag{15}$$

Since, $J_{y,i}^w(x) \in \mathbb{R}$, we have,

$$\left(\sum_{y=1}^C J_{y,i}^w(x)\right)^2 \leq C \sum_{y=1}^C (J_{y,i}^w(x))^2 \tag{16}$$

where C is the total number of output classes. This implies that,

$$\sum_i \left(\sum_{y=1}^C J_{y,i}^w(x)\right)^2 \leq C \sum_i \sum_{y=1}^C (J_{y,i}^w(x))^2 \tag{17}$$

Or,

$$\left\|\frac{\partial \log p_\theta(x)}{\partial x}\right\|_F^2 \leq C \, \|J^w(x)\|_F^2 \tag{18}$$

To summarize, the overall objective we optimize for a mini-batch $B = \{(x, y)\}_{i=1}^{|B|}$ using AMSReg can be expressed as,

$$\mathcal{L}_{\text{Joint}}(B; \theta) = \sum_{(x,y)\in B} \mathcal{L}_{\text{CE}}(x, y; \theta) + \frac{\mu}{2}\left[\frac{1}{|B|}\sum_{x\in B}||J^w(x)||_F^2\right] \tag{19}$$

where $\mathcal{L}_{\text{CE}}(x, y; \theta)$ is the standard cross-entropy loss function and μ is a hyper-parameter that determines the strength of regularization.

Efficient Algorithm for AMSReg. Hoffman et al. [16] introduce an efficient algorithm to calculate the Frobenius norm of the input-output Jacobian $(J_{y,i}(x) = \frac{\partial f_{\theta,y}(x)}{\partial x_i})$ using random projection as,

$$||J(x)||_F^2 = C\mathbb{E}_{\hat{v}\sim S^{C-1}}[||\hat{v}\cdot J||^2] \tag{20}$$

where C is the total number of output classes and $\hat{v}$ is random vector drawn from the $(C-1)$-dimensional unit sphere S^{C-1}. $||J(x)||_F^2$ can then be efficiently approximated by sampling random vectors as,

$$||J(x)||_F^2 \approx \frac{1}{n_{proj}}\sum_{\mu=1}^{n_{proj}}[\frac{\partial(\hat{v}^\mu \cdot z)}{\partial x}]^2 \tag{21}$$

where n_{proj} is the number of random vectors drawn and z is the output vector (see Sect. 2.3 [16] for more details). In AMSReg, we optimize the square of the Frobenius norm of $J^w(x)$ where:

$$J_{y,i}^w(x) = p_\theta(y \mid x)\frac{\partial f_{\theta,y}(x)}{\partial x_i} \tag{22}$$

Since $J^w(x)$ can essentially be expressed as a matrix product of a diagonal matrix of predicted class probabilities and the regular Jacobian matrix, we are able to re-use the efficient random projection algorithm by expressing $||J^w(x)||_F^2$ as,

$$||J^w(x)||_F^2 = C\mathbb{E}_{\hat{v}\sim S^{C-1}}[||\hat{v}^T\text{diag}(p_\theta(c \mid x))J(x)||^2] \tag{23}$$

3.5 Experiments with AMSReg

We run our experiments on the CIFAR-10 and Fashion-MNIST datasets [45]. We compare AMSReg against both Adversarial Training (AT) [24] and Jacobian Regularization (Jacobian Reg) [16]. We perform our experiments across three network architectures – LeNet [20], DDNet [30] and ResNet18 [15]. We train each model for 200 epochs using piecewise learning rate decay (i.e. decay ten-fold after the 100th and 150th epoch). All models are optimized using SGD with a momentum factor of 0.9. Adversarial training models are trained with 10 steps of PGD. When evaluating adversarial robustness, we use an l_∞ bounded 5-steps PGD attack with $\epsilon = 8/255$ and a step size of $2/255$ for models trained on

Table 1. Experiments on CIFAR10: flatness Φ and accuracies(%) on clean and adversarial test samples. PGD attacks are generated with $\epsilon = 8/255$, steps $= 5$ and a step size of $2/255$. AMSReg is able to achieve the highest Φ-Flatness while adversarial training is the most robust defense.

	Defense	Φ (Flatness ↑)	Clean-Acc. ↑ (%)	Adversarial-Acc. ↑ (%) (PGD)
DDNet [30]	No defense	-13035.5	**91.45**	1.7
	AT ($\epsilon = 3/255$)	-2066.4	85.90	**45.71**
	Jacobian Reg. [16] ($\lambda = 0.05$)	-1692.6	86.06	18.31
	AMSReg ($\lambda = 1.0$)	$\mathbf{-768.7}$	85.45	20.78
RN-18 [15]	No Defense	-13974.3	**94.69**	0.1
	AT ($\epsilon = 3/255$)	-1055.0	85.32	**57.48**
	Jacobian Reg. [16] ($\lambda = 0.05$)	-1945.1	82.28	14.73
	AMSReg ($\lambda = 1.0$)	$\mathbf{-659.5}$	86.94	11.07

Table 2. Experiments on Fashion-MNIST: flatness Φ and accuracies(%) on clean and adversarial test samples. PGD attacks are generated with $\epsilon = 25/255$, steps $= 10$ and a step size of $6.25/255$. AMSReg is able to achieve the highest Φ-Flatness while adversarial training is the most robust defense.

	Defense	Φ (Flatness ↑)	Clean-Acc. ↑ (%)	Adversarial-Acc. ↑ (%) (PGD)
LeNet [20]	No defense	-9435.9	**91.53**	0.0
	AT ($\epsilon = 10/255$)	-1504.0	89.35	**62.64**
	Jacobian Reg. [16] ($\lambda = 0.01$)	-1694.6	88.76	34.71
	AMSReg ($\lambda = 0.5$)	$\mathbf{-378.1}$	89.51	34.04
RN-18 [15]	No Defense	-12952.8	**93.75**	0.1
	AT ($\epsilon = 10/255$)	-1257.8	91.64	**65.24**
	Jacobian Reg. [16] ($\lambda = 0.01$)	-821.7	91.04	45.37
	AMSReg ($\lambda = 0.5$)	$\mathbf{-482.9}$	91.29	46.06

CIFAR10 and a 10-step PGD attack with $\epsilon = 25/255$ and a step size of $6.25/255$ for models trained on Fashion-MNIST.

Results and Analysis. Table 1 and Table 2 show the results of AMSReg compared to adversarial training and Jacobian regularization on CIFAR10 and Fashion-MNIST. It is known that there is a trade-off between adversarial robustness and accuracy [41,49]. Therefore, here we only compare models that have similar accuracy on clean test images. As observed in Table 1 and 2, models trained with AMSReg indeed have flatter likelihood landscapes than Jacobian regularization and adversarial training, indicating that the regularization was successful. Furthermore, AMSReg models have comparable adversarial robustness to that of Jacobian regularization. However, they are still significantly susceptible to adversarial attacks than adversarial training.

We conduct further analysis by looking at the flatness along the attack directions (i.e., the direction an FGSM attack chosen to perturb the image) instead of random directions. Figure 5 shows the likelihood landscape projected using one FGSM direction and one random direction. Interestingly enough, both Jacobian and AMS regularization are able to flatten the landscape in the random direction well (with AMS being slightly flatter than Jacobian). However, neither regularization method is able to flatten out the FGSM attack direction nearly as much as adversarial training. This might explain why neither AMS nor Jacobian regularization are able to match the adversarial robustness of adversarial training. We leave the next step of flattening out the worst-case directions (i.e. the attack directions) for future work.

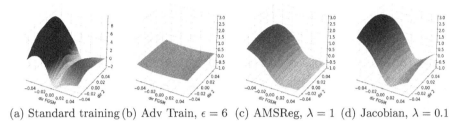

(a) Standard training (b) Adv Train, $\epsilon = 6$ (c) AMSReg, $\lambda = 1$ (d) Jacobian, $\lambda = 0.1$

Fig. 5. Relative likelihood landscapes projected using one FGSM direction (`dir FGSM`) and a random direction (`dir2`). Each visualization uses DDNet trained on CIFAR10. (a) uses standard training with no defense, (b) is adversarially trained with a 10-step PGD attack, (c) uses AMSReg with $\lambda = 1$, and (d) uses Jacobian Regularization with $\lambda = 0.1$. We see that adversarial training is able to flatten the likelihood landscape in the FGSM direction much more than the regularization defenses.

4   Conclusion

In this work, we explore why neural networks are vulnerable to adversarial attacks from a perspective of the marginal distribution of the inputs (images) under the trained model. We first suggest a way to visualize the likelihood landscape of CNNs by leveraging the recently proposed EBM interpretation of a discriminatively learned classifier. Qualitatively, we show that a subset of standard defense techniques such as adversarial training and Jacobian regularization share a common pattern: they induce flat likelihood landscapes. We then quantitatively show this correlation by introducing a measure regarding the flatness of the likelihood landscape in the surrounding area of clean samples. We also explore directly regularizing a term that encourages flat likelihood landscapes, but this results in worse adversarial robustness than adversarial training and is roughly comparable to just Jacobian regularization. After further analysis, we find that adversarial training significantly flattens the likelihood landscape in the directions abused by adversarial attacks. The regularization methods, on the other hand, are better at flattening the landscape in random directions. These findings suggest that flattening the likelihood landscape is important for adversarial robustness, but it's most important for the landscape to be flat in the directions chosen by adversarial attacks.

Overall, our findings in this paper provide a new perspective of adversarial robustness as flattening the likelihood landscape and show how different defenses and regularization techniques address this similar core deficiency. We leave designing a regularizer that flattens the likelihood landscape in attack directions for future investigation.

References

1. Antol, S., et al.: VQA: visual question answering. In: International Conference on Computer Vision (2015)
2. Carlini, N., Wagner, D.: Towards evaluating the robustness of neural networks. In: 2017 IEEE Symposium on Security and Privacy (SP), pp. 39–57. IEEE (2017)
3. Das, A., et al.: Visual dialog. In: Proceedings of the IEEE Conference on Computer Vision and Pattern Recognition (CVPR) (2017)
4. Doersch, C., Gupta, A., Efros, A.A.: Unsupervised visual representation learning by context prediction. In: Proceedings of the IEEE International Conference on Computer Vision, pp. 1422–1430 (2015)
5. Drucker, H., Le Cun, Y.: Improving generalization performance using double back-propagation. IEEE Trans. Neural Netw. **3**(6), 991–997 (1992)
6. Engstrom, L., Ilyas, A., Santurkar, S., Tsipras, D., Tran, B., Madry, A.: Adversarial robustness as a prior for learned representations. arXiv preprint arXiv:1906.00945 (2019)
7. Feinman, R., Curtin, R.R., Shintre, S., Gardner, A.B.: Detecting adversarial samples from artifacts. arXiv preprint arXiv:1703.00410 (2017)
8. Girshick, R.: Fast R-CNN. In: Proceedings of the IEEE International Conference on Computer Vision, pp. 1440–1448 (2015)
9. Goodfellow, I., et al.: Generative adversarial nets. In: Advances in Neural Information Processing Systems, pp. 2672–2680 (2014)
10. Goodfellow, I.J., Shlens, J., Szegedy, C.: Explaining and harnessing adversarial examples. arXiv preprint arXiv:1412.6572 (2014)
11. Goodfellow, I.J., Vinyals, O., Saxe, A.M.: Qualitatively characterizing neural network optimization problems. arXiv preprint arXiv:1412.6544 (2014)
12. Grathwohl, W., Wang, K.C., Jacobsen, J.H., Duvenaud, D., Norouzi, M., Swersky, K.: Your classifier is secretly an energy based model and you should treat it like one. arXiv preprint arXiv:1912.03263 (2019)
13. Gu, S., Rigazio, L.: Towards deep neural network architectures robust to adversarial examples. arXiv preprint arXiv:1412.5068 (2014)
14. Guo, C., Rana, M., Cisse, M., Van Der Maaten, L.: Countering adversarial images using input transformations. arXiv preprint arXiv:1711.00117 (2017)
15. He, K., Zhang, X., Ren, S., Sun, J.: Deep residual learning for image recognition. In: Proceedings of the IEEE Conference on Computer Vision and Pattern Recognition, pp. 770–778 (2016)
16. Hoffman, J., Roberts, D.A., Yaida, S.: Robust learning with Jacobian regularization. arXiv preprint arXiv:1908.02729 (2019)
17. Ilyas, A., Santurkar, S., Tsipras, D., Engstrom, L., Tran, B., Madry, A.: Adversarial examples are not bugs, they are features. In: Advances in Neural Information Processing Systems, pp. 125–136 (2019)
18. Jakubovitz, D., Giryes, R.: Improving DNN robustness to adversarial attacks using Jacobian regularization. In: Ferrari, V., Hebert, M., Sminchisescu, C., Weiss, Y. (eds.) ECCV 2018. LNCS, vol. 11216, pp. 525–541. Springer, Cham (2018). https://doi.org/10.1007/978-3-030-01258-8_32

19. Kurakin, A., Goodfellow, I., Bengio, S.: Adversarial examples in the physical world. arXiv preprint arXiv:1607.02533 (2016)
20. Lecun, Y., Bottou, L., Bengio, Y., Haffner, P.: Gradient-based learning applied to document recognition. Proc. IEEE **86**(11), 2278–2324 (1998). https://doi.org/10.1109/5.726791
21. Lee, K., Lee, K., Lee, H., Shin, J.: A simple unified framework for detecting out-of-distribution samples and adversarial attacks. In: Advances in Neural Information Processing Systems, pp. 7167–7177 (2018)
22. Li, H., Xu, Z., Taylor, G., Studer, C., Goldstein, T.: Visualizing the loss landscape of neural nets. In: Advances in Neural Information Processing Systems, pp. 6389–6399 (2018)
23. Ma, X., et al.: Characterizing adversarial subspaces using local intrinsic dimensionality. arXiv preprint arXiv:1801.02613 (2018)
24. Madry, A., Makelov, A., Schmidt, L., Tsipras, D., Vladu, A.: Towards deep learning models resistant to adversarial attacks. arXiv preprint arXiv:1706.06083 (2017)
25. Moosavi-Dezfooli, S.M., Fawzi, A., Uesato, J., Frossard, P.: Robustness via curvature regularization, and vice versa. In: Proceedings of the IEEE Conference on Computer Vision and Pattern Recognition, pp. 9078–9086 (2019)
26. Pang, T., Du, C., Dong, Y., Zhu, J.: Towards robust detection of adversarial examples. In: Advances in Neural Information Processing Systems, pp. 4579–4589 (2018)
27. Pang, T., Xu, K., Dong, Y., Du, C., Chen, N., Zhu, J.: Rethinking softmax cross-entropy loss for adversarial robustness. arXiv preprint arXiv:1905.10626 (2019)
28. Pang, T., Xu, K., Zhu, J.: Mixup inference: better exploiting mixup to defend adversarial attacks. arXiv preprint arXiv:1909.11515 (2019)
29. Papernot, N., McDaniel, P., Jha, S., Fredrikson, M., Celik, Z.B., Swami, A.: The limitations of deep learning in adversarial settings. In: 2016 IEEE European Symposium on Security and Privacy (EuroS&P), pp. 372–387. IEEE (2016)
30. Papernot, N., McDaniel, P., Wu, X., Jha, S., Swami, A.: Distillation as a defense to adversarial perturbations against deep neural networks. In: 2016 IEEE Symposium on Security and Privacy (SP), pp. 582–597. IEEE (2016)
31. Qin, C., et al.: Adversarial robustness through local linearization. In: Advances in Neural Information Processing Systems, pp. 13824–13833 (2019)
32. Qin, Y., Frosst, N., Sabour, S., Raffel, C., Cottrell, G., Hinton, G.: Detecting and diagnosing adversarial images with class-conditional capsule reconstructions. arXiv preprint arXiv:1907.02957 (2019)
33. Raff, E., Sylvester, J., Forsyth, S., McLean, M.: Barrage of random transforms for adversarially robust defense. In: Proceedings of the IEEE Conference on Computer Vision and Pattern Recognition, pp. 6528–6537 (2019)
34. Ross, A.S., Doshi-Velez, F.: Improving the adversarial robustness and interpretability of deep neural networks by regularizing their input gradients. In: Thirty-Second AAAI Conference on Artificial Intelligence (2018)
35. Samangouei, P., Kabkab, M., Chellappa, R.: Defense-GAN: protecting classifiers against adversarial attacks using generative models. arXiv preprint arXiv:1805.06605 (2018)
36. Shafahi, A., et al.: Adversarial training for free! In: Advances in Neural Information Processing Systems, pp. 3353–3364 (2019)
37. Sokolić, J., Giryes, R., Sapiro, G., Rodrigues, M.R.: Robust large margin deep neural networks. IEEE Trans. Signal Process. **65**(16), 4265–4280 (2017)
38. Svoboda, P., Hradis, M., Barina, D., Zemcik, P.: Compression artifacts removal using convolutional neural networks. arXiv preprint arXiv:1605.00366 (2016)

39. Szegedy, C., et al.: Intriguing properties of neural networks. In: International Conference on Learning Representations (2014). http://arxiv.org/abs/1312.6199

40. Tramèr, F., Boneh, D.: Adversarial training and robustness for multiple perturbations. In: Advances in Neural Information Processing Systems, pp. 5866–5876 (2019)

41. Tsipras, D., Santurkar, S., Engstrom, L., Turner, A., Madry, A.: Robustness may be at odds with accuracy. arXiv preprint arXiv:1805.12152 (2018)

42. Varga, D., Csiszárik, A., Zombori, Z.: Gradient regularization improves accuracy of discriminative models. arXiv preprint arXiv:1712.09936 (2017)

43. Wan, W., Zhong, Y., Li, T., Chen, J.: Rethinking feature distribution for loss functions in image classification. In: Proceedings of the IEEE Conference on Computer Vision and Pattern Recognition, pp. 9117–9126 (2018)

44. Wong, E., Rice, L., Kolter, J.Z.: Fast is better than free: revisiting adversarial training. arXiv preprint arXiv:2001.03994 (2020)

45. Xiao, H., Rasul, K., Vollgraf, R.: Fashion-MNIST: a novel image dataset for benchmarking machine learning algorithms. arXiv preprint arXiv:1708.07747 (2017)

46. Xie, C., Wang, J., Zhang, Z., Ren, Z., Yuille, A.: Mitigating adversarial effects through randomization. arXiv preprint arXiv:1711.01991 (2017)

47. Yu, F., Qin, Z., Liu, C., Zhao, L., Wang, Y., Chen, X.: Interpreting and evaluating neural network robustness. arXiv preprint arXiv:1905.04270 (2019)

48. Zhang, D., Zhang, T., Lu, Y., Zhu, Z., Dong, B.: You only propagate once: accelerating adversarial training via maximal principle. In: Advances in Neural Information Processing Systems, pp. 227–238 (2019)

49. Zhang, H., Yu, Y., Jiao, J., Xing, E.P., Ghaoui, L.E., Jordan, M.I.: Theoretically principled trade-off between robustness and accuracy. arXiv preprint arXiv:1901.08573 (2019)

50. Zheng, Z., Hong, P.: Robust detection of adversarial attacks by modeling the intrinsic properties of deep neural networks. In: Advances in Neural Information Processing Systems, pp. 7913–7922 (2018)

Deep k-NN Defense Against Clean-Label Data Poisoning Attacks

Neehar Peri, Neal Gupta, W. Ronny Huang$^{(\boxtimes)}$, Liam Fowl, Chen Zhu, Soheil Feizi, Tom Goldstein, and John P. Dickerson

Center for Machine Learning, University of Maryland, College Park, USA
wronnyhuang@gmail.com, john@cs.umd.edu

Abstract. Targeted clean-label data poisoning is a type of adversarial attack on machine learning systems in which an adversary injects a few correctly-labeled, minimally-perturbed samples into the training data, causing a model to misclassify a particular test sample during inference. Although defenses have been proposed for general poisoning attacks, no reliable defense for clean-label attacks has been demonstrated, despite the attacks' effectiveness and realistic applications. In this work, we propose a simple, yet highly-effective Deep k-NN defense against both feature collision and convex polytope clean-label attacks on the CIFAR-10 dataset. We demonstrate that our proposed strategy is able to detect over 99% of poisoned examples in both attacks and remove them without compromising model performance. Additionally, through ablation studies, we discover simple guidelines for selecting the value of k as well as for implementing the Deep k-NN defense on real-world datasets with class imbalance. Our proposed defense shows that current clean-label poisoning attack strategies can be annulled, and serves as a strong yet simple-to-implement baseline defense to test future clean-label poisoning attacks. Our code is available on GitHub.

Keywords: Machine learning · Adversarial attacks · Clean label poisoning · Deep k-NN

1 Introduction

Machine-learning-based systems are increasingly being deployed in settings with high societal impact, including hate speech detection on social networks [22], autonomous driving [4], biometric-based applications [29], and malware detection [20]. In these real world applications, a system's robustness to not only noise, but also *adversarial manipulation* is paramount. With an increasing number of machine learning systems trained on data sourced from public and semi-public places such as social networks, collaboratively-edited forums, and multimedia posting services, adversaries can strategically inject training data to manipulate or degrade system performance.

N. Peri, N. Gupta and W. R. Huang—The first three authors contributed equally to this work.

© Springer Nature Switzerland AG 2020
A. Bartoli and A. Fusiello (Eds.): ECCV 2020 Workshops, LNCS 12535, pp. 55–70, 2020.
https://doi.org/10.1007/978-3-030-66415-2_4

Data poisoning attacks on neural networks occur at training time, wherein an adversary places specially-constructed *poisoned examples* into the training data with the intention of manipulating the behavior of the system at test time. Recent work on data poisoning has focused on either (i) an attacker generating a small fraction of training inputs to degrade overall model performance, or (ii) a defender aiming to detect or otherwise mitigate the impact of that attack. In this paper, we focus on *clean-label* data poisoning [25], where an attacker injects a small number of *correctly labeled*, minimally perturbed samples into the training data. In contrast with traditional data poisoning, these samples are crafted to cause a model to misclassify a particular *target* test sample during inference. These attacks are plausible in a wide range of applications, as they do not require the attacker to have control over the labeling function. Many large scale data sets are automatically scraped from the internet without direct supervision, so an adversary need only share their poisoned data online.

Our Contribution: In this paper, we initiate the study of defending against *clean-label* poisoning attacks on neural networks by considering feature collision [25] and convex polytope attacks [35] on the CIFAR-10 dataset. Although poison examples are not easily detected by human annotators, we exploit the property that adversarial examples have different feature distributions than their clean counterparts in higher layers of the network, and that those features often lie near the distribution of the target class. This intuition lends itself to a defense based on k nearest neighbors in feature space, in which the poison examples are detected and removed *prior* to training. Further, the parameter k yields a natural lever for trading off between the number of undetected poisons and number of discarded clean images when filtering the training set.

Our contributions can be outlined as follows.

- We propose a novel Deep k-NN defense for clean-label poisoning attacks. We evaluate it against state-of-the-art clean-label data poisoning attacks, using a slate of architectures and show that our proposed strategy detects 99% of the poison instances without degrading overall performance.
- We reimplement a set of general data poisoning defenses [14], including L_2-Norm Outliers, One-Class SVMs, Random Point Eviction, and Adversarial Training as baselines and show that our proposed Deep k-NN defense is more robust at detection of poisons in the trained victim models.
- From the insights of two ablation studies, we assemble guidelines for implementing Deep k-NN in practice. First we provide instructions for picking an appropriate value for k. Second, we provide a protocol for using the Deep k-NN defense when class imbalance exists in the training set.

2   Overview of Clean-Label Data Poisoning

We briefly describe the how clean-label data poisoning works and the intuition behind a neighborhood conformity defense. Figure 1 shows the feature space representation (i.e. the representations in the penultimate layer of the network) for a targeted poisoning attack that causes a chosen target airplane image (feature representation shown as the dark gray triangle) to be misclassified as a frog during inference. To accomplish this, poison frog images (feature representation shown as dark orange circles) are perturbed to surround the target airplane in feature space. After training on this poisoned data set, the model changes its decision boundary between the two classes in order to accommodate the poison frogs, enveloping them onto the side of the frogs. Inadvertently, the nearby target airplane is also placed on the side of the frogs, leading to misclassification. Under the *feature collision* attack [25], the perturbations are optimized so as to minimize the poison images' distance to the target image in feature space,

$$\boldsymbol{x}_p = \arg\min_{\boldsymbol{x}} \ |\phi(\boldsymbol{x}) - \phi(\boldsymbol{x}_t)|_2^2 + |\boldsymbol{x} - \boldsymbol{x}_b|_2^2,$$

where $\boldsymbol{x}_p$, $\boldsymbol{x}_b$, $\boldsymbol{x}_t$ are the poison, base, and target images, respectively, and ϕ is a feature extractor that propagates input images to the penultimate layer of the network. Alternatively, under the *convex polytope* attack [35], poisoned data points are optimized to form a convex hull of poisons around the target via a more sophisticated loss function. In both cases nonetheless, models fine-tuned on the poisoned dataset will have their decision boundaries adversarially warped and classify the targeted airplane image as a frog at inference time. Though the optimization causes a noticeable change in the feature representations of the images, the poison frogs are perturbed under some small ℓ_2 or ℓ_∞ constraint so that they still appear to be frogs to a human observer.

2.1 Intuition Behind Deep k-NN Defense

As seen in Fig. 1, poisons are surrounded by feature representations of the target class rather than of the base class. For instance, when $k = 3$ and $n_{poison} = 2$, each poison will almost always have a plurality of its neighbors as a non-poison in the target class. Since the plurality label of a poison's neighbors does not match the label of the poison itself, the poison can be removed from the dataset

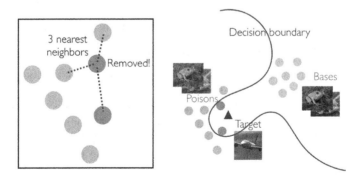

Fig. 1. Proposed Deep k-NN defense ($k = 3$) correctly removing a poisoned example by comparing the class labels of poison with its k neighbors. Since a majority of the k points surrounding the poison do not share the same class label as the poison, it is removed.

or simply not used for training. More generally, if $k > 2n_{poison}$, then we would expect the poisons to be outvoted by members of the target class and be filtered from the training set. Note that by setting $k > 2n_{poison}$, the poisons' label cannot be the majority, but may still be the plurality, or mode, of the Deep k-NN set if the nearest neighbors of the current point are in multiple classes. Empirically, however, we do not observe this to be the case. Extracted features tend to be well-clustered by class; thus there are usually only 2 unique classes in the Deep k-NN neighborhood, base class and target class, with the target class being larger. Therefore, in order to successfully defend against adversarial manipulation, a victim needs only to set a sufficiently large value of k without needing to know exactly how many poisons there are *a-priori*. We further elucidate on the effect of k in Sect. 6.

3   Related Work

We briefly overview related work in the space of defenses to adversarial attacks [2,8], which are categorized into two groups: inference time evasion attacks and train time data poisoning attacks. Most adversarial defenses have focused on mitigating evasion attacks, where inference-time inputs are manipulated to cause misclassification. In neural networks, evasion adversarial examples are perturbed such that the loss on the victim network increases. The search for an optimal perturbation is facilitated by use of the local gradient $\nabla_x \mathcal{L}$ obtained via backpropagation on either a white box network or a surrogate network if the victim network is unknown [16]. Many defenses against evasion attacks leverage the attacker's reliance on gradient information by finding ways to obfuscate gradients, using non-differentiable layers or reshaping the loss surface such that the gradients are highly uncorrelated. [1] showed that obfuscated gradient defenses are insufficient for defending against evasion attacks. Using various strategies to circumvent loss of gradient information, such as replacing non-differentiable layers with differentiable approximations during the backward pass, [1] demonstrates that stronger attacks can reduce inference accuracy to near zero on most gradient-based defenses. Defense strategies that withstand strong attacks are characterized by loss surfaces that are "smooth" with respect to a particular input everywhere in the data manifold. Variants of adversarial training [17,26,33] and linearity or curvature regularizers [18,21] have maintained modest accuracy despite strong multi-iteration PGD attacks [17].

In evasion attacks, Deep k-NN based methods have been used across multiple layers of a neural network to generate confidence estimates of network predictions as a way to detect adversarial examples [19]. Similarly, [27] proposes a white box threat model where an adversary has full access to the training set, and uses prior knowledge of model hyper-parameters, including the value of k used in the Deep k-NN defense when constructing poisons for general attacks. Our Deep k-NN based defense differs in that it identifies and filters poisoned data at training time rather than at test time, using only ground truth labels. Furthermore, a soft nearest neighbor regularizer has been used during training time to improve robustness to evasion examples [6], but its resistance to clean-label poisoning examples has yet to be explored.

Backdoor attacks have recently received attention from the research community as a realistic threat to machine learning models. Backdooring, proposed by [9], can be seen as a subset of data poisoning. In their simplest form, backdoor attacks modify a small number of training examples with a specific *trigger* pattern that is accompanied by a *target* label. These attacks exploit a neural network's ability to over fit to the training set data, and use the trigger at inference time to misclassify an example into the target class. The trigger need not change the ground truth label of the training example, making such attacks clean-label attacks [32]. However, these attacks rely upon the attacker being able to modify data at inference time, an assumption that may not always hold true, and one we do not make in this paper. A number of defenses to backdoor attacks have been proposed, primarily seeking to sanitize training data by detecting and removing poisons. Often, these defenses rely upon the heuristic that backdoor attacks create "shortcuts" in a neural network to induce target misclassification. [28] employed two variants of an ℓ_2 centroid defense, which we adapt in this paper. In one case, data is anomalous if it falls outside of an acceptable radius in feature space. Alternatively, data is first projected onto a line connecting class centroids in feature space and is removed based on its position on this line.

[3] proposed using feature clustering for data sanitation. This defense assumes that naive backdoor triggers will cause poison samples to cluster in feature space. The success for this defense diminishes drastically when exposed to stronger poisoning methods which do not use uniform triggers. Convex polytope attacks [35] create much stronger poisons by surrounding a target image in feature space with a convex hull of poisons. Such attacks will not always result in easily identifiable clusters of poisons. [31] examines spectral signatures as a method for detecting backdoor attacks, stating that all attacks share a set of underlying properties. Spectral signatures are boosted in learned representations, and can be used to identify poisoned images through SVD.

4 Defenses Against Clean-Label Poisoning

In this section, we formally introduce the Deep k-NN defense as well as a set of other baseline defenses against clean-label targeted poisoning attacks. We compare the effectiveness of each defense against both feature collision attacks and convex polytope attacks in Sect. 5.

We use $\boldsymbol{x}_t$ to denote the input space representation of the target image that an adversary tries to misclassify. The target has true label l_t but the attacker seeks to misclassify it as having label l_b. We use $\boldsymbol{x}_b$ to denote a base image having label l_b that is used to build a poison after optimization. We use $\boldsymbol{x}_w$ to denote a base image watermarked with a target image, that is $\gamma \cdot \boldsymbol{x}_t + (1 - \gamma) \cdot \boldsymbol{x}_b$. To a human observer this image will retain the label l_b when γ is sufficiently low. We use $\phi(\boldsymbol{x})$ to denote the activations of the penultimate layer of a neural network. We refer to this as the *feature layer* or *feature space* and $\phi(\boldsymbol{x})$ as *features* of $\boldsymbol{x}$.

Deep k-NN Defense: For each data point in the training set, the Deep k-NN defense takes the plurality vote amongst the labels of that point's k nearest neighbors in feature space. If the point's own label is not the mode amongst labels of its k nearest neighbors, the point is flagged as anomalous, and is not used when training the model. We use Euclidean distance to measure the distance between data points in feature space. See Algorithm 1.

Algorithm 1: Deep k-NN Defense

Result: Filtered training set $X^{train'}$
Let $S_k(x^{(i)})$ denote a set of k points such that for all points $x^{(j)}$ inside the set and points $x^{(l)}$ outside the set, $|\phi(x^{(l)}) - \phi(x^{(i)})|_2 \geq |\phi(x^{(j)}) - \phi(x^{(i)})|_2$
$X^{train'} \leftarrow \{\}$
for *Data points* $\boldsymbol{x}^{(i)} \in X^{train}$ **do**
    Let l denote the label of $x^{(i)}$ and let $l(S_k(x^{(i)}))$ denote the labels of the
    points in $S_k(x^{(i)})$
    if $l \in mode(l(S_k(x^{(i)})))$ **then**
        | $X^{train'} \leftarrow X^{train'} \cup \{x^{(i)}\}$;
    else
        | Omit $x^{(i)}$ from $X^{train'}$;
    end
end

L2-Norm Outlier Defense: The L2 norm outlier defense removes an $\epsilon > 0$ fraction of points that are farthest in feature space from the centroids of their classes. For each class of label $l \in \mathcal{L}$, with size $s_l = |x^{(j)}$ s.t. $l(j) = l|$, we compute the centroid c_l as

$$c_l = \frac{1}{s_l} \sum_{x^{(j)} s.t.l(j)=l} \phi(x^{(j)})$$

and remove $\lfloor \epsilon s_l \rfloor$ points maximizing $|\phi(x^{(j)}) - c_l|_2$. The L2 norm defense relies on the position of the centroid to filter outliers. However, the position of the centroid itself is prone to data poisoning if the per-class data size is small. This defense is adapted from traditional poison defenses not specific to neural networks [14].

One-Class SVM Defense: The one-class SVM defense examines the deep features of each class in isolation by applying the one-class SVM algorithm [24] to identify outliers in feature space for each label in the training set. It utilizes a radial basis kernel and is calibrated with a value $\nu = 0.01$.

Random Point Eviction Defense: The random point eviction defense is a simple experimental control. It filters out a random subset of all training data. We remove 1% of our training data for the feature collision attack and 10% of our training data on the convex polytope attack. If the poisoning attack is sensitive to poisons being removed, the random defense may be successful, at the cost of losing a proportionate amount of the unpoisoned training data.

Adversarial Training Defense: Thus far, we have only considered defenses which filter out examples prior to training. We consider here another defense strategy that does not involve filtering, but rather involves an alternative victim training procedure. Adversarial training, used often to harden networks against evasion attacks [7,17], has been shown to produce neural network feature extractors which are less sensitive to weak features such as norm-bounded adversarial patterns [13]. We explore here whether a victim's use of an adversarially trained feature extractor would yield features that are robust to clean-label data poisoning. Instead of the conventional loss over the training set, adversarial training aims to optimize

$$\min_{\theta} \mathcal{L}_\theta(X + \delta^*), \text{where } \delta^* = \underset{\delta < \epsilon}{\operatorname{argmax}} \mathcal{L}_\theta(X + \delta),$$

where θ, X, and δ are the weights, training input, and adversarial perturbations, respectively, and $\mathcal{L}_\theta$ is some training loss (i.e., cross-entropy). In our experiments, we perform adversarial training following the standard procedure in [17], using an ℓ_∞ PGD adversary of 20 steps and $\epsilon = 8$.

5 Evaluation

In this section, we evaluate the effectiveness of our Deep k-NN defense and baseline defenses against the feature collision [25] and convex polytope [35] attacks on the CIFAR-10 dataset [15]. All model architectures, data splits, and hyperparameters are taken directly from the evaluation setups used in [25,35]. We define the defense success rate as the number of times the poisoning attack fails to cause the target example to be misclassified, divided by the number of attempts. We only consider sets of poisons that lead to successful attacks in the undefended case so by definition the undefended defense success rate is 0%.

5.1 Defense Against Feature Collision Attacks

Attack Procedure. We randomly select 50 images in the base class. For each base image with input representation x_b, we compute the watermark base $x_w \leftarrow \gamma \cdot x_t + (1 - \gamma) \cdot x_b$, then optimize p with initial value w using a forward-backward splitting procedure to solve

$$x_p = \arg \min_{x} |\phi(x) - \phi(x_t)|_2^2 + \beta |x - x_w|_2^2$$

The hyperparameter β is fixed at 0.1. The resulting poisons x_p are both close to the target image x_t in feature space, and close to the watermarked input x_w in image space. To ensure statistical significance, we craft 16 of these collections of 50 poisons and evaluate each collection independently.

Defense Procedure. As in the original setup [25], we first train a modified AlexNet to convergence using only clean data. Next we apply our defenses on the set of clean data plus poisons to obtain a filtered dataset. That filtered dataset is then used to fine tune the pretrained model over 10 epoch with a batch size of 128. We evaluate the performance of all defenses described in Sect. 4 against collections of 50 poisons that successfully cause a targeted misclassification.

Results. As seen in Table 1, the Deep k-NN defense with $k = 5000$, successfully identifies all but one poison across multiple attacks, while filtering just 0.6% of the clean images from the training set. As a result, after victim training, models defended by Deep k-NN have defense success rates of 100%. In contrast, the $L2$-norm defense only identifies roughly half the feature collision poisons using $\epsilon = 0.01$. Both the One-Class SVM and the Random Point Eviction defenses are unable to detect a majority of the feature collision poisons.

Table 1. Comparing the effectiveness of baseline defenses aggregated for all model architectures in Feature Collision Attack

Defense strategy	Poisons removed	Clean images removed (%)	Defense success rate (%)	CIFAR-10 test accuracy (%)
Deep k-NN ($k = 5000$)	**799/800**	**0.6**	**100.0**	**74.6**
$L2$-norm outliers	395/800	1.0	50.0	**74.6**
One-class SVM	168/800	1.0	37.5	74.5
Random point eviction	84/800	10.0	12.5	74.5

5.2 Defense Against Convex Polytope Attacks

Attack Procedure. Following the procedure in [35], the CIFAR-10 dataset is split into 48000 images for pretraining, and 500 images for fine-tuning. The poison base images are taken from the remaining split of 1500 images.

Since the attacker does not know the victim model parameters, they first pre-train their own model to convergence using the same subset of 48000 CIFAR-10 images used for pretraining. Next, an adversary uses this surrogate model to craft 5 poisons using the convex polytope method. To ensure statistical significance, 102 collections of 5 poisons are crafted.

When crafting convex polytope poisons, multiple surrogate models with different architectures are ensembled, so that the generated poisons generalize to victim architectures that the poisons were not crafted on. Our results are based on eight architectures: two of which are not used in crafting the poisons (black box setting), and six which use random initialization (grey box setting). The grey-box architectures are DPN92 [5], GoogLeNet [30], MobileNetV2 [23], ResNet50 [10], ResNeXT29-2x64d [34], and SENet18 [11], while the black-box architectures are DenseNet121 [12] and ResNet18 [10].

Defense Procedure. The victim model is first pretrained to convergence using a random initialization unknown to the attacker on the 48000 pretraining images from CIFAR-10[1]. Our defenses are applied to the 500 fine-tuning images plus poisons to obtain a filtered fine-tuning set[2]. Finally, this filtered dataset is used to fine-tune the victim model.

Again, the performance of all defenses is reported only on collections of poisons that lead to a successful attack in the undefended case. Since the attacker did not have access to the victim architecture or model parameters during crafting of the poisons, the defenses are evaluated independently for each individual victim architecture.

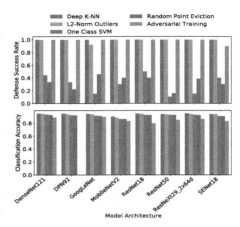

Fig. 2. The Deep k-NN Defense is model-agnostic, achieving high defense success rate and test classification accuracy.

Results. The aggregate results of each defense strategy on all 8 architectures are shown in Table 2. Both the Deep k-NN and $L2$-Norm defense filter out nearly all poisons, while incorrectly removing 4.3% and 9.1% of the clean training examples, respectively. Compared to feature collision poisons, convex polytope poisons trigger more false positive detections (i.e. clean images removed) across all defense methods, leading to fewer remaining clean examples and reduced test accuracy.

Surprisingly, the $L2$-Norm defense is much better able to detect convex polytope poisons compared to feature collision poisons; it detects almost as many as Deep k-NN . However, it has a lower specificity because it removes more clean images, resulting in half-percent lower test accuracy. These results are broken down for each victim architecture in Fig. 2. The Deep k-NN attack is successful on all architectures with perfect defense success rate. $L2$-norm Outliers and Adversarial Training perform almost as well. Other strategies largely fail to be a viable defense.

We evaluate the effectiveness of adversarial training on the Convex Polytope-crafted poisons. In Table 2 and Fig. 2, adversarially trained feature extractors—trained naively to provide resistance against only evasion attacks—do in fact help mitigate poisoning attacks as well. To our knowledge, this is the first time adversarial training has been shown to provide resistance against data poisoning (i.e. training time) attacks and is a direction for future work.

---

[1] We use conventional training loss for all except the adversarial training defense.

[2] There is no filtering in adversarial training.

The defense however significantly hurts test set accuracy (as is common for adversarially trained networks), which drops to 85% on average, compared with 94% on the same architectures without adversarial training. In scenarios when adversarial training for evasion attack robustness is not required, such as in situations when adversaries cannot control test time inputs, the Deep k-NN defense provides the poisoning resistance without the burden of decreased generalization performance.

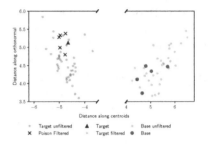

Fig. 3. Feature space visualization of the Deep k-NN defense against a convex polytope attack on DPN92.

Table 2. Comparing the effectiveness of baseline defenses aggregated for all model architectures in convex polytope attack

Defense strategy	Poisons removed	Clean images removed (%)	Defense success rate (%)	CIFAR-10 test accuracy (%)
Deep k-NN ($k = 50$)	**510/510**	**4.3**	**100.0**	**93.9**
$L2$-norm outliers	509/510	9.1	99.0	93.4
One-class SVM	114/510	7.1	29.9	91.7
Random Point Eviction	47/510	10.0	33.2	91.3
Adversarial Training	–	–	98.6	85.2

5.3 Feature Space Visualization

The favorable results of Deep k-NN defense also afford us an opportunity to understand anomaly detection in deep networks more generally via observing the effects in feature representations. A feature space visualization of the penultimate layer of the network is shown in Fig. 3, with both filtered poisons and non-poisons displayed.

Specifically, Fig. 3 shows a projected visualization in the feature space of the fine tuning set in the target (blue) and base (green) classes. Following the projection scheme used in [25], where the x-axis is the direction along the line connecting the centroids of the target and base class features and the y-axis is the component of the parameter vector (i.e. decision boundary) orthogonal to the between-centroids vector, the deep features of the DPN92 network are projected into a two-dimensional plane. The "x" markers denote poisons that are filtered out by the defense and would have otherwise almost formed a convex polytope around the target (blue triangle). The Deep k-NN acts with high specificity: all the poisons are filtered, while only 2 outlying clean points in the target class (not shown) are also filtered. No points in the base class are filtered.

5.4   Limitations of the Deep k-NN Defense

The Deep k-NN defense exploits feature space clustering seen in feature collision and convex polytope attacks. It may not be as effective if this initial condition is not met. We view this as a strong and simple baseline defense for poisoning attacks that shows the need for more sophisticated and adaptive attacks.

6   Ablation Studies and Best Practices

We now turn to ablation studies to gain insight into best practices for using the Deep k-NN defense under realistic situations. All results are reported on the convex polytope attack for CIFAR-10 as described in [35] on all 8 architectures discussed previously. We specifically focus on the convex polytope attack method since it is shown to act as a stronger poison on black-box threat models, and study the transfer learning case to mimic the common practice of using pre-trained feature-extractors trained on large datasets.

We again closely mimick the setup in [35] using the first 4800 images in each class to train a model from scratch and then using the next 50 images of each class (making a fine-tuning set size of 500) to fine-tune the model. The Adam optimizer with a learning rate of 0.1 is used. In both studies, we assign frogs as the target class and ships as the base class. The first 5 ship images from the fine-tuning set are replaced with the 5 poisoned ships. Each set of 5 poisoned ships has an associated target frog image that is neither in the training nor fine-tune set. We use the standard CIFAR-10 test split to measure test accuracy.

6.1   Choosing a Value of k

In our first study, we vary the value of k used in the Deep k-NN defense. Since dataset sizes vary, as well as the number of classes, we normalize k against the number of data points *per class*. Specifically, we measure all metrics against a normalized-k ratio, such that normalized-$k = k/N$ where k is the number of nearest neighbors considered by the Deep k-NN and N is the maximum number of examples for any class in the fine-tune set.

As seen in Fig. 4 (top left and middle left), the defense success rate and MCC begin to reach maximum levels at normalized-$k = 0.2$, corresponding to an (unnormalized) k of twice the number of poisons, $k = 10 = 2n_{poison}$. This confirms our intuition in Sect. 2: when $k > 2n_{poison}$, poisons will be marked anomalous since the poison class cannot be the majority label in the neighborhood and is unlikely to be the plurality because the neighborhood usually only contains two unique classes. Of course, the victim must set a value of k without knowledge of the number of poisons employed by the attacker. Fortunately we observe that defense success rate

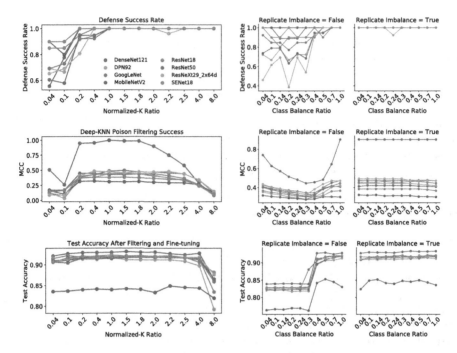

Fig. 4. Ablation studies on the effect of k (left) and class imbalance (right). (Top Left) Defense success rate increases to 100% for all models as normalized-k ratio increases beyond 1.0 for all architectures. (Middle Left) Matthew's correlation coefficient is highest for all models when normalized-k ratio is between 0.4 and 2.0. (Bottom Left) Accuracy on the CIFAR-10 test split drops as normalized-k value increases beyond 4 times the number of examples per class. (Top Right) Defense and performance metrics under class imbalance. Defense success rate is stabilized when the target class training examples are first replicated to match the size of other classes. (Middle Right) Matthews correlation coefficient is also less dependent on the size of the target class when data replication is on. (Bottom Right) Test accuracy is highest when replicating the target examples to match the size of other classes.

remains at 100% as the normalized-k ratio increases beyond $k = 0.2$. Specifically, we see that after a normalized-k value greater than 1.0 ($k = 50$) (i.e. the situation where Deep k-NN considers more neighbors than the per-class number of examples) the convex polytope attack is ineffective on all models. However, there are limitations. Despite successfully detecting all the poisons, an extremely large k could lead to adverse effects on model test performance if too much clean data is removed (i.e. too many false positives).

To take both positives and negatives into account, we again invoke the MCC metric in Fig. 4 (middle left) to measure the trade-off between detecting poisoned images and removing clean images. The maximum correlation coefficient for all models occurs for normalized-k values in the range of 1 and 2. This makes intuitive sense. On one hand, for k smaller than the class size, Deep k-NN could fail to look within a large enough neighborhood around a data point to properly judge its conformity. For example, a poison point may lie within a small, yet very tight cluster of other poison points of the same class and be improperly marked as benign even though the poison cluster itself may lie within a much larger cluster of clean target points. On the other hand, for k larger than 2 times the class size, the neighborhood may be too large and contain too many data points from a competing class. For example, the current target point may lie in a cluster of other target points, but since the neighborhood is so large that it contains all the target points as well as all the points in the nearby poison class cluster, the current target point will be improperly marked as anomalous.

This upper threshold of normalized-$k = 2$ is confirmed by looking at test accuracy performance in Fig. 4 (bottom left). We note that performance is highest in the normalized-k region from 0.2 to 2. It slightly decreases after a normalized-k ratio of 2 and sharply decreases after 4. This shows that a model's ability to generalize suffers when too many legitimate data points are removed under sufficiently large values of k. Based on these experiments, we recommend using a normalized-k value between 1 and 2 for optimal success in defending against poisoning attacks while minimizing false positives.

6.2   Dealing with Class Imbalance

In our second study, we consider the effectiveness of our defense on datasets with an imbalanced number of examples per class. Given an imbalanced dataset, the target class could be either the majority class or a minority class. The easiest case for the defender is when the target is the majority class. In this case, so long as k is set sufficiently large, there will be more than enough target training examples to cause the poisons in their midst (in feature space) to be marked as anomalous after running Deep k-NN . In this section, we will consider the worst case, wherein the target class is the smallest minority class in the dataset. Without applying any protocol to balance out the classes, there may not be enough target class neighbors when running Deep k-NN to know that the poisons clustered in their midst are anomalous.

A typical way to deal with imbalanced classes is to upweight the loss from examples in the minority classes or, equivalently, sample examples from minority classes at a higher rate that is inversely proportional to the fraction of the dataset that their class occupies. We consider a simple and equivalent modification of the latter protocol: given an imbalanced-class dataset, the examples in each class are *replicated* by a factor of N/n, where n is the number of examples in that class and N is the maximum number of examples in any class. After this operation, the dataset will be larger, but once again balanced. We study the effect of this data replication protocol on imbalanced classes. Specifically, we set the number of examples in the target class (frog) to $n < N$ while leaving the number of examples in all other classes as N. We then replicate the frog examples by a factor of N/n such that its size match the size of the other classes. Finally, we plot the defense success rate against the class imbalance ratio n/N in Fig. 4 (top right). The value of normalized-k is fixed at 2 ($k = 100$) for this experiment.

Figure 4 (top right, left panel) shows the defense success rate when no protocol is applied prior to running Deep k-NN : the success rate suffers for class balance ratios below 0.7. When our data replication protocol is applied before the Deep k-NN defense, the defense success rate is near perfect regardless of the class balance ratio. These results show that our minority class replication protocol, combined with the Deep k-NN defense, is very effective at removing poisons in an imbalanced class dataset. Our replication-based balancing protocol normalizes the number of examples considered by the Deep k-NN defense in feature space.

Next, we observe the MCC as a function of class imbalance in the absence of any protocol in Fig. 4 (middle right, left panel). When the ratio is small, then the only thing that can hurt MCC is the misdetection of the targets as being anomalous. On the other hand, when the ratio is large, there is no class imbalance. MCC performs worst when there is a modest underrepresentation of the target class. That is where both the targets and the poisons can cause false negatives and false positives. When the replication protocol is applied in Fig. 4 (middle right), the MCC experiences an improvement, although the relative improvement is small. Interestingly, we observe that data replication stabilizes the MCC against class imbalance; the MCC is essentially a flat curve in Fig. 4 (middle right).

All models experience better test accuracy on the CIFAR-10 test set when replicating target examples as shown in Fig. 4 (bottom right). Despite only having n *unique* points in feature space, replicating them boosts model performance to be similar to the control experiment with a class balance ratio of 1.0. At lower class balance values, replicating data in unbalanced classes improves test accuracy by 8%. Based on these experiments, we recommend the protocol of replicating images of underrepresented classes to match the maximum number of examples in any particular class prior to running Deep k-NN . Defense success rate and model generalizability are both improved and stabilized by this protocol.

7   Conclusion

In summary, we have demonstrated that the simple Deep k-NN approach provides an effective defense against clean-label poisoning attacks with minimal degradation in model performance. With an appropriately selected value of k, the Deep k-NN defense identifies virtually all poisons from two state-of-the-art clean-label data poisoning attacks, while only filtering a small percentage of clean images. The Deep k-NN defense outperforms other data poisoning baselines and provides a strong benchmark on which to measure the efficacy of future defenses.

Acknowledgement. Dickerson and Gupta were supported in part by NSF CAREER Award IIS-1846237, DARPA GARD HR00112020007, DARPA SI3-CMD S4761, DoD WHS Award HQ003420F0035, and a Google Faculty Research Award. Goldstein and his students were supported by the DARPA GARD and DARPA QED4RML programs.

Additional support was provided by the National Science Foundation DMS division, and the JP Morgan Fellowship program.

References

1. Athalye, A., Carlini, N., Wagner, D.: Obfuscated gradients give a false sense of security: circumventing defenses to adversarial examples. arXiv preprint arXiv:1802.00420 (2018)
2. Biggio, B., et al.: Evasion attacks against machine learning at test time. In: Blockeel, H., Kersting, K., Nijssen, S., Železný, F. (eds.) ECML PKDD 2013. LNCS (LNAI), vol. 8190, pp. 387–402. Springer, Heidelberg (2013). https://doi.org/10.1007/978-3-642-40994-3_25
3. Chen, B., et al.: Detecting backdoor attacks on deep neural networks by activation clustering. arXiv preprint arXiv:1811.03728 (2018)
4. Chen, X., Ma, H., Wan, J., Li, B., Xia, T.: Multi-view 3D object detection network for autonomous driving. In: Conference on Computer Vision and Pattern Recognition (CVPR), pp. 1907–1915 (2017)
5. Chen, Y., Li, J., Xiao, H., Jin, X., Yan, S., Feng, J.: Dual path networks. In: Advances in Neural Information Processing Systems, pp. 4467–4475 (2017)
6. Frosst, N., Papernot, N., Hinton, G.: Analyzing and improving representations with the soft nearest neighbor loss. arXiv preprint arXiv:1902.01889 (2019)
7. Goodfellow, I., Shlens, J., Szegedy, C.: Explaining and harnessing adversarial examples. In: International Conference on Learning Representations (2015). http://arxiv.org/abs/1412.6572
8. Goodfellow, I.J., Shlens, J., Szegedy, C.: Explaining and harnessing adversarial examples. arXiv preprint arXiv:1412.6572 (2014)
9. Gu, T., Dolan-Gavitt, B., Garg, S.: BadNets: identifying vulnerabilities in the machine learning model supply chain. arXiv preprint arXiv:1708.06733 (2017)
10. He, K., Zhang, X., Ren, S., Sun, J.: Deep residual learning for image recognition. In: Proceedings of the IEEE Conference on Computer Vision and Pattern Recognition, pp. 770–778 (2016)
11. Hu, J., Shen, L., Sun, G.: Squeeze-and-excitation networks. In: Proceedings of the IEEE Conference on Computer Vision and Pattern Recognition, pp. 7132–7141 (2018)
12. Huang, G., Liu, Z., Van Der Maaten, L., Weinberger, K.Q.: Densely connected convolutional networks. In: IEEE Conference on Computer Vision and Pattern Recognition (CVPR), pp. 2261–2269. IEEE (2017)
13. Ilyas, A., Santurkar, S., Tsipras, D., Engstrom, L., Tran, B., Madry, A.: Adversarial examples are not bugs, they are features. In: Advances in Neural Information Processing Systems, pp. 125–136 (2019)
14. Koh, P.W., Steinhardt, J., Liang, P.: Stronger data poisoning attacks break data sanitization defenses. arXiv preprint arXiv:1811.00741 (2018)
15. Krizhevsky, A., Hinton, G., et al.: Learning multiple layers of features from tiny images. Technical report. Citeseer (2009)
16. Liu, Y., Chen, X., Liu, C., Song, D.: Delving into transferable adversarial examples and black-box attacks. arXiv preprint arXiv:1611.02770 (2016)
17. Madry, A., Makelov, A., Schmidt, L., Tsipras, D., Vladu, A.: Towards deep learning models resistant to adversarial attacks. arXiv preprint arXiv:1706.06083 (2017)

18. Moosavi-Dezfooli, S.M., Fawzi, A., Uesato, J., Frossard, P.: Robustness via curvature regularization, and vice versa. In: Proceedings of the IEEE Conference on Computer Vision and Pattern Recognition, pp. 9078–9086 (2019)
19. Papernot, N., McDaniel, P.: Deep k-nearest neighbors: towards confident, interpretable and robust deep learning. arXiv preprint arXiv:1803.04765 (2018)
20. Pascanu, R., Stokes, J.W., Sanossian, H., Marinescu, M., Thomas, A.: Malware classification with recurrent networks. In: International Conference on Acoustics, Speech and Signal Processing (ICASSP), pp. 1916–1920. IEEE (2015)
21. Qin, C., et al.: Adversarial robustness through local linearization. arXiv preprint arXiv:1907.02610 (2019)
22. Rizoiu, M.A., Wang, T., Ferraro, G., Suominen, H.: Transfer learning for hate speech detection in social media. In: Conference on Artificial Intelligence (AAAI) (2019)
23. Sandler, M., Howard, A., Zhu, M., Zhmoginov, A., Chen, L.C.: MobileNetV2: inverted residuals and linear bottlenecks. arXiv preprint arXiv:1801.04381 (2018)
24. Schölkopf, B., Platt, J.C., Shawe-Taylor, J., Smola, A.J., Williamson, R.C.: Estimating the support of a high-dimensional distribution. Neural Comput. **13**(7), 1443–1471 (2001)
25. Shafahi, A., et al.: Poison frogs! targeted clean-label poisoning attacks on neural networks. In: Advances in Neural Information Processing Systems, pp. 6103–6113 (2018)
26. Shafahi, A., et al.: Adversarial training for free! In: Advances in Neural Information Processing Systems (2019)
27. Sitawarin, C., Wagner, D.: On the robustness of keep k-nearest neighbors (2019)
28. Steinhardt, J., Koh, P.W., Liang, P.: Certified defenses for data poisoning attacks. CoRR abs/1706.03691 (2017). http://arxiv.org/abs/1706.03691
29. Sun, Y., Wang, X., Tang, X.: Deep learning face representation from predicting 10,000 classes. In: Conference on Computer Vision and Pattern Recognition (CVPR), pp. 1891–1898 (2014)
30. Szegedy, C., et al.: Going deeper with convolutions. In: Proceedings of the IEEE Conference on Computer Vision and Pattern Recognition, pp. 1–9 (2015)
31. Tran, B., Li, J., Madry, A.: Spectral signatures in backdoor attacks. In: Advances in Neural Information Processing Systems, pp. 8000–8010 (2018)
32. Turner, A., Tsipras, D., Madry, A.: Clean-label backdoor attacks (2019). https://people.csail.mit.edu/madry/lab/cleanlabel.pdf
33. Xie, C., Wu, Y., Maaten, L.v.d., Yuille, A.L., He, K.: Feature denoising for improving adversarial robustness. In: Proceedings of the IEEE Conference on Computer Vision and Pattern Recognition, pp. 501–509 (2019)
34. Xie, S., Girshick, R., Dollár, P., Tu, Z., He, K.: Aggregated residual transformations for deep neural networks. In: 2017 IEEE Conference on Computer Vision and Pattern Recognition (CVPR), pp. 5987–5995. IEEE (2017)
35. Zhu, C., Huang, W.R., Li, H., Taylor, G., Studer, C., Goldstein, T.: Transferable clean-label poisoning attacks on deep neural nets. In: International Conference on Machine Learning, pp. 7614–7623 (2019)

Ramifications of Approximate Posterior Inference for Bayesian Deep Learning in Adversarial and Out-of-Distribution Settings

John Mitros[(✉)], Arjun Pakrashi, and Brian Mac Namee

School of Computer Science, University College Dublin, Dublin, Ireland
ioannis.mitros@insight-centre.org, {arjun.pakrashi,brian.macnamee}@ucd.ie

Abstract. Deep neural networks have been successful in diverse discriminative classification tasks, although, they are poorly calibrated often
assigning high probability to misclassified predictions. Potential consequences could lead to trustworthiness and accountability of the models when deployed in real applications, where predictions are evaluated
based on their confidence scores. Existing solutions suggest the benefits
attained by combining deep neural networks and Bayesian inference to
quantify uncertainty over the models' predictions for ambiguous data
points. In this work we propose to validate and test the efficacy of likelihood based models in the task of out of distribution detection (OoD).
Across different datasets and metrics we show that Bayesian deep learning models indeed outperform conventional neural networks but in the
event of minimal overlap between in/out distribution classes, even the
best models exhibit a reduction in AUC scores in detecting OoD data.
We hypothesise that the sensitivity of neural networks to unseen inputs
could be a multi-factor phenomenon arising from the different architectural design choices often amplified by the curse of dimensionality. Preliminary investigations indicate the potential inherent role of bias due
to choices of initialisation, architecture or activation functions. Furthermore, we perform an analysis on the effect of adversarial noise resistance
methods regarding in and out-of-distribution performance when combined with Bayesian deep learners.

Keywords: OoD detection · Bayesian deep learning · Uncertainty
quantification · Generalisation · Anomalies · Outliers

1 Introduction

Anomaly detection is concerned with the detection of unexpected events. The
goal is to effectively detect rare or novel patterns which neither comply to the

Electronic supplementary material The online version of this chapter (https://
doi.org/10.1007/978-3-030-66415-2_5) contains supplementary material, which is available to authorized users.

© Springer Nature Switzerland AG 2020
A. Bartoli and A. Fusiello (Eds.): ECCV 2020 Workshops, LNCS 12535, pp. 71–87, 2020.
https://doi.org/10.1007/978-3-030-66415-2_5

norm nor follow a specific trend present in the existing data distribution. The task of anomaly detection is often described as open category classification [30] and anomalies can be referred to as *outliers, novelties, noise deviations, exceptions* or *out-of-distribution* (OoD) examples depending on the context of the application [41]. Anomaly detection raises significant challenges among which some stem from the nature of generated data arising from different sources e.g. multi-modality from different domains such as manufacturing, industrial, health sectors etc. [6,8,22].

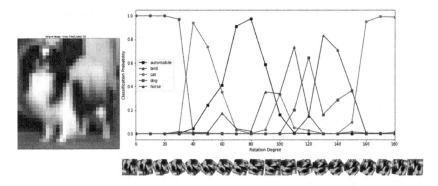

Fig. 1. An example of the OoD detection problem. At each angle $0° \leq \theta \leq 180°$ we record the image classifier's response. The classifier is able to predict the true label, *dog*, with approximately 100% confidence for an angle $\theta \leq 30°$ degrees, while it fails to predict the correct label for any rotation angle $\theta > 30°$. (Inspired by an image in [14].)

In this paper we are concerned with a variant of anomaly detection typically referred to as out-of-distribution (OoD) data detection. Figure 1 shows an illustrative example of an OoD problem. A model is trained to classify images among the following classes: {*horse, dog, cat, bird, automobile*}. The training procedure involved augmentation, including random rotations $\pm 30°$. When a new image $\mathbf{x}$ is provided containing any of these entities, the model should be able to predict the result and assign a probability to each prediction. In the case of Fig. 1 we record the classifier's response $\hat{y} = f_{\mathbf{W}}(\mathbf{x})$ while rotating an image belonging to the class *dog* in the range of $0° \leq \theta \leq 180°$ degrees. From Fig. 1 it is evident that the classifier is able to predict the true label y, which in this instance is *dog*, with approximately 100% confidence for angles up to $\theta \leq 30°$ degrees, while it fails to predict the correct label for any rotation angle $\theta > 30°$. Moreover, in the cases where the model misclassifies the rotated image, it can assign a very high confidence to the incorrect class membership. For example at $\theta = 40°$ and $\theta = 80°$ the model classifies the rotated image to be a *cat* and an *automobile* respectively with very high confidence.

This uncovers two related issues. First, the example demonstrates the incapacity of the model to detect OoD samples previously unseen by it. Second, the classifier misclassifies predictions with high confidence. For instance, the image of the dog was classified as a *cat* or an *automobile* (i.e. black and orange curves in Fig. 1) with high confidence across different rotation angles. This can potentially lead to problems related with trustworthiness, transparency and privacy of the prediction [40]. This phenomenon is not new [42], and since being identified has captivated significant research effort to understand its underpinnings and provide related solutions.

Many of the proposed solutions have their roots in different fields such as differential privacy, information theory, robust high dimensional statistics, robust control theory and robust optimisation. Recently attention has been concentrated to methods that provide a principled approach to quantifying uncertainty through Bayesian deep learning [14, 17, 31, 33]. In spite of their sophistication, questions remain unanswered regarding the effectiveness of these approaches.

In addition to the OoD prediction problem, there is another issue with which deep neural networks struggle: that is adversarial noise [1, 38]. For a trained deep neural network, adversarial noise can be added to an input image causing the model to assign a predicted label to a different class from its true one with high probability. If this noise is chosen carefully, then the image can appear unchanged to a human observer. As many deep learning based image recognition systems are being deployed in real life, such adversarial attacks can result in serious consequences. Therefore, developing strategies to mitigate against them is an important open problem.

In this paper we describe an empirical evaluation of the effectiveness of Bayesian deep learning for OoD detection on a number of different methods using a selection of image classification datasets. This experiment benchmarks the performance of current state-of-the-art methods, and is aligned in principle to the inquisitive nature of [15]. Examining why such powerful models exhibit properties of miscalibration and an incapacity to detect OoD samples will allow us to better understand the major implications in critical domain applications since it implies that their predictions cannot be trusted [2, 18, 25, 28, 36]. Furthermore, we provide discussions into what we consider to be critical components attributing to this phenomenon which suggest directions for future work, inspired by evidence in recent literature. Finally, we investigate the potential benefit of using adversarial noise defence mechanisms for the OoD detection problem as well as the in-sample adversarial attacks for Bayesian neural networks.

The main contributions of this work include:

- A benchmark study demonstrating that Bayesian deep learning methods do indeed outperform point estimate neural networks in OoD detection.
- An empirical validation showing that from the considered approaches in this work, *Stochastic Weight Averaging of Gaussian Samples* (SWAG) [31] is most effective (it achieves the best balance between performance in the OoD detection task and the original image classification task).
- We demonstrate that OoD detection is harder when there exists even minimal overlap between in-distribution and out-of-distribution examples and outlier exposure [24] might not be beneficial in such scenarios.

– We investigate the efficacy of adversarial defence techniques and noise robustness showing that randomised smoothing can be quite effective in combination with Bayesian neural networks.

This paper is structured as follows: Sect. 2 describes related work; Sect. 3 describes an experiment to benchmark the ability of Bayesian neural networks to detect OoD examples; Sect. 4 investigates the effects of adversarial noise defence mechanisms on classification models and the ability of Bayesian neural networks with and without these defences to identify adversarial examples; Sect. 5 concludes the paper.

2   Related Work

The goal of anomaly detection [6], is to detect any inputs that do not conform to in-distribution data—referred to as *out-of-distribution data* (OoD). While there are approaches that are solely targeted at the detection of anomalies [6,11,35], classification models that can identify anomalous inputs as part of the classification process are attractive. For instance, it is desirable that a model trained for image classification (such as that described in Fig. 1) would be capable of not only classifying input images accurately, but also identifying that does not belong to any of the classes that it was trained to recognise. Not only can this be useful for detecting ambiguous inputs [40], but it can also be useful for detecting more insidious adversarial examples [5]. This type of outlier detection as part of the classification process is most commonly referred to as *OoD detection*.

The majority of OoD detection approaches utilise the uncertainty of classifier's outputs as an indicator to identify whether an input instance is OoD. Such methods are mostly based on density estimation techniques utilising generative models or ensembles. For example, Choi et al. [8] presented an ensemble of generative models for density-based OoD detection by estimating epistemic uncertainty of the likelihood, while Hendrycks et al. [21] focused on detecting malformed inputs, $\mathbf{x}$, from their conditional distribution $p(y|\boldsymbol{x})$, assigning OoD inputs lower confidence than normal inputs. Hybrid models for uncertainty quantification are also common. For example, the work of Zhu et al. [52] that describes an end-to-end system using recurrent networks and probabilistic models. The overall architecture is composed of an encoder-decoder recurrent model coupled with a multilayer perceptron (MLP) on top of the encoder allowing to estimate the uncertainty of predictions.

The recent resurgence of interest in combining Bayesian methods with deep neural networks [45] has given rise to different training mechanisms for classification models that output local likelihood estimates of class membership, providing better estimates of classification uncertainty compared to standard deep neural network (DNN) approaches producing point-estimates. Models whose outputs better represent classification uncertainty can be directly applied to the OoD detection problem—intuitively predictions with high uncertainty suggest OoD examples—and recent approaches that use a Bayesian formulation have claimed to be effective at this. One early example is *MC-Dropout* [14] which views the use of dropout at test time as approximate Bayesian inference. Monte Carlo sampling from the model at test time for different subsets of the weights is used to achieve this. MC-Dropout is based on prior work [10] which established a relationship between neural networks with dropout and Gaussian Processes (GP). *Stochastic Weight Averaging of Gaussian Samples* (SWAG) [31], being an extension of *stochastic weight averaging* (SWA) [26], is another example of an approximate posterior inference approach that has been shown to be useful for OoD detection. In SWAG, the weights of a neural network are averaged during different SGD iterations [34] which allows approximates of output distributions to be calculated.

The *Dirichlet Prior Network* (DPN) [33] based on recent work focuses explicitly on detecting OoD instances. While alternative approaches using Bayesian principles aim at constructing an implicit conditional distribution with certain desirable properties—e.g., appropriate choice of prior and inference mechanism—DPN strives to explicitly parameterise a distribution using an arithmetic mixture of Dirichlet distributions. DPN also introduces a separate source of uncertainty through an OoD dataset used at training time as a uniform prior over the OoD data (i.e. similar to outlier exposure [24]). Training a DPN model involves optimising the weights of a neural network by minimising the Kullback-Leibler divergence between in-distribution and out-of-distribution data, where in-distribution data has been modelled with a sharp prior Dirichlet distribution while out-of-distribution data is modelled with a flat prior Dirichlet distribution.

The *Joint Energy Model* (JEM) [17] is an alternative approach that uses generative models to improve the calibration of a discriminative model by reinterpreting an existing neural network architecture as an Energy Based Model (EBM). Specifically they use an EBM [29], which is a type of generative model that parameterises a probability density function using an unnormalised log-density function. It was also demonstrated that JEM is effective on OoD problems [17].

All of these methods perform OoD detection by converting their class membership outputs into a score indicating the likelihood of an out-of-distribution input instance and then applying a threshold to this score. There are four common approaches to calculating these scores:

- *Max probability*: This is the maximum value of the softmax outputs. If $\hat{y} = [y_1, y_3, \ldots, y_k]$ is the prediction, then $\max_k p(y_k|\boldsymbol{x})$ is used as the score.
- *Entropy*: Calculates information entropy $H[Y] = -\sum_{k=1}^{K} p(\hat{y}_k) \log p(\hat{y}_k)$ over the class memberships predicted by a model [33].
- *Mutual Information*: $I(X, Y) = \mathbb{E}_{p(\boldsymbol{x},y)} \left[\frac{p(\boldsymbol{x},y)}{p(\boldsymbol{x})p(y)} \right]$ measures the amount of information obtained about a random variable X by observing some other random variable Y [43]. Here X could describe the entropy of the predictions and Y the entropy of a Dirichlet mixture encompassing the final predictions.
- *Differential Entropy*: Also known as the continuous entropy is defined as follows $h(Y) = -\int p(\hat{y}) \log p(\hat{y}) dx$. This is used to measure distributional uncertainty between in-out distributions.

These techniques can also be applied to the outputs of a simple deep neural network to perform OoD detection in the same manner.

Despite the resurgence of Bayesian methods for deep learning, there are still a number of challenges that need to be addressed before fully harnessing their benefits for OoD detection. Foong et al. [13] draws attention to the poorly understood approximations due to computational tractability upon which Bayesian neural networks rely, indicating that common approximations such as factorised Gaussian assumption and MC-Dropout lead to pathological estimates of predictive uncertainty. Posteriors obtained via mean field variation inference (MFVI) [4] are unable to represent uncertainty between data clusters but have no issue representing uncertainty outside the data clusters. Similarly, MC-Dropout cannot represent uncertainty in-between data clusters, being more confident in the midpoint of the clusters rather than the centres. In addition, Wenzel et al. [46] questioned the efficacy of accurate posterior approximation in Bayesian deep learning, demonstrating through MCMC sampling that the predictions induced by a Bayes posterior systematically produced worse results than point estimates obtained from SGD.

Bias in models has been also observed due to the extreme over-parameterised regime of neural networks. Predictions in the vicinity of the training samples are regularised by the continuity of smoothness constraints, while predictions away from the training samples determine the generalisation performance. The different behaviours reflect the inductive bias of the model and training algorithm. Zhang et al. [51] showed that shallow networks learn to generalise but as their size increases they are more biased into memorising training samples. Increasing the receptive field of convolutional networks introduces bias towards memorising samples, while increasing the width and number of channels does not. This seems to be in accordance with equivalent findings from Azulay et al. [3] identifying that subsampling operations in combination with non-linearities introduce a bias, producing representations that are not shiftable therefore causing the network to lose its invariance properties. Finally, Fetaya et al. [12] showed that conditional generative models have the tendency to assign higher likelihood to interpolated images due to the class-unrelated entropy indicating that likelihood-based density estimation and robust classification might be at odds with each other.

Prior work has focused on investigating the sensitivity of point estimate DNNs for OoD detection on corrupted inputs as well as comparing local vs global methods (i.e. ensembles) [7, 19, 20, 23]. In this work, however, we focus on investigating the effectiveness of local likelihood methods when inputs at test time are considered corrupted with noise by an adversary or simply constitute OoD. In spite of the recent emergence of multiple approaches to the OoD detection problem based on Bayesian methods, and the uncertainty surrounding their effectiveness, no objective benchmark regarding their relative performance at this task currently exists in the literature to best of our knowledge. This paper provides such a benchmark.

To "fool" a deep neural network, examples can be specifically generated in such a way that a specific image is forced to be assigned to a class, from which it does not belong to. This is done by adding noise such that the input image looks identical to the human eye, but it is either i) classified to a specific target class to which the example does not belong to (targetted), or ii) classified to any class other than its original one (untargeted). Such induced noise is called adversarial noise, and can be generated using several methods [1,38]. As deep learning based vision systems are being deployed in real world applications, consequently, research in defending against adversarial noise is increasing in popularity and importance. For this pupose, there have been developed several adversarial noise defence mechanisms in the literature [9,39,47], which we present and evaluate in Sect. 3.

3   Experiment 1: Benchmarking OoD Detection

In this experiment, using a number of well-known image classification datasets, we evaluate the ability of four state-of-the-art methods based on Bayesian deep learning—DPN, MC-Dropout, SWAG, and JEM—to detect OoD examples, and compare this with the performance of a standard deep neural network (DNN).

3.1   Experimental Setup

We use a 28 layers wide and 10 layers deep WideResNet [50] as the DNN model, which is also the same base model used across the other methods for fairness, to avoid biasing results depending on the underlying network architecture. Whenever it was appropriate and possible for specific datasets, we utilised pre-trained models provided by the original authors to avoid any discrepancies in our results. For instance, Grathwohl et al. [17] made available a pre-trained JEM model for the CIFAR-10 dataset, and we use this in our experiments rather than training our own. For the remaining models we trained each for 300 epochs using a validation set for hyper-parameter tuning and rolling back to the best network to avoid overfitting.

The optimiser used during the experiments was Stochastic Gradient Descent (SGD) [16] with momentum set to 0.9 and weight decay in the range $[3e^{-4}, 5e^{-4}]$. Additionally, every dataset was split into three distinct sets {*train, validation, test*} with augmentations such as random rotation, flip, cropping and pixel distortion applied on the training set.

A significant distinction between DPN and the other approaches used is that DPN uses an idea similar to outlier exposure [24] and it requires two datasets during training: one to represent the in-distribution data and the other to simulate out-of-distribution data.

Five well-known image classification datasets are used in this experiment: *CIFAR-10* [27], *CIFAR-100* [27], *SVHN* [37], *FashionMNIST* [48], and *LSUN* [49] (We use only the *bedroom* scene from LSUN in this experiment only as OoD data for testing.)

To comprehensively explore the performance of the different models we perform experiments using each dataset to train a model and use all other datasets to measure the ability of the model to perform OoD detection. The ability of each model to perform the image classification task it was trained for, was first evaluated using the test set associated with each dataset. All datasets have balanced class distributions, therefore we use classification accuracy to measure this performance.

To measure the ability of models to recognise OoD examples we make predictions for the test portion of the dataset used to train the model (i.e. in-distribution data CIFAR-10) and then also make predictions for each test portion of the remaining datasets (i.e. out-of-distribution data, CIFAR-100, SVHN, and LSUN). This means that for each training set we have three different evaluations of OoD detection effectiveness. For example, when SVHN is used as the in-distribution training set, CIFAR-10, LSUN, and CIFAR-100 are used as the out-of-distribution test sets. One of the available out-of-distribution datasets is selected for use at training time for DPN which requires this extra data. Only the training portions of this dataset is used for this purpose, while the test portions are used for evaluation.

The predictions provided by the models are converted into OoD scores using the four alternative approaches described in Sect. 2: *max probability*, *entropy*, *mutual information*, and *differential entropy*. To avoid having to set detection thresholds on these scores we measure the separation between the scores generated for instances on the in-distribution and out-of-distribution test sets using the area under the curve (AUC-ROC) based on the four different approaches for calculating scores. We do this individually for each approach to generate the AUC scores.

3.2   Results and Discussion

Table 1 shows the overall performance of each model for the basic in-distribution image classification tasks they were trained for, measured using classification accuracy. These results indicate that, for most cases, the Bayesian methods are performing comparatively to the DNN baseline (the cases where this is not true will be discussed shortly).

Table 1. Accuracy of models on in-distribution dataset image classification tasks.

Model	CIFAR-10	SVHN	FashionMNIST	CIFAR100
DNN	95.06	96.67	95.27	77.44
DPN	88.10	90.10	93.20	79.34
MC-Dropout	96.22	96.90	95.40	78.39
SWAG	96.53	97.06	93.80	78.61
JEM	92.83	96.13	83.21	77.86

Table 2 shows the results of the OoD detection experiments. The scores represent AUC-ROC scores calculated using entropy of the model outputs for OoD detection. Values inside parenthesis indicate the percentage improvement with respect to DNN, which is treated as a baseline. The ↑ indicates an improvement and the ↓ indicates a degradation. The last row of the table shows the average percentage improvement across the dataset combinations for each approach with respect to the DNN baseline. Similar tables based on OoD scores calculated using max probability, mutual information, and differential entropy are available in the supplementary material.

Table 2. Out-of-distribution experiment results. Scores are Entropy based AUC-ROC in percentage. The values in bracket are % improvement of the corresponding algorithm w.r.t. DNN, taken as a baseline. An ↑ indicates improvement and ↓ degradation wrt. the baseline (DNN). The asterisks (*) next to each dataset indicates out-distribution datasets used to train DPN.

Data		(baseline)	Entropy AUC-ROC score (% gain wrt. baseline)			
In-distribution	OoD	DNN	DPN	MC-Dropout	SWAG	JEM
CIFAR-10	CIFAR-100*	86.27	85.60 (↓0.78%)	89.92 (↑4.23%)	91.89 (↑6.51%)	87.35 (↑1.25%)
	SVHN	89.72	98.90 (↑10.23%)	96.25 (↑7.28%)	98.62 (↑9.92%)	89.22 (↓0.56%)
	LSUN	88.83	83.30 (↓6.23%)	92.04 (↑3.61%)	95.12 (↑7.08%)	89.84 (↑1.14%)
SVHN	CIFAR-100	93.19	99.10 (↑6.34%)	94.33 (↑1.22%)	95.97 (↑2.98%)	92.34 (↓0.91%)
	CIFAR-10*	94.58	99.60 (↑5.31%)	94.97 (↑0.41%)	96.03 (↑1.53%)	92.85 (↓1.83%)
	LSUN	92.97	99.70 (↑7.24%)	93.31 (↑0.37%)	95.71 (↑2.95%)	91.82 (↓1.24%)
FashionMNIST	CIFAR-100	91.20	99.50 (↑9.10%)	93.75 (↑2.80%)	96.19 (↑5.47%)	62.79 (↓31.15%)
	CIFAR-10*	94.59	99.60 (↑5.30%)	96.06 (↑1.55%)	94.28 (↓0.33%)	64.76 (↓31.54%)
	LSUN	93.34	99.80 (↑6.92%)	97.40 (↑4.35%)	99.05 (↑6.12%)	65.38 (↓29.96%)
CIFAR-100	CIFAR-10	78.25	85.15 (↑8.82%)	80.70 (↑3.13%)	84.92 (↑8.52%)	77.64 (↓0.78%)
	SVHN*	81.52	92.64 (↑13.64%)	85.59 (↑4.99%)	94.16 (↑15.51%)	81.22 (↓0.37%)
	LSUN	77.22	86.38 (↑11.86%)	76.58 (↓0.83%)	87.22 (↑12.95%)	77.54 (↑0.41%)
Avg % improvement			(↑6.48%)	(↑2.76%)	(↑6.60%)	(↓7.96%)

Table 3 summarises the results from these tables and presents the average performance increase with respect to the DNN baseline based on AUC-ROC calculated using each OoD scoring method.

It is evident from Table 2 that most of the Bayesian methods—DPN, MC-Dropout and SWAG—almost always improve OoD detection performance over the DNN baseline. Interestingly, in our experiment JEM consistently performed poorly with respect to the DNN baseline. We attribute this mostly to a failure of robust selection of hyper-parameter combinations. Despite our best efforts we were not able to achieve classification accuracy (above 83%) on the FashionM-NIST dataset using this approach, and the OoD performance suffered from this. However, even for the datasets in which the JEM models performed well for the image classification task (SVHN and CIFAR-100) its use did not lead to an

Table 3. Percent performance increase w.r.t. baseline (DPN) for all the evaluation scores. The values in parenthesis are the relative ranks of improvement for the corresponding evaluation score. Last row indicates the improvement ranking of the algorithms, averaged through the evaluation scores.

	DPN	MC-Dropout	SWAG	JEM
Max prob.	5.895% (2)	2.618% (3)	6.067% (1)	−8.350% (4)
Mutual info.	7.262% (1)	6.243% (2)	5.337% (3)	−8.034% (4)
Entropy	6.480% (2)	6.480% (3)	6.601% (1)	−7.960% (4)
Diff. entropy	9.232% (1)	4.386% (3)	5.495% (2)	−10.217% (4)
Avg. rank	(1.50)	(2.75)	(1.75)	(4.00)

increase in OoD detection performance[1]. Overall we found that both JEM and DPN can be quite unstable and extremely sensitive to hyper-parameter choice.

The results in Table 3 indicate that among the different evaluation metrics, DPN performs best for OoD detection, with SWAG following closely behind. MC-Dropout is third, but comparatively worse than the other two. All three Bayesian methods on average increase OoD detection performance with respect to the DNN baseline regardless of the approach used to calculate OoD scores. DPN used in combination with differential entropy is especially effective. For other methods the best performance is achieved using entropy as the score for the OoD detection task. This indicates that overall, Bayesian methods improve the performance on OoD detection, with DPN and SWAG achieving similar performance (although JEM does not achieve improvements over the DNN baseline).

However, it is worth noting from Table 1 that the DPN models do not perform well on the base image classification tasks compared to the other approaches. Therefore, we conclude that of the approaches compared in this experiment SWAG is the most effective as it achieves strong OoD detection performance, without compromising basic in-distribution classification performance. It also does not require the use of an additional OoD dataset at training time. Although, if the objective is achieving the best possible OoD detection performance then DPN should be considered instead.

It is also worth mentioning that the performance of models used in combination with the four approaches to calculating OoD scores (described in Sect. 2), varies slightly. The *max probability, entropy* and *mutual information* approaches are sensitive to class overlap between in-distribution and out-of-distribution data, while *differential entropy* seems to be more sensitive to the target precision

---

[1] Even after suggestions kindly provided by Grathwohl et al. [17] we found that the model diverged instead of closely approximating its original results. There was another instance we noticed a degradation in performance compared to the original results, it was in the case of DPN when trained with CIFAR-10 for in-distribution and CIFAR-100 for OoD [33], due to the overlap between some classes in CIFAR-10 & CIFAR-100.

parameter of the Dirichlet distribution, controlling how it is concentrated over the simplex of possible outputs.

4 Experiment 2: Robustness Analysis

Having identified in the previous experiment that Bayesian neural network models can be utilised for OoD detection, we proceed by verifying whether they can withstand untargeted adversarial attacks. We design a new experiment including the usual DNN acting as a control baseline and two Bayesian models (*MC-Dropout* and *SWAG*), in order to answer the following questions:

1. Are Bayesian neural networks capable of detecting adversarial examples?
2. Could simple defence mechanisms {e.g. *Top-k, Randomised Smoothing, MMLDA*}, outperform BNN or improve their overall performance?

4.1 Experimental Setup

First, we evaluated the ability of each model against untargeted adversarial examples generated with *Projected Gradient Descent (PGD)* [32,38] with the following hyper-parameter values $\epsilon = 0.1$ and $\alpha = 0.01$ for ten iterations. Second, we introduced three recently proposed defence techniques against adversarial examples and evaluated their efficacy on the (i) *clean test set*, (ii) *adversarially corrupted test set*, and finally on the (iii) *OoD detection task*. The defence methods evaluated against adversarial examples are: *Randomised Smoothing* [9], *Sparsify k-winners take all (Top-k)* [47], and *Max-Mahalanobis Linear Discriminant Analysis* (MMLDA) [39]. We use only the CIFAR-10 dataset in these experiments to train the models.

In order to train models with MMLDA and RandSmooth we utilised the same number of hyper-parameters and values as they are depicted in the original papers. For instance, for MMLDA we utilised a variance of value 10 for CIFAR-10 in order to compute the Max-Mahalanobis centres on $\mathbf{z}$, where $\mathbf{z}$ denotes the features regarding the penultimate layer of each model $\mathbf{z} = f(x)$. We should also notice the possibility for randomised smoothing to achieve better results at the expense of increased prediction time. A key combination for obtaining successful results is the number of samples $n = 55$, or randomised copies for each x in the test set, together with the confidence level $\alpha = 0.001$ (i.e. there is a 0.001 probability that the answer will be wrong) and standard deviation $\sigma = 0.56$ of the isotropic Gaussian noise.

4.2 Results and Discussion

We first illustrate the effect of adversarial noise on the ability of different model types to perform the underlying CIFAR-10 classification task in the presence of adversarial noise. Table 4 shows the performance of each model on the basic CIFAR-10 classification problem on a clean test dataset with no adversarial noise. We see that, apart from randomise smoothing, the addition of noise defence strategies does not negatively affect performance at this task. Randomised smoothing had a negative impact on the in-distribution performance across all methods indicating that robustness might be at odds with accuracy [44]. MMLDA and Top-k seem to have a positive effect on DNN and MC-Dropout but not on SWAG.

Table 4. Accuracy on CIFAR-10 clean test set for each defence technique. Values in parenthesis indicate the percentage of abstained predictions.

Model	No defence	Top-k	RandSmooth	MMLDA
DNN	95.06	94.52	86.25 (13.00)	95.18
MC-Dropout	96.22	94.43	86.98 (13.39)	95.21
SWAG	96.53	91.73	79.68 (20.32)	91.30

Table 5 summarises the performance of the same models when adversarial noise is added to the test set. The first column in Table 5 corresponds to the accuracy on the adversarially corrupted test set for each model without applying any adversarial defence mechanism. Lower values indicate that the adversarial attack was successful and managed to force the model to misclassify instances with high confidence. When Randomised smoothing is used, the values in parentheses denote the percentage of abstained predictions on the total number of instances from the test set. The impact of adversarial noise is evident in these results. When there is no defence mechanism used then classification accuracy plummets to less than 2% for all models. The addition of defence techniques, especially MMLDA and randomised smoothing indicates improvement in performance.

Table 5. Accuracy on CIFAR-10 test set corrupted with adversarial noise.

Model	No defence	Top-k	RandSmooth	MMLDA
DNN	1.15	11.41	62.65 (27.75)	45.71
MC-Dropout	1.94	7.28	88.85 (17.98)	47.90
SWAG	0.55	0.79	36.80 (20.97)	47.95

In addition, Table 6, shows that model predictions can be used to identify adversarial examples in the same manner that OoD examples were identified by measuring the entropy over the final predictions. It compares the entropy scores generated for an in-distribution test set versus those generated for adversarial examples and uses AUC-ROC scores to measure the ability of a model to distinguish them. First, notice that in line with the results in the previous section

Table 6. Out-of-distribution detection results for all defence methods on clean vs adversarially corrupted CIFAR-10. Scores represent entropy based AUC-ROC in percentage.

Data		Entropy AUC-ROC score		
In-distribution	OoD	DNN	MC-Dropout	SWAG
CIFAR-10/No defence	CIFAR-10 Adv.	13.67	36.54	73.81
CIFAR-10/Topk	CIFAR-10 Adv.	99.83	99.83	98.90
CIFAR-10/MMLDA	CIFAR-10 Adv.	53.15	85.65	70.38
CIFAR-10/RandSmooth	CIFAR-10 Adv.	62.68	57.06	48.29

Table 7. Out-of-distribution detection results. Scores represent entropy based AUC-ROC in percentage.

Data		Entropy AUC-ROC score		
In-distribution	OoD	DNN	MC-Dropout	SWAG
CIFAR-10/Topk	CIFAR-100	90.59	90.45	84.72
	SVHN	91.20	92.62	94.61
	LSUN	92.42	92.39	89.81
CIFAR-10/MMLDA	CIFAR-100	99.87	99.24	79.78
	SVHN	99.74	99.75	84.56
	LSUN	99.93	99.67	81.72
CIFAR-10/RandSmooth	CIFAR-100	62.91	69.33	60.21
	SVHN	42.62	63.22	66.48
	LSUN	61.94	69.10	61.23

the Bayesian models perform better at this task than the baseline DNN when no defence against adversarial examples is used. SWAG, in particular, seems quite effective. When adversarial defence mechanisms are used, the performance increases dramatically. Top-k in particular seems effective in this occasion.

Finally, Table 7 shows whether the defence techniques provide any improvements in regard to the OoD detection task. The values represent AUC-ROC scores computed based on entropy of the predictions obtained from each model after having introduced the different defence techniques to each model. Again the top-k and MMLDA approaches improve performance over the no defence option.

5   Conclusion

This work investigates if Bayesian methods improve OoD detection for deep neural networks used in image classification. To do so, we investigated four recent Bayesian deep learning methods and compared them against a baseline point estimate neural network on four datasets.

Our findings show that the Bayesian methods overall improve performance at the OoD detection task over the DNN baseline. Of the considered methods examined DPN performed best, with SWAG following closely behind. However, DPN requires additional data during training, is sensitive to hyper-parameter tuning, and led to worse in-distribution performance. Therefore, we conclude that SWAG seems more effective of the approaches examined considering both in and out-distribution performance.

Furthermore we showed that, despite being better than simple DNN models on OoD, Bayesian neural networks do not possess the ability to cope with adversarial examples. Although adversarial defence techniques overall degrade accuracy on the clean test set, at the same time, they robustify against adversarial examples across models and occasionally improve OoD detection. Randomised smoothing performed best, followed by MMLDA. In the case of the OoD detection task it seems that Top-k slightly improved performance for DNN while MMLDA improved performance for DNN and MC-Dropout.

The broader scope of our work is not only to identify whether Bayesian methods could be used for OoD detection and to mitigate against adversarial examples, but also to understand how much of the proposed methods are affected by the inductive bias in the choices of model architecture and objective functions. For our future work we look forward to examine these components and their inherent role regarding the sensitivity of models against outliers.

Acknowledgements. This research was supported by Science Foundation Ireland (SFI) under Grant number SFI/12/RC/2289_P2.

References

1. Akhtar, N., Mian, A.: Threat of adversarial attacks on deep learning in computer vision: a survey. IEEE Access **6**, 14410–14430 (2018)
2. Alemi, A.A., Fischer, I., Dillon, J.V.: Uncertainty in the Variational Information Bottleneck. arXiv:1807.00906 (2018)
3. Azulay, A., Weiss, Y.: Why do deep convolutional networks generalize so poorly to small image transformations? J. Mach. Learn. Res. **20**(184), 1–25 (2019)
4. Blei, D.M., Kucukelbir, A., McAuliffe, J.D.: Variational inference: a review for statisticians. J. Am. Stat. Assoc. **112**(518), 859–877 (2017). https://doi.org/10.1080/01621459.2017.1285773
5. Carlini, N., Wagner, D.: Adversarial Examples Are Not Easily Detected: Bypassing Ten Detection Methods. arXiv:1705.07263 (2017)
6. Chalapathy, R., Chawla, S.: Deep Learning for Anomaly Detection: A Survey. arXiv:190103407 (2019)
7. Chen, J., Wu, X., Liang, Y., Jha, S., et al.: Robust out-of-distribution detection in neural networks. arXiv preprint arXiv:2003.09711 (2020)
8. Choi, H., Jang, E., Alemi, A.A.: WAIC, but Why? Generative Ensembles for Robust Anomaly Detection. arXiv:1810.01392 (2018)
9. Cohen, J.M., Rosenfeld, E., Kolter, J.Z.: Certified adversarial robustness via randomized smoothing (2019)

10. Damianou, A., Lawrence, N.: Deep gaussian processes. In: Carvalho, C.M., Raviku-mar, P. (eds.) Proceedings of the Sixteenth International Conference on Artificial Intelligence and Statistics. Proceedings of Machine Learning Research, vol. 31, pp. 207–215. PMLR, Scottsdale, Arizona, USA, 29 April–01 May 2013

11. Erfani, S.M., Rajasegarar, S., Karunasekera, S., Leckie, C.: High-dimensional and large-scale anomaly detection using a linear one-class SVM with deep learning. Pattern Recognit. **58**, 121–134 (2016). https://doi.org/10.1016/j.patcog.2016.03.028

12. Fetaya, E., Jacobsen, J.H., Grathwohl, W., Zemel, R.: Understanding the limita-tions of conditional generative models. In: International Conference on Learning Representations (2020)

13. Foong, A.Y.K., Burt, D.R., Li, Y., Turner, R.E.: Pathologies of Factorised Gaus-sian and MC Dropout Posteriors in Bayesian Neural Networks. arXiv:1909.00719, September 2019

14. Gal, Y., Ghahramani, Z.: Dropout as a Bayesian Approximation: Representing Model Uncertainty in Deep Learning. arXiv:1506.02142 (2016)

15. Goldblum, M., Geiping, J., Schwarzschild, A., Moeller, M., Goldstein, T.: Truth or backpropaganda? An empirical investigation of deep learning theory. In: Inter-national Conference on Learning Representations (2020)

16. Goodfellow, I., Bengio, Y., Courville, A.: Deep Learning. MIT Press, Cambridge (2016)

17. Grathwohl, W., Wang, K.C., Jacobsen, J.H., Duvenaud, D., Norouzi, M., Swersky, K.: Your classifier is secretly an energy based model and you should treat it like one. arXiv preprint arXiv:1912.03263 (2019)

18. Guo, C., Pleiss, G., Sun, Y., Weinberger, K.Q.: On calibration of modern neural networks. In: ICML (2017)

19. Hendrycks, D., Dietterich, T.: Benchmarking neural network robustness to common corruptions and perturbations. arXiv preprint arXiv:1903.12261 (2019)

20. Hendrycks, D., Gimpel, K.: A baseline for detecting misclassified and out-of-distribution examples in neural networks. arXiv preprint arXiv:1610.02136 (2016)

21. Hendrycks, D., Gimpel, K.: A baseline for detecting misclassified and out-of-distribution examples in neural networks. In: ICLR (2017)

22. Hendrycks, D., Mazeika, M., Dietterich, T.: Deep Anomaly Detection with Outlier Exposure. arXiv:181204606 (2018)

23. Hendrycks, D., Mazeika, M., Dietterich, T.: Deep anomaly detection with outlier exposure. arXiv preprint arXiv:1812.04606 (2018)

24. Hendrycks, D., Mazeika, M., Dietterich, T.: Deep anomaly detection with outlier exposure. In: International Conference on Learning Representations (2019)

25. Hill, B.: Confidence in belief, weight of evidence and uncertainty reporting. In: International Society for Imprecise Probability: Theories and Applications (2019)

26. Izmailov, P., Podoprikhin, D., Garipov, T., Vetrov, D., Wilson, A.G.: Averaging Weights Leads to Wider Optima and Better Generalization. arXiv:1803.05407 [cs, stat], March 2018

27. Krizhevsky, A., Hinton, G., et al.: Learning multiple layers of features from tiny images. Technical report (2009)

28. Kumar, A., Liang, P., Ma, T.: Verified Uncertainty Calibration. arXiv:1909.10155 (2019)

29. LeCun, Y., Chopra, S., Hadsell, R., Ranzato, M., Huang, F.: A tutorial on energy-based learning. Predict. Struct. Data **1** (2006)

30. Liu, S., Garrepalli, R., Dietterich, T., Fern, A., Hendrycks, D.: Open category detection with PAC guarantees. In: Dy, J., Krause, A. (eds.) Proceedings of the 35th International Conference on Machine Learning. Proceedings of Machine Learning Research, vol. 80, pp. 3169–3178. PMLR, Stockholmsmässan, Stockholm Sweden, 10–15 July 2018

31. Maddox, W.J., Izmailov, P., Garipov, T., Vetrov, D.P., Wilson, A.G.: A simple baseline for Bayesian uncertainty in deep learning. In: Wallach, H., Larochelle, H., Beygelzimer, A., d' Alché-Buc, F., Fox, E., Garnett, R. (eds.) Advances in Neural Information Processing Systems 32, pp. 13153–13164. Curran Associates, Inc. (2019)

32. Madry, A., Makelov, A., Schmidt, L., Tsipras, D., Vladu, A.: Towards deep learning models resistant to adversarial attacks. arXiv preprint arXiv:1706.06083 (2017)

33. Malinin, A., Gales, M.: Predictive uncertainty estimation via prior networks. In: Advances in Neural Information Processing Systems, pp. 7047–7058 (2018)

34. Mandt, S., Hoffman, M.D., Blei, D.M.: Stochastic Gradient Descent as Approximate Bayesian Inference. arXiv e-prints, April 2017

35. Marchi, E., Vesperini, F., Weninger, F., Eyben, F., Squartini, S., Schuller, B.: Non-linear prediction with LSTM recurrent neural networks for acoustic novelty detection. In: International Joint Conference on Neural Network (2015). https://doi.org/10.1109/IJCNN.2015.7280757

36. Marin, J.M., Pudlo, P., Robert, C.P., Ryder, R.J.: Approximate Bayesian computational methods. Stat. Comput. **22**(6), 1167–1180 (2012). https://doi.org/10.1007/s11222-011-9288-2

37. Netzer, Y., Wang, T., Coates, A., Bissacco, A., Wu, B., Ng, A.Y.: Reading digits in natural images with unsupervised feature learning. Neural Inf. Process. Syst. (2011)

38. Ozdag, M.: Adversarial attacks and defenses against deep neural networks: a survey. Procedia Comput. Sci. **140**, 152–161 (2018)

39. Pang, T., Xu, K., Dong, Y., Du, C., Chen, N., Zhu, J.: Rethinking softmax cross-entropy loss for adversarial robustness. In: International Conference on Learning Representations (2020). https://openreview.net/forum?id=Byg9A24tvB

40. Schulam, P., Saria, S.: Can you trust this prediction? Auditing pointwise reliability after learning. In: Proceedings of the Artificial Intelligence and Statistics (2019)

41. Shafaei, A., Schmidt, M., Little, J.J.: Does Your Model Know the Digit 6 Is Not a Cat? A Less Biased Evaluation of "Outlier" Detectors. arXiv:1809.04729 (2018)

42. Szegedy, C., Zaremba, W., Sutskever, I., Bruna, J., Erhan, D., Goodfellow, I., Fergus, R.: Intriguing properties of neural networks. arXiv:1312.6199 (2013)

43. Tschannen, M., Djolonga, J., Rubenstein, P.K., Gelly, S., Lucic, M.: On mutual information maximization for representation learning. arXiv preprint arXiv:1907.13625 (2019)

44. Tsipras, D., Santurkar, S., Engstrom, L., Turner, A., Madry, A.: Robustness may be at odds with accuracy. In: International Conference on Learning Representations (2019). https://openreview.net/forum?id=SyxAb30cY7

45. Wang, H., Yeung, D.Y.: Towards bayesian deep learning: a framework and some existing methods. IEEE Trans. Knowl. Data Eng. **28**(12), 3395–3408 (2016). https://doi.org/10.1109/TKDE.2016.2606428

46. Wenzel, F., et al.: How Good is the Bayes Posterior in Deep Neural Networks Really? arXiv:2002.02405, February 2020

47. Xiao, C., Zhong, P., Zheng, C.: Enhancing adversarial defense by k-winners-take-all. In: International Conference on Learning Representations (2020)

48. Xiao, H., Rasul, K., Vollgraf, R.: Fashion-MNIST: a novel image dataset for bench-marking machine learning algorithms. arXiv preprint arXiv:1708.07747 (2017)
49. Yu, F., Zhang, Y., Song, S., Seff, A., Xiao, J.: LSUN: construction of a large-scale image dataset using deep learning with humans in the loop. arXiv preprint arXiv:1506.03365 (2015)
50. Zagoruyko, S., Komodakis, N.: Wide residual networks. arXiv preprint arXiv:1605.07146 (2016)
51. Zhang, C., Bengio, S., Hardt, M., Mozer, M.C., Singer, Y.: Identity crisis: Memorization and generalization under extreme overparameterization. In: International Conference on Learning Representations (2020)
52. Zhu, L., Laptev, N.: Deep and confident prediction for time series at uber. In: 2017 IEEE International Conference on Data Mining Workshop ICDMW, pp. 103–110 (2017). https://doi.org/10.1109/ICDMW.2017.19

Adversarial Shape Perturbations on 3D Point Clouds

Daniel Liu[1], Ronald Yu[2], and Hao Su[2(✉)]

[1] Torrey Pines High School, San Diego, CA, USA
daniel.liu02@gmail.com
[2] University of California San Diego, La Jolla, CA, USA
ronaldyu@ucsd.edu, haosu@eng.ucsd.edu

Abstract. The importance of training robust neural network grows as 3D data is increasingly utilized in deep learning for vision tasks in robotics, drone control, and autonomous driving. One commonly used 3D data type is 3D point clouds, which describe shape information. We examine the problem of creating robust models from the perspective of the attacker, which is necessary in understanding how neural networks can be exploited. We explore two categories of attacks: distributional attacks that involve imperceptible perturbations to the distribution of points, and shape attacks that involve deforming the shape represented by a point cloud. We explore three possible shape attacks for attacking 3D point cloud classification and show that some of them are able to be effective even against preprocessing steps, like the previously proposed point-removal defenses. (Source code available at https://github.com/Daniel-Liu-c0deb0t/Adversarial-point-perturbations-on-3D-objects).

Keywords: Adversarial attacks · Adversarial defenses · 3D point clouds · Robustness · Neural networks · Deep learning · PointNet

1 Introduction

3D data is often used as an input for controlling autonomous driving systems, robotics, drones, and other tasks that rely on deep learning. For example, in tasks like controlling self-driving cars, 3D point cloud data from LiDAR scans, in addition to radar and image data, are being used by companies like Waymo, Uber, Lyft, and NVIDIA. Point cloud data is also produced in many other applications, including LiDAR (Light Detection and Ranging) scans, RGB-D scans (*e.g.*, through Microsoft Kinect), or photogrammetry of 3D objects. However, compared to the bitmap 2D images that are well-studied in many previous works, deep learning with 3D data is still poorly understood. More specifically, there has not been many works studying the robustness of 3D deep learning, despite its importance in safety-critical applications and systems. Since, in general, the robustness of neural networks in adverse scenarios can be measured by looking at adversarial examples (*e.g.*, an adversarial object that looks like a car but the

© Springer Nature Switzerland AG 2020
A. Bartoli and A. Fusiello (Eds.): ECCV 2020 Workshops, LNCS 12535, pp. 88–104, 2020.
https://doi.org/10.1007/978-3-030-66415-2_6

neural network misclassifies it as a person), we study how to generate effective attacks specific to 3D space to evaluate the robustness of 3D deep learning models in this paper. In our experiments, we focus on the task of 3D point cloud classification.

Currently, it is quite easy to create effective adversarial attacks in both 2D and 3D space by maximizing the loss of a neural network to generate an adversarial example that fools the neural network. However, in this paper, we are interested in crafting effective adversarial perturbations that satisfy certain desirable criteria, like morphing the shape of a point cloud to generate certain adversarial features, adding certain adversarial features to an existing point cloud, or being effective even against defenses. We examine two distinct categories of attacks, the well-studied "distributional attacks", which seek to minimize perceptibility, and our new "shape attacks" that modify the surface of a 3D object's point cloud. In total, we propose three new "shape attacks" for exploiting shape data that is native to 3D point clouds data in generating our adversarial examples. We do extensive experimental validation of their effectiveness, and show that some of them are able to break previous point-removal defenses, which represent realistic defenses or preprocessing steps for scanned point clouds.

2   Related Work

2.1   Attacks

[25] and [2] first examined how adversarial perturbations on 2D images could be generated on a neural network by searching for some perturbation that causes a neural network to make a wrong prediction. Later, there were many attacks proposed in [10,14,17], and [13] that were based on constraining the L_p norm of the perturbation, which was a measure of perceptibility. In general, their proposed optimization problems are solved through projected gradient descent, in what is sometimes known as the "iterative gradient attack". There are also many further improvements by [4,8,18–20], and [6] on perturbing 2D images. [3] provides an overview of threat models and 2D attacks.

Only very recently has adversarial attacks been examined in 3D space. For 3D point clouds, [32,33], and [16] all proposed point perturbations based on shifting points using ideas similar to attacks on 2D images. An adversarial attack that shifted the distribution of points on a 3D object's surface was proposed in [16] as an extremely imperceptible attack, by using projected gradient descent.

[32] also proposed an attack that generates clusters of points with different shapes. For this attack, they optimize an objective function that constrains the distance between the large, perceptible generated point clusters and the main benign point cloud.

[35] and [29] both proposed saliency-based techniques for adversarially removing points. This is different from our attacks in this paper since we only consider shifting points. Adversarial attacks on a simulated LiDAR system were proposed in [5] by making the scanning operation that converts an object's surface into a point cloud differentiable. The resulting adversarial examples generated from

a basic box object are large, noticeable shape deformations. We adopt a similar approach in this paper, except our shape deformation are directly performed on raw point clouds and we directly attack point cloud neural networks.

2.2 Defenses

A popular defense for 2D image space is adversarial training, where a neural network is trained with adversarial examples [10]. This is done by adding the loss incurred by adversarial examples to the objective function that is optimized throughout training. Another defensive technique is defensive distillation [21]. However, so far, it is still quite easy to attack and bypass defenses in the 2D image domain. Later work has shown that a neural network can be trained to be provably robust [30] to some degree. For neural networks that classify 3D point clouds, removing points was shown to be more effective than adversarial training. In particular, [16] and [36] both examined removing outlier points, and [16] also examined removing salient points. These defenses were shown to be effective even when the adversarial perturbations were small and imperceptible.

2.3 3D Deep Learning

3D deep learning has recently experienced exponential growth. Many network architectures were proposed for different tasks in 3D deep learning. We specifically examine the case of 3D point cloud classification, using the PointNet [23], PointNet++ [24], and DGCNN [28] architectures. Other work on this task include [22] and [7]. Point cloud networks were previously shown to be robust against small random perturbations and random point dropout. There is also work on handling voxel (quantized point cloud) data [27,31], though this is not examined in this paper.

3   Setting

We have a neural network model $f_\theta : \mathbb{R}^{N \times 3} \mapsto \mathbb{R}^M$, which solves the 3D point cloud classification task by predicting probability vectors for each output class. It takes in x, a set of N 3D points, and outputs a vector $f_\theta(x)$ of length M that contains the probability of the input x being each output possible class. The model f_θ is trained by adjusting its parameters θ to minimize the cross entropy loss, which is denoted by $J(f_\theta(x), y)$, for each sample x and its corresponding one-hot label y.

Threat Model. We assume that the attacker has full access to the architecture and parameters θ of a neural network f_θ (white-box threat model). The attacker also has access to the unstructured 3D point clouds, which they can change before feeding them to the neural network. Therefore, we wish to construct attacks that are purely based on unstructured 3D point sets, without any extra shape or normal information on the structure of the point set. Our experiments occur in a *digital* setting, and we leave experimental evaluations of our attacks on *physical* objects as future work. We will focus on generating *untargeted* attacks, where the attacker is attempting to force a neural network into misclassifying an object of a certain class into a different class. However, our algorithms can be easily extended to targeted attacks.

In our experiments, we also evaluate defensive techniques. After we generate adversarial attacks on a clean/benign point cloud x by perturbing it with $\delta \in \mathbb{R}^{N \times 3}$ to obtain $x^* = x + \delta$, we feed x^* to a defense before it is classified by a neural network. We now discuss some previously proposed defenses that we will be used in our attack evaluation. These defenses will help simulate a realistic scenario, where the scanning operation and subsequent preprocessing, like outlier removal, make generating effective attacks much more difficult.

Random Point Removal. We test basic random point dropout, which randomly selects and removes a set of points from x. We do not expect this to be a very effective defense.

Outlier Removal. This defense proposed in [16] involves first calculating statistical outliers through

$$o[i] = \frac{1}{K} \sum_{j=1}^{K} \mathrm{NN}(x^*[i], j), \quad \forall i \in \{1 \ldots N\} \tag{1}$$

with $\mathrm{NN}(\rho, j)$ returning the j-th nearest neighboring point of the point ρ. Then, we remove points greater than ϵ standard deviations away from the average $o[i]$ across all i:

$$x^{**} = x^* \setminus \left\{ x^*[i] : o[i] > \frac{1}{N} \sum_{j=1}^{N} o[j] + \epsilon \, \mathrm{stddev}(o) \wedge i \in \{1 \ldots N\} \right\} \tag{2}$$

Salient Point Removal. In this defense proposed by [16], salient points are identified and removed by first calculating the saliency

$$s[i] = \max_{j \in \{1 \ldots M\}} \left\| \left(\nabla_{x^*} f_\theta(x^*)[j] \right)[i] \right\|_2, \quad \forall i \in \{1 \ldots N\} \tag{3}$$

and a subset of the points in x^* with the highest saliencies are removed.

4 Attack Types

In this paper, we present two classes of attacks that represent contradictory attack objectives: *distributional* and *shape* attacks. We summarize the benefits and drawbacks of each category of attacks (along with their 2D counterparts) in Table 1. We elaborate on the distributional and shape attacks below:

Table 1. How our new shape attacks compare to some previous types of attacks in both 2D and 3D domains. We mainly focus on *shape attacks* in this paper.

Images	**Small perturbations for all pixels.** L_p norm attacks like [6,8,10,19,25], and other attacks that minimize perceptibility. Small perturbations are easily lost in real photographs or against defenses/preprocessing
	Large perturbations for some pixels. Either strong single-pixel perturbations [20], low frequency attack [11], or large adversarial patches [4] that are visually noticeable and relatively easy to apply. More likely to be effective against preprocessing and in physical domains (taking a photo of the adversarial examples) due to noticeable perturbations
Point clouds	**Distributional attacks.** Small perturbations or imperceptible point insertion or removal, like in [16,29,35], and [32] (some of which are inspired by 2D L_p methods). Similar to its 2D analog, small point perturbations are easily lost due preprocessing/defenses
	Shape attacks. Morphing the shape of the point cloud by changing multiple points in some focused areas of the point cloud, instead of single point perturbations
	Adding disjoint clusters. The attack in [32] works by adding many clusters of smaller point clouds to an existing point cloud. This is extremely noticeable (similar to our shape attacks), and generates disjoint objects

Distributional Attacks. In this paper, we will mainly focus on distributional attacks that attempt to perturb every point by a small, imperceptible amount (for human eyes) on or very near a 3D object. Usually, the objective functions for these distributional attacks will merely involve minimizing perceptibility through some distance function while maximizing the loss of a neural network. However, due to the spread out, low magnitude perturbations generated in these attacks, we show that the adversarial examples will be brittle against some existing defenses and we believe that they are also difficult to realize on real life objects due to the easily ignored perturbations.

Shape Attacks. These attacks create larger perturbations at a select few locations to create more noticeable continuous shape deformations without outliers (as opposed to outlying points created with previous distributional attacks in [16] and [32]) through the use of resampling, etc. We show that this is robust against previous point-removal defenses due to the continuous, outlier-free changes. We believe that the larger perturbations on real objects are also more likely to be picked up by scanning operations, and the clear, focused deformations in a few areas may be easier to physically construct on real life objects than small perturbations for every point in a point cloud. These attacks may make use of more complex objective functions than just minimizing perceptibility.

Constructing Physical Adversarial Examples. Although we do not discuss this in detail in this paper, we note that these shape attacks can be realized through 3D printing with triangulated point clouds [1, 9, 26], forceful shape deformations (*e.g.*, bending, breaking, etc.), or adding new features with flimsy/cheap materials. These can be done by using generated adversarial shape attacks as a guide to modify specific regions of a physical object.

5 Distributional Attacks

We define a distributional attack as any attack that optimizes for imperceptible point perturbations:

$$\underset{\delta}{\text{maximize}} \quad J(f_\theta(x + \delta), y) - \lambda \mathcal{D}(x + \delta, x) \tag{4}$$

where $\mathcal{D}$ is some metric for measuring the perceptibility of an adversarial attack. A variant of this objective is constraining $\mathcal{D}(x + \delta, x) < \epsilon$, for some threshold ϵ. These definitions will include attacks from previous works like [32] and [16] (for example, the iterative gradient L_p attack or projected gradient descent attacks in 3D). Here, we give a basic overview of the previous Chamfer [32] and gradient projection [16] attacks that were previously proposed.

5.1 Chamfer Attack

There are a few ways to measure the distance (and thus, perceptibility) between two point sets. One way is the Chamfer distance, which is given by

$$\mathcal{C}(A, B) = \frac{1}{|A|} \sum_{a \in A} \min_{b \in B} ||b - a||_2 \tag{5}$$

for two sets A and B. With this distance and the L_2 norm to discourage large perturbations, the following optimization problem can be solved using an algorithm like Adam [12] to generate a distributional adversarial example:

$$\underset{\delta}{\text{maximize}} \quad J(f_\theta(x + \delta'), y) - \lambda(\mathcal{C}(x + \delta', x) + \alpha||\delta'||_2) \tag{6}$$

where the L_∞ norm of the perturbation δ' is hard bounded using $\delta' = 0.1 \tanh \delta$ (note that each point cloud is scaled to fit in a unit sphere), so the perturbations cannot become arbitrarily large, which is unfair when comparing to other attacks. We binary search for λ and set a value α for balancing the two distance functions. This is one of the proposed attacks in [32], and in this paper we will refer to it as the *Chamfer attack*. In general, the important of using the Chamfer metric to measure perceptiblity is evident when compared to using a standard L_p norm to measure perceptibility [16], as L_p norms do not encourage perturbations that result in point shifts across the surface of an object (note that the points x lies on the surface of an object).

5.2 Gradient Projection

For our variant of the gradient projection attack [16] that we evaluate, we use a very similar metric to Chamfer distance: the Hausdorff distance

$$\mathcal{H}(A, B) = \max_{a \in A} \min_{b \in B} ||b - a||_2 \tag{7}$$

Although this metric was also used in previous works like [32], our new attack variant involves measuring the amount of perturbation by comparing the perturbed point cloud x^* to the benign surface S of a point cloud, rather than the benign point cloud x. This is less sensitive to the density of points, and imperceptible point shifts along the surface of the object do not increase the Hausdorff distance. Note that in practice, the benign surface S is represented through a triangular mesh t with T triangles. This triangular mesh can be obtained through an algorithm like the alpha shape algorithm [9] using the 3D Delaunay triangulation [15] of each point cloud (other triangulation algorithms like [1] also work).

If we structure our optimization problem for constructing adversarial point clouds to maximize J while bounding the Hausdorff distance with an easy-to-tune parameter τ, then we can craft adversarial perturbations that are imperceptible. The problem can be formulated as

$$\underset{\delta}{\text{maximize}} \quad J(f_\theta(x + \delta), y), \quad \text{subject to} \quad \mathcal{H}(x + \delta, S) \leq \tau \tag{8}$$

This formulation is similar to the optimization problem solved through [17]'s projected gradient descent, but with a different method for measuring perceptibility. Therefore, it can be solved the same way, using n steps of gradient descent (unfortunately, momentum cannot be used) constrained under the L_2 norm by $\frac{\epsilon}{n}$ (for some hyperparameter ϵ) and then projecting onto the set $\{a : \mathcal{H}(\{\rho\}, S) \leq \tau \wedge \rho \in \mathbb{R}^3\}$ for a surface S using a variant of the basic point-to-triangle projection, since S is represented by a set of triangles t. In our implementation of this attack, we use a VP-tree [34] to organize the 3D triangles, which are represented as points, for faster projections by pruning with a calculated query radius in the tree. Note that this algorithm is a generalization of the gradient projection attack proposed in [16], which is just a special case where $\tau = 0$.

6 Shape Attacks

The general idea behind each of our proposed shape attacks is that we try to use point perturbations that are known to be successful in the distributional attacks, and then somehow "connecting" these points with the main point cloud object to create shape deformations, so that the resulting adversarial point cloud is a *single* connected object without any outliers. This is partially inspired by low-frequency attacks on 2D images [11], which generates smooth perturbations over multiple pixels, instead of noisy perturbations for each pixel. In our attacks, we enforce the criteria that the *density* of the sampled points to be as uniform as possible on S^*, the surface of an adversarial object, through methods like resampling.

6.1 Perturbation Resampling

Consider an adversarial object's surface that is represented by the set S^*. We want to ensure that the point distribution of a point cloud $x^* \subset S^*$ is approximately uniform density on this surface by maximizing the following objective:

$$\underset{x^*}{\text{maximize}} \quad \underset{i \in \{1...N\}}{\min} \underset{j \in \{1...N\} \setminus \{i\}}{\min} ||x^*[j] - x^*[i]||_2, \quad \text{subject to} \quad x^* \subset S^* \quad (9)$$

in addition to maximizing the loss, while constraining the perceptibility. The easiest method to accomplish this is through the use of a point resampling algorithm, like farthest point sampling, on the estimated shape S^* (which is obtained using something like the alpha shape algorithm to get a triangulation of the point cloud) after perturbing the point cloud with gradient descent. This simple idea of adding resampling into the optimization process was also examined in a parallel work [26], which also proposed the usage of sampling points on the surface of the estimated shape in the attack.

Overall, the attack procedure can be summarized as the following: during each iteration of the perturbation resampling attack, we first execute one step of gradient descent (without momentum, due to the resampling operation) constrained by the L_2 norm and perturb the point cloud x to get x^*. Then, we approximate the shape of this perturbed point cloud by computing the alpha shape to get S^*. Finally, we use farthest point sampling to resample the κ points with the lowest saliencies (which is defined as the L_2 norm of the gradient of the loss function wrt each point) onto S^*. The $N - \kappa$ points with higher saliencies are automatically considered "picked" from sampling. Intuitively, we are essentially perturbing the underlying shape of the 3D point cloud, while ensuring that points are evenly sampled on the surface of that shape when the surface is stretched due to the perturbations.

6.2 Adversarial Sticks

We can also attempt to create new features on the mesh instead of making small, incremental deformities through the perturbation resampling attack. [32] explores this idea by adding clusters of points and even 3D objects from other classes to attack a benign 3D object. Their attack crafts adversarial examples that can contain many disjoint point clouds. We think that this is unrealistic for a classification task on a single point cloud object, so we attempt to add a very simple new feature (sticks, or line segments) onto a point cloud, where the sticks must originate from the surface of the benign 3D object. This shows that adding only a few new features is an effective method for generating adversarial point clouds.

A straightforward formulation of the problem results in optimizing over the position vectors $\alpha \in \mathbb{R}^{\sigma \times 3}$ and the orientation and length vectors $\beta \in \mathbb{R}^{\sigma \times 3}$ for generating σ sticks:

$$\underset{\alpha, \beta}{\text{maximize}} \quad J(f_\theta(x \cup \mathcal{S}_\kappa(\alpha, \alpha + \beta)), y), \quad \text{subject to} \quad ||\beta||_2 \leq \epsilon, \quad \alpha \subset S \quad (10)$$

Note that $\mathcal{S}_\kappa(\alpha, \alpha + \beta)$ samples κ points across the line segments defined by the points $\alpha[i]$ and $\alpha[i] + \beta[i], \forall i \in \{1 \ldots \sigma\}$. The sampling procedure involves assigning points to each stick with probability proportional to its length, and then all points assigned to a stick are evenly spread out along the stick. These sampled points are added to the point cloud x, and they represent sticks that are connected to the point cloud. However, solving this is difficult due to a number of factors. First, the sampling function $\mathcal{S}$ is not differentiable, which means that the objective function cannot be directly maximized. Second, we need to somehow constrain the position of each stick onto the surface of the benign point cloud x in each step of the attack.

The first problem can be solved by approximating the solution to the optimization problem by perturbing points to get x^* using the saliency of each point (L_2 norm of the gradient of the loss function with respect to each input point) instead of orienting sticks, so $\alpha = \mathcal{P}(x^*)$ and $\beta = x^* - \mathcal{P}(x^*)$. $\mathcal{P}$ is a function that projects each point onto the benign surface S (which is represented as a set of triangles t in practice) of a point cloud:

$$\mathcal{P}(x^*) = \bigcup_{i=0}^{N} \left\{ \underset{\rho \in \text{proj}(x^*[i])}{\arg \min} ||x^*[i] - \rho||_2 \right\}, \quad \text{where}$$

$$\text{proj}(\rho) = \bigcup_{j=0}^{T} \left\{ \text{clip\_to\_tri} \left(\rho - ((\rho - t[j][0]) \cdot n[j])n[j], \ t[j] \right) \right\} \quad (11)$$

Note that $n[j]$ is the unit normal vector of the j-th triangle and clip_to_tri clips an orthogonally projected point to be inside a triangle by examining each edge of the triangle. Additionally, we use iterative gradient L_2 to perturb x to get x^*. In this case, we only keep the subset of σ points with the largest perturbations, so a controlled, small amount of adversarial sticks can be generated. To solve the second problem, we resample low saliency points onto the generated sticks at the end, after directly perturbing points based on the gradient of f_θ. In summary, we use gradient descent constrained with the L_2 norm to attack the point cloud x, pick the σ points with the largest perturbations (saliencies), and sample points on the line segments defined by each perturbed point and its closest point (projection point, using a VP-tree in practice) on the surface of the benign point cloud.

6.3   Adversarial Sinks

Due to resampling and projection operations used in the adversarial sinks attack that make optimization difficult, the natural question is whether it is possible to create new shape features without those operations. Indeed, it is possible to deform the shape of the benign point cloud to generate an adversarial example without inferring its surface. We can guide this deformation through sink points, which pull points in the point cloud in a local spatial region towards them. This means that there will not be outlier points after the attack. In practice, this should lead to smooth shape deformations. The reason for guiding the deformation process with a few points is because we know that perturbing a select few points is effective against some point cloud neural networks due to their max pooling operation that selects a subset of critical points in their prediction process [16,23].

Given a set of σ sink points $s \in \mathbb{R}^{\sigma \times 3}$ that are initialized with $s_0 \in \mathbb{R}^{\sigma \times 3}$ (in practice, this would be the σ points with the highest saliencies), we can deform the point cloud x to create x^* with

$$x^*[i] = x[i] + 0.5 \tanh \Big(\sum_{j=1}^{\sigma} (s[j] - x[i]) \phi_{\mu'}(||s_0[j] - x[i]||_2) \Big), \quad \forall i \in \{1 \dots N\}$$
(12)

where $\phi_{\mu'}$ represents a radial basis function that decreases the influence of each sink point by distance. This means that points in the point cloud that are far away from a sink point's initial position will be influenced less by that sink point as it moves. We use the Gaussian radial basis function which is defined as

$$\phi_{\mu'}(r) = e^{-\left(\frac{r}{\mu'}\right)^2}, \quad \text{for} \quad \mu' = \frac{\mu}{N} \sum_{i=1}^{N} \min_{j \in \{1 \dots N\} \setminus \{i\}} ||x[j] - x[i]||_2 \quad (13)$$

The hyperparameter μ can be tuned to control how the influence of each sink point falls off as distance increases. The influence of each sink point is also scaled by the average distance between point in x and its nearest neighboring point, so it is invariant to the scale of the point cloud x. We choose the Gaussian radial basis function because it is differentiable everywhere and it outputs a value between 0 and 1, which represents a percentage of the distance between each point cloud point and sink point. In essence, Eq. 12 represents shifting each point in the point cloud by the weighted sum of the vectors from that point to each sink point.

To craft an attack, we solve the following optimization problem:

$$\underset{s}{\text{maximize}} \quad J\big(f_\theta(x^*), y\big) - \lambda ||x^* - x||_2 \quad (14)$$

where x^* is constructed according to Eq. 12. The idea is that we are trying to maximize the overall loss by shifting our sink points s, while minimizing the L_2 norm of the perturbation (similar to the Chamfer attack, tanh is used to hard bound the L_∞ norm of the perturbations). λ can be adjusted using binary search to balance the emphasis on maximizing the loss and minimizing the L_2 norm of the perturbation. Since this objective function is fully differentiable, we can directly apply a gradient-based optimization algorithm like Adam [12] to maximize our objective function.

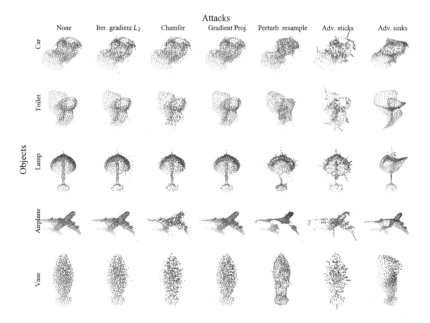

Fig. 1. Visualizations of adversarial examples that are generated on the PointNet architecture, for five different ModelNet40 classes. Notice that the Chamfer, gradient projection, and iterative gradient L_2 attacks produce noisy point clouds, while the perturbation resampling, adversarial sticks, and adversarial sinks attacks change the shape of the point cloud objects. There are some areas that are noticeably missing points in the perturbation resampling and adversarial sticks adversarial examples, which is due to limitations in the triangulation and resampling procedures on a fixed number of points. Other triangulation methods may produce better point clouds, but we leave this as an open question.

7   Results

7.1   Setup

Models and Dataset. We train PointNet [23], PointNet++ [24], and DGCNN [28] with default hyperparameters, but we use a slightly lowered batch size for PointNet++ due to memory constraints. The neural networks are trained on the training split of the ModelNet40 dataset [31]. The test split of the dataset is used for evaluating our adversarial attacks. We use point clouds of size $N = 1,024$ sampled with uniform density from 3D triangular meshes in the ModelNet40 dataset.

Attacks. For the traditional iterative gradient L_2 attack, we use $\epsilon = 2$ and $n = 100$ iterations. For the Chamfer attack, we run the Adam optimizer ($\beta_1 = 0.9$ and $\beta_2 = 0.999$) for $n = 20$ iterations with a learning rate of $\eta = 0.1$, we balance the two distance functions with $\alpha = 0.002$, and we binary search for λ. For the

Table 2. The success rates of untargeted adversarial attacks against different defenses on the PointNet, PointNet++, and DGCNN architectures. The highest attack success rate for each neural network and defense combination is bolded. The last three columns are our new shape attacks.

	Defenses	Attacks						
		None	Iter. grad. L_2	Chamfer	Grad. proj.	Perturb. resample	Adv. sticks	Adv. sinks
PointNet	None	0.0%	99.1%	98.2%	89.8%	97.5%	97.2%	**99.2%**
	Random remove	0.8%	**98.4%**	93.3%	86.3%	96.4%	97.1%	93.3%
	Remove outliers	5.6%	56.9%	56.9%	46.9%	81.6%	**97.0%**	86.2%
	Remove salient	7.7%	51.8%	59.4%	41.6%	82.8%	**96.9%**	87.1%
PointNet++	None	0.0%	**100.0%**	86.2%	98.8%	98.4%	77.8%	95.7%
	Random remove	3.6%	**96.2%**	71.9%	75.9%	95.2%	74.5%	77.0%
	Remove outliers	11.0%	44.4%	47.4%	29.6%	**86.7%**	85.3%	81.6%
	Remove salient	10.2%	46.6%	49.8%	26.8%	**81.0%**	70.7%	76.8%
DGCNN	None	0.0%	42.0%	67.1%	34.5%	76.0%	30.2%	**99.3%**
	Random remove	5.6%	32.4%	45.9%	22.1%	72.8%	37.1%	**78.5%**
	Remove outliers	10.5%	22.6%	31.8%	18.4%	69.1%	63.9%	**84.4%**
	Remove salient	11.9%	24.1%	36.4%	18.0%	65.8%	36.5%	**73.4%**

gradient projection attack, we use $\epsilon = 1$, $\tau = 0.05$, and $n = 20$ iterations (it is relatively more time consuming to run than other attacks). For perturbation resampling, we use $\epsilon = 2$, we resample $\kappa = 500$ points, and we run the attack for $n = 100$ iterations. For adversarial sticks, we use $\epsilon = 2$, $n = 100$, we add $\sigma = 100$ sticks, and we resample $\kappa = 400$ points. For adversarial sinks, we use a learning rate of $\eta = 0.1$ for the Adam optimizer ($\beta_1 = 0.9$ and $\beta_2 = 0.999$), we scale the strength of each sink point by $\mu = 7$, we binary search for λ, and we run the attack for $n = 20$ iterations with $\sigma = 20$ sink points. The hyperparameters were determined through rudimentary grid search, and we will show the effect of changing some of them in our experiments. We use larger ϵ values in our shape attacks, since the perturbations are more focused. Additionally, we use less iterations for the attacks that use the Adam optimizer instead of vanilla gradient descent because Adam is much more efficient. There should be a low number of sink points and adversarial sticks, so they are easy to construct.

Defenses. For random point removal, we drop out 200 random points. For removing outliers, we calculate the average distance from each point to its 10 nearest neighbors, and remove points with an calculated average distance greater than one standard deviation from the average of the average distances across all points. For removing salient points, we remove the 200 points with the highest saliencies. We do not test adversarial training because it is not attack agnostic and it was found to perform worse than the point removal defenses [16].

7.2 Adversarial Attacks

In Table 2, we show the success rates of our attacks on the 2,000+ correctly classified objects from the ModelNet40 dataset, against the PointNet, PointNet++, and DGCNN architectures with different defenses. Adversarial examples are visualized in Fig. 1.

From Table 2, we see that overall, our new attacks are quite successful against the defended PointNet, PointNet++, and DGCNN models, but they are slightly less effective than the traditional iterative gradient L_2 attack when there are no strong defenses. This is due to the extra processes like projection, resampling, and limiting the perturbations to a few select areas, which are used to meet the desired perceptibility or shape deformation criteria. Adversarial sticks does not do as well as the other shape attacks on PointNet++ and DGCNN, probably due to those networks (instead of the defenses) ignoring the focused perturbations in some areas. Adversarial sinks seems to be the most successful, as expected, since its objective function easier to optimize. The effectiveness of the shape attacks against the defenses can be explained through the harder to remove/ignore deformation features, the inability for the defense to rely on unperturbed points with no gradient flow (due to the max pooling operation) for a correct prediction, and the inability for the defense to identify and remove adversarial outliers. We note that the random point removal defenses is very weak compared to the other defenses.

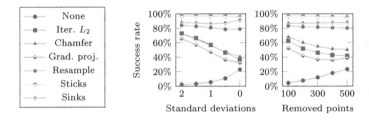

Fig. 2. The success rates of adversarial attacks against the PointNet architecture as we increase the number of points removed with the outlier removal (left) and the salient point removal (right) defenses. For the outlier removal defense, more points are removed as the number of standard deviations from the average decreases.

Stronger Point Removal Defenses. In Fig. 2, as we increase the number of points removed through the outlier removal and salient point removal defenses for stronger attacks, we see that our three shape attacks are able to maintain their high success rates, compared to the distributional attacks. Therefore, strong point removal defenses are relatively ineffective against our shape attacks. In practice, removing such a large amount of points is not recommended, as we see that removing nearly half of the points in each benign point cloud also causes PointNet to misclassify more than 20% of the benign point clouds.

Blending Between a Shape Attack and a Distributional Attack. In Fig. 3, we show the effect of decreasing the number of sampled points in the

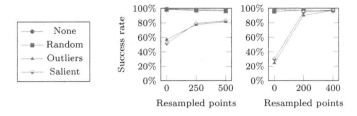

Fig. 3. The success rates of the perturbation resampling (left) adversarial sticks (right) attack against different point removal defenses on the PointNet architecture as the number of resampled points changes. Essentially, we are "blending" between a distributional attack and a shape attack to show the effect of resampling points on the success rates.

adversarial sticks attack. This is essentially blending between a shape attack and a simple "distributional attack" with no resampling procedures. We see that the attacks rely heavily on resampling to be robust against point removal defenses. Additionally, we see that resampling (which only happens for points with low saliencies) does not significantly affect the success rate of an attack against an undefended PointNet. We note that increasing the μ parameter in the adversarial sinks attack will increase its success rate against defenses, similar to resampling more points.

Table 3. The success rates of some adversarial attacks with different ϵ values (magnitude of perturbations) against different defenses on the PointNet and PointNet++ architectures. The highest attack success rate for each model and defense is bolded.

Attacks	ϵ	PointNet				PointNet++			
		None	Random	Outliers	Salient	None	Random	Outliers	Salient
Iter. gradient L_2	1	96.9%	94.5%	46.6%	41.5%	99.7%	80.1%	30.5%	28.1%
	2	99.1%	98.4%	56.9%	51.8%	**100.0%**	96.2%	44.4%	46.6%
	3	**99.4%**	**99.3%**	62.1%	55.8%	**100.0%**	**98.7%**	53.0%	55.5%
Perturb. resample	1	93.2%	91.0%	74.0%	74.7%	92.8%	87.2%	78.4%	71.3%
	2	97.5%	96.4%	81.6%	82.8%	98.4%	95.2%	86.7%	81.0%
	3	98.7%	98.0%	84.3%	84.3%	99.6%	98.1%	**92.3%**	**88.3%**
Adv. sticks	1	91.0%	90.9%	90.7%	90.4%	60.8%	56.8%	78.5%	57.1%
	2	97.2%	97.1%	97.0%	96.9%	77.8%	74.5%	85.3%	70.7%
	3	98.8%	98.7%	**98.5%**	**98.4%**	85.2%	81.8%	88.3%	77.2%

Using Different ϵ Values in the Attacks. For some attacks that requires an ϵ hyperparameter, we show the effect of changing the ϵ values on the success rates of the attacks in Table 3. As expected, the success rate of each attack increases as ϵ increases, which exemplifies the tradeoff between perceptibility and attack success rate.

Perceptibility vs Robustness. There is a definite trade off between the visual perceptibility of the attacks and their robustness against defenses. The shape and distributional attacks represent different extremes of this spectrum. In practice, we think that the shape attacks can be used as *physical* attacks on real-world objects *before* scanning, and distributional attacks can be used as imperceptible *digital* attacks. We believe that effective, but imperceptible physical attacks are very difficult to craft without creating large, focused perturbations in certain areas of an object, as the adversarial features generated by an attack will be easily removed and ignored through the defenses. Small perturbations spread out across an entire shape are too easily masked through point dropout.

8   Conclusion

We examine previous distributional attacks, which optimize for human impercep- tibility, and shape attacks, which are more realistic. We also introduce three new shape attacks for 3D point cloud classifiers, which utilizes shape morphing to be robust against previous defenses or preprocessing steps. These attacks show that morphing the shape of a 3D object or adding a few adversarial features are viable options for generating effective adversarial attacks. Our work is an impor- tant step in understanding the vulnerability of point cloud neural networks and defenses against adversarial attacks. For future work, it is important to create defenses that are robust to our shape attacks.

Acknowledgements. This work was supported in part by NSF awards CNS-1730158, ACI-1540112, ACI-1541349, OAC-1826967, the University of California Office of the President, and the University of California San Diego's California Institute for Telecom- munications and Information Technology/Qualcomm Institute. Thanks to CENIC for the 100Gpbs networks. We want to thank Battista Biggio from the University of Calia- gri for feedback on a draft of this manuscript.

References

1. Bernardini, F., Mittleman, J., Rushmeier, H., Silva, C., Taubin, G.: The ball- pivoting algorithm for surface reconstruction. IEEE Trans. Vis. Comput. Graph. **5**(4), 349–359 (1999)
2. Biggio, B., et al.: Evasion attacks against machine learning at test time. In: Bloc- keel, H., Kersting, K., Nijssen, S., Železný, F. (eds.) ECML PKDD 2013. LNCS (LNAI), vol. 8190, pp. 387–402. Springer, Heidelberg (2013). https://doi.org/10. 1007/978-3-642-40994-3_25
3. Biggio, B., Roli, F.: Wild patterns: ten years after the rise of adversarial machine learning. Pattern Recognit. **84**, 317–331 (2018)
4. Brown, T.B., Mané, D., Roy, A., Abadi, M., Gilmer, J.: Adversarial Patch. arXiv preprint arXiv:1712.09665 (2017)
5. Cao, Y., et al.: Adversarial Objects Against LiDAR-Based Autonomous Driving Systems. arXiv preprint arXiv:1907.05418 (2019)
6. Carlini, N., Wagner, D.: Towards evaluating the robustness of neural networks. In: 2017 IEEE Symposium on Security and Privacy, pp. 39–57. IEEE (2017)

7. Deng, H., Birdal, T., Ilic, S.: PPF-FoldNet: Unsupervised Learning of Rotation Invariant 3D Local Descriptors. arXiv preprint arXiv:1808.10322 (2018)
8. Dong, Y., et al.: Boosting Adversarial Attacks with Momentum. arXiv preprint (2018)
9. Edelsbrunner, H., Kirkpatrick, D., Seidel, R.: On the shape of a set of points in the plane. IEEE Trans. Inf. Theory **29**(4), 551–559 (1983)
10. Goodfellow, I., Shlens, J., Szegedy, C.: Explaining and Harnessing Adversarial Examples. arXiv preprint arXiv:1412.6572 (2014)
11. Guo, C., Frank, J.S., Weinberger, K.Q.: Low frequency adversarial perturbation. arXiv preprint arXiv:1809.08758 (2018)
12. Kingma, D.P., Ba, J.: Adam: A method for stochastic optimization. arXiv preprint arXiv:1412.6980 (2014)
13. Kurakin, A., Goodfellow, I., Bengio, S.: Adversarial Examples in the Physical World. arXiv preprint arXiv:1607.02533 (2016)
14. Kurakin, A., Goodfellow, I., Bengio, S.: Adversarial Machine Learning at Scale. arXiv preprint arXiv:1611.01236 (2016)
15. Lee, D.T., Schachter, B.J.: Two algorithms for constructing a Delaunay triangulation. Int. J. Comput. Inf. Sci. **9**(3), 219–242 (1980)
16. Liu, D., Yu, R., Su, H.: Extending Adversarial Attacks and Defenses to Deep 3D Point Cloud Classifiers. arXiv preprint arXiv:1901.03006 (2019)
17. Madry, A., Makelov, A., Schmidt, L., Tsipras, D., Vladu, A.: Towards Deep Learning Models Resistant to Adversarial Attacks. arXiv preprint arXiv:1706.06083 (2017)
18. Moosavi-Dezfooli, S.M., Fawzi, A., Fawzi, O., Frossard, P.: Universal adversarial perturbations. In: Proceedings of the IEEE Conference on Computer Vision and Pattern Recognition, pp. 1765–1773 (2017)
19. Moosavi-Dezfooli, S.M., Fawzi, A., Frossard, P.: DeepFool: a simple and accurate method to fool deep neural networks. In: Proceedings of the IEEE Conference on Computer Vision and Pattern Recognition, pp. 2574–2582 (2016)
20. Papernot, N., McDaniel, P., Jha, S., Fredrikson, M., Celik, Z.B., Swami, A.: The limitations of deep learning in adversarial settings. In: 2016 IEEE European Symposium on Security and Privacy (EuroS&P), pp. 372–387. IEEE (2016)
21. Papernot, N., McDaniel, P., Wu, X., Jha, S., Swami, A.: Distillation as a defense to adversarial perturbations against deep neural networks. In: 2016 IEEE Symposium on Security and Privacy (SP), pp. 582–597. IEEE (2016)
22. Qi, C.R., Liu, W., Wu, C., Su, H., Guibas, L.J.: Frustum PointNets for 3D Object Detection from RGB-D Data. arXiv preprint arXiv:1711.08488 (2017)
23. Qi, C.R., Su, H., Mo, K., Guibas, L.J.: PointNet: deep learning on point sets for 3D classification and segmentation. In: Proceedings of the IEEE Conference on Computer Vision and Pattern Recognition, vol. 1, no. 2, p. 4 (2017)
24. Qi, C.R., Yi, L., Su, H., Guibas, L.J.: PointNet++: deep hierarchical feature learning on point sets in a metric space. In: Advances in Neural Information Processing Systems, pp. 5099–5108 (2017)
25. Szegedy, C., et al.: Intriguing properties of neural networks. arXiv preprint arXiv:1312.6199 (2013)
26. Tsai, T., Yang, K., Ho, T.Y., Jin, Y.: Robust adversarial objects against deep learning models. In: Proceedings of the AAAI Conference on Artificial Intelligence, vol. 34, pp. 954–962 (2020)
27. Wang, P.S., Liu, Y., Guo, Y.X., Sun, C.Y., Tong, X.: O-CNN: octree-based convolutional neural networks for 3D shape analysis. ACM Trans. Graph. **36**(4), 72 (2017)

28. Wang, Y., Sun, Y., Liu, Z., Sarma, S.E., Bronstein, M.M., Solomon, J.M.: Dynamic graph CNN for learning on point clouds. ACM Trans. Graph. (TOG) (2019)
29. Wicker, M., Kwiatkowska, M.: Robustness of 3D deep learning in an adversarial setting. In: Proceedings of the IEEE Conference on Computer Vision and Pattern Recognition, pp. 11767–11775 (2019)
30. Wong, E., Kolter, J.Z.: Provable Defenses against Adversarial Examples via the Convex Outer Adversarial Polytope. arXiv preprint arXiv:1711.00851 (2017)
31. Wu, Z., et al.: 3D ShapeNets: a deep representation for volumetric shapes. In: Proceedings of the IEEE Conference on Computer Vision and Pattern Recognition, pp. 1912–1920 (2015)
32. Xiang, C., Qi, C.R., Li, B.: Generating 3D Adversarial Point Clouds. arXiv preprint arXiv:1809.07016 (2018)
33. Yang, J., Zhang, Q., Fang, R., Ni, B., Liu, J., Tian, Q.: Adversarial Attack and Defense on Point Sets. arXiv preprint arXiv:1902.10899 (2019)
34. Yianilos, P.N.: Data structures and algorithms for nearest neighbor search in general metric spaces. In: Proceedings of the Fourth Annual ACM-SIAM Symposium on Discrete Algorithms, vol. 93, pp. 311–321 (1993)
35. Zheng, T., Chen, C., Ren, K., et al.: Learning Saliency Maps for Adversarial Point-Cloud Generation. arXiv preprint arXiv:1812.01687 (2018)
36. Zhou, H., Chen, K., Zhang, W., Fang, H., Zhou, W., Yu, N.: Deflecting 3D Adversarial Point Clouds Through Outlier-Guided Removal. arXiv preprint arXiv:1812.11017 (2018)

Jacks of All Trades, Masters of None: Addressing Distributional Shift and Obtrusiveness via Transparent Patch Attacks

Neil Fendley, Max Lennon, I-Jeng Wang, Philippe Burlina,
and Nathan Drenkow[(⊠)]

The Johns Hopkins University Applied Physics Laboratory, Laurel, MD, USA
{neil.fendley,max.lennon,i-jeng.wang,philippe.burlina,
nathan.drenkow}@jhuapl.edu

Abstract. We focus on the development of effective adversarial patch attacks and – for the first time – jointly address the antagonistic objectives of attack success and obtrusiveness via the design of novel semi-transparent patches. This work is motivated by our pursuit of a systematic performance analysis of patch attack robustness with regard to geometric transformations. Specifically, we first elucidate a) key factors underpinning patch attack success and b) the impact of distributional shift between training and testing/deployment when cast under the Expectation over Transformation (EoT) formalism. By focusing our analysis on three principal classes of transformations (rotation, scale, and location), our findings provide quantifiable insights into the design of effective patch attacks and demonstrate that scale, among all factors, significantly impacts patch attack success. Working from these findings, we then focus on addressing how to overcome the principal limitations of scale for the deployment of attacks in real physical settings: namely the obtrusiveness of large patches. Our strategy is to turn to the novel design of irregularly-shaped, semi-transparent partial patches which we construct via a new optimization process that jointly addresses the antagonistic goals of mitigating obtrusiveness and maximizing effectiveness. Our study – we hope – will help encourage more focus in the community on the issues of obtrusiveness, scale, and success in patch attacks.

1 Introduction

Deep learning (DL) has demonstrated performance seemingly on par with humans for various problems such as medical image diagnostics [4], game playing [9], and other tasks [7]. However this apparent success has been met with the prospect [5] that deep networks are very fragile to adversarial attacks, which motivated many studies in the past several years researching methods for attacks and defenses, and the corresponding significant growth of interest in the community in the field of *adversarial machine learning* (AML) specifically applied to

© Springer Nature Switzerland AG 2020
A. Bartoli and A. Fusiello (Eds.): ECCV 2020 Workshops, LNCS 12535, pp. 105–119, 2020.
https://doi.org/10.1007/978-3-030-66415-2_7

DL models. Several approaches to AML as applied to DL have been envisioned, and a simple taxonomy useful to frame the goals for this work consists of distinguishing between a) generating attacks that affect an entire image, are designed to work on a specific image, and are carried out by additive non-perceivable perturbations (such as in [5,10,11,13]), vs. b) attacks that are confined to a specific sub-window of the image and are designed to affect a wide set of images and classes of objects (i.e., a patch attack).

Attacks that are of interest for this work use the approach in (b). These attacks are more suitable to be implemented in the physical domain since they can be printed on contiguous surfaces and placed more easily in a scene – a significant concern for applications such as automotive and robotic autonomy and related areas. The first successful design of such an attack was reported in [2] where it was demonstrated that an adversarial patch can be designed by using a loss function that includes a term that expresses an expectation over geometric transformations including rotation, translation and scale. This was based on work originally reported in [1].

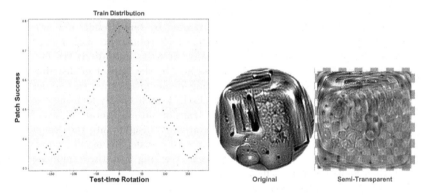

Fig. 1. [**left**] Demonstrating the impact on patch attack success for test-time transformations sampled out-of-distribution. [**right**] A novel method for generating semi-transparent patch attacks to address the balance between patch scale, obtrusiveness, and success. (Toaster patch used above from [2])

Additionally, and to better frame our motivations here, when it comes to patch attacks for physical scenes, one must distinguish between the specific case (where attacks are generated and optimized for a particular scene) and the general case (where attacks are generated without any prior knowledge of the scene/context in which they will be placed). In the former scenario, patch patterns and placement may be optimized more highly given greater knowledge about the attack scene. However, we focus on the latter scenario which makes fewer assumptions about the scene and is thus likely to produce attacks that generalize more effectively. We wish to understand the robustness of deep neural networks to attacks developed for these scenarios and then design the most effective attacks under these assumptions.

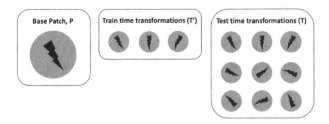

Fig. 2. Constrained EoT limits the range of transformations (e.g., rotation angles, scales, etc.) for optimizing the patch, but increases the range used at test time. Test time transformations are assumed to be unconstrained as this is consistent with many real world use cases.

2   Contributions

The novel contributions of this are as follows: a) we present a principled methodology for the evaluation of patch attacks and the train/test-time factors that impact their success. In particular, we study, under the framework of the expectation over transformation approach, the impact of distributional differences between patch optimization and deployment conditions and their subsequent effect on patch attack success. This study allows new insights into factors leading to attack success and in particular demonstrate that among all – patch scale is a driving factor for success, and that rotation factors suffers from a "jack of all-trades, master of none" pathology. b) Armed with these observations we next consider the question of how to best design effective patches. That is, patches that scale up but still retain desirable factors with regard to deployment and detectability (i.e. unobtrusiveness). To our knowledge we report the first ever design of semi-transparent patches that address these objectives. We also develop a novel c) an optimization framework that results in the design of such patches and d) develop new methods to characterize effectiveness in this new scale/obtrusiveness/success trade space, and, given scale as a key limiting factor, we develop a novel measure for patch obtrusiveness.

3   Analysis of Expectation over Transformation

3.1   Motivation

This paper focuses on patch-based attacks initially inspired by the EoT method first introduced by [1]. Early experiments with EoT produced results as shown in Fig. 1. In particular, it could be observed that when the test-time transformation distribution is unconstrained and mismatched with the train-time distribution, the patch attack severely under-performs. Furthermore, as we seek to optimize new patches, there has been limited examination of how to define the patch transformation distributions used in the optimization.

To dig deeper into these issues, we performed a systematic study of the effects of standard patch transformations on attack success. Specifically, we identify how train and test-time distributions (and mismatches between them) relate to patch effectiveness. Given that EoT is the standard method for generating patch attacks, our intent is not to disprove the method's utility but rather to provide quantifiable evidence and practical insights regarding its usage.

3.2 Expectation over Transformation

In one of the most common formulations of the problem, an adversarial example is generated by perturbing a clean input image x (taken from the set of all possible images X) to produce an adversarial exemplar image, x', such that a classifier will produce a correct classification, y (taken from a set of Y classes) for x, but an incorrect classification, y', for x'. Following [1], the perturbation is determined more formally via optimization:

$$\arg\max_{x'} P(y'|x') \\ \text{subject to } \|x' - x\|_p < \varepsilon \tag{1}$$

In the case of a targeted attack, the optimization is designed to find the x' that maximizes the likelihood of producing the targeted label, y'_t.

As stated in [1], this approach fails to account for real world transformations and was thus extended by including such transformations directly in the optimization process. The result is a modified optimization:

$$\arg\max_{x'} \mathbb{E}_{t \sim T} P(y'|t(x')) \\ \text{subject to } \mathbb{E}_{t \sim T}[d(t(x), t(x'))] < \varepsilon \tag{2}$$

where T is a distribution of transforms, $d()$ is a distance measure, and $t(x)$ is a sampled transform applied to the image x.

This approach was further modified by [2] to produce patch-based attacks (where perturbations are spatially co-located) robust to transformations likely to occur in the physical world. In particular, that work used a patch application operator, $A(p, x)$, which applies transformations and perturbations within a confined spatial region. The result is the following:

$$\arg\max_{p} \mathbb{E}_{x \sim X, t \sim T} P(y'|A(t(p), x)) \tag{3}$$

Without loss of generality, the focus is on transformations T including distributions over location, scale, and rotation.

While the above methods demonstrate the advantage of expanding the optimization to incorporate potential real-world transformations, the trade-off between attack success and transformations considered has not yet been investigated and is the focus of our study.

3.3 Constrained Expectation over Transformation

The main endeavor of this section is to determine how constraints imposed on the distribution of transformations during the patch optimization affect success when transformations are unconstrained at test time. For physical attacks, this is a key question since the attacker can optimize and place the patch in the scene (under certain transformation and distributional assumptions) but has no control over the pose of the system-under-attack, which may present a distributional shift at attack time. In [1] and [2], the transformations were limited to a small range of rotation angles and scales. Here we ask whether there is any trade-off or negative impact on the effectiveness of EoT if the range of transformations is expanded or mismatched between optimization and testing.

We run a series of experiments (described in Sect. 4) with modifications to (2) whereby during the optimization we consider the distribution T (typically, uniform) over a compact support $\mathcal{T}$ (Fig. 2). At test time, we then use the full support of T to assess patch success. We first perform these experiments considering scale and rotation transformations separately, then repeat with scale and rotation jointly optimized.

The motivation for this formulation and set of experiments is three-fold. First, by creating a difference between train and test-time transformation distributions, we can assess whether the added constraint improves or degrades patch success. For instance, we can quantify trade-offs or benefits that might exist for optimizing over the full range of rotation angles relative to a smaller number. Additionally, we can further study in isolation the effects of individual transformations on patch effectiveness. Lastly, we can leverage insights gained to more appropriately constrain patch optimization relative to specific applications. We note that while we limit our experiments to examining patch rotation, scale, and location, this evaluation can be generalized to other relevant transformations under consideration.

4   Experiments

4.1   Base Experiment

All of the following experiments are derived from a common base experiment. Since the primary focus of this section is on providing a quantitative analysis of the effects of testing transformations outside the distribution of training images, we restrict our experiments to a white-box scenario where the model and architecture used for patch optimization is the same as the one that undergoes the patch attack. We view this as a best case scenario in light of the challenges with attack transferability (as discussed in [3,8,12,15]). All experiments in this paper optimize and test patch attacks using a ResNet50 [6] model trained on ImageNet.

In each of the next experiments, we use the following procedure. First, we select the transformation of interest and define its corresponding train time support $\mathcal{T}$ (e.g., rotation angle $\theta \in [0, 45]$). Then, we sample 100 classes uniformly from the standard ImageNet classes to be our target labels and for each class, we optimize a patch attack following [2]. This attack is then tested by sampling a test-time configuration from the full support of the transformation distribution, T, for each image in a set of 512 randomly sampled from the ImageNet test set.

Patch locations during optimization and at attack time are randomized (unless specified otherwise - Sect. 4.4 examines this transformation separately).

4.2 Scale

Question: The scale of patch attacks in the real world may be the most challenging factor to control at attack time. Thus, we ask: *what is the impact of scale on patch success? Will patches optimized at small scales maintain effectiveness at large scales (and vice versa)?*

Approach: These questions address whether the patch optimization process is producing patterns that rely on fine-grained details or embody general yet robust attack patterns. To address the scale questions, we optimize the set of patches following the base experiment setup with patch rotation fixed at $0°$ and location randomized. We perform two variants of the experiment. The first experiment starts at a nominal scale, $s_o \in [s_{min}, s_{max}]$, and bounds the support $\mathcal{T}$ such that scale varies from s_o up to a maximum scale s_{max} (i.e., $[s_o, s_{max}]$). The second experiment starts from a nominal scale and optimizes from the minimum up to nominal (i.e., $[s_{min}, s_o]$). Both experiments allow the full support at test time with scale sampled from the entire distribution $\mathcal{T} \sim U(s_{min}, s_{max})$.

Experiments: Results for both experiments are captured in Fig. 3.

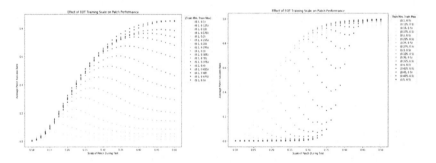

Fig. 3. Results for EoT with variable scale. (left) Shows how patch performance changes when the minimum scale is fixed and the maximum scale varies. (right) Shows performance differences starting from a fixed maximum size and allowing the minimum scale to vary. Shades correspond to the range of scales represented in the expectation. Both plots show attack success rate, so higher is better.

Discussion: Both experiments indicate that patch success is highly dependent on scale. We find that patches optimized at larger scales are more likely to have success at small scales than patches optimized in the reverse direction. We attribute this to the fact that a higher resolution for the patch forces the optimization to place less importance on individual pixels and to reveal patterns which are preserved when forced to lower resolutions. Regardless of the optimization approach, the results clearly indicate significant differences in success rate between large and small patches, which suggests that additional techniques may be necessary to overcome this practical limitation (a topic addressed in Sect. 5).

4.3   Rotation

Question: We next seek to understand how robust patch attacks are to out-of-distribution rotations experienced at test time. We ask the following question: *are patch attacks more robust when the patch is optimized over a larger range of rotations? how does patch success modulate with respect to this range?*

Approach: We again run the base patch optimization procedure now holding scale fixed while allowing location to vary randomly. However, during patch optimization, the rotation distribution is fixed such that angles are selected from a bounded support, $\mathcal{T}$, with angles taken from $[-\theta, \theta]$. We run a series of variants of the base experiment by varying θ at intervals of $\pi/5$.

Experiments: Results of the rotation experiments are captured in Fig. 4.

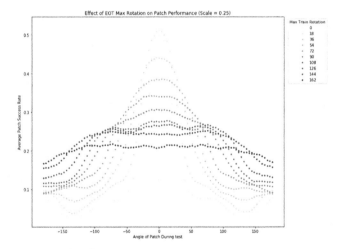

Fig. 4. Results for EoT when only angles are considered. The plot shows patch success (higher is better) at test angles from $[-\pi, \pi]$. Each shade of blue represents the maximum angle allowed during patch optimization (darker shades imply more angles considered).

Discussion: The rotation experiments illustrate a jack-of-all-trades, master-of-none result whereby patches optimized over a wider angle distribution exhibit lower peak performance yet higher average performance. One possible explanation for this result is that the conventional formulation of EoT focuses on finding a patch that maximizes the target class probability in expectation rather than including the rotation transformation directly in the optimization. While the loss is computed over all rotations sampled during a given training batch, the patch is always updated in its canonical orientation. When taking the expectation, it is likely that local patch regions compete with each other at specific angles. As the range of candidate rotations increases, the EoT forces the local regions to become less competitive so that they are more effective over a range of angles. However, the undesired effect is that the local region patterns may have lower performance at specific angles.

4.4   Location

Question: We perform experiments to explore the incidence of the location of adversarial patches in the image on the attack performance. Specifically, *we probe the veracity of the hypothesis that placement of masks at locations of maximum impact results in maximum attack effectiveness (where, as we shall explain next, maximum impact is taken to be maximum saliency).*

Approach: We must first clarify again that we're considering the general case of patch attacks whereby we optimize the patch making assumptions about the types and range of patch transformations, but make no assumptions about the scene or context in which the patch attack is applied. We make this distinction because in the specific attack scenario, a patch and location may be optimized jointly (or conditionally) if certain characteristics of the scene are assumed. In this experiment, we focus on the impact of the presence/absence of systematic biases in the location selection which may impact patch success. We utilize saliency (as determined via white-box methods) to identify how a patch's placement in regions of high/low importance to the deep network affect the patch success.

We compute saliency via conventional computation methods that use gradient descent to find locations in the input image domain that result in maximum softmax output for the image class label. The basic hypothesis driving this choice for location is that, by placing the patch over maximally salient locations in the image, the patch will gain maximum attack potency by both a) occluding the most important cues leading to the correct label and b) replacing them with cues targeting the new targeted class.

Experiments: For EOT-based training and testing, we consider three location strategies, corresponding to minimum and maximum loci in the saliency map as well as random location. We then explore the results of attacks using masks by combining the scenarios of {min, max, random} train location with {min, max, random} test locations. The results are shown in tables given in Fig. 5. We observe a strong dependence between patch success and target label. While we view this as an important relationship to study, for now we simply capture it by comparing patch performance binned according to the top, middle and bottom performing target classes. For each situation we report the mean accuracy when computed on a set of images along with additional statistics (standard deviation, as well as min and max accuracy).

Discussion: Looking at the results we found that the highest effectiveness is obtained in general when selecting the patch placement at test time according to the same rules as during the optimization (e.g., random location in both). There is a slight gain in the patch success when placing it in regions of high saliency, but it is not statistically significant, and the preference on placing the patch in the same type of location as it was trained seems to trump other considerations.

On the other hand, if we train randomly, and place the patch anywhere at attack time, we don't observe much of a performance hit. The fact that maximum placement does not over-perform reveals results that are somewhat counter to what one would hypothesize and suggests an 'easy-does-it' strategy, namely that you don't have to calculate maximum saliency at training time and place patches at optimal locations at attack time. Instead, you can work with random locations at train and attack time and still achieve results that are nearly as good as the maximal location.

In sum, location placement is one area in which – unlike rotation – adopting a lazy approach pays off. While this result was confirmed in this last set of EoT experiments, it helps to justify the choice of random location used for the scale and rotation variants of the base experiment.

5   Improving Patch Attack Effectiveness and Unobtrusiveness

We observed in Sect. 4.2 the dominant effect of scale on patch effectiveness, specifically noting a strong drop-off in performance as the patch image dimension decreases below 30% of the image edge size. This leads us to ask the question: *can we devise a strategy to increase the patch size (and subsequent effectiveness) without simultaneously increasing the observability of the patch itself?*

5.1   Approach

First, we must clarify two things. As before, we make the assumption that the attacker has no control over the camera pose and so cameras can always be positioned to force the patch to be arbitrarily small. However, we aim in this section to devise a method to increase the patch size to improve overall patch success.

Second, while we frame the question as one of observability, we're not asking it from the standpoint of measuring limits of human perception. Instead, we view observability as roughly equated to scene occlusion. To address the motivating question, we devise a measure of *patch obtrusiveness*. Since a primary advantage of patch-based attacks is the ability to print contiguous yet semi-transparent patterns, we can define patch *obtrusiveness* as a measure of the overall opacity of the patch. Fully opaque patches have maximal obtrusiveness as they are most

	Maximum Saliency Test		Random Location Test		Minimum Saliency Test	
Maximum Saliency Train	Mean acc:	0.9604	Mean acc:	0.9448	Mean acc:	0.9068
	Standard dev:	0.0130	Standard dev:	0.0214	Standard dev:	0.0402
	Maximum acc:	"gas mask" : 0.984	Maximum acc:	"football helmet" : 0.969	Maximum acc:	"gas mask" : 0.961
	Minimum acc:	"power drill" : 0.938	Minimum acc:	"power drill" : 0.906	Minimum acc:	"chow chow" : 0.828
Random Location Train	Mean acc:	0.9599	Mean acc:	0.9484	Mean acc:	0.8995
	Standard dev:	0.0215	Standard dev:	0.0237	Standard dev:	0.0509
	Maximum acc:	"gas mask" : 0.984	Maximum acc:	"banjo" : 0.984	Maximum acc:	"gas mask" : 0.969
	Minimum acc:	"power drill" : 0.898	Minimum acc:	"power drill" : 0.891	Minimum acc:	"power drill" : 0.766
Minimum Saliency Train	Mean acc:	0.8859	Mean acc:	0.8396	Mean acc:	0.9615
	Standard dev:	0.0463	Standard dev:	0.0723	Standard dev:	0.0235
	Maximum acc:	ootball helmet" : 0.938	Maximum acc:	"football helmet" : 0.953	Maximum acc:	"football helmet" : 0.992
	Minimum acc:	"banjo" : 0.781	Minimum acc:	"banjo" : 0.656	Minimum acc:	"Sussex spaniel" : 0.906

Middle 15 Target Classes	Maximum Saliency Test		Random Location Test		Minimum Saliency Test	
Maximum Saliency Train	Mean acc:	0.8880	Mean acc:	0.8214	Mean acc:	0.7328
	Standard dev:	0.0273	Standard dev:	0.0387	Standard dev:	0.0466
	Maximum acc:	"brambling" : 0.938	Maximum acc:	"tricycle" : 0.883	Maximum acc:	"guenon monkey" : 0.813
	Minimum acc:	"ibex" : 0.844	Minimum acc:	"stretcher" : 0.758	Minimum acc:	"jaguar" : 0.641
Random Location Train	Mean acc:	0.8974	Mean acc:	0.8417	Mean acc:	0.7667
	Standard dev:	0.0318	Standard dev:	0.0500	Standard dev:	0.0535
	Maximum acc:	"brambling" : 0.953	Maximum acc:	d-backed sandpiper" : 0.93	Maximum acc:	d-backed sandpiper" : 0.867
	Minimum acc:	"ibex" : 0.844	Minimum acc:	"guenon monkey" : 0.758	Minimum acc:	"tench" : 0.641
Minimum Saliency Train	Mean acc:	0.7292	Mean acc:	0.6526	Mean acc:	0.8854
	Standard dev:	0.0820	Standard dev:	0.0915	Standard dev:	0.0324
	Maximum acc:	"tench" : 0.852	Maximum acc:	"tricycle" : 0.781	Maximum acc:	"laptop" : 0.922
	Minimum acc:	"brambling" : 0.523	Minimum acc:	"brambling" : 0.422	Minimum acc:	\and-held computer" : 0.828

Worst 15 Target Classes	Maximum Saliency Test		Random Location Test		Minimum Saliency Test	
Maximum Saliency Train	Mean acc:	0.8880	Mean acc:	0.8214	Mean acc:	0.7328
	Standard dev:	0.0273	Standard dev:	0.0387	Standard dev:	0.0466
	Maximum acc:	"brambling" : 0.938	Maximum acc:	"tricycle" : 0.883	Maximum acc:	"guenon monkey" : 0.813
	Minimum acc:	"ibex" : 0.844	Minimum acc:	"stretcher" : 0.758	Minimum acc:	"jaguar" : 0.641
Random Location Train	Mean acc:	0.8974	Mean acc:	0.8417	Mean acc:	0.7667
	Standard dev:	0.0318	Standard dev:	0.0500	Standard dev:	0.0535
	Maximum acc:	"brambling" : 0.953	Maximum acc:	d-backed sandpiper" : 0.93	Maximum acc:	d-backed sandpiper" : 0.867
	Minimum acc:	"ibex" : 0.844	Minimum acc:	"guenon monkey" : 0.758	Minimum acc:	"tench" : 0.641
Minimum Saliency Train	Mean acc:	0.7292	Mean acc:	0.6526	Mean acc:	0.8854
	Standard dev:	0.0820	Standard dev:	0.0915	Standard dev:	0.0324
	Maximum acc:	"tench" : 0.852	Maximum acc:	"tricycle" : 0.781	Maximum acc:	"laptop" : 0.922
	Minimum acc:	"brambling" : 0.523	Minimum acc:	"brambling" : 0.422	Minimum acc:	\and-held computer" : 0.828

Fig. 5. Results for location: showing the 15 most effective target classes (top), median (middle) and least (bottom).

easily identified at test time relative to a nearly transparent patch. Furthermore, our patch *obtrusiveness* measure creates a trade space whereby scale, opacity, and success rate can be optimized together according to the desired operating point.

More precisely, we measure opacity by first defining a mask, M, which has the same shape (namely $h \times w$ for rectangular patches) as the patch P and where $M_{ij} \in [0, 1]$ for each i, j in M. From here, we define *patch obtrusiveness (PO)* as:

$$PO(M) = \frac{total\ mask\ value}{total\ mask\ area} = \frac{\sum_{i,j} M_{ij}}{hw} \qquad (4)$$

This metric allows us to create some measure of equivalence between small, opaque patches and large, partially transparent patches. When $PO \approx 0$, the patch is essentially transparent and maximally subtle in the image. When $PO = 1$, the patch is fully opaque and is equivalent to the current conventional output from the EoT method. Ultimately, our goal is to side-step the scale limitations observed with EoT (as discussed in Sect. 4.2) and produce larger, more effective patches without increasing observability.

Given our definition of $PO(M)$, we now include the mask M as part of the optimization. Whereas standard EoT applies a sampled transform $t(p)$ to patch p and then applies the transformed patch directly over the image x (i.e., $A(t(p), x)$), we modify the application function $A()$ to blend the patch and the image according to the values of M. Namely, the adversarially attacked image is created as follows: For all i, j in x,

$$
\begin{aligned}
x'_{ij} &= A\left(t(p), x; t(M)\right)_{ij} \\
&= t(M)_{ij} * t(p)_{ij} + (1 - t(M)_{ij}) * x_{ij},
\end{aligned}
\tag{5}
$$

where x is the clean image, M is the optimized mask, p is the optimized patch, and t is the sampled transformation to apply equally to the patch and mask. We assume that $t(p)$ produces an image of the same dimension as x where p has been transformed and placed at a location also valid in x.

Now that we can apply M to the patch to attack a clean image, we optimize the patch p and mask M simultaneously:

$$
\underset{p, PO(M)}{\arg\max} \; \mathbb{E}_{x \sim X, t \sim T} P(y' | A(t(p), x; t(M)))
\tag{6}
$$

For targeted attacks, the optimization loss term for a single attacked image is:

$$
\begin{aligned}
L_{total} &= L_{target} + \gamma \, L_{PO} \\
L_{target} &= CE(y', y_{target}) \\
L_{PO} &= (PO(M))^2
\end{aligned}
\tag{7}
$$

where y' is the label produced by the network for the attacked image x', CE is the standard cross-entropy loss, and $PO(M)$ is the patch obtrusiveness defined in (4). Gradients with respect to the loss term are backpropagated to both the patch and mask pixels so as optimize them jointly. Due to differences in the natural scale of the loss terms, we include γ to allow more control over the competition between patch obtrusiveness and its ability to fool the attacked network.

5.2   Experiment

Following the base experiment approach described in Sect. 4.1, a series of patches (and associated masks) are optimized and tested over a sampling of target labels and test images. We hold rotation fixed and use random locations during patch optimization and evaluation. The rotation assumption can be viewed as a best case given the results in Sect. 4.3 and the location strategy is based on the results in Sect. 4.4.

Given γ is designed to control the trade-off between patch success and obtrusiveness, we employ a curriculum learning-like procedure whereby minimizing patch obtrusiveness is prioritized initially then down-weighted later on to encourage greater patch performance. Inspired by [14], we change the γ value throughout the patch optimization process depending on the value of the loss. The γ term is held fixed until the loss remains below a threshold (e.g., nominally 0.1) for a fixed patience period (e.g., min 5 iterations). After the patience period, the γ value is decreased (according to its schedule) and the process repeats. We observe during training that changes in γ usually produce a spike in the loss, so the patience period promotes a certain degree of stability and is critical to achieving high-performing semi-transparent patches.

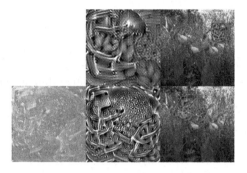

Fig. 6. Example of patch attacks with and without optimization to balance obtrusiveness and success. Top row is the control patch and its application to a test image. The bottom row shows the mask, the full patch, and the application of both to the same test image.

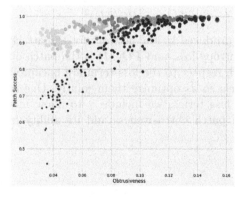

Fig. 7. Partial patch results aggregated over the five top, middle, and bottom target labels each. Light colors imply greater transparency, larger dots imply larger patches. Results demonstrate that the larger, semi-transparent patches can be optimized which perform better than their fully opaque counterpart at the same obtrusiveness operating point.

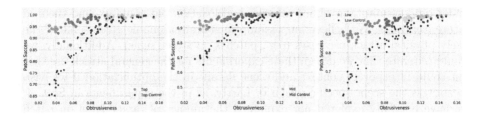

Fig. 8. Partial patch results for the five top (left), middle (center), and bottom (right) performing target labels. Dot size is proportional to patch size; dot transparency is proportional to patch transparency. In all cases, the obtrusiveness-optimized patches demonstrate better performance overall compared to the fully opaque control patches.

Figure 6 illustrates a standard EoT-based patch attack as well as an example produced with our approach showing a sample mask, full patch, and application of both to attack a clean image.

To test our method, we follow the base experiment setup but expanding to 1200 iterations during optimization (i.e., until convergence) and employing a learning rate of 5.0. We fix the scale between [0.4, 0.5] and then run the optimization. At the end of the process, the final obtrusiveness value is recorded and the patch is tested on a random set of 512 held-out images.

For the control experiments, we take the final obtrusiveness score and label for the semi-transparent patch and produce a fully-opaque control patch at a scale that achieves the same obtrusiveness score. Note that because the control patches have no transparency, they are naturally smaller. The control patches are optimized for 500 iterations to achieve target loss convergence (fewer iterations are used due to the absence of the curriculum required for the obtrusiveness term). We then evaluate on the same set of held-out test images and record attack success rates.

Results for these experiments are captured in Figs. 7 and 8.

5.3 Discussion

The results from Sect. 5.2 give us several insights into the trade-off between obtrusiveness and attack success. We demonstrate that we can generate patches that achieve similar or better success rates with lower obtrusiveness values. In particular, we see that for a range of operating points (i.e., obtrusiveness values), we can generate larger patches that achieve higher success but are more subtle relative to the standard patch attacks of the same size. This is the first such demonstration of semi-transparent patches which are capable of achieving this kind of result.

However, we still recognize several potential avenues for improvement. First, while we're the first to define a notion of patch attack obtrusiveness, we have not yet tied it to a true perceptual measure (necessary if we hope to claim to avoid detection by a human). Second, we have constructed these experiments under rigid assumptions about rotation and location but could expand them in future iterations. We believe our assumptions are reasonable given some of the likely physical attack use cases (e.g., patches on street signs). Additionally, we would reiterate that we design these experiments for the general attack use case but could incorporate context and scene details in the optimization if we wish to address the scene-specific scenario. Lastly, as in the experiments in Sect. 4, we have not fully explored nor explained the impact of target label on patch success. We observe a clear impact and can hypothesize about why certain labels do better (e.g., due to scale/deformability/uniformity), but further investigation is necessary to better characterize these relationships.

6   Conclusions

This paper makes two key contributions: (1) we study a method and provide results that characterize the impact of real-world transformations on patch attack optimization and success, (2) we present a novel objective for the design of ideal patches that trade obtrusiveness and success rate, and develop companion approaches for optimization of patch obtrusiveness for producing more subtle patch attacks without sacrificing performance. We have demonstrated that the results of (1) can be directly leveraged to improve patch attacks (e.g., as shown with (2)). We believe that the methodology and results presented in this paper will help provide greater guidance and focus in the community in understanding patch attack capabilities and limitations. Lastly, our obtrusiveness measure led to the production of semi-transparent attacks (for the first time) and opens the door for exploring many other variants of this approach. With these results we want to underscore the importance of generating these attacks (and subsequent defenses) not as a means for defeating visual recognition systems, but rather as a way to improve understanding of the robustness of these systems and gain greater insight into their inner workings and possible defenses, which is left for future endeavors.

Acknowledgements. This work was completed with the support of JHU/APL Internal Research Funding.

References

1. Athalye, A., Engstrom, L., Ilyas, A., Kwok, K.: Synthesizing robust adversarial examples. arXiv preprint arXiv:1707.07397 (2017)
2. Brown, T.B., Mané, D., Roy, A., Abadi, M., Gilmer, J.: Adversarial patch. arXiv preprint arXiv:1712.09665 (2017)
3. Demontis, A., et al.: Why do adversarial attacks transfer? Explaining transferability of evasion and poisoning attacks. In: 28th {USENIX} Security Symposium ({USENIX} Security 19), pp. 321–338 (2019)
4. Esteva, A., et al.: Dermatologist-level classification of skin cancer with deep neural networks. Nature **542**(7639), 115 (2017)

5. Goodfellow, I.J., Shlens, J., Szegedy, C.: Explaining and harnessing adversarial examples. arXiv preprint arXiv:1412.6572 (2014)
6. He, K., Zhang, X., Ren, S., Sun, J.: Deep residual learning for image recognition. In: Proceedings of the IEEE Conference on Computer Vision and Pattern Recognition, pp. 770–778 (2016)
7. LeCun, Y., Bengio, Y., Hinton, G.: Deep learning. Nature **521**(7553), 436 (2015)
8. Liu, Y., Chen, X., Liu, C., Song, D.: Delving into transferable adversarial examples and black-box attacks. arXiv preprint arXiv:1611.02770 (2016)
9. Mnih, V., et al.: Human-level control through deep reinforcement learning. Nature **518**(7540), 529 (2015)
10. Moosavi-Dezfooli, S.M., Fawzi, A., Fawzi, O., Frossard, P.: Universal adversarial perturbations. In: Proceedings of the IEEE Conference on Computer Vision and Pattern Recognition, pp. 1765–1773 (2017)
11. Moosavi-Dezfooli, S.M., Fawzi, A., Frossard, P.: Deepfool: a simple and accurate method to fool deep neural networks. In: Proceedings of the IEEE Conference on Computer Vision and Pattern Recognition, pp. 2574–2582 (2016)
12. Papernot, N., McDaniel, P., Goodfellow, I.: Transferability in machine learning: from phenomena to black-box attacks using adversarial samples. arXiv preprint arXiv:1605.07277 (2016)
13. Szegedy, C., Zaremba, W., Sutskever, I., Bruna, J., Erhan, D., Goodfellow, I., Fergus, R.: Intriguing properties of neural networks. arXiv preprint arXiv:1312.6199 (2013)
14. Wang, B., et al.: Neural cleanse: Identifying and mitigating backdoor attacks in neural networks
15. Zhou, W., et al.: Transferable adversarial perturbations. In: Ferrari, V., Hebert, M., Sminchisescu, C., Weiss, Y. (eds.) ECCV 2018, Part XIV. LNCS, vol. 11218, pp. 471–486. Springer, Cham (2018). https://doi.org/10.1007/978-3-030-01264-9_28

Evaluating Input Perturbation Methods for Interpreting CNNs and Saliency Map Comparison

Lukas Brunke, Prateek Agrawal$^{(\boxtimes)}$, and Nikhil George

Volkswagen Group of America, Belmont, CA 94002, USA
`prateek.agrawal@vw.com`

Abstract. Input perturbation methods occlude parts of an input to a function and measure the change in the function's output. Recently, input perturbation methods have been applied to generate and evaluate saliency maps from convolutional neural networks. In practice, neutral baseline images are used for the occlusion, such that the baseline image's impact on the classification probability is minimal. However, in this paper we show that arguably neutral baseline images still impact the generated saliency maps and their evaluation with input perturbations. We also demonstrate that many choices of hyperparameters lead to the divergence of saliency maps generated by input perturbations. We experimentally reveal inconsistencies among a selection of input perturbation methods and find that they lack robustness for generating saliency maps and for evaluating saliency maps as saliency metrics.

Keywords: Saliency methods · Saliency maps · Saliency metrics · Perturbation methods · Baseline image · RISE · MoRF · LeRF

1 Introduction

Understanding and interpreting convolutional neural networks' (CNN) predictions through saliency methods has become an active field of research in the recent years [3,14,15,17]. Saliency methods create saliency maps, which highlight relevant parts of the input image for a classification task.

Input perturbation methods, which are also referred to as occlusion methods [2], are one of the saliency methods for understanding CNNs. Input perturbation methods follow a simple principle: covering up or masking an important part of an input results in relevant information loss and should reduce the prediction score for a specific class. If an occluded part of the image results in an increase of the prediction score, then this part of the input will be negatively correlated with the target class. Furthermore, there might be parts of the input, which

Electronic supplementary material The online version of this chapter (https:// doi.org/10.1007/978-3-030-66415-2_8) contains supplementary material, which is available to authorized users.

A. Bartoli and A. Fusiello (Eds.): ECCV 2020 Workshops, LNCS 12535, pp. 120–134, 2020.
https://doi.org/10.1007/978-3-030-66415-2_8

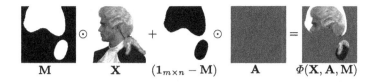

$$\mathbf{M} \qquad \mathbf{X} \qquad (\mathbf{1}_{m\times n} - \mathbf{M}) \qquad \mathbf{A} \qquad \varPhi(\mathbf{X}, \mathbf{A}, \mathbf{M})$$

Fig. 1. Masking an input image $\mathbf{X}$ with a baseline image $\mathbf{A}$. The input is elementwise multiplied with the mask $\mathbf{M}$, keeping all the parts of the input, where the mask has a value of 1, and setting the parts of the input to zero, where the mask has a value of 0. The baseline image is elementwise multiplied with the inverse mask $(\mathbf{1}_{m\times n} - \mathbf{M})$, which is the matrix of ones minus the mask, and added to the masked input, such that the parts, which were set to 0 for the input are set to the corresponding value of the baseline image. The displayed mask shows an intermediate mask from LeRF.

when covered, do not affect the prediction score. Occluding or masking an input is defined as substituting specific elements with a baseline image. The masking of an input image with a given baseline image and a mask is illustrated in Fig. 1. In practice, neutral images are chosen as the baseline image, in such a way that the baseline image's impact on the classification probability is minimal. Furthermore, input perturbation methods are also used to compare and evaluate saliency maps from different saliency methods for the same input image. Input perturbation methods have two major advantages over other saliency methods like gradient-based methods. First, input perturbation methods allow the analysis of black-box models, whose weights and gradients are inaccessible. This is, for example, relevant in the automotive industry for validating black box models received from suppliers. Second, input perturbation methods' saliency maps are easily interpretable by humans unlike gradient-based methods, which often result in noisy or diffuse saliency maps [2,11]. In contrast to these advantages of input perturbation methods, a disadvantage is that they are typically more computationally intensive.

In this paper we experimentally evaluate the robustness of perturbation methods for image classification tasks against different parameters and baseline images. We pick three representative input perturbation methods to illustrate our findings: Randomized Input Sampling for Explanation (RISE) [12], most relevant first (MoRF) [13], and least relevant first (LeRF). However, our results also apply to other input perturbation methods. The goal is to determine the effects of changing the perturbation methods' parameters on the reliability of the perturbation method. We evaluate the generation of saliency maps with RISE by varying its parameters and selecting various baseline images, which are used to occlude an input. MoRF, and LeRF are metrics used to objectively compare different saliency maps for the same input by increasingly perturbing the original input [13]. We also vary the baseline images for MoRF and LeRF. The experiments reveal shortcomings regarding the robustness of input perturbation methods.

The next section introduces the investigated perturbation methods. Section 3 gives an overview of related work. The material used for the experiments is presented in Sect. 4. The experiments on RISE, MoRF, and LeRF are described in detail in Sect. 5 and subsequently discussed in Sect. 6. We conclude our findings in Sect. 7.

2   Input Perturbation Methods

Perturbation methods rely on a baseline image [15] to occlude parts of the original input, see Fig. 1. The application of a baseline image $\mathbf{A} \in \mathbb{R}^{m \times n \times 3}$ with mask $\mathbf{M} \in [0,1]^{m \times n}$ to an input image $\mathbf{X} \in \mathbb{R}^{m \times n \times 3}$ is defined in [7,9] as:

$$\Phi(\mathbf{X}, \mathbf{A}, \mathbf{M}) = \mathbf{M} \odot \mathbf{X} + (\mathbf{1}_{m \times n} - \mathbf{M}) \odot \mathbf{A}, \tag{1}$$

where $\odot$ denotes the Hadamard product and $\mathbf{1}_{m \times n}$ is an m × n-dimensional matrix of ones. The mask can also be of a different dimension, depending on the scaling and cropping operations [12].

No systematic pattern of how to choose a suitable baseline image seems identifiable in the literature. It has been argued to be a neutral input such that $\tilde{f}(\mathbf{X}) \approx \mathbf{0}$, where $\tilde{f} : \mathbb{R}^{m \times n \times 3} \to \mathbb{R}^K$ is a CNN function and K is the number of predicted classes [15]. The function $f = \tilde{f} \circ g$ is the DNN function with preprocessing, where $g : \mathbb{R}^{m \times n \times 3} \to \mathbb{R}^{m \times n \times 3}$ is the preprocessing function applied to the input $\mathbf{X}$ of a DNN. The black image $\mathbf{A} = \mathbf{0}_{m \times n \times 3}$, with the m × n × 3-dimensional matrix of zeros $\mathbf{0}_{m \times n \times 3}$, is the baseline image used in [15]. Samek *et al.* applied baseline images from a uniform distribution, Dirichlet distribution, constant baseline images and blurred baseline images [13]. Petsiuk *et al.* use the baseline image for which $g(\mathbf{A}) = \mathbf{0}_{m \times n \times 3}$ [12]. We refer to this baseline image as the zero baseline image after preprocessing.

2.1   Saliency Map Generation: RISE

Petsiuk *et al.* introduced RISE for creating saliency maps for interpreting CNNs for classification tasks [12]. RISE generates a saliency map by applying N random masks to an input and creating a weighted sum of the N output probabilities for the target class c with the random masks. RISE uses a set of N random masks $\mathcal{M}_N(p, w, h) = \{\mathbf{M}_1(p, w, h), ..., \mathbf{M}_N(p, w, h)\}$, where $\mathbf{M}_i(p, w, h) \in \{0,1\}^{w \times h}, i \in \{1, ..., N\}$ and the probability $p \in [0,1]$ of each element of the mask to be equal to 1 or in this context to be "on". The parameters w and h indicate the size of the low resolution mask before scaling and cropping. In the following $w = h$, since we only use square masks and input images. Each mask is scaled to the input image's size with a transformation $q : \{0,1\}^{w \times h} \to [0,1]^{m \times n}$, which applies a bilinear transformation and a random cropping operation. For a detailed description of the mask generation for RISE we refer the reader to the original paper [12]. The saliency map S is then given by:

$$\mathbf{S} = \frac{1}{\mathbb{E}(\mathcal{M}_N) \cdot N} \sum_{i=1}^{N} \tilde{f}^c(q(\mathbf{M}_i) \odot g(\mathbf{X})), \tag{2}$$

where $\tilde{f}^c$ are the outputs for class c for the DNN function without preprocessing. Directly applying the mask to the preprocessed input is equal to implicitly using the zero after preprocessing baseline image, therefore, using a constant baseline image for the saliency map generation:

$$\tilde{f}^c(q(\mathbf{M}_i) \odot g(\mathbf{X})) = f^c(\Phi(\mathbf{X}, g^{-1}(\mathbf{0}_{m \times n \times 3}), q(\mathbf{M}_i)) \tag{3}$$

with the inverse preprocessing function g^{-1} and f^c, which is the output for class c for the DNN function with preprocessing.

2.2   Saliency Map Comparison: MoRF and LeRF

The MoRF and LeRF metrics quantitatively evaluate the quality of saliency maps [13]. This is relevant for comparing and assessing different saliency methods like RISE [12], Grad-CAM [14], and integrated gradients [15]. MoRF and LeRF measure the effect of occluding pixels from the input image on the target class' output probabilities. MoRF and LeRF replace elements from the input with the corresponding elements from a baseline image. Given a saliency map for an input image as a ranking of importance for individual pixels, MoRF replaces pixels in the input image in decreasing order of the importance (= most relevant first), whereas LeRF replaces pixels in the input image in increasing order of importance (= least relevant first). Replacing pixels in the input image with the corresponding pixels from the baseline image results in a change of the output probabilities. Recording the output probabilities as further elements are occluded results in a curve over the relative number of occluded elements α. The number of additionally occluded pixels r in each step is variable. Here, all presented experiments are run with $r = 1$, occluding only one additional pixel in each step.

MoRF iteratively replaces elements with decreasing importance according to a given ordering by a saliency map. A greater area over the curve (AOC) for MoRF is desirable and suggests a superior saliency map. On the other hand, LeRF iteratively replaces elements with increasing importance and its score is given by the area under the curve (AUC). In this paper we limit the experiments to the MoRF and LeRF metrics, because the MoRF and LeRF metrics can be transformed into the insertion and deletion metrics, which are used in [9,12].

3   Related Work

There has been a limited effort to determine the robustness of input perturbation methods against varying parameters and baseline images. Ancona *et al.* [2] have tested different square sizes for masks for the sliding window method [17]. Increased square sizes exhibited a reduction of details provided in the saliency maps. This inspired our experiments on varying the w and p parameters for RISE, which affect the shape and size of the random masks. Similarly to Ancona *et al.* [2], we find that less fine grained saliency maps are created by less fine grained masks. This corresponds to small values for w in the case of RISE and bigger values for n in the case of Occlusion-n.

Samek *et al.* [13] have investigated various baseline images for the MoRF and LeRF metrics. They determined that the uniform distribution baseline image gives the best MoRF and LeRF scores when averaged over the whole dataset. Samek *et al.* applied MoRF and LeRF with $r = 9 \times 9$ non-overlapping regions and only perturbed up to 15.7% of the input image. Furthermore, they evaluated the blurred baseline image with $\sigma = 3$, which is arguably small and still contains information from the original input.

Fong and Vedaldi [9] generate saliency maps by optimizing perturbation masks, which minimize or maximize the output probabilities. They find that different baseline images yield different saliency maps, which highlights the dependency on the choice of baseline image. Dabkowski and Gal [7] note similar findings and therefore use a set of baseline images to circumvent the dependence on a single baseline image. Petsiuk *et al.* [12] determined through qualitative reasoning for MoRF and through quantitative analysis for LeRF, that a constant gray baseline image and a blurred baseline image work best for their evaluation using MoRF (= deletion) and LeRF (= insertion), respectively. However, our quantitative analysis suggests that this is not the case.

Recently, [16] have investigated the reliability of saliency metrics like MoRF and LeRF. Using two different baseline images, specifically the constant dataset mean baseline image and a uniform noise baseline image, they show that MoRF and LeRF are dependent on the choice of the baseline image. However, since they used a random baseline image for comparison, the baseline image changes for each run of MoRF and LeRF, which lacks consistency and could affect their findings and yield issues regarding reproducibility. In contrast, we use a set of constant baseline images (with the exception of the blurred baseline image). Additionally, they only analyzed the scores for MoRF and LeRF for 100 steps, whereas we run both metrics for all steps.

4   Material

In this work we are exclusively considering CNNs for image classification. Specifically we are using the pre-trained CNN ResNet-50 [10] from Keras [6] in TensorFlow [1]. We process the input in the same way as described by [10], such that the input shape is $224 \times 224 \times 3$. Then we apply a mean shift, which is the preprocessing function used for the pretrained network.

All experiments are run with the validation set from ImageNet [8], which we refer to as dataset $\mathcal{D}_{\text{val}}$. Saliency maps are visualized by normalizing the resulting maps in the range $[0, 1]$ and applying OpenCV's perceptually uniform sequential inferno colormap [4]. This colormap displays parts supporting the prediction with lighter colors and parts opposing the prediction with darker colors.

5   Experiments

This section presents experiments on the robustness of input perturbation methods against different parameter settings and different baseline images. First, we investigate how saliency maps change under varying parameters for the mask generation with RISE. Second, we analyze the impact of using different supposedly uninformative baseline images on the robustness of RISE, MoRF, and LeRF.

5.1   Experiments on RISE

Petsiuk *et al.* apply RISE with $N = 8000$ for ResNet-50, $h = w = 7$, $p = 0.5$ and the zero baseline image after preprocessing [12]. The choice of these parameters is not elaborated on in [12]. Therefore, this section determines the effect of varying the parameters on the convergence of a saliency map and the subjective quality of the saliency map. Throughout this section, saliency maps are generated for their maximum activated class $c_{\max} = \arg\max_c f^c(\mathbf{x})$. In contrast to Petsiuk *et al.* [12], we applied $N_{\max} = 32768$ masks to evaluate the convergence. Unless otherwise stated the experiments are executed with $h = w = 7$, $p = 0.5$, $N = N_{\max}$, and the zero baseline image after preprocessing.

First, the number of masks needed for convergence of a single saliency map is determined. In order to check the convergence RISE is run three times, with three independent random sets of masks $\mathcal{M}^l_{N_{\max}}$, where $l \in \{1, 2, 3\}$ and $N_{\max} = 32768$. Figure 2 displays the resulting saliency maps for an input $\mathbf{X}$ after applying $N_{\max}$ masks. Figure 2 shows that the saliency maps from multiple runs have converged, since they are nearly indistinguishable from each other. Since the visual assessment of convergence for saliency maps is not efficient, the goal is to determine convergence quantitatively. We define a function d that calculates the L^2-distance between two saliency maps S_i and S_j, where $i \neq j, i \in \{1, 2, 3\}$ and $j \in \{1, 2, 3\}$, from different runs and records the L^2-distance for every incremental saliency map. In practice, we only consider saliency maps with a

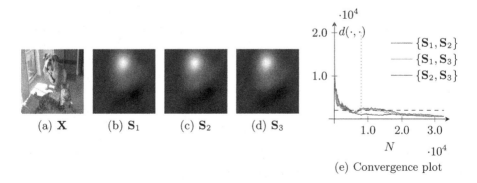

(a) $\mathbf{X}$ (b) $\mathbf{S}_1$ (c) $\mathbf{S}_2$ (d) $\mathbf{S}_3$

(e) Convergence plot

Fig. 2. Convergence for RISE for the class `bull_mastiff`.

maximum distance between saliency maps from independent runs smaller than $d_{\max} = 2000$, which is represented by the dashed horizontal line in Fig. 2e. We chose this threshold, because it indicates the maximum in the histogram in Fig. 3. The histogram shows the L^2-distances for RISE saliency maps from 1000 randomly sampled input images from ImageNet. Manual inspection confirmed that the threshold also yields subjectively good results. Note, that for this input image the use of the suggested number of masks $N = 8000$, which is represented by the dotted vertical line in Fig. 2e, results in $d > d_{\max}$ for some combinations of i and j.

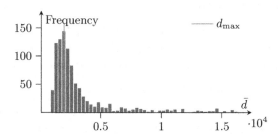

Fig. 3. Selection of $d_{\max} = 2000$.

The input image and saliency maps in Fig. 4 show an example, where the standard parameters for w and p with $N_{\max} = 32768$ masks do not lead to the same saliency map for the three independent runs. The graph in Fig. 4e shows that the L^2-distance between the saliency maps in Fig. 4 never get close to the threshold $d_{\max}$, and therefore do not converge. Doing the same investigation on a random subset of 1000 images from the validation set from ImageNet, yields only 389 converged saliency maps. In the following we investigate the variation of the p and w parameters. We observed, that the example in Fig. 4a-e converged when setting $p = 0.1$, which is shown in Fig. 4f-j. Note, that the threshold $d_{\max}$ is only reached when N approaches $N_{\max}$. Applying different values for w, while p is constant, results in distinct saliency maps, which also impacts the average L^2-distance $\bar{d}$ between saliency maps from different runs at $N_{\max}$. Figure 5 shows the effect of selecting different $w \in \{3, 4, 5, 7\}$. In the two examples, an increasing value for w leads to more defined distinct local optima in the saliency maps, which focus on the individual objects of the target class. While $\bar{d}$ increases with increasing w for the example in Fig. 5a-e, $\bar{d}$ decreases with increasing w for the example in Fig. 5f-j.

Subsequently, the impact of varying p and $w = h$ jointly is investigated. We analyzed the variation of $p \in \{0.1, 0, 3, 0.5, 0.7, 0.9\}$ and $h = w \in \{5, 7, 9, 11\}$, where each set of parameters is run three times to determine convergence. In Fig. 6 we only report the saliency maps, which converged. Saliency maps with $p \in \{0.5, 0.7, 0.9\}$ did not converge for the input image from Fig. 4. For some combinations the maximum number of masks $N_{\max} = 32768$ might not be sufficient for the saliency maps to converge. However, we did not run any experiments

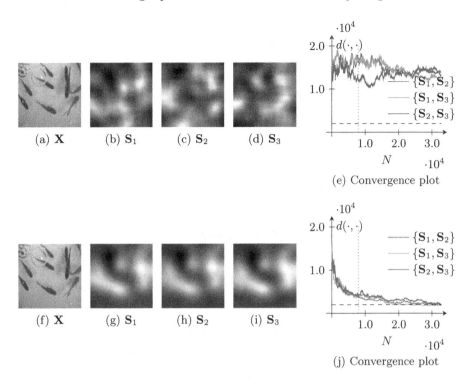

(a) $\mathbf{X}$ (b) $\mathbf{S}_1$ (c) $\mathbf{S}_2$ (d) $\mathbf{S}_3$

(e) Convergence plot

(f) $\mathbf{X}$ (g) $\mathbf{S}_1$ (h) $\mathbf{S}_2$ (i) $\mathbf{S}_3$

(j) Convergence plot

Fig. 4. Convergence for RISE with $p = 0.5$ at the top and $p = 0.1$ at the bottom for the class `goldfish`. The choice of $p = 0.5$ does not lead to convergence, while $p = 0.1$ converges. We found that only 389 of 1000 randomly sampled input images led to convergence with $p = 0.5$.

with $N > N_{max}$, since the application of more masks increases the number of calculations, which leads to computational infeasibility in practical settings.

The systematic choice of a suitable baseline image is still an open research question [7,9,12,15,17]. The proposed condition for the baseline image is, that it is uninformative [15] and therefore different baseline images have been presented in the literature [13]. The following gives an overview over the types of baseline images considered in our work. The constant baseline images are given by $\mathbf{A}_\gamma = \gamma \cdot \mathbf{1}_{m \times n \times 3}, \forall \gamma \in \Gamma \subseteq \{0, ..., 255\}$. Note that $\gamma = 0$, $\gamma = 127$, and $\gamma = 255$ refer to a black, gray and white baseline image, respectively. The zero baseline image after preprocessing, which is used by [12], is $\mathbf{A}_{inv} = g^{-1}(\mathbf{0}_{m \times n \times 3})$. Applying the preprocessing function g to $\mathbf{A}_{inv}$, yields $g(\mathbf{A}_{inv}) = \mathbf{0}_{m \times n \times 3}$. The blurred baseline image is defined as $\mathbf{A}_\sigma = \pi(\mathbf{X}, \sigma), \forall \sigma \in \Sigma$, where π applies a Gaussian blur with standard deviation σ and Σ is the set of standard deviations. In the past, values for the standard deviation of $\sigma = 3$ [13] and $\sigma = 5$ [12] have been chosen, however, applying these standard deviations can result in an output probability for the target class much greater than 0, which is not an uninformative baseline image. Therefore, we follow the choice of $\sigma = 10$ as in [7,9], which yields a baseline image

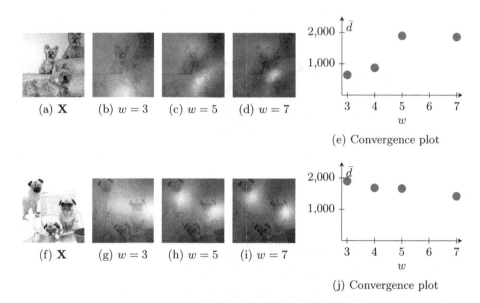

(a) **X**   (b) $w = 3$ (c) $w = 5$ (d) $w = 7$

(e) Convergence plot

(f) **X**   (g) $w = 3$ (h) $w = 5$ (i) $w = 7$

(j) Convergence plot

Fig. 5. Convergence for RISE with varying w for the classes `Yorkshire terrier` and pug, respectively. The choice of w can either reduce or increase $\bar{d}$.

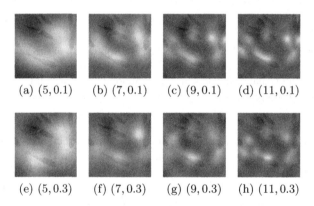

(a) $(5, 0.1)$ (b) $(7, 0.1)$ (c) $(9, 0.1)$ (d) $(11, 0.1)$

(e) $(5, 0.3)$ (f) $(7, 0.3)$ (g) $(9, 0.3)$ (h) $(11, 0.3)$

Fig. 6. RISE with different values for (w, p) for the class `goldfish`. The choice of different hyperparameters yield different saliency maps.

with output probability of approximately 0. Note, that the blurred baseline image is the only baseline image which uses local information from the current input image **X**. We combine the above baseline images in a set of baseline images:

$$\mathcal{A}(\Gamma, \Sigma, g) = \{\mathbf{A}_\gamma, \mathbf{A}_{\text{inv}}, \mathbf{A}_\sigma, |\forall \gamma \in \Gamma, \forall \sigma \in \Sigma\}. \tag{4}$$

Specifically the set $\mathcal{A}_e = \mathcal{A}(\{0, 127, 255\}, \{10\}, g)$ is used in the following experiments. The set could also be extended to include a baseline image sampled from a uniform or Dirichlet distribution, the constant average from the current input, and a constant input image averaged over the dataset. However, the chosen set of baseline images is already sufficient for raising robustness issues for input perturbation methods. We adapt RISE to enable the application of different baseline images:

$$\mathbf{S} = \frac{1}{\mathbb{E}(\mathcal{M}_N) \cdot N} \sum_{i=1}^{N} f^c(\Phi(\mathbf{X}, \mathbf{A}, q(\mathbf{M}_i))), \tag{5}$$

where $\mathbf{A} \in \mathcal{A}_e$. Each saliency map is generated for examples $\mathbf{X} \in \mathcal{D}_{\mathrm{val}}$ for the class c_{max} with the highest output probability for that input image and $N = 16384$. This choice of N is a compromise between accuracy and computational effort. If the L^2-distance among saliency maps for an input image is greater than d_{max}, the input will not be considered for this experiment. Therefore, we only take into account examples, which have converged. An overview of four examples, where different baseline images yield different results, is given in Fig. 7. The top row shows the tested input images and the two bottom rows each display the application of a different baseline image when creating the RISE saliency map for the specific input image. Depending on the baseline image, RISE attributes more or less importance to certain parts of the input. Comparing Figs. 7g and k, fewer or more people are highlighted based on the choice of baseline image. Similarly, in Figs. 7h and l the bird's left wing and the bird's feet are highlighted, respectively. Depending on the selected baseline image the saliency map for RISE can completely change. RISE is therefore highly dependent on the choice of the baseline image.

5.2   Experiments on MoRF and LeRF

MoRF and LeRF provide an objective and automatable metric for assessing different saliency methods. Here, we use MoRF to compare one saliency map generated by RISE with one generated by Grad-CAM [14]. We limit the presented experiments to MoRF and refer the reader to the supplementary material [5] for experiments on LeRF, which yield the same propositions. RISE and Grad-CAM have been chosen for demonstration purposes only. The choice of different saliency methods for the comparison is equally reasonable. The generation of each saliency map uses the standard parameters presented in the original paper for Grad-CAM [14]. We run MoRF with $r = 1$, such that the input images are covered pixel by pixel. The function $m : [0, 1] \times [0, 1]^{224 \times 224} \rightarrow \{0, 1\}^{224 \times 224}$ gives the mask induced by a saliency map S, where the relative occluded area of the input image is given by α.

In the first set of experiments we investigate the behavior of MoRF for different single input images. Similar as in [12], we also only use the zero after preprocessing baseline image for MoRF and the blurred baseline image for LeRF. In the examples in Table 1, we find that comparing the AOC for MoRF for the single input images results in a conflicting assessment on which saliency method is preferable. For example, for the input image on the left side of Table 1, the AOC for the saliency map from Grad-CAM is greater than the AOC for the saliency map from RISE. However, for the input image on the right the AOC for

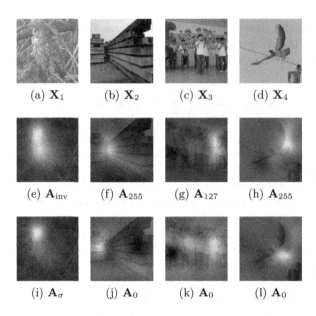

(a) $\mathbf{X}_1$ (b) $\mathbf{X}_2$ (c) $\mathbf{X}_3$ (d) $\mathbf{X}_4$

(e) $\mathbf{A}_{\mathrm{inv}}$ (f) $\mathbf{A}_{255}$ (g) $\mathbf{A}_{127}$ (h) $\mathbf{A}_{255}$

(i) $\mathbf{A}_\sigma$ (j) $\mathbf{A}_0$ (k) $\mathbf{A}_0$ (l) $\mathbf{A}_0$

Fig. 7. RISE with different baseline images.

the saliency map generated by RISE is greater. Note, that the output probability of the original input for the target class is displayed at $\alpha = 0$, when the image is not covered at all. According to the graphs in Table 1 the selected baseline images at $m(1, \mathbf{S}) = \mathbf{A}$ also fulfill the requirement of being uninformative since the output probabilities are close to zero.

The subsequent set of experiments illustrate the dependency of the input perturbation methods on the choice of the baseline image. Again we use MoRF to compare one saliency map generated by RISE with one generated by Grad-CAM. Here only the results for the $\mathbf{A}_\sigma$ and $\mathbf{A}_{\mathrm{inv}}$ baseline images are reported, because these are the baseline images used for the insertion and deletion metrics by [12]. However, the results below can also be observed when applying different baseline images. The results for MoRF on different input images are shown in Table 2. Each graph compares the saliency maps from RISE and Grad-CAM. The table shows that by only changing the baseline image, the result of MoRF inverts. While one saliency method is superior using a specific baseline image, it is inferior using another baseline image. Furthermore, RISE performs better than Grad-CAM with the $\mathbf{A}_\sigma$ baseline image for MoRF but worse than Grad-CAM using the same baseline image for LeRF. Therefore, the results from MoRF and LeRF are highly dependent on the choice of the baseline image. Note, that even though the RISE saliency map was created using the zero after preprocessing baseline image, the AOC for MoRF with the blurred baseline image $\mathbf{A}_\sigma$ is greater than the AOC for MoRF with the zero after preprocessing baseline image. This shows that running MoRF with the same baseline image, that was used to create the RISE saliency map, does not lead to a greater AOC for RISE.

Table 1. MoRF with the zero after preprocessing baseline image for input images of the classes `chimpanzee` (left) and `beacon` (right).

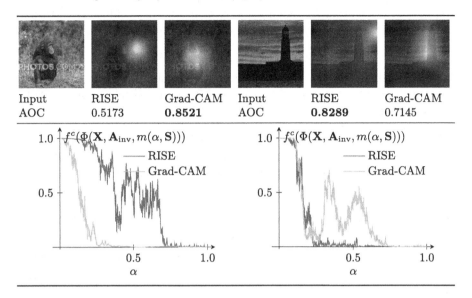

Input	RISE	Grad-CAM	Input	RISE	Grad-CAM
AOC	0.5173	**0.8521**	AOC	**0.8289**	0.7145

Table 2. MoRF with different baseline images for the class `flamingo`.

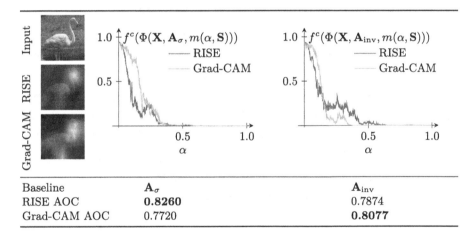

Baseline	$\mathbf{A}_\sigma$	$\mathbf{A}_{\text{inv}}$
RISE AOC	**0.8260**	0.7874
Grad-CAM AOC	0.7720	**0.8077**

6   Discussion

RISE does not converge with the parameters presented by Petsiuk *et al.* [12] for every input image. Therefore, we implemented a convergence check with three independent sets of masks using the L^2-distance. As shown by the experiments, different sets of parameters do not necessarily converge for all input images. Furthermore, the experiments show that varying the parameters for the mask generation can change the resulting saliency map significantly. In the case of Fig. 4 multiple goldfish are visible and RISE does not converge using the standard parameters. In this example, the probability of at least one goldfish being visible is very high when covering half the image ($p = 0.5$). This results in a high prediction score for each masked input image. This could be one possible explanation, why the RISE saliency map is not able to converge in this case. The above example is more likely to converge if fewer goldfish are revealed with each single mask. In practice, a smaller value for p led to convergence, if multiple instances of the same object are visible in the input image. While w can impact convergence, we did not find indicators for a correlation between w and the object size. In general, it is not clear how to decide on the parameters for running RISE on an input image. Determining the set of parameters requires multiple executions of RISE, which increases the computational effort. It is also not obvious how to select one saliency map for assessment, in the case that multiple sets of parameters result in a converged saliency map for an input image. The presented experiments on RISE also raise the issue of reproducibility of saliency maps with RISE.

The application of MoRF and LeRF on single input images when using the same baseline images, shows that the outputs for MoRF and LeRF do not agree for the saliency maps for the different input images in a dataset, which supports the recent findings in [16]. While, [13] average the AOC scores for MoRF or AUC scores for LeRF over the whole dataset, we want to highlight that MoRF and LeRF are not normalized. We argue that this puts input images with a low output probability for the target class at a disadvantage, since they generally exhibit a lower AOC for MoRF or a lower AUC for LeRF. For example, consider MoRF and two input images, which receive an output probability for the target class of 0.4 and 0.9, respectively. The maximum possible AOC for the first image is significantly lower than the maximum possible AOC for the second one.

The results from MoRF, LeRF, and RISE suggest, that these metrics and methods are highly dependent on the choice of the baseline image. Each baseline image introduces a bias into the input image. Even an all zero input can yield activations for the ResNet-50 because of nonzero bias terms. The choice of a neutral baseline image is therefore nontrivial. Creating saliency maps from RISE with different baseline images can result in highlighting contradicting parts of the input image. Consequently, a specific baseline image can lead to an incorrect conclusion when assigning importance to certain elements in the input. Similarly, the experiments on the variation of baseline images display that by changing baseline images it is possible to invert the proposition from MoRF and LeRF on which saliency method is superior.

7   Conclusion

In this paper we investigated input perturbation methods and metrics under changing parameters and baseline images. First, the experiments on RISE revealed, that convergence is not guaranteed for the majority of analyzed input images. Convergence could still be achieved for certain inputs, by varying the parameters for w, h and p. However, different sets of parameters yield significantly distinct results, which might contradict each other by highlighting different parts of the input image. Determining the best parameters by multiple independent runs of RISE with different parameters and checking convergence for each input image increases the amount of required computations. Furthermore, these additional runs and variations still do not guarantee non-contradicting saliency maps from RISE. Second, varying the baseline image has led to notably different saliency maps for RISE and changed the AOC and AUC values for MoRF and LeRF, respectively, such that a reliable evaluation of saliency methods is challenging.

Our results suggest that input perturbation methods are unreliable for understanding CNNs' predictions and the current available methods should not be used in practice. Since RISE, MoRF, and LeRF lack desirable robustness properties, future work will explore the development of robust input perturbation methods along with additional experiments for determining their robustness.

References

1. Abadi, M., et al.: TensorFlow: Large-Scale Machine Learning on Heterogeneous Systems (2015). http://tensorflow.org/
2. Ancona, M., Ceolini, E., Öztireli, A.C., Gross, M.: A unified view of gradient-based attribution methods for deep neural networks. In: NIPS Workshop on Interpreting, Explaining and Visualizing Deep Learning (2017)
3. Bach, S., Binder, A., Montavon, G., Klauschen, F., Müller, K.R., Samek, W.: On pixel-wise explanations for non-linear classifier decisions by layer-wise relevance propagation. PLoS ONE **10**(7), e0130140 (2015)
4. Bradski, G.: The OpenCV library. Dr. Dobb's J. Softw. Tools, article id 2236121 (2000). https://github.com/opencv/opencv/wiki/CiteOpenCV
5. Brunke, L., Agrawal, P., George, N.: Supplementary Material: Evaluating Input Perturbation Methods for Interpreting CNNs and Saliency Map Comparison (2020)
6. Chollet, F., et al.: Keras (2015). https://keras.io
7. Dabkowski, P., Gal, Y.: Real time image saliency for black box classifiers. In: Guyon, I., et al.: (eds.) Advances in Neural Information Processing Systems 30, pp. 6967–6976. Curran Associates, Inc. (2017)
8. Deng, J., Dong, W., Socher, R., Li, L.J., Li, K., Fei-Fei, L.: ImageNet: a large-scale hierarchical image database. In: CVPR09 (2009)
9. Fong, R.C., Vedaldi, A.: Interpretable explanations of black boxes by meaningful perturbation. In: The IEEE International Conference on Computer Vision (ICCV), October 2017
10. He, K., Zhang, X., Ren, S., Sun, J.: Deep residual learning for image recognition. In: The IEEE Conference on Computer Vision and Pattern Recognition (CVPR), June 2016
11. Hooker, S., Erhan, D., Kindermans, P.J., Kim, B.: A benchmark for interpretability methods in deep neural networks. In: Advances in Neural Information Processing Systems 32 (2019)
12. Petsiuk, V., Das, A., Saenko, K.: RISE: randomized input sampling for explanation of black-box models. In: Proceedings of the British Machine Vision Conference (BMVC) (2018)

13. Samek, W., Binder, A., Montavon, G., Lapuschkin, S., Müller, K.: Evaluating the visualization of what a deep neural network has learned. IEEE Trans. Neural Netw. Learn. Syst. **28**(11), 2660–2673 (2017)
14. Selvaraju, R.R., Cogswell, M., Das, A., Vedantam, R., Parikh, D., Batra, D.: Grad-CAM: visual explanations from deep networks via gradient-based localization. In: 2017 IEEE International Conference on Computer Vision (ICCV), pp. 618–626, October 2017
15. Sundararajan, M., Taly, A., Yan, Q.: Axiomatic attribution for deep networks. In: ICML (2017)
16. Tomsett, R., Harborne, D., Chakraborty, S., Gurram, P., Preece, A.: Sanity checks for saliency metrics. In: Proceedings of the AAAI Conference on Artificial Intelligence (2020)
17. Zeiler, M.D., Fergus, R.: Visualizing and understanding convolutional networks. In: Fleet, D., Pajdla, T., Schiele, B., Tuytelaars, T. (eds.) ECCV 2014, Part I. LNCS, vol. 8689, pp. 818–833. Springer, Cham (2014). https://doi.org/10.1007/978-3-319-10590-1_53

Adversarial Robustness of Open-Set Recognition: Face Recognition and Person Re-identification

Xiao Gong[1]($\boxtimes$), Guosheng Hu[2]($\boxtimes$), Timothy Hospedales[3]($\boxtimes$), and Yongxin Yang[3]($\boxtimes$)

[1] Nanjing University, Nanjing, China
gongxiao2020@hotmail.com
[2] AnyVision, Belfast, UK
huguosheng100@gmail.com
[3] Edinburgh University, Edinburgh, UK
{t.hospedales,yongxin.yang}@ed.ac.uk

Abstract. Recent studies show that DNNs are vulnerable to adversarial attacks, in which carefully chosen imperceptible modifications to the inputs lead to incorrect predictions. However most existing attacks focus on closed-set classification, and adversarial attack of open-set recognition has been less investigated. In this paper, we systematically investigate the adversarial robustness of widely used open-set recognition models, namely person re-identification (ReID) and face recognition (FR) models. Specifically, we compare two categories of black-box attacks: transfer-based extensions of standard closed-set attacks and several direct random-search based attacks proposed here. Extensive experiments demonstrate that ReID and FR models are also vulnerable to adversarial attack, and highlight a potential AI trustworthiness problem for these socially important applications.

Keywords: Adversarial attack · Open-world classification · Person re-identification · Face recognition

1 Introduction

In recent years, deep neural networks (DNNs) have been shown to be susceptible to adversarial examples: inputs with perturbations that are imperceptibly small to humans, which are easily misclassified by DNNs. This phenomenon has caused widespread concern about AI security. So far, most existing adversarial attacks [3,11,15,20–22,30] have targeted classification such as object category recognition under the closed-set setting. That is, all testing categories are pre-defined in training set. In contrast, for open-set recognition such as person re-identification and face recognition, the categories of test set are disjoint from the training set. This difference in evaluation metric leads to differing requirements of learning *separable* vs *discriminaive* feature spaces respectively. This paper provides

© Springer Nature Switzerland AG 2020
A. Bartoli and A. Fusiello (Eds.): ECCV 2020 Workshops, LNCS 12535, pp. 135–151, 2020.
https://doi.org/10.1007/978-3-030-66415-2_9

the first systematic study on the adversarial robustness of open-set recognition across multiple major open-set applications.

To study this problem, we conduct extensive experiments to investigate the robustness of widely used open-set recognition models against a variety of adversarial attacks in this paper. We consider the most real-world black-box setting, where an attacker victimizes an online (target) model with hidden parameters and architecture. In this case the attacker proceeds by attacking a surrogate (agent) model whose parameters and structure are known but different from the target model. If adversarial perturbations generated on the agent model can be transferred [11] to the target, the attacker can successfully fool the target model.

In this work we investigate transfer learning attacks based on modifying PGD to the open-set setting (as proposed in [1]) and we analogously extend the C&W attack to the open-set setting. The limited existing work on open-set adversarial robustness has not confirmed whether transfer-learning approaches to black box attack indeed outperform carefully designed gradient-free alternatives applied directly to the target model. We therefore propose four new random-search based black-box attacks (random-PGD-transfer, interior-random-search, bound-random-search and C&W-random-search) that can apply directly to the target model to provide baselines of comparison.

To evaluate open-set attacks concretely on applications that are widely-used and of high social importance, we consider two problems: Person Re-identification (ReID) and Face Recognition (FR) as a case studies. Then, we investigate the robustness of contemporary ReID and FR models against adversarial attacks. Given the typical real-world usage scenarios of ReID, the most relevant attack is an un-targeted *evasion* attack, e.g., to prevent a person of interest from being detected. Meanwhile, to attack FR systems, a targeted *impersonation* attack is also relevant, where an attacker desires to portray them-self as another specific identity, e.g., to fool an access control system.

We comprehensively conduct quantitative and qualitative experiments on adversarial attacks on open-set recognition systems. Our results show that transfer-based methods are more effective than carefully designed random-search-based methods, confirming that existing widely studied and deployed – and socially important – recognition models are also vulnerable to the adversarial attack. Our results can be used as a baseline for further research on adversarial robustness of open-set recognition and inspiring future research in this field. In summary, our main contributions can be summarized as:

- We are the first to *systematically* investigate the adversarial robustness of open-set recognition models in real world black-box conditions.
- To adapt to ReID and FR, we proposed 4 new attacks: random-PGD-transfer, C&W-random-search, interior-random-search and bound-random-search.
- The experimental results demonstrate that the existing recognition models are also vulnerable to adversarial attack. Our results serve as a baseline for further research on open-set recognition attacks and defenses.

2 Related Work

2.1 Adversarial Attack

Since DNNs were first found vulnerable to adversarial examples [30], a large volume of research has contributed various attack methods [3,11,15,20–22,28] against image classification. systems. In this paper, we categorize the various attacks to the following groups:

White-Box and Black-Box Attacks. White-box attacks [3,11,15,21,22,30] require the attackers to have prior knowledge of the target models, and adversarial examples are generated with and tested on the same model. Black-box attacks [6,23,28] do not assume the attacker has access to the structure and parameters of the target model. Under this setting, attackers must resort to random search or gradient-free search [4] for adversarial attacks on the target model, or to attack a surrogate agent model that is trained for the same task and transfer the attack to the target model. This class of methods are called transfer-based because they rely on the transferability of the adversarial examples. In the real world, the structure and parameters of online models are hidden, so the black-box attacks are the most practical and general purpose. We note that the widely used gradient-free and black-box ZOO attack [4] does not directly extend to the open set setting, so we focus on transfer and random search attacks here.

Targeted/Impersonation and Untargeted/Evasion Attacks. In conventional closed-set image classification, untargeted attacks [15,21,28] mean that the attackers attempts to make the target model misclassify the adversarial example to any other category. Meanwhile, targeted attacks [3,6,11,22,23,30] further aim to make the adversarial example be misclassified to a specific alternative category chosen by the attacker.

In ReID [36] and FR [32], untargeted and targeted attacks are often termed evasion and impersonation attacks respectively. The evading attacker attempts to prevent themselves from being identified by the target model. While the impersonation attacker attempts to mislead the model into identifying itself as another specific identity.

Gradient-Based and Optimization-Based Attacks. Adversarial examples aim to satisfy two properties: (1) small perturbation (so as to be imperceptible to humans) and (2) being misclassified by the target model. To achieve these objectives, two categories of attacks have been developed. Gradient-based attacks [11,15,20,22,30] fix the size of perturbation and use the gradient of the loss with respect to the input image to search for the adversarial example in some neighborhood (usually L_p ball) of the original image. While optimization-based attacks [3] minimize the adversarial perturbation based on ensuring the success of the attacks.

2.2 Person Re-identification

Person re-identification (ReID), aims to re-identify people across different camera views, has been significantly advanced by deep learning in recent years. For example, state of the art ReID models [46] have almost solved the Market-1501 benchmark [43] since its introduction a few years earlier. Overall the prior work on ReID is too extensive to review in detail here, and we refer the reader to [16,25–27,33,40–42] for an overview, while just mentioning a few related studies here. Recently, [17] uses adversarial learning in a deep ReID model to synthesise impostors for the gallery, making the model robust to attack from real impostors in the probe set. [46] first introduces the omni-scale deep feature learning and builds a new architecture called OSNet, which can capture discriminative features of either a single scale or a weighted mixture of multiple scales. [39] uses feature uncertainty method to train noise-robust ReID models.

Limited existing work has studied the adversarial robustness of person re-identification. [1] introduces an open-set extension of the PGD attack for attacking Re-ID systems. [36] propose advPattern that formulates the problem of generating adversarial patterns against deep ReID models as an optimization problem of minimizing the similarity scores of the adversary's images across camera views and [9] makes the first attempt to attack ReID models using universal adversarial perturbations. We provide the first systematic study of the black-box robustness of Re-ID systems to both different types of transfer attacks, as well as several different non-transfer alternatives based on random search.

2.3   Face Recognition

Face recognition (FR) is divided into face verification (1 : 1) and face identification (1 : N). Recent works mainly focus on the loss function used for training standard CNN architectures. There are two main categories of loss functions, triplet-ranking loss methods [24] and softmax (classification) loss-based methods [8,18,34,35]. Based on the large-scale training data and the elaborate DNN architectures, both the softmax and triplet loss-based methods achieve excellent performance on face recognition. We refer the reader to [2,19,29,31,44,47] for an overview of this extensive area of research.

Recently, research has also investigated adversarial attacks on FR. [12] investigates image-level and face-level attacks and proposes selective dropout to mitigate the adverse effect of these attacks. However, these attacks are not visually imperceptible. [7,10] introduce GANs to generate adversarial faces. However, these methods are white-box attacks, and we investigate the black-box attack which is more similar to real applications. [32] begins to investigate black-box transfer attacks on FR. We provide the first systematic evaluation of black-box transfer and random search attacks on open-set FR.

3   Black-Box Attacks for Open-Set Problems

In this section, we detail the adversarial attacks for open-set problems. In the real world, the target model parameters and architecture are not accessible, thus we focus on black-box attacks.

3.1 Threat Model

In this work, we investigate two categories of attacks: PGD-like (PGD-transfer, Random-PGD-transfer, interior-random-search and bound-random-search) and C&W-like (C&W-transfer and C&W-random-search). PGD-like attacks are extended based on closed-set PGD attacks [20]. We choose L_∞ ball to be the threat model. C&W-like attacks are based on closed-set C&W attacks [3]. Its threat model is chosen to be L_2 ball.

To attack ReID models, we assume the adversary has no access to the query images. While in FR case, we assume the adversary can hack into the deployed system to get the query images.

Before introducing the various attacks in detail, we first clarify some notations and terminologies. We use f and F denoting our agent and target models respectively. In open-set settings, f and F are both feature extractors, which map images into a discriminative feature space. During identification, a specific distance of the feature space denoted by d is used to measure the similarity between two images. We use x denoting the original test image, x^{adv} denoting the corresponding adversarial example and p denoting the image compared. Particularly, for verification, it requires setting a threshold denoted as T. Then a pair of images whose corresponding distance is less than T is considered to be the same identity.

3.2 Design Objectives

We propose two attack scenarios, evasion attack and impersonation attack, to deceive open-set recognition systems.

Evasion Attacks. An evasion attack is an untargeted attack: open-set models are fooled to match the adversary as an arbitrary person except himself, which looks like that the adversary wears an "invisible cloak". Formally, an open-set model outputs a similarity score $d(f(x), f(p))$ of the test image x and the compared image p. We attempt to find an adversarial example x^{adv} of x by solving the following optimization problem:

$$\text{maximize} \quad d(f(x^{adv}), f(p)) \tag{1}$$
$$s.t. \quad x^{adv} \in \Phi$$

where Φ is the L_∞ ball centered at x for PGD-like attacks and L_2 ball centered at x for C&W-like attacks respectively.

Impersonation Attacks. An impersonation attack is a targeted attack which can be viewed as an extension of evading attack: The adversary attempts to deceive open-set recognition models into mismatching himself as a target person. Given the image x_t of the target identity we want to impersonate, we formulate impersonation attack as the following optimization problem:

$$\text{minimize} \quad d(f(x^{adv}), f(x_t)) \tag{2}$$
$$s.t. \quad x^{adv} \in \Phi$$

where Φ is the L_∞ ball centered at x for PGD-like attacks and L_2 ball centered at x for C&W-like attacks respectively.

In the following, we detail two groups of black-box attacks: (1) transfer-based (2) random-search-based.

3.3   Transfer-Based Attacks

Transfer-based attacks rely on the transferability of adversarial examples. We first train a known agent model on our own dataset, which we can attack with white box methods since we know its structure and parameters. We can then transfer the adversarial examples to the target model. We generate adversarial examples by using open-set extensions of existing closed-set adversarial attacks. We adapt PGD [20] from gradient-based methods and C&W [3] from optimization-based methods to meet the open-set protocol. Details of these attacks are listed below.

PGD-Transfer. The PGD-transfer attack can be formulated as:

$$\begin{cases} x_0^{adv} = x \\ x_{k+1}^{adv} = \text{Clip}_\varepsilon(x_k^{adv} \pm \alpha \cdot \text{sign}(\frac{\partial d(f(x),f(p))}{\partial x_k^{adv}})) \end{cases} \quad (3)$$

where the gray-scale perturbation is ε and one step perturbation as α, and Clip_ε is a clip function that ensures the generated adversarial example is within the ε-ball of the original image. In the case of evasion and impersonation attacks, the sign before α in Eq. (3) is chosen to be positive and negative respectively.

C&W-Transfer. The C&W-transfer attack can be formulated as:

$$\text{minimize} \quad \|\delta\|_2 + c \cdot h(x + \delta, p) \quad (4)$$
$$s.t. \quad x + \delta \in [0,1]^N$$

where $\delta = \frac{1}{2}(tanh(w) + 1) - x$ denotes the perturbation of x in the continuous image space, N denotes the input dimension and $h(x + \delta, p)$ is an equivalent condition of successfully attacking agent model f. Following [3], we optimize over w whose feasible region is $\mathbb{R}^N$.

Random-PGD-Transfer. In order to investigate the effect of the gradients of our agent model, we replace the gradients in PGD-transfer with randomly generated ones, thus introduce a new black-box attack named random-PGD-transfer. If PGD-transfer beats this baseline it confirms the value of exploiting the transferred gradients from the surrogate agent model. The algorithm is shown in Algorithm 1 (Note that resp. is the shorthand for respectively):

Random-Search-based Attacks To contextualize the efficacy of transfer-based attacks, we introduce three new non-transfer attacks based on random-search for comparison. Our main idea is to identify the feasible region for random search. Since the adversarial examples found by the transfer-based attacks also have a high probability of cheating the target model, we use the feasible region

---

Algorithm 1. Random-PGD-Transfer: Evasion (resp. Impersonation)

Input: x, ε, α, iter
Output: x^{adv}

1: Initialization: $x_0^{adv} = x$
2: **for** i=0 : (iter-1) **do**
3:      randomly generate ∇ whose size is equal to x
4:      $x' = \text{Clip}_\varepsilon(x_i^{adv} + (\text{resp.}-)\alpha \cdot \text{sign}\nabla)$
5:      **if** $d(f(x'), f(p)) > (\text{resp.} <)d(f(x_0^{adv}), f(p))$ **then**
6:          $x_{i+1}^{adv} = x'$
7:      **else**
8:          $x_{i+1}^{adv} = x_i^{adv}$
9: **return** $x^{adv} = x_{iter}^{adv}$

---

of the corresponding transfer-based attacks as our random search space. Only if the efficacy of transfer-based attacks is higher than that of random search is it valuable to train an agent model to do transfer-based attacks.

C&W-Random-Search. To compare with the C&W-transfer attack, we first compute the maximal perturbation Δ on the whole dataset. That is,

$$\Delta = \max\{\|x - x^{adv}\|_2 : x \in D\} \tag{5}$$

where D is the test dataset, x^{adv} is the corresponding adversarial example of x. Then we search the adversarial examples of the target model F corresponding to x in $B(x; \Delta) = \{y : \|y - x\|_2 \leq \Delta\}$ since we know that the adversarial examples generated by attacking f also belongs to $B(x; \Delta)$.

Interior-Random-Search. Unlike the C&W attack, the feasible region of the PGD attack is discrete. In our experiments, we scale the gray-scale into $[-1, 1]$. The step size of each iteration is then one gray-scale, that is $\alpha = \frac{2}{255}$. To compare with the PGD-transfer and random-PGD-transfer attack, we search the adversarial examples of the target model F corresponding to x in $S(x; \varepsilon) = \{x - \varepsilon \cdot \alpha, x - (\varepsilon - 1) \cdot \alpha, ..., 0, ..., x + (\varepsilon - 1) \cdot \alpha, x + \varepsilon \cdot \alpha\}^N$, where N is the dimension of the image x. $S(x; \varepsilon)$ is the same domain of the adversarial examples attacking the agent model f.

Bound-Random-Search. Although the adversarial examples generated by PGD are in $S(x; \varepsilon)$, they are usually close to the boundary of the feasible region. So we narrow down the feasible region to the boundary and then search adversarial examples randomly in $\partial S(x; \varepsilon) = [x - \varepsilon \cdot \alpha, x + \varepsilon \cdot \alpha]^N$.

3.4   Open-Set Case Study (ReID and FR)

ReID. In the case of ReID, we make a ranking list according to the distances between the gallery images and the query images. However the query images are unaccessible to the attackers, we can only mislead the model by making the features of the adversarial examples far away from those of the original images. It means we can only conduct evasion attacks. Accordingly, the compared image p is the same as the original image x and the term $h(x + \delta, p)$ in Eq. (4) is $max\{0, T - d(f(x + \delta), f(x))\}$.

FR. In the case of FR, we only investigate face verification (1 vs. 1 face comparison). 3,000 pairs of faces with the same identity denoted by $D_{same} = \{(x_i, p_i)\}_{i=1}^{3000}$ and 3,000 pairs with different identities denoted by $D_{diff} = \{(x_i, p_i)\}_{i=1}^{3000}$ are chosen from LFW dataset [14] as the test datasets, where x_i is used to generate the adversarial example, p_i is used for comparison. On D_{same} we can conduct the evasion attack. The term $h(x + \delta, p)$ in Eq. (4) is $max\{0, T - d(f(x + \delta), f(p))\}$. On D_{diff} we can conduct the impersonation attack, which deserves more attention, since it is more dangerous that an attacker can trick a model into impersonating someone else. In this case, the term $h(x + \delta, p)$ in Eq. (4) is $max\{0, d(f(x + \delta), f(p)) - T\}$.

4 Experiments

In this section, we conduct extensive experiments on the adversarial robustness of the widely used ReID and FR models against the attacks from Section 3.

4.1 Experimental Settings

To make our results reproducible, we detail our experimental settings here.

LFW. LFW (Labled Faces in the Wild) [14] is a commonly used test set for face verification. It has 13,233 face images and 5,749 identities. In our experiment, we randomly selected 3,000 pairs of faces with the same identity and 3,000 pairs with different identities. All images are resized into 112×112.

CASIA-WebFace. CASIA-WebFace [38] is a large face dataset, that is widely used for training contemporary face recognition systems. It includes 10,575 identities and 494,414 face images. In our experiments, we also resize the images into 112×112.

Market-1501. Market-1501 [43] contains 32668 pictures corresponding to 1501 identities taken by 6 cameras. 12936 images of 751 identities are used for training and 19732 images of the rest 750 identities used for testing. During test, 3368 images are used for querying and others used as gallery images.

Experimental Details: Re-ID For ReID experiments, we use pre-trained models in Torchreid [45]. We select ResNet50 [13] trained on MSMT17[37] as the agent model and state of the art osnet-ain-x1-0 [46] trained on MSMT17 as the target model. To evaluate adversarial robustness of the target model, we choose Market-1501 as the test set and investigate the mAP and the rank-k ($k = 1, 5, 10, 20$) of the various adversarial gallery sets.

Experimental Details: Face For FR experiments, we use mobilefacenet [5] as the backbone of the agent model trained with CosFace [35] loss on CASIA-WebFace achieving 97.98% on LFW. While we use IR-SE-ResNet50 [8] as the backbone of the target model trained with ArcFace [8] loss also on CASIA-WebFace achieving 98.99% on LFW. For evaluating the evading attack, we use the 3,000 pairs of faces with the same identity from LFW. For evaluating the impersonation attack, we use the 3,000 pairs of faces with the different identities from LFW.

4.2 Results of ReID

To show our experimental results, we divide the results into two categories: (1) PGD-like: PGD-transfer, random-PGD-transfer, interior-random-search and bound-random-search. (2) C&W-like: C&W-transfer and C&W-random-search. During our experiments, the number of iterations in the transfer-based attacks are set to be 1000. For random-search attacks, we conduct 1000 search times and use the mean of the 1000 results as the final values. The results are as follows:

Quantitative Results. The quantitative results for PGD and C&W-like attacks are shown in Fig. 1 and Table 1 respectively in terms of mAP and accuracy act rank-k. We can see that state of the art ReID systems [46] are indeed vulnerable to black-box adversarial attack. From Fig. 1, among PGD-like attacks PGD transfer attack is best, followed by bound-random-search, interior-random-search and random-PGD-transfer. Similarly, according to Table 1, we see that C&W-transfer is more effective than C&W-random-search. Together the results confirm that transfer attacks outperform carefully designed random search baselines and hence that Re-ID systems are vulnerable to black-box attack by transfer.

Table 1. Comparison of C&W-like attacks on state of the art OSANet [46] with Market-1501 dataset. mAP and rank-k ($k = 1, 5, 10, 20$) performance when setting $d(x, x^{adv}) > 10000$ and maximal adversarial perturbation $\Delta = 24.59$.

Method	Baseline	C&W-transfer	C&W-random-search
mAP	43.3	32.4	39.94
rank1	70.3	60.6	68.02
rank5	83.9	78.7	82.78
rank10	88.5	84.8	87.42
rank20	92.3	89.7	91.36

Visualization of Adversarial Examples. Qualitative results of adversarial attack on ReID are illustrated in Fig. 2. Here, a code $(\varepsilon, d(x^{adv}, p))$ at the top of each sub-figure explains the example where p is the query image and d is the

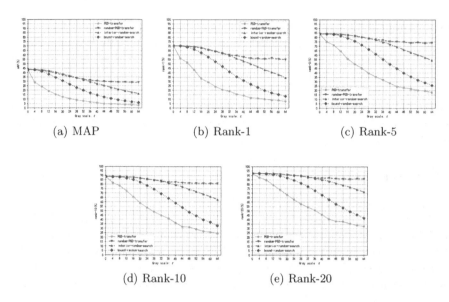

(a) MAP   (b) Rank-1   (c) Rank-5

(d) Rank-10   (e) Rank-20

Fig. 1. Comparison of PGD-like attacks on state of the art OSANet [46] with Market-1501 dataset. mAP and rank-k ($k = 1, 5, 10, 20$) performance when varying the maximum magnitude of adversarial perturbation ε. Transfer attacks from an agent model outperform random search attacks on the target model.

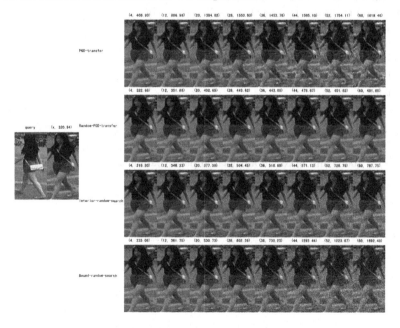

Fig. 2. Visualization of adversarial examples generated by PGD-like attacks by varying the maximum magnitude of adversarial perturbation ε. Each gallery image annotation is $(\varepsilon, d(x^{adv}, p))$ where larger target model distances $d(\cdot, \cdot)$ indicate a more successful attack.

Fig. 3. Visualization of adversarial examples generated by C&W-like attacks. Each gallery image annotation is $d(x^{adv}, p)$, where larger target model distances $d(\cdot, \cdot)$ indicate a more successful attack.

distance according to the target model. The adversarial perturbations generated by the four PGD-like attacks are almost imperceptible to humans when $\varepsilon \leq 12$. In this case, all methods increase the distance to the query image (left vs right example distances), and the distances of the query image and the adversarial examples generated by PGD-transfer are more than twice of the random-search alternatives.

In Fig. 3, the number at the top of each sub-figure is $d(x, p)$, where p is the query image in the first row and d is the distance according to the target model. We can see that the adversarial perturbations of C&W-like attacks are more visually imperceptible than that of PGD-like attacks. Also the distances of the query image and the adversarial examples generated by C&W-transfer are mainly larger than that of C&W-random-search.

4.3   Results on Face Recognition

We next investigate the adversarial robustness of face verification. As discussed in Sect. 4.1, we use 3,000 pairs of faces with the same identity for evaluating the evading attack and 3,000 pairs of faces with the different identities for evaluating the impersonation attack. The thresholds of our agent and target model are 1.474 and 1.315 respectively.

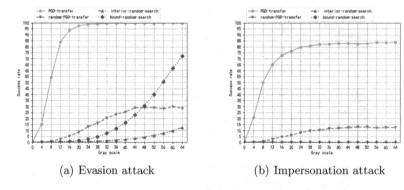

(a) Evasion attack   (b) Impersonation attack

Fig. 4. The success rate of PGD-like attacks against target face recognition model (CosFace+WebFace trained mobilefacenet) when evaluated on LFW benchmark.

Table 2. The success rate of C&W-like attacks against target FR model (Cos-Face+WebFace trained mobilefacenet) when evaluated on LFW benchmark.

Method	Type	Impersonation				Evading					
	T	0.010	0.500	1.000	1.474	1.474	2.000	2.500	3.000	3.500	3.950
	Δ	12.57	6.23	3.73	2.31	2.48	3.27	4.00	4.56	5.67	15.75
CW-random-search		0.00	0.00	0.00	0.00	0.00	0.00	0.00	0.00	0.00	0.00
CW-transfer		25.63	12.23	3.43	0.17	0.07	2.27	7.63	14.83	19.87	22.67

Quantitative results. Quantitative results for PGD and C&W like attacks are shown in Fig. 4 and Table 2. Overall, we can see that transfer-based attacks have overwhelming efficacy over random-search attacks. Random-search attacks are almost completely ineffective in the impersonation case. PGD-transfer attack can achieve high success rate with very small perturbation (93.6% for the evasion attack and 72.8% for the impersonation attack with $\varepsilon = 16$). Unlike ReID, face verification has a threshold to determine whether the pair of faces has the same identity. In this case, random-search attacks generally fail to find samples that cross the threshold. However, the threshold gives a clear target for transfer-based attacks, making it easier to find adversarial examples for the agent model that can be transferred.

Visualization of Adversarial Examples. In Fig. 5, a descriptor at the top of each sub-figure is $(\varepsilon, d(x^{adv}, p))$, where ε is the attack strength, p is the compared image and d is the distance according to the target model. The adversarial perturbations generated by the four PGD-like attacks are almost imperceptible to humans when $\varepsilon \leq 12$. All PGD-like methods increase the perceived distance of the true match to the probe p, but PGD-transfer is clearly the most successful.

In Fig. 6, a descriptor at the top of each sub-figure is $(T, d(x^{adv}, p))$, where T is the threshold in Eq. (4) where p is the compared image and d is the distance according to the target model. We can see that the adversarial perturbations of C&W-like attacks are more visually imperceptible than that of PGD-like attacks.

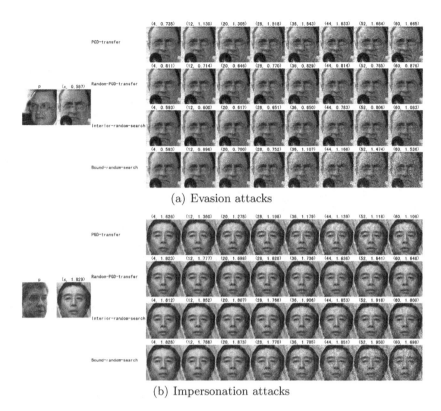

(a) Evasion attacks

(b) Impersonation attacks

Fig. 5. Visualization of adversarial examples generated by PGD-like attacks in face recognition.

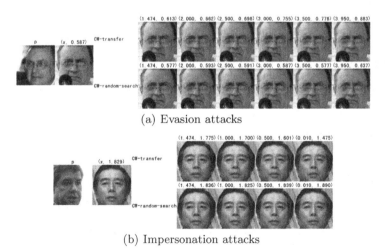

(a) Evasion attacks

(b) Impersonation attacks

Fig. 6. Visualization of adversarial examples generated by C&W-like attacks in face recognition.

Compared with C&W-random-search, the distances of the compared image and the adversarial examples generated by C&W-transfer are better in both cases: larger for dodging case and smaller for impersonation. This confirms the value of transferring attacks from a known surrogate agent model to a black-box target model.

Summary. Overall the results show that PGD-transfer and C&W-transfer attacks can successfully attack state of the art open-set face recognition and ReID systems by generating human-imperceptible perturbations that nevertheless lead to misclassification by the model.

5 Conclusion and Future Work

We investigated the adversarial robustness of state of the art open-set recognition models (ReID and FR) against various attacks in real world black-box conditions. The experimental results demonstrate that these open-set recognition methods are also vulnerable to adversarial attack, which is an import AI-safety concern given the social importance of these applications. Specifically, from the experimental results, we found that the adversarial examples are highly transferable across architectures and datasets under the open-set setting, and such transfer attacks can outperform well designed random search baselines applied directly to the target model. This encourages future work to improve the adversarial robustness of open-set recognition models.

References

1. Bai, S., Li, Y., Zhou, Y., Li, Q., Torr, P.H.S.: Metric attack and defense for person re-identification (2019)
2. Cao, K., Rong, Y., Li, C., Tang, X., Loy, C.C.: Pose-robust face recognition via deep residual equivariant mapping. In: IEEE Conference on Computer Vision and Pattern Recognition (CVPR), pp. 5187–5196 (2018)
3. Carlini, N., Wagner, D.A.: Towards evaluating the robustness of neural networks. In: 2017 IEEE Symposium on Security and Privacy (SP), pp. 39–57 (2017)
4. Chen, P.Y., Zhang, H., Sharma, Y., Yi, J., Hsieh, C.J.: ZOO: zeroth order optimization based black-box attacks to deep neural networks without training substitute models. In: Proceedings of the 10th ACM Workshop on Artificial Intelligence and Security, pp. 15–26 (2017)
5. Chen, S., Liu, Y., Gao, X., Han, Z.: MobileFaceNets: efficient CNNs for accurate real-time face verification on mobile devices. In: Zhou, J., et al. (eds.) CCBR 2018. LNCS, vol. 10996, pp. 428–438. Springer, Cham (2018). https://doi.org/10.1007/978-3-319-97909-0_46
6. Cissé, M., Adi, Y., Neverova, N., Keshet, J.: Houdini: fooling deep structured prediction models. CoRR abs/1707.05373 (2017)
7. Deb, D., Zhang, J., Jain, A.K.: AdvFaces: adversarial face synthesis. CoRR abs/1908.05008 (2019)
8. Deng, J., Guo, J., Xue, N., Zafeiriou, S.: ArcFace: additive angular margin loss for deep face recognition. In: IEEE Conference on Computer Vision and Pattern Recognition (CVPR), pp. 4690–4699 (2019)

9. Ding, W., Wei, X., Hong, X., Ji, R., Gong, Y.: Universal adversarial perturbations against person re-identification. CoRR abs/1910.14184 (2019)
10. Gafni, O., Wolf, L., Taigman, Y.: Live face de-identification in video. In: 2019 IEEE/CVF International Conference on Computer Vision (ICCV), pp. 9377–9386 (2019)
11. Goodfellow, I.J., Shlens, J., Szegedy, C.: Explaining and harnessing adversarial examples. In: 3rd International Conference on Learning Representations (ICLR) (2015)
12. Goswami, G., Ratha, N.K., Agarwal, A., Singh, R., Vatsa, M.: Unravelling robustness of deep learning based face recognition against adversarial attacks. In: Proceedings of the Thirty-Second AAAI Conference on Artificial Intelligence, pp. 6829–6836 (2018)
13. He, K., Zhang, X., Ren, S., Sun, J.: Deep residual learning for image recognition. In: IEEE Conference on Computer Vision and Pattern Recognition (CVPR), pp. 770–778 (2016)
14. Huang, G.B., Ramesh, M., Berg, T., Learned-Miller, E.: Labeled faces in the wild: a database for studying face recognition in unconstrained environments. Technical report (2007)
15. Kurakin, A., Goodfellow, I.J., Bengio, S.: Adversarial examples in the physical world. In: 5th International Conference on Learning Representations (ICLR) (2017)
16. Li, D., Chen, X., Zhang, Z., Huang, K.: Learning deep context-aware features over body and latent parts for person re-identification. In: IEEE Conference on Computer Vision and Pattern Recognition (CVPR), pp. 7398–7407 (2017)
17. Li, X., Wu, A., Zheng, W.-S.: Adversarial open-world person re-identification. In: Ferrari, V., Hebert, M., Sminchisescu, C., Weiss, Y. (eds.) ECCV 2018. LNCS, vol. 11206, pp. 287–303. Springer, Cham (2018). https://doi.org/10.1007/978-3-030-01216-8_18
18. Liu, W., Wen, Y., Yu, Z., Li, M., Raj, B., Song, L.: SphereFace: deep hypersphere embedding for face recognition. In: IEEE Conference on Computer Vision and Pattern Recognition (CVPR), pp. 6738–6746 (2017)
19. Liu, Y., Li, H., Wang, X.: Rethinking feature discrimination and polymerization for large-scale recognition. CoRR abs/1710.00870 (2017)
20. Madry, A., Makelov, A., Schmidt, L., Tsipras, D., Vladu, A.: Towards deep learning models resistant to adversarial attacks. In: 6th International Conference on Learning Representations (ICLR) (2018)
21. Moosavi-Dezfooli, S., Fawzi, A., Frossard, P.: DeepFool: a simple and accurate method to fool deep neural networks. In: IEEE Conference on Computer Vision and Pattern Recognition (CVPR), pp. 2574–2582 (2016)
22. Papernot, N., McDaniel, P.D., Jha, S., Fredrikson, M., Celik, Z.B., Swami, A.: The limitations of deep learning in adversarial settings. In: IEEE European Symposium on Security and Privacy (EuroS&P), pp. 372–387 (2016)
23. Sarkar, S., Bansal, A., Mahbub, U., Chellappa, R.: UPSET and ANGRI : breaking high performance image classifiers. CoRR abs/1707.01159 (2017)
24. Schroff, F., Kalenichenko, D., Philbin, J.: FaceNet: a unified embedding for face recognition and clustering. In: IEEE Conference on Computer Vision and Pattern Recognition (CVPR), pp. 815–823 (2015)
25. Shen, Y., Xiao, T., Li, H., Yi, S., Wang, X.: End-to-end deep Kronecker-product matching for person re-identification. In: IEEE Conference on Computer Vision and Pattern Recognition (CVPR), pp. 6886–6895 (2018)

26. Si, J., Zhang, H., Li, C., Kuen, J., Kong, X., Kot, A.C., Wang, G.: Dual attention matching network for context-aware feature sequence based person re-identification. In: IEEE Conference on Computer Vision and Pattern Recognition (CVPR), pp. 5363–5372 (2018)

27. Song, C., Huang, Y., Ouyang, W., Wang, L.: Mask-guided contrastive attention model for person re-identification. In: IEEE Conference on Computer Vision and Pattern Recognition (CVPR), pp. 1179–1188 (2018)

28. Su, J., Vargas, D.V., Sakurai, K.: One pixel attack for fooling deep neural networks. IEEE Trans. Evol. Comput. **23**(5), 828–841 (2019)

29. Sun, Y., Wang, X., Tang, X.: Deep learning face representation from predicting 10, 000 classes. In: IEEE Conference on Computer Vision and Pattern Recognition (CVPR), pp. 1891–1898 (2014)

30. Szegedy, C., Zaremba, W., Sutskever, I., Bruna, J., Erhan, D., Goodfellow, I.J., Fergus, R.: Intriguing properties of neural networks. In: 2nd International Conference on Learning Representations (ICLR) (2014)

31. Taigman, Y., Yang, M., Ranzato, M., Wolf, L.: DeepFace: closing the gap to human-level performance in face verification. In: IEEE Conference on Computer Vision and Pattern Recognition (CVPR), pp. 1701–1708 (2014)

32. Tang, D., Wang, X., Zhang, K.: Query-free attacks on industry-grade face recognition systems under resource constraints. CoRR abs/1802.09900 (2018)

33. Varior, R.R., Haloi, M., Wang, G.: Gated Siamese convolutional neural network architecture for human re-identification. In: Leibe, B., Matas, J., Sebe, N., Welling, M. (eds.) ECCV 2016. LNCS, vol. 9912, pp. 791–808. Springer, Cham (2016). https://doi.org/10.1007/978-3-319-46484-8_48

34. Wang, F., Xiang, X., Cheng, J., Yuille, A.L.: Normface: L_2 hypersphere embedding for face verification. In: Proceedings of the 2017 ACM on Multimedia Conference, pp. 1041–1049 (2017)

35. Wang, H., Wang, Y., Zhou, Z., Ji, X., Gong, D., Zhou, J., Li, Z., Liu, W.: Cos-Face: large margin cosine loss for deep face recognition. In: IEEE Conference on Computer Vision and Pattern Recognition (CVPR), pp. 5265–5274 (2018)

36. Wang, Z., Zheng, S., Song, M., Wang, Q., Rahimpour, A., Qi, H.: advPattern: physical-world attacks on deep person re-identification via adversarially transformable patterns. In: 2019 IEEE/CVF International Conference on Computer Vision (ICCV), pp. 8340–8349 (2019)

37. Wei, L., Zhang, S., Gao, W., Tian, Q.: Person transfer GAN to bridge domain gap for person re-identification. In: IEEE Conference on Computer Vision and Pattern Recognition (CVPR), pp. 79–88 (2018)

38. Yi, D., Lei, Z., Liao, S., Li, S.Z.: Learning face representation from scratch. CoRR abs/1411.7923 (2014)

39. Yu, T., Li, D., Yang, Y., Hospedales, T.M., Xiang, T.: Robust person re-identification by modelling feature uncertainty. In: IEEE/CVF International Conference on Computer Vision (ICCV), pp. 552–561 (2019)

40. Zhang, X., et al.: AlignedReID: surpassing human-level performance in person re-identification. CoRR abs/1711.08184 (2017)

41. Zhang, Z., Lan, C., Zeng, W., Chen, Z.: Densely semantically aligned person re-identification. In: IEEE Conference on Computer Vision and Pattern Recognition (CVPR), pp. 667–676 (2019)

42. Zhao, L., Li, X., Zhuang, Y., Wang, J.: Deeply-learned part-aligned representations for person re-identification. In: IEEE International Conference on Computer Vision (ICCV), pp. 3239–3248 (2017)

43. Zheng, L., Shen, L., Tian, L., Wang, S., Wang, J., Tian, Q.: Scalable person re-identification: a benchmark. In: IEEE International Conference on Computer Vision (ICCV), pp. 1116–1124 (2015)
44. Zhou, E., Cao, Z., Yin, Q.: Naive-deep face recognition: touching the limit of LFW benchmark or not? CoRR abs/1501.04690 (2015)
45. Zhou, K., Xiang, T.: Torchreid: a library for deep learning person re-identification in Pytorch. CoRR abs/1910.10093 (2019)
46. Zhou, K., Yang, Y., Cavallaro, A., Xiang, T.: Omni-scale feature learning for person re-identification. In: IEEE/CVF International Conference on Computer Vision (ICCV), pp. 3701–3711 (2019)
47. Zhu, Z., Luo, P., Wang, X., Tang, X.: Recover canonical-view faces in the wild with deep neural networks. CoRR abs/1404.3543 (2014)

WaveTransform: Crafting Adversarial Examples via Input Decomposition

Divyam Anshumaan[1], Akshay Agarwal[1,2], Mayank Vatsa[3(✉)], and Richa Singh[3]

[1] IIIT-Delhi, Delhi, India
{divyam17147,akshaya}@iiitd.ac.in
[2] Texas A&M University, Kingsville, USA
[3] IIT Jodhpur, Jodhpur, India
{mvatsa,richa}@iitj.ac.in

Abstract. Frequency spectrum has played a significant role in learning unique and discriminating features for object recognition. Both low and high frequency information present in images have been extracted and learnt by a host of representation learning techniques, including deep learning. Inspired by this observation, we introduce a novel class of adversarial attacks, namely 'WaveTransform', that creates adversarial noise corresponding to low-frequency and high-frequency subbands, separately (or in combination). The frequency subbands are analyzed using wavelet decomposition; the subbands are corrupted and then used to construct an adversarial example. Experiments are performed using multiple databases and CNN models to establish the effectiveness of the proposed WaveTransform attack and analyze the importance of a particular frequency component. The robustness of the proposed attack is also evaluated through its transferability and resiliency against a recent adversarial defense algorithm. Experiments show that the proposed attack is effective against the defense algorithm and is also transferable across CNNs.

Keywords: Transformed domain attacks · Resiliency · Transferability · Wavelet · CNN · Object recognition

1 Introduction

Convolutional neural networks (CNNs) for image classification are known to utilize both high and low frequency information [32,39]. Goodfellow et al. [20] show that the CNN activations are sensitive towards high-frequency information present in an image. It is also shown that some neurons are sensitive towards the upper right stroke, while some are activated for the lower edge. Furthermore, Geirhos et al. [15] have shown that the CNN trained on ImageNet [11] are highly biased towards texture (high-frequency) and shape of the object (low-frequency). We hypothesize that if an attacker can manipulate the frequency information presented in an image, it can fool CNN architectures as well. With this motivation,

© Springer Nature Switzerland AG 2020
A. Bartoli and A. Fusiello (Eds.): ECCV 2020 Workshops, LNCS 12535, pp. 152–168, 2020.
https://doi.org/10.1007/978-3-030-66415-2_10

we propose a novel method of adversarial example generation that utilizes the low-frequency and high-frequency information individually or in combination. To find the texture and shape information, a wavelet-based decomposition is an ideal choice which yields multi-resolution high-frequency and low-frequency images. Therefore, the proposed method incorporates wavelet decomposition to obtain multiple high and low-frequency images and adversarial noise is added to individual or combined wavelet components through gradient descent learning to generate an adversarial example. Since almost every CNN learns these kinds of features; therefore, the attack generated by perturbing the high frequency (edge) information makes it easily transferable to different networks. In brief, the key highlights of this research are:

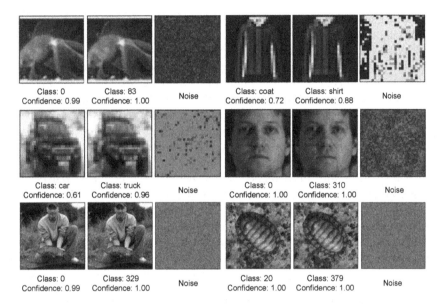

Fig. 1. Fooling of CNN model using the proposed attack on a broad range of databases including object recognition (Tiny ImageNet [45], ImageNet [11], CIFAR-10 [28]), face identification (Multi-PIE [23]), and Fashion data classification (Fashion-MNIST [42]). In each image set, the first image is the clean image, the second is an adversarial image, and the last is the adversarial noise. It can be clearly observed that the proposed attack is able to fool the networks with high confidence score.

- a novel class of adversarial example generation is proposed by decomposing the image into low-frequency and high-frequency information via wavelet transform;
- extensive experiments concerning multiple databases including ImageNet [11], CIFAR-10 [28], and Tiny ImageNet [45],
- multiple CNN models including ResNet [24] and DenseNet [25] are used to showcase the effectiveness of the proposed WaveTransform;

– the robustness of the proposed attack is evaluated against a recent complex adversarial defense.

Figure 1 shows the effectiveness of the proposed attack on multiple databases covering color and gray-scale object images to face images. The proposed attack can fool the network trained on each data type with high confidence. For example, on the color object image (the first image of the top row), the model predicts the correct class (i.e., 0) with confidence 0.99, while, after the attack, the network misclassifies it to the wrong category (i.e., 83) with confidence 1.00.

2   Related Work

Adversarial generation algorithms presented in the literature can be divided into the following categories: (i) gradient-based, (ii) optimization-based, (iii) decision boundary-based, and (iv) universal perturbation.

Goodfellow et al. [20] proposed a fast attack method that calculated the gradient of the image concerning the final output and pushed the image pixels in the direction opposite to the sign of the gradient. The adversarial noise vector can be defined as: $\eta = \epsilon sign(\nabla_x J_\theta(x, l))$, where ϵ controls the magnitude of perturbation, ∇_x represents the gradient of image x with respect to network parameters θ. The perturbation vector η is added in the image to generate the adversarial image. The above process is applied for a single step, which is less effective and can easily be defended [30]. Therefore, several researchers have proposed the variant where the noise is added iteratively [29], [31], and with momentum [13]. Moosavi-Dezfooli et al. [34] have proposed a method that can transfer clean images from their decision boundaries to some other, belonging to a different class. The attacks are performed iteratively using a linear approximation of the non-linear decision boundary. Carlini and Wagner [8] presented attacks by restricting the L_2 norm of an adversarial image. The other variant, such as L_∞ and L_1, are also proposed; however, they are found to be less effective as compared to L_2. Similar to L_2 norm minimization, Chen et al. [10] have proposed the elastic norm optimization attack, which is the combination of L_2 and L_1 norm. Goswami et al. [21,22] presented several black-box attacks to fool the state-of-the-art face recognition algorithms. Later, both adversarial examples detection and mitigation algorithms are also presented in the paper. Agarwal et al. [3] shown the use of filtering operations in generating adversarial noise in a network-agnostic manner.

Other popular adversarial generation algorithms are based on generative networks [41], and EoT [5]. The application of an adversarial attack is not restricted to 2D object recognition but also explored for semantic segmentation [43], 3D recognition [40], audio classification [9], text recognition [14], and reinforcement learning [6]. Goel et al. [19] have developed an adversarial toolbox for the generation of adversarial perturbations and defense against them. The details of the existing algorithms can be found in the survey papers presented by Yuan et al. [47] and Singh et al. [37].

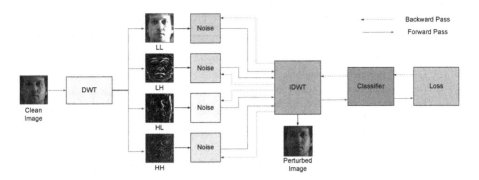

Fig. 2. Schematic diagram of the proposed *'WaveTransform'* adversarial attack algorithm. DWT and IDWT are the forwards and inverse discrete wavelet decomposition. The noise is added to the desired wavelet subband and optimized to increase the loss of the network. LL represents the low pass subband. LH, HL, and HH represent the high-pass subbands in horizontal, vertical, and diagonal directions.

3   Proposed WaveTransform Attack Algorithm

Adversarial attacks generally modify the image in the spatial domain. In this research, we propose a new class of attack termed as WaveTransform where the image is first transformed into the frequency (scale) domain using wavelet decomposition. A digital image is composed of low frequency and high-frequency information, where the role of each frequency component might be different in its spatial representation. With this observation, high and low-frequency bands are perturbed such that the reconstructed image is an adversarial example but visually close to the clean example. The proposed attack can be defined using the following equation:

$$min\ \alpha\|\mathcal{I}_{org} - \mathcal{I}_{pert}\|_\infty + \mathcal{L}(\mathcal{F}(\mathcal{I}_{pert}), t) \tag{1}$$

where, $\mathcal{I}_{org}, \mathcal{I}_{pert} \in [0, 1]$ represent the clean and perturbed images, respectively. α is the loss term trade-off parameter, $\mathcal{L}$ is the classification loss function of the target CNN classifier $\mathcal{F}$, and t is the target label. The aim is to find an adversarial image $\mathcal{I}_{pert}$ that maximizes the classification error for a target label while keeping the noise imperceptible to the human observer.

A discrete wavelet transform (DWT) is applied on $\mathcal{I}_{org}$ to obtain the $LL, LH, HL,$ and HH subbands, using low pass and high pass filters. The LL band contains low frequency information. Whereas LH, HL, and HH contain the high frequency information in horizontal, vertical, and diagonal directions, respectively. These subbands are then modified by taking a step in the direction of the sign of the gradient of the subbands concerning the final output vector. The image is then reconstructed with the modified subbands using an inverse discrete wavelet transform $(IDWT)$, to obtain the desired image $\mathcal{I}_{pert}$. As shown in Fig. 2, the attack is performed iteratively to find an adversarial image with

---

Algorithm 1: Subband Updating (Proposed Adversarial Attack)

Initialization:
- Let the selected subbands be expressed by θ, for a particular image $\mathcal{I}_j$.
- Let the perturbed image be $\mathcal{I}_j' \leftarrow IDWT(\theta)$
- Let l_j be the ground truth label of the image.
- Let r be the number of random restarts taken, k be the number of steps to optimize the objective function
- Let γ be the step size of the update and let n be the minibatch size.
- Let the CNN model be expressed as $\mathcal{F}$.
- Let ϵ be the maximum amount of noise that may be added to $\mathcal{I}_j'$, such that $\mathcal{I}_j' \in [\mathcal{I}_j - \epsilon, \mathcal{I}_j + \epsilon]$

for *restarts in r* **do**

    Initialize $\mathcal{I}_j'$ by adding random noise to $\mathcal{I}_j$ from range $[-\epsilon, \epsilon]$

    for *steps in k* **do**

        Obtain subbands (θ) by decomposing $\mathcal{I}_j'$.

$$\theta_{low}, \theta_{high} \leftarrow DWT(\mathcal{I}_j')$$

        Update subband(s) to maximize classification error by gradient ascent using the term:

$$\theta \leftarrow \theta + \gamma(\, sign(\, \nabla_\theta \mathcal{L}(\, \mathcal{F}(\, IDWT(\theta_{low}, \theta_{high}))), l_j))$$

        $x \leftarrow IDWT(\theta_{low}, \theta_{high})$

        Project x into valid range by clipping pixels and update $\mathcal{I}_j'$;

        $\mathcal{I}_j' \leftarrow x$;

        if $\mathcal{F}(x) \neq l_j$ **then**

            └ Return $\mathcal{I}_j'$

Return $\mathcal{I}_j'$

---

minimal distortion. It is ensured that $\mathcal{I}_{pert}$ remains a valid image after updating its wavelet subbands by projecting the image back onto a L_∞ ball of valid pixel values such that $\mathcal{I}_{pert} \in [0, 1]$. If the noise that can be added or removed, is already limited to ϵ, we add another clipping operation limiting pixel values such that $\mathcal{I}_{pert} \in [\mathcal{I}_{org} - \epsilon, \mathcal{I}_{org} + \epsilon]$. Since, in this setting, there is no need to minimize the added noise explicitly, we also fix the trade-off parameter to $\alpha = 0$. Based on this, we propose our main method called Subband Updating, where particular subbands obtained by the discrete wavelet transform of the image are updated using projected gradient ascent. The proposed 'WaveTransform' adversarial attack algorithm is described in Algorithm 1.

4   Experimental Setup

The experiments are performed using multiple databases and CNN models. This section describes the databases used to generate the adversarial examples, CNN models used to report the results and parameters for adversarial attack and defense algorithms.

Databases: The proposed method is evaluated with databases comprising a wide range of target images: Fashion-MNIST (F-MNIST) [42], CIFAR-10 [28], frontal-image set of Multi-PIE [23], Tiny-ImageNet [45], and ImageNet [11]. Fashion-MNIST comprises low-resolution grayscale images of 10 different apparel categories. CIFAR-10 contains low-resolution RGB images of 10 different object categories. Multi-PIE database has high-resolution RGB images of 337 individuals and Tiny-ImageNet [45] contains 10,000 images from over 200 classes from the ILSVRC challenge [36]. To perform the experiments on ImageNet, the validation set comprising 50,000 images are used. These datasets also vary in color space, CIFAR-10 and Tiny-Imagenet contain color images while F-MNIST contains gray-scale images.

Table 1. Architecture of the custom model used for Fashion-MNIST experiments. [42].

Layer type	Output size	Description
Batch Norm 2D	28×28	Channels 1, affine False
Conv 2D	28×28	5×5, 10, stride 1
Max Pool 2D	24×24	Kernel 2×2
ReLU	23×23	–
Conv 2D	23×23	5×5, 20, stride 1
Max Pool2D	21×21	Kernel 2×2
ReLU	20×20	–
Dropout 2D	20×20	Dropout prob 0.2
Flatten	400×1	Convert to a 1D vector
Linear	400×1	320, 10
Output	10×1	Final logits

CNN Models and Implementation Details: Recent CNN architectures with high classification performance are used for the experiments. For Multi-PIE, we use a ResNet-50 model [24] pretrained on VGG-Face 2 [7] and an InceptionNet-V1 [38] pretrained on CASIA-Webface [46]. For CIFAR-10, pretrained ResNet-50 [24] and DenseNet-121 [25] are used, pretrained on the same. For Fashion-MNIST, a 10-layer custom CNN, as described in Table 1, has been used, and a pretrained ResNet-50 is used with Tiny-ImageNet and ImageNet. The standard models are fine-tuned, replacing the last layer of the network to match the number of classes in the target database and then iterating over the training split of the data for 30 epochs using the Adam [27] optimizer with a learning rate of 0.0001 and batch size of 128. Standard train-validation-test splits are used for CIFAR-10, Fashion-MNIST, and Tiny-ImageNet databases. From the Multi-PIE database [23], 4753 training images, 1690 validation, and 3557 test images are randomly selected. All the models use images in the range [0, 1], and the experimental results are summarized on the test split of the data, except for ImageNet and Tiny-ImageNet, where experimental results are reported on the validation split.

Attack Parameters: In the experiments, each attack follows the same setting unless mentioned. Cross-entropy is used as the classification loss function $\mathcal{L}$. The SGD [26] optimizer is used to calculate the gradient of the subbands concerning the final logits vector used for classification. The experiments are performed using multiple different wavelet filters including Haar, Daubechies (db2 and db3), and Bi-orthogonal. Before computing the discrete wavelet transform, input data is extrapolated by zero-padding. Each attack runs for 20 iterations with 20 restarts, where the adversarial image is initialized with added random noise. This is referred to as random restarts by Madry et al. [31], where the attack algorithm finally returns the first valid adversarial image produced between all restarts. The maximum amount of noise that may be added to (or removed from) a clean image is fixed at $\epsilon = 8.0/255.0$ in terms of L_∞ norm for all the attacks. The step size of the subband update is fixed at $\gamma = 0.05$.

5   Results and Observations

This section describes the results corresponding to original and adversarial images generated via perturbing the individual or combined wavelet components. Extensive analysis has been performed to understand the effect of different filters with wavelet transformation. To demonstrate the effectiveness of the transformed domain attack, we have compared the performance with prevalent pixel-level attacks and recent steganography based attacks. We have also evaluated the transferability of the proposed attack and resiliency against a recent defense algorithm [39].

5.1   Effectiveness of WaveTransform

To evaluate the effectiveness of attacking different subbands, we performed experiments with individual subbands and different combinations of subbands. Figure 3 shows samples of clean and adversarial images corresponding to individual subband from the Multi-PIE [23] and Tiny-ImageNet [45] databases. Individual wavelet components of both image classes help understand the effect of adversarial noise on each frequency information. While the noise in the low-frequency image is quasi subtle, it is visible in the high-frequency components. Among the high-frequency components, the HH component yields the highest amount of distortion. It is interesting to note that the final adversarial images

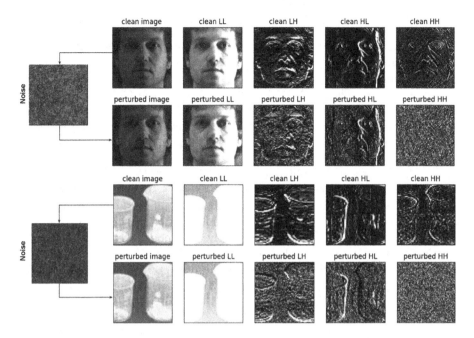

Fig. 3. Illustrating the individual wavelet components of the adversarial images generated using clean images from Multi-PIE [23] and Tiny-ImageNet [45] databases. While the adversarial images are visually close to the clean images; the individual high-frequency components (LH, HL, and HH) clearly show that the noise is injected to fool the system. The wavelet components corresponding to the HH subband show the maximum effect of the adversarial noise.

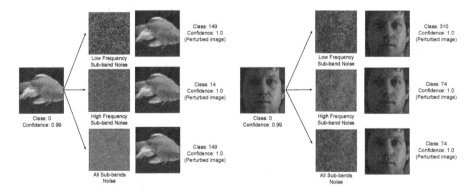

Fig. 4. Adversarial noise generated by attacking different subbands and adding to the clean image to obtain the corresponding adversarial images. Images are taken from Tiny-ImageNet [45] (left) and Multi-PIE [23] (right). It is observed that adversarial images generated using low-frequency or high-frequency or both the components, are effective in fooling the CNN with high confidence.

are close to their clean counterpart. Figure 4 shows adversarial images generated by perturbing the different frequency components of an image. Adversarial image, whether created from low-frequency perturbation or high-frequency, can fool the classifier with high confidence.

Table 2 summarizes the results on each database for the clean as well as the adversarial images. The ResNet-50 model trained on the CIFAR-10 database yields 94.38% object classification accuracy on clean test images. The performance of the model decreases drastically when any of the wavelet frequency band is perturbed. For example, when only the low frequency band is corrupted, the model fails and can classify 3.11% test images only. The performance drops further when all the high subbands (HL, LH, and HH) are perturbed and yields only 1.03% classification accuracy. The results show that each element is essential, and perturbing any component can significantly reduce the network performance. Similarly, on the Tiny-ImageNet [45], the proposed attack can fool the ResNet-50 model almost perfectly. The model, which yields 75.29% object recognition accuracy on clean test images, gives 0.01% accuracy on adversarial images. On the Multi-PIE database, the ResNet-50 model yields 99.41% face identification accuracy, which reduces to 0.06% when both low and high-frequency components are perturbed.

Table 2. Classification rates (%) of the original images and adversarial images generated by attacking different wavelet subbands. The ResNet-50 model is used for CIFAR-10 [28], Multi-PIE [23] and Tiny-ImageNet [45]. The results on F-MNIST [42] are reported using custom CNN (refer Table 1). Bold values represent the best fooling rate achieved by perturbing all subbands, and 'underline' value represents if the fooling rate is the same with all subbands perturbation.

Dataset	CIFAR-10	F-MNIST	Multi-PIE	Tiny-ImageNet
Original accuracy	**94.38**	**87.88**	**99.41**	**75.29**
LL subband attack	3.11	59.04	0.08	<u>0.01</u>
LH subband attack	7.10	78.51	<u>0.06</u>	<u>0.01</u>
HL subband attack	6.56	72.73	0.10	<u>0.01</u>
HH subband attack	13.77	80.56	0.10	0.54
High subbands attack	1.03	70.04	0.08	<u>0.01</u>
All subbands attack	**0.16**	**58.36**	**0.06**	**0.01**

On the Fashion-MNIST [42] database, the proposed attack reduces the model accuracy from 87.88% to 58.36%. In comparison to other databases, the drop on the F-MNIST database is low, which can be attributed to the lack of high textural and object shape information. It is also interesting to note that the model used on F-MNIST is much shallower as compared to the models used for other databases. While the deeper models give higher recognition accuracy as compared to the shallow model; they also find more sensitivity against adversarial perturbations in comparison to the shallow model [30]. The results reported in Table 2 corresponds to a *'white-box'* scenario where an attacker has complete access to the classification network.

Importance of Filter: A filter is a critical part of DWT; therefore, to understand which types of filters are useful in crafting the proposed attack, we have performed experiments with multiple types of filters: Haar, Daubechies (db2 and db3), and Bi-orthogonal. Across the experiments on each database and CNN models, it is observed that 'Haar' is more effective in comparison to other filters in reducing the classification performance. For example, on the F-MNIST [42] database, the Haar filter reduces the model accuracy to 58.36% from 87.88%, which is at least 1.61% better than Daubechies and Bi-orthogonal.

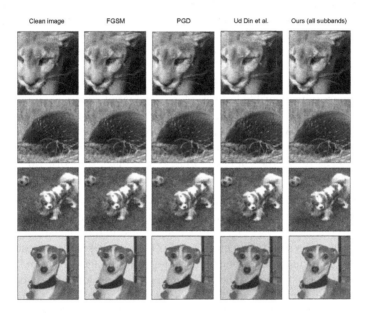

Fig. 5. Comparison of the adversarial images generated using the proposed and existing attacks including FGSM, PGD and Ud Din et al. [12] on the ImageNet. Images used have been resized and center-cropped to make them of size $224 \times 224 \times 3$.

5.2 Comparison with Existing Attack Algorithms

We next compare the performance of WaveTransform with pixel-level attacks and recent wavelet based attacks in literature. Figure 5 shows the adversarial images generated using the proposed, existing pixel level attacks FGSM and PGD, and steganography attack by Ud Din et al. [12].

Pixel-Level Attacks: While most of the existing adversarial attack algorithms work at the pixel level, i.e., in the image space only; the proposed attack works at the transformation level. Therefore, we have also compared the performance of the proposed attack with popular methods such as Projected Gradient Descent (PGD) [31] and Fast Gradient Sign Method (FGSM) [20] with $\epsilon = 0.03$ in terms of accuracy and image degradation. Image degradation metrics such as Universal Image Quality Index (UIQI) [50] is a useful measure for attack quality. An adversarial example with a higher UIQI (with the maximum being 100, for the original image), is perceptually harder to distinguish from the clean image. On the CIFAR-10 database, while the proposed attack with perturbation on both low and high-frequency subbands reduces the performance of ResNet-50 to 0.16% from 94.38%, existing PGD and FGSM reduce the performance to 0.06% and 46.27%, respectively. Similarly, on the ImageNet validation database, the proposed attack reduces the performance of ResNet-50 to 0.05% from 76.13%. On the contrary, the existing PGD and FGSM attacks reduce the recognition accuracy to 0.01% and 8.2%, respectively. The experiments show that the proposed attack can either surpass the existing attack or perform comparably on both databases.

While the perturbation strength both in the existing and proposed attacks is fixed to quasi imperceptible level, we have evaluated the image quality of the adversarial examples. The average UIQI computed from the adversarial examples computed on the CIFAR-10 and ImageNet databases show a value of more than 99. The higher value (close to maximum, i.e., 100) shows that both existing and proposed attacks retain the quality of images and make the noise imperceptible to humans.

Comparison with Recent Attack: The closest attack to the proposed attack is recently proposed by Yahya et al. [44] and Ud Din et al. [12]. These attacks are based on the concept of steganography, where a watermark image referred to as a secret image is embedded in the clean images using wavelet decomposition. The performance of the model is dependent on the secret image. To make the attack highly successful, i.e., to reduce the CNN's recognition performance, a compelling steganography image is selected based on its fooling rate on the target CNN. However, the proposed approach has no requirement of an additional watermark image and learns the noise vector from the network itself. Since Yahya et al. [44] have shown the effectiveness of the attack on the simple MNIST database only, we have compared the performance with Ud Din et al. [12]. They have evaluated their method on a validation set of ImageNet.

To maintain consistency, the experiments are performed on a validation set of ImageNet with ResNet-50. Along with visual comparison, the results are also compared using fooling ratio as the evaluation metric, which is defined as

$$\psi = \frac{|\{f(x_i + \eta) \neq f(x_i)\}|}{M}, \forall i \in \{1, 2, ..., M\} \qquad (2)$$

where f is a trained classifier, x_i is a clean image from the database, M is the total number of samples, and η is the adversarial noise. Using the best steganography image, the attack by Ud Din et al. [12] on a pretrained ResNet-50 achieves a fooling ratio of 84.77% whereas, the proposed attack achieves a fooling ratio of 99.95%.

DenseNet to ResNet

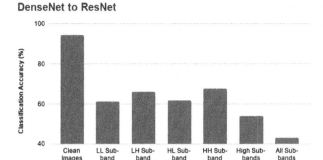

ResNet to DenseNet

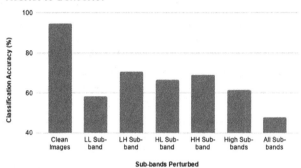

Fig. 6. Illustrating transfer capability of the proposed attack using CIFAR-10 database [28]. The graph on the right shows the results of adversarial images generated on ResNet-50 being tested on DenseNet-121. The plot on the left shows the results of adversarial images generated on DenseNet-121 being tested on ResNet-50. The performance of ResNet-50 and DenseNet-121 are degraded upto 43.02% from 94.38% and 47.82% from 94.76%, respectively.

5.3   Transferability and Resiliency of WaveTransform

Finally, we evaluate the transferability and resiliency of the proposed attack on multiple databases.

Transferability: In the real-world settings, the attacker might not know the target CNN model, which he/she wants to fool. In such a scenario, to make the attack more practical, it is necessary to evaluate its effectiveness with an unseen testing network - the adversarial images generated using one model are used to fool another unseen model. The scenario refers to 'black-box' setting in adversarial attack literature [4] where an attacker does not have access to the target model. The experiments are performed on the CIFAR-10 [28] and Multi-PIE [23] databases.

Table 3. Classification rates (%) for the original and adversarial images generated by attacking different wavelet subbands in the presence of kernel defense [39]. ResNet-50 model is used for CIFAR-10 [28] and the results on F-MNIST [42] are reported using the custom CNN.

Database		Original	Wavelet subbands					
			LL	LH	HL	HH	High	All
CIFAR-10	Before defense	**94.38**	3.11	7.10	6.56	13.77	1.03	**0.16**
	After defense	**91.92**	2.42	5.73	5.03	10.05	0.65	**0.11**
F-MNIST	Before defense	**87.88**	59.04	78.51	72.73	80.56	70.04	**58.36**
	After defense	**81.29**	57.99	72.74	69.05	74.84	66.76	**57.84**

For CIFAR-10 [28], two state-of-the-art CNN models are used, i.e., ResNet-50 and DenseNet-121, and the results are summarized in Fig. 6. The ResNet model yields 94.38% accuracy on clean images of CIFAR-10 [28]; on the other hand, DenseNet gives 94.76% classification accuracy. When the adversarial images generated using the ResNet model are used for classification, the performance of the DenseNet model reduces to 46.94%. Similar performance reduction can be observed on the performance of the ResNet model when the adversarial images generated using the DenseNet model are used. The adversarial images generated by perturbing all the high-frequency wavelet bands reduce the classification accuracy up to 51.36%. The sensitivity of the network against the unseen attack generated models shows the practicality of the proposed attack. Other than that, when adversarial examples are generated using the ResNet on Multi-PIE [23] and used for classification by InceptionNet [38], the performance of the network reduces by 26.77%. The perturbation of low-frequency components hurts the performance most in comparison to the modification of high-frequency components. The highest reduction in accuracy across the unseen testing network is observed when both low and high-frequency components are perturbed.

WaveTransform works by corrupting low frequency and high frequency information contained in the image. It is well understood that the low frequency information corresponds to the high level features learned in the deeper layers of the network [15,32]. Moosavi-Dezfooli et al. [33] have shown that the high level features learned by different models tend to be similar. We assert that since the proposed method perturbs low frequency information that is used across models, it shows good transferability.

Adversarial Resiliency: With the advancement in the adversarial attack domain, researchers have proposed several defense algorithms [1,2,16–18,35]. We next evaluate the resiliency of the attack images generated using the proposed WaveTransform against the recently proposed defense algorithm by Wang et al. [39][1]. The concept of the defense algorithm is close to the proposed attack, thus

---

[1] Original codes provided by the authors are used to perform the experiment.

making it a perfect fit for evaluation. The defense algorithm performs smoothing of the CNN neurons at earlier layers to reduce the effect of adversarial noise.

Table 3 summarizes the results with the defense algorithm on CIFAR10 and F-MNIST databases. Interestingly, we observe that in the presence of the defense algorithm, the performance of the network is further reduced. We hypothesize that, while the proposed attack is perturbing the frequency components, the kernel smoothing further attenuates the noise and yields a higher fooling rate. This phenomenon can also be seen from the accuracy of clean images. For example, the ResNet model without defense yields 94.38% accuracy on CIFAR-10, which reduces to 91.92% after defense incorporation. The proposed attack can fool the defense algorithm on each database. For example, on the CIFAR-10 database, the proposed attack reduces the accuracy up to 0.16% and it further reduces to 0.11% after the defense. Similar resiliency is observed on the F-MNIST database as well.

The PGD attack, which shows a similar reduction in the performance on the CIFAR-10 database, is found less resilient against the defense proposed by Wang et al. [39]. The defense algorithm can successfully boost the recognition performance of ResNet-50 by 30%; whereas, the proposed attack is found to be resilient against the defense.

The robustness of the proposed attack is also evaluated against state-of-the-art defense methods such as Madry et al. [31] and Zhang et al. [49] on the CIFAR-10 database. The defense model presented by Madry et al. [31] utilizes the ResNet-50 model, which yields 87.03% accuracy on clean images of the database, but the accuracy significantly reduces to 60.81% when the proposed attack is applied. Similarly, the accuracy of the defended WideResNet [48] model by Zhang et al. [49] reduces to 62.73% from 84.92%.

6   Conclusion

High and low frequency components present in an image play a vital role when they are processed by deep learning models. Several recent research works have also highlighted that CNN models are highly sensitive towards high and low-frequency components. The attack generation algorithms in the literature generally learn the additive noise vector without considering the individual frequency components. In this research, intending to understand the role of different frequencies, we have proposed a novel attack by decomposing the images using discrete wavelet transform and adding learned adversarial noise in different frequency subbands. The experiments using multiple databases and deep learning models show that the proposed attack poses a significant challenge to the classification models. The proposed attack is further evaluated under unseen network training-testing settings to showcase its real-world application. Other than that, the proposed WaveTransform attack is found to be challenging to mitigate/defend.

Acknowledgements. A. Agarwal was partly supported by the Visvesvaraya PhD Fellowship. R. Singh and M. Vatsa are partially supported through a research grant from MHA, India. M. Vatsa is also partially supported through Swarnajayanti Fellowship by the Government of India.

References

1. Agarwal, A., Singh, R., Vatsa, M., Ratha, N.: Are image-agnostic universal adversarial perturbations for face recognition difficult to detect? In: IEEE BTAS, pp. 1–7 (2018)
2. Agarwal, A., Vatsa, M., Singh, R.: The role of sign and direction of gradient on the performance of CNN. In: IEEE CVPRW (2020)
3. Agarwal, A., Vatsa, M., Singh, R., Ratha, N.: Noise is inside me! Generating adversarial perturbations with noise derived from natural filters. In: IEEE CVPRW (2020)
4. Akhtar, N., Mian, A.: Threat of adversarial attacks on deep learning in computer vision: a survey. IEEE Access **6**, 14410–14430 (2018)
5. Athalye, A., Engstrom, L., Ilyas, A., Kwok, K.: Synthesizing robust adversarial examples. In: ICML, pp. 284–293 (2018)
6. Behzadan, V., Munir, A.: Vulnerability of deep reinforcement learning to policy induction attacks. In: Perner, P. (ed.) MLDM 2017. LNCS (LNAI), vol. 10358, pp. 262–275. Springer, Cham (2017). https://doi.org/10.1007/978-3-319-62416-7_19
7. Cao, Q., Shen, L., Xie, W., Parkhi, O.M., Zisserman, A.: VGGFace2: a dataset for recognising faces across pose and age. In: IEEE FG, pp. 67–74 (2018)
8. Carlini, N., Wagner, D.: Towards evaluating the robustness of neural networks. In: IEEE S&P, pp. 39–57 (2017)
9. Carlini, N., Wagner, D.: Audio adversarial examples: targeted attacks on speech-to-text. In: IEEE S&PW, pp. 1–7 (2018)
10. Chen, P.Y., Sharma, Y., Zhang, H., Yi, J., Hsieh, C.J.: EAD: elastic-net attacks to deep neural networks via adversarial examples. In: AAAI (2018)
11. Deng, J., Dong, W., Socher, R., Li, L.: ImageNet: a large-scale hierarchical image database. In: IEEE CVPR, pp. 710–719 (2009)
12. Din, S.U., Akhtar, N., Younis, S., Shafait, F., Mansoor, A., Shafique, M.: Steganographic universal adversarial perturbations. Pattern Recogn. Lett. **135**, 146–152 (2020)
13. Dong, Y., et al.: Boosting adversarial attacks with momentum. In: IEEE CVPR, pp. 9185–9193 (2018)
14. Gao, J., Lanchantin, J., Soffa, M.L., Qi, Y.: Black-box generation of adversarial text sequences to evade deep learning classifiers. In: IEEE S&PW, pp. 50–56 (2018)
15. Geirhos, R., Rubisch, P., Michaelis, C., Bethge, M., Wichmann, F.A., Brendel, W.: ImageNet-trained CNNs are biased towards texture; increasing shape bias improves accuracy and robustness. In: ICLR (2019)
16. Goel, A., Agarwal, A., Vatsa, M., Singh, R., Ratha, N.: DeepRing: protecting deep neural network with blockchain. In: IEEE CVPRW (2019)
17. Goel, A., Agarwal, A., Vatsa, M., Singh, R., Ratha, N.: Securing CNN model and biometric template using blockchain. In: IEEE BTAS, pp. 1–6 (2019)
18. Goel, A., Agarwal, A., Vatsa, M., Singh, R., Ratha, N.: DNDNet: reconfiguring CNN for adversarial robustness. In: IEEE CVPRW (2020)

19. Goel, A., Singh, A., Agarwal, A., Vatsa, M., Singh, R.: SmartBox: benchmarking adversarial detection and mitigation algorithms for face recognition. In: IEEE BTAS (2018)

20. Goodfellow, I.J., Shlens, J., Szegedy, C.: Explaining and harnessing adversarial examples. arXiv preprint arXiv:1412.6572 (2014)

21. Goswami, G., Ratha, N., Agarwal, A., Singh, R., Vatsa, M.: Unravelling robustness of deep learning based face recognition against adversarial attacks. In: AAAI, pp. 6829–6836 (2018)

22. Goswami, G., Agarwal, A., Ratha, N., Singh, R., Vatsa, M.: Detecting and mitigating adversarial perturbations for robust face recognition. Int. J. Comput. Vision **127**, 719–742 (2019). https://doi.org/10.1007/s11263-019-01160-w

23. Gross, R., Matthews, I., Cohn, J., Kanade, T., Baker, S.: Multi-pie. I&V Comp. **28**(5), 807–813 (2010)

24. He, K., Zhang, X., Ren, S., Sun, J.: Deep residual learning for image recognition. In: IEEE CVPR, pp. 770–778 (2016)

25. Huang, G., Liu, Z., Van Der Maaten, L., Weinberger, K.Q.: Densely connected convolutional networks. In: IEEE CVPR, pp. 4700–4708 (2017)

26. Kiefer, J., Wolfowitz, J., et al.: Stochastic estimation of the maximum of a regression function. Ann. Math. Stat. **23**(3), 462–466 (1952)

27. Kingma, D.P., Ba, J.: Adam: a method for stochastic optimization. In: ICLR (2015)

28. Krizhevsky, A.: Learning multiple layers of features from tiny images (2009)

29. Kurakin, A., Goodfellow, I., Bengio, S.: Adversarial examples in the physical world. In: ICLR-W (2017)

30. Kurakin, A., Goodfellow, I., Bengio, S.: Adversarial machine learning at scale. In: ICLR (2017)

31. Madry, A., Makelov, A., Schmidt, L., Tsipras, D., Vladu, A.: Towards deep learning models resistant to adversarial attacks. In: ICLR (2018)

32. Matthew, D., Fergus, R.: Visualizing and understanding convolutional neural networks. In: ECCV, pp. 6–12 (2014)

33. Moosavi-Dezfooli, S.M., Fawzi, A., Fawzi, O., Frossard, P.: Universal adversarial perturbations. In: IEEE CVPR, pp. 1765–1773 (2017)

34. Moosavi-Dezfooli, S.M., Fawzi, A., Frossard, P.: DeepFool: a simple and accurate method to fool deep neural networks. In: IEEE CVPR, pp. 2574–2582 (2016)

35. Ren, K., Zheng, T., Qin, Z., Liu, X.: Adversarial attacks and defenses in deep learning. Engineering 1–15 (2020)

36. Russakovsky, O., et al.: ImageNet large scale visual recognition challenge. IJCV **115**(3), 211–252 (2015)

37. Singh, R., Agarwal, A., Singh, M., Nagpal, S., Vatsa, M.: On the robustness of face recognition algorithms against attacks and bias. In: AAAI SMT (2020)

38. Szegedy, C., et al.: Going deeper with convolutions. In: IEEE CVPR, pp. 1–9 (2015)

39. Wang, H., Wu, X., Huang, Z., Xing, E.P.: High-frequency component helps explain the generalization of convolutional neural networks. In: IEEE/CVF CVPR, pp. 8684–8694 (2020)

40. Xiang, C., Qi, C.R., Li, B.: Generating 3D adversarial point clouds. In: IEEE CVPR, pp. 9136–9144 (2019)

41. Xiao, C., Li, B., Zhu, J.Y., He, W., Liu, M., Song, D.: Generating adversarial examples with adversarial networks. arXiv preprint arXiv:1801.02610 (2018)

42. Xiao, H., Rasul, K., Vollgraf, R.: Fashion-MNIST: a novel image dataset for benchmarking machine learning algorithms. arXiv preprint arXiv:1708.07747 (2017)

43. Xie, C., Wang, J., Zhang, Z., Zhou, Y., Xie, L., Yuille, A.: Adversarial examples for semantic segmentation and object detection. In: IEEE ICCV, pp. 1369–1378 (2017)
44. Yahya, Z., Hassan, M., Younis, S., Shafique, M.: Probabilistic analysis of targeted attacks using transform-domain adversarial examples. IEEE Access **8**, 33855–33869 (2020)
45. Yao, L., Miller, J.: Tiny imagenet classification with convolutional neural networks. CS 231N **2**(5), 8 (2015)
46. Yi, D., Lei, Z., Liao, S., Li, S.Z.: Learning face representation from scratch. arXiv preprint arXiv:1411.7923 (2014)
47. Yuan, X., He, P., Zhu, Q., Li, X.: Adversarial examples: attacks and defenses for deep learning. IEEE TNNLS **30**(9), 2805–2824 (2019)
48. Zagoruyko, S., Komodakis, N.: Wide residual networks. arXiv preprint arXiv:1605.07146 (2016)
49. Zhang, H., Yu, Y., Jiao, J., Xing, E.P., Ghaoui, L.E., Jordan, M.I.: Theoretically principled trade-off between robustness and accuracy. In: ICML (2019)
50. Wang, Z., Bovik, A.C.: A universal image quality index. IEEE Signal Process. Lett. **9**(3), 81–84 (2002)

Robust Super-Resolution of Real Faces Using Smooth Features

Saurabh Goswami$^{(\boxtimes)}$, Aakanksha, and A. N. Rajagopalan

Indian Institute of Technology, Madras, India
{ee18s003,ee18d405}@smail.iitm.ac.in, raju@ee.iitm.ac.in

Abstract. Real low-resolution (LR) face images contain degradations which are too varied and complex to be captured by known downsampling kernels and signal-independent noises. So, in order to successfully super-resolve real faces, a method needs to be robust to a wide range of noise, blur, compression artifacts etc. Some of the recent works attempt to model these degradations from a dataset of real images using a Generative Adversarial Network (GAN). They generate synthetically degraded LR images and use them with corresponding real high-resolution (HR) image to train a super-resolution (SR) network using a combination of a pixel-wise loss and an adversarial loss. In this paper, we propose a two module super-resolution network where the feature extractor module extracts robust features from the LR image, and the SR module generates an HR estimate using only these robust features. We train a degradation GAN to convert bicubically downsampled clean images to real degraded images, and interpolate between the obtained degraded LR image and its clean LR counterpart. This interpolated LR image is then used along with it's corresponding HR counterpart to train the super-resolution network from end to end. Entropy Regularized Wasserstein Divergence is used to force the encoded features learnt from the clean and degraded images to closely resemble those extracted from the interpolated image to ensure robustness.

1 Introduction

Face Super-Resolution (SR) is an important preprocessing step for high-level vision tasks like facial detection and recognition. Robustness to real degradations like noise, blur, compression artifacts, etc. is one of the key aspects of the human visual system and hence highly desirable in machine vision applications as well. Incorporating this robustness in the Super-Resolution stage itself would ease all the downstream tasks. Unfortunately, most of the face SR methods are trained with a fixed degradation model (downsampling with a known kernel and adding noise) that is unable to capture the complexity and diversity of real degradations

Electronic supplementary material The online version of this chapter (https://doi.org/10.1007/978-3-030-66415-2_11) contains supplementary material, which is available to authorized users.

© Springer Nature Switzerland AG 2020
A. Bartoli and A. Fusiello (Eds.): ECCV 2020 Workshops, LNCS 12535, pp. 169–185, 2020.
https://doi.org/10.1007/978-3-030-66415-2_11

and hence performs poorly when applied on real degraded face images. This problem becomes more pronounced when the image is extremely small. Since most of the useful information is degraded, it further increases the ambiguity in reconstruction process. Previous methods such as [3, 32, 34] use facial heatmaps and facial landmarks as priors to reduce ambiguity. [30, 33] leverage autoencoders to build networks which are robust to synthetic noise and [18] leverage wavelet transform to train a network which is robust to gaussian noise. However, none of the above methods have been proven to be robust to real degradation except [3]. In [4], a Generative Adversarial Network (GAN) was trained to generate realistically degraded Low-Resolution (LR) versions of clean High-Resolution (HR) face images and another GAN was trained to super-resolve the synthetic degraded images to their corresponding clean HR counterparts. To the best of our knowledge, this is the only previous work which super-resolves real degraded faces without the aid of any facial priors. However, we observed that [4] produces visually different outputs for different degradations. This can be attributed to the fact that the network sees every degraded image independently and there is no explicit constraint to extract the same features from different degraded versions of the same image.

In this paper, we focus on incorporating robustness to degradations in the task of tiny face super-resolution without the need of a face specific prior and without a dataset of *degraded LR-clean HR* image pairs. Premised upon the observation that humans are remarkably adept at registering different degrdaded versions of the same image as visually similar images, we prepend a smooth feature extractor module to our Super-Resolution (SR) module. Since our feature extractor is smooth with respect to real degradations, its output does not vary wildly when we move from clean images to degraded images. The SR module which produces clean HR images from features extracted by the smooth feature extractor, thus, produce similar images regardless of the degradation. Features which remain smooth under degradations are also features that are common between clean and degraded LR. So, our network, in essence, learns to look at features which are similar between clean and degraded LR.

Following [4], we train a GAN to convert clean LR images to corresponding degraded LR images. One training iteration of our network involves two back-propagations. During the first backpropagation, we update parameters of both modules of our network to learn a super-resolution mapping from an interpolated LR (by combining clean and degraded LR) to its corresponding clean HR. The interpolation is carried out to avoid having the network overfit one of two LR domains (clean and degraded). During the second backpropagation, we minimize the Entropy Regularized Wasserstein Distance between features extracted from clean as well as degraded LR and those extracted from interpolated LR. The interpolation also helps in ensuring smoothness of the feature extractor.

During test time, we put an image (clean or degraded) through the feature extractor module first and then feed the extracted features to the SR Module to get the corresponding super-resolved image. Since the extracted features do not change significantly between clean and degraded images, the super-resolution output for a degraded image does not change significantly from that of a clean image. We perform tests to visualise the robustness of our network as well as smoothness of the features extracted by our feature extractor.

The main contributions of our work are as follows:

- We propose a new approach for unpaired face SR where the SR network relies on features that are common between corresponding clean and degraded images.
- To the best of our knowledge, ours is the first work that handles robustness separately from the task of super-resolution. This enables us to explicitly enforce robustness constraints on the network.

2   Related Works

Single Image Super-Resolution (SISR) is a highly ill-posed inverse problem. Traditional methods mostly impose handcrafted constraints as priors to restrict the space of solutions. With the availability of Large-scale Image Datasets and the consistent success of Convolutional Neural Networks (CNNs), learning (rather than handcrafting) a prior from a set of natural images became a possibility. Many such approaches have been explored subsequently.

2.1   Deep Single Image Super-Resolution

We classify all the deep Single Image Super-Resolution (SISR) methods in two broad categories - (i) deep Paired SISR and (ii) deep Unpaired SISR. In paired SISR, corresponding pairs of LR and HR images are available and the network is evaluated on its ability to estimate an HR image given its LR counterpart. Most of the available deep paired SISR networks are trained under a setting where LR images are generated by downsampling HR images (from datasets such as Set5, Set14, DIV2K [1], BSD100 [2] etc.) using a known kernel (often bicubic). These networks are trained using either a pixel wise Mean Squared Error (MSE) loss e.g. [13,21,26], L_1 loss e.g. [36], Charbonnier loss e.g. [22] or a combination of pixel-wise L_1 loss, perceptual loss [20] and adversarial loss [16] e.g. [10,23,29]. Even though these networks perform really well in terms of PSNR and SSIM, and the GAN based ones produce images that are highly realistic, these networks often fail when they are applied on real images with unseen degradations such as realistic noise and blur. To address this, RealSR [6] dataset was introduced in NTIRE 2019 Challenge [5] containing images taken at two different focal lengths of a camera. Networks like [14,15,19] were trained on this dataset and are therefore robust to real degradations.

On the other hand, in unpaired SISR, only the LR images are available in the dataset. In [35], a CycleGAN [37] was trained to denoise the input image and another one to finetune a pretrained super-resolution network. In [27], a CycleGAN was trained to generate degraded versions of clean images and a super-resolution network was then trained using pairs of synthetically degraded

LR and clean HR images.

However, all these networks are meant for natural scenes and not faces in particular. Humans are highly sensitive to even the subtlest changes when it comes to human faces, making the task of perceptually super-resolving human faces a challenging and interesting one.

2.2   Deep Face SISR

General SR networks as the ones mentioned above, often produce undesired artifacts when applied on faces. Hence, paired face SR networks often rely on face-specific prior information to subdue the artifacts and make the network focus on important features.

Networks like [3,10,32,34] rely on facial landmarks and heatmaps to impose additional constraints on the output whereas [12] leverage HR exemplars to produce high-quality HR outputs. On the other hand, networks like [9,18] rely on pairs of LR and HR face images to perceptually super-resolve faces. Even though the above methods are somewhat robust to noise and occlusion, they are not equipped well enough to handle noises which are as complex and as diverse as those in real images. [30,33] leverage capsule networks and transformative autoencoders to class-specifically super-resolve noisy faces but the noises are synthetic. As of yet, there seems to be no dataset with paired examples of degraded LR and clean HR images of faces available. As a result, in recent years, there has been a shift in face SISR methods from paired to unpaired. Recently, with the release of *Widerface* [31] dataset of real low-resolution faces and the wide availability of high resolution face recognition datasets such *AFLW*[28], *VGGFace2*[7] and *CelebAMask-HQ* [24], Bulat et al. [4] propose a training strategy where a High-to-Low GAN is trained to convert instances from clean HR face images to corresponding degraded LR images and a Low-to-High GAN is then trained using synthetically degraded LR images and their clean HR counterparts. This method is highly effective since it does not require facial landmarks or heatmaps for faces (as they are not available for real face images captured in the wild).

However, despite producing sharp outputs, it is not very robust as different outputs are obtained for different degradations in the LR images. In order to explicitly impose robustness, we introduce a smooth feature extractor module to extract similar features from a degraded LR image and its clean LR counterpart. This enabled us to get features that are more representative of the actual face in the image and is significantly less affected by the degradations in the input.

2.3   Robust Feature Learning

Our work builds on the existing methods in robust feature learning. Haoliang et al. [25] extract robust features from multiple datasets of similar semantic contents by minimizing Maximum Mean Discrepancy (MMD) between features extracted from these datasets. Cemgil et al. [8], achieve robustness by forcing Entropy Regularized Wasserstein Distance to be low between features extracted from clean images and their noisy counterparts. None of these works handle

Super-Resolution where rigorous compression using an autoencoder may hurt the reconstruction quality. We propose a method of incorporating robust feature learning in super-resolution without requiring any face specific prior information.

3   Proposed Method

3.1   Motivation

Super-Resolution networks which are meant to be used on real facial images need to satisfy two criteria: (i) they need to be robust under real degradations, (ii) they should preserve the identity and pose of a face. Deep state-of-the-art super-resolution networks usually derive the LR images by bicubically downsampling HR images. Hence, an SR network trained on pairs of LR and HR images used for training fail to meet the first criterion. On the other hand, SR networks trained with real degradations fail to satisfy the second criterion. Noting the fact that the face recognition ability of us humans does not change very significantly with reasonably high degradation in images, it should be possible to find features that remain invariant under significant degradation and train a super-resolution network that would rely only on these features. Now, features which are robust to degradations would also be smooth under the said degradations. So, by enforcing explicit smoothness constraints on the extracted features, we can ensure robustness.

3.2   Overall Pipeline

We have a clean High-Resolution dataset Y_c and a degraded Low-Resolution dataset X_d. We obtain clean Low-Resolution dataset, X_c, corresponding to Y_c, by downsampling every image in Y_c with a bicubic downsampling kernel. So every x_c in X_c is a downsampled version of some y_c in Y_c, using the equation

$$x_c = (y_c * k)_{\downarrow s} \tag{1}$$

where, k is the bicubic downsampling kernel and s is the scale factor. Following [4], we train a Degradation GAN, G_d to convert clean samples from X_c to look like they have been drawn from the degraded LR dataset X_d. We call this synthetic degraded LR dataset $\widehat{X_d}$ and samples in this dataset $\widehat{x_d}$. So,

$$\widehat{x_d} = G_d(x_c, z) \in \widehat{X_d} \quad \forall \quad x_c \in X_c \tag{2}$$

where $z \in Z$ is an additional vector input which is sampled from a distribution Z to capture the one-to-many relation between HR and degraded LR images.

Our network basically comprises 2 modules - (i) Feature Extractor Module (f) and (ii) Super-Resolution Module (g). During training, we first sample an x_c from X_c and generate one of its degraded counterparts $\widehat{x_d} = G_d(x_c, z)$ using G_d. We then combine these two LR images with a mixing coefficient α

$$x_{in} = \alpha x_c + (1 - \alpha)\widehat{x_d} \tag{3}$$

where $0 < \alpha < 1$. We, then, put x_{in} through the convolutional feature extractor $f(x)$ and the SR module $g(h)$ to estimate the corresponding clean HR output $\widehat{y}_c$ and do a backpropagation.

$$h_{in} = f(x_{in}), \quad \widehat{y}_c = g(h_{in}) \tag{4}$$

To ensure smoothness of f under real degradations, we extract features h_c and h_d from x_c and $\widehat{x_d}$

$$h_c = f(x_c) \quad h_d = f(\widehat{x_d}) \tag{5}$$

and minimize the Entropy Regularized Wasserstein Distance (Sinkhorn distance) between (h_c, h_{in}) and (h_d, h_{in}) through another backpropagation. We recalculate h_{in} during this operation as well. Figure 1 shows a schematic diagram of our approach.

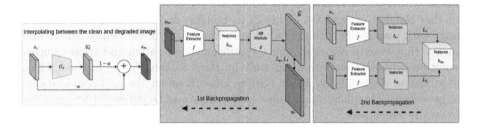

Fig. 1. The proposed approach.

Here, if we use $\alpha = 0$, since the entire network, during the first backpropagation, would be trained using pairs of synthetically degraded LR and clean HR samples, it may end up learning a mapping that would fail to preserve the identity of a face. However, if we take $\alpha = 1$, the network may exhibit preference to the domain of clean LR images. So, we needed an input LR image which is not as sharp as x_c but not as degraded as $\widehat{x_d}$ either. Since the edges in x_c are much sharper than those in $\widehat{x_d}$, x_{in} continues to appear reasonably clean even when $\alpha < 0.5$. This is why we do not sample α from a distribution since that might end up giving one domain advantage over the other and keep it fixed at 0.3 since $\alpha = 0.3$ appears to us to have struck the right balance between the two LR domains visually.

Also, using $0 < \alpha < 1$, enables us to apply the smoothness constraint between (h_c, h_{in}) and (h_d, h_{in}) which is a better way to ensure smoothness than imposing smoothness constraint on pairs of (h_c, h_d).

3.3   Modeling Degradations with Degradation GAN

Owing to the complex and diverse nature of real degradations, it is extremely difficult to mathematically model them by hand. So, following previous works [4,27], we train a GAN (termed Degradation GAN) to model real degradations.

Generator. Our Degradation GAN Generator, shown in Fig. 3, G_d, has 3 down-sampling blocks, each consisting of a ResNet block followed by a 3×3 convolutional with *stride* = 2, and 3 upsampling blocks each comprising ResNet blocks followed a Nearest Neighbour Upsampling layer and a 3×3 convolutional block with *stride* = 1. The downsampling and upsampling paths are connected through skip connections. All the ResNet blocks used in Generator follow the structure described in Fig. 2. Our Generator takes a bicubic downsampled image x_c and an n dimensional random vector z sampled from a normal distribution. We expand each of the n dimensions of the random vector into a channel of size $H \times W$ (filled with a single value) where H and W are the height and width of every image. We concatenate the expanded volume with the image and feed it to the generator.

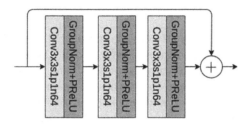

Fig. 2. ResNet block used in Degradation GAN Generator.

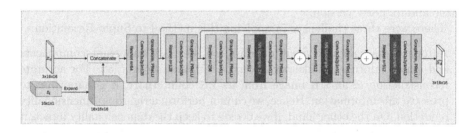

Fig. 3. The overall architecture of the Degradation GAN Generator G_d.

Critic. We use the same discriminator used in [23]. Since we train the degradation GAN as Wasserstein GAN [17], we replace the Batch Normalization layers with Group Normalization and remove the last Sigmoid layer. Following the nomenclature, we call it critic instead of discriminator.

Loss Functions. We train the degradation GAN as a Wasserstein GAN with Gradient Penalty (WGAN-GP) [17]. So, the critic is trained by minimizing the following loss function:

$$L_D = (\mathbb{E}_{x \in \widehat{X_d}}[D(x)] - \mathbb{E}_{x \in X_d}[D(x)]) + \lambda \mathbb{E}_{\widehat{x} \sim \mathbb{P}_{\widehat{x}}}[(\|\nabla_{\widehat{x}} D(x)\|_2 - 1)^2] \quad (6)$$

where, as in [17], the first term is the original critic loss and the second term is the gradient-penalty.

To maintain the correspondence between inputs and outputs of the generator, we add a Mean Square Loss (MSE loss) term to the WGAN loss in the objective function L_G of the generator.:

$$L_G = \lambda_{WGAN} L_{WGAN} + \lambda_{MSE} L_{MSE} \tag{7}$$

where,

$$L_{WGAN} = -\mathbb{E}_{(x_c,z) \in (X_c, Z)}[G_d(x_c, z)] \tag{8}$$

and

$$L_{MSE} = \|x_c - G_d(x_c, z)\|^2 \tag{9}$$

3.4   Super-Resolution Using Smooth Features

The main objective of our work is to design a robust SR network the performance of which does not deteriorate under real degradation. Our network has two modules (a) a fully-convolutional feature extractor f and (b) a fully-convolutional SR module g. The way we achieve robustness is by making the feature extractor smooth under degradations and making the SR module g rely solely on the features extracted by f. In [8], Cemgil et al. proposed a method to enforce robustness on the representations learnt by Variational Autoencoders (VAEs). They trained a VAE to reconstruct clean images and minimized the Entropy Regularized Wasserstein Distance between representations derived from a clean image and its noisy version.

There were three challenges in applying this method to Super-Resolution:

1. Autoencoders compress an input down to its most important components and ignore information like occlusion, background objects, etc. For accurate reconstruction of an HR image from its LR counterpart, it is important to preserve this information. Hence, we cannot perform a rigorous dimensionality reduction. On the other hand, if we decide to keep the dimensionality intact, it will make it harder to achieve robustness since there are too many distractors. So it is important to choose a reduction factor that will achieve the best trade-off between reconstruction and robustness.
2. They train their network for synthetic noise. However, real degradation involves signal-dependent noise, blur and a variety of other artifacts. So, we need a mechanism to realistically degrade images.
3. As we show in the supplementary material, despite smoothness constraint and despite the network being reasonably robust, naively applying their method on SR still leaves a gap between its performance on clean and degraded images. So, we need a better training strategy.

To address (1), we try a number of different dimensionality reduction choices $(1\times, 4\times, 16\times)$ for the features extracted by f and we observed that $4\times$ dimensionality reduction attains the best trade-off. To address (2), we train a degradation GAN to realistically degrade clean images. To address (3), we interpolate between a clean image (x_c) and one of its synthetically degraded counterpart ($G_d(x_c, z)$) using a mixing coefficient α as shown in Eq. 3. We call this x_{in}.

Feature Extractor f: Our feature extractor consists of 4 Residual Channel Attention (RCA) downsampling and 2 upsampling blocks. As shown in Fig. 4, there are 2 skip connections. It is a fully convolutional module which takes an LR image of dimension $3 \times 16 \times 16$ at the input and produces a feature volume of dimension $64 \times 4 \times 4$. In Fig. 4, 'RCA, n64' denotes an RCA block with 64 output channels and 'Conv3x3, s2 p1 n64' denotes a 3×3 convolutional layer with $stride = 2$, $padding = 1$ and 64 output channels.

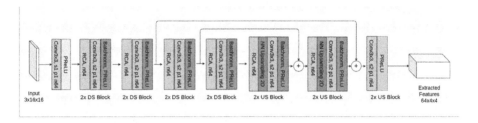

Fig. 4. Feature extractor f.

Super-Resolution Module g: Our Super-Resolution module consists of 6 upsampling blocks and 2 DenseBlocks as shown in Fig. 5. The upsampling blocks comprise a Pixel-Shuffle layer, a convolution layer, a Batch-Normalization layer and a PReLU layer. The DenseBlocks contain a number of Residual Channel Attention (RCA) blocks and Residual Channel Attention Back-Projection (RCABP) blocks connected in a dense fashion as in [19]. In Fig. 5, 'Pixel Shuffle (2)' denotes 2x pixel-shuffle upsampling layer and 'RCABP, n64' stands for an RCABP block with 64 heatmaps at the output.

During one forward pass, we pass a minibatch of x_{in} through our feature extractor f to produce the feature volume h_{in}. We put h_{in} through our Super-Resolution module g to produce a high resolution estimate $\hat{y}_c$ and do a back propagation through both g and f. This ensures that the features are useful for SR. Since x_{in} is neither as clean as x_c nor as severely degraded as $\widehat{x_d}$, the possibility of our SR network being biased to any one of the domains is eliminated.

After the first backpropagation, we put one minibatch each of $x_c, \widehat{x_d}$ and x_{in} (again) through f, as shown in Eq. 5, and calculate the Sinkhorn Distance [11] (which calculates the Entropy Regularized Wasserstein Divergence) between (h_c, h_{in}) and (h_d, h_{in}),

$$L_c = Sinkhorn(h_c, h_{in}), \quad L_d = Sinkhorn(h_d, h_{in}) \tag{10}$$

Using a combination of L_c and L_d as a loss function, we backpropagate through f one more time to enforce smoothness under degradations.

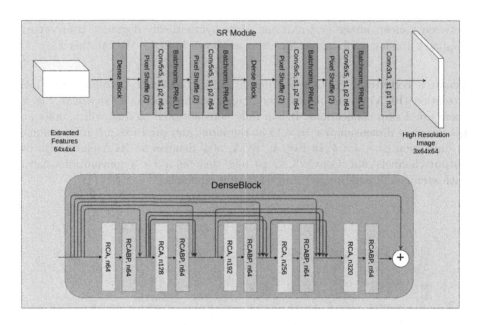

Fig. 5. Architecture of SR module g and DenseBlock.

Like our Degradation GAN, we train our robust super-resolution network (during the first back propagation) like a Wasserstein GAN. So, the objective function here is a combination of adversarial loss (L_{adv}), pixel-level L_1 loss (L_p) and a perceptual loss [20] (L_f) computed between features extracted from the estimated ($\widehat{y}_c$) and ground-truth (y_c) HR images through a subset of VGG16 network. Hence, the overall objective function optimized during the first back propagation is

$$L_{sr} = \lambda_p L_p + \lambda_f L_f + \lambda_{adv} L_{adv} \tag{11}$$

where,

$$L_p = \|y_c - \widehat{y}_c\|_1 \tag{12}$$

$$L_f = \|f_{vgg}(y_c) - f_{vgg}(\widehat{y}_c)\|_1 \tag{13}$$

$$L_{adv} = -\mathbb{E}_{x_{in} \sim \widehat{\mathbb{P}}_x}[D_{sr}(g(f(x_{in})))] \tag{14}$$

with f_{vgg} being a subset of VGG16 network, $\mathbb{P}_x$ being the distribution described by x_{in} and D_{sr} being the critic comparing the generated HR images with the ground-truth HR images. The architecture of D_{sr} is same as the critic of degradation GAN and it is trained with the following loss function:

$$L_{DSR} = (\mathbb{E}_{\widehat{y}_c \sim \mathbb{P}_y}[D_{sr}(\widehat{y}_c)] - \mathbb{E}_{y_c \in Y_c}[D(y_c)]) + \lambda \mathbb{E}_{\widehat{y} \sim \mathbb{P}_{\widehat{y}}}[(\|\nabla_{\widehat{y}} D(\widehat{y})\|_2 - 1)^2] \tag{15}$$

where $\mathbb{P}_y$ is the distribution generated by the outputs of our network and $\mathbb{P}_{\widehat{y}}$ is the distribution of samples interpolated between $\widehat{y}_c$ and y_c.

For the second back propagation, we optimize a combination of the Sinkhorn Distances mentioned earlier

$$L_{robust} = \lambda_c L_c + \lambda_d L_d \tag{16}$$

Since the second backpropagation is only through f, it does not directly affect the mapping learnt by g and only makes f smooth under degradations.

4 Experiments

4.1 Training Details

We use two-time step update for both our Degradation GAN and Robust Super-Resolution Network. For both D and D_{sr}, we start with a learning rate of 4×10^{-4} and decrease them by a factor of 0.5 after every 10000 iterations. For all the other networks (G_d, f, g) we set the initial training at 10^{-4} and decay it by a factor of 0.5 after every 10000 iterations.

For all networks, we use Adam Optimizer with $\beta_1 = 0.0$ and $\beta_2 = 0.9$. For every 5 updates of discriminators, we update the corresponding generator networks once. We try out a number of different values of λ and the ones that worked best for us are $[\lambda_{WGAN} = 0.05, \lambda_{MSE} = 1, \lambda_p = 1, \lambda_f = 0.5, \lambda_{adv} = 0.05, \lambda_c = 0.3, \lambda_d = 0.7]$. For G_d, we sample z from a $16-$dimensional multivariate normal distribution with zero mean and unit standard deviation.

4.2 Datasets

We train our network for $4\times$ super-resolution $(s = 4)$. However, our robustness strategy is not scale dependent. For training our network, we used two datasets: one with degraded images and the other with clean images. To make the degraded image dataset, we randomly sample 153446 images from the *Widerface* [31] dataset. This dataset contains face images with a wide range of degradations such as: varying degrees of noise, extreme poses and expressions, occlusions, skew, non-uniform blur etc. We use 138446 of these images for training and 15000 for testing. While compiling the clean dataset, to make sure it is diverse enough in terms of poses, occlusions, skin colours and expressions, we combined the entire *AFLW* [28] dataset with 60000 images from *CelebAMask-HQ* [24] dataset and 100000 images from *VGGFace2* [7] dataset. To obtain clean LR images, we simply downsample images from the clean dataset.

4.3 Results

To assess the accuracy as well as robustness of our work, we test our network on 3 different datasets - (i) Bicubically-Degraded Dataset, (ii) Synthetically-Degraded Dataset and (iii) Real-Degraded Dataset.

1. **Bicubically-Degraded Dataset:** To compile this dataset, we randomly sample 4000 HR images from the clean Facial Recognition Datasets as mentioned above. We bicubically downsample them to obtain paired LR-HR images. Evaluation on this dataset tells us about the reconstruction accuracy of our SR network.

   As shown in Fig. 6a, ESRGAN [29] performs best on this dataset since it was trained on bicubic downsampled images. Interestingly, the results of [4] appear to be a little different from the HR ground truth in terms of identity. We observe this in all our experiments. Also, their outputs contain a lot of undesired artifacts. Our outputs are faithful to the ground-truth HR and contain less artifacts. However, our method performs a little poorly in terms of PSNR and SSIM as shown in Table 1. However, since we focus primarily on the robustness part of the problem, the strength of our network becomes evident with the evaluation of robustness.

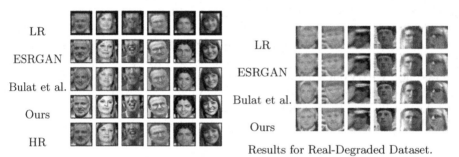

Results for Bicubic-Degraded Dataset.

Results for Real-Degraded Dataset.

Fig. 6. Comparison of results.

Table 1. Comparison of PSNR/SSIM on Bicubic-Degraded Dataset

Method	PSNR	SSIM
ESRGAN [29]	**25.25**	**0.351**
Bulat et al. [4]	24.55	0.220
Ours	24.48	0.218

2. **Synthetically-Degraded Dataset:** This dataset contains the same HR images as the Bicubically-Degraded Dataset but we obtain 5 different synthetically degraded LR versions of each HR image using our degradation GAN. We perform two tests on this dataset:
   - **Robustness Test:** Here, for each HR image, we put all 5 degraded LR images through our SR network. This test shows us how the output changes for different degradation. The similarity between the outputs will tell us how robust our network is to realistic degradations which is the focus of our work.

LR

ESR
GAN
[29]

Bulat et
al. [4]

Ours

HR

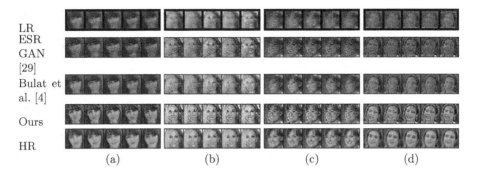

(a)   (b)   (c)   (d)

Fig. 7. Comparison of robustness.

As shown in Fig. 7, these images are extremely degraded. ESRGAN [29] gives the worst performance on this dataset. [4] produces slightly different-looking faces for different degradations. Our method, however, produces outputs that look similar for all these degradations. This shows that our network is robust to realistic degradations.

– **Smoothness Test:** This experiment enables us to visualise the smoothness of our feature extractor (f). Here, we combine every degraded LR image in the dataset with their bicubically downsampled counterparts using 5 different values of α_i such as $[0.0, 0.2, 0.4, 0.8, 1.0]$ to create a set of 5 different images $(\{x_{mix}\})$. Since α is the coefficient we use to mix clean and corresponding degraded images, by gradually varing α from 0 to 1 and noting the output, we get an idea of how adept our network is at maintaining its output as we gradually move from a clean image, through increasingly degraded images, to one of its realistically degraded versions. If our network manages to maintain its output without altering its overall appearance (changing pose, identity, etc.), it would mean that the learnt features are smooth and robust to degradations.

$$x^i_{mix} = \alpha_i x_c + (1 - \alpha_i)\widehat{x_d} \qquad (17)$$

Figure 8 shows a comparison of the output of our network with those of [4] and [29]. The outputs of ESRGAN [29] becomes increasingly worse as α decreases. The outputs of [4] changes significantly as α goes from 0 to 1, sometimes even producing different faces. The output of our network does not undergo any visually significant changes. This establishes the features learnt by the feature extractor are smooth under realistic degradation. Figure 7 shows that ESRGAN [29] consistently performs poorly in terms of robustness than the other two methods. This is expected since it was trained with bicubically downsampled LR images only. The behavior of [4] is interesting. In Fig. 7(a), (b) and (g), it is generating additional facial components that are unrelated to the content of the input. The performance of our network, as shown in (f) and (g), drops a little when the input is heavily degraded but the recognizable features do not change much.

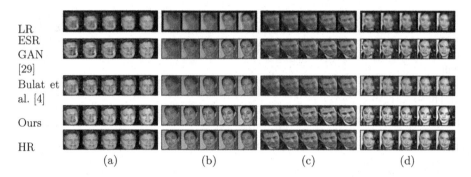

LR

ESR
GAN
[29]

Bulat et
al. [4]

Ours

HR

(a)   (b)   (c)   (d)

Fig. 8. Visualizing smoothness for $\alpha = [0, 0.2, 0.4, 0.8, 1.0]$.

3. **Real-Degraded Dataset:** This dataset contains 15000 images from the *Widerface* Dataset. Performance on this dataset will dictate how effective our method is in super-resolving real degraded facial images.

   As shown in Fig. 6b, our method is able to super-resolve real degraded faces. The outputs of [4] contain undesired artifacts and sometimes exhibit identity discrepancy as well. ESRGAN [29] is able to maintain the identity but the outputs are not sharp. Since we do not have ground-truth HR images for these LR images, we can not compute PSNR/SSIM. So, we use Fretchet Inception Distance (FID) as a metric to assess how close the output is to the target distribution of sharp images. Table 2 shows the FIDs of [4, 29] and our method computed over 15000 images. Lower FID denotes better adherence to target distribution and hence sharper output.

Table 2. Comparison of FID.

Method	FID
ESRGAN [29]	139.2599
Bulat et al. [4]	**74.2798**
Ours	77.1359

As shown in Table 2, our method performs very close to [4] in terms of realness of the output and at the same time, maintains a fixed output under varying degradations. So, our method is robust and at the same time, effective on real degraded faces.

5   Conclusion

We propose a robust super-resolution network that would give consistent output under a wide range of degradations. We train a feature extractor that is able to extract similar features from both bicubically downsampled images and their corresponding realistically degraded counterparts. We perform robustness test to put our claim of robustness to test and smoothness test to visualize the variation in extracted features as we gradually move from a clean to a degraded LR image. There is still room to improve our network for better performance in terms of PSNR/SSIM. In our future works, we will attempt to address this.

References

1. Agustsson, E., Timofte, R.: Ntire 2017 challenge on single image super-resolution: dataset and study. In: The IEEE Conference on Computer Vision and Pattern Recognition (CVPR) Workshops, July 2017
2. Arbelaez, P., Maire, M., Fowlkes, C., Malik, J.: Contour detection and hierarchical image segmentation. IEEE Trans. Pattern Anal. Mach. Intell. **33**(5), 898–916 (2011). https://doi.org/10.1109/TPAMI.2010.161
3. Bulat, A., Tzimiropoulos, G.: Super-fan: integrated facial landmark localization and super-resolution of real-world low resolution faces in arbitrary poses with GANs. CoRR abs/1712.02765 (2017). http://arxiv.org/abs/1712.02765
4. Bulat, A., Yang, J., Tzimiropoulos, G.: To learn image super-resolution, use a GAN to learn how to do image degradation first. In: Ferrari, V., Hebert, M., Sminchisescu, C., Weiss, Y. (eds.) ECCV 2018. LNCS, vol. 11210, pp. 187–202. Springer, Cham (2018). https://doi.org/10.1007/978-3-030-01231-1_12
5. Cai, J., Gu, S., Timofte, R., Zhang, L.: Ntire 2019 challenge on real image super-resolution: methods and results. In: Proceedings of the IEEE Conference on Computer Vision and Pattern Recognition Workshops (2019)
6. Cai, J., Zeng, H., Yong, H., Cao, Z., Zhang, L.: Toward real-world single image super-resolution: a new benchmark and a new model. In: Proceedings of the IEEE International Conference on Computer Vision (2019)
7. Cao, Q., Shen, L., Xie, W., Parkhi, O.M., Zisserman, A.: VGGFace2: a dataset for recognising faces across pose and age. In: International Conference on Automatic Face and Gesture Recognition (2018)
8. Cemgil, T., Ghaisas, S., Dvijotham, K.D., Kohli, P.: Adversarially robust representations with smooth encoders. In: International Conference on Learning Representations (2020). https://openreview.net/forum?id=H1gfFaEYDS
9. Chen, X., Wang, X., Lu, Y., Li, W., Wang, Z., Huang, Z.: RBPNET: an asymptotic residual back-projection network for super-resolution of very low-resolution face image. Neurocomputing **376**, 119–127 (2020). https://doi.org/10.1016/j.neucom.2019.09.079. http://www.sciencedirect.com/science/article/pii/S0925231219313530
10. Chen, Y., Tai, Y., Liu, X., Shen, C., Yang, J.: FSRNet: end-to-end learning face super-resolution with facial priors. CoRR abs/1711.10703 (2017). http://arxiv.org/abs/1711.10703
11. Cuturi, M.: Sinkhorn distances: Lightspeed computation of optimal transportation distances (2013)
12. Dogan, B., Gu, S., Timofte, R.: Exemplar guided face image super-resolution without facial landmarks. CoRR abs/1906.07078 (2019). http://arxiv.org/abs/1906.07078
13. Dong, C., Loy, C.C., He, K., Tang, X.: Image super-resolution using deep convolutional networks. CoRR abs/1501.00092 (2015). http://arxiv.org/abs/1501.00092

14. Du, C., Zewei, H., Anshun, S., Jiangxin, Y., Yanlong, C., Yanpeng, C., Siliang, T., Ying Yang, M.: Orientation-aware deep neural network for real image super-resolution. In: The IEEE Conference on Computer Vision and Pattern Recognition (CVPR) Workshops, June 2019

15. Feng, R., Gu, J., Qiao, Y., Dong, C.: Suppressing model overfitting for image super-resolution networks. In: The IEEE Conference on Computer Vision and Pattern Recognition (CVPR) Workshops, June 2019

16. Goodfellow, I.J., et al.: Generative adversarial networks (2014)

17. Gulrajani, I., Ahmed, F., Arjovsky, M., Dumoulin, V., Courville, A.C.: Improved training of Wasserstein GANs. CoRR abs/1704.00028 (2017). http://arxiv.org/abs/1704.00028

18. Huang, H., He, R., Sun, Z., Tan, T.: Wavelet-SRNET: a wavelet-based CNN for multi-scale face super resolution. In: 2017 IEEE International Conference on Computer Vision (ICCV), pp. 1698–1706 (2017)

19. Jang, D., Park, R.: DenseNet with deep residual channel-attention blocks for single image super resolution. In: 2019 IEEE/CVF Conference on Computer Vision and Pattern Recognition Workshops (CVPRW), pp. 1795–1803 (2019)

20. Johnson, J., Alahi, A., Fei-Fei, L.: Perceptual losses for real-time style transfer and super-resolution (2016)

21. Kim, J., Lee, J.K., Lee, K.M.: Accurate image super-resolution using very deep convolutional networks. CoRR abs/1511.04587 (2015). http://arxiv.org/abs/1511.04587

22. Lai, W., Huang, J., Ahuja, N., Yang, M.: Fast and accurate image super-resolution with deep Laplacian pyramid networks. CoRR abs/1710.01992 (2017). http://arxiv.org/abs/1710.01992

23. Ledig, C., et al.: Photo-realistic single image super-resolution using a generative adversarial network. CoRR abs/1609.04802 (2016). http://arxiv.org/abs/1609.04802

24. Lee, C.H., Liu, Z., Wu, L., Luo, P.: MaskGAN: towards diverse and interactive facial image manipulation. arXiv preprint arXiv:1907.11922 (2019)

25. Li, H., Jialin Pan, S., Wang, S., Kot, A.C.: Domain generalization with adversarial feature learning. In: The IEEE Conference on Computer Vision and Pattern Recognition (CVPR), June 2018

26. Lim, B., Son, S., Kim, H., Nah, S., Lee, K.M.: Enhanced deep residual networks for single image super-resolution. CoRR abs/1707.02921 (2017). http://arxiv.org/abs/1707.02921

27. Lugmayr, A., Danelljan, M., Timofte, R.: Unsupervised learning for real-world super-resolution (2019)

28. Martin Koestinger, Paul Wohlhart, P.M.R., Bischof, H.: Annotated facial landmarks in the wild: a large-scale, real-world database for facial landmark localization. In: proceedings of the First IEEE International Workshop on Benchmarking Facial Image Analysis Technologies (2011)

29. Wang, X., et al.: ESRGAN: enhanced super-resolution generative adversarial networks. In: Leal-Taixé, L., Roth, S. (eds.) ECCV 2018. LNCS, vol. 11133, pp. 63–79. Springer, Cham (2019). https://doi.org/10.1007/978-3-030-11021-5_5

30. Xin, J., Wang, N., Jiang, X., Li, J., Gao, X., Li, Z.: Facial attribute capsules for noise face super resolution (2020)

31. Yang, S., Luo, P., Loy, C.C., Tang, X.: Wider face: a face detection benchmark. In: IEEE Conference on Computer Vision and Pattern Recognition (CVPR) (2016)

32. Yu, X., Fernando, B., Hartley, R., Porikli, F.: Super-resolving very low-resolution face images with supplementary attributes. In: 2018 IEEE/CVF Conference on Computer Vision and Pattern Recognition, pp. 908–917 (2018)
33. Yu, X., Porikli, F.: Hallucinating very low-resolution unaligned and noisy face images by transformative discriminative autoencoders. In: 2017 IEEE Conference on Computer Vision and Pattern Recognition (CVPR), pp. 5367–5375 (2017)
34. Yu, X., Fernando, B., Ghanem, B., Porikli, F., Hartley, R.: Face super-resolution guided by facial component heatmaps. In: Ferrari, V., Hebert, M., Sminchisescu, C., Weiss, Y. (eds.) ECCV 2018. LNCS, vol. 11213, pp. 219–235. Springer, Cham (2018). https://doi.org/10.1007/978-3-030-01240-3_14
35. Yuan, Y., Liu, S., Zhang, J., Zhang, Y., Dong, C., Lin, L.: Unsupervised image super-resolution using cycle-in-cycle generative adversarial networks. In: 2018 IEEE/CVF Conference on Computer Vision and Pattern Recognition Workshops (CVPRW), pp. 814–81409 (2018)
36. Zhang, Y., Li, K., Li, K., Wang, L., Zhong, B., Fu, Y.: Image super-resolution using very deep residual channel attention networks. CoRR abs/1807.02758 (2018). http://arxiv.org/abs/1807.02758
37. Zhu, J., Park, T., Isola, P., Efros, A.A.: Unpaired image-to-image translation using cycle-consistent adversarial networks. CoRR abs/1703.10593 (2017). http://arxiv.org/abs/1703.10593

Improved Robustness to Open Set Inputs via Tempered Mixup

Ryne Roady[1(✉)], Tyler L. Hayes[1], and Christopher Kanan[1,2,3]

[1] Rochester Institute of Technology, Rochester, NY 14623, USA
{rpr3697,tlh6792,kanan}@rit.edu
[2] Paige, New York, NY 10044, USA
[3] Cornell Tech, New York, NY 10044, USA

Abstract. Supervised classification methods often assume that evaluation data is drawn from the same distribution as training data and that all classes are present for training. However, real-world classifiers must handle inputs that are far from the training distribution including samples from unknown classes. Open set robustness refers to the ability to properly label samples from previously unseen categories as novel and avoid high-confidence, incorrect predictions. Existing approaches have focused on either novel inference methods, unique training architectures, or supplementing the training data with additional background samples. Here, we propose a simple regularization technique easily applied to existing convolutional neural network architectures that improves open set robustness without a background dataset. Our method achieves state-of-the-art results on open set classification baselines and easily scales to large-scale open set classification problems.

1 Introduction

Modern supervised classification methods often assume train and test data are drawn from the same distribution and all classes in the test set are present for training. However, deployed models will undoubtedly be exposed to out-of-distribution inputs that do not resemble training samples and these models are expected to robustly handle these novel samples. Performance in this 'open-world' setting is often hidden by current computer vision benchmarks in which the train and test sets have the same classes and the data is sampled from the same underlying sources. One solution to this problem is to develop open set classifiers which have the ability to identify novel inputs that do not belong to any training classes so that they are not assigned an incorrect label [23]. This capability is especially important for the development of safety-critical systems (e.g., medical applications, self-driving cars) and lifelong learning agents that automatically learn during deployment [19].

There are two major paradigms for enabling open set classification in existing convolutional neural network (CNN) architectures. The first paradigm replaces

© Springer Nature Switzerland AG 2020
A. Bartoli and A. Fusiello (Eds.): ECCV 2020 Workshops, LNCS 12535, pp. 186–201, 2020.
https://doi.org/10.1007/978-3-030-66415-2_12

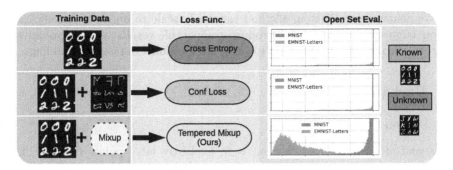

Fig. 1. TEMPERED MIXUP. Traditional methods for building model robustness to unknown classes involves training with a background set, which is problematic for large-scale datasets. Tempered Mixup provides state-of-the-art robustness to novel inputs without the requirement of a representative background set. The model confidence of a LeNet++ model trained to classify MNIST digits is shown for both known and unknown classes here represented by the Extended-MNIST-Letters dataset. Confidence is calculated using only the softmax of the model output and does not rely on a computationally expensive open set inference method to separate known from unknown classes.

the standard closed-set prediction scheme with a new inference mechanism [2, 16,17,24]. The second paradigm is to regularize the classification model during training to enable it to better separate known classes from potential unknowns [8, 15]. For the latter approach, the most effective methods involve training with a large set of background images to penalize an overconfident prediction when a sample from a novel or unknown class is encountered. This approach has been shown to excel at open set classification, but existing results have been limited to small-scale datasets. In part, this is because it is increasingly difficult to construct an *effective background dataset that does not semantically overlap with a large-scale training dataset.

In this paper, we overcome this limitation by proposing a new training approach that penalizes overconfident predictions on samples outside of the known training classes without access to a set of background/unknown inputs. Instead our method hallucinates invalid images using a novel form of the Mixup data augmentation technique [29] that we combine with a unique auxiliary loss function (Fig. 1). **This paper makes the following contributions:**

1. We propose a novel end-to-end training algorithm for regularizing existing CNN architectures for open set classification that does not require the use of an explicit background dataset.
2. We propose a new loss function specifically designed to train a model to be less confident towards samples from unknown classes.
3. We show that our method exceeds methods that use explicit background datasets on standard small-scale benchmark settings for open set classification (e.g., MNIST and CIFAR).

4. We demonstrate that our method can easily scale to open set classification on large-scale datasets and produce comparable results to background set regularization without having to build an additional dataset for training.

2 Background

2.1 Open Set Classification

The goal of open set classification is to explicitly label points that are 'far' from known training classes as an unknown class instead of arbitrarily assigning one of the known categories. Open set classification methods discriminate among K known categories seen during training and reject samples that do not belong to the known categories. Formally, given a training set $D_{train} = \{(X_1, y_1), (X_2, y_2), \ldots, (X_n, y_n)\}$, where X_i is the i-th training input tensor and $y_i \in C_{train} = \{1, 2, \ldots, K\}$ is its corresponding class label, the goal is to learn a classifier $F(X) = (f_1, \ldots, f_k)$, that correctly identifies the label of a known class and separates known from unknown examples:

$$\hat{y} = \begin{cases} \text{argmax}_k\, F(X) & \text{if } S(X) \geq \delta \\ K + 1 & \text{if } S(X) < \delta \end{cases} \tag{1}$$

where $S(X)$ is an acceptance score function that determines whether the input belongs to the training data distribution, δ is a user-defined threshold, and $K + 1$ indicates 'unknown class.' For proper open set testing, the evaluation set contains samples from both the set of classes seen during training and additional unseen classes, i.e., $D_{test} = \{(X_1, y_1), (X_2, y_2), \ldots, (X_n, y_n)\}$, where $y_i \in (C_{train} \bigcup C_{unk})$ and C_{unk} contains classes that are not observed during training.

2.2 Inference Methods for Open Set Classification

Inference methods incorporate open set classification abilities into a pre-trained CNN by creating a unique acceptance score function and threshold that rejects novel inputs [2,3,23]. Current state-of-the-art inference methods often rely on features from multiple layers of the CNN [1,16] and many methods use multiple forward and backward passes through the CNN to improve performance [16,17]. These approaches significantly increase computational and memory requirements during inference [21], which may be sub-optimal for deployed models. We instead focus on building better open set classification performance through model regularization so that, during inference time, a much simpler and computationally efficient method such as confidence thresholding can be used to detect unknown classes.

2.3 Confidence Loss Training for Open Set Classification

One approach for improving open set robustness is training with a confidence loss penalty [8,15], which improves detection of unknown classes by penalizing overconfident predictions on samples that are outside the training classes. As previously observed [8,21], inputs from unknown classes tend to be centered around the origin and have a smaller magnitude than samples from known classes in the deep feature space of a well-trained CNN.

Given an effective background dataset that is representative of unknowns expected to be seen during deployment, confidence loss training [8] collapses novel samples toward the origin of the deep feature space, resulting in lower confidence model outputs. The regularization penalty is a simple addition to standard cross-entropy loss during training, i.e.,

$$\mathcal{L}_{conf} = \begin{cases} -\log S_k(F(X)) & \text{if } X \in D_{train} \\ -\frac{1}{K}\sum_{k=1}^{K} \log S_k(F(X)) & \text{if } X \in D_{bkg} \end{cases} \tag{2}$$

where the first term is a standard cross-entropy loss for known classes (D_{train}) and the second loss term forces the model to push samples from a representative background class (D_{bkg}) toward a uniform posterior distribution.

Multiple approaches have been developed for producing open set images including using alternative datasets which are distinct from the training set [8,20] and even using generative methods to produce images outside of the manifold defined by the training set [15,18]. Naturally, the performance of the open set classifier trained with open set images is tied to the "representativeness" of the background dataset and its similarity to the known training set. The difficulty for even small-scale classification is that the effectiveness of certain background sets cannot be known a priori and seemingly representative background datasets can sometimes fail to produce better

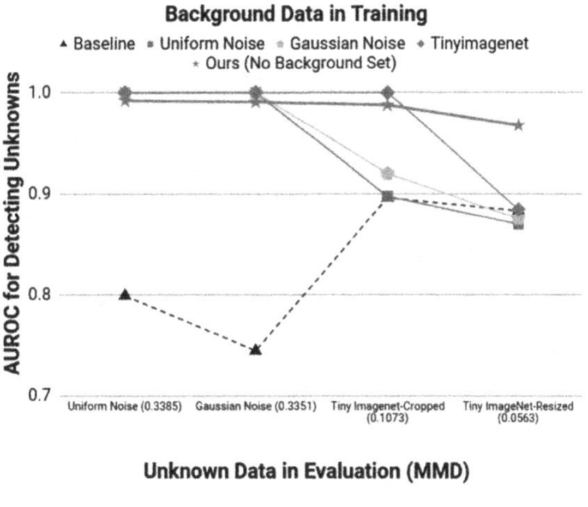

Fig. 2. NOVELTY DETECTION VS BACKGROUND CLASS SELECTION The performance of confidence loss for improving open set performance is dependent on the selected background class. As the background class becomes more dissimilar from the in-distribution data, it becomes less useful for improving model performance on difficult open set classification problems. Our method does not require a background class for training and thus is robust across a wide-range of different input types.

open set robustness. We demonstrate this phenomenon using the CIFAR-10 dataset in Fig. 2. To overcome this limitation, we designed our method around drawing novel samples for confidence loss using data augmentation instead of relying on an explicit background set.

2.4 Mixup Training

Mixup is a regularization approach that combines two separate images from the training set into a single example by forming an elementwise convex combination of the two input samples [29]. Mixup can improve model accuracy [29], model calibration [26], and model robustness to certain types of image corruptions [5]. However, Mixup has not been shown to be beneficial for open set classification, and in [5] Mixup resulted in a 50% reduction in detecting unknown classes versus baseline cross-entropy training.

Mixup is based on the principle of Vicinal Risk Minimization (VRM) [4] where a classification model is trained not only on the direct samples in the training set, but also in the *vicinity* of each training sample to better sample the training distribution. In Mixup, these vicinal training samples $(\tilde{x}, \tilde{y})$ are generated as a simple convex combination of two randomly selected input samples, x_i and x_j:

$$\tilde{x} = \lambda x_i + (1 - \lambda)x_j$$
$$\tilde{y} = \lambda y_i + (1 - \lambda)y_j \tag{3}$$

where y_i and y_j are the associated targets for the selected input samples. The linear interpolation factor $\lambda \in [0, 1]$ is drawn from a symmetric $Beta(\alpha, \alpha)$ distribution where the shape of the distribution is determined by the hyper-parameter, α which trades off training with mostly unmixed examples versus training with averaged inputs and labels. By training with standard cross-entropy loss on these vicinal examples, the model learns to vary the strength of its output between class manifolds. The effect of this training is a substantial improvement in model calibration and accuracy on large-scale image classification tasks [5,26,30]. Our approach uses a variant of Mixup to overcome the need for an explicit background set with confidence loss training.

3 Mixup and Open Set Classification

3.1 Re-balancing Class Targets

As originally proposed, Mixup is a form of model regularization that trains on linear combinations of inputs and targets, and encourages a model to learn smoother decision boundaries between class manifolds. While smoother decision boundaries promote better generalization, they also benefit open set classification. This is because smoother boundaries reduce the likelihood of a model producing a confident but wrong prediction with an input that does not lie on the class manifolds learned from the training set. Mixup essentially turns a single label classification problem into a multi-label problem by creating additional samples through the linear mixture of inputs and feature space embeddings. The question then becomes: is a linear combination of the targets appropriate for a linear combination of features in training a model to produce accurate uncertainty estimates in the space between class manifolds?

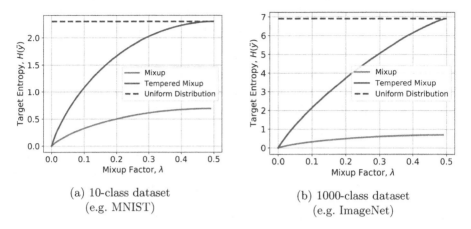

(a) 10-class dataset
(e.g. MNIST)

(b) 1000-class dataset
(e.g. ImageNet)

Fig. 3. TARGET REBALANCING: Ideally, for unknown inputs a classifier's output probabilities will approach a uniform distribution. Using Mixup, samples that are more mixed should have higher entropy, but this does not occur to the extent necessary in the original formulation. Instead, Tempered Mixup uses a novel formulation that ensures the target entropy for mixed examples approaches that of a uniform distribution. This figure demonstrates this in both the 10 class and 500 class setting.

We answer this question by looking at how the target entropy of a model trained via cross-entropy loss changes as a function of the mixing factor λ. As shown in Fig. 3, using a linear combination of targets, Eq. 3, to mix the labels does not capture the increase in uncertainty that we desire for examples that are off of the class manifold, e.g., the highly mixed examples. Instead we can re-balance the target labels with an additional label smoothing term modulated by the interpolation factor, λ, with an adjusted target mixing scheme as follows:

$$\tilde{y}_{tm} = |2\lambda - 1|\tilde{y} + \frac{1 - |2\lambda - 1|}{K} \ , \qquad (4)$$

where $\tilde{y}$ is the normal linear mixing from Eq. 3 and K is the number of known target classes. Using this novel re-balancing approach, we can temper the model confidence for highly mixed samples such that they approach a uniform distribution prediction. As shown in Fig. 3, this approach assigns a much higher target entropy to highly mixed up samples, including when they are mixed from the same class. This tempering effect is magnified as the number of known classes increases.

3.2 Tempered Mixup

Tempered Mixup is an open set classification training method that overcomes the need for a background training set by using a modified form of Mixup with a novel variant of confidence loss regularization. Using Mixup enables the creation of *off-manifold* samples based on the training input distribution, and it enables control over how similar the simulated outliers are to the known classes. This allows the CNN to learn features that are robust to open set classes through a number of different specialized inference methods, including baseline confidence thresholding.

Instead of training with standard cross-entropy (softmax) loss with labels drawn from the convex combination of two randomly selected images as prescribed in the standard Mixup algorithm, Tempered Mixup uses the same mixed up input but a modified version of the auxiliary confidence loss function to regularize how the model maps these between-class inputs in deep feature space. To do this, we apply the Mixup coefficient drawn per sample from a symmetric Beta distribution to a modified confidence loss equation. This allows us to simultaneously minimize the loss for misclassifying samples from the known classes and map unknown samples that are far from the known classes to the origin. The Tempered Mixup loss is given by:

$$\mathcal{L}_{TempMix} = -|2\lambda - 1| \sum_{k=1}^{K} \tilde{y}_k \log \sigma_S(F(\tilde{X}))_k - \zeta \frac{1 - |2\lambda - 1|}{K} \sum_{k=1}^{K} \log \sigma_S(F(\tilde{X}))_k,$$
(5)

where σ_S is the softmax function applied to the vector $F(\tilde{X})$, λ is the sampled mixing interpolation factor, $\tilde{y}$ denotes the linearly mixed targets, K is the number of known classes, and ζ weights the amount of confidence loss applied to highly mixed up samples. Tempered Mixup is a straight-forward extension of traditional mini-batch stochastic gradient descent with cross-entropy loss training for deep neural network models.

3.3   Visualizations of Deep CNN Feature Space to Unknowns

To visually illustrate the benefit of Tempered Mixup in separating known and unknown samples, we trained a simple CNN model (LeNet++ architecture [14]) to classify MNIST digits as known classes and Extended MNIST Letters as unknown classes (overlaid as black points). The CNN architecture has a bottlenecked two dimensional feature space to allow the visualization of the resulting embeddings.

As shown in Fig. 4, the Tempered Mixup model collapses the embedding of samples from the unknown classes towards the origin, thus reducing the overlap (and confusion with) known classes. This is a dramatic improvement over common supervised training methods that improve model robustness including Label Smoothing [25] and Center Loss [27]. Tempered Mixup even improves on methods trained with an explicit background set such as Entropic Open Set and Objectosphere [8].

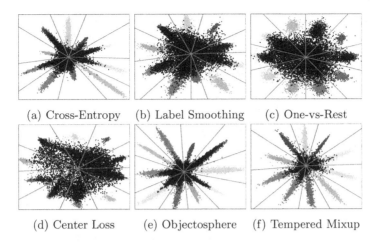

(a) Cross-Entropy   (b) Label Smoothing   (c) One-vs-Rest

(d) Center Loss   (e) Objectosphere   (f) Tempered Mixup

Fig. 4. 2-D visualization of the effect of the different feature space regularization strategies on separating in-distribution and outlier inputs. The in-distribution training set is MNIST, while the unknown samples are from the **Extended-MNIST-Letters** dataset [6]. For Entropic Open Set and Objectosphere, the Omniglot dataset is used as a source for background samples.

4   Experiments

To evaluate the open set robustness of our method against current state-of-the-art techniques, we first compare against baselines established using the MNIST [14] and CIFAR-10 [12] datasets as known classes and samples drawn from similar, but distinct, datasets as unknown classes. We then compare the performance of Tempered Mixup to standard Mixup training and other forms of VRM data augmentation. Finally, we extend our small-scale experiments to show how our method scales to large-scale open set classification problems.

4.1   Open Set Performance Assessment

An open set classifier needs to correctly classify samples from known classes and identify samples from unknown classes. This makes evaluation more complex than out-of-distribution detection, which simplifies the detection task to a binary in/out classification problem.

Our primary metric is the area under the open set classification (AUOSC) curve. It measures the correct classification rate among known classes versus the false positive rate for accepting an open set sample, and has been used as a standard metric for open set classification [8]. The correct classification rate can be viewed as the difference between normal model accuracy and the false negative rate for rejection. Intuitively, AUOSC takes into account whether true positive samples are actually classified as the correct class and thus rewards methods which reject incorrectly classified positive samples before rejecting samples that are correctly classified. In addition to reporting the area under the curve, we also calculate the correct classification rate at a specific false positive rate of 10^{-1}. Finally, we also report the area under the Receiver Operating Characteristic (AUROC) for identifying unknown classes as a measure of pure out-of-distribution detection performance.

4.2 Comparison Methods

We compare our method against the following approaches:

1 **Cross-entropy**: As a baseline, we train each network with standard cross-entropy loss to represent a common feature space for CNN-based models.
2 **One-vs-Rest** [24]: The one-vs-rest training strategy was implemented by substituting a sigmoid activation layer for the typical softmax activation and using a binary cross-entropy loss function. In this paradigm, every image is a negative example for every category it is not assigned to. This creates a much larger number of negative training examples for each class than positive examples.
3 **Label Smoothing** [25]: By smoothing target predictions during cross-entropy training, the model learns to regularize overconfident predictions and produce less confident and more calibrated predictions.
4 **CenterLoss** [27]: A form of model regularization that increases the robustness of the class-conditional feature representation by encouraging tightly grouped class clusters. This is achieved by penalizing the Euclidean distance between samples and their class-mean (inter-class variance). By reducing the inter-class variance, a more precise rejection threshold can theoretically be established for separating known from unknown samples.
5 **Entropic Open Set** [8]: The Entropic Open Set method applies the confidence loss formulation (Eq. 2) using a background class to train the model to reduce model confidence on unknown classes.
6 **Objectosphere** [8]: Objectosphere separates known from unknown classes by training with a background class and reducing the magnitude of learned features for unknown classes. Instead of using the confidence loss formulation, it uses a margin based hinge-loss centered around the origin in deep feature space. This loss reduces the magnitude of features from the background set and increases the magnitude of features for samples from known classes to be larger than a user defined margin. This method was shown to better separate known from unknown samples in network architectures with higher dimensional feature spaces.

For all methods, identification of unknown classes is done by thresholding the maximum class posterior found by passing the model's output through a softmax activation.

4.3 Open Set Baselines

We first study Tempered Mixup on common open set benchmarks. Following the protocol established in [8], the first baseline uses the LeNet++ CNN architecture [14] with MNIST [14] for known classes and a subset of the Extended-MNIST-Letters dataset [6] for unknown classes.

The second benchmark uses a 32-layer Pre-Activation ResNet architecture [10] with CIFAR-10 [12] as known classes and 178 classes from TinyImageNet [13] as unknown classes. TinyImageNet images were all resized to 32×32 images. We removed 22 classes from the original TinyImageNet dataset because they contained semantic overlap with CIFAR-10 classes based on hypernym or hyponyms, which were determined using the Wordnet lexical database [9].

Table 1. OPEN-SET CLASSIFICATION BASELINES. The correct classification rate at a false positive rate for open set classification of 10^{-1} and the areas under the resulting OSC and ROC curves. In these experiments, the positive class used samples from the known set of classes seen during training and the negative class used samples from the unknown classes. Best performance for each experiment and metric is in **bold**.

Experiment	Algorithm	CCR @ FPR 10^{-1}	AUOSC	AUROC
MNIST Unknown: EMNIST-Letters Arch: LeNet++	Baseline	0.7259	0.9066	0.9103
	One-vs-rest	0.9556	0.9654	0.9814
	Label Smoothing	0.8543	0.9315	0.9443
	CenterLoss	0.9633	0.9695	0.9877
	Entropic Open-Set	0.9712	0.9797	0.9892
	Objectosphere	0.9570	0.9739	0.9801
	Tempered Mixup (Ours)	**0.9761**	**0.9821**	**0.9924**
CIFAR-10 Unknown: TinyImageNet Arch: Pre-ResNet-32	Baseline	0.5211	0.7694	0.8105
	One-vs-rest	0.2363	0.7064	0.7559
	Label Smoothing	0.0920	0.6841	0.7283
	CenterLoss	0.4930	0.7613	0.8038
	Entropic Open-Set	0.6766	0.7880	0.8344
	Objectosphere	0.6720	0.8045	**0.8584**
	Tempered Mixup (Ours)	**0.6923**	**0.8099**	0.8503

For both benchmarks, we train all open set classification methods on the known classes. Methods that use a background class (i.e., Entropic Open Set and Objectosphere) are additionally trained on the first 13 classes in the Extended-MNIST-Letters dataset. For the CIFAR-10 baseline, a background class training set is drawn from non-overlapping classes in the CIFAR-100 dataset. All methods are then evaluated on an even split of samples drawn from known and unknown classes. For MNIST, 10000 samples are used as a source of unknowns drawn from the final 13 classes in the Extended-MNIST-Letters dataset. For CIFAR-10, 10000 samples are randomly selected from the TinyImageNet dataset as a source of unknowns.

Results. Tempered Mixup achieves state-of-the-art-results in open set classification without the use of an explicit background class (see Table 1). For the MNIST experiment, all methods show an improvement over the cross-entropy baseline. Tempered Mixup surpasses cross-entropy by 8% in terms of AUOSC and even surpasses confidence loss methods that use a background set that is derived from the same dataset as the unknowns. For CIFAR-10, which uses a modern ResNet v2 architecture, Tempered Mixup achieves more than a 5% improvement in terms of AUOSC over all methods that do not require a background set for training and is state-of-the-art in terms of AUOSC over all evaluated methods including those that train with an additional background set.

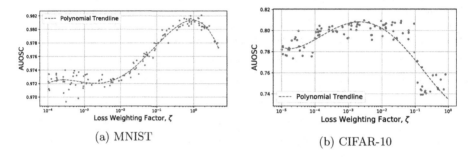

(a) MNIST

(b) CIFAR-10

Fig. 5. EFFECT OF TEMPERED LABELS: The effect of the tempering label smoothing loss term is varied to demonstrate the overall benefit on MNIST (left) and CIFAR-10 (right) Open Set Robustness.

4.4   Additional Evaluations

Performance Improvement from Target Rebalancing. We seek to understand the benefit from target rebalancing towards improving the open set robustness gained from normal Mixup training. To model this effect, we varied the weight applied to the label smoothing term using the loss weighting factor (ζ in Eq. 5). This allowed us to see the performance difference as the confidence loss portion of our formulation is emphasized over standard Mixup training (which is equivalent as $\zeta \to 0$). As Fig. 5 shows, performance improves as we increase the weight of the confidence loss until a point when the confidence loss prevents the model from achieving a high closed set accuracy, thus reducing the overall open set performance.

Alternate Data Augmentation Schemes. Mixup is part of a family of data augmentation approaches which work on the VRM principle. As an alternative to the standard Empirical Risk Minimization formulation, VRM attempts to enlarge the support of the empirical training distribution by creating *virtual* examples through various data augmentation schemes. Other VRM data augmentation schemes have been proposed recently which have demonstrated increased robustness to certain forms of input corruption [5]; however, few have explicitly tested for open set robustness.

The Cutmix strategy [28] overlays a random path from a separate training image and adjusts the target labels during training based on the ratio of the patch area to the original image. Cutout [7] is a similar variation, however the patch is made up of black (zero-valued) pixels. We evaluated our Tempered Mixup formulation against these competing VRM schemes both in their normal formulation and with a tempered target label set where the entropy of the target distribution is adjusted based on the interpolation factor, λ. As results show in Table 2, Tempered Mixup is superior to both Cutmix and Cutout and their tempered variants.

Table 2. ALTERNATE VRM COMPARISON. AUOSC performance of our method versus other VRM data augmentation techniques and their tempered variants. Experiments use either MNIST or CIFAR-10 as known dataset and three different unknown datasets that vary in similarity to the known dataset.

Experiment	Algorithm	Gaussian Noise	FMNIST	EMNIST-Letters
MNIST	Baseline	**0.9878**	0.9848	0.9066
	Cutmix	0.8837	0.8751	0.8172
	Cutout	0.9830	0.9813	0.9028
	Mixup	0.9874	**0.9875**	0.9737
	Tempered Cutmix	0.9805	0.9780	0.9249
	Tempered Cutout	0.9844	0.9829	0.9121
	Tempered Mixup (Ours)	0.9846	**0.9875**	**0.9821**
Experiment	Algorithm	SVHN	LSUN	Tiny ImageNet
CIFAR-10	Baseline	0.8271	0.7934	0.7694
	Cutmix	0.6249	0.7956	0.7697
	Cutout	0.8174	0.7425	0.7414
	Mixup	0.8193	0.7966	0.7886
	Tempered Cutmix	0.7274	0.7803	0.7572
	Tempered Cutout	0.7783	0.6963	0.7150
	Tempered Mixup (Ours)	**0.8340**	**0.8062**	**0.8099**

4.5   Large-Scale Open Set Classification

Deployed systems typically operate on images with far higher resolution and many more categories than the open set baselines previously established for model regularization techniques. It is necessary to understand how well these systems work with higher resolution images and when the number of categories exceeds 100. As the number of categories increases, it can become increasingly difficult to identify a suitable set of background images for background regularization methods.

To study open set classification for large-scale problems, we use the ImageNet Large-scale Visual Recognition Challenge 2012 dataset (ImageNet) [22]. ImageNet has 1.28 million training images (732–1300 per class) and 50000 labeled validation images (50 per class), which we use for evaluation. All methods use an 18-layer ResNet CNN for classification [11] with an input image resolution of 224 × 224. In our experimental setup, the known set of classes consists of 500 classes from ImageNet. Following [2], unknown images for open set evaluation are drawn from categories of the 2010 ImageNet challenge that were not subsequently used and do not have semantic overlap with the 2012 ImageNet dataset. In total the open set dataset consisted of 16950 images drawn from the 339 categories.

We compare Tempered Mixup against the baseline cross-entropy method, Objectosphere, and a combination that incorporates both Objectosphere and Tempered Mixup training. Objectosphere was chosen because it is the best method for using a background dataset based on our previous experiments. To our knowledge, Objectosphere has not been previously evaluated on large-scale problems, thus for a background training set, we use 1300 images from the Places scene understanding validation dataset [31]. We again ensure that all classes do not have semantic overlap with any ImageNet category in either the known or unknown evaluation set, as verified by hypernym and hyponym relationship lookup in the Wordnet lexical database. All models are trained using SGD with a mini-batch size of 256, momentum weighting of 0.9, and weight decay penalty factor of 0.0001 for 90 epochs, starting with a learning rate of 0.1 that is decayed by a factor of 10 every 30 epochs. The baseline cross-entropy trained model for the 500 class partition achieves 78.04% top-1 (94.10% top-5) accuracy.

Results. The results from our ImageNet experiments are shown in Fig. 6. We compute the AUOSC metric using the top-1 correct classification rate and report AUROC as a measure of OOD detection capability. For this large-scale experiment, Tempered Mixup shows roughly the same open set robustness as compared to Objectosphere without having to train with an additional dataset of background samples. To try and gain even better open set performance we augmented our Tempered Mixup formulation with the same background samples used in the Objectosphere training and a uniform distribution target among the known classes for these samples. In this way, our model trains on multiple combinations of mixed up samples, including combinations of known and unknown classes. The resulting hybrid model achieved the best open set performance over either the Tempered Mixup or the Objectosphere methods alone.

5   Discussion

The results from both small-scale and large-scale open set classification problems are evidence that Tempered Mixup is an effective means of improving open set robustness through feature space regularization without having to train with a source of representative unknown samples. We additionally have shown that when a representative background class is available, samples can easily be added into the training pipeline to gain additional robustness.

Our ImageNet open set classification results show that our Tempered Mixup formulation is as effective alone as training with a background class without the additional overhead. When an effective background class is available, we have also demonstrated that our formulation can take advantage of this additional training data to further improve open set robustness. In this case when the nature of the unknown classes to be rejected is known to a degree that an effective background class can be procured, then this is equivalent to hard negative mining for training with a confidence loss framework to reduce the network activation towards these unknown samples.

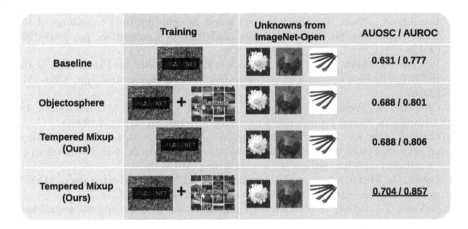

	Training	Unknowns from ImageNet-Open	AUOSC / AUROC
Baseline			0.631 / 0.777
Objectosphere	+		0.688 / 0.801
Tempered Mixup (Ours)			0.688 / 0.806
Tempered Mixup (Ours)	+		0.704 / 0.857

Fig. 6. LARGE-SCALE OPEN SET CLASSIFICATION. Training data is made up of either ImageNet only or ImageNet plus background images from the Places dataset. Unknowns for evaluation are drawn from the ImageNet-Open dataset.

While we have demonstrated that our training paradigm builds a robust feature representation that improves model robustness in detecting novel classes unseen during training, this property is only tested in our work with a baseline confidence thresholding inference method. More advanced inference methods could easily be applied to our models to yield even better open set performance.

6   Conclusion

In this paper we developed a novel technique for improving the feature space of a deep CNN classifier to enable better robustness towards samples from unknown classes. We combined the concept of Mixup augmentation with a novel formulation of confidence loss to train a CNN to produce less confident predictions for samples off of the input distribution defined by the training set. Experimental evidence shows that this formulation performs favorably against current state-of-the-art methods including confidence loss regularization with a background class. This strategy could be especially useful when an appropriate background set is not available in large-scale, real-world classification environments.

Acknowledgements. This work was supported in part by NSF award #1909696, the DARPA/MTO Lifelong Learning Machines program [W911NF-18-2-0263], and AFOSR grant [FA9550-18-1-0121]. The views and conclusions contained herein are those of the authors and should not be interpreted as representing the official policies or endorsements of any sponsor.

References

1. Abdelzad, V., Czarnecki, K., Salay, R., Denounden, T., Vernekar, S., Phan, B.: Detecting out-of-distribution inputs in deep neural networks using an early-layer output. arXiv preprint arXiv:1910.10307 (2019)
2. Bendale, A., Boult, T.: Towards open world recognition. In: Proceedings of the IEEE Conference on Computer Vision and Pattern Recognition, pp. 1893–1902 (2015)
3. Bendale, A., Boult, T.E.: Towards open set deep networks. In: Proceedings of the IEEE Conference on Computer Vision and Pattern Recognition, pp. 1563–1572 (2016)
4. Chapelle, O., Weston, J., Bottou, L., Vapnik, V.: Vicinal risk minimization. In: Advances in Neural Information Processing Systems, pp. 416–422 (2001)
5. Chun, S., Oh, S.J., Yun, S., Han, D., Choe, J., Yoo, Y.: An empirical evaluation on robustness and uncertainty of regularization methods. In: ICML Workshop on Uncertainty and Robustness in Deep Learning (2019)
6. Cohen, G., Afshar, S., Tapson, J., van Schaik, A.: EMNIST: an extension of MNIST to handwritten letters. arXiv preprint arXiv:1702.05373 (2017)
7. DeVries, T., Taylor, G.W.: Improved regularization of convolutional neural networks with cutout. arXiv preprint arXiv:1708.04552 (2017)
8. Dhamija, A.R., Günther, M., Boult, T.: Reducing network agnostophobia. In: Advances in Neural Information Processing Systems, pp. 9157–9168 (2018)
9. Fellbaum, C.: Wordnet. In: Poli, R., Healy, M., Kameas, A. (eds.) Theory and Applications of Ontology: Computer Applications, pp. 231–243. Springer, Dordrecht (2010). https://doi.org/10.1007/978-90-481-8847-5_10
10. He, K., Zhang, X., Ren, S., Sun, J.: Deep residual learning for image recognition. In: Proceedings of the IEEE Conference on Computer Vision and Pattern Recognition, pp. 770–778 (2016)
11. He, K., Zhang, X., Ren, S., Sun, J.: Deep residual learning for image recognition. In: IEEE Conference on Computer Vision and Pattern Recognition (CVPR), June 2016
12. Krizhevsky, A., Hinton, G.: Learning multiple layers of features from tiny images. Technical report, Citeseer (2009)
13. Le, Y., Yang, X.: Tiny imagenet visual recognition challenge. CS 231N (2015)
14. LeCun, Y., Bottou, L., Bengio, Y., Haffner, P.: Gradient-based learning applied to document recognition. Proc. IEEE **86**(11), 2278–2324 (1998)
15. Lee, K., Lee, H., Lee, K., Shin, J.: Training confidence-calibrated classifiers for detecting out-of-distribution samples. arXiv preprint arXiv:1711.09325 (2017)
16. Lee, K., Lee, K., Lee, H., Shin, J.: A simple unified framework for detecting out-of-distribution samples and adversarial attacks. In: Advances in Neural Information Processing Systems, pp. 7167–7177 (2018)
17. Liang, S., Li, Y., Srikant, R.: Enhancing the reliability of out-of-distribution image detection in neural networks. In: International Conference on Learning Representations (2018). https://openreview.net/forum?id=H1VGkIxRZ

18. Neal, L., Olson, M., Fern, X., Wong, W.K., Li, F.: Open set learning with counter-factual images. In: Proceedings of the European Conference on Computer Vision (ECCV), pp. 613–628 (2018)
19. Parisi, G.I., Kemker, R., Part, J.L., Kanan, C., Wermter, S.: Continual lifelong learning with neural networks: a review. Neural Netw. **113**, 54–71 (2019)
20. Perera, P., Patel, V.M.: Learning deep features for one-class classification. IEEE Trans. Image Process. **28**, 5450–5463 (2019)
21. Roady, R., Hayes, T.L., Kemker, R., Gonzales, A., Kanan, C.: Are out-of-distribution detection methods effective on large-scale datasets? arXiv preprint arXiv:1910.14034 (2019)
22. Russakovsky, O., et al.: ImageNet large scale visual recognition challenge. IJCV **115**(3), 211–252 (2015). https://doi.org/10.1007/s11263-015-0816-y
23. Scheirer, W.J., Rocha, A., Sapkota, A., Boult, T.E.: Towards open set recognition. IEEE Trans. Pattern Anal. Mach. Intell. (T-PAMI) **35**, 1757–1772 (2013)
24. Shu, L., Xu, H., Liu, B.: DOC: deep open classification of text documents. In: Proceedings of the 2017 Conference on Empirical Methods in Natural Language Processing, pp. 2911–2916 (2017)
25. Szegedy, C., Vanhoucke, V., Ioffe, S., Shlens, J., Wojna, Z.: Rethinking the inception architecture for computer vision. In: Proceedings of the IEEE Conference on Computer Vision and Pattern Recognition, pp. 2818–2826 (2016)
26. Thulasidasan, S., Chennupati, G., Bilmes, J.A., Bhattacharya, T., Michalak, S.: On mixup training: Improved calibration and predictive uncertainty for deep neural networks. In: Advances in Neural Information Processing Systems, pp. 13888–13899 (2019)
27. Wen, Y., Zhang, K., Li, Z., Qiao, Yu.: A discriminative feature learning approach for deep face recognition. In: Leibe, B., Matas, J., Sebe, N., Welling, M. (eds.) ECCV 2016. LNCS, vol. 9911, pp. 499–515. Springer, Cham (2016). https://doi.org/10.1007/978-3-319-46478-7_31
28. Yun, S., Han, D., Oh, S.J., Chun, S., Choe, J., Yoo, Y.: CutMix: regularization strategy to train strong classifiers with localizable features. In: Proceedings of the IEEE International Conference on Computer Vision, pp. 6023–6032 (2019)
29. Zhang, H., Cisse, M., Dauphin, Y.N., Lopez-Paz, D.: mixup: beyond empirical risk minimization. arXiv preprint arXiv:1710.09412 (2017)
30. Zhang, H., Cisse, M., Dauphin, Y.N., Lopez-Paz, D.: mixup: beyond empirical risk minimization. In: International Conference on Learning Representations (ICLR) (2018). https://openreview.net/forum?id=r1Ddp1-Rb
31. Zhou, B., Lapedriza, A., Khosla, A., Oliva, A., Torralba, A.: Places: a 10 million image database for scene recognition. IEEE Trans. Pattern Anal. Mach. Intell. **40**, 1452–1464 (2017)

Defenses Against Multi-sticker Physical Domain Attacks on Classifiers

Xinwei Zhao$^{(\boxtimes)}$ (iD) and Matthew C. Stamm (iD)

Drexel University, Philadelphia, PA, USA
{xz355,mstamm}@drexel.edu

Abstract. Recently, physical domain adversarial attacks have drawn significant attention from the machine learning community. One important attack proposed by Eykholt et al. can fool a classifier by placing black and white stickers on an object such as a road sign. While this attack may pose a significant threat to visual classifiers, there are currently no defenses designed to protect against this attack. In this paper, we propose new defenses that can protect against multi-sticker attacks. We present defensive strategies capable of operating when the defender has full, partial, and no prior information about the attack. By conducting extensive experiments, we show that our proposed defenses can outperform existing defenses against physical attacks when presented with a multi-sticker attack.

Keywords: Real-world adversarial attacks · Defenses · Classifiers · Deep learning

1 Introduction

Deep neural networks have been widely used for many visual classification systems, such as autonomous vehicles [13,35] and robots [38].However, deep neural networks are vulnerable to adversarial attacks [5,6,14,17,18,21,22,24,25,27,28,33,34]. By modifying the pixel values of an image, many classifiers can be fooled.

Recently, attacks that can operate in the physical world have started to attract increasing attention [1,4,11]. While some physical domain attacks require crafting a new object [1,11,19], other attacks can fool the classifiers by adding one or a few physical perturbations, such as printable patches [4,11] on or next to an object. The adversarial patch attack creates one universal patch that can be used to attack an arbitrary object once it is trained, regardless of scale, location and orientation [4]. The camouflage art attack uses black and white stickers that are applied to an object such as a traffic sign to make a classifier believe it is a different object. [11] Since these physical perturbations are very concentrated and confined to small regions, it is easy for attackers to craft these physical perturbations and put the attack in practice in the real world.

Previous research shows that defenses against digital domain attacks [2,3,7–10,12,15,16,20,22,26,29,30,32,36] may not be able to defend against physical domain attacks, such as the camouflage art attack, because physical

© Springer Nature Switzerland AG 2020
A. Bartoli and A. Fusiello (Eds.): ECCV 2020 Workshops, LNCS 12535, pp. 202–219, 2020.
https://doi.org/10.1007/978-3-030-66415-2_13

perturbations are usually stronger than those produced by digital domain attacks. Some recent research has been done to defend against physical domain attacks [7, 16, 20, 26, 37].

Fig. 1. Attacked signs (a) & (d) as well as their Grad-CAM activation maps before attack (b) & (e) and after attack (c) & (f).

Existing research, however, focuses on defending against adversarial patches, and does not translate to defend against other physical attacks like the camouflage art attack (i.e white and black sticker attack). For example, one approach to defend against the adversarial patch attack is to first locate the perturbed area using an attention-based or gradient-based model, and then remove or diminish these areas [16, 26, 37]. The perturbations produced by multi-sticker attacks like the camouflage art attack, however, cannot be detected the same way due to several reasons. First, the black and white stickers produced camouflage art attack are not highly textured, and hence are unlikely to be detected via gradient-based methods. Second, the camouflage art attack works in conjunction with the scene content to redirect the classifiers decision instead of hijacking its attention like the adversarial patch does. As a result, multi-sticker attacks are unlikely to be identified using attention-base models.

An example of this phenomenon can be seen in Fig. 1, which shows activation maps produced by Grad-CAM [31] when presented with images before and after a multi-sticker attack. When examining the activation maps of the pedestrian crossing sign before an attack shown in Fig. 1(b) and after the attack shown in Fig. 1(c), we can see that the attack has shifted the classifier's attention off of attacked sign. Defenses that operate by removing or altering these regions will have no effect on the attack. Alternatively, from examining the activation maps of an unattacked speed limit sign in Fig. 1(e) and it's attacked counterpart in Fig. 1(f), the classifier is paying attention to nearly the entire sign. Defenses that operate by removing or distorting these regions will degrade the image so severely that the classifier will be unable to operate.

Furthermore, it is important for defenses against physical domain attacks to be evaluated on real images of physically attacked objects. Digital simulations of physical attacks are sometimes used for evaluation due to the ease of creating a dataset, for example, digitally adding perturbations that simulate a physical attack into an image. However, these digital simulations do not capture many effects that occur during imaging, such as lighting conditions, the curvature of surfaces, focus blur, sampling effects, etc. In practice, phenomena such as these can impact how a camera captures physical domain perturbations, and can potentially affect the success of defenses. Defenses that are highly tuned to features of "pristine" digital simulations of attacks may be less successful when confronted with real images of physically attacked objects or scenes.

In this paper, we propose a new defense strategy that does not rely on attention models to identify attacked image regions and can successfully defend against multi-sticker attacks, like the camouflage art attack. Our proposed defense operates by first creating defensive masks that can maximize the likelihood of guessing the location of the perturbations, then mitigates the effect of the perturbations through targeted modifications, and eventually make a final decision based on defended images.

Our Contributions:

- We propose a set of new defenses that can protect against multi-sticker physical domain attacks such as the camouflage art attack by Ekyholt et al. [11]. To the best of our knowledge, no existing defenses are designed to defend against such attacks.
- We present practical defenses that can be utilized depending on whether the defender has full knowledge of the attack (non-blind), partial information about the attack (semi-blind), or no information regarding the attack (blind).
- We create a new database of front-facing photos of 90 physically attacked signs using camouflage art attack and use this database to assess our defense.
- We demonstrate that our proposed defenses outperform other state-of-the-art defenses against physical attacks, such as the digital watermark defense [16], when presented with multi-sticker attacks.

2   Additive Physical Domain Attacks

Adversarial attacks pose an important threat against deep neural networks [1,4–6,11,14,17–19,21,22,24,25,27,28,33,34]. Some physical domain attacks, like the adversarial patch [4] and the camouflage art attack [11], have shown that adding perceptible but localized patches to an object can make a classifier identify it as a different object. We now briefly describe how these two physical domain attacks are launched at a classifier C using attack target class t'.

Adversarial Patch: To generate an adversarial patch A', the authors of [4] use an operator $O(I, A, \theta_l, \theta_t)$ to transform a given patch A, then apply it to an image I at location θ_l. Similarly to an Expectation over Transformation attack (EoT) [1], the adversarial patch can be obtained by optimizing over sampled transformation and locations,

$$A' = \max_A \mathbb{E}_{I \sim \mathcal{I}, \theta_l \sim \Theta_L, \theta_t \sim \Theta_T} C(t' | O(I, A, \theta_l, \theta_t)) \tag{1}$$

where $\mathcal{I}$ denotes the training image dataset, Θ_T denotes the distribution of transformation and Θ_L denotes the distribution of the location. Once the patch is trained, it can universally attack any object.

Camouflage Art Attack: Launching the camouflage art attack involves finding a single set of perturbations P that are capable of fooling a classifier under different physical conditions. This attack, which produces perturbations for a given pairing of source and target class, was demonstrated by using it to fool a classifier trained to distinguish between different US traffic signs. Let H^v denote the distribution of the image of an object under both digital and physical transformations, and h_i denote each sample from this distribution. The attack perturbations can be obtained via optimizing,

$$\operatorname*{argmin}_{P} \lambda||M_h, P||_p + \mathbb{E}_{h_i \sim H^v} J(C(h_i + G(M_h, P), t') \tag{2}$$

where M_h is the mask that applies spatial constraints to the perturbation (i.e ensures the perturbation is within the surface area of the object), λ is a hyperparameter that regularize the distortion, $J(\cdot)$ is the loss function that measures the difference between the classifier's prediction of the attacked object and the target class, $G(\cdot)$ is the alignment function that maps transformations on the object to transformations on the perturbation, $|| \cdot ||_p$ denotes ℓ_p norm.

3   Problem Formulation

We assume that the system under attack wishes to analyze some scene $S(x, y)$ containing an object to be classified. To do this, the system will capture a digital image $I(x, y)$ of the scene, which will then be provided to a pre-trained classifier $C(\cdot)$ which maps the image into one of N classes $t \in \mathcal{T}$. For the purposes of this work, we assume that if no adversarial attack is launched, then the image provided to the classifier is $I = S$.

An attacker, may attempt to fool the classifier by launching a physical domain attack $\alpha(\cdot)$. This corresponds to physically modifying an object within the scene by adding adversarial perturbations P to it. Since these perturbations must be physically added to the scene, we assume that they will be spatially localized to one or more regions of the object under attack. These regions can be specified by a spatial mask M, where $M(x, y) = 1$ corresponds to a perturbation being present at spatial location (x, y) and $M(x, y) = 0$ corresponds to no perturbation occurring at (x, y). As a result, we can express a physically attacked scene $\alpha(S)$

$$\alpha(S(x, y)) = (1 - M(x, y))S(x, y) + M(x, y)P(x, y). \tag{3}$$

In this paper, we assume that the adversarial perturbations will take the form of black and white stickers added to an object as proposed by Eykholt et al. [11], i.e. $P(x, y) = \{black, white\}$. Other physical domain attacks, such as the adversarial patch [4] can still be modeled using (3) by allowing $P(x, y)$ to correspond to the full range of color values. Since the majority of the defenses proposed in this paper do not rely on knowledge of the color values of P, it is likely that these defenses can be used against other physical domain attacks such as the adversarial patch. We note that this work only addresses physical domain attacks that involve modifying an existing physical object, and not attacks that involve the creation of a new physical object such as synthesized 3D objects [1] and printed photos or posters [11,19].

4 Knowledge Scenarios

To defend a classifier, we first assume that the defender has full access to the classifier and implicitly knows the N classes that it is trained to distinguish between. We examine three scenarios corresponding to different levels of knowledge available to the defender.

Non-blind: We assume that defender knows if an object is attacked or not, the perturbation masks M that indicates the perturbation areas and the perturbations P. Therefore, locations of perturbations can be directly located.

Semi-blind: We assume that the defender does not know if the object is attacked or not. We also assume that if the object was attacked, the defender does not know the perturbation masks M. However, the defender knows the attack method $\alpha(\cdot)$. Therefore, for any source A and target B pairing, the defender can obtain a perturbation mask $M_{A,\,B}$ via launching the attack.

Blind: We assume that defender has zero knowledge. Specifically, the defender does not know whether an object is attacked or not. We also assume that if the object was attacked, the defender does not know the perturbation regions. Additionally, the defender does not know the attack method.

5 Proposed Defenses

To defend against a physical domain attack, we propose a set of defenses based on the amount of knowledge available to the defender. These defenses attempt to interfere with or remove adversarial multi-sticker perturbations to mitigate their effects. If the defender is able to leverage information about the potential locations of these perturbations, defenses are guided to these regions. Otherwise, our defenses are designed with the intuition that adversarial perturbations are more fragile to distortions than the underlying object that they are attacking.

Our defensive strategy is composed of three major steps. First, we obtain a defensive mask R or set of defensive masks $\mathcal{R}$ indicating candidate areas to apply defenses. Second, we launch a local defense in regions indicated by a defensive mask to produce a defended image δ. When our first step results in a set of defensive masks, local defenses can either be sequentially applied in conjunction with each mask to produce a single defended image, or they can be applied in parallel to produce a set of defended images. In the third step, the defended image or images are provided to the classifier. If a set of defended images are produced by the second step, a fusion strategy is employed to produce a single classification decision. In what follows, we discuss each step of our proposed defenses in detail.

5.1   Defensive Mask Selection

The goal of each defensive mask is to ensure that defensive distortions are only applied to small regions of the image, since each perturbation produced by the multi-sticker attack is still confined to a small region. We do not want to change the ground truth object. Let $R(x, y) \in \{0, 1\}$ denote a defensive mask, where 1 indicates the area need to be defended, 0 indicates the area of the ground truth content. Now we discuss the acquisition of defensive masks.

Oracle Approach: If the perturbation mask M is known, such as in the non-blind scenario, we simply let $R = M$.

Estimated Defensive Mask Sets: In semi-blind scenarios, the defender may know the potential attack method α, but not perturbation masks or the potential attack mask if the attack was launched. They can, however, leverage knowledge of α to create a set of estimated defensive masks.

To do this, first we assume that I is an image of an attacked scene. The attack's target class $\hat{t}$ can be inferred by using C to classify the image such that $\hat{t} = C(I)$. Next, the defender can create their own implementation of α and use it to recreate an attack aimed to move true class j to target class $\hat{t}$. The attack's perturbation mask can then be used as the estimated defensive mask $R_{j,\hat{t}}$ for source j and target t. This process can be repeated for all $j \in \mathcal{T}$ such that $j \neq \hat{t}$ to produce the set of estimated masks $\mathcal{R}_{\hat{t}} = \{R_{1,\hat{t}}, \ldots, R_{\hat{t}-1,\hat{t}}, R_{\hat{t}+1,\hat{t}}, \ldots, R_{N,\hat{t}}\}$. To reduce computational costs while launching the defense, the set $\mathcal{R}_{\hat{t}}$ can be precomputed for each target class. With increasing number of classes, the computational cost may become high for constructing sets of estimated set of defense masks and launching the defense. To solve this problem, defender can use a subset of defensive masks instead of every single mask. We propose two methods to form these subsets.

Ranked Selection: The defender can utilize class activations to guide the selection of the subset of defensive masks to use. Since physical attacks operate by constraining perturbations to small areas to avoid suspicion, it is reasonable to assume these perturbation push the object just across the boundary of its true class. Therefore, the true class of an attacked image most likely shows up in the top few activated classes. To guide the selection of defensive masks, first we assume that a scene is always under attack (regardless of whether this is true or not) and treat the class with the highest activation as the target class. The true class then lies among the remaining classes, which are ranked according to their activation scores. The subset of k defensive masks is then chosen as the set of masks created using the assumed target class (i.e. the class with the highest activation) and the k top candidates for the true source class (i.e. the classes with the second highest through $k + 1$ highest activations). By doing this, the defender can control the computation cost of the defense while increasing the chance that the most useful defensive masks are utilized.

Random Selection: A heuristic way to form a subset of defensive masks is through random selection. Since each of the selected mask is related to the target class, each selected mask can be used to defend a partial of the image. By grouping several defensive masks, it may increase the chance for a successful defense.

Randomly Chosen Regions: In blind scenarios, the defender cannot leverage any prior information about the attack or possible perturbation locations. In these situations, we create a set of defensive masks made by randomly choosing defensive regions. Our intuition is that if we use many random defensive masks, several of them will interfere with the adversarial perturbations. Each mask is made by randomly selecting m different $w \times w$ windows to apply localized defenses. We use two different approaches for randomly choosing these regions:

Overlapping: The locations of each window are chosen uniformly at random from throughout the image area. As a result, some windows may overlap with one another.

Non-overlapping: In this approach, we ensure that defensive regions are spread throughout the region by disallowing overlaps. We do this by first dividing the defensive mask into non-overlapping $w \times w$ blocks, then randomly choosing m of these blocks as defensive regions.

5.2   Local Defense Strategies

After the defensive masks are obtained, we can apply local defenses to image regions specified by these masks. To make it clear, we first show how to obtain the defended image using single defensive mask, then we adapt the proposed defenses to accommodate multiple defensive masks.

Given one defensive mask, we propose two methods to defend against the attack.

Targeted Perturbation Remapping: This idea is to interfere the perturbations instead of removing it. Specifically, we can using remapping functions to destroy the spatial correlation between perturbed regions. Let $\phi(\cdot)$ be the remapping function, then a single defended image can be expressed as,

$$\delta(x, y) = \begin{cases} I(x, y) & R(x, y) = 0 \\ \phi(I(x, y)) & R(x, y) = 1 \end{cases} \qquad (4)$$

In this work, we consider three mapping functions:

RemapW: Change pixels to white.

RemapB: Change pixels to black.

RemapT: Pick a threshold τ, change pixels to black if the luminance value is above the threshold and to white if below the threshold.

Localized Region Reconstruction: The idea is to diminish or remove the effects of that perturbation by reconstructing perturbed local regions of input image on the basis of other parts of the image. Since the perturbations are confined to a small region, we can use the inpainting algorithm to reconstruct the image.

The defenses discussed above can be easily adapted for multiple defensive masks. Let $\psi(\cdot)$ denote the defense. For a set of defensive masks $\mathcal{R}$ that comprises k mask, $\mathcal{R} = \{R_1, R_2, ..., R_k\}$, we can either obtain one single final defended image via sequential defense, or obtain a sequence of individually defended images via parallel defense and then fuse the results. Now we discuss sequential and parallel defense individually.

Sequential Defense: We attempt to make the defense stronger by recursively applying the defense and obtain a single defended image. For iteration ℓ, the defense $\psi(\cdot)$ is applied to the output of the previous step using ℓ^{th} defensive mask, $\psi_\ell(\cdot) = \psi(\delta_{\ell-1}, R_\ell)$. The final defended image is obtained by sequential applying the defense using each of k individual defensive mask via,

$$\delta = (\psi_k \circ \psi_{k-1} \circ \ldots \circ \psi_1)(I) \tag{5}$$

Parallel Defense: The idea is to generate many copies of defended image with each copy being able to defend one part of input image. Using ℓ^{th} defensive mask, we define $\ell^t h$ defended image as $\delta_\ell = \psi(I, R_\ell)$, then using k defensive masks we get k individual defended images, $\{\delta_1, \delta_2, \ldots, \delta_k\}$.

5.3   Defensive Classification

After applying local defenses, we need to use the classifier to make a final decision on the defended image or images. We propose two decision making strategies.

Single Defended Image: After the sequential defense, the defender will obtain a single defended image. We simple use the classifier to classify the defended image, $t = C(\delta)$.

Multiple Defended Images: The parallel defense will result in a sequence of defended images. The defender can use the classifier to get a fused decision by combining the decisions of the individually defended images. We propose two fusion strategies.

Majority vote (MV): Use the classifier to make a decision with each individual defended image, $t_\ell = C(\delta_\ell)$, then take a majority votes of all decisions

$$t = \underset{n \in N}{\operatorname{argmax}} \sum_{\ell=1}^{k} \mathbb{1}(C(\delta_\ell) = t_n) \tag{6}$$

where $\mathbb{1}$ is the indicator function.

Softmax fusion (SF): Let $\mathbf{v}^{(\ell)}$ denote the softmax output of the classifier for the ℓth defended image, $\mathbf{v}^{(\ell)} = C_{softmax}(\delta_\ell)$, next add the softmax output of each of the k defended images to form a single vector $\mathbf{v}$,

$$\mathbf{v} = \sum_{\ell=1}^{k} \mathbf{v}^{(\ell)} \tag{7}$$

then take the class corresponding to the largest value in $\mathbf{v}$ as the final decision,

$$t = \operatorname*{argmax}_{n \in N} v_n \tag{8}$$

where v_n is the nth element in the vector $\mathbf{v}$.

6 Evaluation Metrics

When formulating our evaluation metrics, we let t^* denote the ground truth class of a scene. Additionally, we let π_A denote the a priori probability that an attack is launched against a scene.

Classifier: To evaluate the baseline performance of the classifier $C(\cdot)$, we calculate the classification accuracy as the probability that the image of a scene being correctly classified as its ground true class,

$$\mathrm{CA} = Pr(C(I) = t^* | I = S) \tag{9}$$

Attack: To evaluate the baseline performance of the attack, we calculate the targeted attack success rate (T-ASR) and the untargeted attack success rate(U-ASR).

T-ASR is defined as the probability that the image of an attacked scene is classified as the target class,

$$\mathrm{T\text{-}ASR} = Pr(C(I)) = t' | I = \alpha(S)) \tag{10}$$

U-ASR is defined as the probability that the image of an attacked scene is classified as any other class than the true class,

$$\mathrm{U\text{-}ASR} = Pr(C(I)) \neq t^* | I = \alpha(S)) \tag{11}$$

Defense: To evaluate the performance of our proposed defenses, we calculate the Defense Rate (DR) for an attacked scene, the Classification Drop (CD) for an unattacked scene, and the Post-Defense Accuracy (PDA) for any scene.

DR is defined as the probability that the defended image of a scene is classified as true class, given it is an attacked scene and its image was not classified as the true class before the defense,

$$\mathrm{DR} = Pr(C(D(I)) = t^* | I = \alpha(S), C(I) \neq t^*) \tag{12}$$

CD is defined as the probability that the image of an unattacked scene get misclassified after applying the defense.

$$CD = CA - Pr(C(D(I)) = t^*|I = S) \qquad (13)$$

PDA is defined as the probability that the image of any scene is correctly classified as the true class after the defense,

$$PDA = (1 - \pi_A)Pr(C(D(I)) = t^*|I = S)$$
$$+ \pi_A Pr(C(D(I)) = t^*|I = \alpha(S)) \qquad (14)$$

When U-ASR=1, using Eq. 11, 12 and 13, Eq. 14 can be expressed as,

$$PDA = (1 - \pi_A)(CA - CD) + \pi_A DR \qquad (15)$$

7   Experimental Results

To evaluate the performance of our proposed defenses, we conducted a series of experiments. The physical attack we attempt to defend against is the camouflage art attack proposed by Eykholt et al. [11]. The classifier we used to evaluate the proposed defense was trained to differentiate 17 common US traffic signs using LISA traffic sign database [23] (a US traffic sign database). The classifier was reported to achieve 91% classification accuracy in their paper. We started by making a dataset composed of photos of unattacked ground truth source signs and physical attacked signs. Then we demonstrated the effectiveness of the proposed defense method under the three scenarios we discussed in Sect. 4. We assume $\pi_A = 0.5$ in all scenarios.

7.1   Dataset

To the best of our knowledge, there exists no database that made specifically for physical attack, especially using camouflage art attack. A physical attack database should be constructed with the photos of the physically attacked objects. This is because empirically we found that defenses against physical perturbations are very different from the digital simulation. One reason is that the many effects introduced during capturing images of physically attacked objects, such as the curvature of surfaces, focus blur, sampling effects, sensor noise, will result in significant discrepancies between physical perturbations and digital approximation. Therefore, it is important to create a new database to fill this gap and benefit future research in the community.

To make the database, we first purchased six US road signs which were included among the 16 which classes the LISA-CNN is trained to distinguish between. These six signs are indicated above in Table 1 as 'source' signs.

To create training data for the attack, and assess the baseline performance of the LISA-CNN, we first captured a set of images of the six unattacked signs in our possession. This was done by photographing each sign at angles running

Table 1. Source and target traffic signs. S denotes "source" and T denotes "target".

Category	Sign name	Category	Sign name
S & T	crossing	T	added lane
S & T	stop	T	keep right
S & T	yield	T	lane ends
S & T	signal ahead	T	stop ahead
S & T	speed limit 25	T	turn right
S & T	speed limit 45	T	school/limit 25
T	merge	T	speed limit 30
T	school	T	speed limit 35

Table 2. Non-blind evaluation of our proposed defenses.

Proposed defense	DR	CD	PDA
RemapW	0.4339	0.0000	0.7170
RemapB	0.4556	0.0000	0.7283
RemapT	**0.9222**	0.0000	**0.9611**
Reconst	0.6778	0.0000	0.8389

from −50 to +50 degrees in an increments of 10 degrees, to create a set of 66 images of unattacked signs.

Next, we launched a series of multi-sticker attacks against the six signs in our possession, using each of the 15 remaining classes listed in Table 1 as the attack's target. This was done by following the attack protocol described in [11]. For each pair of source and target signs, we first created a digital copy of the attacked sign. This digital copy was projected onto the corresponding physical copy of the source sign, then black and white stickers were placed on the sign in regions indicated by the digitally attacked version. Front facing images of all of the attacked signs were captured, then cropped to approximately 340 × 340 pixels and saved as PNG files. This resulted in a set of 90 images of physically attacked signs, each with a different source-target class pairing. The database is publicly available at https://drive.google.com/drive/folders/1qOmSubSOVY8JzB3KfXhDQ38ihoY5GExK?usp=sharing.

7.2   Baseline Evaluation of the Classifier and Attack

To assess the baseline classification accuracy of the LISA-CNN classifier trained by Eykholt et al., we evaluated its performance on the unattacked signs captured as part of our database. In this evaluation, the LISA-CNN achieved 100% classification accuracy. We note that Eykholt et al. reported a 91% classification accuracy during their evaluation of this trained classifier. In this paper, when reporting metrics that depend on classification accuracy, we use the value that we obtained since this classification accuracy is measured on the same set of road signs in the attack set. Furthermore, this corresponds to more challenging test conditions for our defense, since perfect performance would need to bring the defense rate equal to this higher classification accuracy.

Next, we measured the baseline performance of the attack by using the LISA-CNN to classify the images of physically attacked signs in our database. Our implementation of the camouflage art attack achieved a 0.9556 targeted attack success rate (T-ASR) and a 1.0000 untargeted attack success rate (U-ASR). This result verifies that we were able to reproduce the attack, and that this attack can successfully fool the classifier.

7.3   Non-Blind

In our first set of experiments, we evaluated our defenses' performance in the non-blind scenario. We used the digital versions of the perturbation masks obtained while training the attack as the oracle defensive masks known to the defender. While these digital masks are not perfect ground truth locations of the actual perturbations they are sufficiently close to evaluate our experiment.

Using these oracle masks, we evaluated the three perturbation remapping defenses remap to white (RemapB), black (RemapW), and threshold (RemapT) as well as the targeted region reconstruction (Reconst) defense. We note that the classification drop is always zero in this experiment because the defender always knows if an attack is present and can choose when not to apply the defense.

Table 2 shows the performance of our defenses in the non-blind scenario. Thresholded perturbation achieved strongest performance with the highest defense rate of 0.9222 and post-defense accuracy of 0.9611. Since both the remap-to-white and remap-to-black strategies will only affect approximately half of the stickers added to an object, it is reasonable to expect that the thresholded perturbation remapping approach outperforms these approaches. Reconstruction approach achieved second highest performance. We believe that lower defense rate is predominantly due to the slight misalignment between the ideal digital perturbation masks and the true locations of the physical perturbations in the attacked images.

7.4   Semi-blind

To evaluate our defenses in the semi-blind scenario, we created a set of estimated defensive masks for each of the 15 possible target classes. Each set of defensive masks contained six pairings of source and target sign, i.e. one for each source sign in our database that an attack could be launched against.

Next, we used these sets of defensive masks to evaluate our relevant defensive strategies. The results of these experiments are shown in Table 3. We adopt the notation Par and Seq to denote that a defense was applied either in parallel or sequentially, and (k) to denote the number of defensive masks used for defense. When defenses were applied in parallel, we use the notation MV to denote majority vote fusion and SF to denote softmax fusion. We use Rand and

Table 3. Evaluation of proposed defenses in semi-blind scenario

Defense strategies	DR	CD	PDA	Defense strategies	DR	CD	PDA
RemapW-Par(6) + MV	0.3989	0.3333	0.5328	RemapW-Par(6) + SF	0.4186	0.1667	0.6260
RemapB-Par(6) + MV	0.0794	0.1667	0.4563	RemapB-Par(6) + SF	0.0690	0.0000	0.5348
RemapT-Par(6) + MV	0.5174	0.3333	0.5921	RemapT-Par(6) + SF	0.6453	0.1667	0.7393
Reconst-Par(6) + MV	0.3560	0.0000	0.6780	Reconst-Par(6) + SF	0.3514	0.0000	0.6757
Reconst-Seq-Rand(1)	0.2815	0.0556	0.6130	Reconst-Seq-Rand4)	0.6200	0.1112	0.7544
Reconst-Seq-Rand(2)	0.4237	0.0556	0.6840	Reconst-Seq-Rand(5)	0.6648	0.1389	0.7630
Reconst-Seq-Rand(3)	0.5350	0.0834	0.7250	Reconst-Seq(6)	0.7000	0.1667	0.7667
Reconst-Seq-Rank(1)	0.3780	0.0000	0.6890	Reconst-Seq-Gtd(1)	0.6778	0.0000	0.8389
Reconst-Seq-Rank(2)	0.6336	0.0000	0.8168	Reconst-Seq-Gtd(2)	0.6623	0.0333	0.8145
Reconst-Seq-Rank(3)	**0.7001**	**0.0000**	**0.8501**	Reconst-Seq-Gtd(3)	0.6855	0.0667	0.8094
Reconst-Seq-Rank(4)	0.6667	0.0000	0.8333	Reconst-Seq-Gtd(4)	0.7022	0.1000	0.8011
Reconst-Seq-Rank(5)	0.7000	0.0000	0.8500	Reconst-Seq-Gtd(5)	0.7044	0.1333	0.7856
Other Methods	**DR**	**CD**	**PDA**	**Other Methods**	**DR**	**CD**	**PDA**
DW [16]	0.2222	0.0000	0.6111	Median Filter (kernel=7) [36]	0.3777	0.3333	0.5222
JPEG (QF=10) [10]	0.1333	0.0000	0.5667	Local Smooth [26]	0.0000	0.0000	0.5000

Rank to denote the random or ranked mask selection strategy. Additionally, we use Gtd to denote a special "Guaranteed scenario" in which the defensive mask with the correct source-target pair was always included and the remaining masks were randomly chosen.

Results in Table 3 show that for any mask selection strategy, sequential reconstruction outperforms both parallel reconstruction and perturbation remapping. The defense using three defensive masks selected using ranked activation (Reconst-Seq-Rank(3)) outperforms all other strategies and achieved the highest defense rate of 0.7001, highest post-defense accuracy of 0.8501, and zero classification drop on unattacked images. We note that Reconst-Seq-Rank(3) is statistically the same performance as Reconst-Rank(5), but it is more computationally efficient using less masks.

Comparisons of Defensive Mask Selection Strategies: For all values of k, the ranked selection strategy achieved a higher defense rate and post-defense accuracy than the random selection strategy. This shows that using a well chosen subset of defensive masks improves our system's performance. Additionally, it reinforces our observation that important information about the true source class and attack target class can be observed in the top few class activations.

To explore impacts from the correct source and target defensive mask, we ran another set of experiments for the "Guaranteed scenario". Compared to the Reconst-Seq-Rand(k) strategy, the Reconst-Seq-Gtd(k) strategy always achieved higher defense rate, post-defense accuracy, and lower classification drop for the same k. These results imply that the inclusion of the estimated mask for the correct source-target class pair can significantly improve the performance of defenses.

Comparing the Ranked strategy with the "Guaranteed scenario", Ranked results are in a higher post-defense accuracy for $k \geq 2$. The main reason is that

Ranked produces a significantly lower classification drop. These results suggest that using the ranked selection strategy not only can pick out the "best" subset of defensive masks to use, but can also exclude those that deteriorate the classification accuracy of unattacked images.

This is reinforced by examining the classification drop as k increases. For both Reconst-Seq-Gtd(k) and Reconst-Seq-Rand(k), the classification drop increases as k increases, thus hurting the overall post-defense accuracy. This is likely because some masks that negatively effect the overall performance are included. By contrast, Reconst-Seq-Rank(k) does not suffer from the same decrease in classification drop because unlikely defensive masks that may hurt performance are excluded.

Comparisons with Related Defenses: We compared the performance of our proposed defenses with several existing defenses against physical domain attacks. These include distortions that are universally applied to an image such as JPEG compression and median filtering, as well as the more sophisticated digital watermarking (DW) defensive method and the local smooth approach. While we evaluated the performance of the JPEG defense using multiple quality factors and the median filtering defense using multiple kernel sizes, we report only the strongest results in the interest of space.

The results in Table 3 show that all of our proposed strategies with the reconstruction defense can significantly outperform each of these existing defenses. The digital watermarking defense proved to be the strongest performing existing defense, with a defense rate of 0.2222 and a post-defense accuracy of 0.6111. However, even when only one randomly chosen estimated defensive mask is used, our region reconstruction defense outperforms this approach. Our best performance achieved more than three times higher in defense rate and about 40% more in post-defense accuracy than this approach. The relatively poor performance of these existing defenses likely occurs because they are targeted to defend against the adversarial patch attack. Since the multi-sticker camouflage art attack works in a different manner and exhibits different visual properties, these defenses are not as well suited to protect against this and similar attacks.

7.5 Blind

To evaluate our defenses in the blind scenario, we created randomly chosen defensive masks using both the overlapping (OL) and non-overlapping (NOL) strategies.

Table 4 shows the results of these experiments. In each experiment, we identified the optimal window size and number of windows for use in these masks through a grid search. We chose window size w vary from 2, 4, 8 16 pixels. Next we controlled the number of windows m by randomly selecting a ratio of total number of windows based on the given window sizes. The results reported in Table 4 correspond to the pairing of w and ratio that achieved the highest post-defense accuracy. A detailed examination of the choice of w and ratio is provided later in this section.

From Table 4, we can see the strongest performance in terms of all evaluation metrics was achieved using targeted region reconstruction applied in parallel using 100 random masks with non-overlapping windows in conjunction with majority vote decision fusion (NOL-Reconst-Par(100) + SF). Even though no information regarding the attack could be leveraged, this defense was still able to achieve a defense rate of 0.4102 with a corresponding classification drop of 0.0017 and a post-defense accuracy of 0.7043. Though performance is worse than in the semi-blind scenario, we are still able to outperform existing defenses in all evaluation metrics. We note that the local region reconstruction defense uniformly outperformed the targeted perturbation remapping strategy, and for targeted region reconstruction, applying the defense in parallel outperformed sequential application of the defense.

Creating defensive masks using the non-overlapping strategy significantly improves our defense's performance over the overlapping approach (i.e. choosing window locations uniformly at random). Furthermore, we note that performance increases as the number of randomly chosen masks increases. While this comes at the price of additional computation costs, in practice we found that our proposed defense takes 0.4 seconds on average using 100 masks without any attempt at execution time optimization.

Effect of Size and Number of Windows: To understand the effect that the window size and number of windows (or ratio) in each randomly chosen defensive mask has on our defense, we provide detailed results of our search over these parameters in Table 5. The symbol $*$ means when window size was 16 and ratio was 0.625, the computed number of windows was not an integer. However, it equals to when ratio was 0.5 if rounded down, and equals to when ratio was 0.75 when rounded up.

The results show that the defense rate increases as the ratio (i.e the number of windows) increases. After a certain point, the classification drop also increases, resulting in a negative effect on the post-defense accuracy. We also find that increasing the window size increases the defense rate up to a certain point, after which the defense rate begins to decrease. Additionally, after a certain point, increasing the window size also leads to an increase in the classification drop and a decrease in the post-defense accuracy. In our experiments, we found that the optimal window size was 8 pixels and ratio was 0.625. More importantly, when choosing the window size and the ratio (i.e the number of windows), the defender must balance the trade-off between interfering with the attack and interfering with the unattacked scene content used by the classifier.

Comparisons with Related Defenses: From Table 4, we can see that applying region reconstruction in parallel using non-overlapping masks outperforms the existing defenses that were also considered in the semi-blind scenario (the performance of these defenses do not change in the blind scenario). This result holds true even when only six randomly generated non-overlapping masks are used.

Table 4. Defense performance in the blind-scenario.

Defense strategies	DR	CD	PDA	Defense strategies	DR	CD	PDA
OL-RemapT-Par(6) + MV	0.1210	0.2367	0.4422	NOL-RemapT-Par(6) + MV	0.1236	0.1283	0.4976
OL-RemapT-Par(6) + SF	0.1271	0.1650	0.4811	NOL-RemapT-Par(6) + SF	0.1255	0.1267	0.4994
OL-RemapT-Par(100) + MV	0.0713	0.0333	0.5190	NOL-RemapT-Par(100) + MV	0.0482	0.0500	0.4991
OL-RemapT-Parallel(100) + SF	0.0778	0.0333	0.5222	NOL-RemapT-Par(100) + SF	0.0737	0.0667	0.5035
OL-Reconst-Seq(6)	0.2352	0.1782	0.5286	NOL-Reconst-Seq(6)	0.1942	0.0775	0.5584
OL-Reconst-Seq(100)	0.1588	0.8450	0.1569	NOL-Reconst-Seq(100)	0.2553	0.6300	0.3027
OL-Reconst-Par(6) + MV	0.1836	0.0717	0.5560	NOL-Reconst-Par(6) + MV	0.3444	0.0467	0.6488
OL-Reconst-Par(6) + SF	0.1762	0.0350	0.5706	NOL-Reconst-Par(6) + SF	0.3501	0.0317	0.6593
OL-Reconst-Par(100) + MV	0.1362	0.0000	0.5681	NOL-Reconst-Par(100) + MV	0.4129	0.0067	0.7031
OL-Reconst-Par(100) + SF	0.2415	0.0167	0.6124	NOL-Reconst-Par(100) + SF	**0.4102**	**0.0017**	**0.7043**
Other Methods	**DR**	**CD**	**PDA**	**Other Methods**	**DR**	**CD**	**PDA**
DW [16]	0.2222	0.0000	0.6111	Median Filter (kernel=7) [36]	0.3777	0.3333	0.5222
JPEG (QF=10) [10]	0.1333	0.0000	0.5667	Local Smooth [26]	0.0000	0.0000	0.5000

Table 5. Local region reconstruction using 100 parallel masks with non-overlapping windows and softmax fusion. * For window size 16, ratio 0.625 results in a non-integer number of windows.

	Ratio = 0.25			Ratio = 0.5			Ratio = 0.625			Ratio = 0.75		
	DR	CD	PDA	DR	CD	PDA	DR	CD	PDA	DR	CD	PDA
$w = 2$	0.0546	0.0000	0.5273	0.1528	0.0000	0.5764	0.2027	0.0033	0.5997	0.2770	0.1283	0.5744
$w = 4$	0.0537	0.0000	0.5269	0.1818	0.0000	0.5909	0.2395	0.0000	0.6198	0.3153	0.0233	0.6460
$w = 8$	0.1648	0.0000	0.5824	0.3268	0.0000	0.6634	**0.4102**	**0.0017**	**0.7043**	0.4701	0.2500	0.6101
$w = 16$	0.0183	0.0000	0.5092	0.1073	0.1400	0.4837	*			0.1353	0.3333	0.401

8   Conclusions

In this paper, we proposed new defense strategies against physical domain attacks with a special focus on the multi-sticker attacks, like camouflage art attack. Our proposed methods attempt to maximize the likelihood of diminishing the effect of the physical perturbation, given the defender's different levels of knowledge. We conducted an extensive amount of experiments to show that our proposed defense can successfully defend against the camouflage art attack under many scenarios with small classification drops on unattacked objects. Additionally, we built a new database using the camouflage art attack that contains photos of 90 physical attacked traffic signs and six source signs. This database may benefit future research in the community.

References

1. Athalye, A., Engstrom, L., Ilyas, A., Kwok, K.: Synthesizing robust adversarial examples. arXiv preprint arXiv:1707.07397 (2017)
2. Bastani, O., Ioannou, Y., Lampropoulos, L., Vytiniotis, D., Nori, A., Criminisi, A.: Measuring neural net robustness with constraints. In: Advances in Neural Information Processing Systems, pp. 2613–2621 (2016)

3. Bhagoji, A.N., Cullina, D., Mittal, P.: Dimensionality reduction as a defense against evasion attacks on machine learning classifiers 2. arXiv preprint arXiv:1704.02654 (2017)
4. Brown, T.B., Mane, D., Roy, A., Abadi, M., Gilmer, J.: Adversarial patch (2017)
5. Carlini, N., Wagner, D.: Towards evaluating the robustness of neural networks. In: 2017 IEEE Symposium on Security and Privacy (sp), pp. 39–57. IEEE (2017)
6. Chen, P.Y., Sharma, Y., Zhang, H., Yi, J., Hsieh, C.J.: Ead: elastic-net attacks to deep neural networks via adversarial examples. In: Thirty-Second AAAI Conference on Artificial intelligence (2018)
7. Chiang, P.Y., Ni, R., Abdelkader, A., Zhu, C., Studor, C., Goldstein, T.: Certified defenses for adversarial patches. arXiv preprint arXiv:2003.06693 (2020)
8. Das, N., et al.: Shield: fast, practical defense and vaccination for deep learning using jpeg compression. In: Proceedings of the 24th ACM SIGKDD International Conference on Knowledge Discovery & Data Mining, pp. 196–204 (2018)
9. Dhillon, G.S., et al.: Stochastic activation pruning for robust adversarial defense. arXiv preprint arXiv:1803.01442 (2018)
10. Dziugaite, G.K., Ghahramani, Z., Roy, D.M.: A study of the effect of jpg compression on adversarial images. arXiv preprint arXiv:1608.00853 (2016)
11. Eykholt, K., et al.: Robust physical-world attacks on deep learning visual classification. In: Proceedings of the IEEE Conference on Computer Vision and Pattern Recognition, pp. 1625–1634 (2018)
12. Feinman, R., Curtin, R.R., Shintre, S., Gardner, A.B.: Detecting adversarial samples from artifacts. arXiv preprint arXiv:1703.00410 (2017)
13. Geiger, A., Lenz, P., Urtasun, R.: Are we ready for autonomous driving? the kitti vision benchmark suite. In: 2012 IEEE Conference on Computer Vision and Pattern Recognition, pp. 3354–3361. IEEE (2012)
14. Goodfellow, I.J., Shlens, J., Szegedy, C.: Explaining and harnessing adversarial examples. arXiv preprint arXiv:1412.6572 (2014)
15. Guo, C., Rana, M., Cisse, M., Van Der Maaten, L.: Countering adversarial images using input transformations. arXiv preprint arXiv:1711.00117 (2017)
16. Hayes, J.: On visible adversarial perturbations & digital watermarking. In: Proceedings of the IEEE Conference on Computer Vision and Pattern Recognition Workshops, pp. 1597–1604 (2018)
17. Karmon, D., Zoran, D., Goldberg, Y.: Lavan: Localized and visible adversarial noise. arXiv preprint arXiv:1801.02608 (2018)
18. Kos, J., Fischer, I., Song, D.: Adversarial examples for generative models. In: 2018 IEEE Security and Privacy Workshops (SPW), pp. 36–42. IEEE (2018)
19. Kurakin, A., Goodfellow, I., Bengio, S.: Adversarial examples in the physical world. arXiv preprint arXiv:1607.02533 (2016)
20. Levine, A., Feizi, S.: (de) randomized smoothing for certifiable defense against patch attacks. arXiv preprint arXiv:2002.10733 (2020)
21. Liu, Y., Chen, X., Liu, C., Song, D.: Delving into transferable adversarial examples and black-box attacks. arXiv preprint arXiv:1611.02770 (2016)
22. Madry, A., Makelov, A., Schmidt, L., Tsipras, D., Vladu, A.: Towards deep learning models resistant to adversarial attacks. arXiv preprint arXiv:1706.06083 (2017)
23. Mogelmose, A., Trivedi, M.M., Moeslund, T.B.: Vision-based traffic sign detection and analysis for intelligent driver assistance systems: perspectives and survey. IEEE Trans. Intell. Transp. Syst. 13(4), 1484–1497 (2012)
24. Moosavi-Dezfooli, S.M., Fawzi, A., Fawzi, O., Frossard, P.: Universal adversarial perturbations. In: Proceedings of the IEEE Conference on Computer Vision and Pattern Recognition, pp. 1765–1773 (2017)

25. Moosavi-Dezfooli, S.M., Fawzi, A., Frossard, P.: Deepfool: a simple and accurate method to fool deep neural networks. In: The IEEE Conference on Computer Vision and Pattern Recognition (2016)
26. Naseer, M., Khan, S., Porikli, F.: Local gradients smoothing: defense against localized adversarial attacks. In: 2019 IEEE Winter Conference on Applications of Computer Vision (WACV), pp. 1300–1307. IEEE (2019)
27. Nguyen, A., Yosinski, J., Clune, J.: Deep neural networks are easily fooled: high confidence predictions for unrecognizable images. In: Proceedings of the IEEE Conference on Computer Vision and Pattern Recognition, pp. 427–436 (2015)
28. Papernot, N., McDaniel, P., Jha, S., Fredrikson, M., Celik, Z.B., Swami, A.: The limitations of deep learning in adversarial settings. In: 2016 IEEE European Symposium on Security and Privacy (EuroS&P), pp. 372–387. IEEE (2016)
29. Papernot, N., McDaniel, P., Wu, X., Jha, S., Swami, A.: Distillation as a defense to adversarial perturbations against deep neural networks. In: 2016 IEEE Symposium on Security and Privacy (SP), pp. 582–597. IEEE (2016)
30. Raghunathan, A., Steinhardt, J., Liang, P.: Certified defenses against adversarial examples. arXiv preprint arXiv:1801.09344 (2018)
31. Selvaraju, R.R., Cogswell, M., Das, A., Vedantam, R., Parikh, D., Batra, D.: Gradcam: visual explanations from deep networks via gradient-based localization. In: The IEEE International Conference on Computer Vision (ICCV) (2017)
32. Shaham, U., et al.: Defending against adversarial images using basis functions transformations. arXiv preprint arXiv:1803.10840 (2018)
33. Su, J., Vargas, D.V., Sakurai, K.: One pixel attack for fooling deep neural networks. IEEE Trans. Evol. Comput. 23(5), 828–841 (2019)
34. Szegedy, C., et al.: Intriguing properties of neural networks (2013)
35. Urmson, C., et al.: Autonomous driving in urban environments: boss and the urban challenge. J. Field Rob. 25(8), 425–466 (2008)
36. Xu, W., Evans, D., Qi, Y.: Feature squeezing: Detecting adversarial examples in deep neural networks. arXiv preprint arXiv:1704.01155 (2017)
37. Xu, Z., Yu, F., Chen, X.: Lance: A comprehensive and lightweight CNN defense methodology against physical adversarial attacks on embedded multimedia applications (2019)
38. Zhang, F., Leitner, J., Milford, M., Upcroft, B., Corke, P.: Towards vision-based deep reinforcement learning for robotic motion control. arXiv preprint arXiv:1511.03791 (2015)

Adversarial Attack on Deepfake Detection Using RL Based Texture Patches

Steven Lawrence Fernandes[1(✉)] and Sumit Kumar Jha[2]

[1] Creighton University, Omaha, NE 68131, USA
stevenfernandes@creighton.edu
[2] University of Texas at San Antonio, San Antonio, TX 78249, USA
sumit.jha@utsa.edu

Abstract. The advancements in GANs have made creating deepfake videos a relatively easy task. Considering the threat that deepfake videos pose for manipulating political opinion, recent research has focused on ways to better detect deepfake videos. Even though researchers have had some success in detecting deepfake videos, it has been found that these detection systems can be attacked.

The key contributions of this paper are (a) a deepfake dataset created using a commercial website, (b) validation of the efficacy of DeepExplainer and heart rate detection from the face for differentiating real faces from adversarial attacks, and (c) the proposal of an attack on the FaceForensics++ deepfake detection system using a state-of-the-art reinforcement learning-based texture patch attack. To the best of our knowledge, we are the first to successfully attack FaceForensics++ on our commercial deepfake dataset and DeepfakeTIMIT dataset.

Keywords: Adversarial attack · Deepfake · FaceForensics++

1 Introduction

Deepfake videos are manipulated videos produced by sophisticated generative adversarial networks [15] that yield realistic images and videos that are very difficult to identify as manipulated by the human eye [26]. They include automated manipulations of faces in a video while preserving the other original facial features, such as nose size and distance between the lips and nose [51]. This technology can be misused to generate pornographic videos [41] and manipulate political opinions [34], among other activities. Generative adversarial networks (GANs) are widely used in such videos, audios and image generations. One of the most popular deepfake video applications is Snapchat [6]. It captures a 3D model of users' faces, swaps the face instantaneously and then generates fake videos and images. Other similar deepfake applications include Zao, FaceApp, and FaceSwap, among others. These applications are built using GANs.

Recent research has focused on the detection of deepfakes using spatiotemporal convolutional networks [31], visual heartbeat rhythms [43], optical flow [8],

© Springer Nature Switzerland AG 2020
A. Bartoli and A. Fusiello (Eds.): ECCV 2020 Workshops, LNCS 12535, pp. 220–235, 2020.
https://doi.org/10.1007/978-3-030-66415-2_14

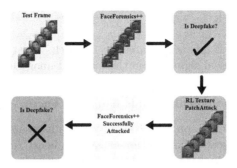

Fig. 1. Overview of the RL based texture patch attack on FaceForensics++.

heart rates [19], attribution-based confidence [18], face weighting [37], variational autoencoders [24], analysis of convolutional traces [20], and the widely used FaceForensics++ [44]. In this paper, we attack FaceForensics++ using a state-of-the-art reinforcement learning (RL)-based texture patch attack [53]. The test video frames are given to FaceForensics++ to be classified as real or fake. The correctly classified real/fake frames are then submitted to RL-based texture patch attack. If the test video is correctly classified by FaceForensics++ as being fake, we apply the RL-based texture patch attack to it and classify the label changes as real. Similarly, if the test video is correctly classified by FaceForensics++ as being real, we apply the RL-based texture patch attack to it and classify the label changes as fake. The overview of our proposed system is shown in Fig. 1.

Apart from proposing the RL-based texture patch attack for FaceForensics++, we also validated the efficacy of using DeepExplainer [33] and heart rates obtained from fakes to detect deepfakes [2]. Using commercial website [1], we created a new deepfake dataset. The commercial website [1] is easily accessible by anyone. It is therefore necessary to develop approaches that produce excellent results. To create a deepfake dataset, we considered original and donor videos from YouTube. Three hundred minutes GPU time [1] was used to generate deepfake video. The dataset will be made publicly available for research.

The significant contributions of this work include:

1. A new deepfake dataset created using donor and original videos from YouTube.
2. Validation of the efficacy of DeepExplainer and heart rate detection from the face for differentiating real faces from and adversarial attacks.
3. The proposal of an attack on the FaceForensics++ deepfake detection system using a state-of-the-art reinforcement learning-based texture patch attack.

To the best of our knowledge, we are the first to attack FaceForensics++ using a state-of-the-art reinforcement learning-based texture patch attack.

2   Related Work

Detecting deepfakes is an active area of research. Most deepfake detection approaches use biological signals from the face and image forensics.

2.1   Extracting Biological Signals from the Face

Remote photoplethysmography (rPPG) [45] and ballistocardiograms (BCGs) [9] are two commonly used techniques for extracting heart rate from face videos. Between the two approaches, rPPG is found to produce heart rate values closer to the ground truthed values. To extract heart rate data from faces, color variations and subtle motion are observed in videos [52]. The commonly used rPPG techniques include applying Kalman filters [42], analyzing optical properties [17] and extracting heart rates from specific positions in the face [54].

Recently, researchers have approximated heart rates according to the face [13]. They exploited temporal consistency and spatial coherence to detect deepfakes using heart rates obtained from a face video. Another approach was proposed by [43]. They inferred that heart rates obtained from original face videos will disrupt deepfake videos. This is done using spatially and temporally based attention to dynamically adapt to the changes that occur between the original and deepfake faces. Neural ordinary differential equations (Neuro-ODE) were also used to detect heart rate variability in the original and deepfake videos by Fernandes et al. [19]. The major drawback of using biological signals from the face to detect deepfakes is that there are few datasets with ground truth physiological signals. The commonly used datasets include the UBFC-RPPG [10] and LGI-PPGI [40] databases.

2.2   Image Forensics

The first actual expression transmission for facial re-enactment was demonstrated by Thies et al. [48]. They used a customer-level RGB-D camera, reconstructed the footage, and tracked the 3D model of the actor both at the source and at the target. The distortions they came across were applied to the target actor. Kim et al. [25] converted computer graphic versions of faces to actual images using a network that learns image-to-image translation. Neural Textures is better than pure image-to-image translation networks, as it optimizes neural texture in combination with a network to calculate the re-enactment result. Neural Textures gives sharper results than Deep Video Portraits [25] in the mouth area. Face aging is another approach applied by [7] to generate good results. Deep feature interpolation [50] is found to work well in changing face qualities such as age, moustaches, and laughing, among others. Other similar methods used to detect deepfakes are reported by [5,11,21,22,32,36,39].

Among the various deepfake detection approaches, FaceForensics++ [44] is widely used because it was tested on 1.8 million altered images. The authors used the approach reported by Thies et al. [50] to track the faces in the videos and extract only the face region in the image.

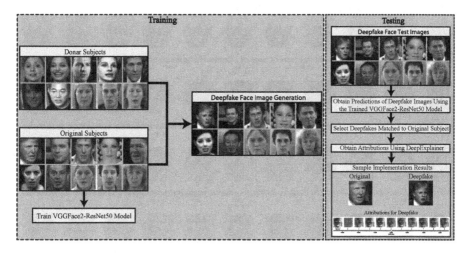

Fig. 2. Training using VGGFace2-ResNet50 on original videos and applying DeepExplainer to obtain attributions of deepfake videos matched to the original subjects.

3 Proposed Approach

To create deepfake videos, donor and original subject videos were taken from YouTube and submitted to a commercial website [1]. The original subject videos were used to train the VGGFace2-ResNet50 model. During testing, the deepfake videos obtained from the commercial website [1] were given to the trained model. The test video frames that did not match the original subject were discarded, and those that matched the original subject were given to DeepExplainer to obtain the attributions. The Fig. 2 shows the training and testing process.

3.1 DeepExplainer Attributions for Deepfake Videos

DeepExplainer uses game theory to explain the prediction of the VGGFace2-ResNet50 model. All features are considered contributors. It uses DeepLIFT (Deep Learning Important FeaTures) [47] and Shapley [46] values. DeepLIFT manages the contribution score of each neuron by using the linear, rescale, and RevealCancel rules.

Consider t is the output neuron and $x_1, x_2, x_3, \cdots, x_n$ the intermediate layers of the deep learning network that are sufficient to calculate t. Let t_1 be the reference activation of t. Then, Δt is the difference between t and t_1. The contribution score is represented by $C_{\Delta x_i, \Delta t}$. DeepLIFT assigns this score to δx_i.

In the raw gradient method, the importance of the feature to the prediction of the model might become saturated. When this happens, changing the value of the feature does not cause a noticeable change in the prediction of the model. The feature takes up a value that has a maximum impact on model prediction. The gradient method therefore fails to capture maximal impacting values of the feature. Unlike the gradient method, the contribution score approach presented in

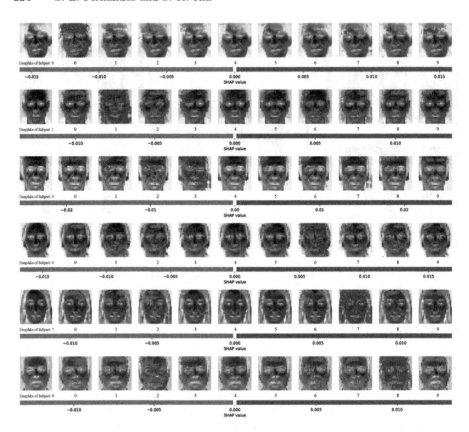

Fig. 3. Attributions obtained using DeepExplainer for deepfake videos frames that matched the original subjects from the YouTube dataset.

DeepExplainer will be nonzero during saturation. The rules and constraints are defined so that it is possible to overcome the saturation problem. Furthermore, sudden jumps and gaps in the importance score cannot cause the contribution score to suffer and compromise the prediction value of the network.

In Fig. 3, the first column is the deepfake video frame that is matched to the corresponding original subject. The attributions are obtained using DeepExplainer for deepfake video frames that are matched to the original subjects. The deepfake videos are obtained by uploading the donor and original videos from the YouTube dataset to the commercial website [1].

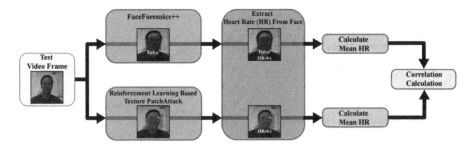

Fig. 4. Extraction of heart rate from the face before and after using the texture patch attack.

3.2 Heart Rate Extraction from the Face

Heart rate [13] and visual heartbeat rhythms [43] are used to detect deepfakes. We therefore extracted the heart rate from the face using the approach presented in [2] before and after the texture patch attack [53]. We found the mean heart rates and calculated the correlation between them as shown in Fig. 4. Algorithm 1 explains the process in detail.

3.3 Detecting Deepfake Using FaceForensics++

The test video frame was given to FaceForensics++ [44] to detect whether it is real or fake. The predicted output from FaceForensics++ was compared with the ground truthed data. Predictions for the video frames that matched the ground truthed data were submitted to the RL-based texture patch attack, and the accuracy was calculated.

Algorithm 1. Heart rate extraction from the face video using this approach [2]

Input:

→ Video V	→ Each frame of video F^N
→ beats per minute/heart rate bpm	→ size of the data buffer buffer_size
→ size of the data buffer buffer_size	→ buffer to store data data_buffer
→ Fourier transform on data fft	→ current time to
→ frame after patch attack F^P	→ Frames index to be attacked: indexes
→ current frame F	

Output: Heart (bpm) rate of each of the frame F of the video V.

→ bpms_afterattack = [ ]	→ Sample = [ ]
→ buffer_size = 100	→ data_buffer = [ ]
→ fft = [ ]	→ bpm = 0
→ bpms = [ ]	→ peaks = [ ]
→ to = current_time	

```
 1: for i in range(2) do
 2:     for f_i in indexes do
 3:         if i == 0 then
 4:             F = F^N[f_i]
 5:         else
 6:             F = F^p[f_i]
 7:             face = detectFace(F)                          ▷ detect face from image
 8:             face_align = alignface(face)                       ▷ align the face
 9:             ROI_1, ROI_2 = detectchecks(face_align)      ▷ detect checks from face
10:             mask = face_remap(face_align)            ▷ get keypoints from face
11:             g_1 = mean(ROI_1)                ▷ mean value of right check (ROI1)
12:             g_2 = mean(ROI_2)                 ▷ mean value of left check (ROI2)
13:             L = len(data_buffer)
14:             g = mean(g_1, g_2)
15:             if abs(g - mean(data_buffer)) > 10 and L>99 then     ▷ sudden spikes
16:                 g =data_buffer[-1]        ▷ set mean to most recent data buffer value
17:                 times.append(current_time_to)
18:                 if L>buffer_size then
19:                     data_duffer = data_buffer[-buffer_size]
20:                     bpms = bpms[-buffer_size//2:]   ▷ Keep last n/2 values for bpms
21:                     L = buffer_size
22:                     times = times[-buffer_size:]              ▷ Keep last n values
23:                 end if
24:                 if L==buffer_size then   ▷ Minimum frames to measure heart rate
25:                     Fps = L/times[-1] - times[0]     ▷ Calculate FPS of the processor
26:                     even_time = generatetime(times[0], times[1], L)
27:                     processed = dctrend(processed)            ▷ detrend the signals
28:                     interpolated = interpolate(even_times,times,processed)
29:                     interpolated = hamming(L)*interpolated ▷ signal made periodic
30:                     norm = Norm(interpolated)
31:                     raw = rfft(norm*30)                    ▷ Fourier transformation
32:                     freqs = fps/L * (0,1, ..., L/2+1)
33:                     freqs = 60 * freqs
34:                     fft = abs(raw)**2                    ▷ get amplitude spectrum
35:                     idx = freq > 50 and freq < 180                    ▷ HR range
36:                     Fft = fft[idx]
37:                     Freq = freq[idx] bpm = freq[argmax(fft)]          ▷ Heart Rate
38:                     bpms·append(bpm)
39:                     processed = band_pass_filter(processed, 0.8, 3, fps, order=3)
40:                 end if
41:             end if
42:         end if
43:         if i == 0 then
44:             bpms.append(bpm)
45:         else
46:             bpms_afterattack·append(bpm)
47:         end if
48:         processed = band_pass_filter(processed, 0.8, 3, fps, order = 3)
49:     end for
50: end for
51: print(corr(bpms, bpms_afterattack))
return
```

Algorithm 2. Applied the RL based patch attack [53] on FaceForensics++

1: $S = (U_1^1, V_1^2, i_1^3, U_1^4, V_1^5, \cdots, U_c^1, V_c^2, i_c^3, U_c^4, V_c^5)$ ▷ U_c^1, V_c^2 determines patch position
        in x ▷ i_c^3 is the texture image   ▷ U_c^4, V_c^5 determines patch position in i_c^3 texture
2: **for** C in range(maximumpatches **do**
3:     **for** t in range(total_iterations **do**
4:       $a = A(\Theta_A)$ ▷ $A(\theta_A) = P(a_t | a_1, \cdots, a_{t-1}, F(\cdot, \theta), x)$
5:       $r = \ln y^* - A(a)/\sigma^2$
6:       $L = -r \cdot \ln P$
7:       $y = f(g(x); \theta_A)$ ▷ $\epsilon = J_t^1(U_c^1, U_c^2)$
8:       **if** $y \cdot$ argmax() $== y^*$ **then**   ▷ $Tx_{U;V} = J_t^2(i_c, V_c^4, V_c^5)$
9:          print("successful")   ▷ J_t^1 patch crop
10:         return ymax(), "success"   ▷ P probability seq of actions
11:       **end if**
12:     **end for**
13: **end for**
return

3.4 RL Based Texture Patch Attack on FaceForensics++

Adversarial patch attacks are designed to fool deep learning models. They are the most practical threats against real-world computer vision techniques. The test video frame is given to FaceForensics++ [44] to detect if it is real or fake. The predicted output from FaceForensics++ is compared with the ground truthed data. To the correctly predicted frames, we apply the RL-based texture patch attack. The RL-based texture patch attack proposed by [53] is currently the state-of-the-art black-box patch attack. The authors have shown that its procedures have high accuracy for nontargeted and targeted attacks. The procedure performs a search over the position and shape of the adversarial patches using reinforcement learning. In addition, the authors showed that attacks with monochrome patches cannot succeed as targeted attacks. This is because monochrome patches can neither remove any information from the images nor add any information. This makes it critical to fool the model in a targeted manner. Hence, the authors applied texture to the adversarial patches to overcome this limitation.

The two classes that we have considered in our approach are real and fake. If the output of FaceForensics++ was real and the ground truth data were real, then the correctly classified video frames were submitted to the RL-based texture patch attack. Similarly, if the output of FaceForensics++ was fake and the ground truth data were fake, then the correctly classified video frames were submitted to the RL-based texture patch attack. The mathematical formulations are presented in Algorithm 2. This method shows that even small textured patches are able to break deep neural networks. The attack process is modeled on RL with an agent trained to superimpose patches onto the images to induce mis-classifications as shown in Fig. 5.

Fig. 5. FaceForensics++ attacked using the RL based texture patch attack.

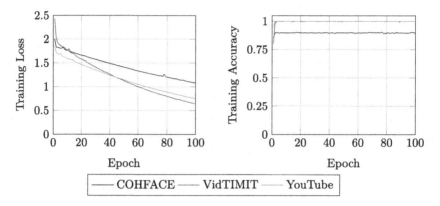

Fig. 6. (Left) Training Loss. (Right) Training Accuracy. Obtained for three original datasets: COHFACE, VidTIMIT, and YouTube.

4   Results and Discussion

We validated our approach by using three deepfake datasets. Two were created by uploading the original videos from the YouTube and COHFACE datasets to a commercial website [1]. The third deepfake dataset was DeepfakeTIMIT [26]; its original videos were from VidTIMIT. We used DeepExplainer to obtain the attributions of deepfake videos. Heart rate was extracted from the face of the original/fake videos after applying FaceForensics++ and the RL-based texture patch attack. The accuracy of the RL-based texture patch attack was calculated using all three of the datasets.

4.1   DeepExplainer Attributions for Deepfake Videos

We trained the VGGFace2-ResNet50 model for 100 epochs on only the original videos using Google Colab Pro with 64 GB RAM and TPUs. The training loss and accuracy for the YouTube, COHFACE and VidTIMIT datasets are shown Fig. 6. The trained model was tested on deepfake videos; the frames that did not match the original subject were discarded, and the attributions for the others were obtained using DeepExplainer. The DeepExplainer results obtained for YouTube, COHFACE and DeepfakeTIMIT are shown in Fig. 3, 8, and 10, respectively.

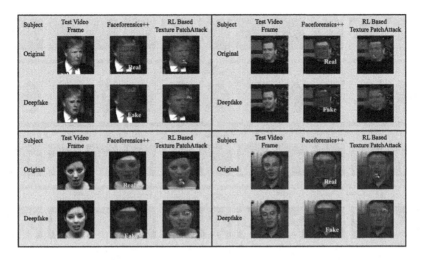

Fig. 7. Original video from the YouTube dataset and deepfake videos were applied to FaceForensics++ for prediction and then attacked using the RL based texture patches.

4.2   Heart Rate Extraction from the Face

The test video, which can be an original or a deepfake video, is converted to frames and applied to FaceForensics++. The heart rate is then extracted using Algorithm 1. The frames are also applied to the RL-based texture patch attack, and the heart rate is again extracted using the algorithm presented in Sect. 3.3. Mean heart rate is obtained for each subject, and the correlation between these values is calculated. The correlation values are obtained for the original and fake videos with all 3 datasets.

4.3   RL Based Texture Patch Attack on FaceForensics++

Prior work on attacking deepfake detectors included manipulating the face images proposed by [38] using well-known adversarial noise [49] on basic CNN models XceptionNet [12] and MesoNet [5]. However, there were several defenses against such adversarial attacks [14,16,23,27–29,35]. In this paper, we used a state-of-the-art RL-based texture patch attack [53] to expose the vulnerability of deepfake detectors. The original videos from the YouTube, COHFACE and VidTIMIT datasets were applied to FaceForensics++ and compared to the ground truthed data. If they were correctly predicted, they were then attacked using the RL-based texture patch attack. Similarly, the deepfake videos from the three datasets were applied to FaceForensics++ and compared to the ground truthed data. If they were correctly predicted, they were then attacked using the RL-based texture patch attack. The accuracy of the attack was calculated. The results obtained for the YouTube, COHFACE and DeepfakeTIMIT datasets are shown in Fig. 7, 9, and 11, respectively.

Fig. 8. Attributions obtained using deepexplainer for deepfake videos frames that are matched to the original subjects from COHFACE dataset

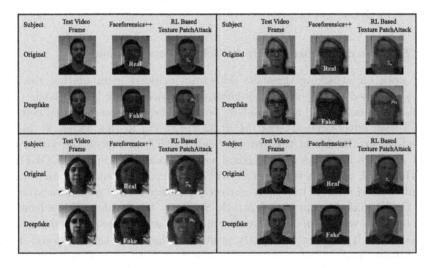

Fig. 9. Original video from the COHFACE dataset and deepfake videos applied to FaceForensics++ for prediction and then attacked using the RL-based texture patches.

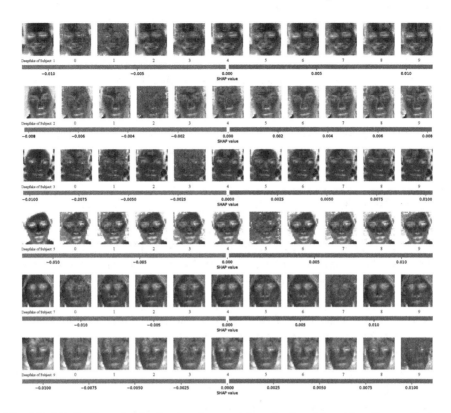

Fig. 10. Attributions obtained using deepexplainer for deepfake videos frames that are matched to the original subjects from VidTIMIT dataset

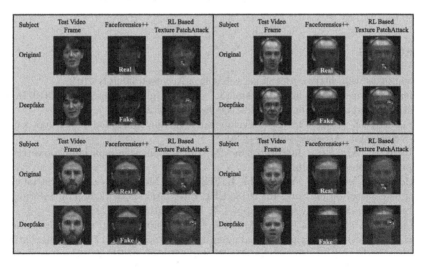

Fig. 11. Original video from the VidTIMIT dataset and deepfake videos were applied to FaceForensics++ for prediction and then attacked using the RL based texture patches.

Table 1. Accuracy of the texture patch attack calculated for the three datasets.

Database	Total no. of frames	No. of frames correctly classified by FaceForensics++	No. of frames successfully attacked by RL texture patches	Accuracy
COHFACE	6263	2212	1865	84.31%
VidTIMIT	1137	244	211	86.40%
YouTube	885	563	548	97.34%

5   Conclusion and Future Work

In this paper, we used state-of-the-art RL-based texture patches to attack Face-Forensics++ with three datasets. The heart rate correlation values range from -0.3 to 0.8. The RL-based texture patch attack accuracy is tabulated in Table 1.

The key contributions of our paper are:

1. The creation of deepfake datasets using a commercial website.
2. Validation of the efficacy of DeepExplainer and heart rate detection from the face for differentiating real faces from adversarial attacks.
3. The attack on FaceForensics++ using the RL-based texture patch attack, obtaining an accuracy of 84.31%, 86.40%, 97.34% for real video frames obtained from COHFACE, VidTIMIT and YouTube datasets.

To the best of our knowledge, we were the first to attack FaceForensics++ using the RL-based texture patch attack. In the future, we will implement the proposed approach using the latest embedded devices, including the Jetson AGX Xavier [3] and OpenCV AI Kits [4] on the Celeb-DF [30] dataset.

References

1. Deepfakes web. https://deepfakesweb.com/. Accessed 20 July 2019
2. Heart-rate-measurement-using-camera. https://github.com/habom2310/Heart-rate-measurement-using-camera. Accessed 30 Apr 2020
3. Jetson AGX Xavier. https://developer.nvidia.com/embedded/jetson-agx-xavier-developer-kit. Accessed 30 Mar 2020
4. OpenCV AI Kit. https://opencv.org/introducing-oak-spatial-ai-powered-by-opencv. Accessed 30 June 2020
5. Afchar, D., Nozick, V., Yamagishi, J., Echizen, I.: MesoNet: a compact facial video forgery detection network. In: 2018 IEEE International Workshop on Information Forensics and Security (WIFS), pp. 1–7. IEEE (2018)
6. Agarwal, A., Singh, R., Vatsa, M., Noore, A.: Swapped! Digital face presentation attack detection via weighted local magnitude pattern. In: 2017 IEEE International Joint Conference on Biometrics (IJCB), pp. 659–665. IEEE (2017)
7. Amerini, I., Ballan, L., Caldelli, R., Del Bimbo, A., Serra, G.: A sift-based forensic method for copy-move attack detection and transformation recovery. IEEE Trans. Inf. Forensics Secur. **6**(3), 1099–1110 (2011)

8. Amerini, I., Galteri, L., Caldelli, R., Del Bimbo, A.: Deepfake video detection through optical flow based CNN. In: Proceedings of the IEEE/CVF International Conference on Computer Vision (ICCV) Workshops, October 2019

9. Balakrishnan, G., Durand, F., Guttag, J.: Detecting pulse from head motions in video. In: Proceedings of the IEEE Conference on Computer Vision and Pattern Recognition, pp. 3430–3437 (2013)

10. Bobbia, S., Macwan, R., Benezeth, Y., Mansouri, A., Dubois, J.: Unsupervised skin tissue segmentation for remote photoplethysmography. Pattern Recogn. Lett. **124**, 82–90 (2019)

11. Choi, Y., Choi, M., Kim, M., Ha, J.W., Kim, S., Choo, J.: StarGAN: unified generative adversarial networks for multi-domain image-to-image translation. In: Proceedings of the IEEE Conference on Computer Vision and Pattern Recognition, pp. 8789–8797 (2018)

12. Chollet, F.: Xception: deep learning with depthwise separable convolutions. In: Proceedings of the IEEE Conference on Computer Vision and Pattern Recognition, pp. 1251–1258 (2017)

13. Ciftci, U.A., Demir, I.: Fakecatcher: detection of synthetic portrait videos using biological signals. arXiv preprint arXiv:1901.02212 (2019)

14. Cohen, J.M., Rosenfeld, E., Kolter, J.Z.: Certified adversarial robustness via randomized smoothing. arXiv preprint arXiv:1902.02918 (2019)

15. Creswell, A., White, T., Dumoulin, V., Arulkumaran, K., Sengupta, B., Bharath, A.A.: Generative adversarial networks: an overview. IEEE Signal Process. Mag. **35**(1), 53–65 (2018)

16. Erichson, N.B., Yao, Z., Mahoney, M.W.: JumpReLU: a retrofit defense strategy for adversarial attacks. CoRR abs/1904.03750 (2019)

17. Feng, L., Po, L.M., Xu, X., Li, Y., Ma, R.: Motion-resistant remote imaging photoplethysmography based on the optical properties of skin. IEEE Trans. Circuits Syst. Video Technol. **25**(5), 879–891 (2014)

18. Fernandes, S., et al.: Detecting deepfake videos using attribution-based confidence metric. In: Proceedings of the IEEE/CVF Conference on Computer Vision and Pattern Recognition (CVPR) Workshops, June 2020

19. Fernandes, S., et al.: Predicting heart rate variations of deepfake videos using neural ode. In: Proceedings of the IEEE International Conference on Computer Vision Workshops (2019)

20. Guarnera, L., Giudice, O., Battiato, S.: Deepfake detection by analyzing convolutional traces. In: Proceedings of the IEEE/CVF Conference on Computer Vision and Pattern Recognition (CVPR) Workshops, June 2020

21. He, Z., Zuo, W., Kan, M., Shan, S., Chen, X.: AttGAN: facial attribute editing by only changing what you want. IEEE Trans. Image Process. **28**(11), 5464–5478 (2019)

22. Hochreiter, S., Schmidhuber, J.: Long short-term memory. Neural Comput. **9**(8), 1735–1780 (1997)

23. Hu, S., Yu, T., Guo, C., Chao, W.L., Weinberger, K.Q.: A new defense against adversarial images: turning a weakness into a strength. In: Advances in Neural Information Processing Systems, pp. 1635–1646 (2019)

24. Khalid, H., Woo, S.S.: OC-FakeDect: classifying deepfakes using one-class variational autoencoder. In: Proceedings of the IEEE/CVF Conference on Computer Vision and Pattern Recognition (CVPR) Workshops, June 2020

25. Khodabakhsh, A., Ramachandra, R., Raja, K., Wasnik, P., Busch, C.: Fake face detection methods: can they be generalized? In: 2018 International Conference of the Biometrics Special Interest Group (BIOSIG), pp. 1–6. IEEE (2018)

26. Korshunov, P., Marcel, S.: DeepFakes: a new threat to face recognition? assessment and detection. CoRR abs/1812.08685. arXiv preprint arXiv:1812.08685 (2018)
27. Kurakin, A., Goodfellow, I., Bengio, S.: Adversarial machine learning at scale. arXiv preprint arXiv:1611.01236 (2016)
28. Lecuyer, M., Atlidakis, V., Geambasu, R., Hsu, D., Jana, S.: Certified robustness to adversarial examples with differential privacy. In: 2019 IEEE Symposium on Security and Privacy (SP), pp. 656–672. IEEE (2019)
29. Li, B., Chen, C., Wang, W., Carin, L.: Certified adversarial robustness with additive noise. In: Wallach, H., Larochelle, H., Beygelzimer, A., d' Alché-Buc, F., Fox, E., Garnett, R. (eds.) Advances in Neural Information Processing Systems, vol. 32, pp. 9464–9474. Curran Associates, Inc. (2019). http://papers.nips.cc/paper/9143-certified-adversarial-robustness-with-additive-noise.pdf
30. Li, Y., Yang, X., Sun, P., Qi, H., Lyu, S.: Celeb-DF: a large-scale challenging dataset for deepfake forensics. In: Proceedings of the IEEE/CVF Conference on Computer Vision and Pattern Recognition, pp. 3207–3216 (2020)
31. de Lima, O., Franklin, S., Basu, S., Karwoski, B., George, A.: Deepfake detection using spatiotemporal convolutional networks. arXiv preprint arXiv:2006.14749 (2020)
32. Liu, M., et al.: StGAN: a unified selective transfer network for arbitrary image attribute editing. In: Proceedings of the IEEE Conference on Computer Vision and Pattern Recognition, pp. 3673–3682 (2019)
33. Lundberg, S.M., Lee, S.I.: A unified approach to interpreting model predictions. In: Advances in Neural Information Processing Systems, pp. 4765–4774 (2017)
34. Maddocks, S.: 'A deepfake porn plot intended to silence me': exploring continuities between pornographic and political deep fakes. Porn Stud. 1–9 (2020)
35. Madry, A., Makelov, A., Schmidt, L., Tsipras, D., Vladu, A.: Towards deep learning models resistant to adversarial attacks. arXiv preprint arXiv:1706.06083 (2017)
36. Matern, F., Riess, C., Stamminger, M.: Exploiting visual artifacts to expose deepfakes and face manipulations. In: 2019 IEEE Winter Applications of Computer Vision Workshops (WACVW), pp. 83–92. IEEE (2019)
37. Montserrat, D.M., et al.: Deepfakes detection with automatic face weighting. In: Proceedings of the IEEE/CVF Conference on Computer Vision and Pattern Recognition (CVPR) Workshops, June 2020
38. Neekhara, P., Hussain, S., Jere, M., Koushanfar, F., McAuley, J.: Adversarial deepfakes: evaluating vulnerability of deepfake detectors to adversarial examples. arXiv preprint arXiv:2002.12749 (2020)
39. Nguyen, H.H., Fang, F., Yamagishi, J., Echizen, I.: Multi-task learning for detecting and segmenting manipulated facial images and videos. arXiv preprint arXiv:1906.06876 (2019)
40. Pilz, C.S., Zaunseder, S., Krajewski, J., Blazek, V.: Local group invariance for heart rate estimation from face videos in the wild. In: Proceedings of the IEEE Conference on Computer Vision and Pattern Recognition Workshops, pp. 1254–1262 (2018)
41. Popova, M.: Reading out of context: pornographic deepfakes, celebrity and intimacy. Porn Stud. 1–15 (2019)
42. Prakash, S.K.A., Tucker, C.S.: Bounded Kalman filter method for motion-robust, non-contact heart rate estimation. Biomed. Opt. Express $9(2)$, 873–897 (2018)
43. Qi, H., et al.: DeepRhythm: exposing deepfakes with attentional visual heartbeat rhythms. arXiv preprint arXiv:2006.07634 (2020)

44. Rossler, A., Cozzolino, D., Verdoliva, L., Riess, C., Thies, J., Nießner, M.: Face-Forensics++: learning to detect manipulated facial images. In: Proceedings of the IEEE International Conference on Computer Vision, pp. 1–11 (2019)

45. Rouast, P.V., Adam, M.T., Chiong, R., Cornforth, D., Lux, E.: Remote heart rate measurement using low-cost RGB face video: a technical literature review. Front. Comput. Sci. **12**(5), 858–872 (2018)

46. Shapley, L.S.: A value for n-person games. In: Contributions to the Theory of Games vol. 2, no. 28, pp. 307–317 (1953)

47. Shrikumar, A., Greenside, P., Kundaje, A.: Learning important features through propagating activation differences. arXiv preprint arXiv:1704.02685 (2017)

48. Suwajanakorn, S., Seitz, S.M., Kemelmacher-Shlizerman, I.: Synthesizing Obama: learning lip sync from audio. ACM Trans. Graph. (TOG) **36**(4), 1–13 (2017)

49. Szegedy, C., et al.: Going deeper with convolutions. In: Proceedings of the IEEE Conference on Computer Vision and Pattern Recognition, pp. 1–9 (2015)

50. Thies, J., Zollhofer, M., Stamminger, M., Theobalt, C., Nießner, M.: Face2Face: real-time face capture and reenactment of RGB videos. In: Proceedings of the IEEE Conference on Computer Vision and Pattern Recognition, pp. 2387–2395 (2016)

51. Tolosana, R., Vera-Rodriguez, R., Fierrez, J., Morales, A., Ortega-Garcia, J.: Deep-Fakes and beyond: a survey of face manipulation and fake detection. arXiv preprint arXiv:2001.00179 (2020)

52. Wu, H.Y., Rubinstein, M., Shih, E., Guttag, J., Durand, F., Freeman, W.: Eulerian video magnification for revealing subtle changes in the world (2012)

53. Yang, C., Kortylewski, A., Xie, C., Cao, Y., Yuille, A.: PatchAttack: a black-box texture-based attack with reinforcement learning. arXiv preprint arXiv:2004.05682 (2020)

54. Zhao, C., Lin, C.L., Chen, W., Li, Z.: A novel framework for remote photoplethysmography pulse extraction on compressed videos. In: Proceedings of the IEEE Conference on Computer Vision and Pattern Recognition Workshops, pp. 1299–1308 (2018)

W02 - BioImage Computation

W02 - BioImage Computation

BioImage Computing (BIC) is a rapidly growing field at the interface of computer vision and the life sciences. Advanced light microscopy can deliver 2D and 3D image sequences of living cells with unprecedented image quality and ever increasing resolution in space and time. The emergence of novel and diverse microscopy modalities has provided biologists with unprecedented means to explore cellular mechanisms, embryogenesis, or neural development, to mention only a few areas of research. Electron microscopy provides information on the cellular structure at nanometer resolution. Here, correlating light microscopy and electron microscopy at the subcellular level, and relating both to animal behavior at the macroscopic level, is of paramount importance. The enormous size and complexity of these data sets, which can exceed multiple TB per volume or video, requires not just accurate, but also efficient computer vision methods.

Since 2015, the BIC workshop brings the latest challenges in bioimage computing to the computer vision community. It showcases the specificities of bioimage computing and its current achievements, including issues related to image modeling, denoising, super-resolution, multi-scale instance and semantic segmentation, motion estimation, image registration, tracking, classification, and event detection.

This year's BIC workshop was held virtually on August 23 at ECCV 2020 and was organized by Jan Funke (HHMI Janelia, USA), Dagmar Kainmueller (BIH and MDC Berlin, Germany), Florian Jug (CSBD and MPI-CBG, Dresden, Germany), Anna Kreshuk (EMBL Heidelberg, Germany), Peter Bajcsy (NIST, USA), Martin Weigert (EPFL, Switzerland), Patrick Bouthemy (INRIA, France), and Erik Meijering (UNSW Sydney, Australia).

Following an open review process through OpenReview (https://openreview.net/group?id=thecvf.com/ECCV/2020/Workshop/BIC), the organizers selected 17 out of 25 submissions for presentation at the workshop. Those submissions are representative of the wide range of exciting computer vision challenges we encounter in the life sciences, e.g., computational imaging, denoising of microscopy images, bioimage segmentation and classification, multi-modal registration, and tracking of cells and animals in videos.

August 2020

Jan Funke
Dagmar Kainmueller
Florian Jug
Anna Kreshuk
Peter Bajcsy
Martin Weigert
Patrick Bouthemy
Erik Meijering

A Subpixel Residual U-Net and Feature Fusion Preprocessing for Retinal Vessel Segmentation

Sohom Dey[✉]

Kalinga Institute of Industrial Technology, Bhubaneswar, India
sohom21d@gmail.com

Abstract. Retinal Image analysis allows medical professionals to inspect the morphology of the retinal vessels for the diagnosis of vascular diseases. Automated extraction of the vessels is vital for computer-aided diagnostic systems to provide a speedy and precise diagnosis. This paper introduces SpruNet, a Subpixel Convolution based Residual U-Net architecture which re-purposes subpixel convolutions as down-sampling and up-sampling method. The proposed subpixel convolution based down-sampling and up-sampling strategy efficiently minimizes the information loss during the encoding and decoding process which in turn increases the sensitivity of the model without hurting the specificity. A feature fusion technique of combining two types of image enhancement algorithms is also introduced. The model is trained and evaluated on three mainstream public benchmark datasets, and detailed analysis and comparison of the results are provided which shows that the model achieves state-of-the-art results with less complexity. The model can make inference on 512×512 pixel full image in $0.5\,s$.

Keywords: Medical image analysis · Retinal Vessel Segmentation · Subpixel convolution · Residual network · U-Net

1 Introduction

Retinal Vessel Analysis is a non-invasive method of examining the retinal vasculature comprising of a complicate and elaborate network of arteries, veins and capillaries. The morphology of the retinal vasculature is an important biomarker for diseases like Diabetic retinopathy, Hypertensive retinopathy, Retinal vein occlusion, Retinal artery occlusion, etc., which affects the retinal blood vessels. Retinal examination allows medical professionals dealing with vascular diseases to get a unique perspective, allowing them to directly inspect the morphology and draw conclusions about the health of a patient's micro-vasculature anywhere in the body. The retinal vasculature is also adopted as the most stable feature for multi modal retinal image registration and retinal mosaic. They are also being used for bio-metric identification.

© Springer Nature Switzerland AG 2020
A. Bartoli and A. Fusiello (Eds.): ECCV 2020 Workshops, LNCS 12535, pp. 239–250, 2020.
https://doi.org/10.1007/978-3-030-66415-2_15

Quantitative analysis of the retinal blood vessels, requires the vascular tree to be extracted so that morphological features like length, width, branching, angle, etc. can be calculated. Manual segmentation of the blood vessel is a difficult and time consuming task and requires expertise. Automating the task of vessel segmentation has gained importance and has been accepted as a vital and challenging step for retinal image analysis in computer aided diagnostic systems for ophthalmic and cardiovascular diseases.

Retinal vessels vary in shape, size and intensity level locally. The vessel width may range anywhere between 1 to 20 pixels depending on the anatomical vessel width and image resolution. Segmentation using artificially designed features or conventional image-processing based segmentation algorithms is quite difficult because of the presence of vessel crossing, overlap, branching and center-line reflex. The Segmentation can be further complicated due to the presence of pathologies in the form of lesions and exudates.

Deep learning based supervised segmentation models have proved to perform much better than the classical unsupervised methods which depend on hand crafted features. While most of them uses an encoder-decoder architecture, the image is down-sampled and up-sampled many times in the encoder and decoder respectively. These two processes are most commonly performed by a max-pooling operation and transposed convolution operation respectively and in the process, some amount of important information or features are lost which could have been beneficial for the segmentation algorithm. This paper introduces a Subpixel Residual U-Net architecture or SpruNet, based on the U-Net [11] framework, which re-purposes the subpixel convolutions [13] to perform image down-sampling and up-sampling in the U-shaped encoder-decoder architecture. This approach preserves information and provides better accuracy than the state-of-the-art algorithms. Also, this architecture is simpler, faster and has fewer parameters ($\sim$20 M) than the previous state-of-the-art algorithms.

The main problem with retinal images is the varying brightness and contrasts. To tackle this, a feature fusion technique is proposed to increase the accuracy and robustness of the model a bit more. Contrast Limited adaptive histogram equalization (CLAHE) [10] is combined with Ben Graham's [2] preprocesing method of subtracting the local average colour. This increases the clarity of the vessels in most of the images as compared to using either of the algorithms alone. But after some experiments its seen that in some scenarios, specially on the Chase dataset the combination of CLAHE and Ben Graham's algorithm performs slightly lower than the standalone CLAHE. To solve this, a feature fusion approach of concatenating the standalone CLAHE preprocessed image with the combined CLAHE and Ben Graham's preprocessed image is used, which tops all of the experiments done on all three datasets.

2   Related Work

In the last couple of years significant improvement has been seen in the field of Retinal Image Analysis, especially for the task of vessel extraction. This section gives a brief overview of the latest high-performance supervised approaches.

Roychowdhury S. et al. [12] in 2015 proposed a unsupervised method for Retinal Vessel Segmentation where they used iterative adaptive thresholding to identify vessel pixels from vessel enhanced image which is generated by top-hat reconstruction of the negative green plane image.

Liskowski P. and Krawiec K. [7] in 2016 proposed a supervised deep neural network model for Retinal Vessel Segmentation and preprocessed the images with global contrast normalization, zero-phase whitening, and used different augmentations like geometric transformations and gamma corrections.

Orlando J.I. et al. [9] in 2017 proposed a conditional random field model for Retinal Vessel Segmentation which enables real time inference of fully connected models. It uses a structured output support vector machine to learn the parameters of the method automatically.

Alom M.Z. et al. [1] in 2018 proposed their R2Unet which harnessed the power of Recurrent Convolutional Neural Networks (RCNN). The residual blocks allowed the network to train faster while the recurrent convolution layers allowed feature accumulation which helped in better feature representation.

Wang B. et al. [14] in 2019 proposed their Dual Encoding U-Net (DEU-Net) architecture, having two encoding paths: a spatial path with large kernels and a context path with multi-scale convolution blocks. This allowed the network to preserve the spatial information and encode contextual information via the two pathways.

Wu Y. et al. [3] in 2019 proposed their Vessel-Net, where they used inception inspired residual convolution blocks in the encoder part of a U-like encoder-decoder architecture and introduced four supervision paths to preserve the rich and multi-scale deep features.

Jin Q. et al. [5] in 2019 proposed their DUNet architecture based on the U-Net which used deformable convolution blocks in place of few of the standard convolution blocks. Deformable convolution blocks allow the network to adaptively adjust the receptive fields, thus enabling this architecture to classify the retinal vessels at various scales.

Yan Z. et al. [15] in 2019 proposed a three stage model for Retinal Vessel Segmentation, which segments thick vessels and thin vessels separately. The separate segmentation approach helps in learning better discriminative features. The final stage refines the results and identifies the non vessel pixels. This way the overall vessel thickness consistency is improved.

Zang S. et al. [16] in 2019 proposed their Attention Guided Network for Retinal Vessel Segmentation which uses a guided filter as a structure sensitive expanding path and an attention block which exclude noise. Their method preserves structural information.

Li L. et al. [6] in 2020 proposed their IterNet. They used a standard U-Net followed by an iteration of mini U-Nets with weight-sharing and skip connections. This allowed them to pass the output features of the standard U-Net through a number of light-weight intermediate networks to fix any kind of defects in the results.

3   Proposed Method

3.1   Dataset

The following three prominent benchmark datasets are used for experiment:

- The **Drive** dataset has 40 colored images of the retinal fundus along with the pixel-level segmentation mask. The size of the images is 584×565 pixels. The dataset already has a predefined 20–20 train-test split.
- The **Chase** dataset has 28 colored images of the retinal fundus along with the pixel-level segmentation mask. The size of the images is 999×960 pixels. The dataset has no predefined split so a random 20-8 train-test split is used.
- The **Stare** dataset has 20 colored images of the retinal fundus along with the pixel-level segmentation mask. The size of the images is 700×605 pixels. The dataset has no predefined split so a random 16-4 train-test split is used.

3.2   Preprocessing and Augmentation

The retinal images come with different lighting conditions and vary a lot in terms of brightness and contrast. Also different methods of image acquisition and presence of diseases makes it hard to build a robust segmentation algorithm. Image enhancement algorithms are used to increase the clarity of the images as much as possible before passing them to the model. The most popular retinal image enhancement algorithm used in the literature is the Contrast Limited Adaptive Histogram Equalization (CLAHE) algorithm. It increases the contrast of local regions to enhance the visibility of local details. It is quite robust and performs very well in most scenarios. In 2015 Ben Graham used a preprocessing technique in the Kaggle Diabetic Retinopathy competition, where he subtracted the local average colour from the image. In this method the vessels popped out quite distinctively from the background. In this paper we further enhance this method with a two step preprocessing. First CLAHE is used to increase the contrast, and then the local average colour is subtracted out from it, resulting in better clarity as shown in Fig. 1.

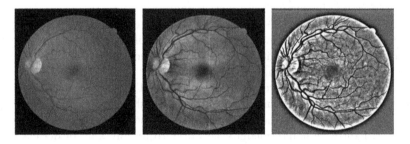

Fig. 1. From left to right: original images, CLAHE, CLAHE + Local Mean Colour Subtraction

Though this method performed better than the standalone CLAHE method on both the Drive and Stare dataset, it performed slightly worse on the Chase dataset. To tackle this problem, a feature fusion approach is taken, and a three step preprocessing is done as explained in the following steps.

1. Perform CLAHE on original image.
2. Perform Ben Graham's preprocessing of subtracting the local average colour from the CLAHE preprocessed image from step 1.
3. Concatenate the preprocessed images from step 1 and step 2 along the channels.

This method allows the model to use the all the information available in the CLAHE preprocessed image from step 1 along with the local average colour subtracted image from step 2. A diagram is shown in Fig. 2. A detailed ablation study is shown later which shows the improvement achieved with the Feature Fusion preprocessing method.

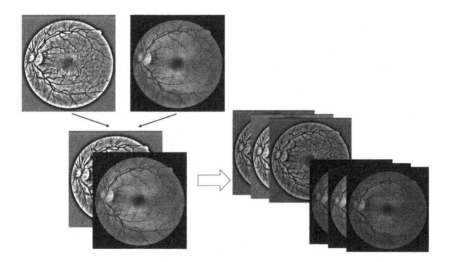

Fig. 2. Concatenating the preprocessed images

There are only a few public retinal datasets which includes vessel segmentation ground truths and all of them contains very few images. Since only around 20 images from each of the above-mentioned datasets are used for training, heavy data augmentation is used to increment the sample count to avoid overfitting. This includes: random horizontal and vertical flips, random rotations, minor RGB shifts for few images, and random brightness, contrast and gamma. Spatial level transformation like Grid distortion, Elastic transform and Optical distortion is also used to prevent over-fitting.

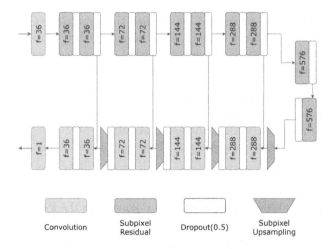

Fig. 3. Architecture

3.3 Architecture

The proposed architecture - SpruNet or Subpixel Residual U-Net (shown in Fig. 3) is adapted from the encoder-decoder architecture of the U-Net. The model is a fully convolutional model [8] with a contraction and expansion path. The encoder is a 25-layer novel ResNet [4] architecture which uses subpixel residual blocks instead of standard residual convolution blocks. The decoder is a stack of 13-layers with two convolution layers following every subpixel convolutional up-sampling blocks. Skip connections from encoder to decoder facilitates feature transfer from earlier layers to later layers which helps in the fine-grained segmentation. Each convolution layer is followed by a batch-normalization layer. Relu activation is used in all the layers except in the convolution layer in the subpixel convolution block.

Subpixel Convolution is re-purposed as a down-sampling and up-sampling method for semantic segmentation in the encoder and decoder respectively. Subpixel Convolution is just a standard convolution followed by a pixel reshuffle. Normally a Subpixel convolution is used for up-sampling process, but we adapt it to be used for both tasks as it preserves the data unlike any other methods commonly used. Figure 4 shows how the image dimension changes in the subpixel down-sampling block and the subpixel up-sampling block. Figure 5 shows how the residual convolution block is adapted to use subpixel convolution.

– **Subpixel down-sampling**: In this, an array of dimension (H, W, C) is converted to (H/2, W/2, 2C). For this we first pass it through a 1×1 2D convolution layer to reduce the depth to (H, W, C/2), then use pixel shuffling to change the dimension from (H, W, C/2) to (H/2, W/2, 2C). This process proves to be more efficient in preserving the spatial information as we down-sample the image but at the same time encode sufficient semantic information for efficient pixel-wise classification.

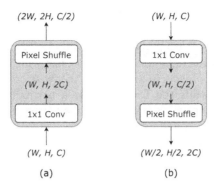

Fig. 4. (a) Subpixel up-sampling and (b) Subpixel down-sampling

- **Subpixel up-sampling**: In this, an array of dimension (H, W, C) is converted to (2H, 2W, C/2). For this we first pass it through a 1×1 2D convolution layer to increase the depth to (H, W, 2C), then use pixel shuffling to change the dimension from (H, W, 2C) to (2H, 2W, C/2).

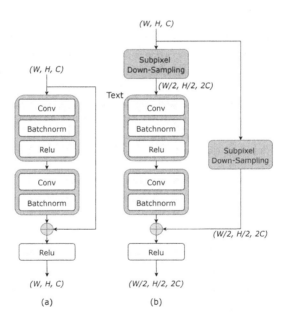

Fig. 5. (a) Identity block and (b) Subpixel residual block

3.4   Loss Function

We used BCE-Dice loss, a combination of pixel-wise Binary Cross-entropy loss which compares each pixel individually and Dice loss which measures the amount of overlap between two objects in an image. Pixel-wise cross-entropy loss suffers from class imbalance while the Dice loss has a normalizing effect and is not affected by class imbalance. This combination gave better segmentation accuracy than any one of them used individually.

$$BCE\text{-}Loss = y(log(p) + (1 - y)log(1 - p)$$ (1)

$$Dice\text{-}Loss = 2 \cdot \frac{A \cap B}{A \cup B}$$ (2)

$$BCE\text{-}Dice\text{-}Loss = BCE\text{-}Loss + Dice\text{-}Loss$$ (3)

3.5   Metrics

We evaluated our model on 5 evaluation metrics to provide a good comparison with the other methods: Accuracy, F1-score, Sensitivity, Specificity and ROC-AUC. The F1 score is the harmonic mean of the precision and recall. Since there is a large class imbalance, the F1 score is a better metric than the Accuracy. The higher the F1 score the better. Sensitivity is the ability of the model to correctly identify vessel pixels (true positive rate), whereas Specificity is the ability of the model to correctly identify non-vessel pixels (true negative rate). ROC - curve is a graph which plots the true positive rate against the true negative rate at various thresholds. AUC is the area under the ROC curve. The higher the AUC the better the model is at distinguishing vessel vs non-vessel pixels (Tables 1, 2 and 3).

$$F1\text{-}score = 2 \cdot \frac{Precision \cdot Recall}{Precision + Recall}$$ (4)

$$Sensitivity = \frac{TP}{TP + FN}$$ (5)

$$Specificity = \frac{TN}{TN + FP}$$ (6)

Table 1. Ablation table for Drive dataset

Method	F-1	SE	SP	AC	AUC
CLAHE + Subpixel	0.8445	0.8296	0.9843	0.9682	0.9868
CLAHE + Ben Graham's + Subpixel	0.8514	0.8394	0.9848	0.9701	0.9874
Information Fusion + No-Subpixel	0.8477	0.8287	0.9850	0.9694	0.9858
Information Fusion + Subpixel	**0.8533**	**0.8401**	**0.9852**	**0.9703**	**0.9888**

Table 2. Ablation table for Chase dataset

Method	F-1	SE	SP	AC	AUC
CLAHE + Subpixel	0.8576	0.8452	0.9878	0.9746	0.9912
CLAHE + Ben Graham's + Subpixel	0.8561	0.8370	0.9879	0.9740	0.9906
Information fusion + No-Subpixel	0.8528	0.8308	0.9875	0.9736	0.9896
Information fusion + Subpixel	**0.8591**	**0.8472**	**0.9880**	**0.9747**	**0.9913**

Table 3. Ablation table for Stare dataset

Method	F-1	SE	SP	AC	AUC
CLAHE + Subpixel	0.8427	0.7842	0.9919	0.9721	0.9899
CLAHE + Ben Graham's + Subpixel	0.8682	0.8220	0.9925	0.9763	0.9945
Information fusion + No-Subpixel	0.8677	0.8216	0.9924	0.9766	0.9941
Information fusion + Subpixel	**0.8686**	**0.8240**	**0.9926**	**0.9768**	**0.9945**

4   Experimental Evaluation

This section provides the implementation details of the proposed method and a detailed analysis for a number of experiments performed for a robust evaluation of the method. The experiments are performed on a single 16GB NVIDIA Tesla P100. Instead of patch-based training we used full images resized to 512×512 resolution for both training and testing. A batch size of 5 is used, keeping in mind the hardware limitation. We used ADAM optimizer with an initial learning rate of 0.001. The learning rate is dynamically reduced by a factor 0.1 when the validation loss reaches a plateau. We used early stopping to stop the training when the validation loss remains stable for 10 consecutive epochs. Our model takes an hour and a half on average on the specified hardware to train on 1000 augmented full images of 512×512 resolution. Inference can be done within half a second on a 512×512 resolution image. The results of the experiments and comparisons are provided in the following tables (Tables 4, 5 and 6) (Fig. 6).

Table 4. Results and comparisons on Drive dataset

Method	Year	F-1	SE	SP	AC	AUC
U-Net	2018	0.8174	0.7822	0.9808	0.9555	0.9752
Dense Block U-Net	2018	0.8146	0.7928	0.9776	0.9541	0.9756
DUNet	2019	0.8237	0.7963	0.9800	0.9566	0.9802
DE-UNet	2019	0.8270	0.7940	0.9816	0.9567	0.9772
Vessel-Net	2019	–	0.8038	0.9802	0.9578	0.9821
IterNet	2020	0.8205	0.7735	0.9838	0.9573	0.9816
Proposed method	2020	**0.8533**	**0.8401**	**0.9852**	**0.9703**	**0.9888**

Table 5. Results and comparisons on Chase dataset

Method	Year	F-1	SE	SP	AC	AUC
U-Net	2018	0.7993	0.7841	0.9823	0.9643	0.9812
Dense Block U-Net	2018	0.8006	0.8178	0.9775	0.9631	0.9826
DUNet	2019	0.7883	0.8155	0.9752	0.9610	0.9804
DE-UNet	2019	0.8037	0.8074	0.9821	0.9661	0.9812
Vessel-Net	2019	–	0.8132	0.9814	0.9661	0.9860
IterNet	2020	0.8073	0.7970	0.9823	0.9655	0.9851
Proposed method	2020	**0.8591**	**0.8472**	**0.9880**	**0.9745**	**0.9913**

Table 6. Results and comparisons on Stare dataset

Method	Year	F-1	SE	SP	AC	AUC
U-Net	2018	0.7595	0.6681	0.9915	0.9639	0.9710
Dense Block U-Net	2018	0.7691	0.6807	0.9916	0.9651	0.9755
DUNet	2019	0.8143	0.7595	0.9878	0.9641	0.9832
IterNet	2020	0.8146	0.7715	0.9886	0.9701	0.9881
Proposed method	2020	**0.8686**	**0.8240**	**0.9923**	**0.9762**	**0.9945**

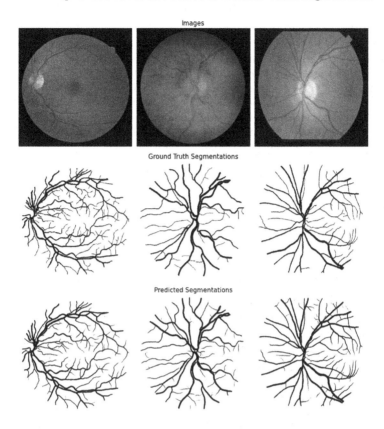

Fig. 6. From top to bottom: original images, ground-truth segmentations, predicted segmentations. From left to right: Drive, Chase, Stare

5   Conclusion

In this experiment we showed that Subpixel Convolutions are very efficient in preserving information and are an effective way of changing image sizes during encoding and decoding of an image. Residual convolutional blocks can be adapted to use subpixel convolutions in place of max-pooling and thus a residual network can be improved to encode as much information as possible, both spatial and contextual with minimal loss of information during down-sampling. Similarly, subpixel based upsampling increases the spatial dimension of an image in a learnable way, preserving the information in an efficient manner. The proposed model SpruNet achieves AUC of 0.9888, 0.9913 and 0.9945 on the Drive, Chase and Stare datasets respectively. Thus beating the state-of-the-art models with a much simpler architecture with lesser parameter ($\sim$20 M) with a fast inference speed of 0.5 s on a 512×512 full image.

References

1. Alom, M.Z., Hasan, M., Yakopcic, C., Taha, T.M., Asari, V.K.: Recurrent residual convo-lutional neural network based on U-Net (R2U-Net) for medical image segmentation. arXiv preprint. arXiv:1802.06955 (2018)
2. Graham, B.: Kaggle diabetic retinopathy detection competition report (2015)
3. Wu, Y., et al.: Vessel-Net: retinal vessel segmentation under multi-path supervision. In: Shen, D., et al. (eds.) MICCAI 2019. LNCS, vol. 11764, pp. 264–272. Springer, Cham (2019). https://doi.org/10.1007/978-3-030-32239-7_30
4. He, K., Zhang, X., Ren, S., Sun, J.: Deep residual learning for image recognition. 2016 IEEE Conference on Computer Vision and Pattern Recognition (CVPR) (2016)
5. Jin, Q., Meng, Z., Pham, T.D., Chen, Q., Wei, L., Su, R.: DUNet: a deformable network for retinal vessel segmentation. Knowl. Based Syst. **178**, 149–162 (2019)
6. Li, L., Verma, M., Nakashima, Y., Nagahara, H., Kawasaki, R.: IterNet: retinal image segmentation utilizing structural redundancy in vessel networks. 2020 IEEE Winter Conference on Applications of Computer Vision (WACV) (2020)
7. Liskowski, P., Krawiec, K.: Segmenting retinal blood vessels with deep neural networks. IEEE Trans. Med. Imaging **35**(11), 2369–2380 (2016)
8. Long, J., Shelhamer, E., Darrell, T.: Fully convolutional networks for semantic segmentation. 2015 IEEE Conference on Computer Vision and Pattern Recognition (CVPR) (2015)
9. Orlando, J.I., Prokofyeva, E., Blaschko, M.B.: Discriminatively trained fully connected conditional random field model for blood vessel segmentation in fundus images. IEEE Trans. Biomed. Eng. **64**(1), 16–27 (2017)
10. Pizer, S.M., et al.: Adaptive histogram equalization and its variations. Comput. Vis. Graph. Image Process. **39**, 355–368 (1987)
11. Ronneberger, O., Fischer, P., Brox, T.: U-Net: convolutional networks for biomedical image segmentation. In: Navab, N., Hornegger, J., Wells, W.M., Frangi, A.F. (eds.) MICCAI 2015. LNCS, vol. 9351, pp. 234–241. Springer, Cham (2015). https://doi.org/10.1007/978-3-319-24574-4_28
12. Roychowdhury, S., Koozekanani, D.D., Parhi, K.K.: Iterative vessel segmentation of fundus images. IEEE Trans. Biomed. Eng. **62**(7), 1738–1749 (2015)
13. Shi, W., et al.: Real-time single image and video super-resolution using an efficient sub-pixel convolutional neural network. In: 2016 IEEE Conference on Computer Vision and Pattern Recognition (CVPR), pp. 1874–1883 (2016)
14. Wang, B., Qiu, S., He, H.: Dual encoding U-Net for retinal vessel segmentation. In: Shen, D., et al. (eds.) MICCAI 2019. LNCS, vol. 11764, pp. 84–92. Springer, Cham (2019). https://doi.org/10.1007/978-3-030-32239-7_10
15. Yan, Z., Yang, X., Cheng, K.T.: A three-stage deep learning model for accurate retinal vessel segmentation. IEEE J. Biomed. Health Inf. **23**(4), 1427–1436 (2019)
16. Zhang, S., et al.: Attention guided network for retinal image segmentation. In: Shen, D., et al. (eds.) MICCAI 2019. LNCS, vol. 11764, pp. 797–805. Springer, Cham (2019). https://doi.org/10.1007/978-3-030-32239-7_88

Attention Deeplabv3+: Multi-level Context Attention Mechanism for Skin Lesion Segmentation

Reza Azad[1] , Maryam Asadi-Aghbolaghi[2](✉) , Mahmood Fathy[2] ,
and Sergio Escalera[3]

[1] Computer Engineering Department, Sharif University of Technology, Tehran, Iran
rezazad68@gmail.com
[2] Computer Science School, Institute for Research in Fundamental Science (IPM),
Tehran, Iran
{masadi,mahfathy}@ipm.ir
[3] Universitat de Barcelona and Computer Vision Center, Barcelona, Spain
sergio@maia.ub.es

Abstract. Skin lesion segmentation is a challenging task due to the large variation of anatomy across different cases. In the last few years, deep learning frameworks have shown high performance in image segmentation. In this paper, we propose Attention Deeplabv3+, an extended version of Deeplabv3+ for skin lesion segmentation by employing the idea of attention mechanism in two stages. We first capture the relationship between the channels of a set of feature maps by assigning a weight for each channel (i.e., channels attention). Channel attention allows the network to emphasize more on the informative and meaningful channels by a context gating mechanism. We also exploit the second level attention strategy to integrate different layers of the atrous convolution. It helps the network to focus on the more relevant field of view to the target. The proposed model is evaluated on three datasets ISIC 2017, ISIC 2018, and PH^2, achieving state-of-the-art performance.

Keywords: Medical image segmentation · Deeplabv3+ · Attention mechanism

1 Introduction

With the development of computer vision, medical image segmentation has become an important part of computer-aided diagnosis. These years the computer-aided diagnosis (CAD) systems are required to assist the experts by providing accurate interpretation of medical images. Among many medical image processing tasks, automatic image segmentation is an important and effective step toward the analysis phase. Medical image segmentation is included in a

---

R. Azad and M. Asadi-Aghbolaghi—Contributed equally to this work.

© Springer Nature Switzerland AG 2020
A. Bartoli and A. Fusiello (Eds.): ECCV 2020 Workshops, LNCS 12535, pp. 251–266, 2020.
https://doi.org/10.1007/978-3-030-66415-2_16

large number of application domains like skin cancer segmentation. Skin cancer is one of the most widespread and deadly forms of cancer. The human skin includes three types of tissues, i.e., dermis, epidermis, and hypodermis. The epidermis has melanocytes and under some conditions (like the strong ultraviolet radiation from sunshine), it produces melanin at a greatly unusual rate. A lethal type of skin cancer is melanoma which is the result of unusual growth of melanocytes [14]. With a mortality rate of 1.62%, melanoma is reported as the most lethal skin cancer [32]. In 2019, the American Cancer Society reported there are approximately 96,480 new cases of melanoma and about 7230 will die from this cancer [30]. The non-melanoma cancers are also the reason for a large number of deaths. The World Health Organization reported that between 2 and 3 million non-melanoma skin cancers and 132,000 melanoma skin cancers are recorded every year in the world [1].

Although the dermatologists detect melanoma in medical images, their detection may be inaccurate. On the other hand, early diagnosis of the melanoma is critical in terms of treatment. The early detection and diagnosis of melanoma helps to have the proper treatment and ensure a complete recovery. It is reported that early detection increases the five-year relative survival rate to 92% [29]. As a result, skin lesion segmentation (Fig. 1) has a critical role in the early and accurate diagnosis of skin cancer by computerized systems. During the last decade, automatic detection of skin cancer has been significantly taken into account. It is applied for different kinds of skin problems like three main types of abnormal skin cells are noticed i.e. Basic cell carcinoma, Squamous cell carcinoma and Melanoma.

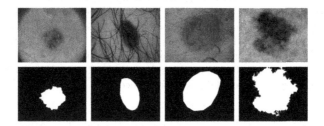

Fig. 1. Some samples of skin lesion segmentation.

Skin cancer segmentation is a challenging task due to several factors like low contrast of the images, differences in texture, position, color, and size of skin lesions in medical images. Moreover, the existence of air bubbles, hair, ebony frames, ruler marks, blood vessels, and color illumination make the lesion segmentation task extremely difficult. Various approaches have been proposed for the skin lesion segmentation. Different types of deep convolutional neural networks have been proposed for image segmentation, and like other fields of research in computer vision, deep learning approaches have achieved outstanding results in this field. Fully convolutional neural network (FCN) [21] was one of the first deep networks proposed for segmentation. This network was then extended to U-Net [27], consisting of an encoding and a decoding path. U-Net achieved good segmentation performance alongside leveraging the need of a large amount of training data.

CNN allows to learn increasingly abstract data representation which yields the network robust to local image transformation. Abstraction of spatial information may be undesirable for semantic segmentation. To solve that, Deeplab [7] was proposed by utilizing "atrous spatial pyramid pooling" (ASPP). Deeplabv3 [8] was then proposed to capture contextual information at multiple scales by employing several parallel ASPPs. Chen et al. [9] proposed DeepLabv3+ by combing the idea of both U-Net and Deeplabv3, as an extension of Deeplabv3 by adding a decoder module to recover the object boundaries. Recently, attention-based networks have been widely utilized in different tasks of computer vision [31]. The attention strategy helps the network by avoiding the use of multiple similar feature maps and focusing on the most salient and informative features for a given task without additional supervision. It has been proved that attention enhances the result of semantic segmentation networks [25,31,35].

In this paper, we propose Att-Deeplabv3+ *(attention Deeplabv3+)* for skin lesion segmentation. In particular, we improve Deeplabv3+ by inserting two attention modules in the atrous convolution. The first attention module is a kind of channel wise attention and it is employed to recalibrate the feature map in each layer of the atrous convolution to pay more attention to more informative channels. That is to say, the network concentrates more on channel features with more useful information (based on their relationship) by assigning different weights to various channels of feature maps.

The second attention module is exploited to aggregate the features extracted by different layers of the atrous convolution through a multi-scale attention mechanism. Different layers of an atrous convolution extract features from the input features map at different sizes of field of view. Utilizing a small field of view for extracting features results in lower-level resolution which encodes local information. The local representation is important to discriminate the input image details. On the other hand, by enlarging the field of view, a larger image context is taken into account for extracting features, and therefore, more global information is taken to account. Especially, with large scales of receptive fields, the layer extracts long range features. The utilized scale attention module yields the network to emphasize the extracted features of the layers which are more relevant to the scale of the targets. We evaluate the proposed network on three datasets ISIC 2017, ISIC 2018, and Ph^2. The experimental results demonstrate that the proposed network achieves superior performance than existing alternatives. The main contributions of the paper are as follows:

- We propose an extended version of Deeplabv3+, Att-Deeplabv3+ for skin lesion segmentation.
- A channel wise attention module is utilized in each layer of the atrous convolution of Deeplabv3+ to focus on the more informative channel features.

- A multi-scale attention module is employed to aggregate the information of all layers of the atrous convolution.
- The proposed network achieves superior performance than existing alternatives on three datasets ISIC 2017, ISIC 2018, and PH^2.

The rest of the paper is organized as follows. Section 2 reviews related work. The proposed network is presented in Sect. 3. The experimental results are described in Sect. 4. Finally, Sect. 5 concludes the paper.

2   Related Work

Semantic segmentation is one of the most important tasks in medical imaging. Before the revolution of deep learning in computer vision, traditional hand-crafted features have been utilized in semantic segmentation. During the last few years, deep learning-based approaches have achieved outstanding results in image segmentation. These networks can be divided into two main groups, i.e., encoder-decoder structures and the models using spatial pyramid pooling [9].

2.1   Encoder-Decoder

The encoder-decoder networks have been successfully utilized for semantic segmentation. These networks include encoding and decoding paths. The encoder generates a large number of feature maps with reduced dimensionality, and the decoder produces segmentation maps by recovering the spatial information. U-Net [27] is one of the most popular encoder-decoder networks. A main advantage of this network is that the network works well with few training samples by employing the global location and context information at the same time. Many methods [4,5,20,26] demonstrate the effectiveness of encoder-decoder structure for image segmentation in different applications.

Different extensions of the U-Net have been proposed for image segmentation. V-Net [23] is proposed to predict segmentation of a given volume. For segmentation of 3D volumes (e.g., MRI volumes), Çiçek et al. propose 3D U-Net [11]. BCDU-Net [5] improves the performance of the U-Net by utilizing ConvL-STM to combine the feature maps extracted from the corresponding encoding path and the previous decoding up-convolutional layer in a non-linear way. In that network, densely connected convolutions are also inserted in the last convolutional layer of the encoder to strengthen feature propagation. Inspired by the effectiveness of the recently proposed squeeze and excitation modules, Asadi et al. propose MCGU-Net [4] by inserting these modules in decoder for medical image segmentation.

Alom et al. [3] improve U-Net by using recurrent convolution network and recurrent residual convolutional network, and propose RU-Net and R2U-Net for medical image segmentation. While the U-Net uses skip connections to concatenate features from the encoder, the SegNet [6] passes pooling indices to the decoder to reduce computational complexities of weight concatenation. SegNet uses VGG-16-like network structure in the encoder and it works better than U-Net when the image content becomes complicated. Ghiasi et al. [15] propose a multi-resolution reconstruction architecture based on a Laplacian pyramid. In that network, the segmented boundaries of the low-resolution maps are reconstructed by the skip connections from the higher resolution maps. U-Net has been also improved by utilizing different attention-based modules in both encoder and decoder [25, 28].

2.2 Spatial Pyramid Pooling

The usual deep convolutional networks (DCNNs) have some drawbacks for semantic image segmentation. The spatial feature resolution is considerably decreased due to a consecutive max-pooling and down-sampling functions in the network. Moreover, objects (e.g., skin lesions) can have different scales in the images [7]. To mitigate this problem, spatial pyramid pooling models capture rich contextual information by pooling features at different resolutions. PSP-Net [34] is a pyramid scene parsing network to embed difficult scenery context features. In an FCN based pixel prediction framework, global context information is captured by using different region-based context aggregation through a pyramid pooling module. Atrous spatial pyramid pooling (ASPP) at several grid scales is performed in deeplab [7, 8]. In that network, parallel atrous convolution layers with different rates capture multi-scale information. Multi-scale information helps these models to improve the performance of the network on several segmentation benchmarks.

Hesamian et al. [16] utilize the Atrous convolution which increases the field of view of the filters and helps to improve the performance of the network for segmentation of Lung Nodule. Recurrent neural networks with LSTM have been also exploited in several methods [19, 33] to aggregate global context information for semantic segmentation. Chen et al. [9] take into account the advantages of both encoder-decoder networks and pyramid spatial pooling and introduce Deeplabv3+ shown in Fig. 2. In that network, Deeplabv3 is extended by adding a decoder module to recover the object boundaries. Atrous convolution extracts rich semantic information in the encoding path and controls the density of the decoder features. The decoder module helps the network to recover the object boundaries. Many attention mechanisms have been utilized to improve the performance of the U-Net [25, 28].

In this paper, we improve the performance of the Deeplabv3+ by inserting two kinds of attentions, channel-wise attention and multi-scale attention, in the atrous convolutions. The squeeze and excitation blocks [17] have been successfully inserted into convolutional networks. These blocks improved the performance by applying more attention on the most informative channel features. In this paper, we use the attention mechanism to focus on the scales with informative features in atrous convolutions.

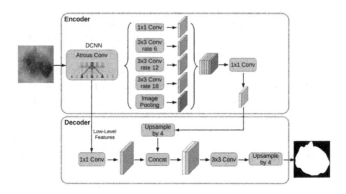

Fig. 2. Deeplabv3+ [9].

3   Proposed Method

We propose Att-Deeplabv3+, an attention-based Deeplabv3+ extension, (Fig. 3) for skin lesion segmentation. The network utilizes the strengths of attention mechanism to focus on the features with more informative data, and avoid the use of multiple similar feature maps. We highlight different parts of the proposed network in details in the following subsections.

3.1   Encoder

Xception model [10] works well in object classification with fast computation. This model has been adopted for the task of image segmentation [9], and achieved promising results. Therefore, we utilize the modified version of Xception model [9] as the encoder of the proposed network. In that method, a deeper Xception model is utilized, and extra batch normalization and ReLU functions have been used after the 3×3 depth-wise convolution. Moreover, all the max pooling layers are replaced by depth-wise separable convolution with striding. By considering this idea, atrous separable convolution can be applied at an arbitrary resolution to extract features. This model consists of three parts, i.e., entry flow, middle flow, and exit flow. The entry flow includes 2 convolutional and 9 depth-wise separable convolutional layers in four blocks. The middle flow contains one block consisting of three depth-wise separable convolutional layers which is repeated 16 times. The exit flow includes 6 depth-wise separable convolutional layers in 2 blocks. We refer the readers to [9] for more details.

3.2   Attention-Based Atrous Convolution

In usual CNNs, the spatial feature resolution is considerably decreased due to a set of consecutive max-pooling and down-sampling functions in the network. Moreover, objects can have different scales in the images. To mitigate this problem, atrous Convolutions with multiple rates [7] have been introduced. In atrous

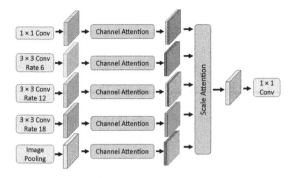

Fig. 3. Multi-level attention Deeplabv3+.

Convolutions, from the last few max pooling layers, the down-sampling operations have been removed while the filters have been up-sampled in the subsequent convolutional layers. To up-sample the filters, the full resolution image is convolved with filters with holes, i.e., zeros have been inserted between filter's values. Since non-zero filters' values are only considered in the calculations, the number of parameters stay constant. The atrous convolution yields the method to control the spatial resolution of the feature responses. Moreover, we can enlarge the field of view of the filters to compute feature responses at any layer which results in incorporating larger context information. For one-dimensional signal, the atrous convolution [7] is calculated as

$$y[i] = \sum_k x[i + r.k]w[k] \tag{1}$$

where y is the output feature map, i is a spatial location on y, x is the input feature map, and w is a convolution filter. Moreover, the atrous rate r determines the stride with which we sample the input signal. As it can be seen in Fig. 2, the atrous convolution consists of five layers. Each feature map includes a number of channels. In Deeplabv3+, the output of these layers are concatenated and passed to the next block of the network. In other words, all channels of a set of feature map are processed in the network with the same attention while some of these channels may be informative. Inspired by the squeeze and excitation network [17], we employ a channel-based attention on the output of each layer of the atrous convolution to capture explicit relationship between channels of the convolutional layers (shown in Fig. 4). This strategy helps the network to selectively empathize informative features and suppress less useful ones by utilizing the global information of the input data. That is to say, it encodes feature maps by assigning a weight for each channel (i.e. channel attention).

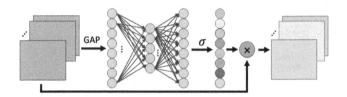

Fig. 4. Channel-wise attention for each layer of the atrous convolution.

The model exploits the global context information of the input features to produce the weight for each input channel. To do that, the global average pooling is calculated for each channel as

$$z_f = \frac{1}{H \times W} \sum_{i}^{H} \sum_{j}^{W} x_f(i, j) \tag{2}$$

where x_f is the fth channel, $H \times W$ is the size of the channel, and z_f is the output of the global average pooling. In the next step, we learn non-mutually-exclusive relationship and nonlinear interaction between channels. Inspired by [17], two fully connected layers are then employed to capture the channel-wise dependencies (Fig. 4). The output of these layers is calculated as

$$s_f = \sigma\left(\mathbf{W}_2 \delta(\mathbf{W}_1 z_f)\right) \tag{3}$$

where $\mathbf{W}_1$ and $\mathbf{W}_2$ are the parameters of the two fc layers, δ is ReLU, and the σ refers to the sigmoid activation, and s_f is the learnt scale factor for the f_{th} channel. The final output of each layer of the attention-based atrous convolution is $X = [\tilde{x}_1, \tilde{x}_2, ..., \tilde{x}_F]$ where $\tilde{x}_f = s_f . x_f$ is a channel-wise multiplication between the fth channel x_f, and its corresponding scale factor s_f.

3.3 Multi-scale Attention-Based Depth-Wise Aggregation in Atrous Convolution

In Deeplabv3+, the output of all layers of atrous convolution are concatenated and then passed to the depth-wise separable convolution. Depth-wise separable convolution factorizes the standard convolution into depth-wise convolutions followed by a pointwise convolution. In particular, a spatial convolution is independently employed for each input channel, and a pointwise convolution is then performed to combine the result of depth-wise convolution for all of the input channels. Atrous convolution produces multi-scale features with different resolutions containing different semantic information. Lower-level features encode more information about local representation while higher-level features focus on global representation. In particular, different layers of a multi-scale strategy contain different semantic information, i.e., the layers with lower resolution are responsible for smaller objects and layers with higher resolution are responsible for objects with a larger scale. In the original Deeplabv3+, these features are simply concatenated with the same weight. Since we have objects with different sizes, it is to process the features of these layers with a different attention scale. To mitigate this problem, instead of a simple concatenation of multi-scale features, we integrate multi-scale features by employing a multi-scale depth-wise attention strategy (Fig. 5). Especially, the depth-wise aggregation performs the integration independently for the corresponding channels of all five layers of atrous convolution.

Assume that X_s^f is the fth ($f \in \{1, 2, ..., 256\}$) channel of the sth ($s \in \{1, 2, ..., 5\}$) layer (scale) of the atrous convolution. The output of the original Deeplabv3+ is the concatenation of feature maps from all layers, i.e, $Y = [X_1, X_2, ...X_5]$. In Att-Deeplabv3+, these feature maps are integrated with a multi-scale attention depth-wise method. We learn a weight for each scale, and apply a non-linear depth-wise aggregation to combine these features. The output of the atrous convolution is calculated as

$$Y_f = \sigma \left(\sum_{s=1}^{5} w_s X_s^f \right) \tag{4}$$

where the X_s^f is the fth channel of the sth scale, w_s is the learnt weight for the sth scale, σ is the sigmoid function, and Y_f is the fth output channel of the atrous convolution.

3.4   Decoder

We utilize the same decoder as Deeplabv3+. To reduce the number of channels, a 1×1 convolution is applied on both encoder and decoder features. The encoder features are bilinearly upsampled and are then concatenated with the corresponding features (with the same spatial resolution) from the network backbone. To refine the features, the concatenated features are then passed to a few 3×3 convolutions. At the end, another bilinear upsampling is performed on the features.

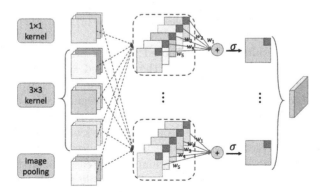

Fig. 5. Attention-based depth-wise aggregation in atrous convolution.

4 Experimental Results

The proposed method is evaluated on three datasets ISIC 2017, ISIC 2018, and PH^2. Several performance metrics have been employed for the experimental comparative, including accuracy (AC), sensitivity (SE), specificity (SP), F1-Score, Jaccard similarity (JS), and area under the curve (AUC). We stop the training of the network when the validation loss remains the same in 10 consecutive epochs. For all three datasets, we finetune the network from a pre-trained model which is learnt on PASCAL VOC dataset.

Table 1. Performance comparison of the proposed network and other methods on ISIC 2017.

Methods	F1-Score	Sensitivity	Specificity	Accuracy	Jaccard similarity
U-net [27]	0.8682	0.9479	0.9263	0.9314	0.9314
Melanoma det. [13]	–	–	–	0.9340	–
Lesion analysis [18]	–	0.8250	0.9750	0.9340	–
R2U-net [3]	0.8920	**0.9414**	0.9425	0.9424	0.9421
BCDU-Net [5]	0.8810	0.8647	0.9751	0.9528	0.9528
MCGU-Net [4]	0.8927	0.8502	0.9855	0.9570	0.9570
Deeplabv3+ [9]	0.9162	0.8733	**0.9921**	0.9691	0.9691
Att-Deeplabv3+	**0.9190**	0.8851	0.9901	**0.9698**	**0.9698**

4.1 ISIC 2017

The ISIC 2017 dataset [13] is designed for skin cancer segmentation published in 2017. Att-Deeplabv3+ is evaluated on the provided data for skin lesion segmentation. The dataset consists of 2000 skin lesion images with masks (including cancer or non-cancer lesions). Like other approaches [3], we use the standard evaluation setting for this dataset, i.e, 1250 samples for training, 150 samples for validation, and the other 600 samples for test. The original size of each sample is 576×767. We resize images to 256×256.

In Table 1, the results of the proposed network are compared with other approaches. By comparing the result of the proposed attention-based network with the original Deeplabv3+, we can see that the multi-level attention mechanisms enhance the performance of the network. The segmentation outputs of the proposed network for some samples from this dataset are depicted in Fig. 6. By comparing visually the results of the proposed network and Deeplabv3+, we can conclude that the two level attention mechanism utilized in the network helps to extract finer boundaries for the melanoma. The reason behind this result is the additional attention on the most informative scale and channel features in atrous convolution.

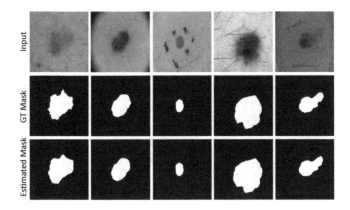

Fig. 6. Segmentation results of Att-Deeplabv3+ on ISIC 2017 dataset.

Table 2. Performance comparison of the proposed network and other methods on ISIC 2018.

Methods	F1-Score	Sensitivity	Specificity	Accuracy	Jaccard similarity
U-net [27]	0.647	0.708	0.964	0.890	0.549
Att U-net [25]	0.665	0.717	0.967	0.897	0.566
R2U-net [3]	0.679	0.792	0.928	0.880	0.581
Att R2U-Net [3]	0.691	0.726	0.971	0.904	0.592
BCDU-Net [5]	0.851	0.785	0.982	0.937	0.937
MCGU-Net [4]	0.895	0.848	0.986	0.955	0.955
Deeplab v3+ [9]	0.882	0.856	0.977	0.951	0.951
Att-Deeplab v3+	**0.912**	**0.875**	**0.988**	**0.964**	**0.964**

4.2   ISIC 2018

The International Skin Imaging Collaboration (ISIC) published the ISIC 2018 dataset [12] as a large-scale dataset of dermoscopy images in 2018. This dataset contains 2594 dermoscopy images. For each sample the original image and corresponding ground truth annotation (containing cancer or non-cancer lesions) are available. Like other approaches, we use the standard evaluation setting, and utilize 1815 images for training, 259 for validation and 520 for testing. We resize images to 256 × 256.

In Table 2, the performance of the proposed method is compared with other approaches. The proposed Att-Deeplabv3+ outperforms state-of-the-art methods. It can be seen there is a high gap between the result of the proposed method and the original Deeplabv3+. We have also compute the threshold Jaccard index for comparing the performance of two networks of Att-Deeplabv3+ and the original Deeplabv3+. For Deeplabv3+, the threshold Jaccard index is 0.9404, and for the proposed network is 0.9599. Figure 7 shows some segmentation outputs of the proposed network on ISIC 2018 dataset. Like ISIC 2017, we can see that

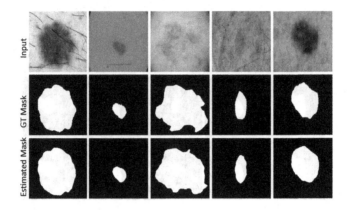

Fig. 7. Segmentation results of Att-Deeplabv3+ on ISIC 2018 dataset.

Att-Deeplabv3+ results in segmenting melanoma with more boundary details. In other words, the attention mechanism improves the performance by emphasizing features with more detained information.

Table 3. Performance comparison of the proposed network and other methods on PH^2.

Methods	F1-Score	Sensitivity	Specificity	Accuracy	Jaccard similarity
FCN [24]	0.8903	0.9030	0.9402	0.9282	0.8022
U-net [27]	0.8761	0.8163	0.9776	0.9255	0.7795
SegNet [6]	0.8936	0.8653	0.9661	0.9336	0.8077
FrCN [2]	0.9177	**0.9372**	0.9565	0.9508	0.8479
Deeplab v3+ [9]	0.9202	0.8818	0.9832	0.9503	0.9503
Att-Deeplab v3+	**0.9456**	0.9161	**0.9896**	**0.9657**	**0.9657**

4.3 PH^2

The PH^2 dataset is a dermoscopic image database proposed for segmentation and classification [22]. The total number of samples for this dataset is 200 melanocytic lesions, including 80 common nevi, 80 atypical nevi, and 40 melanomas. The manual segmentations of the skin lesions are available as the ground truth. Each input image is a 8-bit RGB color images with the resolution of 768 × 560 pixels. There is not a standard evaluation setting (test and train sets) for this dataset. We use the same setting as [4] and randomly split the dataset into two sets of 100 images, and then use one set as the test data, 80% of the other set for the train, and the remained data for the validation.

Table 3 lists the quantitative results obtained by other methods and the proposed network on PH^2 dataset. It is shown that the Att-Deeplabv3+ outperforms state-of-the-art approaches. Moreover, the proposed method surpasses the original Deeplabv3+. Some precise and promising segmentation results of the proposed network for this dataset are shown in Fig. 8. In this Figure, we can see similar results to other datasets, i.e., improving the boundary segmentation. That is to say, the proposed network utilizes discriminative information with the attention to segment the input data.

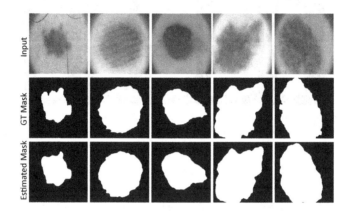

Fig. 8. Segmentation results of Att-Deeplabv3+ on PH^2 dataset

4.4   Discussion

In this paper, we employ multi-level attention mechanism to improve the performance of the Deeplabv3+. In Fig. 9, the output segmentation results of some samples for both Att-deeplabv3+ and deeplabv3 networks are compared. It can be seen that Att-Deeplabv3+ performs better than the original network and its output results are more precious. The boundary of the segmented lesion of the Att-deeplabv3+ is finer, i.e., the output of the Att-Deeplabv3+ includes more local details in boundary of the object. In medical image segmentation, the precise and true boundaries of skin lesions are vital to locate the melanoma accurately in dermoscopic images. The proposed approach is able to segment finer boundaries rather than the original Deeplabv3+ by extracting most useful and informative features from the input, and focusing more on the features with more discriminative and informative data. In the proposed network, a channel wise attention is first applied to all the layers of the atrous convolution. This attention mechanism yields the network to focus on the more informative channels. The second level of the attention is used as a multi-scale attention method for aggregating all the layers of the atrous convolution. The multi-scale attention allows the network to focus on more scale relevant feature to the target.

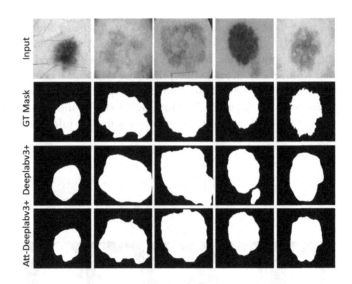

Fig. 9. Visual effect of the multi-level attention mechanism in the proposed network.

5   Conclusion

We proposed Att-Deeplabv3+ for skin lesion segmentation. It has been shown that by including two level attention based mechanism in Deeplabv3+, the network is able to learn more discriminative information. Compared to the original deeplabv3+, the proposed network has more precise segmentation results. The experimental results on three public skin cancer benchmark datasets showed high gain in semantic segmentation in relation to state-of-the-art alternatives.[1].

Acknowledgment. This work has been partially supported by the Spanish project PID2019-105093GB-I00 (MINECO/FEDER, UE) and CERCA Programme/Generalitat de Catalunya, and ICREA under the ICREA Academia programme. We gratefully acknowledge the support of NVIDIA Corporation with the donation of the GPU used for this research.

References

1. who.int/uv/faq/skincancer/en/index1.html
2. Al-Masni, M.A., Al-antari, M.A., Choi, M.T., Han, S.M., Kim, T.S.: Skin lesion segmentation in dermoscopy images via deep full resolution convolutional networks. Comput. Methods Programs Biomed. **162**, 221–231 (2018)
3. Alom, M.Z., Hasan, M., Yakopcic, C., Taha, T.M., Asari, V.K.: Recurrent residual convolutional neural network based on U-Net (R2U-Net) for medical image segmentation. arXiv preprint arXiv:1802.06955 (2018)

---

[1] Source code is available on https://github.com/rezazad68/AttentionDeeplabv3p.

4. Asadi-Aghbolaghi, M., Azad, R., Fathy, M., Escalera, S.: Multi-level context gating of embedded collective knowledge for medical image segmentation. arXiv preprint arXiv:2003.05056 (2020)
5. Azad, R., Asadi-Aghbolaghi, M., Fathy, M., Escalera, S.: Bi-directional ConvL-STM U-Net with Densley connected convolutions. In: Proceedings of the IEEE International Conference on Computer Vision Workshops (2019)
6. Badrinarayanan, V., Kendall, A., Cipolla, R.: SegNet: a deep convolutional encoder-decoder architecture for image segmentation. IEEE Trans. Pattern Anal. Mach. Intell. **39**(12), 2481–2495 (2017)
7. Chen, L.C., Papandreou, G., Kokkinos, I., Murphy, K., Yuille, A.L.: DeepLab: semantic image segmentation with deep convolutional nets, atrous convolution, and fully connected Ds. IEEE Trans. Pattern Anal. Mach. Intell. **40**(4), 834–848 (2017)
8. Chen, L.C., Papandreou, G., Schroff, F., Adam, H.: Rethinking atrous convolution for semantic image segmentation. arXiv preprint. arXiv:1706.05587 (2017)
9. Chen, L.C., Zhu, Y., Papandreou, G., Schroff, F., Adam, H.: Encoder-decoder with atrous separable convolution for semantic image segmentation. In: Proceedings of the European conference on computer vision (ECCV), pp. 801–818 (2018)
10. Chollet, F.: Xception: deep learning with depthwise separable convolutions. In: Proceedings of the IEEE Conference on Computer Vision and Pattern Recognition, pp. 1251–1258 (2017)
11. Çiçek, Ö., Abdulkadir, A., Lienkamp, S.S., Brox, T., Ronneberger, O.: 3D U-Net: learning dense volumetric segmentation from sparse annotation. In: Ourselin, S., Joskowicz, L., Sabuncu, M.R., Unal, G., Wells, W. (eds.) MICCAI 2016. LNCS, vol. 9901, pp. 424–432. Springer, Cham (2016). https://doi.org/10.1007/978-3-319-46723-8_49
12. Codella, N., et al.: Skin lesion analysis toward melanoma detection 2018: a challenge hosted by the international skin imaging collaboration (ISIC). arXiv preprint arXiv:1902.03368 (2019)
13. Codella, N.C., et al.: Skin lesion analysis toward melanoma detection: a challenge at the 2017 international symposium on biomedical imaging (ISBI), hosted by the international skin imaging collaboration (ISIC). In: 2018 IEEE 15th ISBI 2018, pp. 168–172. IEEE (2018)
14. Feng, J., Isern, N.G., Burton, S.D., Hu, J.Z.: Studies of secondary melanoma on C57BL/6J mouse liver using 1H NMR metabolomics. Metabolites **3**(4), 1011–1035 (2013)
15. Ghiasi, G., Fowlkes, C.C.: Laplacian pyramid reconstruction and refinement for semantic segmentation. In: Leibe, B., Matas, J., Sebe, N., Welling, M. (eds.) ECCV 2016. LNCS, vol. 9907, pp. 519–534. Springer, Cham (2016). https://doi.org/10.1007/978-3-319-46487-9_32
16. Hesamian, M.H., Jia, W., He, X., Kennedy, P.J.: Atrous convolution for binary semantic segmentation of lung nodule. In: ICASSP 2019–2019 IEEE International Conference on Acoustics, Speech and Signal Processing (ICASSP), pp. 1015–1019, May 2019
17. Hu, J., Shen, L., Sun, G.: Squeeze-and-excitation networks. In: Proceedings of the CVPR, pp. 7132–7141 (2018)
18. Li, Y., Shen, L.: Skin lesion analysis towards melanoma detection using deep learning network. Sensors **18**(2), 556 (2018)
19. Liang, X., Shen, X., Xiang, D., Feng, J., Lin, L., Yan, S.: Semantic object parsing with local-global long short-term memory. In: Proceedings of the IEEE Conference on Computer Vision and Pattern Recognition, pp. 3185–3193 (2016)

20. Lin, G., Milan, A., Shen, C., Reid, I.: RefineNet: multi-path refinement networks for high-resolution semantic segmentation. In: Proceedings of the IEEE Conference on Computer Vision and Pattern Recognition, pp. 1925–1934 (2017)
21. Long, J., Shelhamer, E., Darrell, T.: Fully convolutional networks for semantic segmentation. In: Proceedings of the IEEE Conference on Computer Vision and Pattern Recognition, pp. 3431–3440 (2015)
22. Mendonça, T., Ferreira, P.M., Marques, J.S., Marcal, A.R., Rozeira, J.: PH 2-a dermoscopic image database for research and benchmarking. In: 2013 35th Annual International Conference of the IEEE Engineering in Medicine and Biology Society (EMBC), pp. 5437–5440. IEEE (2013)
23. Milletari, F., Navab, N., Ahmadi, S.A.: V-Net: fully convolutional neural networks for volumetric medical image segmentation. In: 2016 Fourth International Conference on 3D Vision (3DV), pp. 565–571. IEEE (2016)
24. Noh, H., Hong, S., Han, B.: Learning deconvolution network for semantic segmentation. In: Proceedings of the CVPR, pp. 1520–1528 (2015)
25. Oktay, O., et al.: Attention U-Net: learning where to look for the pancreas. arXiv preprint arXiv:1804.03999 (2018)
26. Peng, C., Zhang, X., Yu, G., Luo, G., Sun, J.: Large kernel matters-improve semantic segmentation by global convolutional network. In: Proceedings of the IEEE Conference on Computer Vision and pattern recognition, pp. 4353–4361 (2017)
27. Ronneberger, O., Fischer, P., Brox, T.: U-Net: convolutional networks for biomedical image segmentation. In: Navab, N., Hornegger, J., Wells, W.M., Frangi, A.F. (eds.) MICCAI 2015. LNCS, vol. 9351, pp. 234–241. Springer, Cham (2015). https://doi.org/10.1007/978-3-319-24574-4_28
28. Schlemper, J., et al.: Attention gated networks: learning to leverage salient regions in medical images. Med. Image Anal. **53**, 197–207 (2019)
29. Siegel, R.L., Miller, K.D., Jemal, A.: Cancer statistics, 2018. CA Cancer J. Clin. **68**(1), 7–30 (2018)
30. Siegel, R.L., Miller, K.D., Jemal, A.: Cancer statistics, 2019. CA: Cancer J. Clin. **69**(1), 7–34 (2019)
31. Sinha, A., Dolz, J.: Multi-scale guided attention for medical image segmentation. arXiv preprint arXiv:1906.02849 (2019)
32. Tarver, T.: American cancer society. Cancer facts and figures 2014. J. Consum. Health Internet **16**, 366–367 (2012)
33. Wang, G., Luo, P., Lin, L., Wang, X.: Learning object interactions and descriptions for semantic image segmentation. In: Proceedings of the IEEE Conference on Computer Vision and Pattern Recognition, pp. 5859–5867 (2017)
34. Zhao, H., Shi, J., Qi, X., Wang, X., Jia, J.: Pyramid scene parsing network. In: Proceedings of the IEEE Conference on Computer Vision and Pattern Recognition, pp. 2881–2890 (2017)
35. Zhao, H., et al.: PSANet: point-wise spatial attention network for scene parsing. In: Proceedings of the European Conference on Computer Vision (ECCV), pp. 267–283 (2018)

Automated Assessment of the Curliness of Collagen Fiber in Breast Cancer

David Paredes[1(✉)], Prateek Prasanna[2], Christina Preece[2], Rajarsi Gupta[2], Farzad Fereidouni[3], Dimitris Samaras[1], Tahsin Kurc[2], Richard M. Levenson[3], Patricia Thompson-Carino[2], Joel Saltz[2], and Chao Chen[2]

[1] Department of Computer Science, Stony Brook University, New York, USA
dparedesmeri@cs.stonybrook.edu
[2] Department of Biomedical Informatics, Stony Brook University, New York, USA
[3] Department of Pathology and Laboratory Medicine, UC Davis Medical Center, Sacramento, USA

Abstract. The growth and spread of breast cancer are influenced by the composition and structural properties of collagen in the extracellular matrix of tumors. Straight alignment of collagen has been attributed to tumor cell migration, which is correlated with tumor progression and metastasis in breast cancer. Thus, there is a need to characterize collagen alignment to study its value as a prognostic biomarker. We present a framework to characterize the curliness of collagen fibers in breast cancer images from DUET (DUal-mode Emission and Transmission) studies on hematoxylin and eosin (H&E) stained tissue samples. Our novel approach highlights the characteristic fiber gradients using a standard ridge detection method before feeding into the convolutional neural network. Experiments were performed on patches of breast cancer images containing straight or curly collagen. The proposed approach outperforms in terms of area under the curve against transfer learning methods trained directly on the original patches. We also explore a feature fusion strategy to combine feature representations of both the original patches and their ridge filter responses.

Keywords: Collagen fiber · Deep learning · Ridge detection · Digital pathology

1 Introduction

Collagen is an abundant structural protein in the extracellular matrix of tissues. In addition to providing structural support, collagen also plays a major regulatory role in other tissue functions, including cell adhesion, migration, and proliferation. Recent experimental and clinical studies have shown how the structural integrity of collagen directly influences the behavior of breast cancer (BCa) cells and their capacity for metastasis.

Compelling evidence has emerged to show how the structure and alignment of collagen is correlated with BCa progression and metastasis. Our goal is to use

© Springer Nature Switzerland AG 2020
A. Bartoli and A. Fusiello (Eds.): ECCV 2020 Workshops, LNCS 12535, pp. 267–279, 2020.
https://doi.org/10.1007/978-3-030-66415-2_17

deep learning to characterize these architectural features by classifying whether collagen appears straight or curly in BCa images. Therefore, we utilized the DUET imaging technique (DUal-mode Emission and Transmission) [11], which focuses on color (spectral) differences in H&E fluorescence to directly highlight collagen distributions in H&E slides. Compared to existing techniques, DUET uses readily available H&E slides with the potential to be a better collagen detection tool in clinical settings due to low cost, simplicity, rapid imaging speed, and non-destructive effects in tissue samples. See Fig. 1 for examples. Quantitative methods to identify collagen structure orientation patterns in DUET images have not been explored.

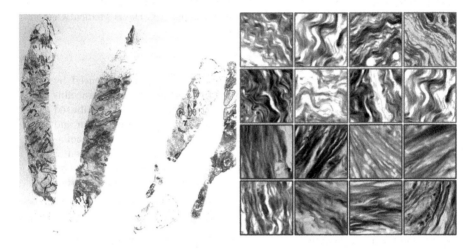

Fig. 1. Example DUET images capturing different collagen fiber patterns. <u>Left</u>: red and blue contours highlight regions of interest (ROIs). Preliminary annotations are given to these ROIs. Red circles enclose regions which are dominantly curly. Blue circles enclose regions which are dominantly straight. <u>Right</u>: Patches were sampled from these ROIs and are individually annotated by expert annotators. Red boxes enclose curly patches and blue boxes enclose straight patches. Annotations of individual patches are correlated to but are not fully consistent with the annotations of ROIs. For example, a dominantly curly region may contain straight patches. (Color figure online)

We present a deep learning-based automatic method to classify BCa DUET image patches into two categories, curly and straight, based on given annotations by our domain expert (Fig. 1). There are two challenges for the direct application of deep neural networks: First, the amount of training data is relatively limited. It is insufficient for training a deep neural network classifier from scratch. Second, the distinct geometry and textural patterns of collagen fiber patches makes it hard to directly use neural networks trained on other pathology images (mostly H&E). On the other hand, classic image filters are well designed to characterize object flow and detect these collagen fiber structures of interests.

Our novel approach combines the strengths of both deep neural networks and classic image filters. In particular, we leverage classic second-order-derivatives-based filters that are well-designed for identification of collagen fiber structures. The identified structures are combined with deep neural nets pretrained on public datasets for the best classification of curly vs. straight fiber patterns. We also explore different existing approaches for feature fusion in order to find the best strategy for our application. Usage of the classic filters that are well-suited to the structure of interest alleviates the demand of training data from deep networks. Meanwhile, incorporating these filters with pretrained deep nets ensures the structural features can be fully leveraged in a data-driven manner. Inspired by the analogy of the collagen images with "flow" images and the use of second-order derivative filters for optical flow computation, we experiment with two stream networks analogous to those successfully applied in video analysis tasks, where spatial (frames) and temporal information (optical flow) are aggregated [10,26, 30].

In summary, our main contributions are:

1. We present the first deep learning approach to capture orientation patterns of collagen fibers;
2. We incorporate sophisticated image filters (i.e., ridge detectors) that are very appropriate for highlighting fiber structures in a two-stream deep learning framework.
3. We evaluate our methods on a unique cohort of DUET collagen images. We demonstrate that incorporating these additional structural features will significantly improve the generalization performance of the classifier, especially when there is insufficient training data.

2   Background

Importance of Collagen in Breast Cancer: Brabrand et al. [1] studied the structure of collagen fibers in intratumoral, juxtatumoral, and extratumoral regions and classified collagen fibers as (1) curly or straight and (2) parallel or not parallel. They observed that collagen fibers appeared more straight and aligned at tumor boundaries while tumor cells invade surrounding tissues. In another study, Carpino et al. [3] observed correlations between the reorganization of collagen and progression. Early in tumorigenesis and during the initial stages of tumor progression, collagen appears to be mostly curly and dense. As tumorigenesis progresses, collagen architecture becomes straight and aligned in parallel to the tumor boundary. In more advanced stages of tumor progression, collagen fibers are oriented perpendicular to the tumor boundary. These and other studies indicate that the structure and alignment of collagen in close proximity of BCa can be a critical indicator of tumor progression [4,6,27,28,31].

The association between collagen alignment with breast tumor progression and metastasis underlies the need for quantitative methods to extract and characterize useful information from collagen-tumor relationships. A variety of techniques have been proposed to capture images of collagen fibers in tissue specimens [2,7,8,24]. Collagen fiber orientation may be quantified using Spatial Light

Interference Microscopy (SLIM). Majeed et al. [20] have used SLIM images to extract prognostic information from BCa collagen fibers. Second harmonic generation (SHG) microscopy techniques have been used to localize patterns associated with collagen fiber extracellular matrix without the need for specialized stains. Collagen alignment has been characterized by wavelet transform type methods [18,19]. These techniques are rather expensive; they require specialized equipment and are restricted to specific subtypes of collagen fibers. In the clinical setting, trichrome staining is a common collagen staining approach. However, it is subject to high staining inconsistencies across institutions and even across different subjects from the same cohort. Furthermore, trichrome stains not only collagen, but also other undesired tissues [5,16,32].

Image Analysis and Machine Learning Methods: Hand-crafted ridge detectors have been used for several applications, such as for neurite in fluorescence microscopy images [22], for enhancement of vessel structures [12,29], and for detecting wrinkles on human skin [25]. Hidayat et al. [14] have leveraged ridge detection for real-time texture boundary detection on natural images. There have been previous attempts to characterize structure of collagen from a machine learning perspective using hand-crafted features on spectroscopy images. Mayerich et al. [21] have classified several types of tissues, including collagen, on a breast histology dataset of mid-infrared spectroscopy images. This method uses random forests and features from principal component analysis. [15,23] train a support vector machine (SVM) classifier using texture features from multiphoton microscopy images and second-harmonic generation microscopy images, respectively. The collagen fiber orientation in human arteries in light microscopic images was explored in [9]. Here, robust clustering and morphological operations in the color space were first used to identify fiber regions. This was followed by ridge and valley detections to calculate fiber orientations. Subsequently, region growing was used to obtain areas of homogenous fiber orientation. However, none of the previous works have utilized deep learning based methods to characterize fiber orientations. This is partly owing to the lack of well annotated pathology datasets. The ability of flow-based approaches to enhance image characteristics provides us the unique opportunity to leverage such methods in the context of our problem by fusing them with deep learning methods in a multi-stream setting.

Two-stream networks have been successfully applied in video analytics tasks, where spatial (frames) and temporal information (optical flow) are aggregated as shown in [10,26,30]. Particularly, [26] explores multiple fusion methods of intermediate outputs for human action recognition. Lin et al. [17] have introduced a new application of two-stream networks, where the architecture can model local pairwise feature interactions in a translationally invariant manner which is particularly useful for fine-grained categorization. Our methods leverage such two-stream networks in conjunction with gradient-based operators to build a curliness classifier on DUET images.

3   Methodology

We first introduce the filtering operations using the ridge filter. Next, we explain how to combine the ridge filter with deep neural networks in a seamless fashion in order to train a robust collagen fiber classifier.

3.1   Ridge Filter for Collagen Fiber Structure Detection

Hessian-based filters are well-suited for detection of curvelinear structures in various biomedical contexts, such as vessels [12], neurons [22], etc. In particular, we employ a ridge detector. In a 2D image, ridges are the collection of curves on which the function value remains almost constant along the tangent direction, and are local maxima along the normal direction. See Fig. 2 for sample input DUET patches and the ridge detection results. The ridge detector successfully highlights the fiber structures so that they can be used to distinguish curly vs. straight patterns.

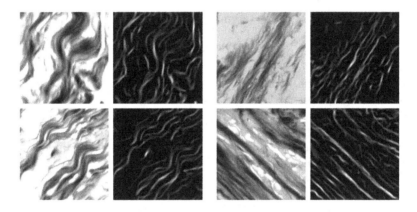

Fig. 2. A ridge detector is applied to curly and straight patches. From left to right: the DUET patches of curly structures, their corresponding ridge filter response, the DUET patches of straight structures, their ridge filter responses. Note the original image is inverted before the ridge filer being applied.

A ridge detector relies on the Hessian matrix of a gray-scale image. At any point of the image domain, the eigenvalues and eigenvectors of the Hessian correspond to the two principle curvatures and their corresponding directions (called principle directions). On a point in a ridge, one of the two Hessian eigenvalues will be close to zero (curvature along the tangent direction), the other will be negative (curvature along the normal direction). Ridge detectors identify ridge points based on this principle. Our method is particularly inspired by the one by Meijering et al. [22].

Our algorithm for ridge detection is illustrated in Fig. 3. At first, the original DUET image (A) is inverted so that the white pixels correspond to collagen fibre. Next, the image is smoothed with a Gaussian filter to ensure numerical stability when computing the second-degree derivatives (B). Next, the Hessian matrix $\mathcal{H}$ is calculated for each pixel in the smoothed image (denoted as $\mathcal{I}$) as follows

$$\mathcal{H}_{i,j} = \frac{\partial^2 \mathcal{I}}{\partial x_i \partial x_j} \tag{1}$$

where x_i and x_j are the two dimensions of the image domain. Since the Hessian matrix is symmetric, its eigenvalues can be computed as

$$\lambda_1 = \frac{1}{2}\left[\mathcal{H}_{1,1} + \mathcal{H}_{2,2} + (\sqrt{4\mathcal{H}_{1,2}^2 + (\mathcal{H}_{1,1} - \mathcal{H}_{2,2})^2})\right], \tag{2}$$

$$\lambda_2 = \frac{1}{2}\left[\mathcal{H}_{1,1} + \mathcal{H}_{2,2} - (\sqrt{4\mathcal{H}_{1,2}^2 + (\mathcal{H}_{1,1} - \mathcal{H}_{2,2})^2})\right]. \tag{3}$$

We rank the two eigenvalues based on their absolute magnitudes. We call λ_H and λ_L the high and low eigenvalues with respect to the absolute magnitude, or in short, the *high* and *low* eigenvalues. For each pixel, we have

$$\lambda_H = \text{argmax}_{\lambda \in \{\lambda_1, \lambda_2\}}|\lambda| \tag{4}$$

$$\lambda_L = \text{argmax}_{\lambda \in \{\lambda_1, \lambda_2\}}|\lambda| \tag{5}$$

Subfigures C.1 and C.2 of Fig. 3 visualize the high and low eigenvalues respectively. We note that the high eigenvalues can be either positive and negative. The desired ridge points are the locations whose λ_H has high absolute magnitude yet negative sign. We normalize λ_H by dividing with its minimum value. In this way, the range of the results can be up to +1. Subfigure D of Fig. 3 shows that the ridge points have values close to 1. They correspond to points whose λ_H has high absolute magnitude yet negative sign. Finally, we remove irrelevant responses by setting all negative values to zero as follows

$$\mathcal{R} = \max(0, \lambda_H / \min(\lambda_H)) \tag{6}$$

where $\mathcal{R}$ is the final filter response. See subfigure E of Fig. 3 for an example.

3.2   Convolutional Neural Networks

We explore 3 different settings of training CNNs with ridge filter response as shown in Fig. 4. All experiments use ResNet [13] models with pre-trained weights using ImageNet. The backbone network has frozen weights and its output is an 1D-vector of size 512. In the first setting, the input to the CNN are the original DUET patches. We apply transfer learning by fine-tuning the final layers for two classes (curly and straight). In the second setting, the ridge detector is applied to the original DUET patches and is then provided to the CNN for transferred learning. In the third setting, the original DUET patch and the ridge filter are both use as input for a pretrained network. The output of the network are high-level feature representations of the original image and the ridge response. We fuse these features for the final binary classifier training.

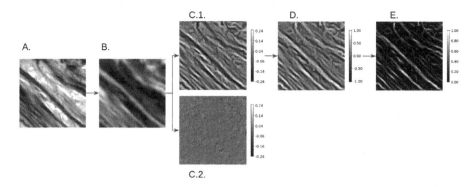

Fig. 3. <u>A.</u> Raw input patch. <u>B.</u> Input is inverted and smoothed with a Gaussian filter. <u>C.1</u> High eigenvalues λ_L of Eq. 4. <u>C.2</u> Low eigenvalues λ_H of Eq. 5. <u>D.</u> High eigenvalues λ_H are divided by its minimum value. <u>E.</u> Final ridge filter response after the negative values of the previous step are set to zero as stated in Eq. 6.

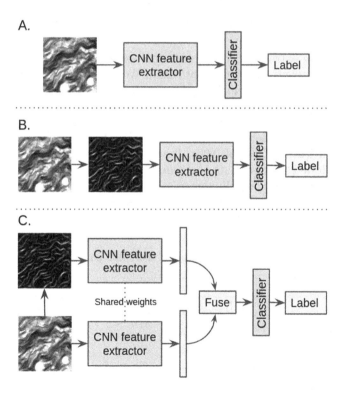

Fig. 4. <u>A.</u> The input is the raw patch of collagen content. <u>B.</u> The ridge detector is applied to the raw patch before feeding into the CNN. <u>C.</u> Features of the raw patch and the ridge detector response are fused for the classification.

We investigate 4 types of feature fusion strategies. Element-wise product (Elt-wise prod) is the element-wise multiplication of the elements of two 1D input vectors and its output has the same size. Similarly, Addition outputs a 1D vector of same size as the inputs and its elements are the sum of the elements of the two 1D input vectors. Another method is the concatenation of both 1D input vectors. Finally, we use the bilinear operation for two-stream networks, introduced in [17]. It computes the L2-normalized outer product of the two features as the input feature for the final classifier.

4 Experiments

Dataset and Settings. Six whole slide images (WSIs) of DUET images of BCa tissue are used to evaluate the proposed method. Our study comprised images from aggressive early stage breast cancer patients with known 5-year distant metastasis status. For each WSI, our domain expert selected regions of interests (ROIs) with strong collagen structures (see the left part of Fig. 1). Preliminary binary labels are also given to these ROIs (as curly or straight). Next, non-overlapping patches were sampled from inside these ROIs. The size of the patches is 400 × 400 pixels with a physical resolution of 0.22 microns per pixel.

Each of these extracted patches are annotated by the domain annotator as curly, straight, mixed, and insufficient signal. We discard patches of mixed type or insufficient signal. The remaining ones are used for training and validation. 5 WSIs and 1 WSI were used to create the training set and validation set, respectively. The distribution of the patches is shown in Table 1.

The ResNet-family networks were trained using the Adam optimizer for 100 epochs with a learning rate of 0.001. Input patches were resized to 224 × 224 pixels. We followed standard data augmentation regimes such as a 0.5 probability of horizontally and vertically flipping as well as random rotation of the input by up to 20 degrees. For the ridge detection step, we use a σ value of 5 as smoothing factor in the Gaussian filter before computing the Hessian matrix.

Table 1. Distribution of 400 × 400 patches

	Training	Validation
Curly	1037	314
Straight	1018	418

Quantitative Results. For our binary classification problem, we evaluated the area under the ROC Curve (AUC). We compare the 3 pipelines of Fig. 4 and the results are presented in Table 2. We observe that when the networks were trained with the output of the ridge detector, they achieve higher AUC than when the networks were trained directly on raw patches. Furthermore, we explore 4 different methods to aggregate the outputs of as shown in part C. of Fig. 4. In this setting, the addition of the outputs of the CNN is the only method that can boost the AUC performance on most of the networks.

Although the feature fusion methods do not seem to achieve significantly higher accuracy, we suspect it is due to the insufficient training data. When scaling up to a larger cohort, we expect the feature fusion network will achieve higher accuracy and become the best choice in practice.

Table 2. Results (AUC) after fine-tuning the ResNet models using raw patch only, ridge response only, and feature fusion result. The latter case has 4 ways of combining the intermediate outputs of the CNN.

Network	Raw	Ridge $\sigma = 5$	Raw + Ridge			
			Elt-wise prod	Bilinear	Concatenation	Addition
ResNet-18	0.7365	**0.8467**	0.7027	0.7545	0.8055	0.8148
ResNet-34	0.7415	0.8235	0.8321	0.8100	0.8129	**0.8423**
ResNet-50	0.7400	0.8352	0.8391	0.7582	0.8446	**0.8624**
ResNet-101	0.7774	0.8432	0.7677	0.7988	0.8199	**0.8463**
Average	0.7489	0.8372	0.7854	0.7804	0.8207	**0.8415**

Table 3 shows how the smoothing factor (σ) of the ridge filter affects the AUC performance of the ResNet models that were fine-tuned on patches with a ridge filter applied before feeding into the network as shown in part B. of Fig. 4. On average, the ResNet networks benefit the most when σ was 5, so we used this value for our experiments.

Table 3. Results (AUC) of the fine-tuned ResNet models when the input of the CNN is the output of the ridge detector. σ is the smoothing factor of the ridge detector.

	Ridge filter			
Network	$\sigma = 1$	$\sigma = 3$	$\sigma = 5$	$\sigma = 7$
ResNet-18	0.6968	0.8151	**0.8467**	0.8402
ResNet-34	0.7202	0.7934	**0.8235**	0.8231
ResNet-50	0.7589	0.8289	0.8352	**0.8354**
ResNet-101	0.7383	0.8179	**0.8432**	0.8203
Average	0.7286	0.8138	**0.8372**	0.8298

Qualitative Results. The network with highest AUC from the previous section is used to predict the degree of curliness as a heatmap on the annotated regions of interest. We run inference on the non-overlapping patches from inside the regions of interest and we save the output of the softmax layer for the "curly" class. On the left side of Fig. 5, preliminary annotations of the regions of interest are depicted for a WSI. The red boundary regions correspond to areas with dominating curliness and the blue boundary regions are dominantly straight. The predicted heatmap is shown on the right side of Fig. 5 and each pixel represents the likelihood of curliness of a patch of 400×400 pixels. We observe high consistency between patchwise prediction of our model and the preliminary labels of ROIs. At the bottom of Fig. 5, we also included a magnification of a sub-region of the WSI.

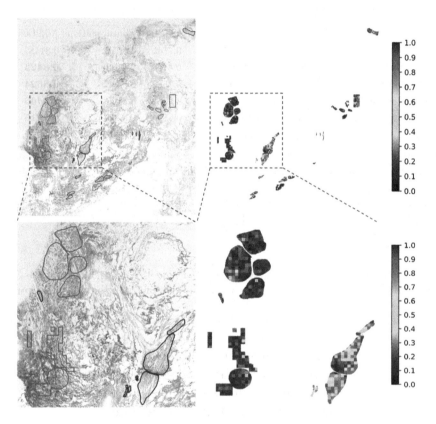

Fig. 5. Visualization of prediction results of our method. A high value on the heatmap represents a high likelihood of curliness. Left: ROIs on a WSI. Colors of the boundaries correspond to preliminary annotation by our experts. Red boundaries enclose regions that are dominantly curly. Blue boundaries enclose dominantly straight regions. Right: The likelihood map of curliness predicted by our method. Top: Full size WSI. Bottom: Magnification of a sub-region. (Color figure online)

5   Conclusion

In an effort to optimize the research and clinical opportunities of tumor associated collagen as a predictive and prognostic biomarker, it is imperative to first localize the collagen content and then characterize the degree of curliness to further elucidate the relationship of collagen structure with tumor progression and metastasis. We present a quantitative approach that identifies characteristic collagen architecture in a unique dataset of BCa DUET images. Rather than learning to estimate collagen patterns using standard deep learning approaches directly on the DUET images, we enhance the images using ridge detectors and incorporate the filter responses using a feature-fusion strategy. Our work provides a novel and promising direction for research in digital pathology that can both further our understanding of BCa biology and meaningfully impact clinical management. Our results show that certain primitives can be enhanced in the pre-analytical steps, which can then help in providing additional architectural information about tissue samples. This can then be further integrated with other types of digital pathology analyses like automated tumor and TIL detection and characterization of different immunohistochemical (IHC) stains.

Acknowledgements. This work was supported in part by National Cancer Institute grants 3U24CA215109, 1UG3CA225021, 1U24CA180924 and its supplement 3U24CA180924-05S2, as well as the grant R33CA202881 and its supplement 3R33CA202881-02S1. Partial support for this effort was also funded through the generosity of Bob Beals and Betsy Barton.

References

1. Brabrand, A., et al.: Alterations in collagen fibre patterns in breast cancer. A premise for tumour invasiveness? Apmis **123**(1), 1–8 (2015)
2. Burke, K., Tang, P., Brown, E.: Second harmonic generation reveals matrix alterations during breast tumor progression. J. Biomed. Opt. **18**(3), 031106–031106 (2013)
3. Carpino, G., et al.: Matrisome analysis of intrahepatic cholangiocarcinoma unveils a peculiar cancer-associated extracellular matrix structure. Clin. Proteomics **16**(1), 1–12 (2019)
4. Case, A., et al.: Identification of prognostic collagen signatures and potential therapeutic stromal targets in canine mammary gland carcinoma. PLOS One **12**(7), e0180448 (2017)
5. Cason, J.E.: A rapid one-step Mallory-Heidenhain stain for connective tissue. Stain Technol. **25**(4), 225–226 (1950)
6. Martins Cavaco, A.C., Dâmaso, S., Casimiro, S., Costa, L.: Collagen biology making inroads into prognosis and treatment of cancer progression and metastasis. Cancer Metastasis Rev. **39**(3), 603–623 (2020). https://doi.org/10.1007/s10555-020-09888-5
7. Dolber, P., Spach, M.: Conventional and confocal fluorescence microscopy of collagen fibers in the heart. J. Histochem. Cytochem. **41**(3), 465–469 (1993)
8. Drifka, C.R., et al.: Highly aligned stromal collagen is a negative prognostic factor following pancreatic ductal adenocarcinoma resection. Oncotarget **7**, 76197 (2016)

9. Elbischger, P., Bischof, H., Regitnig, P., Holzapfel, G.: Automatic analysis of collagen fiber orientation in the outermost layer of human arteries. Pattern Anal. Appl. **7**(3), 269–284 (2004). https://doi.org/10.1007/BF02683993

10. Feichtenhofer, C., Pinz, A., Zisserman, A.: Convolutional two-stream network fusion for video action recognition. In: Proceedings of the IEEE Conference on Computer Vision and Pattern Recognition, pp. 1933–1941 (2016)

11. Fereidouni, F., et al.: Dual-mode emission and transmission microscopy for virtual histochemistry using hematoxylin-and eosin-stained tissue sections. Biomed. Opt. Express **10**(12), 6516–6530 (2019)

12. Frangi, A.F., Niessen, W.J., Vincken, K.L., Viergever, M.A.: Multiscale vessel enhancement filtering. In: Wells, W.M., Colchester, A., Delp, S. (eds.) MICCAI 1998. LNCS, vol. 1496, pp. 130–137. Springer, Heidelberg (1998). https://doi.org/10.1007/BFb0056195

13. He, K., Zhang, X., Ren, S., Sun, J.: Deep residual learning for image recognition. In: Proceedings of the IEEE Conference on Computer Vision and Pattern Recognition, pp. 770–778 (2016)

14. Hidayat, R., Green, R.D.: Real-time texture boundary detection from ridges in the standard deviation space. In: BMVC, pp. 1–10 (2009)

15. Kistenev, Y.V., Vrazhnov, D.A., Nikolaev, V.V., Sandykova, E.A., Krivova, N.A.: Analysis of collagen spatial structure using multiphoton microscopy and machine learning methods. Biochemistry (Moscow) **84**(1), 108–123 (2019). https://doi.org/10.1134/S0006297919140074

16. Lillie, R., Miller, G.: Histochemical acylation of hydroxyl and amino groups. Effect on the periodic acid Schiff reaction, anionic and cationic dye and Van Gieson collagen stains. J. Histochem. Cytochem.**12**(11), 821–841 (1964)

17. Lin, T.Y., RoyChowdhury, A., Maji, S.: Bilinear CNN models for fine-grained visual recognition. In: Proceedings of the IEEE International Conference on Computer Vision, pp. 1449–1457 (2015)

18. Liu, Y., Keikhosravi, A., Mehta, G.S., Drifka, C.R., Eliceiri, K.W.: Methods for quantifying fibrillar collagen alignment. In: Rittié, L. (ed.) Fibrosis. MMB, vol. 1627, pp. 429–451. Springer, New York (2017). https://doi.org/10.1007/978-1-4939-7113-8_28

19. Liu, Y., et al.: Fibrillar collagen quantification with curvelet transform based computational methods. Front. Bioeng. Biotechnol. **8**, 198 (2020)

20. Majeed, H., Okoro, C., Kajdacsy-Balla, A., Toussaint, K.C., Popescu, G.: Quantifying collagen fiber orientation in breast cancer using quantitative phase imaging. J. Biomed. Opt. **22**(4), 046004 (2017)

21. Mayerich, D.M., Walsh, M., Kadjacsy-Balla, A., Mittal, S., Bhargava, R.: Breast histopathology using random decision forests-based classification of infrared spectroscopic imaging data. In: Medical Imaging 2014: Digital Pathology, vol. 9041, p. 904107. International Society for Optics and Photonics (2014)

22. Meijering, E., Jacob, M., Sarria, J.C., Steiner, P., Hirling, H., Unser, M.: Design and validation of a tool for neurite tracing and analysis in fluorescence microscopy images. Cytometry Part A J. Int. Soc. Anal. Cytol. **58**(2), 167–176 (2004)

23. Mostaço-Guidolin, L.B., et al.: Collagen morphology and texture analysis: from statistics to classification. Sci. Rep. **3**(1), 1–10 (2013)

24. Natal, R.A., et al.: Collagen analysis by second-harmonic generation microscopy predicts outcome of luminal breast cancer. Tumor Biol. **40**(4), 1010428318770953 (2018)

25. Ng, C.-C., Yap, M.H., Costen, N., Li, B.: Automatic wrinkle detection using hybrid Hessian filter. In: Cremers, D., Reid, I., Saito, H., Yang, M.-H. (eds.) ACCV 2014. LNCS, vol. 9005, pp. 609–622. Springer, Cham (2015). https://doi.org/10.1007/978-3-319-16811-1_40

26. Park, E., Han, X., Berg, T.L., Berg, A.C.: Combining multiple sources of knowledge in deep CNNs for action recognition. In: 2016 IEEE Winter Conference on Applications of Computer Vision (WACV), pp. 1–8. IEEE (2016)

27. Provenzano, P.P., Eliceiri, K.W., Campbell, J.M., Inman, D.R., White, J.G., Keely, P.J.: Collagen reorganization at the tumor-stromal interface facilitates local invasion. BMC Med. **4**(1), 1–15 (2006)

28. Provenzano, P.P., et al.: Collagen density promotes mammary tumor initiation and progression. BMC Med. **6**(1), 1–15 (2008)

29. Sato, Y., et al.: Three-dimensional multi-scale line filter for segmentation and visualization of curvilinear structures in medical images. Med. Image Anal. **2**(2), 143–168 (1998)

30. Simonyan, K., Zisserman, A.: Two-stream convolutional networks for action recognition in videos. In: Advances in Neural Information Processing Systems, pp. 568–576 (2014)

31. Velez, D., et al.: 3D collagen architecture induces a conserved migratory and transcriptional response linked to vasculogenic mimicry. Nat. Commun. **8**(1), 1–12 (2017)

32. Whittaker, P., Kloner, R., Boughner, D., Pickering, J.: Quantitative assessment of myocardial collagen with picrosirius red staining and circularly polarized light. Basic Res. Cardiol. **89**(5), 397–410 (1994). https://doi.org/10.1007/BF00788278

Bionic Tracking: Using Eye Tracking to Track Biological Cells in Virtual Reality

Ulrik Günther[1,2,3($\boxtimes$)] (ID), Kyle I.S. Harrington[6,7] (ID), Raimund Dachselt[4,5] (ID), and Ivo F. Sbalzarini[2,3,4,5] (ID)

[1] Center for Advanced Systems Understanding, Görlitz, Germany
ulrik.guenther@hzdr.de
[2] Center for Systems Biology Dresden, Dresden, Germany
[3] Max Planck Institute of Molecular Cell Biology and Genetics, Dresden, Germany
[4] Faculty of Computer Science, Technische Universität Dresden, Dresden, Germany
[5] Cluster of Excellence Physics of Life, Technische Universität Dresden, Dresden, Germany
[6] Virtual Technology and Design, University of Idaho, Moscow, ID, USA
[7] HHMI Janelia Research Campus, Ashburn, VA, USA

Abstract. We present Bionic Tracking, a novel method for solving biological cell tracking problems with eye tracking in virtual reality using commodity hardware. Using gaze data, and especially smooth pursuit eye movements, we are able to track cells in time series of 3D volumetric datasets. The problem of tracking cells is ubiquitous in developmental biology, where large volumetric microscopy datasets are acquired on a daily basis, often comprising hundreds or thousands of time points that span hours or days. The image data, however, is only a means to an end, and scientists are often interested in the reconstruction of cell trajectories and cell lineage trees. Reliably tracking cells in crowded three-dimensional space over many time points remains an open problem, and many current approaches rely on tedious manual annotation or curation. In the Bionic Tracking approach, we substitute the usual 2D point-and-click interface for annotation or curation with eye tracking in a virtual reality headset, where users follow cells with their eyes in 3D space in order to track them. We detail the interaction design of our approach and explain the graph-based algorithm used to connect different time points, also taking occlusion and user distraction into account. We demonstrate Bionic Tracking using examples from two different biological datasets. Finally, we report on a user study with seven cell tracking experts, highlighting the benefits and limitations of Bionic Tracking compared to point-and-click interfaces.

Electronic supplementary material The online version of this chapter (https://doi.org/10.1007/978-3-030-66415-2_18) contains supplementary material, which is available to authorized users.

A. Bartoli and A. Fusiello (Eds.): ECCV 2020 Workshops, LNCS 12535, pp. 280–297, 2020.
https://doi.org/10.1007/978-3-030-66415-2_18

1   Introduction

In cell and developmental biology, the image data generated by fluorescence microscopy is often a means to an end: Many biological studies require information about the positions of cells during development, or their ancestral history, the cell lineage tree. Both the creation of such a tree using cell tracking, and tracking of single cells, are difficult and cannot always be fully automated. Therefore, cell tracks and lineage trees often are curated in a tedious manual process using point-and-click user interfaces. The same point-and-click user interfaces are also used to generate ground-truth or training data for machine-learning approaches to cell tracking (see [23] for a review).

This typically is a tedious task, as users have to go through each time point and 2D cutting plane in order to connect cells in 3D+time. Tracking a single cell through 100 time points with this manual process takes 5 to 30 min, depending on complexity of the images. Tracking an entire developmental dataset with many 3D images can thus take weeks of manual curation effort.

Part of this difficulty is due to fluorescence microscopy images not having well-defined intensity scales. Intensities might vary strongly even within single cells. Cells also move, divide, change their shape – sometimes drastically – or die Cells might also not appear isolated, and may move through densely-populated tissue, making it difficult to tell one cell apart from another. These difficulties make cell tracking in developmental systems hard. Further aggravating the problem, advances in fluorescence microscopy, such light-sheet microscopy [10], have led to an enormous increase in size and number of images, with datasets ranging from a gigabyte to several terabytes for time-lapse images [27]. Given these data volumes, manual point-and-click curation is not a scalable long-term solution.

Here, we propose a way to improve the efficiency of curation and generation of cell-tracking data from fluorescence microscopy videos. We introduce *Bionic Tracking*, a method that uses smooth pursuit eye movements as detected by eye trackers inside a virtual reality head-mounted display (HMD) to render 3D cell tracking and track curation tasks easier, faster, and more ergonomic. To track a cell using Bionic Tracking, users just have to follow the cell with their eyes in Virtual Reality (VR). The main contributions we present to this end are:

- A system for interactively tracking cells by following them with the eyes in a 3D volume rendering using a virtual reality headset equipped with commodity eye tracking devices;
- An iterative, graph-based algorithm to connect gaze samples over time, addressing the problems of cell occlusion and user distraction;
- A user study evaluating the setup and the workflow with seven cell-tracking experts.

2   Related Work

The main problem we address in this paper is the manual curation or tracking step, which is necessary for: validation, handling cases where automatic tracking produces incorrect or no results, or producing training data for machine learning-based tracking.

In the past, software for solving cell-tracking problems was typically developed for a specific model organism, such as for the roundworm *Caenorhabditis elegans*, the fruitfly *Drosophila melanogaster*, or the zebrafish *Danio rerio*—all highly studied animals in biology—and often leveraged on stereotypical developmental dynamics to successfully track cells. However, these tools can fail or produces unreliable results for other organisms, or for organisms whose development is not stereotyped. For this reason, (semi-)automated approaches have been developed that are independent of the model organism and can track large numbers of cells. These, however, often require manual tracking of at least a subset of the cells in a dataset, or manual curation. Examples of such frameworks include:

- *TGMM*, Tracking by Gaussian Mixture Models [1,2], an offline tracking solution that works by generating over-segmented supervoxels from the original image data, then fit cell nuclei with a Gaussian Mixture Model and evolve that through time, and finally use the temporal context of a cell track to create the lineage tree;
- *TrackMate* [33], a plugin for Fiji [28] that provides automatic, semi-automatic, and manual tracking of objects in image datasets. TrackMate can be extended with custom spot detection and tracking algorithms.
- *MaMuT*, the Massive MultiView Tracker [38], is another plugin for Fiji that allows the user to manually track cells in large datasets, often originating from multi-view light-sheet microscopes. MaMuT's viewer is based on Big-DataViewer [24] and is able to handle terabyte-sized videos.

Manual tracking and curation in these approaches is usually done with mouse-and-keyboard interaction to select a cell and create a track, often while viewing a single slice of a 3D time point. In Bionic Tracking, we replace this interaction by leveraging the user's gaze in a VR headset, while the user can move around freely in the dataset. Gaze has been used in human-computer interaction for various purposes: It has been used as an additional input modality in conjunction with touch interaction [31] or pedaling [15], and for building user interfaces, e.g., for text entry [20]. Important for a gaze-based user interface is the type of eye movements that are to be used.

The particular kind of eye movements we exploit for Bionic Tracking are the so-called *smooth pursuits*. During smooth pursuit, the eyes follow a stimulus in a smooth, continuous manner. This type of eye movement is not typically used in 3D or VR applications. Instead, smooth pursuit interfaces can be found mainly in 2D interfaces, such as in applications that evaluate cognitive load [16] or for item selection [36]. In the context VR, we are only aware of two previous works: *Radial Pursuit*, a technique where the user can select an object in a 3D scene by tracking it with her/his eyes [26], and [14], where smooth pursuits are explored for selection of mesh targets in VR. In addition, [14] provides some design recommendations for smooth pursuit VR interfaces.

All previous works focused on navigation or selection tasks in structured, geometric scenes. In Bionic Tracking however, we use smooth pursuits to track cells in unstructured, volumetric data that cannot simply be queried for the objects contained or their positions.

This is the typical context in biomedical image analysis, where VR has previously been applied with success, e.g., for virtual colonoscopy [22] or for neuron tracing [35]. In the latter, the user traces neurons in VR using a handheld controller. The authors state that this resulted in faster and better-quality annotations. Tracking cells using handheld VR controllers is an alternative to gaze, but could increase physical strain on the user.

3   The Bionic Tracking Approach

For Bionic Tracking, we exploit smooth pursuit eye movements. Smooth pursuits are the only smooth movements performed by our eyes. They occur when following an optical stimulus, and cannot be triggered without one [6]. Instead of using a regular 2D screen and however, we perform the cell tracking process in VR, since VR provides the user with improved navigation and situational awareness compared to 2D when exploring a complex 3D+time dataset [30].

In addition, the HMD orientation data can be used to impose constraints on the data acquired from the eye trackers. In order to remove outliers from gaze data, one can calculate the quaternion distance between eyeball rotation and head rotation, which is physiologically limited: more than a 90-degree angle between eye direction and head direction is not possible, and head movement tends to follow eye movement through the vestibo-ocular reflex.

We believe that the combination of both a VR HMD and an integrated eye tracking device is necessary, for the following reasons:

- *Without eye tracking*, the head orientation from the HMD could still be used as a cursor. However, following small and smooth movements with the head is not something humans are used to doing. The eyes always lead the way, and the head follows. This would therefore provide an unnatural user interface.
- *Without virtual reality*, the effective space in which the user can follow cells is restricted to the rather small part of the visual field a regular screen occupies. The user furthermore loses the ability to move around without an additional input modality, e.g., to avoid temporary occlusion of the tracked cell. As an alternative to HMDs, a system using large screens or projectors, such as a PowerWall or a CAVE, could be used, at the cost of increased technical complexity.

3.1   Hardware Selection

We have chosen the HTC Vive as HMD, since it is comfortable to wear, provides good resolution, and comes with a tracking system for room-scale VR. Furthermore, it is compatible with the SteamVR/OpenVR software API, which we use.

For eye tracking, we have selected the *Pupil* eye tracker produced by Pupil Labs [12], since it provides LGPL-licensed open-source software and competitively-priced hardware that is simple to integrate into off-the-shelf HMDs. We used the cable-bound version of the HTC Vive, as the wireless kit does not provide the additional USB port required by the eye tracker.

In addition to being open-source, the *Pupil* software makes the gaze data and image frames available to external applications via a simple ZeroMQ- and MessagePack-based communication protocol, which is essential for using these data in our external tracking application, possibly even remotely over a network.

Alternatives, such as the HTC Vive Pro Eye, or an HTC Vive with integrated Tobii eye tracker, were either not available at the time this project started, or were costlier.

3.2   Software Framework

We have implemented Bionic Tracking using the visualization framework *scenery* [8]. Scenery supports simultaneous rendering of mesh data and volumetric data in VR. This is essential for Bionic Tracking, where the images that contain the cells to be tracked are volumetric data, and the overlaid tracks are mesh data. Scenery supports rendering to all SteamVR/OpenVR-supported VR HMDs and also supports the Pupil eye tracking devices out-of-the-box. In addition, scenery runs on the Java VM and is therefore compatible with the popular bio-image analysis software Fiji and the existing tracking tools *TrackMate* and *MaMuT* (see Sect. 2).

3.3   Rendering

We use simple, alpha-blending volume rendering to display the image data in the VR headset using scenery's Vulkan backend. While more advanced volume rendering (e.g., Metropolis Light Transport [17]) would provide higher visual quality, achieving a high and consistent frame rate is important for Bionic Tracking. For this work, we have only used in-core rendering, but the framework in principle also supports out-of-core volume rendering for larger datasets. In addition to the image data, we also display a gray, unobtrusive box around the volume for spatial anchoring (see supplementary video).

4   Tracking Cells with Bionic Tracking

Using Bionic Tracking to track cells or nuclei in 3D+time microscopy videos requires a brief preparation phase, followed by the actual tracking task.

4.1   Preparation

After putting on the VR HMD, making sure the eye tracker cameras can see the user's eyes, and launching the application, a calibration routine is run in order to establish a mapping between the user's gaze and world-space positions in the VR scene. For calibration, we show the user a total of 18 white spheres, with 5 of them layered on three concentric circles 1 m apart (distances in the VR scene are the same as in the physical world). Using three circles of increasing radii achieves good coverage of the field of view. In addition to the spheres on the circles, three spheres are shown at the centers of the three circles in order to calibrate the center of the field of view. During calibration, only a single sphere is shown to the user at a time for 500 ms, after that, the next one is shown. Since the calibration targets follow the head movements of the user, the user does not need to stand still while doing so. At the end of the calibration, the user is notified of success or failure, and can repeat the calibration process if necessary.

Calibration typically needs to be done only once per session, and can then be used to track as many cells as the user likes. Calibration needs to be repeated if the HMD significantly slips, or if the HMD is removed and remounted. Our calibration routine is similar to the one used in *Pupil's* HMDeyes Unity example[1].

Body movement in VR can be performed physically, but the handheld controllers additionally allow control over the following functions (handedness can be swapped, default bindings shown in Suppl. Fig. 1):

- move the dataset by holding the left-hand trigger and moving the controller,
- use the directional pad on the left-hand controller to move the observer (forward, backward, left, or right – with respect to the direction the user is looking to),
- start and stop tracking by pressing the right-hand side trigger,
- deleting the most recently created track by pressing the right-side button, and confirming within three seconds with another press of the same button,
- play and pause the dataset over time by pressing the right-hand menu button,
- play the dataset faster or slower in time by pressing the right-hand directional pad up or down, and
- stepping through the timepoints of the dataset one by one, forward or backward, by pressing the right-hand directional pad left or right.

When the dataset is not playing, the user can use the directional pad on the right-hand controller to scale the dataset. Initially the dataset is scaled to fit within a height of 2 m.

4.2   Tracking

After preparation, the user can position herself/himself freely in space. To track a cell, the user performs the following steps:

1. Find the timepoint and cell with which the track should start, adjust playback speed between one and 20 volumes/second, and start looking at the cell or object of interest.

---

[1] https://github.com/pupil-labs/hmd-eyes

2. Start playback of the multi-timepoint dataset, while continuing to follow the cell by looking at it, and maybe moving physically to follow the cell around occlusions.
3. End or pause the track at the final timepoint. Tracking will stop automatically when playback has reached the end of the dataset, and the dataset will play again from the beginning.

In order to minimize user strain in smooth-pursuit VR interactions, design guidelines have been suggested [14]: large trajectory sizes, clear instructions what the user has to look at, and relatively short selection times. The controls available to the user in Bionic Tracking enable free positioning and zooming. The selection time, here the tracking time, depends on the individual cell to be tracked, but as the tracking can be paused, and the playback speed adjusted, the user is free to choose both a comfortable length and speed.

During the tracking procedure, we collect the following data for each time point:

– the entry and exit points of the gaze ray through the volume in normalised volume-local coordinates, i.e., as a vector $\in [0.0, 1.0]^3$,
– the confidence rating – calculated by the *Pupil* software – of the gaze ray,
– the user's head orientation and position,
– the time point of the volume, and
– a list of sampling points with uniform spacing along the gaze ray through the volume and the actual image intensity sample values at these points, calculated by tri-linear interpolation from the voxel data.

A *spine* is a single gaze ray including the above metadata. A *hedgehog* is the set of all spines for a single track over time (Suppl. Fig. 2). By collecting the spines through the volume, we are able to transform each time point's 3D cell localization problem into a 1D problem along the spine. This results in the tracking algorithm presented below.

5   Tracking Algorithm

In previous applications of smooth-pursuit eye tracking [26,36], the tracked objects were geometric and not volumetric in nature, and therefore well-defined in space with their extents and shapes fully known. Here, in contrast, we use the indirect information about the objects contained in the spines and hedgehogs to find the tracked object in the unstructured volumetric data and follow it.

After a hedgehog has been collected to create a new cell track, all further processing relies on the data contained in this hedgehog. To illustrate the analysis, it is useful to visualize a hedgehog in 2D by laying out all spines in a plane next to each other (see Fig. 1). In this plane, time advances along the X axis and depth through the volume along a given spine is on the Y axis. Each line parallel to the Y axis represents one spine and therefore one gaze sample, of which we collect up to 60 per second. In Fig. 1, this amounts to 1614 spines in total with

16 spines per time point on average collected within 30 s. In the figure, we have highlighted the first local intensity maximum along each spine in red. The track of the cell the user was following is then mostly visible.

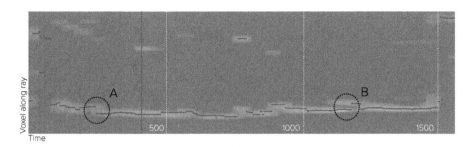

Fig. 1. A hedgehog visualized in 2D, with nearest local intensity maxima in red. Each vertical line is one spine of the hedgehog with the observer at the bottom. On the X axis, time runs from left to right, and is counted in number of gaze samples. After every 500 spines, a dotted white line is shown for orientation. The gray line before spine 500 is the spine whose profile is shown in Suppl. Fig. 3. The discontinuities in the track at A and B have different origins: At A, the user seems to have moved further away, resulting in a gap, while at B, another cell appeared closely behind the tracked one and might have distracted the user. See main text for details. (Color figure online)

5.1   World-Space Distance Weighted Temporal Linking

Movements of the user and temporary occlusion of the tracked cell by other cells or objects need to be accounted for when extracting a space-time trajectory from the information in the hedgehog. In order to reliably link cell detections across time points, we apply an iterative graph-based algorithm to all spines that have local maxima in their sample values. The intensity profile of an exemplary spine through a volume is shown in Suppl. Fig. 3. In that figure, the distance from the observer in voxels along the spine is shown on the X axis, while the Y axis shows the intensity value of the volume data at the sample points along the spine. To initialize the algorithm, we assume that when starting a track the user looks at a non-occluded cell that is manifest as the nearest local maximum along the spine. In Suppl. Fig. 3 that would be the leftmost local maximum.

Depending on the playback speed chosen by the user, the number of spines collected for each time point of the data varies. In our tests, it varied between 0 and 120. Zero spines are obtained if the user closes her/his eyes, or if no detection is possible for other reasons; 120 Hz is the maximum frame rate of the eye trackers used. More than 120 samples might be obtained if the playback speed is below one volume/second.

In order to track a cell across spines over time after the initial seed point on the first spine has been determined, we step through the spines in the hedgehog one by one, performing the following weighted nearest-neighbor linking, as illustrated in Fig. 2:

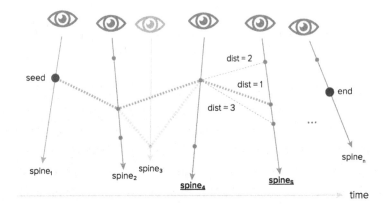

Fig. 2. Illustration of the iterative linking algorithm used to extract cell tracks from a hedgehog. Time runs along the X axis from left to right. $spine_1$ contains the initial seed point where to start tracking. The algorithm is currently at $spine_4$, determining how to proceed to $spine_5$. In this example, the middle track with dist $= 1$ wins, as it is the shortest world-space distance away from the current point. The algorithm continues the path search until it reaches the last spine, $spine_n$. In this manner, the algorithm closes gaps in the tracks and leaves out detected cells further along the individual rays. $spine_3$ is connected initially, but removed in the final statistical pruning step. It is therefore grayed out. See main text for details.

1. advance to the next spine in the hedgehog,
2. find the indices of all local intensity maxima along the spine, ordered by world-space distance to the selected point from the previous spine,
3. connect the selected point from the previous spine with the closest (in world-space distance) local maximum in the current spine,
4. calculate the world-space position of the new selected point, and
5. add the selected point to the set of points for the current track.

World-space distance weighting bridges gaps in the track (e.g., discontinuity A in Fig. 1). Indeed, this approach is a variant of *dynamic fringe-saving A** search on a grid [32] with all rays extended to the maximum length in the entire hedgehog along the Y axis, and time increasing along the X axis.

This algorithm constructs one cell track from the spines of each hedgehog. Extracting a track typically takes less than a second. Each track is then visualized in the VR scene right away, enabling the user to immediately decide whether to keep it, or to discard it.

5.2   Handling User Distraction and Object Occlusion

In some cases, world-space distance weighting is not sufficient, as a *Midas touch* problem [11] remains: When the user briefly looks elsewhere than at the cell of interest, and another local intensity maximum is detected there, then that local maximum may indeed have the smallest world-space distance and win. This would introduce a wrong link in the track. Usually, such problems are avoided by resorting to multi-modal input [21,31]. Here, we aim to avoid the Midas touch problem without burdening the user with additional modalities of interaction. We instead use statistics: for each segment distance (the distance between successive timepoints) d, we calculate the z-score $Z(d) = (d - \mu_{dist})/\sigma_{dist}$, where μ_{dist} is the mean distance in the entire hedgehog and σ_{dist} is the standard deviation of all distances in the entire hedgehog. We then prune all candidate track vertices with a z-score higher than 2.0. This corresponds to segment distances larger than double the standard deviation of all segment distances in the hedgehog. Pruning and linking are repeated iteratively until no vertices with a z-score higher than 2.0 remain, effectively filtering out discontinuities like B in Fig. 1.

6   Proof of Concept

We demonstrate the applicability of Bionic Tracking to two different datasets:

- A developmental 101-timepoint dataset of a *Platynereis dumerilii* embryo, an ocean-dwelling ringworm, acquired using a custom-built OpenSPIM [25] light-sheet microscope, with cell nuclei tagged with the fluorescent GFP protein (16-bit stacks, $700 \times 660 \times 113$ pixel, 100 MB/timepoint, 9.8 GByte dataset size),
- A 12-timepoint dataset of *MDA231* human breast cancer cells, embedded in a collagen matrix and infected with viruses tagged with the fluorescent GFP protein, acquired using a commercial Olympus FluoView F1000 confocal microscope (dataset from the Cell Tracking Challenge [34], 16-bit stacks, $512 \times 512 \times 30$ pixel, 15 MB/timepoint, 180 MByte total size.

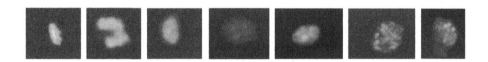

Fig. 3. Some example nucleus shapes encountered in the *Platynereis* test dataset.

The *Platynereis* dataset is chosen as a test case because it poses a current research challenge, with all tested semi-automatic algorithms failing on this dataset due to the diverse nuclei shapes and cell movements. Examples of shapes encountered in the dataset are shown in Fig. 3. The MDA231 dataset is chosen because it currently has the worst success scores for automatic tracking methods on the http://celltrackingchallenge.net. For Bionic Tracking, both datasets are rendered in VR at their full resolution with a typical playback frame rate of 60...90 fps.

For the *Platynereis* dataset, we are able to obtain high-quality cell tracks using the Bionic Tracking prototype system. A visualization of one such cell track is shown in Suppl. Fig. 4. In the supplementary video, we show both the gaze tracking process to create the track and a visualization of all spines used to generate the track.

For the MDA231 dataset, we are able to obtain tracks for six cells in the dataset in about 10 min. A visualization of these tracks is shown in Suppl. Fig. 5; see the supplementary video for a part of the tracking process. This example also demonstrates that the Bionic Tracking approach is useful even on "nearly 2D" microscopy images, as this dataset only has 30 Z slices, for a lateral resolution of 512×512.

7   Evaluation

We evaluate Bionic Tracking by performing a user study, to gain insight into user acceptance and practicability, and by comparing the tracks created by Bionic Tracking with manually annotated ground truth. Together, these evaluations provide an initial characterization of the accuracy and performance of Bionic Tracking.

7.1   User Study

We recruited seven cell-tracking experts who were either proficient with manual cell tracking tasks in biology, proficient in using or developing automated tracking solutions, or both (median age 36, s.d. 7.23, 1 female, 6 male) to take part in the study. All users were tasked with tracking cells in the *Platynereis* dataset already used in Sect. 6. One of the users was already familiar with this particular dataset. The study was conducted on a Dell Precision Tower 7910 workstation (Intel Xeon E5-2630v3 CPU, 8 cores, 64 GB RAM, GeForce GTX 1080Ti GPU) running Windows 10, build 1909.

Before starting to use Bionic Tracking, all users were informed of the goals and potential risks (e.g., simulator sickness) of the study. With a questionnaire, they were asked for presence of any visual or motor impairments (apart from needing to wear glasses or contact lenses, none were reported), about previous VR experience and physical wellbeing. After having performed the task, users were again asked about their physical wellbeing, and they had to judge their experience using the NASA Task Load Index (TLX, [9]) and the Simulator Sickness Questionnaire (SSQ, [13]). In addition, they were asked qualitative and quantitative questions about the software based on both the User Experience Questionnaire [18] and the System Usability Scale [3]. We concluded the study for each participant with a short interview, where users were asked to state areas of improvement, and what they liked about Bionic Tracking. The full questionnaire used in the study is available in the supplementary material.

After filling in the pre-task part of the questionnaire, users were given a brief introduction to the controls in the software. Upon ensuring a good fit of the HMD on the user's head, the inter-pupillary distance (IPD) of the HMD was adjusted to the user's eyes, as were the ROIs of the eye tracking cameras. The users then ran the calibration routine on their own. If the calibration was found to not be sufficiently accurate, we re-adjusted HMD fit and camera ROIs, and ran the calibration routine again. Then, users took time to freely explore the dataset before starting to track the cells in the *Platynereis* dataset. Users were then able to create cell tracks on their own, creating up to 32 complete tracks in 10...29 min.

All participants in the study had no or very limited experience with using VR interfaces (5-point scale, 0 means no experience, 4 daily use: mean 0.43, s.d. 0.53), and only one user had previously used an eye-tracking user interface (same 5-point scale: mean 0.14, s.d. 0.37).

7.2   User Study Results

The average SSQ score was 25.6 ± 29.8 s.d. (median 14.9), which is on par with other VR applications that have been evaluated using SSQ [29]. From TLX, we used all categories (mental demand, physical demand, temporal demand, success, effort, insecurity) on a 7-point scale where $0 =$ Very Low and $6 =$ Very High for the demand metrics, and $0 =$ Perfect, $6 =$ Failure for the performance metrics. Users reported medium scores for mental demand (2.71 ± 1.70) and for effort (2.86 ± 1.68), while reporting low scores for physical demand (1.86 ± 1.95), temporal demand (1.57 ± 0.98), and insecurity (1.14 ± 1.68). Importantly, the participants judged themselves to have been rather successful with the tracking tasks (1.71 ± 0.75).

All questions related to software usability and acceptance are summarized in Fig. 4a. The users estimated that the Bionic Tracking method would yield a speedup of a factor 2...10 (3.33 ± 6.25) compared to tracking cells with a 2D interface, and they expressed interest in using Bionic Tracking for their own tracking tasks (3.43 ± 0.53; 5-point scale here and for the following: $0 =$ No agreement, $4 =$ Full agreement), as the tracks created were judged to look reasonable (2.57 ± 0.98), Bionic Tracking would provide an improvement over their current methods (3.14 ± 0.90), and they could create new cell tracks not only with reasonable confidence (2.86 ± 0.69), but much faster (3.29 ± 0.76). Users found the software to be relatively intuitive (2.43 ± 0.98) and did not need long to learn how to use it (0.59 ± 0.79), which they also remarked in the follow-up interviews:

> "It was so relaxing, actually, looking at this [cell] and just looking." (P2, the user remarked further after the interview that the technique might prevent carpal tunnel issues often encountered when tracking using mouse and keyboard.)

> "I figured this could be like a super quick way to generate the [cell] tracks." (P7)

The user study also showed that users adjust playback speed more often than image size in VR. After exploring different settings – users could choose speeds from 1...20 timepoints/second – all users independently settled on a playback speed of 4...5 timepoints/second for tracking, corresponding to 200...250 ms of viewing time per timepoint, which coincides with the onset delay of smooth-pursuit eye movements [6]. Despite having no or limited previous VR experience, the users did not at all feel irritated by the environment (0.00 ± 0.00), nor by the use of eye tracking (0.29 ± 0.49).

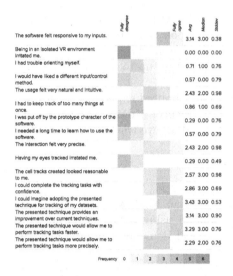

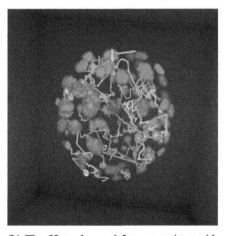

(b) The 52 tracks used for comparison with manual tracking, visualized together with the volumetric data of one time point. This is the same view a user had, taken from within the VR headset. See the supplementary video for a dynamic visualization over time.

(a) Results of usability and acceptance question from the user study. Note that the questions are formulated both positively and negatively.

Fig. 4. User study results and exemplary tracks created

7.3   Comparison with Manual Tracking Results

To quantify the accuracy and quality of the tracks generated by Bionic Tracking, we compare them with manually curated ground-truth tracks. The purpose of this comparison is to assess the capacity of Bionic Tracking to substitute for manual tracking and curation approaches. We compare 52 tracks created by an expert annotator using Bionic Tracking (see Fig. 4b) on the *Platynereis* dataset with their respective best-matching ground truth tracks. We find that 25 of the 52 tracks have a distance score [34] of less than 1 cell diameter, suggesting that these tracks will, on average, link the correct cells.

8   Discussion

We have shown that gaze in VR can be used to help reconstruct tracks of biological cells in 3D microscopy. Not only does Bionic Tracking accelerate the tracking process, but it makes it easier and less physically demanding than manual point-and-click interfaces. Although our expert-based user study was rather small, limiting its statistical power, we believe that it provides a first indication that the use of Bionic Tracking can improve the user experience and productivity in cell tracking tasks, and that developing it further is worthwhile.

Even though all users had very limited previous VR experience, they were quickly able to independently create cell tracks with confidence. Multiple users complimented the ergonomics of the technique, although it remains to be seen whether this would still be the case for longer (1 h+) tracking sessions. With the projected speedups, however, it might not even be necessary to have such long sessions anymore. Users reported that for manual tracking they would not do sessions longer than 3 to 4 h, which, with the estimated speedups, would reduce to 20...90 min when using Bionic Tracking.

However, for tracking large lineages comprising thousands of cells, Bionic Tracking on it own is going to be cumbersome for combinatorial reasons. It can, however, augment existing techniques for parts of the lineaging process, e.g., to track cells during early stages of development, where they tend to have less well-defined shapes. It may also provide constraints and training data for machine-learning algorithms for automated tracking. Furthermore, Bionic Tracking could be used in conjunction with any automatic tracking algorithm that provides uncertainty scores in order to restrict gaze input to regions where the algorithm cannot perform below a given uncertainty threshold. This could be done, e.g., by superimposing a heatmap on the volume rendering to indicate to the user areas that need additional curation. Hybrid semi-automated/manual approaches are already among the most popular tools for challenging biological datasets [37].

9   Future Work and Limitations

In the future, we would like to integrate Bionic Tracking into existing tracking software [38] and user-facing visualization tools [7], such that it can be used by a broader audience. Unfortunately, eye-tracking HMDs are not yet commonly available, but according to current product announcements, this is likely to change soon. Current developments in eye-tracking hardware and VR HMDs also indicate falling prices in the near future, such that these devices might soon become more common. An immediate possibility is to supply one or two eye-tracking HMDs as an institute, making them available to users in a bookable item-facility manner. Furthermore, the calibration of the eye trackers can still be problematic, but this is likely to improve in the future, too, with machine learning approaches expected to render the process faster, more reliable, and more user-friendly.

In order for Bionic Tracking to become a tool that can be routinely used for biological research, it will also be necessary to implement interactions that allow the user to record certain events, like cell divisions or cell death. Such an interaction could, e.g., include the user pressing a certain button whenever she/he is looking at a cell division occurring, and then track until the next cell division. In such a way, the user can track from cell division to cell division, applying divide-and-conquer for tracking (a part of) a cell lineage tree. The design and evaluation of algorithms to detect and track entire cell lineage trees is currently an active research area in the bio-imaging community [34]. In this study, we have used comparison algorithms from the Particle Tracking Challenge (PTC) [5], which were designed to compare single tracks. There are limitations when applying the PTC metric to compare cell tracking annotations. However, until additional tracking events—such as the aforementioned cell divisions—can be recorded with Bionic Tracking, PTC is the only metric that can be applied.

In our evaluation, we have still seen some spurious detections, which led to wrong tracks. This calls for more evaluation in crowded environments: While Bionic Tracking seems well suited for crowded scenes in principle – as users can, e.g., move around corners as tracked by the HMD – it is not clear whether eye tracking is precise enough in such situations.

A potential improvement to the user interface would include using head-tracking data from the HMD to highlight the region of the volumetric dataset the user is looking toward, e.g., by dimming the surrounding areas similar to foveated rendering techniques [4, 19]. We have not yet explored foveation, but could imagine it might improve tracking accuracy, rendering throughput, and cognitive load.

10   Conclusion

We have presented *Bionic Tracking*, a method for object tracking over time in volumetric image datasets, leveraging gaze data and virtual reality displays for biological cell tracking problems. Bionic Tracking can augment manual parts of cell tracking tasks by rendering them faster, more ergonomic, and more enjoyable for the user.

As part of Bionic Tracking, we have introduced an algorithm for iterative graph-based temporal tracking, which robustly connects gaze samples with cell or object detections in volumetric data over time.

The results from our research prototype are encouraging. Users judged that Bionic Tracking accelerated their cell tracking tasks 2 to 10-fold, they were confident using the system, felt no irritation, and generally enjoyed this way of interacting with 3D+time fluorescence microscopy images. We therefore plan to continue this line of research with further studies, extending the evaluation to more datasets and users, and adding a more detailed evaluation of the accuracy of the created cell tracks on datasets with known ground truth. We would also like to include Bionic Tracking into a pipeline where the gaze-determined cell tracks are used to train machine-learning algorithms to improve automatic tracking.

A prototype implementation of Bionic Tracking is available as open source software at github.com/scenerygraphics/bionic-tracking.

Acknowledgements. The authors thank all participants of the user study. Thanks to Mette Handberg-Thorsager for providing the *Platynereis* dataset and for feedback on the manuscript. Thanks to Vladimir Ulman and Jean-Yves Tinevez for helpful discussions regarding track comparison. Thanks to Bevan Cheeseman, Aryaman Gupta, and Stefanie Schmidt for helpful discussions. Thanks to Pupil Labs for help with the eye tracking calibration.

This work was partially funded by the Center for Advanced Systems Understanding (CASUS), financed by Germany's Federal Ministry of Education and Research (BMBF) and by the Saxon Ministry for Science, Culture and Tourism (SMWK) with tax funds on the basis of the budget approved by the Saxon State Parliament.

References

1. Amat, F., Höckendorf, B., Wan, Y., Lemon, W.C., McDole, K., Keller, P.J.: Efficient processing and analysis of large-scale light-sheet microscopy data. Nat. Protoc. **10**(11) (2015). https://doi.org/10.1038/nprot.2015.111

2. Amat, F., et al.: Fast, accurate reconstruction of cell lineages from large-scale fluorescence microscopy data. Nat. Methods **11**(9) (2014). https://doi.org/10.1038/nmeth.3036

3. Brooke, J.: SUS - a quick and dirty usability scale. In: Usability Evaluation in Industry, p. 7. CRC Press, June 1996

4. Bruder, V., Schulz, C., Bauer, R., Frey, S., Weiskopf, D., Ertl, T.: Voronoi-based foveated volume rendering. In: EUROVIS 2019, Porto, Portugal (2019)

5. Chenouard, N., et al.: Objective comparison of particle tracking methods. Nat. Methods **11**(3), 281–289 (2014). https://doi.org/10.1038/nmeth.2808

6. Duchowski, A.T.: Eye Tracking Methodology: Theory and Practice, 3rd edn. Springer, Cham (2017). https://doi.org/10.1007/978-3-319-57883-5

7. Günther, U., Harrington, K.I.S.: Tales from the trenches: developing sciview, a new 3D viewer for the ImageJ community. In: VisGap - The Gap between Visualization Research and Visualization Software at EuroGraphics/EuroVis 2020, p. 7 (2020). https://doi.org/10.2312/VISGAP.20201112

8. Gunther, U., et al.: Scenery: flexible virtual reality visualization on the Java VM. In: 2019 IEEE Visualization Conference (VIS), Vancouver, BC, Canada, pp. 1–5. IEEE, October 2019. https://doi.org/10.1109/VISUAL.2019.8933605

9. Hart, S.G., Staveland, L.E.: Development of NASA-TLX (task load index): results of empirical and theoretical research. Adv. Psychol. **52** (1988). https://doi.org/10.1016/s0166-4115(08)62386--9

10. Huisken, J.: Optical sectioning deep inside live embryos by selective plane illumination microscopy. Science **305**(5686) (2004). https://doi.org/10.1126/science.1100035

11. Jacob, R.J.K.: Eye tracking in advanced interface design. In: Virtual Environments and Advanced Interface Design, pp. 258–290 (1995)

12. Kassner, M., Patera, W., Bulling, A.: Pupil: an open source platform for pervasive eye tracking and mobile gaze-based interaction. In: Proceedings of the 2014 ACM International Joint Conference on Pervasive and Ubiquitous Computing, Seattle, Washington, pp. 1151–1160. ACM Press (2014). https://doi.org/10.1145/2638728.2641695

13. Kennedy, R.S., Lane, N.E., Berbaum, K.S., Lilienthal, M.G.: Simulator sickness questionnaire: an enhanced method for quantifying simulator sickness. Int. J. Aviat. Psychol. **3**(3) (1993). https://doi.org/10.1207/s15327108ijap0303_3

14. Khamis, M., Oechsner, C., Alt, F., Bulling, A.: VRpursuits: interaction in virtual reality using smooth pursuit eye movements. In: Proceedings of the 2018 International Conference on Advanced Visual Interfaces - AVI 2018, Castiglione della Pescaia, Grosseto, Italy, pp. 1–8. ACM Press (2018). https://doi.org/10.1145/3206505.3206522

15. Klamka, K., Siegel, A., Vogt, S., Göbel, F., Stellmach, S., Dachselt, R.: Look & pedal: hands-free navigation in zoomable information spaces through gaze-supported foot input. In: Proceedings of the 2015 ACM on International Conference on Multimodal Interaction - ICMI 2015, Seattle, Washington, USA, pp. 123–130. ACM Press (2015). https://doi.org/10.1145/2818346.2820751

16. Kosch, T., Hassib, M., Woźniak, P.W., Buschek, D., Alt, F.: Your eyes tell: leveraging smooth pursuit for assessing cognitive workload. In: Proceedings of the 2018 CHI Conference on Human Factors in Computing Systems - CHI 2018, Montreal QC, Canada, pp. 1–13. ACM Press (2018). https://doi.org/10.1145/3173574.3174010

17. Kroes, T., Post, F.H., Botha, C.P.: Exposure render: an interactive photo-realistic volume rendering framework. PLoS ONE 7(7) (2012). https://doi.org/10.1371/journal.pone.0038586

18. Laugwitz, B., Held, T., Schrepp, M.: Construction and evaluation of a user experience questionnaire. In: Holzinger, A. (ed.) USAB 2008. LNCS, vol. 5298, pp. 63–76. Springer, Heidelberg (2008). https://doi.org/10.1007/978-3-540-89350-9_6

19. Levoy, M., Whitaker, R.: Gaze-directed volume rendering. ACM SIGGRAPH Comput. Graph. 24(2) (1990). https://doi.org/10.1145/91385.91449

20. Lutz, O.H.-M., Venjakob, A.C., Ruff, S.: SMOOVS: towards calibration-free text entry by gaze using smooth pursuit movements. J. Eye Mov. Res. 8(1) (2015). https://doi.org/10.16910/jemr.8.1.2

21. Meena, Y.K., Cecotti, H., Wong-Lin, K., Prasad, G.: A multimodal interface to resolve the Midas-Touch problem in gaze controlled wheelchair. In: Conference Proceedings : ... Annual International Conference of the IEEE Engineering in Medicine and Biology Society, IEEE Engineering in Medicine and Biology Society, Annual Conference 2017 (2017). https://doi.org/10.1109/embc.2017.8036971

22. Mirhosseini, S., Gutenko, I., Ojal, S., Marino, J., Kaufman, A.: Immersive virtual colonoscopy. IEEE Trans. Visual. Comput. Graph. 25(5) (2019). https://doi.org/10.1109/tvcg.2019.2898763

23. Moen, E., Bannon, D., Kudo, T., Graf, W., Covert, M., Van Valen, D.: Deep learning for cellular image analysis. Nat. Methods 16(12), 1233–1246 (2019). https://doi.org/10.1038/s41592-019-0403-1

24. Pietzsch, T., Saalfeld, S., Preibisch, S., Tomancak, P.: BigDataViewer: visualization and processing for large image data sets. 12(6) (2015). https://doi.org/10.1038/nmeth.3392

25. Pitrone, P.G., et al.: OpenSPIM: an open-access light-sheet microscopy platform. Nat. Methods 10(7) (2013). https://doi.org/10.1038/nmeth.2507

26. Piumsomboon, T., Lee, G., Lindeman, R.W., Billinghurst, M.: Exploring natural eye-gaze-based interaction for immersive virtual reality. In: 2017 IEEE Symposium on 3D User Interfaces (3DUI), Los Angeles, CA, USA, pp. 36–39. IEEE (2017). https://doi.org/10.1109/3DUI.2017.7893315

27. Reynaud, E.G., Peychl, J., Huisken, J., Tomancak, P.: Guide to light-sheet microscopy for adventurous biologists. Nat. Methods 12(1) (2014). https://doi.org/10.1038/nmeth.3222

28. Schindelin, J., et al.: Fiji: an open-source platform for biological-image analysis. Nat. Methods 9(7) (2012). https://doi.org/10.1038/nmeth.2019

29. Singla, A., Fremerey, S., Robitza, W., Raake, A.: Measuring and comparing QoE and simulator sickness of omnidirectional videos in different head mounted displays. In: 2017 Ninth International Conference on Quality of Multimedia Experience (QoMEX), pp. 1–6, May 2017. https://doi.org/10.1109/QoMEX.2017.7965658

30. Slater, M., Sanchez-Vives, M.V.: Enhancing our lives with immersive virtual reality. Front. Robot. AI **3** (2016). https://doi.org/10.3389/frobt.2016.00074

31. Stellmach, S., Dachselt, R.: Look & touch: gaze-supported target acquisition. In: Proceedings of the 2012 ACM Annual Conference on Human Factors in Computing Systems - CHI 2012, Austin, Texas, USA, p. 2981. ACM Press (2012). https://doi.org/10.1145/2207676.2208709

32. Sun, X., Yeoh, W., Koenig, S.: Dynamic fringe-saving A*. In: Proceedings of the 8th International Conference on Autonomous Agents and Multiagent Systems, vol. 2, pp. 891–898. International Foundation for Autonomous Agents and Multiagent Systems, Richland, SC (2009)

33. Tinevez, J.-Y., et al.: TrackMate: an open and extensible platform for single-particle tracking. Methods **115** (2017). https://doi.org/10.1016/j.ymeth.2016.09.016. (IEEE Signal Proc. Mag. 23 3 2006)

34. Ulman, V., et al.: An objective comparison of cell-tracking algorithms. Nat. Methods **14**(12), 1141–1152 (2017). https://doi.org/10.1038/nmeth.4473

35. Usher, W., et al.: A virtual reality visualization tool for neuron tracing. IEEE Trans. Visual. Comput. Graph. **24**(1) (2017). https://doi.org/10.1109/tvcg.2017.2744079

36. Vidal, M., Bulling, A., Gellersen, H.: Pursuits: spontaneous interaction with displays based on smooth pursuit eye movement and moving targets. In: Proceedings of the 2013 ACM International Joint Conference on Pervasive and Ubiquitous Computing - UbiComp 2013, Zurich, Switzerland, p. 439. ACM Press (2013). https://doi.org/10.1145/2493432.2493477

37. Winnubst, J., et al.: Reconstruction of 1,000 projection neurons reveals new cell types and organization of long-range connectivity in the mouse brain. Cell **179**(1), 268–281.e13 (2019). https://doi.org/10.1016/j.cell.2019.07.042

38. Wolff, C., et al.: Multi-view light-sheet imaging and tracking with the MaMuT software reveals the cell lineage of a direct developing arthropod limb. eLife **7** (2018). https://doi.org/10.7554/elife.34410

Cardiac MR Image Sequence Segmentation with Temporal Motion Encoding

Pengxiang Wu[1]([⊠]) [ID], Qiaoying Huang[1], Jingru Yi[1], Hui Qu[1], Meng Ye[1],
Leon Axel[2], and Dimitris Metaxas[1]

[1] Rutgers University, New Brunswick, NJ 08854, USA
{pw241,qh55,jy486,hui.qu,meng.ye,dnm}@rutgers.edu
[2] New York University, New York, NY 10016, USA
Leon.Axel@nyulangone.org

Abstract. The segmentation of cardiac magnetic resonance (MR) images is a critical step for the accurate assessment of cardiac function and the diagnosis of cardiovascular diseases. In this work, we propose a novel segmentation method that is able to effectively leverage the temporal information in cardiac MR image sequences. Specifically, we construct a Temporal Aggregation Module (TAM) to incorporate the temporal image-based features into a backbone spatial segmentation network (such as a 2D U-Net) with negligible extra computation cost. In addition, we also introduce a novel Motion Encoding Module (MEM) to explicitly encode the motion features of the heart. Experimental results demonstrate that each of the two modules enables clear improvements upon the base spatial network, and their combination leads to further enhanced performance. The proposed method outperforms the previous methods significantly, demonstrating the effectiveness of our design.

Keywords: Cardiac MRI · LV segmentation · Temporal · Motion

1 Introduction

Cardiac magnetic resonance imaging (MRI) is one of the major imaging modalities that can be used for the quantitative spatio-temporal analysis of cardiac function and disease diagnosis. Accurate assessment of cardiac function is essential for both diagnosis and treatment of cardiovascular diseases. Recent developments in machine learning methods promise to enable the design of automatic cardiac analysis tools, thereby significantly reducing the manual effort currently required by clinicians. In particular, the automatic segmentation of left ventricle (LV) contours is an important first step to enable the accurate quantification of regional cardiac function, including ejection fraction, temporal changes in ventricular volumes and strain analysis of myocardium [16,20]. However, accurate LV boundary segmentation is non-trivial due to LV shape variability, imaging artifacts, and poor LV boundary delineation. Such complexities make this

© Springer Nature Switzerland AG 2020
A. Bartoli and A. Fusiello (Eds.): ECCV 2020 Workshops, LNCS 12535, pp. 298–309, 2020.
https://doi.org/10.1007/978-3-030-66415-2_19

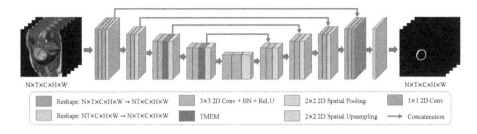

Fig. 1. Overview of the proposed method. The input is a volumetric sequence of cardiac images, and the output is the corresponding LV segmentation results. The overall structure of our method is based on (but not limited to) a 2D U-Net, where we insert two Temporal Motion Encoding Modules (TMEM) to effectively exploit the temporal information for cardiac MR image segmentation. N: the batch size; T: the number of frames in a sequence; H and W: the spatial size of feature maps; $C = 1$: the channel number for grayscale image.

task still an open problem despite the existence of important works for several decades.

Recent methods for cardiac LV segmentation are mainly based on deep neural networks, given their superior performance. One representative of the recent development is the 2D U-Net [13], which has proven one of the most effective methods in image-based segmentation since it learns and combines multi-scale features. This design has inspired many follow-up methods. However, this type of methods segment the slices individually without considering their spatial and temporal correlations. To address this issue, the 3D U-Net [3] extends the 2D U-Net by replacing the 2D convolutions with 3D ones in order to capture long-range dependencies between different slices. While achieving improved accuracy, 3D U-Net inevitably increases the computational cost and tends to cause over-fitting. Such weaknesses can be alleviated by the recurrent U-Net [10], which employs ConvGRU [2] to connect the slices but still suffers from computational inefficiency. Recent works [11,18] have sought to exploit optical flow for capturing cardiac dynamic features and enforcing temporal coherence. However, the extraction of optical flow is non-trivial, expensive and prone to significant errors, making these methods often inaccurate and difficult to deploy in real applications. Other attempts at improving U-Nets include adopting a hybrid solution [19] or integrating the attention mechanism into the network for feature refinement [9]. However, these methods are difficult to train, and fail to explicitly and efficiently model temporal relationships.

To address the above limitations, we propose a novel method for the spatio-temporal segmentation of cardiac MR image sequences. Our method is based on a 2D U-Net, and aggregates the temporal features with only 1D and 2D convolutions, thereby eliminating the heavy computation and massive parameters of 3D and recurrent convolutions. Specifically, we construct a Temporal Aggregation Module (TAM) to capture the inter-slice temporal features. TAM reformulates

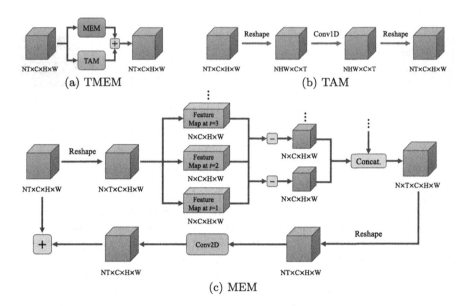

Fig. 2. Illustration of the different modules. (a) Temporal Motion Encoding Module (TMEM). (b) Temporal Aggregation Module (TAM). (c) Motion Encoding Module (MEM), with $\delta = 1$. These modules are computationally efficient, and enable to significantly improve the performance of base network with small extra overhead. "+" and "−" denote element-wise addition and subtraction, respectively. The Conv2D block consists of one 3×3 2D convolution layer, followed by BatchNorm and ReLU.

the input feature map as a 1D signal, and utilizes 1D convolution for temporal feature learning with small extra computation and parameter overhead. In addition, we introduce a Motion Encoding Module (MEM), which explicitly models the cardiac motion features using 2D convolutions without relying on optical flow. By integrating such dynamic information into a 2D U-Net, MEM is able to guide and regulate the segmentation. Finally, we integrate these two modules in a Temporal Motion Encoding Module (TMEM), which feeds the network with complementary temporal and motion information while preserving the simplicity and efficiency of 2D U-Net (see Fig. 1 and Fig. 2). Experimental results on two cardiac MR image datasets demonstrate the effectiveness and superiority of the proposed method compared to the state-of-the-arts.

2   Method

The overall architecture of our method is illustrated in Fig. 1, where we integrate two Temporal Motion Encoding Modules into the classic 2D U-Net. Below we give the details of our module design and the corresponding motivation.

2.1   Temporal Aggregation Module

The Temporal Aggregation Module (TAM) aims to extract the temporal features with very limited extra overhead. As shown in Fig. 2(b), given a 2D feature map $\mathbf{F} \in \mathbb{R}^{NT \times C \times H \times W}$, we first reshape it into a 1D signal $\mathbf{F}' \in \mathbb{R}^{NHW \times C \times T}$. Then, we apply 1D convolution to $\mathbf{F}'$ along the dimension of T to aggregate temporal features. Compared to the 3D convolution, this design discards the 2D component of the 3D kernel, and has the benefit that the temporal information is propagated among different slices with only 1D convolution, which requires a very small number of parameters and computational cost. In our implementation of TAM, we use a 1D convolution with kernel size 3.

2.2   Motion Encoding Module

Existing works, such as [11,18], have indicated that the optical-flow motion information is able to regulate the network and thereby significantly improve the segmentation performance. However, the computation or learning of optical flow in their methods is non-trivial, computationally expensive and error-prone, which hinders their practical applications. To address this issue, we design a Motion Encoding Module (MEM), which aims to capture the motion features efficiently rather than recover the exact motion patterns as in optical flow. To be specific, the motion information from MEM is at the feature level, and can be computed efficiently and used to improve the segmentation.

Specifically, given a feature map $\mathbf{F} \in \mathbb{R}^{NT \times C \times H \times W}$, we first reshape it to expose the temporal dimension, obtaining $\mathbf{F}' \in \mathbb{R}^{N \times T \times C \times H \times W}$. Then we split $\mathbf{F}'$ into a set of feature maps $\mathbf{F}'_1, \ldots, \mathbf{F}'_T$, where $\mathbf{F}'_t \in \mathbb{R}^{N \times C \times H \times W}, t \in [1, T]$. Afterwards, the motion information is extracted from every pair of feature maps $\mathbf{F}'_t$ and $\mathbf{F}'_{t'}$, where $t' = (t + \delta) \bmod T$ (i.e., the temporal interval between $\mathbf{F}'_t$ and $\mathbf{F}'_{t'}$ is δ). Formally,

$$\widetilde{\mathbf{F}}_t = f(\mathbf{F}'_t - \mathbf{F}'_{(t+\delta) \bmod T}), \quad t \in [1, T], \tag{1}$$

where $\widetilde{\mathbf{F}}_t \in \mathbb{R}^{N \times C \times H \times W}$ is the captured motion information, and f denotes a nonlinear function, which in our case is a 2D convolution followed by BatchNorm and ReLU. Then, all the generated motion features $\widetilde{\mathbf{F}}_t$ are stacked along the temporal dimension, providing the feature map $\widetilde{\mathbf{F}} \in \mathbb{R}^{N \times T \times C \times H \times W}$. To make the size of $\widetilde{\mathbf{F}}$ consistent with the input feature $\mathbf{F}$, we reshape $\widetilde{\mathbf{F}}$ into an output feature map $\widehat{\mathbf{F}} \in \mathbb{R}^{NT \times C \times H \times W}$. Finally, $\widehat{\mathbf{F}}$ is added back to the input feature $\mathbf{F}$, a step we find helpful in stabilizing the network training. In Fig. 2(c) we show an instance of MEM with $\delta = 1$. Note that, the use of modulo operation in MEM follows the cyclic property of cardiac image sequences.

As is illustrated, the proposed MEM is simple and only relies on 2D convolution, and thus it is more efficient than the 3D and recurrent counterparts. In the experiments, we will show that MEM is able to largely improve the segmentation performance of a basic 2D U-Net.

Multi-scale Motion Features. For a given δ, it actually defines the "receptive field" or "scale" where the motion information is extracted. Inspired by the widely adopted principle that multi-scale feature learning is critical for the segmentation task [21], we propose to aggregate the motion features at different scales, thereby giving a multi-scale MEM. Specifically, we compute the motion features $\widehat{\mathbf{F}}_{\delta_i}$ for different temporal intervals δ_i, and then sum them up to produce the final motion feature $\widehat{\mathbf{F}}$, i.e., $\widehat{\mathbf{F}} = \sum_i \widehat{\mathbf{F}}_{\delta_i}$. This multi-scale MEM further improves the performance, but at the cost of increased computational complexity, as shown in the experiment.

2.3 Temporal Motion Encoding Module

TAM and MEM extract the temporal features from two different perspectives. To combine their strengths we design a Temporal Motion Encoding Module (TMEM), which consists TAM and MEM (see Fig. 2(a)). TMEM is able to fuse the temporal and motion features together and can be integrated into any layer of a 2D U-Net. In practice, we empirically observe that placing TMEM within the third and fourth Conv2D blocks of the encoder gives the best results. This observation is in accordance with the findings of [17], which suggest temporal representation learning on high-level semantic features is more useful.

3   Experiments and Results

We evaluate the proposed method on two cardiac MR image datasets. (1) DYS, a dataset which contains 24 subjects, of which the patients are with heart failure due to dyssynchrony. The number of phases is 25 for each cardiac cycle. In total, there are around 4000 2D short-axis (SAX) slices. The LV myocardium contours of these SAX images over different spatial locations and cardiac phases are manually annotated based on consensus of three medical experts. The sizes of images vary between 224×204 and 240×198 pixels, and their in-plane resolutions vary from $1.17\,\mathrm{mm}$ to $1.43\,\mathrm{mm}$. We use 3-fold cross validation in our experiments, and make sure both the training and test sets contain the normal subjects and patients. (2) CAP, a publicly available dataset consisting of steady-state free precession (SSFP) cine MR images from Cardiac Atlas database [4,7,14]. CAP involves 100 patients with coronary artery disease and prior myocardial infarction. The ground-truth myocardium annotations are generated by various raters with consensus. There exists large variability within this dataset: the data are generated from different MRI scanner systems, the image size varies from 138×192 to 512×512 and the cardiac phases range from 18 to 35. These factors make CAP more challenging than DYS. In our experiments, we perform cross validation with 3 different partitions of the dataset. In each particular partition, we select 70 subjects for training, 15 for validation and 15 for testing.

During training, for both datasets, we crop the regions around the LV to generate training images of size 144×144. Data augmentation, including random flip and rotation, is adopted to improve the model robustness. To train the models, we use cross entropy loss and optimize the network parameters with Adam optimizer. We set the learning rate to 0.0005, and decay it by 0.5 after every 15 training epochs. The batch size is 8 (i.e., 8 cardiac sequences), each of which is padded/subsampled to contain 32 slices (i.e., the temporal dimension $T = 32$). The weight decay is 0.0001, and the number of training epochs is chosen to be 75 to ensure convergence. For all the 2D convolutions of our method, we set their kernel sizes to 3×3. The training of our model takes around 0.5–1.5 h on a single NVIDIA RTX 4000 GPU, and the inference takes about 0.01 s for a sequence of slices.

Table 1. Evaluation of segmentation accuracy for different methods in terms of Dice and Jaccard metrics, as well as Hausdorff distance (HD) in pixels. We report the mean and standard deviation over different folds. For MEM/TMEM, we compute the single-scale motion features and set $\delta = 4$.

Dataset	Method	Dice	Jaccard	HD
DYS	2D U-Net	0.7854 ± 0.0384	0.6633 ± 0.0465	4.6204 ± 2.8586
	3D U-Net	0.7566 ± 0.0143	0.6229 ± 0.0175	10.159 ± 3.4157
	Att. U-Net	0.7984 ± 0.0269	0.6801 ± 0.0303	5.9199 ± 1.3563
	RFCN	0.7936 ± 0.0316	0.6721 ± 0.0391	4.5095 ± 1.3524
	S3D U-Net	0.7912 ± 0.0348	0.6719 ± 0.0393	5.1291 ± 1.6801
	Ours + TAM	0.8101 ± 0.0254	0.6944 ± 0.0288	3.8381 ± 0.7819
	Ours + MEM	0.8118 ± 0.0238	0.6990 ± 0.0253	3.7334 ± 1.0233
	Ours + TMEM	$\mathbf{0.8201 \pm 0.0166}$	$\mathbf{0.7074 \pm 0.0184}$	$\mathbf{3.2332 \pm 0.9241}$
CAP	2D U-Net	0.7158 ± 0.0243	0.6234 ± 0.0271	4.4257 ± 0.4978
	3D U-Net	0.7434 ± 0.0084	0.6501 ± 0.0051	4.2577 ± 0.3889
	Att. U-Net	0.7341 ± 0.0124	0.6445 ± 0.0127	4.6684 ± 0.1695
	RFCN	0.7172 ± 0.0297	0.6259 ± 0.0278	4.6394 ± 0.3271
	S3D U-Net	0.7191 ± 0.0118	0.6276 ± 0.0101	5.3890 ± 0.3459
	Ours + TAM	0.7653 ± 0.0175	0.6766 ± 0.0188	4.3039 ± 0.4061
	Ours + MEM	0.7727 ± 0.0165	0.6862 ± 0.0159	4.0223 ± 0.3886
	Ours + TMEM	$\mathbf{0.7865 \pm 0.0126}$	$\mathbf{0.7013 \pm 0.0119}$	$\mathbf{3.9852 \pm 0.2541}$

We compare our method with several representative works, including 2D U-Net [13], 3D U-Net [3], recurrent U-Net (RFCN) [10], and Attention U-Net [9]. In particular, the 3D U-Net was originally developed to capture the spatial relationships among the slices of 3D volumetric images (i.e., stacks of images). Contrary to its original application, here we apply the 3D U-Net to a sequence of SAX images from a cardiac cycle, and set its channel numbers equal to its 2D counterpart. Similarly, RFCN was designed for processing the slices of a single 3D volumetric cardiac image, and here we apply its recurrent unit to the temporal dimension. The Attention U-Net aims to improve the classic U-Net, and generates attention maps from higher-level features to help the network focus on important regions. Apart from the above methods, we also build another

U-Net variant which is inspired by [17]. Specifically, we replace the TMEM in our method with a separable 3D convolution, which decomposes the traditional $3 \times 3 \times 3$ 3D convolution into two separate ones: a 3D convolution with kernel size $1 \times 3 \times 3$ followed by another one with kernel size $3 \times 1 \times 1$. We term this model as S3D U-Net, which only requires a small extra parameter and computation overhead while being able to capture the temporal information. Finally, to validate the effectiveness of our two modules, we also conduct ablation studies. In particular, we remove MEM from our model, obtaining a network with

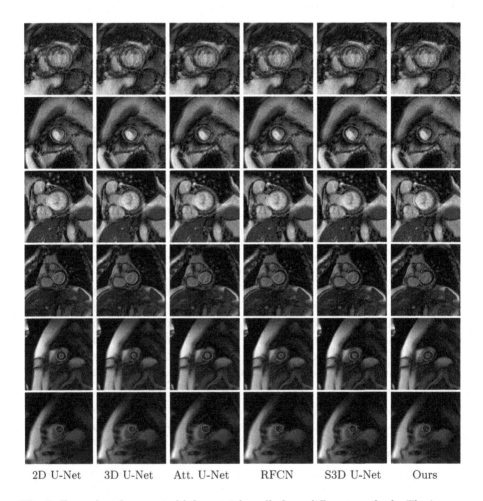

2D U-Net   3D U-Net   Att. U-Net   RFCN   S3D U-Net   Ours

Fig. 3. Examples of segmented left ventricle walls from different methods. The images are from the CAP dataset, and from different LV locations: middle (row 1–2), base (row 3–4) and apex (row 5–6). These images show the cases of myocardial infarction; our method overall achieves the best performance. Green contours represent the ground truth and red contours are the model predictions. "Ours" refers to the model using TMEM (with single-scale motion feature). (Color figure online)

TAM only (i.e., Ours + TAM in Table 1). Similarly, we remove TAM, leading to a model with MEM only (i.e., Ours + MEM in Table 1). In the experiment, we train all these approaches with the same data augmentation strategies as our method, and tune their hyperparameters with the validation set.

Table 1 lists the segmentation results for different methods. As is shown, both the TAM and MEM are able to significantly boost the performance of vanilla 2D U-Net. This demonstrates the importance of leveraging temporal information in cardiac image sequence segmentation, as well as the effectiveness of the proposed modules on exploiting temporal features. In addition, when combining TAM and MEM into a single module, we observe a further improved segmentation accuracy, which indicates that the temporal and motion features are complementary to each other. From Table 1 it can also be observed that our method outperforms RFCN, 3D U-Net and S3D U-Net, even when only one of the proposed modules is employed. This validates the advantages of our temporal feature encoding over the recurrent 2D and vanilla/separable 3D convolutions. In Table 2 we also report the model complexities of different methods. It can be observed that our modules only introduce a small extra parameter and computation overhead while bringing a clear performance gain.

In Table 3 we investigate the influence of temporal interval δ on the model performance. As is shown, when using single-scale motion features, setting $\delta = 4$ gives the best overall results. The reason could be that the motion between two

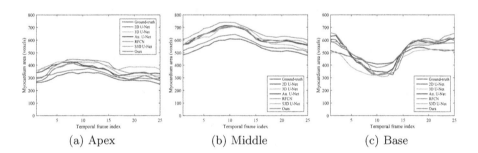

Fig. 4. Representative examples of LV myocardium area over a cardiac cycle for different methods, at the apex, middle and base, respectively. The data sample is from the CAP dataset, and shows a case of myocardial infarction. "Ours" refers to the model using TMEM (with single-scale motion feature).

Table 2. The model complexities of different methods. FLOPs are calculated over a sequence of 32 images, with size 144×144. We·use single-scale motion for MEM/TMEM.

Method	#Parameter	FLOPs	Method	#Parameter	FLOPs
2D U-net	7.9 M	143 G	Att. U-Net	8.5 M	150 G
3D U-Net	23.5 M	183 G	S3D U-Net	8.8 M	159 G
RFCN	22.0 M	179 G	Ours + TAM	8.1 M	146 G
Ours + MEM	8.6 M	154 G	Ours + TMEM	8.8 M	158 G

temporally-close frames is not salient, making the motion features less effective. Similarly, when the two frames are temporally far apart, it would be difficult to learn the motion features due to the disparity of frames. In Table 3 we also show the performance when using multi-scale motion features. It can be observed that multi-scale motion features further boost the performance, but incur increased computational cost (with parameter number 9.5M, and FLOPs 169G).

Table 3. The effects of δ on the model performance. The multi-scale model computes the motion features at scales $\delta = 1$ and $\delta = 4$. We report the segmentation accuracy for models equipped with TMEM.

Dataset	Method	Dice	Jaccard	HD
DYS	$\delta = 1$	0.8173 ± 0.0170	0.7036 ± 0.0197	3.4863 ± 1.2752
	$\delta = 2$	0.8171 ± 0.0282	0.7039 ± 0.0176	3.3761 ± 1.2229
	$\delta = 3$	0.8165 ± 0.0230	0.7029 ± 0.0259	3.4145 ± 1.1859
	$\delta = 4$	0.8201 ± 0.0166	0.7074 ± 0.0184	$\mathbf{3.2332 \pm 0.9241}$
	$\delta = 5$	0.8139 ± 0.0223	0.6972 ± 0.0293	3.7832 ± 1.4499
	Multi-scale	$\mathbf{0.8252 \pm 0.0263}$	$\mathbf{0.7153 \pm 0.0317}$	3.9343 ± 1.4710
CAP	$\delta = 1$	0.7782 ± 0.0154	0.6916 ± 0.0148	4.2917 ± 0.4339
	$\delta = 2$	0.7817 ± 0.0094	0.6965 ± 0.0061	4.2235 ± 0.2736
	$\delta = 3$	0.7807 ± 0.0143	0.6953 ± 0.0136	4.0481 ± 0.3136
	$\delta = 4$	0.7865 ± 0.0126	0.7013 ± 0.0119	3.9852 ± 0.2541
	$\delta = 5$	0.7849 ± 0.0163	0.7011 ± 0.0146	3.8311 ± 0.3143
	Multi-scale	$\mathbf{0.7874 \pm 0.0130}$	$\mathbf{0.7015 \pm 0.0131}$	$\mathbf{3.8236 \pm 0.4159}$

Figure 3 shows several segmentation results for different methods. We can observe that our method is able to delineate the ventricular walls accurately, especially at the base and the apex which are challenging (see row 3–6). Moreover, the proposed method is able to generate smoother outcomes while preserving the topological shape properties of the myocardium walls (i.e., a closed loop). In contrast, the original 2D U-Net fails to accurately localize the boundary at the base and apex, and leads to disconnected segmented shapes. This demonstrates the effectiveness of our method on leveraging temporal and motion information for the segmentation of cardiac MR image sequences.

In Fig. 4, we plot the myocardium area over a cardiac cycle for different methods. We observe that, compared to the baselines, the results by our method are smoother and closer to the ground-truth. In particular, our method largely improves the 2D U-Net, thanks to the explicit modeling of temporal information.

4   Related Work

Cardiac segmentation has been studied for several decades. Here we briefly review some recent works that also attempt to leverage the temporal and motion information for more robust segmentation. In [10], a recurrent neural network (RNN) is employed to capture the temporal information, and thereby improves the segmentation performance upon classic 2D U-Net. Similarly, in [1] the RNN is utilized to facilitate temporal coherence under the setting of semi-supervised learning. Instead of relying on RNN, another direction seeks to model the motion features with optical flow. For example, in [18], the optical flow between temporal frames is computed and then employed to warp the feature maps. Such a strategy enables the deep network to aggregate features from other frames. In a similar spirit, in [11] a joint learning framework is developed to simultaneously learn the optical flow and segmentation masks. Finally, the benefits of RNN and optical flow can be combined for further improved performance, as shown in [5].

Beyond cardiac image sequence segmentation, the learning of temporal features plays an important role in other video-related tasks, such as action recognition and video classification. Recent works on this topic attempts to reduce the computational cost while ensuring the performance. Some representative works include [6,8,12,15,17]. Our method is inspired by these approaches, yet with a focus on image segmentation rather than classification.

5   Conclusions

In this work, we proposed a new method for the segmentation of cardiac MR image sequences, based on the use of a 2D U-Net. The key elements of our method are two new modules, which are able to leverage the temporal and motion information volumetrically in cardiac image sequences. The proposed modules work collaboratively and enable us to improve the feature learning of the base network in a computationally efficient manner. Experimental results on two cardiac MR image datasets demonstrate the effectiveness of our method. As a future work, we plan to apply the proposed modules to 4D cardiac segmentation.

References

1. Bai, W., et al.: Recurrent neural networks for aortic image sequence segmentation with sparse annotations. In: Frangi, A.F., Schnabel, J.A., Davatzikos, C., Alberola-López, C., Fichtinger, G. (eds.) MICCAI 2018. LNCS, vol. 11073, pp. 586–594. Springer, Cham (2018). https://doi.org/10.1007/978-3-030-00937-3_67
2. Ballas, N., Yao, L., Pal, C., Courville, A.: Delving deeper into convolutional networks for learning video representations. In: International Conference on Learning Representations (2016)
3. Çiçek, Ö., Abdulkadir, A., Lienkamp, S.S., Brox, T., Ronneberger, O.: 3D U-Net: learning dense volumetric segmentation from sparse annotation. In: Ourselin, S., Joskowicz, L., Sabuncu, M.R., Unal, G., Wells, W. (eds.) MICCAI 2016. LNCS, vol. 9901, pp. 424–432. Springer, Cham (2016). https://doi.org/10.1007/978-3-319-46723-8_49
4. Fonseca, C.G., et al.: The cardiac atlas project' an imaging database for computational modeling and statistical atlases of the heart. Bioinformatics 27(16), 2288–2295 (2011)

5. Jafari, M.H., et al.: A unified framework integrating recurrent fully-convolutional networks and optical flow for segmentation of the left ventricle in echocardiography data. In: Stoyanov, D., et al. (eds.) DLMIA/ML-CDS -2018. LNCS, vol. 11045, pp. 29–37. Springer, Cham (2018). https://doi.org/10.1007/978-3-030-00889-5_4

6. Jiang, B., Wang, M., Gan, W., Wu, W., Yan, J.: STM: spatiotemporal and motion encoding for action recognition. In: Proceedings of the IEEE conference on computer vision and pattern recognition, pp. 2000–2009 (2019)

7. Kadish, A.H., et al.: Rationale and design for the defibrillators to reduce risk by magnetic resonance imaging evaluation (DETERMINE) trial. J. Cardiovasc. Electrophysiol. **20**(9), 982–987 (2009)

8. Li, Y., Ji, B., Shi, X., Zhang, J., Kang, B., Wang, L.: TEA: temporal excitation and aggregation for action recognition. In: Proceedings of the IEEE/CVF Conference on Computer Vision and Pattern Recognition, pp. 909–918 (2020)

9. Oktay, O., et al.: Attention U-Net: learning where to look for the pancreas. In: Medical Imaging with Deep Learning (2018)

10. Poudel, R.P.K., Lamata, P., Montana, G.: Recurrent fully convolutional neural networks for multi-slice MRI cardiac segmentation. In: Zuluaga, M.A., Bhatia, K., Kainz, B., Moghari, M.H., Pace, D.F. (eds.) RAMBO/HVSMR -2016. LNCS, vol. 10129, pp. 83–94. Springer, Cham (2017). https://doi.org/10.1007/978-3-319-52280-7_8

11. Qin, C., et al.: Joint learning of motion estimation and segmentation for cardiac MR image sequences. In: Frangi, A.F., Schnabel, J.A., Davatzikos, C., Alberola-López, C., Fichtinger, G. (eds.) MICCAI 2018. LNCS, vol. 11071, pp. 472–480. Springer, Cham (2018). https://doi.org/10.1007/978-3-030-00934-2_53

12. Qiu, Z., Yao, T., Mei, T.: Learning spatio-temporal representation with pseudo-3d residual networks. In: Proceedings of the IEEE conference on computer vision and pattern recognition, pp. 5533–5541 (2017)

13. Ronneberger, O., Fischer, P., Brox, T.: U-Net: convolutional networks for biomedical image segmentation. In: Navab, N., Hornegger, J., Wells, W.M., Frangi, A.F. (eds.) MICCAI 2015. LNCS, vol. 9351, pp. 234–241. Springer, Cham (2015). https://doi.org/10.1007/978-3-319-24574-4_28

14. Suinesiaputra, A., et al.: A collaborative resource to build consensus for automated left ventricular segmentation of cardiac MR images. Med. Image Anal. **18**(1), 50–62 (2014)

15. Tran, D., Wang, H., Torresani, L., Feiszli, M.: Video classification with channel-separated convolutional networks. In: Proceedings of the IEEE International Conference on Computer Vision (2019)

16. Wu, P., et al.: Optimal topological cycles and their application in cardiac trabeculae restoration. In: Niethammer, M., et al. (eds.) IPMI 2017. LNCS, vol. 10265, pp. 80–92. Springer, Cham (2017). https://doi.org/10.1007/978-3-319-59050-9_7

17. Xie, S., Sun, C., Huang, J., Tu, Z., Murphy, K.: Rethinking spatiotemporal feature learning: Speed-accuracy trade-offs in video classification. In: Proceedings of the European Conference on Computer Vision (ECCV), pp. 305–321 (2018)

18. Yan, W., Wang, Y., Li, Z., van der Geest, R.J., Tao, Q.: Left ventricle segmentation via optical-flow-net from short-axis cine MRI: preserving the temporal coherence of cardiac motion. In: Frangi, A.F., Schnabel, J.A., Davatzikos, C., Alberola-López, C., Fichtinger, G. (eds.) MICCAI 2018. LNCS, vol. 11073, pp. 613–621. Springer, Cham (2018). https://doi.org/10.1007/978-3-030-00937-3_70

19. Yang, D., Huang, Q., Axel, L., Metaxas, D.: Multi-component deformable models coupled with 2d–3d u-net for automated probabilistic segmentation of cardiac walls and blood. In: 2018 IEEE 15th International Symposium on Biomedical Imaging (ISBI 2018). pp. 479–483. IEEE (2018)
20. Yang, D., Wu, P., Tan, C., Pohl, K.M., Axel, L., Metaxas, D.: 3D motion modeling and reconstruction of left ventricle wall in cardiac MRI. In: Pop, M., Wright, G.A. (eds.) FIMH 2017. LNCS, vol. 10263, pp. 481–492. Springer, Cham (2017). https://doi.org/10.1007/978-3-319-59448-4_46
21. Zhao, H., Shi, J., Qi, X., Wang, X., Jia, J.: Pyramid scene parsing network. In: Proceedings of the IEEE Conference on Computer Vision and Pattern Recognition, pp. 2881–2890 (2017)

Classifying Nuclei Shape Heterogeneity in Breast Tumors with Skeletons

Brian Falkenstein[1]([envelope]) [iD], Adriana Kovashka[1] [iD], Seong Jae Hwang[1] [iD],
and S. Chakra Chennubhotla[2,3] [iD]

[1] School of Computing and Information, University of Pittsburgh, Pittsburgh, PA
15260, USA
briankfalkenstein@gmail.com
[2] Department of Computational and Systems Biology, University of Pittsburgh,
Pittsburgh, PA 15260, USA
[3] Spintellx, Inc., Pittsburgh, PA 15203, USA
https://www.spintellx.com

Abstract. In this study, we demonstrate the efficacy of scoring statistics derived from a medial axis transform, for differentiating tumor and non-tumor nuclei, in malignant breast tumor histopathology images. Characterizing nuclei shape is a crucial part of diagnosing breast tumors for human doctors, and these scoring metrics may be integrated into machine perception algorithms which aggregate nuclei information across a region to label whole breast lesions. In particular, we present a low-dimensional representation capturing characteristics of a skeleton extracted from nuclei. We show that this representation outperforms both prior morphological features, as well as CNN features, for classification of tumors. Nuclei and region scoring algorithms such as the one presented here can aid pathologists in the diagnosis of breast tumors.

Keywords: Medial axis transform · Breast cancer · Digital
pathology · Computer vision

1 Introduction

Recent advancements in whole-slide imaging technology have paved the way for what has been termed Digital Pathology, the digitization of pathology data. Once a biopsy sample has been processed, it can then be scanned into a digital format for viewing by the pathologist. This allows for viewing at any time, while not affecting diagnostic accuracy [23]. Additional benefits of Digital Pathology include the ability to store large amounts of cases for teaching and research purposes, as well as facilitating telepathology, allowing multiple pathologists to view cases simultaneously and remotely. However, what we are most interested in is how the digitization of pathology data allows for the use of computational algorithms to analyze the data and provide useful information to the pathologist.

Digitization proceeds as follows. Slices of the extracted tissue sample are scanned into whole slide images (WSI's). These images are far too large and

© Springer Nature Switzerland AG 2020
A. Bartoli and A. Fusiello (Eds.): ECCV 2020 Workshops, LNCS 12535, pp. 310–323, 2020.
https://doi.org/10.1007/978-3-030-66415-2_20

high resolution (around 5 gigapixels) to perform most analytical computation on, so regions of interest (ROI's) may be extracted via a plethora of methods [15]. Regions are often extracted via texture analysis, where regions of high texture variance are more likely to contain useful information, like a breast duct. These smaller sub-regions of the whole slide (usually around 1M pixels) are likely to contain information useful to the diagnosis, while ignoring large parts of the image which are not useful (such as white space). Most computational analysis is done on the region level. With these manageable ROI's, we can employ a variety of machine learning and computer vision techniques to guide pathologists in the decision making process.

This study focuses on breast tumor analysis in particular for several reasons. Breast cancer is the second most common form of cancer found in women in the US, with an estimated 12% of women developing invasive breast cancer over the course of their life [5]. Annually, roughly 1.6 million breast biopsies are conducted, with around 25% of them showing malignancy [9]. The remaining cases may be classified along a spectrum of labels, from fully benign up to a very high risk of becoming malignant in the future. The differences in appearance between these classes of pre-malignant tumors can be incredibly subtle, which is reflected in the high rates of discordant diagnoses among pathologists for those high-risk cases [9]. In particular, discordance rates are 52% for high-risk benign lesions, compared to just 4% for distinguishing between invasive and non-invasive cancer [9]. Despite the challenges, accurate diagnosis of these lesions is crucial in providing optimal treatment, and early detection and treatment reduces the risk of future malignancy [7,13,14].

Differentiating between a benign lesion and a low or high-risk lesion requires analysis of subtle structural and shape changes in the breast tissue. Complex, high-level structural changes occur in the tumor region as it progresses along the spectrum from benign to malignant [12]. High-risk lesions often show breast ducts being crowded with nuclei in rigid patterns around the lumen (interior opening of the duct). Even the smallest structures present in the tissue, the cell nuclei, demonstrate a *change in shape* as the disease progresses along the spectrum from benign to malignant.

In particular, *higher-risk* lesions, like flat epithelial atypia (FEA) have *rounder* epithelial nuclei than lower risk lesions like columnar cell change (CCC) [31]. Atypical ductal hyperplasia (ADH) is one of the highest risk tumors. Epithelial nuclei in ADH tumors are typically completely rounded [17]. Further, to get a full picture of the tumor environment in order to make an accurate prediction, one must consider not only the distribution of structures across the image (how densely the nuclei are crowded and in what pattern), but also the shape of the structures themselves, in particular the nuclei. This is the focus of our study, to find an accurate and robust way of characterizing the shape of epithelial nuclei in breast tumors.

In this study, we apply the Medial Axis Transform (MAT) [4], a skeletonization algorithm, to the task of classifying nuclei as tumor or non-tumor based solely on their shape. We propose a variety of features to aggregate MAT per-pixel scores into region scores, including novel features measuring branching, bend, and eccentricity. Using our features extracted from the skeleton, we show considerable separation of the two classes of nuclei. In particular, we show that our MAT-based features allow more accurate cancer classification, compared to prior morphological features and those extracted from a CNN applied on a color image or one applied on a 3-channel skeleton image.

We highlight the importance of model interpretability in medical studies such as this one. Many deep learning models exist which obtain impressive classification results in this domain, such as in [2] where they achieve almost 98% accuracy on the BreakHis dataset (benign vs. malignant tumors). However, one thing missing from deep learning models is the ability to directly interpret results. Methods exist for probing deep nets to obtain some result interpretability, such as LIME [21], but these require further processing. This is especially important when considering who will be using the software, primarily pathologists, who are not experts in data science. If our aim is to improve the field of pathology by providing pathologists with tools that will aid them in their diagnostic process, a crucial step is gaining the trust of pathologists, and that can be accomplished through model interpretability. Using hand-picked features which can be mapped to real world descriptions (such as a nucleus being rounded) and simpler decision models leads to better potential interpretability and more trust in the model. Our method thus explores a shape representation that achieves competitive results but is also interpretable. Further work could include adding interpretability into the proposed feature extraction framework.

2   Related Works

Although classification of *whole* breast tumors is a more common problem, several studies have been conducted on classifying *individual* nuclei within the tumor images. Although the end goal is to be able to classify whole tumor images, studying the individual nuclei in the image can be a useful sub-task, as tumor nuclei change shape as a tumor progresses. Being able to highlight individual potentially problematic nuclei in a tumor image would also lend to a more interpretable model. Models that seek to classify from the whole image without special attention to the nuclei, as well as other biological structures, are ignoring biological precedent for the classification. The classic Wisconsin Breast Cancer dataset [27], which features nuclei labeled as either malignant or benign, along with 10 hand-computed features for each nucleus (e.g. radius, perimeter, texture), was used to train simple classifiers with near 100% accuracy on a small test set, such as in [28]. However, a model trained on these parameters could never be used in a clinical setting, as it would require the manual measurement of each nuclei in the biopsy scans, a task more expensive and time consuming than having a pathologist look at it manually.

Many methods exist which seek to replicate the above results in a more scalable fashion, with the nuclei features being computed by an algorithm. Once these explicitly defined features are extracted, simple classifiers can be trained on them with reasonable success. In [26], the authors extracted four simple shape features and 138 textural features. They then classified the nuclei into benign and malignant using an SVM, achieving 96% accuracy. Similarly, in [30], the authors extracted 32 shape and texture features and classified the nuclei into high and low risk categories using an SVM with an accuracy of 90%. One downside of taking this approach to nuclei classification is the requirement of manually obtaining or computing nuclei segmentations, or boundaries around each nuclei, in order to do the shape and texture analysis.

Methods which exploit deep learning do not strictly require a segmentation. For example, Raza et al. [25] use a CNN to classify colon tumor nuclei into four classes with state-of-the-art results. Many works exist which study various CNN frameworks on classifying whole tumor images, such as in [3,11,32], all of which show very promising results. However, the main issue with deep learning approaches is a lack of inherent interpretability. Pathologists may be reluctant to incorporate such black-box methods into their workflow. Although methods exist which allow us to interpret results regardless of the model (such as LIME [21]), there is motivation to keep models more simple and inherently interpretable.

3   Approach

We first describe the data we use, and how we pre-process it. Next, we describe our approach that leverages the well-known Medial Axis Transform (MAT) [4]. Finally, we relate our full set of novel features that we extract from the MAT skeleton.

3.1   Data

We use the *BreCaHAD* dataset of [1], which contains labeled tumor and non-tumor nuclei (i.e. malignant nuclei present inside a malignant tumor and nuclei outside of the tumor, in the stroma, etc.). We leave studying the finer-grained nuances in the middle of the spectrum (i.e. classifying high-risk benign lesions) as future work, because no such annotated dataset is currently available. *BreCaHAD* consists of 162 breast lesion images, all of which show varying degrees of malignancy. The centroid of each nuclei in the images is labeled with one of the following six classes: mitosis, apoptosis, tumor nuclei, non-tumor nuclei, tubule, and non-tubule. For our purposes, we are just interested in differentiating the tumor and non-tumor nuclei. Theoretically, if we can design a framework to differentiate tumor and non-tumor nuclei in malignant cases based on shape, the same framework should be successful at analyzing high-risk benign tumor nuclei, the more challenging task (for which no annotated datasets currently exist).

The dataset presents some challenges. Most importantly, all of the images show malignant tissue samples. Thus, at the nuclei (rather than full tissue sample) level, there is a dramatic class imbalance in the data, with almost 92% of nuclei labeled as 'tumor'. Additionally, the ground truth labels (centroid

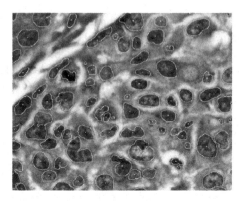

Fig. 1. Segmentation results using the method described in [6]. Overall the output is fairly accurate, but some of the nuclei are segmented together, into a larger blob.

and class label) are not enough for our purposes since nuclei need to be segmented first for further processing (feature extraction and classification). We next describe how we handle both challenges.

Nuclei Segmentation: If we wish to analyze the shape of the nuclei present in the biopsy images, we need some definition of the boundaries of the nuclei. Ideally, this would come with the dataset, and be a ground truth value. However, in our case, we are only given centroid locations of the nuclei, and thus must produce our own segmentation. We use the segmentation algorithm proposed by Chen et al. [6], which is a fully convolutional network based on U-Net [22], with additional smoothing applied after the upsampling. A segmentation on a sample from the dataset using this method is shown in Fig. 1. Once the segmentation is computed, the labeled centroids can be matched to their corresponding nuclei boundary by determining which (masked) boundary the centroid falls within. The labels in the dataset could then be associated with a nuclei boundary. Using this segmentation scheme, 80.18% of the ground truth nuclei in the dataset were found (18149 nuclei). Of these, 92% were positive (16689). Crops of size $(81, 81)$ pixels centered at each nuclei were taken from the original images, and used as the training and testing data.

Class Imbalance: To account for the class imbalance, training and testing sets were obtained by randomly sampling with replacement from the over-represented class (tumor) until there was an equal number of samples for both classes. Without this step, the model would tend to demonstrate very low specificity (high rate of false positives), while still maintaining a low loss. There were 1460 negative (non-tumor) nuclei after segmentation, so the same number of positive nuclei were randomly selected from the initial set of 16689. This is not ideal, as many positives are thrown out, however it is preferable to an overfitted model.

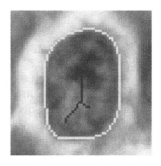

Fig. 2. Medial Axis Transform on a binary image of a horse (left) [18]. A nuclei segmentation (cyan) and skeletonization (red) on a tumor nucleus is shown on the right. (Color figure online)

3.2 Medial Axis Transform

We propose a novel representation for nuclei shape analysis, based on the medial axis transform (MAT) [4]. The MAT is a skeletonization algorithm wherein a closed binary shape is reduced down to a 1-pixel-wide branching structure wholly contained within the original shape. This may be conceptualized as a thinning process, where the border of the shape becomes thinner, or erodes, until opposite sides of the shape come together, defining the skeleton. The skeleton points are center points of maximally inscribed discs, i.e. circles with more than one point on their surface tangent to the shape boundary. This property makes the MAT skeletonization an appropriate candidate for deriving a shape descriptor. We show an example in Fig. 2.

The skeleton itself does not contain enough information on its own to derive a shape descriptor, as it only provides us with a set of pixels which define the skeleton. However, it can be used to derive several scoring statistics. Rezanejad et al. [19] derive three scores from the MAT: ribbon, taper, and separation. They also show that these scores can be successfully used by a CNN to perform scene classification from line images (no color or texture). The scene classification task is significantly different from the nuclei classification one, as scene images feature many contours, some of which may not be closed shapes. Further, a nuclei image will contain far fewer contours compared to a dense scene. However, the results in [19] imply that these scores capture contour shape accurately, which applies to nuclei shape analysis, where we are hoping to capture how rounded the nuclei are.

Let p be a point on the skeleton defined by its pixel location in the image, $R(p)$ be the radius function for that point (shortest distance from the skeleton point to the boundary), and $[\alpha, \beta]$ be a range of skeleton points of size k. Finally, let $p \in [\alpha, \beta]$. Then the three score metrics are defined as follows, based on the first and second derivative of $R(p)$:

$$S_{ribbon}(p) = R'(p) \tag{3.1}$$

$$S_{taper}(p) = R''(p) \tag{3.2}$$

$$S_{separation}(p) = 1 - \left(\int_{\alpha}^{\beta} \frac{1}{R(p)} dp\right)/k \tag{3.3}$$

Because pixels are discrete, and thus so are the intervals we are integrating and deriving over, we must use numerical gradients. A small value of k was used, as the nuclei are fairly small (roughly 100 pixel perimeter), and the scoring metrics are designed to capture local symmetry. For all of the tests, we used a value of k=8. The skeletonization method was adapted from [8,18,20,24].

The Rezanejad algorithm defines three scores for each skeleton point. Thus, it could be treated as a 3-channel color image, where all non-skeletal pixels have a value of 0 and skeleton points are described with 3 scores. The ribbon score $(S_{ribbon}(p))$ captures the degree of parallelity of the surrounding contours, and increases as they become more parallel. The taper score $(S_{taper}(p))$ is designed to increase as contours resemble the shape of a funnel or railroad tracks, but also has a high value for parallel contours. Separation $(S_{separation}(p))$ captures the degree of separation between the contours, and increases with distance.

Table 1. Proposed skeleton features

Feature	Description
Min rib/tap/sep	Minimum of each of the 3 scores for all pixels in the skeleton (ribbon, taper, separation)
Max rib/tap/sep	Maximum of each of the 3 scores
Mean rib/tap/sep	Mean of each of the 3 scores
Deviation rib/tap/sep	Standard deviation of each of the 3 scores
Max branch length	The length (in pixels) of the longest branch of the skeleton
Avg branch length	Average length (in pixels) of the branches of the skeleton
Num branches	The number of branches in the skeleton
Bend	Angle between the line connecting the furthest two points on the skeleton and the skeleton's major axis
Major/minor axis len	Length (in pixels) of the major and minor axes of the ellipse with the same normalized second central moments as the skeleton
Eccentricity	Eccentricity of the ellipse that has the same second-moment as the skeleton, the ratio of the distance between the foci of the ellipse and its major axis length
Solidity	Proportion of pixels in convex hull that are also in the skeleton

3.3   Final Representation

We propose a set of 20 features, described in Table 1. Three of those features rely on the per-pixel ribbon, taper and separation features described above. The per-pixel features are aggregated on the whole nuclei level by taking their min, max, mean, and standard deviation.

The following additional skeleton structural features were extracted: number of branches, average and max branch length, and bend (angle between line connecting furthest 2 skeleton points and major axis). Branches were isolated by computing junction points using [10], then setting junction points to zero and applying connected component analysis on the separated branches.

Another interesting method for extracting a feature from the skeleton data would be to define the skeleton as a graph, with node features being the 3 scores, and apply graph convolutions or graph kernels to do the classification. Deep learning on graph structured data has shown promise in recent years [29] and could prove useful in this task. However, the graph analysis is outside of the scope of this study currently.

4   Experimental Validation

We describe the methods compared and metrics used. We next show quantitative results, and finally some examples of the features that our method relies on.

4.1   Methods Compared

In order to show the efficacy of the proposed method for capturing nuclei shape, the tests will be conducted on features with no information of texture or color in the nuclei. For this reason, we test our method against another method which does not incorporate texture, but has shown to be successful. Yamamoto et al. [30] obtained an accuracy of 85% using only a small number of hand-picked morphological features and an SVM. We were able to replicate their accuracy results on the balanced test set using the following set of features originally proposed in [30]: area, eccentricity, extend, major axis, minor axis, convex area, circularity, equivalent diameter, filled area, perimeter, and solidity. Note again how none of these features incorporate any color or texture information. In total, 19 morphological features were extracted, similar to the dimensionality of the skeleton feature vector we will be comparing it to. We refer to this method as **Morphological features** in Table 2.

However, color and texture features have also been used with success [26]. Thus, we also compare the proposed shape-based method to a standard convolutional neural network which includes color, namely AlexNet [16], pre-trained on ImageNet. We used this model to extract features from the last fully-connected layer ($D = 1000$). The CNN was tested both on the full color crops, as well as the 3-channel skeleton score image (see below). We refer to these methods as **CNN on full color crop** and **CNN on 3-channel skeleton image** in Table 2, respectively.

Table 2. BreCaHAD classification: main results

Method	Accuracy	Sensitivity	Specificity
Morphological features	0.846 +/− 0.001	0.831 +/− 0.001	**0.861 +/− 0.001**
CNN on full color crop	0.632 +/− 0.001	0.477 +/− 0.001	0.798 +/− 0.001
CNN on 3-channel skeleton image	0.654 +/− 0.002	0.623 +/− 0.001	0.687 +/− 0.001
MAT skeleton features (Ours)	**0.850 +/− 0.004**	**0.844 +/− 0.004**	0.852 +/− 0.003

Our method uses the 20-dimensional proposed feature vector described in Sect. 3.3, and is referred to as **MAT skeleton features**.

4.2 Metrics and Setup

We evaluate model performance using accuracy, sensitivity, and specificity, with special attention paid to sensitivity. In medical tests, maximizing sensitivity, or minimizing false negatives, is an important goal, as we do not want to under interpret a sample and cause a patient to not receive necessary treatment. The dataset was split into 70% for training, and 30% for testing, giving sizes of 2044 and 876 nuclei respectively. Classification for all tests was done using a linear SVM with RBF kernel. We use an equal class representation, so a random algorithm would achieve an accuracy of roughly 0.5. Scores were averaged over 20 runs, with a different random train/test split for each run.

4.3 Quantitative Results

Our results are shown in Table 2. Our proposed method achieves the highest accuracy. Further, the results are statistically significant. We reject the null hypothesis that the distributions of accuracies for the skeleton features and the morphological features come from different distributions at significance level 5%.

Additionally, our MAT skeleton features achieve the highest sensitivity of the methods testing. Sensitivity is especially important in a medical setting, where a false negative (missed disease diagnosis) is much more costly than a false positive. This also highlights the failure of the color CNN feature, where its passable accuracy score was achieved by outputting mostly negative labels, resulting in many false negatives. It also demonstrates that the 3-channel skeleton image CNN feature was actually significantly better than the CNN on the full color image, as is seen in the difference in sensitivity. Note that we could have tried a larger CNN for extracting the features, but the sizable gap in the CNN methods' accuracy vs. ours and the morphological features, indicated that further exploration of CNN features may not work well.

Finally, we compare performance of the SVM on the MAT skeleton scores feature which closely follows [19] (min/max/mean/deviation of the ribbon, taper,

Table 3. BreCaHAD classification: ablation results

Method	Accuracy	Sensitivity	Specificity
Min/max/avg/dev skeleton features	0.820 +/− 0.003	0.789 +/− 0.011	**0.853 +/− 0.006**
Other skeleton features	0.692 +/− 0.002	0.680 +/− 0.004	0.705 +/− 0.009
All skeleton features (Ours)	**0.850 +/− 0.004**	**0.844 +/− 0.004**	0.852 +/− 0.003

Fig. 3. The nuclei with the lowest maximum separation score (left), a non-tumor nuclei, and that with the highest maximum separation score (right), a tumor nuclei.

and separation scores) and the new skeleton features that we propose (max branch length, number of branches, etc.). The results can be seen in Table 3. While the MAT score-derived features are more successful in the classification of the nuclei shape in isolation, the other features we propose do significantly contribute to performance.

Scores could likely be improved for both the morphological and skeleton features by taking several steps. First, color features could be incorporated into the skeleton and morphological features. These could be extracted from a CNN, or just be basic statistical measures such as in [26]. However, with these results we have shown the efficacy of the MAT skeleton feature as a means of distinguishing nuclei shapes.

4.4 Qualitative Results

Analysis of the linear predictor coefficients of the SVM with all MAT skeleton features in Table 4 shows that the maximum separation score feature was by far the most discriminant. Recall that the separation score measures the level of separation between the skeleton and its boundary. Intuitively, this makes sense, as we know tumor nuclei to be more engorged and rounded (and thus have high separation), and non-tumor nuclei to be thinner and more elongated (resulting in a smaller separation). This intuition is validated by viewing the data samples which exhibit the highest and lowest max separation scores for the respective classes. The sample which exhibited the highest max separation score was from the positive class, and the sample which exhibited the lowest

Table 4. SVM Beta values for the top 5 features

Feature	Beta value
Max separation	4.082
Major axis length	0.844
Average branch length	0.582
Number of branches	0.580
Minor axis length	0.519

separation was negative, both of which are illustrated in Fig. 3. Other important features include our new axis and branch features.

5 Conclusion

The primary aim of the study was not to achieve the best classification metrics and outperform all other state of the art models, but rather to define a robust shape descriptor that accurately captures the change of shape of the nucleus. The MAT skeleton feature showed improvement over the baseline morphological features. Additionally, features extracted with a CNN on the weighted skeleton image performed better than similar features extracted from the full color image. The MAT skeleton scoring algorithm outlined in this paper is thus a useful shape descriptor in regards to tracking nuclei shape heterogeneity in breast tumors.

Although here the MAT skeleton feature was applied to distinguishing tumor and non-tumor nuclei in malignant breast tumors, it may be robust enough to tackle harder challenges, such as the classification of high-risk tumors, or be applied to other modalities and cancers, such as prostate cancer (which also uses stained slide images for diagnosis). Further work will focus on applying this technique to more challenging tasks and datasets.

We can also apply the nucleus scoring algorithm, where each nuclei receives a set of features, to classify a whole region by considering all contained nuclei. One approach is to take a majority vote: if most nuclei in the region are cancerous, then predict the image as cancerous. Another method could be normalized mean statistics being taken over the graph to return a 1-dimensional vector. A more interesting way to combine the nuclei features across the image could be to define a graph over the image, where the nodes in the graph are the nuclei, and each node inherits the features describing the nuclei. This would maintain the structure and distribution of nuclei in the original image, while allowing for various graph learning methods to be applied, which can consider both the individual nuclei features, as well as the overall structure of the tumor environment. This is a topic for further study.

References

1. Aksac, A., Demetrick, D.J., Ozyer, T., Alhajj, R.: Brecahad: a dataset for breast cancer histopathological annotation and diagnosis. BMC Res. Notes **12**(1), 82 (2019). https://doi.org/10.1186/s13104-019-4121-7

2. Alom, M.Z., Yakopcic, C., Taha, T.M., Asari, V.K.: Breast cancer classification from histopathological images with inception recurrent residual convolutional neural network. CoRR abs/1811.04241 (2018). http://arxiv.org/abs/1811.04241

3. Araújo, T., et al.: Classification of breast cancer histology images using convolutional neural networks. PLOS ONE **12**(6), 1–14 (06 2017). https://doi.org/10.1371/journal.pone.0177544

4. Blum, H.: A transformation for extracting new descriptors of shape. In: Wathen-Dunn, W. (ed.) Models for the Perception of Speech and Visual Form, pp. 362–380. MIT Press, Cambridge (1967)

5. BreastCancerOrg: U.S. Breast Cancer Statistics, January 2020. https://www.breastcancer.org/symptoms/understand_bc/statistics?gclid=CjwKCAjwvtX0BRA FEiwAGWJyZJpuc9fByPVUbO4838EcKnRJ4uXckwegdsznOarkrF8EJ0z_2fQYG hoCbWQQAvD_Bw. Accessed Mar 2020

6. Chen, K., Zhang, N., Powers, L., Roveda, J.: Cell nuclei detection and segmentation for computational pathology using deep learning. In: MSM 2019: Proceedings of the Modeling and Simulation in Medicine Symposium, p. 12 (2019). https://doi.org/10.22360/springsim.2019.msm.012

7. Collins, L.C., Laronga, C., Wong, J.S.: Ductal carcinoma in situ: Treatment and prognosis, January 2020. https://www.uptodate.com/contents/ductal-carcinoma-in-situ-treatment-and-prognosis?search=ductal-carcinoma-in-situ-treatment-and-prognosi&source=search_result&selectedTitle=1~54&usage_type=default& display_rank=1

8. Dimitrov, P., Damon, J.N., Siddiqi, K.: Flux invariants for shape. In: 2003 IEEE Computer Society Conference on Computer Vision and Pattern Recognition. Proceedings, vol. 1, pp. I–I. IEEE (2003)

9. Elmore, J.G., et al.: Diagnostic concordance among pathologists interpreting breast biopsy specimens. JAMA **313**(11), 1122–1132 (2015). https://doi.org/10.1001/jama.2015.1405

10. Fong, A.: Skeleton intersection detection, December 2003. https://www.mathworks.com/matlabcentral/fileexchange/4252-skeleton-intersection-detection

11. Gao, F., et al.: SD-CNN: A shallow-deep CNN for improved breast cancer diagnosis. Comput. Med. Imaging Graph. **70**, 53–62 (2018). https://doi.org/10.1016/j.compmedimag.2018.09.004, http://www.sciencedirect.com/science/article/pii/S0895611118302349

12. Halls, S.: Micropapillary breast cancer, May 2019. https://breast-cancer.ca/micropap/

13. Hartmann, L.C., Degnim, A.C., Santen, R.J., Dupont, W.D., Ghosh, K.: Atypical hyperplasia of the breast - risk assessment and management options. New Engl. J. Med. **372**(1), 78–89 (2015). https://doi.org/10.1056/NEJMsr1407164. pMID: 25551530

14. Kader, T., Hill, P., Rakha, E.A., Campbell, I.G., Gorringe, K.L.: Atypical ductal hyperplasia: update on diagnosis, management, and molecular landscape. Breast Cancer Res. **20**(1), 39 (2018). https://doi.org/10.1186/s13058-018-0967-1

15. Kothari, S., Phan, J., Stokes, T., Wang, M.: Pathology imaging informatics for quantitative analysis of whole-slide images. J. Am. Med. Inf. Assoc. JAMIA **20**, 1099–1108 (2013). https://doi.org/10.1136/amiajnl-2012-001540

16. Krizhevsky, A., Sutskever, I., Hinton, G.E.: ImageNet classification with deep convolutional neural networks. In: Pereira, F., Burges, C.J.C., Bottou, L., Weinberger, K.Q. (eds.) Advances in Neural Information Processing Systems, vol. 25, pp. 1097–1105. Curran Associates, Inc. (2012). http://papers.nips.cc/paper/4824-imagenet-classification-with-deep-convolutional-neural-networks.pdf
17. Page, D.L., Rogers, L.W.: Combined histologic and cytologic criteria for the diagnosis of mammary atypical ductal hyperplasia. Hum. Pathol. **23**(10), 1095–1097 (1992). https://doi.org/10.1016/0046-8177(92)90026-Y, http://www.sciencedirect.com/science/article/pii/004681779290026Y
18. Rezanejad: mrezanejad/aofskeletons, June 2019. https://github.com/mrezanejad/AOFSkeletons
19. Rezanejad, M., et al.: Scene categorization from contours: medial axis based salience measures. CoRR abs/1811.10524 (2018). http://arxiv.org/abs/1811.10524
20. Rezanejad, M., Siddiqi, K.: Flux graphs for 2D shape analysis. In: Dickinson, S., Pizlo Z. (eds.) Shape Perception in Human and Computer Vision. Advances in Computer Vision and Pattern Recognition, pp. 41–54. Springer, London. https://doi.org/10.1007/978-1-4471-5195-1_3
21. Ribeiro, M.T., Singh, S., Guestrin, C.: "Why should I trust you?": explaining the predictions of any classifier. In: Proceedings of the 22nd ACM SIGKDD International Conference on Knowledge Discovery and Data Mining, pp. 1135–1144. KDD 2016. Association for Computing Machinery, New York (2016). https://doi.org/10.1145/2939672.2939778
22. Ronneberger, O., Fischer, P., Brox, T.: U-Net: convolutional networks for biomedical image segmentation. CoRR abs/1505.04597 (2015). http://arxiv.org/abs/1505.04597
23. Shah, K.K., Lehman, J.S., Gibson, L.E., Lohse, C.M., Comfere, N.I., Wieland, C.N.: Validation of diagnostic accuracy with whole-slide imaging compared with glass slide review in dermatopathology. J. Am. Acad. Dermatol. **75**(6), 1229–1237 (2016). https://doi.org/10.1016/j.jaad.2016.08.024
24. Siddiqi, K., Bouix, S., Tannenbaum, A., Zucker, S.W.: Hamilton-Jacobi skeletons. Int. J. Comput. Vision **48**(3), 215–231 (2002). https://doi.org/10.1023/A:1016376116653
25. Sirinukunwattana, K., Raza, S.E.A., Tsang, Y., Snead, D.R.J., Cree, I.A., Rajpoot, N.M.: Locality sensitive deep learning for detection and classification of nuclei in routine colon cancer histology images. IEEE Trans. Med. Imaging **35**(5), 1196–1206 (2016)
26. Wang, P., Hu, X., Li, Y., Liu, Q., Zhu, X.: Automatic cell nuclei segmentation and classification of breast cancer histopathology images. Signal Process. **122**, 1–13 (2016). https://doi.org/10.1016/j.sigpro.2015.11.011, http://www.sciencedirect.com/science/article/pii/S0165168415003916
27. Wolberg, W.H., Street, N., Mangasarian, O.L.: Breast cancer Wisconsin (diagnostic) data set (1995). http://archive.ics.uci.edu/ml/datasets/breastcancerwisconsin(diagnostic)
28. Wolberg, W.H., Street, W., Mangasarian, O.: Machine learning techniques to diagnose breast cancer from image-processed nuclear features of fine needle aspirates. Cancer Lett. **77**(2), 163–171 (1994). https://doi.org/10.1016/0304-3835(94)90099-X, http://www.sciencedirect.com/science/article/pii/030438359490099X. Computer applications for early detection and staging of cancer
29. Wu, Z., Pan, S., Chen, F., Long, G., Zhang, C., Yu, P.S.: A comprehensive survey on graph neural networks. IEEE Trans. Neural Netw. Learn. Syst. 1–21 (2020). https://doi.org/10.1109/tnnls.2020.2978386

30. Yamamoto, Y., et al.: Quantitative diagnosis of breast tumors by morphometric classification of microenvironmental myoepithelial cells using a machine learning approach. Sci. Rep. **7**, 46732–46732 (2017). https://doi.org/10.1038/srep46732, https://pubmed.ncbi.nlm.nih.gov/28440283. 28440283[pmid]

31. Yamashita, Y., Ichihara, S., Moritani, S., Yoon, H.S., Yamaguchi, M.: Does flat epithelial atypia have rounder nuclei than columnar cell change/hyperplasia?: A morphometric approach to columnar cell lesions of the breast. Virchows Arch. **468**(6), 663–673 (2016). https://doi.org/10.1007/s00428-016-1923-z

32. Zuluaga-Gomez, J., Masry, Z.A., Benaggoune, K., Meraghni, S., Zerhouni, N.: A CNN-based methodology for breast cancer diagnosis using thermal images (2019)

DenoiSeg: Joint Denoising and Segmentation

Tim-Oliver Buchholz[1,2], Mangal Prakash[1,2], Deborah Schmidt[1,2],
Alexander Krull[1,2,3], and Florian Jug[1,2,4]

[1] Center for Systems Biology, Dresden, Germany
[2] Max Planck Institute of Molecular Cell Biology and Genetics, Dresden, Germany
jug@mpi-cbg.de
[3] Max Planck Institute for Physics of Complex Systems, Dresden, Germany
[4] Fondazione Human Technopole, Milano, Italy
florian.jug@fht.org

Abstract. Microscopy image analysis often requires the segmentation of objects, but training data for this task is typically scarce and hard to obtain. Here we propose DenoiSeg, a new method that can be trained end-to-end on only a few annotated ground truth segmentations. We achieve this by extending Noise2Void, a self-supervised denoising scheme that can be trained on noisy images alone, to also predict dense 3-class segmentations. The reason for the success of our method is that segmentation can profit from denoising, especially when performed jointly within the same network. The network becomes a denoising expert by seeing all available raw data, while co-learning to segment, even if only a few segmentation labels are available. This hypothesis is additionally fueled by our observation that the best segmentation results on high quality (very low noise) raw data are obtained when moderate amounts of synthetic noise are added. This renders the denoising-task non-trivial and unleashes the desired co-learning effect. We believe that DenoiSeg offers a viable way to circumvent the tremendous hunger for high quality training data and effectively enables learning of dense segmentations when only very limited amounts of segmentation labels are available.

Keywords: Segmentation · Denoising · Co-learning

1 Introduction

The advent of modern microscopy techniques has enabled the routine investigation of biological processes at sub-cellular resolution. The growing amount of microscopy image data necessitates the development of automated analysis methods, with object segmentation often being one of the desired analyses. Over the years, a sheer endless array of methods have been proposed for segmentation [9], but deep learning (DL) based approaches are currently best performing [3,14,18]. Still, even the best existing methods offer plenty of scope for improvements, motivating further research in this field [7,22,24].

T.-O. Buchholz and M. Prakash—Equal contribution (alphabetical order).

© Springer Nature Switzerland AG 2020
A. Bartoli and A. Fusiello (Eds.): ECCV 2020 Workshops, LNCS 12535, pp. 324–337, 2020.
https://doi.org/10.1007/978-3-030-66415-2_21

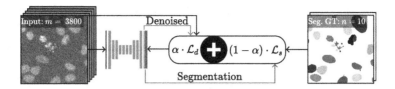

Fig. 1. The proposed DENOISEG training scheme. A U-NET is trained with a joint self-supervised denoising loss ($\mathcal{L}_d$) and a classical segmentation loss ($\mathcal{L}_s$). Both losses are weighted with respect to each other by a hyperparameter α. In this example, $\mathcal{L}_d$ can be computed on all $m = 3800$ training patches, while $\mathcal{L}_s$ can only be computed on the $n = 10$ annotated ground truth patches that are available for segmentation.

A trait common to virtually all DL-based segmentation methods is their requirement for tremendous amounts of labeled ground truth (GT) training data, the creation of which is extraordinarily time consuming. In order to make the most out of a given amount of segmentation training data, data augmentation [23,28] is used in most cases. Another way to increase the amount of available training data for segmentation is to synthetically generate it, *e.g.* by using Generative Adversarial Networks (GANs) [8,15,20]. However, the generated training data needs to capture all statistical properties of the real data and the respective generated labels, thereby making this approach cumbersome in its own right.

For other image processing tasks, such as denoising [2,12,27], the annotation problem has been addressed via self-supervised training [1,10,11,17]. While previous denoising approaches [27] require pairs of noisy and clean ground truth training images, self-supervised methods can be trained directly on the noisy raw data that is to be denoised.

Very recently, Prakash *et al.* [16] demonstrated on various microscopy datasets that self-supervised denoising [10] prior to object segmentation leads to greatly improved segmentation results, especially when only small numbers of segmentation GT images are available for training. The advantage of this approach stems from the fact that the self-supervised denoising module can be trained on the full body of available microscopy data. In this way, the subsequent segmentation module receives images that are easier to interpret, leading to an overall gain in segmentation quality even without having a lot of GT data to train on. In the context of natural images, a similar combination of denoising and segmentation was proposed by Liu *et al.* [13] and Wang *et al.* [26]. However, both methods lean heavily on the availability of paired low- and high-quality image pairs for training their respective denoising module. Additionally, their cascaded denoising and segmentation networks make the training comparatively computationally expensive.

Here, we present DENOISEG, a novel training scheme that leverages denoising for object segmentation (see Fig. 1). Like Prakash *et al.*, we employ self-supervised NOISE2VOID [10] for denoising. However, while Prakash *et al.* rely on two sequential steps for denoising and segmentation, we propose to use a single network to jointly predict the denoised image and the desired object segmentation. We use a simple U-NET [19] architecture, making training fast and accessible on moderately priced consumer hardware. Our network is trained on noisy microscopy data and requires only a small fraction of images to be annotated with GT segmentations. We evaluate our method on different datasets and with different amounts of annotated training images. When only small amounts of annotated training data are available, our method consistently outperforms not only networks trained purely for segmentation [4,6], but also the currently best performing training schemes proposed by Prakash *et al.* [16].

2   Methods

We propose to jointly train a single U-NET for segmentation and denoising tasks. While for segmentation only a small amount of annotated GT labels are available, the self-supervised denoising module benefits from all available raw images. In the following we will first discuss how these tasks can be addressed separately and then introduce a joint loss function combining the two.

Segmentation. We see segmentation as a 3-class pixel classification problem [4,6,16] and train a U-NET to classify each pixel as foreground, background or border (this yields superior results compared to a simple classification into foreground and background [22]). Our network uses three output channels to predict each pixel's probability of belonging to the respective class. We train it using the standard cross-entropy loss, which will be denoted as $\mathcal{L}_s\big(\boldsymbol{y}_i, f(\boldsymbol{x}_i)\big)$, where $\boldsymbol{x}_i$ is the i-th training image, $\boldsymbol{y}_i$ is the ground truth 3-class segmentation, and $f(\boldsymbol{x}_i)$ is the network output.

Self-Supervised Denoising. We use the NOISE2VOID setup described in [10] as our self-supervised denoiser of choice. We extend the above mentioned 3-class segmentation U-NET by adding a fourth output channel, which is used for denoising and trained using the NOISE2VOID scheme. NOISE2VOID uses a Mean Squared Error (MSE) loss, which is calculated over a randomly selected subset of blind spot pixels that are masked in the input image. Since the method is self-supervised and does not require ground truth, this loss $\mathcal{L}_d\big(\boldsymbol{x}_i, f(\boldsymbol{x}_i)\big)$ can be calculated as a function of the input image $\boldsymbol{x}_i$ and the network output $f(\boldsymbol{x}_i)$.

Joint-Loss. To jointly train our network for denoising and segmentation we use a combined loss. For a given training batch $(\boldsymbol{x}_1, \boldsymbol{y}_1, \ldots, \boldsymbol{x}_m, \boldsymbol{y}_m)$ of m images, we assume that GT segmentation is available only for a subset of $n \ll m$ raw images. We define $\boldsymbol{y}_i = \boldsymbol{0}$ for images where no segmentation GT is present. The loss over a batch is calculated as

$$\mathcal{L} = \frac{1}{m} \sum_{i=1}^{m} \alpha \cdot \mathcal{L}_d\big(\boldsymbol{x}_i, f(\boldsymbol{x}_i)\big) + (1-\alpha) \cdot \mathcal{L}_s\big(\boldsymbol{y}_i, f(\boldsymbol{x}_i)\big), \tag{1}$$

where $0 \leq \alpha \leq 1$ is a tunable hyperparameter that determines the relative weight of denoising and segmentation during training. Note that the NOISE2VOID loss is self-supervised, therefore it can be calculated for all raw images in the batch. The cross-entropy loss however requires GT segmentation and can only be evaluated on a subset of images, where this information is available. For images where no GT segmentation is available we define $\mathcal{L}_s\big(\boldsymbol{y}_i = \mathbf{0}, f(\boldsymbol{x}_i)\big) = 0$.

In the setup described above, setting $\alpha = 1$ corresponds to pure NOISE2VOID denoising. However, setting $\alpha = 0$ does not exactly correspond to the vanilla 3-class segmentation, due to two reasons. Firstly, only some of the images are annotated but in Eq. 1 the loss is divided by the constant batch size m. This effectively corresponds to a reduced batch size and learning rate, compared to the vanilla method. Secondly, our method applies NOISE2VOID masking of blind spot pixels in the input image.

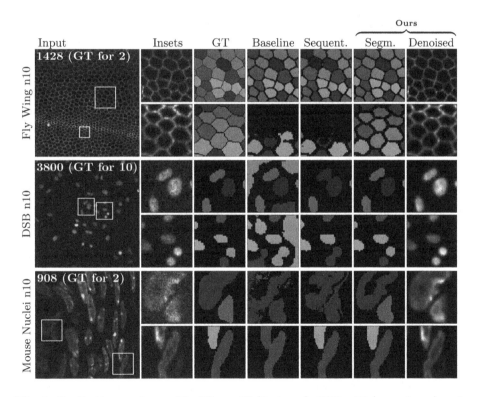

Fig. 2. Qualitative results on Fly Wing n10 (first row), DSB n10 (second row) and Mouse Nuclei n10 (third row). The first column shows an example test image. Numbers indicate how many noisy input and annotated ground truth (GT) patches were used for training. Note that segmentation GT was only available for at most 10 images, accounting for less than 0.27% of the available raw data. Other columns show depicted inset regions, from left to right showing: raw input, segmentation GT, results of two baseline methods, and our DENOISEG segmentation and denoising results.

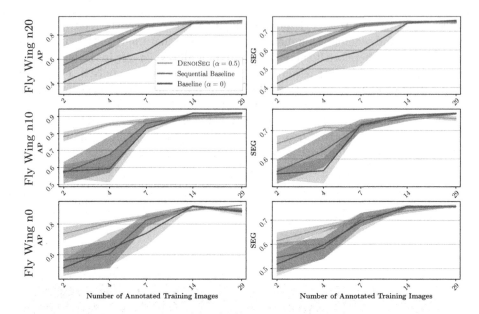

Fig. 3. Results for Fly Wing n0, n10 and n20, evaluated with Average Precision (AP) [22] and SEG-Score [25]. DENOISEG outperforms both baseline methods, mainly when only limited segmentation ground truth is available.

Implementation Details. Our DENOISEG implementation is publicly available[1]. The proposed network produces four output channels corresponding to denoised images, foreground, background and border segmentation. For all our experiments we use a U-NET architecture of depth 4, convolution kernel size of 3, a linear activation function in the last layer, 32 initial feature maps, and batch normalization during training. All networks are trained for 200 epochs with an initial learning rate of 0.0004. The learning rate is reduced if the validation loss is not decreasing over ten epochs. For training we use 8-fold data augmentation by adding 90° rotated and flipped versions of all images.

Additionally, we also provide a DENOISEG plugin for Fiji [21], a popular Java-based image processing software. The plugin does not require computational expertise to be used and provides comprehensive visual feedback during training. Trained models can then be applied to new data and shared with other users for use in either Python or Fiji.

3   Experiments and Results

We use three publicly available datasets for which GT annotations are available (data available at DENOISEG-Wiki[2]). For each dataset we generate noisy versions

---

[1] https://github.com/juglab/DenoiSeg, https://imagej.net/DenoiSeg.

[2] https://github.com/juglab/DenoiSeg/wiki.

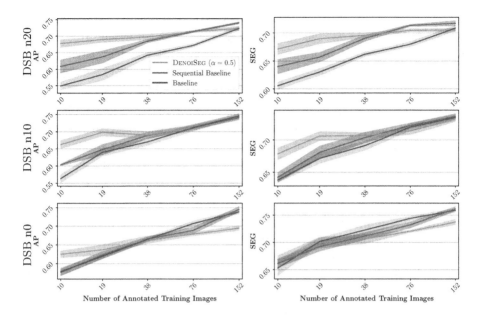

Fig. 4. Results for DSB n0, n10 and n20, evaluated with Average Precision (AP) [22] and SEG-Score [25]. DENOISEG outperforms both baseline methods, mainly when only limited segmentation ground truth is available. Note that the advantage of our proposed method is at least partially compromised when the image data is not noisy (row 3).

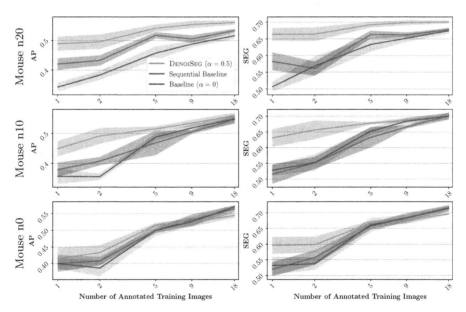

Fig. 5. Results for Mouse nuclei n0, n10 and n20, evaluated with Average Precision (AP) [22] and SEG-Score [25]. DENOISEG outperforms both baseline methods, mainly when only limited segmentation ground truth is available.

by adding pixel-wise independent Gaussian noise with zero-mean and standard deviations of 10 and 20. The dataset names are extended by n0, n10, and n20 to indicate the respective additional noise. For network training, patches of size 128×128 are extracted and randomly split into training (85%) and validation (15%) sets.

- **Fly Wing.** This dataset from our collaborators consist of 1428 training and 252 validation patches of size 128×128 showing a membrane labeled fly wing. The test set is comprised of 50 additional images of size 512×512.
- **DSB.** From the Kaggle 2018 Data Science Bowl challenge, we take the same images as used by [16]. The training and validation sets consist of 3800 and 670 patches respectively of size 128×128, while the test set counts 50 images of different sizes.
- **Mouse Nuclei.** Finally, we choose a challenging dataset depicting diverse and non-uniformly clustered nuclei in the mouse skull, consisting of 908 training and 160 validation patches of size 128×128. The test set counts 67 additional images of size 256×256.

For each dataset, we train DENOISEG and compare it to two different competing methods: DENOISEG trained purely for segmentation with $\alpha = 0$ (referred to as *Baseline*), and a sequential scheme based on [16] that first trains a denoiser and then the aforementioned baseline (referred to as *Sequential*). We chose our network with $\alpha = 0$ as baseline to mitigate the effect of batch normalization on the learning rate as described in Sect. 2. A comparison of our baseline to a vanilla 3-class U-NET with the same hyperparameters leads to very similar results and can be found in Appendix B. Furthermore, we investigate DENOISEG performance when trained with different amounts of available GT segmentation images. This is done by picking random subsets of various sizes from the available GT annotations. Note that the self-supervised denoising task still has access to all raw input images. A qualitative comparison of DENOISEG results with other baselines (see Fig. 2) indicates the effectiveness of our method.

As evaluation metrics, we use Average Precision (AP) [5] and SEG [25] scores. The AP metric measures both instance detection and segmentation accuracy while SEG captures the degree of overlap between instance segmentations and GT. To compute the scores, the predicted foreground channel is thresholded and connected components are interpreted as instance segmentations. The threshold values are optimized for each measure on the validation data. All conducted experiments were repeated 5 times and the mean scores along with ± 1 standard error of the mean are reported in Fig. 8.

Performance with Varying Quantities of GT Data and Noise. Fig. 3 shows the results of DENOISEG with $\alpha = 0.5$ (equally weighting denoising and segmentation losses) for Fly Wing n0, n10 and n20 datasets. For low numbers of GT training images, DENOISEG outperforms all other methods. Similar results are seen for the other two datasets (see Fig. 4 and Fig. 5). Note that, as the number of GT training images reaches full coverage, the performance of the baselines are similar to DenoiSeg. Appendix A shows this for all datasets. Results for all performed experiments showing overall similar trends can be found on the DenoiSeg-Wiki.

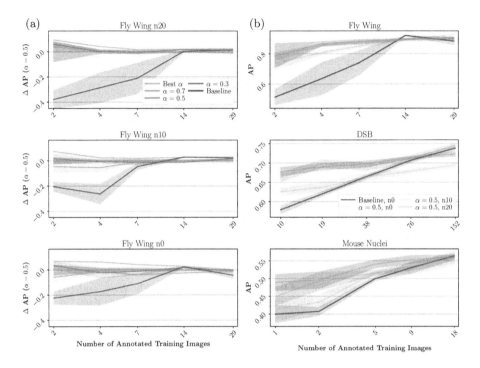

Fig. 6. In (**a**), we show that DenoiSeg consistently improves results over the baseline for a broad range of hyperparameter α values by looking at the difference Δ of AP to $\alpha = 0.5$. The results come close to what would be achievable by choosing the best possible α (see main text). In (**b**), we show that adding synthetic noise can lead to improved DenoiSeg performance. For the Fly Wing, DSB, and Mouse Nuclei data, we compare baseline results with DenoiSeg results on the same data (n0) and with added synthetic noise (n10 and n20, see main text).

Importance of α. We further investigated the sensitivity of our results to the hyperparameter α. In Fig. 6(a) we look at the segmentation performance for different values of hyperparameter α. We compare the results of $\alpha = 0.3$ and $\alpha = 0.7$ by computing the difference (Δ **AP**). Δ **AP** is the difference of the obtained AP score and the AP score obtained with the default $\alpha = 0.5$. Additionally we also compare to the Baseline and results that use (the a priori unknown) best α. The best α for each trained network is found by a grid search for $\alpha \in \{0.1, 0.2, \ldots, 0.9\}$. Figure 6(a) shows that our proposed method is robust with respect to the choice of α. Results for the other datasets showing similar trends are illustrated in Fig. 7.

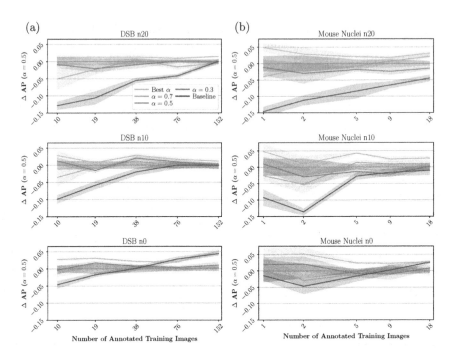

Fig. 7. We show the sensitivity of hyperparameter α for **(a)** Fly Wing and **(b)** Mouse Nuclei datasets by looking at the difference Δ of AP to $\alpha = 0.5$. Note that DENOISEG consistently improves results over the baseline for a broad range of hyperparameter α values.

Noisy Inputs Lead to Elevated Segmentation Performance. Here we want to elaborate on the interesting observation we made in Fig. 3: when additional noise is synthetically added to the raw data, the segmentation performance reaches higher AP and SEG scores, even though segmentation should be more difficult in the presence of noise. We investigate this phenomenon in Fig. 6(b). We believe that in the absence of noise the denoising task can be solved trivially, preventing the regularizing effect that allows DENOISEG to cope with small amounts of training data.

Evaluation of Denoising Performance. Although we are not training DENOISEG networks for their denoising capabilities, it is interesting to know how their denoising predictions compare to dedicated denoising networks. Table 1 compares our denoising results with results obtained by NOISE2VOID [10]. It can be seen that co-learning segmentation is only marginally impeding the network's ability to denoise its inputs.

Table 1. Comparing the denoising performance of DENOISEG and NOISE2VOID. Mean Peak Signal-to-Noise Ratio values (with ± 1 SEM over 5 runs) are shown. Similar tables for DENOISEG results when more segmentation GT was available can be found online in the DENOISEG-Wiki.

	DSB (GT for 10)		Fly Wing (GT for 2)		Mouse N. (GT for 1)	
Noise	DENOISEG	NOISE2VOID	DENOISEG	NOISE2VOID	DENOISEG	NOISE2VOID
n10	37.57 ± 0.07	38.01 ± 0.05	33.12 ± 0.01	33.16 ± 0.01	37.42 ± 0.10	37.86 ± 0.01
n20	35.38 ± 0.08	35.53 ± 0.02	30.45 ± 0.20	30.72 ± 0.01	34.21 ± 0.19	34.59 ± 0.01

4   Discussion

Here we have shown that (i) joint segmentation and self-supervised denoising leads to improved segmentation quality when only limited amounts of segmentation ground truth is available (Figs. 2, 3, 4 and 5), (ii) the hyperparameter α is modulating the quality of segmentation results but leads to similarly good solutions for a broad range of values (Figs. 6(a), 7), and (iii) results on input data that are subject to a certain amount of intrinsic or synthetically added noise lead to better segmentations than DENOISEG trained on essentially noise-free raw data (Fig. 6(b)).

We reason that the success of our proposed method originates from the fact that similar "skills" are required for denoising and segmentation. The segmentation task can profit from denoising, and compared to [16], performs even better when jointly trained within the same network. When a low number of annotated images are available, denoising is guiding the training and the features learned from this task, in turn, facilitate segmentation.

Since DENOISEG is crucially depending on NOISE2VOID, we also inherit the limited applicability from it. More concretely, DENOISEG will perform best when the noise in the input data is conditionally pixel-independent [10].

Users of DENOISEG should further be aware that the used segmentation labels need to sample all structures of interest. While we show that only a few labeled images can suffice, for more diverse structures and datasets it might be needed to choose the labeled images wisely (manually or even with automated approaches [29]).

We believe that DENOISEG offers a viable way to enable the learning of dense segmentations when only a very limited amount of segmentation labels are available, effectively making DENOISEG applicable in cases where other methods are not. We also show that the amount of required training data can be so little, even ad-hoc label generation by human users is a valid possibility, expanding the practical applicability of our proposed method manyfold.

Acknowledgment. The authors would like to acknowledge Romina Piscitello and Suzanne Eaton from MPI-CBG for fly wing data and Diana Afonso and Jacqueline Tabler from MPI-CBG for mouse nuclei data. We also acknowledge the Scientific Computing Facility at MPI-CBG for giving us access to their HPC cluster.

Appendices

A   DSB Results with Increasingly Many GT labels

The DSB dataset we have used offers the possibility to run DENOISEG with arbitrary many segmentation GT labels. While DENOISEG is intended in cases where the amount of such labels is very limited, in Fig. 8 we plot the segmentation results of DENOISEG, the sequential baseline, as well as the baseline as defined in the main text.

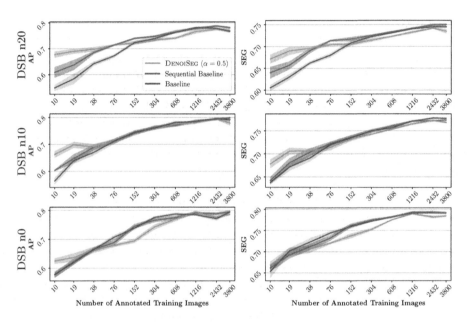

Fig. 8. Extended version of Fig. 4. Results for DSB n0, n10 and n20, evaluated with Average Precision (AP) [22] and SEG-Score [25]. DENOISEG outperforms both baseline methods, mainly when only limited segmentation ground truth is available. Note that the advantage of our proposed method for this dataset is at least partially compromised when the image data is not noisy (row 3).

As expected, With additional labels, the advantage of also seeing noisy images decreases, leading to similarly good results for all compared methods. It is still reassuring to see that the performance of DENOISEG is still essentially on par with the results of a vanilla U-NET that does not perform the joint training we propose.

B   Our Baseline *vs* Vanilla 3-class U-Net

The baseline method we used in this work is, as explained in the main text, a DENOISEG network with α being set to 0. This is, in fact, very similar to using a vanilla 3-class U-NET. While we are still feeding noisy images, we are not backpropagating any denoising loss, meaning that only the data for which segmentation labels exist will contribute to the training. The one difference is, that some of the hyperparameters (number of epochs, adaptation of learning rate, *etc.*) will slightly diverge in these two baseline setups. Figure 9 shows that these subtle differences are in fact not making any practical differences.

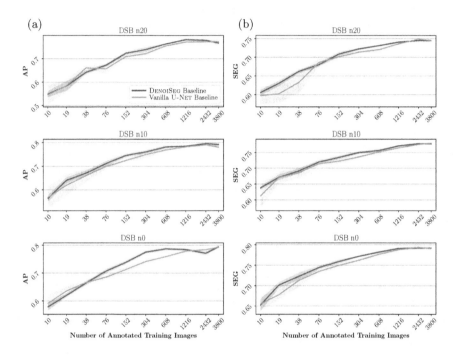

Fig. 9. Comparison of vanilla U-Net with our DENOISEG $\alpha = 0$ baseline for DSB datasets. Our DENOISEG $\alpha = 0$ baseline is at least as good or better than the vanilla U-net baseline both in terms of Average Precision (AP) [22] and SEG-Score [25] metrics. Hence, we establish a stronger baseline with DENOISEG $\alpha = 0$ and measure our performance against this baseline (see Fig. 8, Fig. 3 and Fig. 5.)

References

1. Batson, J., Royer, L.: Noise2Self: blind denoising by self-supervision (2019)

2. Buchholz, T.O., Jordan, M., Pigino, G., Jug, F.: Cryo-CARE: Content-aware image restoration for cryo-transmission electron microscopy data. In: 2019 IEEE 16th International Symposium on Biomedical Imaging (ISBI 2019), pp. 502–506. IEEE (2019)
3. Caicedo, J.C., et al.: Evaluation of deep learning strategies for nucleus segmentation in fluorescence images. Cytometry Part A **95**(9), 952–965 (2019)
4. Chen, H., Qi, X., Yu, L., Heng, P.A.: DCAN: deep contour-aware networks for accurate gland segmentation. In: Proceedings of the IEEE Conference on Computer Vision and Pattern Recognition, pp. 2487–2496 (2016)
5. Everingham, M., Van Gool, L., Williams, C.K., Winn, J., Zisserman, A.: The pascal visual object classes (VOC) challenge. Int. J. Comput. Vis. **88**(2), 303–338 (2010)
6. Guerrero-Pena, F.A., Fernandez, P.D.M., Ren, T.I., Yui, M., Rothenberg, E., Cunha, A.: Multiclass weighted loss for instance segmentation of cluttered cells. In: 2018 25th IEEE International Conference on Image Processing (ICIP), pp. 2451–2455. IEEE (2018)
7. Hirsch, P., Mais, L., Kainmueller, D.: Patchperpix for instance segmentation. arXiv preprint arXiv:2001.07626 (2020)
8. Ihle, S.J., et al.: Unsupervised data to content transformation with histogram-matching cycle-consistent generative adversarial networks. Nat. Mach. Intell. **1**(10), 461–470 (2019)
9. Jug, F., Pietzsch, T., Preibisch, S., Tomancak, P.: Bioimage informatics in the context of drosophila research. Methods **68**(1), 60–73 (2014)
10. Krull, A., Buchholz, T.O., Jug, F.: Noise2Void-learning denoising from single noisy images. In: Proceedings of the IEEE Conference on Computer Vision and Pattern Recognition, pp. 2129–2137 (2019)
11. Krull, A., Vicar, T., Jug, F.: Probabilistic Noise2Void: unsupervised content-aware denoising (2019). https://www.frontiersin.org/articles/10.3389/fcomp.2020.00005/full
12. Lehtinen, J., et al.: Noise2Noise: learning image restoration without clean data. arXiv preprint arXiv:1803.04189 (2018)
13. Liu, D., Wen, B., Liu, X., Wang, Z., Huang, T.S.: When image denoising meets high-level vision tasks: a deep learning approach. arXiv preprint arXiv:1706.04284 (2017)
14. Moen, E., Bannon, D., Kudo, T., Graf, W., Covert, M., Van Valen, D.: Deep learning for cellular image analysis. Nat. Methods **16**, 1–14 (2019)
15. Osokin, A., Chessel, A., Carazo Salas, R.E., Vaggi, F.: GANs for biological image synthesis. In: Proceedings of the IEEE International Conference on Computer Vision, pp. 2233–2242 (2017)
16. Prakash, M., Buchholz, T.O., Lalit, M., Tomancak, P., Jug, F., Krull, A.: Leveraging self-supervised denoising for image segmentation (2019)
17. Prakash, M., Lalit, M., Tomancak, P., Krull, A., Jug, F.: Fully unsupervised probabilistic Noise2Void (2019). https://ieeexplore.ieee.org/document/9098612
18. Razzak, M.I., Naz, S., Zaib, A.: Deep learning for medical image processing: overview, challenges and the future. In: Dey, N., Ashour, A.S., Borra, S. (eds.) Classification in BioApps. LNCVB, vol. 26, pp. 323–350. Springer, Cham (2018). https://doi.org/10.1007/978-3-319-65981-7_12
19. Ronneberger, O., Fischer, P., Brox, T.: U-Net: convolutional networks for biomedical image segmentation. In: Navab, N., Hornegger, J., Wells, W.M., Frangi, A.F. (eds.) MICCAI 2015. LNCS, vol. 9351, pp. 234–241. Springer, Cham (2015). https://doi.org/10.1007/978-3-319-24574-4_28

20. Sandfort, V., Yan, K., Pickhardt, P.J., Summers, R.M.: Data augmentation using generative adversarial networks (cyclegan) to improve generalizability in ct segmentation tasks. Sci. rep. **9**(1), 1–9 (2019)
21. Schindelin, J., et al.: Fiji: an open-source platform for biological-image analysis. Nat. Methods **9**(7), 676–682 (2012)
22. Schmidt, U., Weigert, M., Broaddus, C., Myers, G.: Cell detection with star-convex polygons. In: Frangi, A.F., Schnabel, J.A., Davatzikos, C., Alberola-López, C., Fichtinger, G. (eds.) MICCAI 2018. LNCS, vol. 11071, pp. 265–273. Springer, Cham (2018). https://doi.org/10.1007/978-3-030-00934-2_30
23. Shorten, C., Khoshgoftaar, T.M.: A survey on image data augmentation for deep learning. J. Big Data **6**(1), 60 (2019)
24. Stringer, C., Michaelos, M., Pachitariu, M.: Cellpose: a generalist algorithm for cellular segmentation. bioRxiv (2020)
25. Ulman, V., et al.: An objective comparison of cell-tracking algorithms. Nat. Methods **14**(12), 1141 (2017)
26. Wang, S., Wen, B., Wu, J., Tao, D., Wang, Z.: Segmentation-aware image denoising without knowing true segmentation. arXiv preprint arXiv:1905.08965 (2019)
27. Weigert, M., et al.: Content-aware image restoration: pushing the limits of fluorescence microscopy. Nat. Methods **15**(12), 1090–1097 (2018)
28. Zhao, A., Balakrishnan, G., Durand, F., Guttag, J.V., Dalca, A.V.: Data augmentation using learned transformations for one-shot medical image segmentation. In: Proceedings of the IEEE Conference on Computer Vision and Pattern Recognition, pp. 8543–8553 (2019)
29. Zheng, H., et al.: Biomedical image segmentation via representative annotation. In: Proceedings of the AAAI Conference on Artificial Intelligence, vol. 33, pp. 5901–5908 (2019)

DoubleU-Net: Colorectal Cancer Diagnosis and Gland Instance Segmentation with Text-Guided Feature Control

Pei Wang$^{(\boxtimes)}$(iD) and Albert C. S. Chung(iD)

The Hong Kong University of Science and Technology, Kowloon, Hong Kong SAR
{pwangai,achung}@ust.hk

Abstract. With the rapid therapeutic advancement in personalized medicine, the role of pathologists for colorectal cancer has greatly expanded from morphologists to clinical consultants. In addition to cancer diagnosis, pathologists are responsible for multiple assessments based on glandular morphology statistics, like selecting appropriate tissue sections for mutation analysis [6]. Therefore, we propose DoubleU-Net that determines the initial gland segmentation and diagnoses the histologic grades simultaneously, and then incorporates the diagnosis text data to produce more accurate final segmentation. Our DoubleU-Net shows three advantages: (1) Besides the initial segmentation, it offers histologic grade diagnosis and enhanced segmentation for full-scale assistance. (2) The textual features extracted from diagnosis data provide high-level guidance related to gland morphology, and boost the performance of challenging cases with seriously deformed glands. (3) It can be extended to segmentation tasks with text data like key clinical phrases or pathology descriptions. The model is evaluated on two public colon gland datasets and achieves state-of-the-art performance.

Keywords: Cancer diagnosis · Gland segmentation · Multi-domain learning · Morphological feature guidance

1 Introduction

Colorectal cancer is among the leading causes of mortality and morbidity in the world. It is the third most common cancer worldwide (following tumors of the lung and breast), and the fourth most common cause of oncological death [22]. More than 90% of the colorectal cancers are adenocarcinomas, which are malignant tumors originating from glandular epithelium. It is determined by pathologists on Hematoxylin and Eosin (H&E) stained tissue specimens. The morphological information of intestinal glands, like architectural appearance and gland formation, is one of the primary features to inform prognosis and plan the treatment [3]. Therefore, the automated segmentation methods that extract

© Springer Nature Switzerland AG 2020
A. Bartoli and A. Fusiello (Eds.): ECCV 2020 Workshops, LNCS 12535, pp. 338–354, 2020.
https://doi.org/10.1007/978-3-030-66415-2_22

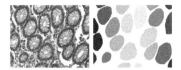 

Patient 15: benign, healthy   Patient 7: malignant, moderately differentiated

Fig. 1. Examples of gland instance segmentation with different histologic grades (i.e. benign or malignant) and differentiation levels. Malignant cases usually show great morphological changes. Glands are denoted by different colors. (Color figure online)

quantitative features with morphological statistics are essential in clinical practice to boost assessing efficiency and reliability, reducing inter- and intra-observer variability, and handling the ever-increasing image quantity and variety.

While previous approaches to address this problem focus on hand-crafted features and prior knowledges of objects [1,5,7,9,14,19–21], recent convolutional neural networks have promoted this area by learning semantic features [4,17,24–26]. Besides glandular objects, the most advanced approaches focus more on capturing the boundary of the gland with different network architecture or loss function [8,25,26,29,30]. These methods show advancement in identifying clustered and touching gland objects through detailed features, but fail to cope with the morphological variance. The glandular morphology shows great diversity for different histologic grades and differentiation levels, as shown in Fig. 1. The significant deformation in architectural appearance and glandular formation could undermine the robustness of methods that focus on detailed local features.

Moreover, the morphological statistics of glandular objects play an increasingly important role in clinical practice for colorectal cancer treatment, which raises higher requirements to segmentation accuracy. With the rapid development in personalized medicine, the role of pathologists has greatly expanded from traditional morphologists to clinical consultants (for gastroenterologists, colorectal surgeons, oncologists, and medical geneticists) [6]. Therefore, besides providing accurate histopathologic diagnosis as a very first step, pathologists are responsible for accurately assessing pathologic staging, analyzing surgical margins, selecting appropriate tissue section for microsatellite instability (MSI) testing and mutation analysis, searching for prognostic parameters, and assessing therapeutic effect, with the aid of quantitative features of gland [6].

To address these challenges, we propose DoubleU-Net that diagnose colorectal cancer and segment gland instance simultaneously, and utilize the diagnosis text which emphasizes the high-level features and overall structural information for more accurate gland instance segmentation. Our DoubleU-Net achieves the state-of-the-art performances in segmentation on public dataset GlaS and CRAG. Our major contributions of DoubleU-Net are listed as follows.

1. Besides the initial gland segmentation, we offer cancer diagnosis and greatly improved segmentation results as full-scale assistance for pathologists according to the clinical routine.

2. It emphasizes high-level features including the structural and morphological appearance of objects, and the significant improvement on shape similarity validates the effectiveness of textual feature guidance.
3. DoubleU-Net can well incorporate text and image from different domains for medical image segmentation. The extracted textual features are applied directly to image other than the text related tasks (e.g. label/report generation). It is also designed in a generalized way for segmentation with key clinical phrases or pathology descriptions as supplementary text input.

2   Related Work

Gland Instance Segmentation. In the last few years, various methods have been proposed for gland segmentation. Pixel-based methods [5,14,19,21] and structure-based methods [1,7,9,20] make full use of the hand-crafted features and prior knowledge of glandular structures. These methods achieved satisfying results on benign objects, but not for the adenoma cases with diversity in size and appearance. Recently, deep learning methods have shown remarkable performance for histological image analysis. The popular U-Net [17] is a U-shaped network with a symmetric encoder and decoder branch. Chen et al. proposed DCAN [4] that focuses on both glandular object and boundary to separate clustered glands, and won the 2015 MICCAI GlaS challenge [18]. Based on DCAN, MILD-Net [8] introduced a loss function to retain maximal information during feature extraction in training. Xu et al. [25,26] applied a multi-channel multi-task network for foreground segmentation, edge detection, and object detection respectively. Besides, Yan et al. [29] proposed a shape-preserving loss function to regularize the glandular boundary for the gland instance. Furthermore, several methods aim at utilizing less manual annotations or computational expenses for gland instance segmentation. For example, Yang et al. [30] presented a deep active learning approach using suggestive annotation, and Zheng et al. [33] proposed a representative annotation (RA) framework. Quantization performed on FCNs [10] reduces computational requirement and overfitting while maintaining segmentation accuracy [24]. Unannotated image data can be utilized effectively by the proposed deep adversarial network [32] for considerably better results.

Text Related Image Processing. There exist close relationships between language and visual attention in psychology according to recent studies [12]. This suggests that spoken or text language associated with a scene provides useful information for visual attention in the scene [13]. Mu et al. [13] introduce a text-guided model for image captioning, which learns visual attention from associated exemplar captions for a given image, referred to as guidance captions, and enables to generate proper and fine-grained captions for the image. TieNet [23] incorporates chest X-ray image with the associated medical report for disease classification and reporting. MULAN [28] is a multi-task model for lesion detection, tagging, and segmentation. To mine training labels, Yan tokenizes the sentences in the radiological report and then match and filter the tags

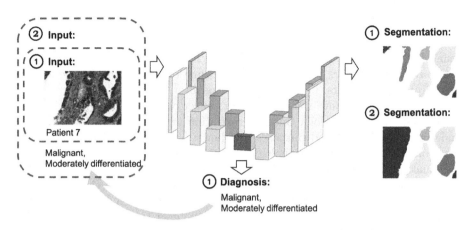

Fig. 2. Pipeline of the DoubleU-Net: (1) Colorectal cancer diagnosis and initial gland segmentation are predicted, and then (2) the diagnosis text data is fed into the text-to-feature encoder of DoubleU-Net, which provides morphological feature guidance and yields largely improved segmentation results.

in the sentences using a text mining module. Similarly, the training labels are text-mined from radiology reports and further utilized in the training of lesion annotation in LesaNet [27].

It is worth noticing that the text data in these approaches are not directly applied to images, they are processed for image captioning, tag extraction, and label selection, which are related to text output or label generation. In this paper, it is novel that the features from diagnosis text data directly control the high-level visual features and the gland segmentation, which serve as the guidance on gland morphology in the learning process.

3   Method

Our proposed DoubleU-Net consists of two encoders for feature extraction of histological images and the corresponding text information, and two decoders [4] for pixel-wise prediction of glandular region and boundary. The architecture is similar to the shape of two Us connected at the bottom, therefore we name it DoubleU-Net. It works in the following two steps: (1) diagnose histologic grades and differentiation levels while performing initial gland segmentation, (2) the diagnosis data is fed into a text-to-feature encoder for higher level feature guidance, and the final gland instance segmentation results are obtained based on the features from different information domain.

3.1   Colorectal Cancer Grading

In clinical practice, the pathologists initially determine the histologic grade and differentiation level based on the glandular morphology. Similarly, DoubleU-Net performs colorectal cancer diagnosis by utilizing the visual features, and the initial segmentation results are provided as well. In Fig. 3 the grading classifier is connected to the bottom of DoubleU-Net, which consists of average pooling layers, convolutional layers, and a linear transformation layer. Besides, the feature maps from two encoders (i.e. all features maps after the max-pooling layers) are concatenated and fed into the classifier for cancer diagnosis.

The automated histologic and differentiation grading in model achieves its underlying advantages and purposes as follows: (1) Directly offer final diagnosis results with glandular morphology statistics to pathologists as reference and assistance for cancer grading. (2) The joint supervision of prediction and segmentation increases the learning ability of the model and alleviate the over-fitting during training. (3) The predicted diagnosis data (or revised data by pathologists, if possible) can further control the abstract higher-level features to improve segmentation accuracy for other treatments by pathologists, like selecting appropriate tissue sections for MSI testing and mutation analysis [6].

Besides, the gland segmentation is usually performed on image patches extracted from the original digitalized H&E stain slides (up to 10000^2 pixels) to focus on the region of interests, and current datasets also contain image patches instead of the whole slides. Therefore, the model is required to maintain classification consistency for different images from the same patient, which is also crucial in clinical routine to perform automated segmentation and prediction on image level. Instead of appending an algorithm to unify the output grades, the patient ID of corresponding histological images can be further utilized. With patient ID as an additional input, the model is able to learn the similarity among images from the same patient, and gradually exploit the interdependency among glandular features, diagnosis grades, and patients. As shown in Table 3, text encoder with additional input significantly promotes the diagnosis accuracy in GlaS dataset. Then the predicted cancer grades and patient ID together served as text input to further refine the gland segmentation, and the grade classifier is deactivated in the remaining process.

3.2   Text-Guided Feature Control

The colorectal cancer is diagnosed based on the morphological statistics of colon gland instance by pathologists, and DoubleU-Net also predicts the cancer grades based on the visual features of glands. Since the diagnosis data like *moderately differentiated* suggests changes in the gland structure and appearance, we could further utilize these diagnosis data for more accurate segmentation.

Word Embedding. Diagnosis text data is fed into the network as an additional input to dominate high-level features from corresponding histological images, and word embedding is the very first step to represent these words as low dimensional vectors. Given limited text information for each image, the simple one-hot encoding is a straightforward option. However, it has the following limitations: (1) Most values of the one-hot encoding will be zero, which is inefficient and fails

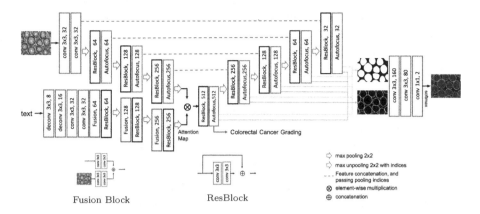

Fig. 3. Overall architecture of DoubleU-Net. It consists of two encoders for feature extraction of histological images and corresponding text information, and two decoders [4] for pixel-wise prediction of glandular region and boundary.

to boost the learning ability of the network for text data. (2) It is not sufficient enough to reveal the semantic relationship between different phrases, because the vectors obtained by one-hot encoding are orthogonal to each other. This independent representation is suitable for histologic grade *benign* and *malignant* that have opposite meanings, but it fails to demonstrate the similarity among *benign*, *moderately differentiated* and *moderately-to-poorly differentiated*. (3) Furthermore, word embedding is a more generalized approach that can be used for a different and larger amount of text data like pathology descriptions of histological slides.

To fully capture the semantic, syntactic similarity of the word, and relation with other words, we adopt popular word2vec [11] from Google for distributed representation. This model is trained on a huge text corpus (one billion words) to construct vocabulary and learns the vector representation of words. Besides the excellent representation capability, the other advantage of word2vec is that the encoded vectors can be meaningfully combined using simple vector summation. We now formulate the word embedding of clinical data as follows:

$$v_i = \sum_j \mathcal{G}(m_{ij}; \boldsymbol{\theta}), \tag{1}$$

$$v = v_0 \parallel v_1 \parallel v_2 \parallel ... \parallel v_n, \tag{2}$$

where $\mathcal{G}$ denotes the word2vec model with parameter $\boldsymbol{\theta}$. m_{ij} represents the jth word in ith sentence (or phrase in our case). v_i is the corresponding vector representation of the phrase (summation of word vector representation), and all n vectors are concatenated (denoted by $\parallel$) to form the textual feature v. In Table 1, we show the cosine distance of vectors encoded by word2vec, where histologic and differentiation grades are from several different histology images.

Table 1. Semantic similarity of encoded diagnosis text from GlaS dataset.

Semantic similarity	Healthy	Adenomatous	Moderately differentiated	Moderately-to-poorly differentiated
Benign	0.7113	0.5578	0.5341	0.5273

Visual and Textual Feature Fusion. For each histological image, textual features are extracted from the corresponding encoded text data. To fully utilize the high-level clinical information for dense pixel-wise prediction, it is necessary to well incorporate the textual features with the visual ones. Therefore, the original input image is concatenated with the textual features after max-pooling and convolutional layer, as formulated in the equation below:

$$y = \mathcal{F}(v; \theta_\alpha) \parallel \mathcal{F}(x; \theta_\beta), \tag{3}$$

where $\mathcal{F}$ denotes the convolution with corresponding weights θ. x represents the down-sample original input image, and v is the encoded text feature.

We explain the fusion block from the following three aspects. First, image and text data are from different information domains and deliver features of different levels. Without thorough incorporation, text data contains no object localization could fail to guide the visual features on the pixel level. Our multiple fusion blocks adaptively combine spatial features with appearance features from clinical text information, where local details and overall glandular morphology are gradually fused and balanced. Second, the feature integration strengthens the interdependency of glandular morphology and histologic grades, and improve the capability of the network to learn and distinguish the visual features for different cancer grades. Furthermore, combining feature maps and original input images, the network preserves detailed information that may be lost during the feature extraction.

Feature Guidance and Attention. The high-level features extracted from images are closely related to the glandular structure and morphology, which can be considered as a response to the histologic grades. On the other hand, the clinical diagnosis is also determined based on the architectural appearance and glandular formation. Therefore, we guide the extracted features to accurately focus on the gland morphology and structure during training, with the aid of features from known clinical text data. The emphasis on global structures and shapes enables the network to distinguish the gland instance from different cancer differentiation grades.

To provide structural information for the local feature extraction process, we employ feature maps from the text encoder as attention maps. Given a feature $u \in \mathbb{R}^{C \times H \times W}$ from the text encoder, we perform two-dimensional softmax to the each channel of the feature maps, and calculate the attention weight for each spatial location $(channels, h, w)$. Therefore, we are able to control the local feature with spatial attention map by element-wise multiplication. The feature guidance by attention unit can be formulated as below:

$$y_{k,l} = \mathcal{F}(x_{k,l}a_{k,l}; \boldsymbol{\theta}), \tag{4}$$

$$a_{k,l} = \frac{exp(u_{k,l}/\tau)}{\sum_{k=1}^{H}\sum_{l=1}^{W} exp(u_{k,l}/\tau)}, \tag{5}$$

where $\boldsymbol{x} = \{x_{k,l}\}$ denotes feature from the image encoder, and $\boldsymbol{u} = \{u_{l,l}\}$ from the text branch. $\boldsymbol{a} = \{a_{k,l}\}$ represents corresponding attention probability distribution over each channel of feature map, and followed by the convolutional layer $\mathcal{F}$ with weight $\boldsymbol{\theta}$. τ is a temperature parameter. For high temperatures ($\tau \to \infty$), all actions have nearly the same probability. For lower temperatures ($\tau \to 0^+$), the probability of the action with the highest expected reward tends to 1, which may cause gradient issues in training.

As shown in Table 4, the increasing amount of text input effectively promotes the gland segmentation performance and especially the object-level Hausdorff distance, which validates that the clinical text data controls and emphasizes the glandular structure and morphology.

3.3   Loss

The task of our model varies with the two training phases of DoubleU-Net. Initially, it performs gland instance segmentation and cancer grade classification simultaneously with total loss $\mathcal{L}_{seg} + \mathcal{L}_{grading}$. Secondly, we improve the segmentation task with the incorporation of clinical text input, with loss $\mathcal{L}_{seg}$ only. The loss function of cancer grading (denoted as g) is an image level cross-entropy:

$$\mathcal{L}_{grading} = -\sum_{i \in \mathcal{I}} \sum_{m \in \mathcal{M}} \lambda_g \log p_{g,m}(i, y_{g,m}(i); \boldsymbol{\theta}_g), \tag{6}$$

where λ_g is the weight of task g to the total loss. For any input image i in the dataset $\mathcal{I}$, $p_{g,m}(i, y_{g,m}(i))$ represents the image-based softmax classification of true labels $y_{g,c}(i)$ for class m in $\mathcal{M}$. $\boldsymbol{\theta}_g$ is the weight parameters of grading task.

As for segmentation, two decoders are applied for the glandular region and boundary independently order to separate the clustered glands. Unlike DCAN [4] that merged the final output of gland region and boundary manually, we integrate the combination step into the network by performing additional convolutional layers and softmax function. The total loss function for segmentation has three cross-entropy for different sub-tasks: gland foreground, gland boundary, and the final combined segmentation task (denoted as f, b, c respectively).

The loss function is written as:

$$\mathcal{L}_{seg} = \sum_{k \in \{f,b,c\}} \lambda_k \boldsymbol{w} \cdot \mathcal{L}_k \tag{7}$$

$$= - \sum_{k \in \{f,b,c\}} \sum_{x \in \mathcal{X}} \lambda_k w(x) \log p_k(x, y_k(x); \boldsymbol{\theta}_k) \tag{8}$$

$$w(x) = w_0 \cdot \frac{1}{max(dist(x), d_0) + \mu}, \tag{9}$$

where $\mathcal{L}_k$, $\boldsymbol{\theta}_k$ and λ_k are the loss function, corresponding weight parameters, and coefficient of task k. $p_k(x, y_k(x); \boldsymbol{\theta}_k)$ represents the pixel-based softmax classification at task k for true labels $y_k(x)$. x denotes a given input pixel in image space $\mathcal{X}$. To better identify the clustered gland instance, a weight map $\boldsymbol{w}$ is constructed in pixel-wise fashion and performed to emphasize the boundary in the segmentation task. In Eq. 9, given the input pixel x, $dist(x)$ is the Euclidean distance from x to the nearest gland. d_0 is the maximum distance that a pixel within d_0 range of the boundary will be considered. In our experiments, w_0 is set to 3, d_0 to 15 and μ equals to 1. The loss coefficient is [1, 1, 1.5, 2] for task g, f, b, c respectively.

Additionally, to address the variation of gland size and obtain effective receptive fields, we adopt Autofocus block [16] in our model. It consists of parallel dilated convolutional [31] branches with different rates and combined with learnable weights. In our experiments, we implement 4 branches with rates 2, 6, 10, and 14 respectively.

4   Experiments

We evaluated DoubleU-Net on two publicly available datasets of colon histological images: the Gland Segmentation (GlaS) dataset [18] from MICCAI challenge, and an independent colorectal adenocarcinoma gland (CRAG) dataset [8] originally used by Awan et al. [2]. Both datasets are from the University Hospitals Coventry and Warwickshire (UHCW) NHS Trust in Coventry, United Kingdom.

Datasets and Pre-processing. GlaS dataset is composed of 165 histological images from the H&E stained slides of 16 patients with a wide range of cancer grades (2 histologic grades, 5 differentiation levels). It consists of 85 training (37 benign (BN) and 48 malignant (MT) from 15 patients) and 80 test images (33 BN and 27 MT in Part A, 4 BN and 16 MT in Part B, from 12 patients). Most images are of size 775×522, and all the histological images are associated with instance-level annotation, corresponding patient ID, histologic grade and differentiation level. In CRAG dataset, 213 H&E images from different cancer grades are split into 173 training and 40 testing images. All images are associated with instance-level annotation and are mostly of size 1512×1516, and no text information is provided. For both datasets, we split 20% of the training images for validation during training to adjust the hyperparameters. According

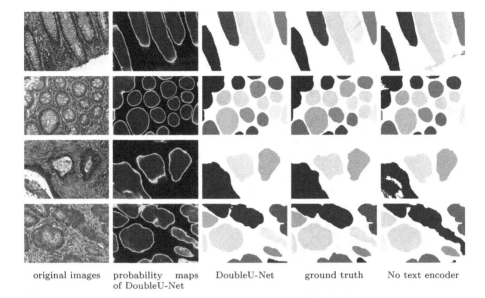

original images   probability   maps   DoubleU-Net   ground truth   No text encoder
of DoubleU-Net

Fig. 4. Segmentation examples of benign (top two cases) and malignant cases (bottom two cases) on GlaS dataset, and gland morphology is well recognized.

Table 2. Performance on GlaS dataset in comparison with other methods.

Method	F1 Score		Object dice		Object hausdorff	
	Part A	Part B	Part A	Part B	Part A	Part B
DoubleU-Net	**0.935**	**0.871**	**0.929**	**0.875**	**27.835**	**76.045**
Mild-Net [8]	0.914	0.844	0.913	0.836	41.540	105.890
Quantization [24]	0.930	0.862	0.914	0.859	41.783	97.390
Shape loss [29]	0.924	0.844	0.902	0.840	49.881	106.075
Suggestive annotation [30]	0.921	0.855	0.904	0.858	44.736	96.976
Multichannel [26]	0.893	0.843	0.908	0.833	44.129	116.821
DCAN [4]	0.912	0.716	0.897	0.781	45.418	160.347

to Graham et al. [8], both training and testing images are from different cancer grades. In our experiments, we augmented the images on the fly by performing elastic transformation, random rotation, random flip, Gaussian blur and color distortion. Eventually, we randomly cropped patches of size 480×480 as input.

Evaluation Criteria. We evaluated the performance of DoubleU-Net by the metrics used in MICCAI GlaS challenge from different aspects: F1 score for object detection, *object-level* Dice index for instance segmentation, and *object-level* Hausdorff distance for glandular shape similarity. All these evaluation metrics are conducted on gland instance instead of image level. For example, the

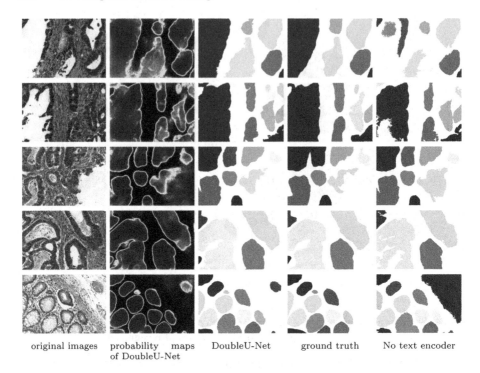

original images   probability   maps   DoubleU-Net   ground truth   No text encoder
                   of DoubleU-Net

Fig. 5. Segmentation results of DoubleU-Net on challenging cases with major improvement on GlaS dataset. DoubleU-Net with text encoder captures the structure and shape of seriously deformed glands and recognizes the misleading background. Top four rows: malignant. Bottom row: benign.

object-level Dice index is a weighted summation of the Dice index for all the glandular objects in this image, where the weights are determined by the relative area of the glands. Equations and details can be found in the challenge paper [18].

Implementation Details. Our method is implemented with PyTorch [15]. We adopt Gaussian distribution ($\mu = 0, \sigma = 0.01$) for weight initialization and train with Adam optimization of initial learning rate 8×10^{-4}, with batch size of 2. We choose skip-gram architecture for the word2vec with hierarchical softmax training approach. The dimensionality of each word vector is 300 and then reshaped as 15×20. For cancer grading, our model infers the histologic grade based on the predicted differentiation levels, because each differentiation level is mapped to exactly one histologic grade. There is no text data in CRAG dataset (and there is no other public gland dataset with text, to the best of our knowledge), we diagnose colorectal cancer by the text encoder trained on GlaS dataset and then utilize the data for gland instance segmentation. The model is trained and tested on an NVIDIA Titan X Pascal for both datasets.

Table 3. Accuracy of colorectal cancer diagnosis on GlaS dataset.

Method	Histologic grade	Differentiation level
DoubleU-Net	**1.000**	**0.925**
Without text encoder	0.950	0.725

Table 4. Ablation study of text encoder on GlaS dataset.

Method	F1 Score		Object dice		Object hausdorff	
	Part A	Part B	Part A	Part B	Part A	Part B
Without text encoder	0.926	0.857	0.916	0.859	41.454	96.473
Text encoder + ID	0.926	0.859	0.918	0.858	41.198	96.074
Text encoder + grade 1	0.928	0.866	0.922	0.865	34.456	86.325
Text encoder + grade 2	0.930	0.862	0.924	0.864	32.645	82.654
Text encoder + ID & grade 1	0.928	0.868	0.925	0.866	33.325	85.435
Text encoder + ID & grade 2	**0.935**	0.863	0.925	0.869	31.754	81.446
Text encoder + grade 1 & 2	0.934	0.867	**0.929**	0.871	28.723	78.943
DoubleU-Net	**0.935**	**0.871**	**0.929**	**0.875**	**27.835**	**76.045**

4.1   Results

Our DoubleU-Net achieves the best performances compared to the state-of-the-art methods on two public datasets. The morphological feature guidance from text encoder largely promotes the overall cancer diagnosis accuracy and segmentation results for seriously deformed cases.

GlaS Dataset. Table 2 shows the segmentation performances on GlaS dataset in comparison with the state-of-the-art methods. Among these approaches, the quantization model [24] and Mild-Net [8] currently achieve the best results, and DCAN [4] won the MICCAI GlaS challenge 2015. Other methods focus on different aspects of gland segmentation as reviewed previously [26,29,30]. As shown in Table 2, our DoubleU-Net achieves the *best* results on F1, the Dice index Hausdorff distance on object-level. More importantly, DoubleU-Net outperforms other approaches by a large margin in glandular shape similarity, which validates the effectiveness of high-level feature guidance on glandular morphology and structure by abstract information from text domain. Figure 4 shows exemplary visual results from DoubleU-Net on the GlaS test images for benign and malignant cases, where glands with various structures, shapes and texture can be well-identified. The corresponding probability maps with clear glandular boundaries indicate the local details are well preserved while the high-level features are emphasized.

Table 5. Performance on CRAG dataset in comparison with other methods.

Method	F1 Score	Object dice	Object hausdorff
DoubleU-Net	**0.835**	**0.890**	**117.25**
Mild-Net [8]	0.825	0.875	160.14
DCAN [4]	0.736	0.794	218.76

Seriously Deformed Cases. Figure 5 shows the challenging cases that are seriously deformed. The center of the deformed malignant glands is very similar to the white background in other cases. Our model without text encoder has failed to identify these white areas, and some papers presented similar results and even list these as failure cases [26,29]. The possible reason is that the network fails to balance the local details and high-level features from a wider contextual range, and focuses more on the boundary than the overall gland structure. Our DoubleU-Net emphasizes exactly on the glandular morphology and appearance, and successfully identify these misleading areas and deformed glands with high confidence as shown in the probability maps. Besides, based on our analysis of the experimental results, the success on extremely complex cases (like Fig. 5) contribute a major improvement to the overall results of DoubleU-Net, especially the object-level Hausdorff distance.

Colorectal Cancer Grading. Table 3 presents classification results on histologic grade and differentiation level on GlaS dataset. With the text encoder and additional patient ID, the model achieves 100% accuracy on the histologic grade and 92.5% on the differentiation level. Therefore, DoubleU-Net is aware of the interdependencies among images from the same histological slide and maintain classification consistency for images from the same patient. Besides, due to the number of classes, it is more challenging for both methods to identify the differentiation level than the binary histologic grade.

Ablation Study. Table 4 shows the effectiveness of text encoder on improving the gland segmentation on GlaS dataset. Based on the network architecture of Double-Net, we gradually feed more text data and evaluate the performances. In Table 4, ID denotes the 16 patient IDs (i.e. 1 to 16) of histological slides; grade 1 means the 2 histologic grades (i.e. *benign* and *malignant*); grade 2 represents the 5 differentiation levels (i.e. *healthy, adenomatous, moderately differentiated, moderately-to-poorly differentiated, and poorly differentiated*). Comparing to DoubleU-Net without text encoder, the attention maps from the textual features significantly boosts the performance in all three evaluation metrics. We find out the patient ID brings more improvement on the classification task (Table 3) over segmentation, which by maintaining consistency and building interdependency for the images of the same slide. Both two diagnosis grades associated with gland appearance promote the segmentation results evidently, and especially on

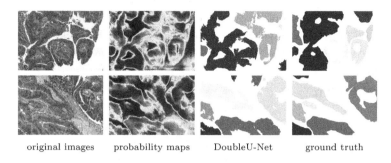

original images   probability maps   DoubleU-Net   ground truth

Fig. 6. Failure segmentation results by DoubleU-Net on GlaS dataset.

Hausdorff distance that measures the shape similarity. The slight advantage of grade 2 over grade 1 is probably because of the more detailed classification criteria brings more information into appearance features. With text encoder and all text data involved, we achieve the best gland segmentation performance with remarkable improvement.

GRAG Dataset. Table 5 shows the segmentation performance on CRAG dataset. The dataset is recently released by Graham et al. [8], and we report their segmentation results together with DCAN [4]. Similarly, DoubleU-Net achieves the best results on the F1 score and object-level Dice index, and a major advancement in Hausdorff distance because of the control on the overall glandular structure. As mentioned earlier, there is no text description or diagnosis information in CRAG dataset, and we use the text encoder trained on GlaS dataset to determine the histologic grades and then utilize them as feature guidance. Despite all this, DoubleU-Net still manages to promote the glandular shape similarity.

In Fig. 6 we also investigate some poorly segmented cases by DoubleU-Net. Besides the extreme complexity of these gland structures, there are limited similar training samples in the dataset, which could be the major reason for the failure. In addition, our DoubleU-Net outputs some unclear or rough glandular boundaries, and some connecting glands for some challenging cases, appropriate post-processing or regularization on the edges can be further considered.

Discussions. The pathologists diagnose cancer based on the morphology of the gland. Since the diagnosis data like *moderately differentiated* suggest changes in the gland structure and appearance, we could further utilize the diagnosis data for more accurate segmentation. The diagnosis accuracy and segmentation performance validate the effectiveness of textual features. Without given clinical text data, our model also demonstrates its robustness by enhancing the segmentation performance based on predicted cancer grades in CRAG dataset. This work is an attempt to incorporate information from different domains for segmentation task (current methods utilize text for label/report generation that is not directly applied to images), and achieves the best performances because of the significant improvement in very complicated and misleading cases. We hope to see more approaches that make sufficient use of clinical text data or even other types of information for accurate medical image segmentation in future.

5   Conclusion

We establish a pipeline to offer the pathologists initial segmentation statistics, histologic grades, differentiation levels, and greatly improved segmentation results for full-scale assistance in the clinical assessment routine of colorectal cancer. Applying features extracted from diagnosis text data to visual cues directly, our proposed DoubleU-Net effectively guides and controls the extracted high-level features to the precise gland structure and morphology against the histological changes. We achieve the best segmentation performances among the state-of-the-art methods on the two publicly available colon gland datasets, with a major improvement on the extremely deformed glandular cases.

References

1. Altunbay, D., Cigir, C., Sokmensuer, C., Gunduz-Demir, C.: Color graphs for automated cancer diagnosis and grading. IEEE Trans. Biomed. Eng. **57**(3), 665–674 (2009)
2. Awan, R., et al.: Glandular morphometrics for objective grading of colorectal adenocarcinoma histology images. Sci. rep. **7**(1), 16852 (2017)
3. Bosman, F.T., Carneiro, F., Hruban, R.H., Theise, N.D., et al.: WHO classification of tumours of the digestive system. edn. 4. World Health Organization (2010)
4. Chen, H., Qi, X., Yu, L., Heng, P.A.: DCAN: deep contour-aware networks for accurate gland segmentation. In: Proceedings of the IEEE Conference on Computer Vision and Pattern Recognition, pp. 2487–2496 (2016)
5. Doyle, S., Madabhushi, A., Feldman, M., Tomaszeweski, J.: A boosting cascade for automated detection of prostate cancer from digitized histology. In: Larsen, R., Nielsen, M., Sporring, J. (eds.) MICCAI 2006. LNCS, vol. 4191, pp. 504–511. Springer, Heidelberg (2006). https://doi.org/10.1007/11866763_62
6. Fleming, M., Ravula, S., Tatishchev, S.F., Wang, H.L.: Colorectal carcinoma: pathologic aspects. J. Gastrointest. Oncol. **3**(3), 153 (2012)
7. Fu, H., Qiu, G., Shu, J., Ilyas, M.: A novel polar space random field model for the detection of glandular structures. IEEE Trans. Med. Imaging **33**(3), 764–776 (2014)
8. Graham, S., et al.: MILD-net: minimal information loss dilated network for gland instance segmentation in colon histology images. Med. Image Anal. **52**, 199–211 (2019)
9. Gunduz-Demir, C., Kandemir, M., Tosun, A.B., Sokmensuer, C.: Automatic segmentation of colon glands using object-graphs. Med. Image Anal. **14**(1), 1–12 (2010)
10. Long, J., Shelhamer, E., Darrell, T.: Fully convolutional networks for semantic segmentation. In: Proceedings of the IEEE Conference on Computer Vision and Pattern Recognition, pp. 3431–3440 (2015)

11. Mikolov, T., Sutskever, I., Chen, K., Corrado, G.S., Dean, J.: Distributed representations of words and phrases and their compositionality. In: Advances in Neural Information Processing Systems, pp. 3111–3119 (2013)
12. Mishra, R.K.: Interaction Between Attention and Language Systems in Humans. Springer, New Delhi (2015). https://doi.org/10.1007/978-81-322-2592-8
13. Mun, J., Cho, M., Han, B.: Text-guided attention model for image captioning. In: Thirty-First AAAI Conference on Artificial Intelligence (2017)
14. Nguyen, K., Sarkar, A., Jain, A.K.: Structure and context in prostatic gland segmentation and classification. In: Ayache, N., Delingette, H., Golland, P., Mori, K. (eds.) MICCAI 2012. LNCS, vol. 7510, pp. 115–123. Springer, Heidelberg (2012). https://doi.org/10.1007/978-3-642-33415-3_15
15. Paszke, A., et al.: PyTorch: an imperative style, high-performance deep learning library. In: Advances in Neural Information Processing Systems, pp. 8026–8037 (2019)
16. Qin, Y., et al.: Autofocus layer for semantic segmentation. In: Frangi, A.F., Schnabel, J.A., Davatzikos, C., Alberola-López, C., Fichtinger, G. (eds.) MICCAI 2018. LNCS, vol. 11072, pp. 603–611. Springer, Cham (2018). https://doi.org/10.1007/978-3-030-00931-1_69
17. Ronneberger, O., Fischer, P., Brox, T.: U-Net: convolutional networks for biomedical image segmentation. In: Navab, N., Hornegger, J., Wells, W.M., Frangi, A.F. (eds.) MICCAI 2015. LNCS, vol. 9351, pp. 234–241. Springer, Cham (2015). https://doi.org/10.1007/978-3-319-24574-4_28
18. Sirinukunwattana, K., et al.: Gland segmentation in colon histology images: the glas challenge contest. Med. Image Anal. 35, 489–502 (2017)
19. Sirinukunwattana, K., Snead, D.R., Rajpoot, N.M.: A novel texture descriptor for detection of glandular structures in colon histology images. In: Medical Imaging 2015: Digital Pathology, vol. 9420, p. 94200S. International Society for Optics and Photonics (2015)
20. Sirinukunwattana, K., Snead, D.R., Rajpoot, N.M.: A stochastic polygons model for glandular structures in colon histology images. IEEE Trans. Med. Imaging 34(11), 2366–2378 (2015)
21. Tabesh, A., et al.: Multifeature prostate cancer diagnosis and gleason grading of histological images. IEEE Trans. Med. Imaging 26(10), 1366–1378 (2007)
22. Torre, L.A., Ayangnd Bray, F., Siegel, R.L., Ferlay, J., Lortet-Tieulent, J., Jemal, A.: Global cancer statistics, 2012. CA Cancer J. Clin. 65(2), 87–108 (2015)
23. Wang, X., Peng, Y., Lu, L., Lu, Z., Summers, R.M.: TieNet: text-image embedding network for common thorax disease classification and reporting in chest x-rays. In: Proceedings of the IEEE Conference on Computer Vision and Pattern Recognition (2018)
24. Xu, X., et al.: Quantization of fully convolutional networks for accurate biomedical image segmentation. In: Proceedings of the IEEE Conference on Computer Vision and Pattern Recognition, pp. 8300–8308 (2018)
25. Xu, Y., et al.: Gland instance segmentation by deep multichannel side supervision. In: Ourselin, S., Joskowicz, L., Sabuncu, M.R., Unal, G., Wells, W. (eds.) MICCAI 2016. LNCS, vol. 9901, pp. 496–504. Springer, Cham (2016). https://doi.org/10.1007/978-3-319-46723-8_57
26. Xu, Y., et al.: Gland instance segmentation using deep multichannel neural networks. IEEE Trans. Biomed. Eng. 64(12), 2901–2912 (2017)

27. Yan, K., Peng, Y., Sandfort, V., Bagheri, M., Lu, Z., Summers, R.M.: Holistic and comprehensive annotation of clinically significant findings on diverse CT images: learning from radiology reports and label ontology. In: Proceedings of the IEEE Conference on Computer Vision and Pattern Recognition (2019)

28. Yan, K., et al.: MULAN: multitask universal lesion analysis network for joint lesion detection, tagging and segmentation. In: Proceedings of the IEEE Conference on Computer Vision and Pattern Recognition (2019)

29. Yan, Z., Yang, X., Cheng, K.-T.T.: A deep model with shape-preserving loss for gland instance segmentation. In: Frangi, A.F., Schnabel, J.A., Davatzikos, C., Alberola-López, C., Fichtinger, G. (eds.) MICCAI 2018. LNCS, vol. 11071, pp. 138–146. Springer, Cham (2018). https://doi.org/10.1007/978-3-030-00934-2_16

30. Yang, L., Zhang, Y., Chen, J., Zhang, S., Chen, D.Z.: Suggestive annotation: a deep active learning framework for biomedical image segmentation. In: Descoteaux, M., Maier-Hein, L., Franz, A., Jannin, P., Collins, D.L., Duchesne, S. (eds.) MICCAI 2017. LNCS, vol. 10435, pp. 399–407. Springer, Cham (2017). https://doi.org/10.1007/978-3-319-66179-7_46

31. Yu, F., Koltun, V.: Multi-scale context aggregation by dilated convolutions. arXiv preprint arXiv:1511.07122 (2015)

32. Zhang, Y., Yang, L., Chen, J., Fredericksen, M., Hughes, D.P., Chen, D.Z.: Deep adversarial networks for biomedical image segmentation utilizing unannotated images. In: Descoteaux, M., Maier-Hein, L., Franz, A., Jannin, P., Collins, D.L., Duchesne, S. (eds.) MICCAI 2017. LNCS, vol. 10435, pp. 408–416. Springer, Cham (2017). https://doi.org/10.1007/978-3-319-66179-7_47

33. Zheng, H., et al.: Biomedical image segmentation via representative annotation. In: Proceedings of the AAAI Conference on Artificial Intelligence, vol. 33, pp. 5901–5908 (2019)

Dynamic Image for 3D MRI Image Alzheimer's Disease Classification

Xin Xing$^{(\boxtimes)}$, Gongbo Liang, Hunter Blanton, Muhammad Usman Rafique, Chris Wang, Ai-Ling Lin, and Nathan Jacobs

University of Kentucky, Lexington, KY 40506, USA
{xxi242,gli238}@g.uky.edu

Abstract. We propose to apply a 2D CNN architecture to 3D MRI image Alzheimer's disease classification. Training a 3D convolutional neural network (CNN) is time-consuming and computationally expensive. We make use of approximate rank pooling to transform the 3D MRI image volume into a 2D image to use as input to a 2D CNN. We show our proposed CNN model achieves 9.5% better Alzheimer's disease classification accuracy than the baseline 3D models. We also show that our method allows for efficient training, requiring only 20% of the training time compared to 3D CNN models. The code is available online: https://github.com/UkyVision/alzheimer-project.

Keywords: Dynamic image · 2D CNN · MRI image · Alzheimer's disease

1 Introduction

Alzheimer's disease (AD) is the sixth leading cause of death in the U.S. [1]. It heavily affects the patients' families and U.S. health care system due to medical payments, social welfare cost, and salary loss. Since AD is irreversible, early stage diagnosis is crucial for helping slow down disease progression. Currently, researchers are using advanced neuroimaging techniques, such as magnetic resonance imaging (MRI), to identify AD. MRI technology produces a 3D image, which has millions of voxels. Figure 1 shows example slices of Cognitive Unimpaired (CU) and Alzheimer's disease (AD) MRI images.

With the promising performance of deep learning in natural image classification, convolutional neural networks (CNNs) show tremendous potential in medical image diagnosis. Due to the volumetric nature of MRI images, the natural deep learning model is a 3D convolutional neural network (3D CNN) [10]. Compared to 2D CNN models, 3D CNN models are more computationally expensive and time consuming to train due to the high dimensionality of the input. Another issue is that most current medical datasets are relatively small. The limited data makes it difficult to train a deep network that generalizes to high

---

X. Xing, G. Liang—authors show equal contribution.

A. Bartoli and A. Fusiello (Eds.): ECCV 2020 Workshops, LNCS 12535, pp. 355–364, 2020.
https://doi.org/10.1007/978-3-030-66415-2_23

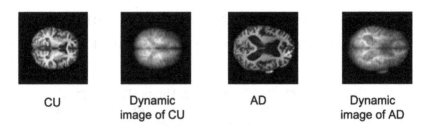

CU   Dynamic   AD   Dynamic
       image of CU   image of AD

Fig. 1. The MRI sample slices of the CU and AD participants and the corresponding dynamic images.

accuracy on unseen data. To overcome the problem of limited medical image training data, transfer learning is an attractive approach for feature extraction. However, pre-trained CNN models are mainly trained on 2D image datasets. There are few suitable pre-trained 3D CNN models. In our paper, We propose to apply approximate rank pooling [3] to convert a 3D MRI volume into a 2D image over the height dimension. Thus, we can use a 2D CNN architecture for 3D MRI image classification. The main contributions of our work are following:

- We propose to apply a CNN model that transforms the 3D MRI volume image into 2D dynamic image as the input of 2D CNN. Incorporating with an attention mechanism, the proposed model significantly boosts the accuracy of the Alzheimer's Disease MRI diagnosis.
- We analyze the effect of skull MRI images on the dynamic image method, showing that the applied dynamic image method is sensitive to the noise introduced by the skull. Skull striping is necessary before using the dynamic image technology.

2 Related Work

Learning-based Alzheimer's disease (AD) research can be mainly divided into two branches based on the type of input: (1) manually selected region of interest (ROI) input and (2) whole image input. With ROI models [6,14], manual region selection is needed to extract the interest region of the original brain image as the input to the CNN model, which is a time consuming task. It is more straightforward and desirable to use the whole image as input. Korolev et al. [11] propose two 3D CNN architectures based on VGGNet and ResNet, which is the first study to prove the manual feature extraction step for Brain MRI image classification is unnecessary. Their 3D models are called 3D-VGG and 3D-ResNet, and are widely used for 3D medical image classification study. Cheng et al. [4] proposes to use multiple 3D CNN models trained on MRI images for AD classification in an ensemble learning strategy. They separate the original MRI 3D images into many patches (n = 27), then forward each patch to an independent 3D CNN for feature extraction. Afterward, the extracted features are concatenated for classification. The performance is satisfactory, but the computation cost and training time overhead are very expensive. Yang et al. [18] uses the 3D-CNN models of Korolev et al. [11] as a backbone for studying the explainability of AD classification in MRI images by extending class activation mapping (CAM) [20] and gradient-based CAM [16] on 3D images. In our work, we use the whole brain MRI image as input and use 3D-VGG and 3D-ResNet as our baseline models. Dynamic images where first applied to medical imagery by Liang et al. [13] for breast cancer diagnosis. The authors use the dynamic image method to convert 3D digital breast tomosynthesis images into dynamic images and combined them with 2D mammography images for breast cancer classification. In our work, we propose to combine dynamic images with an attention mechanism for 3D MRI image classification.

3 Approach

We provide a detailed discussion of our method. First, we summarize the high-level network architecture. Second, we provide detailed information about the approximate rank pooling method. Next, we show our classifier structure and attention mechanism. Finally, we discuss the loss function used for training.

3.1 Model Architecture

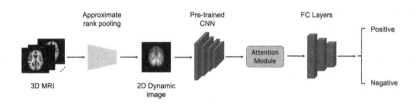

Fig. 2. The architecture of our 2D CNN model.

Figure 2 illustrates the architecture of our model. The 3D MRI image is passed to the approximate rank pooling module to transform the 3D MRI image volume into a 2D dynamic image. We apply transfer learning for feature extraction with the dynamic image as the input. We leveraged a pre-trained CNN as the backbone feature extractor. The feature extraction model is pre-trained with the ImageNet dataset [5]. Because we use a lower input resolution than the resolution used for ImageNet training, we use only a portion of the pre-trained CNN. The extracted features are finally sent to a small classifier for diagnosis prediction. The attention mechanism, which is widely used in computer vision community, can boost CNN model performance, so we embed the attention module in our classifier.

3.2   Dynamic Image

The temporal rank pooling [3,7] was originally proposed for video action recognition. For a video with T frames $I_1, ..., I_T$, the method compresses the whole video into one frame by temporal rank pooling. The compressed frame is called a dynamic image. The construction of the dynamic image is based on Fernando et al. [7]. The authors use a ranking function to represent the video. $\psi(I_t) \in \Re^d$ is a feature representation of the individual frame I_t of the video. $V_t = \frac{1}{t} \sum_{\tau=1}^{t} \psi(I_\tau)$ is the temporal average of the feature up to time t. V_t is measured by a ranking score $S(t|d) =< d, V_t >$, where $d \in \Re^m$ is a learned parameter. By accumulating more frames for the average, the later times are associated with larger scores, e.g. $q > t \rightarrow S(q|d) > S(t|d)$, which are constraints for the ranking problem. So the whole problem can be formulated as a convex problem using RankSVM:

$$d^* = \rho(I_1, ..., I_t; \tau) = \underset{d}{\operatorname{argmin}} E(d) \tag{1}$$

$$E(d) = \frac{\lambda}{2}||d||^2 + \frac{2}{T(T-1)} \times \sum_{q>t} \max\{0, 1 - S(q|d) + S(t|d)\} \tag{2}$$

In Eq. (2), the first term is a quadratic regularization used in SVMs, the second term is a hinge-loss counting incorrect rankings for the pairs $q > t$.

The RankSVM formulation can be used for dynamic image generation, but the operations are computationally expensive. Bilen et al. [3] proposed a fast approximate rank pooling for dynamic images:

$$\hat{\rho}(I_1, ..., I_t; \psi) = \sum_{t=1}^{T} \alpha_t \cdot \psi(I_t) \tag{3}$$

where, $\psi(I_t) = \frac{1}{t} \sum_{\tau=1}^{t} I_\tau$ is the temporal average of frames up to time t, and $\alpha_t = 2t - T - 1$ is the coefficient associated to frame $\psi(I_t)$. We take this approximate rank pooling strategy in our work for 3D MRI volume to 2D image transformation. In our implementation, the z-dimension of 3D MRI image is equal to temporal dimension of the video.

3.3   Classifier with Attention Mechanism

The classifier is a combination of an attention mechanism module and a basic classifier. Figure 3 depicts the structure of attention mechanism, which includes four 1×1 convolutional layers. The first three activation functions of convolutional layers are ReLU, the last convolutional layer is attached with softmax activation function. The input feature maps $A \in R^{H \times W \times C}$ are passed through the four convolutional layers to calculate attention mask $S \in R^{H \times W \times 1}$. We apply element-wise multiplication between the attention mask and input feature maps to get the final output feature map $O \in R^{H \times W \times C}$. Our basic classifier contains three fully connected (FC) layers. The output dimensions of the three FC layers are 512, 64, and 2. Dropout layers are used after the first two layers with dropout probability 0.5.

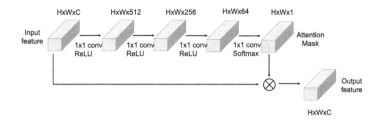

Fig. 3. The attention mechanism structure in our CNN model.

3.4 Loss Function

In previous AD classification studies, researchers mainly concentrated on binary classification. In our work, we do the same for ease of comparison. The overall loss function is binary cross-entropy. For a 3D image V with label l and probability prediction $p(l|V)$, the loss function is:

$$loss(l, V) = -[l \cdot log(p(l|V)) + (1 - l) \cdot log(1 - p(l|V))] \tag{4}$$

where the label $l = 0$ indicates a negative sample and $l = 1$ indicates a positive sample.

4   Evaluation

We use the publicly available dataset from the Alzheimer's Disease Neuroimaging Initiative (ADNI) [2] for our work. Specifically, we trained CNNs with the data from the "spatially normalized, masked, and N3-corrected T1 images" category. The brain MRI image size is $110 \times 110 \times 110$. Since a subject may have multiple MRI scans in the database, we use the first scan of each subject to avoid data leakage. The total number of data samples is 100, containing 51 CU samples and 49 AD samples.

The CNNs are implemented in PyTorch. We use five-fold cross validation to better evaluate model performance. The batch size used for our model is 16. The batch size of the baseline models is 8, which is the maximum batch size of the 3D CNN model trained on the single GTX-1080ti GPU. We use the Adam optimizer with $beta_1 = 0.9$ and $beta_2 = 0.999$. The learning rate is 0.0001. We train for 150 epochs. To evaluate the performance of our model, we use accuracy (Acc), the area under the curve of Receiver Operating Characteristics (ROC), F1 score (F1), Precision, Recall and Average Precision (AP) as our evaluation metrics.

4.1 Quantitive Results

High quality feature extraction is crucial for the final prediction. Different pre-trained CNN models can output different features in terms of size and effective receptive field. We test different pre-trained CNNs to find out which CNN models perform best as our feature extractor. Table 1 shows various CNN models and the corresponding output feature size.

Table 1. The different pre-trained CNN model as feature extractors and the output feature sizes

CNN model	Output feature size
AlexNet [12]	$256 \times 5 \times 5$
VggNet11 [17]	$512 \times 6 \times 6$
ResNet18 [8]	$512 \times 7 \times 7$
MobileNet_v2 [15]	$1280 \times 4 \times 4$

Since our dynamic image resolution is $110 \times 110 \times 3$, which is much smaller than the ImageNet dataset resolution: $256 \times 256 \times 3$, we use only part of the pre-trained CNN as the feature extractor. Directly using the whole pre-trained CNN model as feature extractor will cause the output feature size to be too small, which decreases the classification performance. In the implementation, we get rid of the maxpooling layer of each pre-trained model except for the MobileNet_v2 [15], which contains no maxpooling layer. Also, because there is a domain gap between the natural image and medical image we set the pre-trained CNN models' parameters trainable, so that we can fine tune the models for better performance.

Table 2. The performance results of different backbone models with dynamic image as input

Model	Acc	ROC	F1	Precision	Recall	AP
AlexNet	0.87	0.90	0.86	0.89	0.83	0.82
ResNet18	0.85	0.84	0.84	0.86	0.81	0.79
MobileNet_v2	0.88	0.89	0.87	0.89	0.85	0.83
VggNet11	0.91	0.92	0.91	0.88	0.93	0.86

When analyzing MRI images using computer-aided detectors (CADs), it is common to strip out the skulls from the brain images. Thus, we first test the proposed method using the MRI with the skull stripped. Our proposed model takes dynamic images (Dyn) as input, VGG11 as feature extractor, and a classifier with the attention mechanism: $Dyn + VGG11 + Att$. The whole experiment can be divided into three sections: the backbone and attention section, the baseline model section, and the pooling section. In the backbone and attention section, we use 4 different pre-trained models and test the selected backbone with and without the attention mechanism. Based on the performance shown in Table 2, we

Table 3. The performance results of different 2D and 3D CNN models

Model	Acc	ROC	F1	Precision	Recall	AP
3D-VGG [11]	0.80	0.78	0.78	0.82	0.75	0.74
3D-ResNet [11]	0.84	0.82	0.82	0.86	0.79	0.78
Max. + VGG11	0.80	0.77	0.80	0.78	0.81	0.73
Avg. + VGG11	0.86	0.84	0.86	0.83	0.89	0.79
Max. + VGG11 + Att	0.82	0.76	0.82	0.80	0.83	0.75
Avg. + VGG11 + Att	0.88	0.89	0.88	0.85	**0.91**	0.82
Ours	**0.92**	**0.95**	**0.91**	**0.97**	0.85	**0.90**

choose VGG11 as the backbone model. In the baseline model section, we compare our method with two baselines, namely 3D-VGG and 3D-ResNet. Table 3 shows the performance under different CNN models. The proposed model achieves 9.52% improvement in accuracy and 15.20% better ROC over the 3D-ResNet. In the pooling section: we construct two baselines by replacing the dynamic image module with the average pooling (Avg.) layer or max pooling (Max.) layer. The pooling layer processes the input 3D image over the z-dimension and outputs the same size as the dynamic image. Comparing with the different 3D-to-2D conversion methods under the same configuration, the dynamic image outperforms the two pooling methods.

4.2   Pre-processing Importance Evaluation

Table 4. The performance results of different 2D and 3D CNN models on the MRI image with skull.

Model	Acc	ROC	F1	Precision	Recall	AP
3D-VGG [11]	0.78	0.62	0.77	0.80	0.75	0.72
Ours	0.63	0.52	0.63	0.62	0.64	0.57

In this section, we show results using the raw MRI image (including skull) as input. We perform experiments on the same patients' raw brain MRI image with the skull included to test the performance of our model. The raw MRI image category is "MT1, GradWarp, N3m". The image size of the raw MRI image is "176 × 256 × 256". Figure 4 illustrates the dynamic images of different participants' MRI brain images with the skull. The dynamic images are blurrier than the images under skull striping processing. This is because the skull variance can be treated as noise in the dynamic image. Table 4 shows the significant performance decrease when using 3D Brain MRI images with skull. Figure 4 shows a visual representation of how the dynamic images are affected by including the

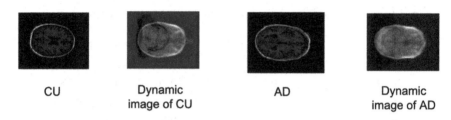

CU   Dynamic   AD   Dynamic
          image of CU   image of AD

Fig. 4. The MRI sample slices with skull of the CU and AD participants and the corresponding dynamic images.

skull in the image. In this scenario, the model can not sufficiently diagnose the different groups. A potential cause of this decrease in performance is that the dynamic image module is a pre-processing step, and the module is not trainable. We believe an end-to-end, learnable rank pooling module would improve performance.

4.3   Models Training Time

Table 5. The total 150 epochs training time of different CNN models.

	Training time(s)
3D-VGG [11]	2359
3D-ResNet [11]	3916
Ours	414

Another advantage of the proposed model is faster training. We train all of our CNN models for 150 epochs on the same input dataset. Table 5 shows the total training time of the different 2D and 3D CNN models. Compared with the 3D-CNN networks, the proposed model trains in about 20% of the time. Also, due to the higher dimension of the 3D convolutional layer, the number of parameters of the 3D convolutional layer is naturally higher than the 2D convolutional layer. By applying the MobileNet [9] or ShuffleNet [19] in medical image diagnosis, there is potential for mobile applications. We used MobileNet for our experiments. We used the MobileNet v1 architecture as the feature extractor and obtained 84.84% accuracy, which is similar in accuracy to the 3D ResNet.

5   Conclusions

We proposed to apply the dynamic image method to convert 3D Brain MRI images into 2D dynamic images as the inputs for a pre-trained 2D CNN. The proposed model outperforms a 3D CNN with much less training time and improves

9.5% better performance than the baselines. We trained and evaluated on MRI brain imagery and found out that brain skull striping pre-processing is useful before applying the dynamic image conversion. We used an offline approximate rank pooling module in our experiments, but we believe it would be interesting to explore a learnable temporal rank pooling module in the future.

Acknowledgement. This work is supported by NIH/NIA R01AG054459.

References

1. https://www.nia.nih.gov/
2. http://adni.loni.usc.edu/
3. Bilen, H., Fernando, B., Gavves, E., Vedaldi, A., Gould, S.: Dynamic image networks for action recognition. In: 2016 IEEE Conference on Computer Vision and Pattern Recognition (CVPR), pp. 3034–3042 (2016)
4. Cheng, D., Liu, M., Fu, J., Wang, Y.: Classification of MR brain images by combination of multi-CNNs for ad diagnosis. In: Proceedings of Ninth International Conference on Digital Image Processing (ICDIP) (2017)
5. Deng, J., Dong, W., Socher, R., Li, L.J., Li, K., Fei-Fei, L.: ImageNet: a large-scale hierarchical image database. In: CVPR09 (2009)
6. Duraisamy, B., Venkatraman, J., Annamala, S.J.: Alzheimer disease detection from structural MR images using FCM based weighted probabilistic neural network. Brain Imag. Behav. **13**(1), 87–110 (2019)
7. Fernando, B., Gavves, E., José Oramas, M., Ghodrati, A., Tuytelaars, T.: Modeling video evolution for action recognition. In: 2015 IEEE Conference on Computer Vision and Pattern Recognition (CVPR), pp. 5378–5387 (2015)
8. He, K., Zhang, X., Ren, S., Sun, J.: Deep residual learning for image recognition. arXiv preprint arXiv:1512.03385 (2015)
9. Howard, A.G., et al.: MobileNets: efficient convolutional neural networks for mobile vision applications (2017). http://arxiv.org/abs/1704.04861
10. Ji, S., Xu, W., Yang, M., Yu, K.: 3D convolutional neural networks for human action recognition (2013)
11. Korolev, S., Safiullin, A., Belyaev, M., Dodonova, Y.: Residual and plain convolutional neural networks for 3D brain MRI classification. In: 2017 IEEE 14th International Symposium on Biomedical Imaging (ISBI 2017), pp. 835–838 (2017)
12. Krizhevsky, A., Sutskever, I., Hinton, G.E.: Imagenet classification with deep convolutional neural networks. Commun. ACM **60**(6), 84–90 (2017)
13. Liang, G., et al.: Joint 2D–3D breast cancer classification. In: 2019 IEEE International Conference on Bioinformatics and Biomedicine (BIBM), pp. 692–696 (2019)
14. Rondina, J., et al.: Selecting the most relevant brain regions to discriminate alzheimer's disease patients from healthy controls using multiple kernel learning: a comparison across functional and structural imaging modalities and atlases. NeuroImage. Clin. **17**, 628–641 (2017)
15. Sandler, M., Howard, A., Zhu, M., Zhmoginov, A., Chen, L.C.: MobileNetv 2: inverted residuals and linear bottlenecks. In: Proceedings of the IEEE Conference on Computer Vision and Pattern Recognition (CVPR), June 2018
16. Selvaraju, R.R., Cogswell, M., Das, A., Vedantam, R., Parikh, D., Batra, D.: Grad-CAM: visual explanations from deep networks via gradient-based localization. In: 2017 IEEE International Conference on Computer Vision (ICCV), pp. 618–626 (2017)

17. Simonyan, K., Zisserman, A.: Very deep convolutional networks for large-scale image recognition. In: International Conference on Learning Representations (2015)
18. Yang, C., Rangarajan, A., Ranka, S.: Visual explanations from deep 3D convolutional neural networks for alzheimer's disease classification. In: AMIA Annual Symposium proceedings. AMIA Symposium, vol. 2018, pp. 1571–1580 (2018)
19. Zhang, X., Zhou, X., Lin, M., Sun, J.: ShuffleNet: an extremely efficient convolutional neural network for mobile devices (2017)
20. Zhou, B., Khosla, A., Lapedriza, A., Oliva, A., Torralba, A.: Learning deep features for discriminative localization. In: Computer Vision and Pattern Recognition (2016)

Feedback Attention for Cell Image Segmentation

Hiroki Tsuda[(✉)], Eisuke Shibuya, and Kazuhiro Hotta

Meijo University, 1-501 Shiogamaguchi, Tempaku-ku, 468-8502 Nagoya, Japan
{193427019,160442066}@ccalumni.meiji-u.ac.jp, kazuhotta@meiji-u.ac.jp
http://www1.meiji-u.ac.jp/~kazuhotta/cms_new/

Abstract. In this paper, we address cell image segmentation task by Feedback Attention mechanism like feedback processing. Unlike conventional neural network models of feedforward processing, we focused on the feedback processing in human brain and assumed that the network learns like a human by connecting feature maps from deep layers to shallow layers. We propose some Feedback Attentions which imitate human brain and feeds back the feature maps of output layer to close layer to the input. U-Net with Feedback Attention showed better result than the conventional methods using only feedforward processing.

Keywords: Cell image · Semantic segmentation · Attention mechanism · Feedback mechanism

1 Introduction

Deep neural networks has achieved state-of-the-art performance in image classification [19], segmentation [22], detection [25], and tracking [3]. Since the advent of AlexNet [19], several Convolutional Neural Network (CNN) [20] has been proposed such as VGG [28], ResNet [13], Deeplabv3+ [5], Faster R-CNN [25], and Siamese FC [3]. These networks are feedfoward processing. Neural network is mathematical model of neurons [34] that imitate the structure of human brain. Human brain performs not only feedfoward processing from shallow layers to deep layers of neurons, but also feedback processing from deep layers to shallow layers. However, conventional neural networks consist of only feedfoward processing from shallow layers to deep layers, and do not use feedback processing to connect from deep layers to shallow layers. Therefore, in this paper, we propose some Feedback Attention methods using position attention mechanism and feedback process.

Semantic segmentation assigns class labels to all pixels in an image. The study of this task can be applied to various fields such as automatic driving [4,6], cartography [11,23] and cell biology [10,17,26]. In particular, cell image segmentation requires better results in order to ensure that cell biologists can perform many experiments at the same time.

© Springer Nature Switzerland AG 2020
A. Bartoli and A. Fusiello (Eds.): ECCV 2020 Workshops, LNCS 12535, pp. 365–379, 2020.
https://doi.org/10.1007/978-3-030-66415-2_24

In addition, overall time and cost savings are expected to be achieved by automated processing without human involvement to reduce human error. Manual segmentation by human experts is slow to process and burdensome, and there is a significant demand for algorithms that can do the segmentation quickly and accurately without human. However, cell image segmentation is a difficult task because the number of supervised images is smaller and there is not regularity compared to the other datasets such as automatic driving. A large number of supervised images requires expert labeling which takes a lot of effort, cost and time. Therefore, it is necessary to enhance the segmentation ability for pixel-level recognition with small number of training images.

Most of the semantic segmentation approaches are based on Fully Convolutional Network (FCN) [22]. FCN is composed of some convolutional layers and some pooling layers, which does not require some fully connected layers. Convolutional layer and pooling layer reproduce the workings of neurons in the visual cortex. These are proposed in Neocognitron [9] which is the predecessor of CNN. Convolutional layer which is called S-cell extracts local features of the input. Pooling layer which is called C-cell compresses the information to enable downsampling to obtain position invariance. Thus, by repeating the feature extraction by convolutional layer and the local position invariance by pooling layer, robust pattern recognition is possible because it can react only to the difference of shape without much influence of misalignment and size change of the input pattern. Only the difference between CNN and Neocognitron is the optimization method, and the basic elements of both are same structure.

We focused on the relationship between the feature map close to the input and output of the semantic segmentation, and considered that it is possible to extract effective features by using between the same size and number of channels in the feature maps close to the input and output. In this paper, we create an attention map based on the relationship between these different feature maps, and a new attention mechanism is used to generate segmentation results. We can put long-range dependent spatial information from the output into the feature map of the input. The attention mechanism is fed back into the feature map of the input to create a model that can be reconsidered in based on the output.

In experiments, we evaluate the proposed method on a cell image datasets [10]. We confirmed that the proposed method gave higher accuracy than conventional method. We evaluate our method by some ablation studies and show the effectiveness of our method.

This paper is organized as follows. In Sect. 2, we describe related works. The details of the proposed method are explained in Sect. 3. In Sect. 4, we evaluate our proposed method on segmentation of cell images. Finally, we describe conclusion and future works in Sect. 5.

2 Related Works

2.1 Semantic Segmentation

FCNs [22] based methods have achieved significant results for semantic segmentation. The original FCN used stride convolutions and pooling to gradually downsize the feature map, and finally created high-dimensional feature map with low-resolution. This feature map has semantic information but fine information such as fine objects and correct location are lost. Thus, if the upsampling is used at the final layer, the accuracy is not sufficient. Therefore, encoder-decoder structure is usually used in semantic segmentation to obtain a final feature map with high-resolution. It consists of encoder network that extracts features from input image using convolutional layers, pooling layers, and batch normalization layers, and decoder network that classifies the extracted feature map by upsampling, convolutional layers, and batch normalization layers. Decoder restores the low-resolution semantic feature map extracted by encoder and middle-level features to the original image to compensate for the lost spatial information, and obtains a feature map with high resolution semantic information.

SegNet [2] is a typical network of encoder-decoder structures. Encoder uses 13 layers of VGG16 [28], and decoder receives some indexes selected by max pooling of encoder. In this way, decoder complements the positional information when upsampling and accelerates the calculation by unpooling, which requires no training.

Another famous encoder-decoder structural model is U-net [26]. The most important characteristic of U-Net is skip connection between encoder and decoder. The feature map with the spatial information of encoder is connected to the restored feature map of the decoder. This complements the high-resolution information and improves the resolution so that labels can be assigned more accurately to each pixel. In addition, deconvolution is used for up-sampling in decoder.

2.2 Attention Mechanism

Attention mechanism is an application of the human attention mechanism to machine learning. It has been used in computer vision and natural language processing. In the field of image recognition, important parts or channels are emphasized.

Residual Attention Network [31] introduced a stack network structure composed of multiple attention components, and attention residual learning applied residual learning [13] to the attention mechanism. Squeeze-and-Excitation Network (SENet) [15] introduced an attention mechanism that adaptively emphasizes important channels in feature maps. Accuracy booster blocks [29] and efficient channel attention module [32] made further improvements by changing the fully-connected layer in SENet. Attention Branch Network [8] is Class Activation Mapping (CAM) [39] based structure to build visual attention maps for image classification. Transformer [30] performed language translation only with the attention mechanism. There are Self-Attention that uses the same tensor, and Source-Target-Attention that uses two different tensors. Several networks have been proposed that use Self-Attention to learn the similarity between pixels in feature maps [7, 16, 24, 33, 37].

2.3 Feedback Mechanism Using Recurrent Neural Networks

Feedback is a fundamental mechanism of the human perceptual system and is expected to develop in the computer vision in the future. There have been several approaches to feedback using recurrent neural networks (RNNs) [1, 12, 36].

Feedback Network [36] uses convLSTM [35] to acquire hidden states with high-level information and provide feedback with the input image. However, this is intended to solve the image classification task and is not directly applicable to the segmentation task.

RU-Net [1] consists of a U-Net [26] and a recurrent neural network, where each convolutional layer is replaced by recurrent convolutional layer [21]. The accumulation of feature information at each scale by the recurrent convolutional layer gives better results than the standard convolutional layer. However, this is not strictly feedback but the deepening of network.

Feedback U-Net [27] is the segmentation method using convLSTM [35]. The probability for segmentation at final layer is used as the input image for segmentation at the second round, while the first feature map is used as the hidden state for the second segmentation to provide feedback.

Since RNNs is a neural network that contains loop connections, it can be easily used for feedback mechanisms. However, the problem with RNNs is that the amount of operations increases drastically and a lot of memory is consumed, which makes processing difficult and often results in the phenomenon that information is not transmitted. Thus, we applied RNNs-free feedback mechanism to U-Net, and excellent performance is shown by the feedback attention mechanism on the segmentation task.

3   Proposed Method

This section describes the details of the proposed method. Section 3.1 outlines the network of our method. In Sect. 3.2, we describe the details of the proposed attention mechanism.

3.1   Network Structure Details

The proposed method is based on U-Net [26], which is used as a standard in medical and cell images. Figure 1 shows the detail network structure of our proposed method using U-net. We design to do segmentation twice using U-Net in order to use the feature maps in input and output. Since the proposed method uses the feature maps of input and output, we use the model twice with shared weights. First, we perform segmentation by U-Net to obtain high-resolution important

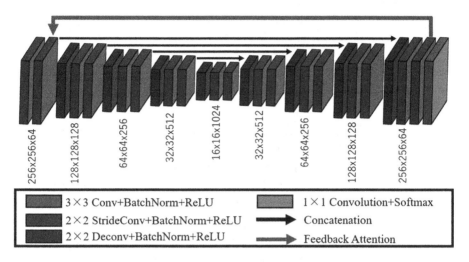

Fig. 1. Network structure of the proposed method using Feedback Attention

feature maps at the final layer. Then, we connect to Feedback Attention to a feature map that is close to the input with the same size and number of channels as this feature map. In this case, we use the input feature map that was processed two times by convolution.

The reason is that a feature map convolved twice can extract more advanced features than a feature map convolved once. The details of Feedback Attention is explained in Sect. 3.2. By applying Attention between the feature maps of input and output, we can obtain an input that takes the output into account as feedback control. In training, U-Net is updated by using only the gradients at the second round using feedback attention. In addition, the loss function is trained using Softmax cross-entropy loss.

3.2 Feedback Attention

We propose two kinds of Feedback Attentions to aggregate feature maps with the shape of $C \times H \times W$. Figure 2 (a) shows the Source-Target-Attention method that directly aggregates similar features between the feature maps of input and output. Figure 2 (b) shows the self-attention method that performs self-attention for output feature map and finally adds it to the feature map of input. Both Feedback Attentions are explained in the following subsections.

Feedback Attention Using Source-Target-Attention. We use Source-Target-Attention to aggregate the correlation between feature maps based on the relationship between input and output. Since the feature map in the final layer close to the output contains all the information for judging, it can be fed back using attention and effectively extract features again from the shallow input layer. We elaborate the process to aggregate each feature map.

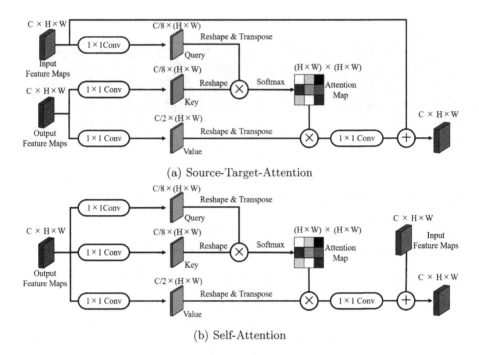

(a) Source-Target-Attention

(b) Self-Attention

Fig. 2. Feedback attention

As shown in Fig. 2 (a), we feed the feature maps of input or output into 1×1 convolutions and batch normalization to generate two new feature maps **Query** and **Key**, respectively, we are inspired by Self-Attention GAN (SAGAN) [37] to reduce the channel number to $C/8$ for memory efficiency. Then, we reshape them to $C/8 \times (H \times W)$. After we perform a matrix multiplication between the transpose of **Query** and **Key**, and we use a softmax function to calculate the attention map. Attention map in vector form is as follows.

$$w_{ij} = \frac{1}{Z_i} \exp(Query_i^T \ Key_j), \tag{1}$$

where w_{ij} measures the ith **Query**'s impact on j^{th} **Key**. Z_i is the sum of similarity scores as

$$Z_i = \sum_{j=1}^{H \times W} \exp(Query_i^T \ Key_j), \tag{2}$$

where $H \times W$ is the total number of pixels in **Query**. By increasing the correlation between two locations, we can create an attention map that takes into account output's feature map.

On the other hand, we feed the feature map of output into 1×1 convolution and batch normalization to generate a new feature map **Value** and reshape it to $C/2 \times (H \times W)$. Then, we perform a matrix multiplication between attention map and the transpose of **Value** and reshape the result to $C/2 \times H \times W$. In addition, we feed the new feature map into 1×1 convolution and batch normalization to generate feature map the same size as the feature map of input $C \times H \times W$. Finally, we multiply it by a scale parameter α and perform a element-wise sum operation with the input feature map to obtain the final output as follows.

$$A_i = \alpha \sum_{j=1}^{H \times W} (w_{ij} \, Value_j^T)^T + F_i, \tag{3}$$

where α is initialized as 0 and gradually learns to assign more weight [37]. A_i indicates the feedbacked output and F_i indicates the feature map of the input. By adding $\alpha \sum_{j=1}^{H \times W} (w_{ij} \, Value_j^T)^T$ to the feature map close to input, we can get the feature map considering feature map of output. The new feature map A_i is fed into the network again, and we obtain the segmentation result.

From Eq. (3), it can be inferred that the output A_i is the weighted sum of all positions in output and the feature map of input. Therefore, the segmentation accuracy is improved by transmitting the information of the output to the input.

Feedback Attention Using Self-Attention. In Source-Target-Attention, the feature map between input and output is aggregated. Thus, the relationship between each feature map can be emphasized. However, the feature map of the input may not extract enough information and therefore may result in poorly relational coordination. We construct Feedback Attention using Self-Attention that aggregates only the feature map of output.

The structure is shown in Fig. 2 (b). We feed the feature maps of output into 1×1 convolution and batch normalization to generate new feature maps **Query**, **Key** and **Value**. This is similar to Source-Target-Attention. We also reshape **Query** and **Key** to $C/8 \times (H \times W)$. Then, we perform a matrix multiplication between the transpose of **Query** and **Key**, and use a softmax function to calculate the attention map. Attention map in vector form is as follows.

$$w_{pq} = \frac{\exp(Query_p^T \, Key_q)}{\sum_{q=1}^{H \times W} \exp(Query_p^T \, Key_q)}, \tag{4}$$

where w_{pq} measures the pth **Query**'s impact on q^{th} **Key**.

We reshape **Value** to $C/2 \times (H \times W)$. Then, we perform a matrix multiplication between attention map and the transpose of **Value** and reshape the result to $C \times H \times W$ after 1×1 convolution. Finally, we multiply it by a scale parameter β and perform a element-wise sum operation with the feature maps of input to obtain the final output as follows.

$$A_p = \beta \sum_{q=1}^{H \times W} (w_{pq} \, Value_q^T)^T + F_p, \tag{5}$$

where β is initialized as 0 and gradually learns to assign more weight [37]. A_p indicates the output, F_p indicates the feature map of input. New feature map A_p is fed into the network again, and we obtain the segmentation result.

Unlike Eq. (3), Eq. (5) calculates the similarity using only the information of output. In addition, consistency can be improved because information can be selectively passed to the input by the scale parameter.

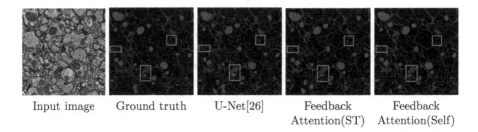

Input image   Ground truth   U-Net[26]   Feedback   Feedback
                                            Attention(ST)   Attention(Self)

Fig. 3. Examples of segmentation results on ssTEM dataset. ST indicates Source-Target-Attention, Self indicates Self-Attention.

Table 1. Segmentation accuracy (IoU and mIoU) on ssTEM Dataset. ST indicates Source-Target-Attention, Self indicates Self-Attention.

Method	Membrane	Mitochondria	Synapse	Cytoplasm	Mean IoU%
U-Net [26]	74.24	71.01	43.08	92.03	70.09
RU-Net [1]	75.81	74.39	43.26	92.25	71.43
Feedback U-Net [27]	76.44	75.20	42.30	92.43	71.59
Feedback attention (ST)	**76.65**	**78.27**	**43.32**	**92.64**	**72.72**
Feedback attention (Self)	76.94	79.52	45.29	92.80	73.64

4   Experiments

This section shows evaluation results by the proposed method. We explain the datasets used in experiments in Sect. 4.1. Experimental results are shown in Sect. 4.2. Finally, Sect. 4.3 describes Ablation studies to demonstrate the effectiveness of the proposed method.

4.1   Dataset

In experiments, we evaluated all methods 15 times with 5-fold cross-validation using three kinds of initial values on the Drosophila cell image data set [10].

We use Intersection over Union (IoU) as evaluation measure. Average IoU of 15 times evaluations is used as final measure.

This dataset shows neural tissue from a Drosophila larva ventral nerve cord and was acquired using serial section Transmission Electron Microscopy at HHMI Janelia Research Campus [10]. This dataset is called ssTEM dataset. There are 20 images of 1024×1024 pixels and ground truth. In this experiment, semantic segmentation is performed for four classes; membrane, mitochondria, synapses and cytoplasm. We augmented 20 images to 320 images by cropping 16 regions of 256×256 pixels without overlap from an image. We divided those images into 192 training, 48 validation and 80 test images.

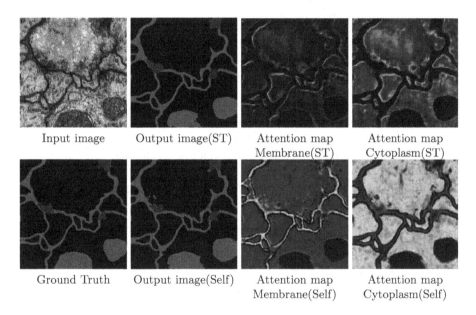

| Input image | Output image(ST) | Attention map Membrane(ST) | Attention map Cytoplasm(ST) |

| Ground Truth | Output image(Self) | Attention map Membrane(Self) | Attention map Cytoplasm(Self) |

Fig. 4. Visualization results of Attention Map on ssTEM dataset. ST indicates Source-Target-Attention, Self indicates Self-Attention.

4.2 Experimental Results

Table 1 shows the accuracy on ssTEM dataset, and Fig. 3 shows the segmentation results. Bold red letters in the Table represent the best IoU and black bold letters represent the second best IoU. Table 1 shows that our proposed Feedback Attention improved the accuracy of all classes compared to conventional U-Net [26]. We also evaluated two feedback methods using RNNs; RU-Net [1] with recurrent convolution applied to U-Net and Feedback U-Net [27] with feedback segmentation applied to U-Net. The result shows that the proposed method gave high accuracy in all classes. In addition, we can see that Self-Attention, which calculates the similarity in the output, is more accurate than Source-Target-Attention which calculates the similarity from the relationship between the input and the output. This indicates that the feature map of the input does not extract enough features and therefore the similarity representation between the input and the output does not work well.

From the yellow frames in Fig. 3, our method using Feedback Attention can identify mitochondria that were detected excessively by conventional methods. In the conventional methods, cell membranes were interrupted, but in our proposed method, we confirm that cell membranes are segmented in such a way that they are cleanly connected. Experimental results show that cell membrane and the mitochondria have been successfully identified even in places where it is difficult to detect by conventional methods.

We visualize some attention maps in Fig. 4 to understand our two kinds of Feedback Attentions. White indicates similarity and black indicates dissimilarity. We find that Self-Attention maps has many similar pixels but Source-Target-Attention maps has fewer pixels. This is because Source-Target-Attention uses the feature maps of input and output, and the feature map near input is different from that of output, so the number of similar pixels are smaller than Self-Attention map. However, the membranes and cytoplasm have different values in the attention map. This means that they are emphasized as different objects. On the other hand, Self-Attention generates attention maps from only the feature map of output. Therefore, as shown in the Fig. 4, when cell membrane and cytoplasm are selected, they are highlighted as similar pixels.

Table 2. Comparison of different feedback connections.

Method	Membrane	Mitochondria	Synapse	Cytoplasm	Mean IoU%
Add	75.56	77.36	41.84	92.46	71.81
1×1 Conv	75.22	**78.39**	**43.46**	92.49	72.39
SE-Net [15]	75.89	77.31	42.92	92.49	72.15
Light Attention [14]	76.20	78.27	43.18	92.57	72.56
Feedback attention (ST)	**76.65**	78.27	43.32	**92.64**	**72.72**
Feedback attention (Self)	76.94	79.52	45.29	92.80	73.64

4.3 Ablation Studies

We performed three ablation studies to show the effectiveness of the proposed method. The first ablation study evaluated the different connection methods. The second ablation study confirmed the effectiveness of connection location from the output to the input. The last ablation study confirmed the effectiveness of before and after Feedback Attention was used.

Comparison of Difference Feedback Connection. The effectiveness of the other feedback connection methods from the output to the input was experimentally confirmed. We compare four methods. We compare two methods that do not use the attention mechanism. The first one is that we simply add the feature map in the output to the input. The second one is that we feed the feature map in the output to 1×1 convolution and then add it to the feature map in the input. Both methods use scale parameter as our propose method.

In addition, we compare two methods using attention mechanism. The first one is that we apply SE-Net [15], which suppresses and emphasizes the feature map between channels, to the output feature map, and add it to the input feature map. The second one is that we apply Light Attention [14], which suppresses and emphasizes the important locations and channels in feature map by 3×3 convolutional processing, to the output feature map and adding it to the input feature map.

From Table 2, we can see that the above four methods improve the accuracy from U-Net [26] because the feedback mechanism is effective. However, our proposed method is more accurate than those four methods. This shows that our proposed Feedback Attention can use the output's information effectively in the input.

Table 3. Comparison between different connection locations.

Method	Membrane	Mitochondria	Synapse	Cytoplasm	Mean IoU%
Feedback attention using source-target-attention					
One conv	76.54	77.39	43.06	91.96	72.24
Two conv (Ours)	76.65	78.27	43.32	92.64	72.72
Feedback attention using self-attention					
One conv	**76.69**	**78.73**	**45.23**	**92.66**	**73.33**
Two conv (Ours)	76.94	79.52	45.29	92.80	73.64

Table 4. Comparison before and after feedback attention.

Method	Membrane	Mitochondria	Synapse	Cytoplasm	Mean IoU%
Feedback attention using source-target-attention					
First output	76.07	76.76	41.28	92.39	71.62
Second output (Ours)	**76.65**	**78.27**	**43.32**	**92.64**	**72.72**
Feedback attention using self-attention					
First output	75.49	74.29	41.57	92.03	70.84
Second output (Ours)	76.94	79.52	45.29	92.80	73.64

Comparison Between Different Connection Locations. We experimentally evaluated the location of the input feature map which is the destination of feedback. Since the size of feature map should be the same as final layer, the candidates are only two layers close to input. The first one is the feature map closest to the input which is obtained by only one convolution process. The other one is the feature map obtained after convolution is performed two times. We compared the two feature map locations that we use Feedback Attention.

Table 3 shows that the Feedback Attention to the feature map after two convolution process is better for both Source-Target-Attention and Self-Attention. This indicates that only one convolution process does not extract good features than two convolution processes.

Comparison Before and After Feedback Attention. When we use Feedback Attention, the output of network is feedback to input as attention. Thus, we get the outputs twice. Although we use the output using Feedback Attention at the second round is used as final result, we compare the results of the outputs at the first and second rounds to show the effectiveness of Feedback Attention. From Table 4, the output using Feedback Attention as the second round is better than that at the first round. This demonstrates that the accuracy was improved through the feedback mechanism.

5   Conclusions

In this paper, we have proposed two Feedback Attention for cell image segmentation. Feedback Attention allows us to take advantage of the feature map information of the output and improve the accuracy of the segmentation, and segmentation accuracy is improved in comparison with conventional feedforward network, RU-Net [1] which uses local feedback at each convolutional layer and Feedback U-Net [27] which uses global feedback between input and output. Ablation studies show that Feedback Attention can obtain accurate segmentation results by choosing the location and attention mechanism that conveys the output information.

In the future, we aim to develop a top-down attention mechanism that directly utilizes ground truth, such as self-distillation [38]. Feedback networks are also categorized as a kind of top-down networks, because the representation of feature extraction will be expanded if the ground truth can be used for direct learning in the middle layer as well. In addition, Reformer [18] using Locality Sensitive Hashing has been proposed in recent years. Since Transformer-based Attention uses a lot of memory, Reformer will work well in our Feedback Attention. These are subjects for future works.

References

1. Alom, M.Z., Hasan, M., Yakopcic, C., Taha, T.M., Asari, V.K.: Recurrent residual convolutional neural network based on u-net (R2U-Net) for medical image segmentation. arXiv preprint arXiv:1802.06955 (2018)

2. Badrinarayanan, V., Kendall, A., Cipolla, R.: SegNet: a deep convolutional encoder-decoder architecture for image segmentation. IEEE Trans. Pattern Anal. Mach. Intell. **39**(12), 2481–2495 (2017)
3. Bertinetto, L., Valmadre, J., Henriques, J.F., Vedaldi, A., Torr, P.H.S.: Fully-convolutional siamese networks for object tracking. In: Hua, G., Jégou, H. (eds.) ECCV 2016. LNCS, vol. 9914, pp. 850–865. Springer, Cham (2016). https://doi.org/10.1007/978-3-319-48881-3_56
4. Brostow, G.J., Fauqueur, J., Cipolla, R.: Semantic object classes in video: a high-definition ground truth database. Pattern Recogn. Lett. **30**(2), 88–97 (2009)
5. Chen, L.-C., Zhu, Y., Papandreou, G., Schroff, F., Adam, H.: Encoder-decoder with atrous separable convolution for semantic image segmentation. In: Ferrari, V., Hebert, M., Sminchisescu, C., Weiss, Y. (eds.) ECCV 2018. LNCS, vol. 11211, pp. 833–851. Springer, Cham (2018). https://doi.org/10.1007/978-3-030-01234-2_49
6. Cordts, M., et al.: The cityscapes dataset for semantic urban scene understanding. In: Proceedings of the IEEE Conference on Computer Vision and Pattern Recognition, pp. 3213–3223 (2016)
7. Fu, J., et al.: Dual attention network for scene segmentation. In: Proceedings of the IEEE Conference on Computer Vision and Pattern Recognition, pp. 3146–3154 (2019)
8. Fukui, H., Hirakawa, T., Yamashita, T., Fujiyoshi, H.: Attention branch network: learning of attention mechanism for visual explanation. In: Proceedings of the IEEE Conference on Computer Vision and Pattern Recognition, pp. 10705–10714 (2019)
9. Fukushima, K., Miyake, S.: Neocognitron: a new algorithm for pattern recognition tolerant of deformations and shifts in position. Pattern Recogn. **15**(6), 455–469 (1982)
10. Gerhard, S., Funke, J., Martel, J., Cardona, A., Fetter, R.: Segmented anisotropic ssTEM dataset of neural tissue, November 2013. https://doi.org/10.6084/m9.figshare.856713.v1. https://figshare.com/articles/Segmented_anisotropic_ssTEM_dataset_of_neural_tissue/856713
11. Ghamisi, P., Benediktsson, J.A.: Feature selection based on hybridization of genetic algorithm and particle swarm optimization. IEEE Geosci. Remote Sens. Lett. **12**(2), 309–313 (2014)
12. Han, W., Chang, S., Liu, D., Yu, M., Witbrock, M., Huang, T.S.: Image super-resolution via dual-state recurrent networks. In: Proceedings of the IEEE Conference on Computer Vision and Pattern Recognition, pp. 1654–1663 (2018)
13. He, K., Zhang, X., Ren, S., Sun, J.: Deep residual learning for image recognition. In: Proceedings of the IEEE Conference on Computer Vision and Pattern Recognition, pp. 770–778 (2016)
14. Hiramatsu, Y., Hotta, K.: Semantic segmentation using light attention mechanism. In: Proceedings of the 15th International Joint Conference on Computer Vision, Imaging and Computer Graphics Theory and Applications, pp. 622–625 (2020)
15. Hu, J., Shen, L., Sun, G.: Squeeze-and-excitation networks. In: Proceedings of the IEEE Conference on Computer Vision and Pattern Recognition, pp. 7132–7141 (2018)
16. Huang, Z., Wang, X., Huang, L., Huang, C., Wei, Y., Liu, W.: CCNet: criss-cross attention for semantic segmentation. In: Proceedings of the IEEE International Conference on Computer Vision, pp. 603–612 (2019)
17. Imanishi, A., et al.: A novel morphological marker for the analysis of molecular activities at the single-cell level. Cell Struct. Func. **43**(2), 129–140 (2018)
18. Kitaev, N., Kaiser, Ł., Levskaya, A.: Reformer: the efficient transformer. In: International Conference on Learning Representations (2020)

19. Krizhevsky, A., Sutskever, I., Hinton, G.E.: ImageNet classification with deep convolutional neural networks. In: Advances in Neural Information Processing Systems, pp. 1097–1105 (2012)
20. LeCun, Y., Bottou, L., Bengio, Y., Haffner, P.: Gradient-based learning applied to document recognition. Proc. IEEE **86**(11), 2278–2324 (1998)
21. Liang, M., Hu, X.: Recurrent convolutional neural network for object recognition. In: Proceedings of the IEEE Conference on Computer Vision and Pattern Recognition, pp. 3367–3375 (2015)
22. Long, J., Shelhamer, E., Darrell, T.: Fully convolutional networks for semantic segmentation. In: Proceedings of the IEEE Conference on Computer Vision and Pattern Recognition, pp. 3431–3440 (2015)
23. Maggiori, E., Tarabalka, Y., Charpiat, G., Alliez, P.: Convolutional neural networks for large-scale remote-sensing image classification. IEEE Trans. Geosci. Remote Sens. **55**(2), 645–657 (2016)
24. Ramachandran, P., Parmar, N., Vaswani, A., Bello, I., Levskaya, A., Shlens, J.: Stand-alone self-attention in vision models. In: Advances in Neural Information Processing Systems, pp. 68–80 (2019)
25. Ren, S., He, K., Girshick, R., Sun, J.: Faster R-CNN: towards real-time object detection with region proposal networks. In: Advances in Neural Information Processing Systems, pp. 91–99 (2015)
26. Ronneberger, O., Fischer, P., Brox, T.: U-Net: convolutional networks for biomedical image segmentation. In: Navab, N., Hornegger, J., Wells, W.M., Frangi, A.F. (eds.) MICCAI 2015. LNCS, vol. 9351, pp. 234–241. Springer, Cham (2015). https://doi.org/10.1007/978-3-319-24574-4_28
27. Shibuya, E., Hotta, K.: Feedback U-Net for cell image segmentation. In: Proceedings of the IEEE Conference on Computer Vision and Pattern Recognition Workshops (2020)
28. Simonyan, K., Zisserman, A.: Very deep convolutional networks for large-scale image recognition. arXiv preprint arXiv:1409.1556 (2014)
29. Singh, P., Mazumder, P., Namboodiri, V.P.: Accuracy booster: performance boosting using feature map re-calibration. arXiv preprint arXiv:1903.04407 (2019)
30. Vaswani, A., et al.: Attention is all you need. In: Advances in Neural Information Processing Systems, pp. 5998–6008 (2017)
31. Wang, F., et al.: Residual attention network for image classification. In: Proceedings of the IEEE Conference on Computer Vision and Pattern Recognition, pp. 3156–3164 (2017)
32. Wang, Q., Wu, B., Zhu, P., Li, P., Zuo, W., Hu, Q.: ECA-Net: efficient channel attention for deep convolutional neural networks. arXiv preprint arXiv:1910.03151 (2019)
33. Wang, X., Girshick, R., Gupta, A., He, K.: Non-local neural networks. In: Proceedings of the IEEE Conference on Computer Vision and Pattern Recognition, pp. 7794–7803 (2018)
34. Widrow, B., Lehr, M.A.: Perceptrons, adalines, and backpropagation. In: The Handbook of Brain Theory and Neural Networks, pp. 719–724. MIT Press (1998)
35. Xingjian, S., Chen, Z., Wang, H., Yeung, D.Y., Wong, W.K., Woo, W.C.: Convolutional LSTM network: a machine learning approach for precipitation nowcasting. In: Advances in Neural Information Processing Systems, pp. 802–810 (2015)
36. Zamir, A.R., et al.: Feedback networks. In: Proceedings of the IEEE Conference on Computer Vision and Pattern Recognition, pp. 1308–1317 (2017)

37. Zhang, H., Goodfellow, I., Metaxas, D., Odena, A.: Self-attention generative adversarial networks. In: International Conference on Machine Learning, pp. 7354–7363 (2019)
38. Zhang, L., Song, J., Gao, A., Chen, J., Bao, C., Ma, K.: Be your own teacher: improve the performance of convolutional neural networks via self distillation. In: Proceedings of the IEEE International Conference on Computer Vision, pp. 3713–3722 (2019)
39. Zhou, B., Khosla, A., Lapedriza, A., Oliva, A., Torralba, A.: Learning deep features for discriminative localization. In: Proceedings of the IEEE Conference on Computer Vision and Pattern Recognition, pp. 2921–2929 (2016)

Improving Blind Spot Denoising
for Microscopy

Anna S. Goncharova[1,2] ![ORCID], Alf Honigmann[1] ![ORCID], Florian Jug[1,2,3(✉)] ![ORCID],
and Alexander Krull[1,2,4] ![ORCID]

[1] Max Planck Institute of Molecular Cell Biology and Genetics, Dresden, Germany
{jug,krull}@mpi-cbg.de
[2] Center for Systems Biology Dresden (CSBD), Dresden, Germany
[3] Fondazione Human Technopole, Milan, Italy
[4] Max Planck Institute for the Physics of Complex Systems, Dresden, Germany

Abstract. Many microscopy applications are limited by the total amount of usable light and are consequently challenged by the resulting levels of noise in the acquired images. This problem is often addressed via (supervised) deep learning based denoising. Recently, by making assumptions about the noise statistics, self-supervised methods have emerged. Such methods are trained directly on the images that are to be denoised and do not require additional paired training data. While achieving remarkable results, self-supervised methods can produce high-frequency artifacts and achieve inferior results compared to supervised approaches. Here we present a novel way to improve the quality of self-supervised denoising. Considering that light microscopy images are usually diffraction-limited, we propose to include this knowledge in the denoising process. We assume the clean image to be the result of a convolution with a point spread function (PSF) and explicitly include this operation at the end of our neural network. As a consequence, we are able to eliminate high-frequency artifacts and achieve self-supervised results that are very close to the ones achieved with traditional supervised methods.

Keywords: Denoising · CNN · Light microscopy · Deconvolution

1 Introduction

For most microscopy applications, finding the right exposure and light intensity to be used involves a trade-off between maximizing the signal to noise ratio and minimizing undesired effects such as phototoxicity. As a consequence, researchers often have to cope with considerable amounts of noise. To mitigate this issue, denoising plays an essential role in many data analysis pipelines, enabling otherwise impossible experiments [2].

Currently, deep learning based denoising, also known as content-aware image restoration (CARE) [24], achieves the highest quality results. CARE methods

© Springer Nature Switzerland AG 2020
A. Bartoli and A. Fusiello (Eds.): ECCV 2020 Workshops, LNCS 12535, pp. 380–393, 2020.
https://doi.org/10.1007/978-3-030-66415-2_25

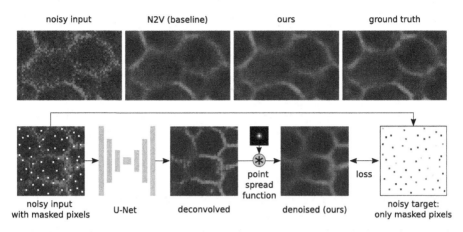

Fig. 1. Improved Denoising for Diffraction-Limited Data. Top: Given a noisy input, self-supervised methods like NOISE2VOID (N2V) [9] often produce high-frequency artifacts that do not occur in diffraction-limited data. Based on the assumption that the true signal must be the product of a convolution with a *point spread function* (PSF), our method is able to considerably improve denoising quality and remove these artifacts. **Bottom:** Our method is based on the NOISE2VOID masking scheme. Unpaired training images simultaneously serve as input and target. The loss is only calculated for a randomly selected set of pixels, which are masked in the input image. Our contribution is to convolve the output of the network with the PSF in order to produce a denoising result that is guaranteed to be consistent with diffraction-limited imaging. The output of the network before the convolution operation can be interpreted as a deconvolution result, which is a byproduct of our method. Our system can be trained in an end-to-end fashion, calculating the loss between our denoising result and the selected pixel set of the input image.

learn a mapping from noisy to clean images. Before being applied, they must be trained with pairs of corresponding noisy and clean training data.

In practice, this dependence on training pairs can be a bottleneck. While noisy images can usually be produced in abundance, recording their clean counterparts is difficult or impossible.

Over the last years, various solutions to the problem have been proposed. Lehtinen *et al.* showed that a network can be trained for denoising using only pairs of corresponding noisy images. This method is known as NOISE2NOISE [12].

The first self-supervised approaches NOISE2VOID [9] and NOISE2SELF [1] were introduced soon after this. These methods can be trained on unpaired noisy image data. In fact, they can be trained on the very same data that is to be denoised in the first place. The underlying approach relies on the assumption that (given the true signal) the noise in an image is generated independently for each pixel, as is indeed the case for the dominant sources of noise in light microscopy (Poisson shot noise and Gaussian readout noise) [13,25]. Both methods employ so-called *blind spot* training, in which random pixels are masked in the input image with the network trying to predict their value from the surrounding patch.

Unfortunately, the original self-supervised methods typically produce visible high-frequency artifacts (see Fig. 1) and can often not reach the quality achieved by supervised CARE training. It is worth noting that the high-frequency artifacts produced by these self-supervised methods never occur in the real fluorescence signal. Since the image is diffraction-limited and oversampled, the true signal has to be smooth to some degree.

Multiple extensions of NOISE2VOID and NOISE2SELF have been proposed [6,10,11,17]. All of them improve results by explicitly modeling the noise distribution.

Here, we propose an alternate and novel route to high-quality self-supervised denoising. Instead of making additional assumptions about the noise, we show that the result can be improved by including additional knowledge about the structure of our signal. We believe that our approach might ultimately complement existing methods that are based on noise modeling, to further improve denoising quality.

We assume that the true signal is the product of a convolution of an unknown *phantom image* and an approximately known point spread function (PSF) – a common assumption in established deconvolution approaches [20]. We use a U-NET [21] to predict the phantom image and then explicitly perform the convolution to produce the final denoised result (see Fig. 1). We follow [1,9] and use a blind spot masking scheme allowing us to train our network in an end-to-end fashion from unpaired noisy data.

We demonstrate that our method achieves denoising quality close to supervised methods on a variety of real and publicly available datasets. Our approach is generally on-par with modern noise model based methods [10,16], while relying on a much simpler pipeline.

As a byproduct, our method outputs the predicted phantom image, which can be interpreted as a deconvolution result. While we focus on the denoising task in this paper, we find that we can produce visually convincing deconvolved images by including a positivity constraint for the deconvolved output.

2   Related Work

In the following, we will discuss related work on self-supervised blind spot denoising and other unsupervised denoising methods. We will focus on deep learning-based methods and omit the more traditional approaches that directly operate on individual images without training. Finally, we will briefly discuss concurrent work that tries to jointly solve denoising and inverse problems such as deconvolution.

2.1   Self-Supervised Blind Spot Denoising

By now, there is a variety of different blind spot based methods. While the first self-supervised methods (NOISE2VOID and NOISE2SELF) use a masking scheme to implement blind spot training, Laine *et al.* [11] suggest an alternative approach. Instead of masking, the authors present a specific network architecture that directly implements the blind spot receptive field. Additionally, the authors proposed a way to improve denoising quality by including a simple pixel-wise Gaussian based noise model. In parallel, Krull *et al.* [10] introduced a similar noise model based technique for improving denoising quality, this time using the pixel masking approach. Instead of Gaussians, Krull *et al.* use histogram-based noise models together with a sampling scheme. Follow-up work additionally introduces parametric noise models and demonstrates how they can be bootstrapped (estimated) directly from the raw data [17].

All mentioned methods improve denoising quality by modeling the imaging noise. We, In contrast, are the first to show how blind spot denoising can be improved by including additional knowledge of the signal itself, namely the fact that it is diffraction-limited and oversampled.

While the blind spot architecture introduced in [11] is computationally cheaper than the masking scheme from [6,9], it is unfortunately incompatible with our setup (see Fig. 1). Applying a convolution after a blind spot network would break the blind spot structure of the overall architecture. We thus stick with the original masking scheme, which is architecture-independent and can directly be applied for end-to-end training.

2.2   Other Unsupervised Denoising Approaches

An important alternative route is based on the theoretical work known as *Stein's unbiased risk estimator* (SURE) [22]. Given noisy observation, such as an image corrupted by additive Gaussian noise, Stein's 1981 theoretical work enables us to calculate the expected mean-squared error of an estimator that tries to predict the underlying signal without requiring access to the true signal. The approach was put to use for conventional (non-deep-learning-based) denoising in [18] and later applied to derive a loss function for neural networks [15]. While it has been shown that the same principle can theoretically be applied for other noise models beyond additive Gaussian noise [19], this has to our knowledge not yet been used to build a general unsupervised deep learning based denoiser.

In a very recent work called DivNoising [16] unsupervised denoising was achieved by training a variational autoencoder (VAE) [7] as a generative model of the data. Once trained, the VAE can produce samples from an approximate posterior of clean images given a noisy input, allowing the authors to provide multiple diverse solutions or to combine them to a single estimate.

Like the previously discussed extensions of blind spot denoising [6,10,11,17] all methods based on SURE as well as DivNoising rely on a known noise model or on estimating an approximation. We, in contrast, do not model the noise distribution in any way (except assuming it is zero centered and applied at the pixel level) and achieve improved results.

A radically different path that does not rely on modeling the noise distribution was described by Ulyanov *et al.* [23]. This technique, known as *deep image prior*, trains a network using a fixed pattern of random inputs and the noisy image as a target. If trained until convergence, the network will simply produce the noisy image as output. However, by stopping the training early (at an adequate time) this setup can produce high-quality denoising results. Like our self-supervised method, deep image prior does not require additional training data to be applied. However, it is fundamentally different in that it is trained and applied separately for each image that is to be denoised, while our method can, once it is trained, be readily applied to previously unseen data.

2.3 Concurrent Work on Denoising and Inverse Problems

Kobayashi et al. [8] developed a similar approach in parallel to ours. They provide a mathematical framework on how inverse problems such as deconvolution can be tackled using a blind spot approach. However, while we use a comparable setup, our perspective is quite different. Instead of deconvolution, we focus on the benefits for the denoising task and show that the quality of the results on real data can be dramatically improved.

Yet another alternative approach was developed by Hendriksen et al. [5]. However, this technique is limited to well-conditioned inverse problems like computer tomography reconstruction and is not directly applicable to the type of microscopy data we consider here.

3 Methods

In the following, we first describe our model of the image formation process, which is the foundation of our method, and then formally describe the denoising task. Before finally describing our method for blind spot denoising with diffraction-limited data, we include a brief recap of the original NOISE2VOID method described in [9].

3.1 Image Formation

We think of the observed noisy image $\mathbf{x}$ recorded by the microscope, as being created in a two-stage process. Light originates from the excited fluorophores in the sample. We will refer to the unknown distribution of excited fluorophores as the *phantom image* and denote it as $\mathbf{z}$. The phantom image is mapped through the optics of the microscope to form a distorted image $\mathbf{s}$ on the detector, which we will refer to as *signal*. We assume the signal is the result of a convolution $\mathbf{s} = \mathbf{z} * \mathbf{h}$ between the phantom image $\mathbf{z}$ and a known PSF $\mathbf{h}$ [20].

Finally, the signal is subject to different forms of imaging noise, resulting in the noisy observation $\mathbf{x}$. We think of $\mathbf{x}$ as being drawn from a distribution $\mathbf{x} \sim p_{\mathrm{NM}}(\mathbf{x}|\mathbf{s})$, which we call the *noise model*. Assuming that (given a signal $\mathbf{s}$) the noise is occurring independently for each pixel, we can factorize the noise model as

$$p_{\mathrm{NM}}(\mathbf{x}|\mathbf{s}) = \prod_i^N p_{\mathrm{NM}}(x_i, s_i), \tag{1}$$

where $p_{\mathrm{NM}}(x_i, s_i)$ is the unknown probability distribution, describing how likely it is to measure the noisy value x_i at pixel i given an underlying signal s_i. Note that such a noise model that factorizes over pixels can describe the most dominant sources of noise in fluorescent microscopy, the Poisson shot noise and readout noise [4,25]. Here, the particular shape of the noise model does not have to be known. The only additional assumption we make (following the original NOISE2VOID [9]) is that the added noise is centered around zero, that is the expected value of the noisy observations at a pixel is equal to the signal $\mathbb{E}_{p_{\mathrm{NM}}(x_i, s_i)}[x_i] = s_i$.

3.2 Denoising Task

Given an observed noisy image $\mathbf{x}$, the denoising task as we consider it in this paper is to find a suitable estimate $\hat{\mathbf{s}} \approx \mathbf{s}$. Note that this is different from the deconvolution task, attempting to find an estimate $\hat{\mathbf{z}} \approx \mathbf{z}$ for the original phantom image.

3.3 Blind Spot Denoising Recap

In the originally proposed NOISE2VOID, the network is seen as implementing a function $\hat{s}_i = f(\mathbf{x}_i^{\mathrm{RF}}; \theta)$, that predicts an estimate for each pixel's signal $\hat{s}_i$ from its surrounding patch $\mathbf{x}_i^{\mathrm{RF}}$, which includes the noisy pixel values in a neighborhood around the pixel i but excludes the value x_i at the pixel itself. We use θ to denote the network parameters.

The authors of [9] refer to $\mathbf{x}_i^{\mathrm{RF}}$ as a *blind spot receptive field*. It allows us to train the network using unpaired noisy training images x, with the training loss computed as a sum over pixels comparing the predicted results directly to the corresponding values of the noisy observation

$$\sum_i (\hat{s}_i - x_i)^2 . \tag{2}$$

Note that the blind spot receptive field is necessary for this construction, as a standard network, in which each pixel prediction is also based on the value at the pixel itself would simply learn the identity transformation when trained using the same image as input and as target.

To implement a network with a blind spot receptive field NOISE2VOID uses a standard U-NET [21] together with a masking scheme during training. The loss is only computed for a randomly selected subset of pixels M. These pixels are *masked* in the input image, replacing their value with a random pixel value from a local neighborhood. A network trained in this way acts as if it had a blind spot receptive field, enabling the network to denoise images once it has been trained on unpaired noisy observations.

3.4   Blind Spot Denoising for Diffraction-Limited Data

While the self-supervised NOISE2VOID method [9] can be readily applied to the data **x** with the goal of directly producing an estimate $\hat{s} \approx s$, this is a sub-optimal strategy in our setting.

Considering the above-described process of image formation, we know that, since **s** is the result of a convolution with a PSF, high-frequencies must be drastically reduced or completely removed. It is thus extremely unlikely that the true signal would include high-frequency features as they are *e.g.* visible in the NOISE2VOID result in Fig. 1. While a network might in principle learn this from data, we find that blind spot methods usually fail at this and produce high-frequency artifacts.

To avoid this problem, we propose to add a convolution with the PSF after the U-NET (see Fig. 1). When we now interpret the final output after the convolution as an estimate of the signal $\hat{s} \approx s$, we can be sure that this output is consistent with our model of image formation and can *e.g.* not contain unrealistic high-frequency artifacts.

In addition, we can view the direct output before the convolution as an estimate of the phantom image $\hat{z} \approx z$, *i.e.* an attempt at deconvolution.

To train our model using unpaired noisy data, we adhere to the same masking scheme and training loss (Eq. 2) as in NOISE2VOID. The only difference being that our signal is produced using the additional convolution, thus enforcing the adequate dampening of high-frequencies in the final denoising estimate.

3.5   A Positivity Constraint for the Deconvolved Image

Considering that the predicted deconvolved phantom image $\hat{z}$ describes the distribution of excited fluorophores in our sample (see Sect. 3.1), we know that it cannot take negative values. After all, a negative fluorophore concentration can never occur in a physical sample.

We propose to enforce this constraint using an additional loss component, linearly punishing negative values. Together with the original NOISE2VOID loss our loss is computed as

$$\frac{1}{|M|} \sum_{i \in M} (\hat{s}_i - x_i)^2 + \lambda \frac{1}{N} \sum_{i=1}^{N} \max(0, -\hat{z}_i), \tag{3}$$

where N is the number of pixels and λ is a hyperparameter controlling the influence of the positivity constraint. Note that the new positivity term can be evaluated at each pixel in the image, while the NOISE2VOID component can only be computed at the masked pixels.

4   Experiments and Results

In the following, we evaluate the denoising performance of our method comparing it to various baselines. Additionally, we investigate the effect of the positivity constraint (see Sect. 3.5). Finally, we describe an experiment on the role of the PSF used for reconstruction.

4.1   Datasets

Fluorescence Microscopy Data with Real Noise. We used 6 fluorescence microscopy datasets with real noise.

The *Convallaria* [10,17] and *Mouse actin* [10,17] datasets each consist of a set of 100 noisy images of 1024×1024 pixels showing a static sample. The *Mouse skull nuclei* [10,17] consist of a set of 200 images of 512×512 pixels. In all 3 datasets, the ground truth is derived by averaging all images. We use all 5 images in each dataset for validation and the rest for training. The authors of [10,17] define a region of each image that is to be used for testing, while the whole image can be used for training of self-supervised methods. We adhere to this procedure.

We additionally use data from [26], which provides 3 channels with training and test sets each consisting of 80 and 40, respectively. We use 15% of the training data for validation. Images are 512×512 pixels in size. Note that like [16] we use the raw data made available to us by the authors as the provided normalized data is not suitable for our purpose. The dataset provides 5 different versions of each image with different levels of noise. In this work, we use only the version with the minimum and maximum amount of noise. We will refer to them as *W2S avg1* and *W2S avg16* respectively, as they are created by averaging different numbers of raw images.

Fluorescence Microscopy Data with Synthetic Noise. Additionally, we use 2 fluorescence microscopy datasets from [3] and added synthetic noise. We will refer to them as *Mouse (DenoiSeg)* and *Flywing (DenoiSeg)*. While the original data contains almost no noise, we add pixel-wise Gaussian noise with standard deviation 20 and 70 for *Mouse (DenoiSeg)* and *Flywing (DenoiSeg)*, respectively. Both datasets are split into a training, validation, and test fraction. The *Mouse* dataset, provides 908 images of 128×128 pixels for training, 160 images of the same size as a validation set, and 67 images of 256×256 as a test set. The *Flywing* dataset, provides 1428 images size 128×128 as a training set, 252 images for validation (same size), and also 42 images size 512×512 as test set. As our method does not require ground truth, we follow [16] and add the test fraction to the training data in order to achieve a fair comparison.

Synthetic Data. While the above-mentioned datasets are highly realistic, we do not know the true PSF that produced the images. To investigate the effect of a mismatch between the true PSF and the PSF used in the training of our method, we used the clean rendered text data from the book *The beetle* [14] previously introduced in [16], synthetically convolved it using a Gaussian PSF with a standard deviation of 1 pixel width. Finally, we added pixel-wise Gaussian noise with a standard deviation of 100. The resulting data consists of 40800 small images of 128×128 pixels in size. We split off a validation fraction of 15%.

4.2 Implementation Details and Training

Our implementation is based on the *pytorch* NOISE2VOID implementation from [10]. We use the exact same network architecture, with the only difference being the added convolution with the PSF at the end of the network.

In all our experiments, we use the same network parameters: A 3-depth U-NET with 1 input channel and 64 channels in the first layer. All networks were trained for 200 epochs, with 10 steps per epoch. We set the initial learning rate to 0.001 and used Adam optimizer, batch size = 1, virtual batch size = 20, and patch size = 100. We mask 3.125% (the default) of pixels in each patch. We use the positivity constraint with $\lambda = 1$ (see Sect. 3.5).

4.3 Denoising Performance

We report the results for all fluorescence microscopy datasets in Table 1. The performance we can achieve in our denoising task is measured quantitatively by calculation of the average peak signal-to-noise ratio (**PSNR**). Qualitative results can be found in Fig. 2.

We run our method using a Gaussian PSF with a standard deviation of 1 pixel width for all datasets. Figure 2 shows examples of denoising results on different datasets.

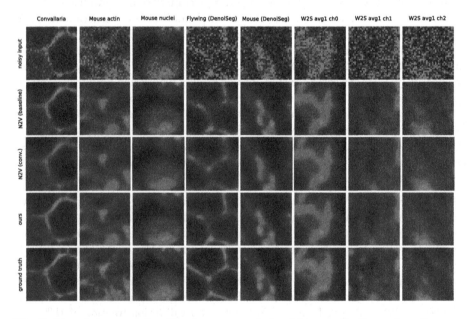

Fig. 2. Denoising results. We show cropped denoising results for various fluorescence microscopy datasets. Our method achieves considerable visual improvements for all datasets compared to NOISE2VOID. The *N2V (conv.)* baseline corresponds to the NOISE2VOID result convolved with the same PSF we use for our proposed method.

Table 1. Quantitative Denoising Results. We report the average peak signal to noise ratio for each dataset and method. Here, *ours+* and *ours−* correspond to our method with ($\lambda = 1$) and without positivity constraint ($\lambda = 0$), see Sect. 3.5 for details. The best results among self-supervised methods without noise model are highlighted in bold. The best results overall are underlined. Here *DivN.* is short for DivNoising [16].

Dataset/network	Raw data	Self-supervised						Superv.
		No noise model				Noise model		
		N2V	N2V conv.	ours−,	ours+	PN2V	DivN.	CARE
Convallaria	28.98	35.85	32.86	**36.39**	36.26	36.47	<u>36.94</u>	36.71
Mouse actin	23.71	33.35	33.48	33.94	**34.04**	33.86	33.98	<u>34.20</u>
Mouse nuclei	28.10	35.86	34.59	**36.34**	36.27	36.35	36.31	<u>36.58</u>
Flywing (DenoiSeg)	11.15	23.62	23.51	24.10	**24.30**	24.85	25.10	<u>25.60</u>
Mouse (DenoiSeg)	20.84	33.61	32.27	**33.91**	33.83	34.19	34.03	<u>34.63</u>
W2S avg1 ch0	21.86	34.30	34.38	**34.90**	34.24	-	34.13	34.30
W2S avg1 ch1	19.35	31.80	32.23	**32.31**	32.24	-	32.28	32.11
W2S avg1 ch2	20.43	34.65	**35.19**	35.03	35.09	32.48	35.18	34.73
W2S avg16 ch0	33.20	38.80	38.73	**39.17**	37.84	39.19	39.62	<u>41.94</u>
W2S avg16 ch1	31.24	37.81	37.49	**38.33**	38.19	38.24	38.37	<u>39.09</u>
W2S avg16 ch2	32.35	40.19	40.32	40.60	**40.74**	40.49	40.52	<u>40.88</u>

To assess the denoising quality of our method we compare its results to various baselines. We compared our method to Noise2Void, noise model based self-supervised methods (PN2V [10], DivNoising [16]), as well as the well-known supervised CARE [24] approach. While we run Noise2Void ourselves, the PSNR values for all other methods were taken from [16].

We created a simple additional baseline by convolving the Noise2Void result with the same PSF used in our own method. This baseline is referred to as *N2V (conv.)*.

4.4 Effect of the Positivity Constraint

Here we want to discuss the effect of the positivity constraint (see Sect. 3.5) on the denoising and deconvolution results. We compare our method without positivity constraint ($\lambda = 0$, see Eq. 3) and with positivity constraint ($\lambda = 1$). Choosing different values for λ did not have a noticeable effect.

We find that the constraint does not provide a systematic advantage or disadvantage with respect to denoising quality (see Table 1). In Fig. 3 we compare the results visually. While it is difficult to make out any differences in the denoising results, we see a stunning visual improvement for the deconvolution result when the positivity constraint is used. While the deconvolution result without positivity constraint contains various artifacts such as random repeating structures and grid patterns, these problems largely disappear when the positivity constraint is used.

We find it is an interesting observation that such different predicted phantom images can lead to virtually indistinguishable denoising results after convolution with the PSF, demonstrating how ill-posed the unsupervised deconvolution problem really is.

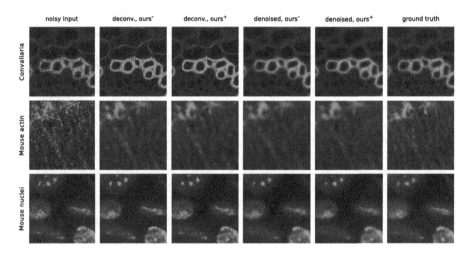

Fig. 3. Effect of the proposed Positivity Constraint. We show cropped denoising and deconvolution results from various datasets with (*ours+*) and without positivity constraint (*ours−*), see Sect. 3.5 for details. While the denoising results are almost indistinguishable, the deconvolution results show a drastic reduction of artifacts when the positivity constraint is used.

4.5 Effect of the Point Spread Function

Here we want to discuss an additional experiment on the role of the PSF used in the reconstruction and the effect of a mismatch with respect to the PSF that actually produced the data.

We use our synthetic *The beetle* dataset (see Sect. 4.1) that has been convolved with a Gaussian PSF with a standard deviation of $\sigma = 1$ pixel width and was subject to Gaussian noise of standard deviation 100. We train our method on this data using different Gaussian PSFs with standard deviations between $\sigma = 0$ and $\sigma = 2$. We used an active positivity constraint with $\lambda = 1$. The results of the experiment can be found in Fig. 4.

We find that the true PSF of $\sigma = 1$ gives the best results. While lower values lead to increased artifacts, similar to those produced by NOISE2VOID, larger values lead to an overly smooth result.

Fig. 4. Effects of Point Spread Function Mismatch. We use synthetic data to investigate how the choice of PSF influences the resulting denoising quality. The data was generated by convolving rendered text with a Gaussian PSF of standard deviation $\sigma = 1$ (highlighted in red) and subsequently adding noise. Here, we show the results of our method when trained using Gaussian PSFs of various sizes. We achieve the best results by using the true PSF. Smaller PSFs produce high-frequency artifacts. Larger PSFs produce overly smooth images.

5   Discussion and Outlook

Here, we have proposed a novel way of improving self-supervised denoising for microscopy, making use of the fact that images are typically diffraction-limited. While our method can be easily applied, results are often on-par with more sophisticated second-generation self-supervised methods [10,16]. We believe that the simplicity and general applicability of our method will facilitate fast and widespread use in fluorescence microscopy where oversampled and diffraction-limited data is the default. While the standard deviation of the PSF is currently a parameter that has to be set by the user, we believe that future work can optimize it as a part of the training procedure. This would provide the user with an *de facto* parameter-free turn-key system that could readily be applied to unpaired noisy raw data and achieve results very close to supervised training.

In addition to providing a denoising result, our method outputs a deconvolved image as well. Even though deconvolution is not the focus of this work, we find that including a positivity constraint in our loss enables us to predict visually plausible results. However, the fact that dramatically different predicted deconvolved images give rise to virtually indistinguishable denoising results (see Fig. 3) illustrates just how underconstrained the deconvolution task is. Hence, further regularization might be required to achieve deconvolution results of optimal quality. In concurrent work, Kobayashi *et al.* [8] have generated deconvolution results in a similar fashion and achieved encouraging results in their evaluation. We expect that future work will quantify to what degree the positivity constraint and other regularization terms can further improve self-supervised deconvolution methods.

We believe that the use of a convolution after the network output to account for diffraction-limited imaging will in the future be combined with noise model based techniques, such as the self-supervised [10,11] or with novel techniques like

DivNoising. In the latter case, this might even enable us to produce diverse deconvolution results and allow us to tackle uncertainty introduced by the under-constrained nature of the deconvolution problem in a systematic way.

Code Availability. Our code is available at https://github.com/juglab/DecoNoising.

Acknowledgments. We thank the Scientific Computing Facility at MPI-CBG for giving us access to their HPC cluster.

References

1. Batson, J., Royer, L.: Noise2Self: blind denoising by self-supervision (2019)
2. Belthangady, C., Royer, L.A.: Applications, promises, and pitfalls of deep learning for fluorescence image reconstruction. Nat. Methods **16**, 1–11 (2019)
3. Buchholz, T.O., Prakash, M., Krull, A., Jug, F.: DenoiSeg: joint denoising and segmentation. arXiv preprint arXiv:2005.02987 (2020)
4. Foi, A., Trimeche, M., Katkovnik, V., Egiazarian, K.: Practical Poissonian-Gaussian noise modeling and fitting for single-image raw-data. IEEE Trans. Image Process. **17**(10), 1737–1754 (2008)
5. Hendriksen, A.A., Pelt, D.M., Batenburg, K.J.: Noise2Inverse: self-supervised deep convolutional denoising for linear inverse problems in imaging (2020)
6. Khademi, W., Rao, S., Minnerath, C., Hagen, G., Ventura, J.: Self-supervised Poisson-Gaussian denoising. arXiv preprint arXiv:2002.09558 (2020)
7. Kingma, D.P., Welling, M.: Auto-encoding variational bayes. In: Bengio, Y., LeCun, Y. (eds.) 2nd International Conference on Learning Representations, ICLR 2014, Banff, AB, Canada, 14–16 April 2014, Conference Track Proceedings (2014). http://arxiv.org/abs/1312.6114
8. Kobayashi, H., Solak, A.C., Batson, J., Royer, L.A.: Image deconvolution via noise-tolerant self-supervised inversion. arXiv preprint arXiv:2006.06156 (2020)
9. Krull, A., Buchholz, T.O., Jug, F.: Noise2Void-learning denoising from single noisy images. In: Proceedings of the IEEE Conference on Computer Vision and Pattern Recognition, pp. 2129–2137 (2019)
10. Krull, A., Vicar, T., Prakash, M., Lalit, M., Jug, F.: Probabilistic Noise2Void: unsupervised content-aware denoising. Front. Comput. Sci. **2**, 60 (2020)
11. Laine, S., Karras, T., Lehtinen, J., Aila, T.: High-quality self-supervised deep image denoising. In: Advances in Neural Information Processing Systems, pp. 6968–6978 (2019)
12. Lehtinen, J., et al.: Noise2noise: learning image restoration without clean data. In: International Conference on Machine Learning, pp. 2965–2974 (2018)
13. Luisier, F., Blu, T., Unser, M.: Image denoising in mixed Poisson-Gaussian noise. IEEE Trans. Image Process. **20**(3), 696–708 (2010)
14. Marsh, R.: The Beetle. Broadview Press (2004)
15. Metzler, C.A., Mousavi, A., Heckel, R., Baraniuk, R.G.: Unsupervised learning with stein's unbiased risk estimator. arXiv preprint arXiv:1805.10531 (2018)
16. Prakash, M., Krull, A., Jug, F.: Divnoising: diversity denoising with fully convolutional variational autoencoders. arXiv preprint arXiv:2006.06072 (2020)

17. Prakash, M., Lalit, M., Tomancak, P., Krull, A., Jug, F.: Fully unsupervised probabilistic Noise2Void. In: 2020 IEEE 17th International Symposium on Biomedical Imaging (ISBI), pp. 154–158. Iowa City, IA, USA (2020). https://doi.org/10.1109/ISBI45749.2020.9098612

18. Ramani, S., Blu, T., Unser, M.: Monte-Carlo sure: a black-box optimization of regularization parameters for general denoising algorithms. IEEE Trans. Image Process. **17**(9), 1540–1554 (2008)

19. Raphan, M., Simoncelli, E.P.: Learning to be Bayesian without supervision. In: Advances in Neural Information Processing Systems, pp. 1145–1152 (2007)

20. Richardson, W.H.: Bayesian-based iterative method of image restoration. JoSA **62**(1), 55–59 (1972)

21. Ronneberger, O., Fischer, P., Brox, T.: U-Net: convolutional networks for biomedical image segmentation. In: Navab, N., Hornegger, J., Wells, W.M., Frangi, A.F. (eds.) MICCAI 2015. LNCS, vol. 9351, pp. 234–241. Springer, Cham (2015). https://doi.org/10.1007/978-3-319-24574-4_28

22. Stein, C.M.: Estimation of the mean of a multivariate normal distribution. Ann. Stat. **9**(6), 1135–1151 (1981)

23. Ulyanov, D., Vedaldi, A., Lempitsky, V.: Deep image prior. In: CVPR (2018)

24. Weigert, M., et al.: Content-aware image restoration: pushing the limits of fluorescence microscopy. Nat. Methods **15**(12), 1090–1097 (2018)

25. Zhang, Y., et al.: A Poisson-Gaussian denoising dataset with real fluorescence microscopy images. In: CVPR (2019)

26. Zhou, R., Helou, M.E., Sage, D., Laroche, T., Seitz, A., Süsstrunk, S.: W2S: a joint denoising and super-resolution dataset. arXiv preprint arXiv:2003.05961 (2020)

Learning to Restore ssTEM Images from Deformation and Corruption

Wei Huang, Chang Chen, Zhiwei Xiong$^{(\boxtimes)}$, Yueyi Zhang, Dong Liu, and Feng Wu

University of Science and Technology of China, Hefei, China
zwxiong@ustc.edu.cn

Abstract. Serial section transmission electron microscopy (ssTEM) plays an important role in biological research. Due to the imperfect sample preparation, however, ssTEM images suffer from inevitable artifacts that pose huge challenges for the subsequent analysis and visualization. In this paper, we propose a novel strategy for modeling the main type of degradation, *i.e.*, Support Film Folds (SFF), by characterizing this degradation process as a combination of content deformation and corruption. Relying on that, we then synthesize a sufficient amount of paired samples (degraded/groundtruth), which enables the training of a tailored deep restoration network. To the best of our knowledge, this is the first learning-based framework for ssTEM image restoration. Experiments on both synthetic and real test data demonstrate the superior performance of our proposed method over existing solutions, in terms of both image restoration quality and neuron segmentation accuracy.

Keywords: Degradation modeling · Image restoration · Deep learning

1 Introduction

The development of electron microscopy (EM) at synapse resolution has greatly promoted the study of neuron morphology and connectomics, which is essential for understanding the working principle of intelligence. Among different EM imaging techniques [2,6,11,15,31,34], serial section transmission electron microscopy (ssTEM) has a clear advantage in the imaging speed and resolution [9,11,34]. It is thus widely adopted in analyzing the connectivity in volumetric samples of brain tissue by imaging many thin sections in sequence. Recently, relying on the ssTEM technique, researchers have imaged the first complete EM volume of the brain of adult *Drosophila* melanogaster [34], which is regarded as a milestone in brain science.

Due to the imperfect sample preparation, however, ssTEM images suffer from inevitable artifacts. Among these artifacts, one type of degradation is most common and has a significant influence on the image quality, *i.e.*, Support Film Folds

Electronic supplementary material The online version of this chapter (https://doi.org/10.1007/978-3-030-66415-2_26) contains supplementary material, which is available to authorized users.

© Springer Nature Switzerland AG 2020
A. Bartoli and A. Fusiello (Eds.): ECCV 2020 Workshops, LNCS 12535, pp. 394–410, 2020.
https://doi.org/10.1007/978-3-030-66415-2_26

(SFF, accounting for 3.2% of samples) [9]. SFF is caused by imperfect cutting in a complicated way. Figure 1(a) visualizes this process with an intuitive example, where a 2D cartoon image is folded along a certain line and content around this line disappears in the deformed image. Figure 1(b) shows three consecutive images with the middle one suffering from SFF degradation. As can be seen, the folding effect not only results in content corruption (in a form of dark line), but also introduces severe deformation of surrounding regions which thus have large misalignments with the adjacent images.

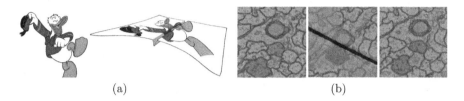

(a)   (b)

Fig. 1. (a) A cartoon image as an intuitive example to demonstrate SFF degradation. (b) Three consecutive ssTEM images where the middle one is with SFF artifacts.

The quality of ssTEM images is greatly reduced with the aforementioned SFF degradation, which severely hinders the subsequent tasks, such as alignment [25] and segmentation [8,16]. For example, in the neuron segmentation task, one image with artifacts could interrupt many neuron structures, resulting in erroneous reconstruction (see Fig. 7 for exemplar cases). Li *et al.* [16] also indicated that neuron segments are interrupted at multiple sections due to SFF degradation. However, it is challenging to restore ssTEM images from SFF degradation. Compared with degradations that generally occur on 2D natural images, SFF exhibits drastically different characteristics, making classic restoration methods difficult to apply. On the other hand, since there is no corresponding groundtruth for the degraded ssTEM image, it is also difficult to directly leverage on the power of deep learning based restoration. To the best of our knowledge, ssTEM image restoration from SFF degradation still remains an open problem.

To fill this gap, we propose to model SFF degradation as a combination of content deformation and corruption. Backed up by a statistic analysis from abundant real samples, we design an algorithm to synthesize degraded images from artifacts-free ones, where the latter are adopted as the corresponding groundtruth. Leveraging on the above obtained image pairs for training, we then propose a tailored deep learning framework, which consists of three modules, *i.e.*, interpolation, unfolding, and fusion. Specifically, the interpolation module utilizes the adjacent images to obtain an interpolated image, which is then used as a reference for the unfolding and fusion modules to address content deformation and corruption, respectively.

As demonstrated by comprehensive experimental results, our method significantly outperforms existing solutions for SFF restoration in ssTEM images, both quantitatively and qualitatively. With the generalizability of the degradation modeling and the scalability of the restoration network, the proposed method could play an essential role in bridging the gap between ssTEM images acquisition and subsequent analysis tasks, which thus facilitates the research of neuron morphology and connectomics.

The main contributions of this work are summarized as follows:

- We propose the first learning-based framework for ssTEM image restoration from SFF artifacts.
- We conduct a comprehensive analysis on the statistics of SFF. Based on the degradation modeling derived from the analysis, we propose a synthesis algorithm to generate degraded/groundtruth image pairs for training the deep restoration network.
- Experiments on both synthetic and real test data demonstrate the advantage of our proposed method, in terms of both image restoration quality and neuron segmentation accuracy.

2   Related Work

EM Image Restoration. Mainstream EM imaging techniques can be divided into two groups: transmission EM (TEM) [7,11,29,34] (with ssTEM as a representative) and scanning EM (SEM) [4,6,15,31]. These two types of techniques have different imaging advantages, and the acquired EM images suffer from different artifacts due to the flaws of respective imaging principles. Previously, EM image restoration is largely investigated within the scope of denoising [26], where the target degradation is relatively simple. Recently, a few works are reported to restore other types of artifacts in EM images. Khalilian-Gourtani et al. [13] proposed a method to detect and correct the striping artifacts in SEM images by solving a variational formulation of the image reconstruction problem. Maraghechi et al. [18] corrected three dominant types of SEM artifacts, i.e., spatial distortion, drift distortion, and scan line shifts under an integrated digital image correlation framework. Meanwhile, Minh-Quan et al. [19] presented an asymmetrically cyclic adversarial network to remove the blob-like artifacts alongside Gaussian noise-like corruption in SEM images and background noise in TEM images. There have been other works addressing axial deformation [28] and varying axial slice thickness artifacts [10]. However, the dominant type of degradation in ssTEM images are seldom investigated, since SFF is much more complicated and challenging compared with the above mentioned ones.

Image Inpainting. Restoration from SFF degradation is kind of similar to the image inpainting task which aims to fill in the corrupted areas in an image. Recently, a number of deep learning based approaches are proposed to accomplish this task, which generate promising results in certain challenging conditions [17,20,24,32,33]. However, directly applying existing inpainting methods would not work for SFF restoration. Take Fig. 1(b) for example, only filling in the dark line cannot fully recover the corrupted content which actually has a larger area than the dark line. Even the dark line is repaired in this way, the resultant image would have large misalignments with the adjacent images.

Substitution/Interpolation. In the neuron segmentation task [16], Li *et al.* completely discarded the degraded image and replaced it with a copy of the adjacent image. This substitution strategy reduces interrupted neurons to a certain extent, but may still lead to inaccurate reconstruction. Another straightforward solution for SFF restoration would be video frame interpolation. However, it is difficult to recover sufficient details since the content of adjacent images is less similar in ssTEM compared to that in common video due to low imaging resolution along the axial direction. Although SFF artifacts are accompanied by content deformation, there is still useful information in the degraded image that should not be discarded along with artifacts. To this end, we propose to synthesize degraded/groundtruth image pairs for training a restoration network, which enables the usage of information in the degraded image.

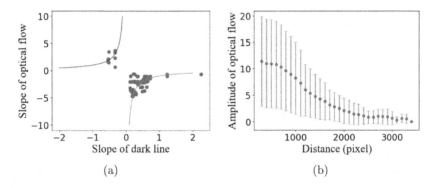

(a)   (b)

Fig. 2. Statistical analysis of SFF degradation from 50 real samples. (a) illustrates the relationship between the slope of optical flow and the slope of dark line. Red points represent the slope values on real samples. The curve is the inverse function: $y = -1/x$. Within the error range, this curve can fit these red points well, which indicates that the main orientation of deformation is roughly perpendicular to the dark line. (b) illustrates the relationship between the amplitude of optical flow and the radial distance of dark line. The horizontal axis is the radial distance from each point in the image to the dark line. Blue points represent the mean of amplitude and gray lines represent variance. It can be seen that the degree of deformation gradually attenuates away from the corruption. (Color figure online)

3   SFF Modeling

3.1   Statistical Analysis

In order to have a better understanding on the characteristics of SFF degradation, we collect 50 images with SFF artifacts from the Full Adult Fly Brain

(FAFB) data [34]. Based on these samples, we statistically analyze the relationship between content deformation and dark line corruption, which is important for SFF simulation. We adopt optical flow to describe the deformation. A straightforward solution is to extract optical flow by using one of the adjacent images as reference. However, such a solution will suffer from intrinsic estimation error, due to low imaging resolution along the axial direction in ssTEM. Therefore, we first use the video frame interpolation method [22] to obtain an interpolation result from the adjacent images, and then extract optical flow between the degraded image and the interpolated one through a block-matching method [30]. The obtained results are then utilized for statistical analysis.

We take the dark line as the boundary and divide the degraded image into two parts for analysis, since the orientations of optical flows in the two parts are opposite. In order to alleviate the influence of estimation error, we only consider the main orientation, *i.e.*, the major component in the orientation histogram. We adopt the concept of slope in mathematics to describe the main orientation in a 2D coordinate system. We observe that the main orientations of the dark line and the extracted optical flow are roughly perpendicular to each other. In other words, the product of their slopes should be close to -1. Analysis on the 50 samples verify this observation, as shown in Fig. 2(a). In addition to the orientation information, we also analyze the relationship between the amplitude of optical flow and the radial distance to the dark line. As shown in Fig. 2(b), we find that the degree of deformation gradually attenuates away from the dark line. The above two observations on orientation and amplitude of content deformation serve as the basis of SFF simulation.

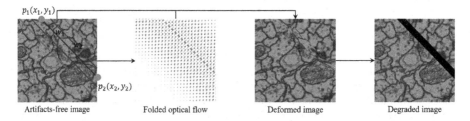

Artifacts-free image   Folded optical flow   Deformed image   Degraded image

Fig. 3. Pipeline to simulate the SFF degradation process. Firstly, we generate the optical flow by using parameters randomly selected in fixed ranges, including two endpoints $p_1(x_1, y_1)$ and $p_2(x_2, y_2)$, the width of dark line w_1, and the width of actually corrupted region w_2. Then, based on the optical flow, we warp the artifacts-free image to obtain the deformed image. Finally, we add a straight black line on top of the deformed image to get the degraded result. (Color figure online)

3.2 SFF Simulation

Realistic artifacts are generated during the acquisition process of ssTEM images, yet we cannot get the groundtruth corresponding to these degraded samples. It is

---

Algorithm 1: Folded optical flow synthesis.

Input: I: Artifacts-free image

$p_1(x_1, y_1)$ and $p_2(x_2, y_2)$: Two endpoints on two different boundaries of image I

w_1: Width of dark line

w_2: Width of actually corrupted region $(w_2 \geq w_1)$

α: Amplitude decay factor of optical flow

H: Height of image I

W: Width of image I

1 $p_1(x_1, y_1)$ and $p_2(x_2, y_2)$ determine the linear function of dark line l:
$y = k_l * x + b_l, k_l = (y_2 - y_1)/(x_2 - x_1), b_l = y_1 - k * x_1$.

2 The slope of optical flow: $k_f = -1/k_l$.

3 **for** $i \in [1, H]$ **do**

4   **for** $j \in [1, W]$ **do**

5   The distance from point (i, j) to the line l:
      $d(i, j) = |(k_l * i - j + b_l)/\sqrt{k_l^2 + 1}|$

6   The amplitude of optical flow:
      $A(i, j) = \alpha * d(i, j) + b_f, b_f = (w_2 - w_1) - \alpha * w_2$

7   **if** $A(i, j) < 0$ **then**

8           $A(i, j) = 0$

9   **end**

10        $F(i, j, 1) = A(i, j) * \cos(\arctan(k_f))$

11        $F(i, j, 2) = A(i, j) * \sin(\arctan(k_f))$

12   **endfor**

13 **endfor**

Output: $F[H, W, 2]$: Folded optical flow.

---

thus difficult to directly train a deep restoration network which generally requires a large amount of paired samples (degraded/groundtruth). To address this issue, we propose an effective modeling strategy considering the aforementioned characteristics of SFF degradation, relying on which we can then synthesize a sufficient amount of paired samples to train a deep restoration network.

As demonstrated in Fig. 1(b), SFF degradation consists of two kinds of artifacts: corruption in the dark line and deformation of surrounding regions. Ideally, if we can unfold the dark line in the correct way, the deformation issue can be addressed and then filling up the corrupted content would not be difficult with the assistance of adjacent images. The challenge is that, since the folding effect during sample preparation is highly non-rigid, the deformation could vary in different regions and the corrupted content is unpredictable. Therefore, accurately modeling SFF degradation in an analytical way is difficult. Instead, we simulate SFF degradation using the following strategy as shown in Fig. 3.

First, we collect a number of artifacts-free images from FAFB. According to the characteristics analyzed in Sect. 3.1, we produce simulated optical flows to deform the artifacts-free images (see Algorithm 1 for detailed implementation). Specifically, two endpoints $p_1(x_1, y_1)$ and $p_2(x_2, y_2)$ are randomly generated on two different boundaries of the image, which determine the slope of the dark line and thus the slope of optical flow (since the two are orthogonal). For the amplitude of the optical flow, we randomly assign the width of the dark line (w_1) and the width of actually corrupted region ($w_2, w_2 \geq w_1$). For simplicity, we only consider a linear decay relationship between the amplitude of the optical flow and the radial distance to the dark line.

Based on the above synthesized optical flows, the artifacts-free images are deformed correspondingly and the content falling in the corrupted region with width w_2 is masked by a black line with width w_1, as shown in Fig. 3. In practice, however, the realistic SFF artifacts may not satisfy the ideal corruption in the simulation process, *i.e.*, with a straight line shape and zero intensity values. To address this issue, we propose to cover a straight black line (with a slightly larger width) on top of the realistic dark line for regularization during the inference stage, which contributes to the restoration performance (see ablation study in Sect. 5.5). In addition to the optical flow for simulating the deformation process, we also generate the inverse optical flow for the unfolding purpose, which serves as the groundtruth in the unfolding module, as detailed in Sect. 4.2.

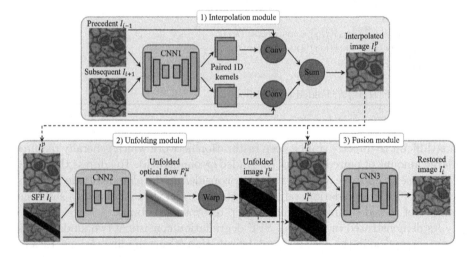

Fig. 4. The proposed restoration framework consists of three modules: interpolation, unfolding and fusion. Interpolation module takes the two adjacent images as input to obtain the interpolated result, which is used as the reference image in the next two modules. Unfolding module aims to address the content deformation in the degraded image. Fusion module is designed to fill up the corrupted content in the unfolded image. Detailed specifications of each module can be found in the supplementary material.

4 SFF Restoration

4.1 Artifacts Detection

In practice, automatic detection of SFF artifacts is highly desired, especially when handling a vast amount of samples from the whole brain. In this work, we adopt a simple yet effective line detector, *i.e.*, Hough transformation [5], to achieve automatic detection of the dark line corruption caused by SFF degradation [9,25]. Specifically, we perform Hough transformation in the binarized version of each image. For the binarization process, we traverse different threshold settings and find a suitable one in a validation dataset. Hough transformation can also localize the dark line, which enables regularization of corruption for a better restoration, as mentioned in Sect. 3.2.

4.2 Restoration Framework

The proposed restoration framework is composed of three modules, *i.e.*, interpolation, unfolding, and fusion, as shown in Fig. 4. On the one hand, the interpolation result is adopted as the reference for the optical-flow based unfolding, which addresses the deformation caused by SFF degradation. On the other hand, the interpolated image is fused with the unfolded image, which fills up the corrupted content in the dark line. For the implementation of each module, advanced network design may lead to a better performance. Yet, as the first attempt of learning-based restoration, we focus more on the realization of functions.

Interpolation Module. Our interpolation module is built upon the kernel prediction network (KPN) [21], which is originally designed for the video frame interpolation task. Here, we view two adjacent images (I_{i-1} and I_{i+1}) as input and predict the corresponding per-pixel kernels to interpolate I_i^p in between. Specifically, the predicted kernels are applied to input images in a convolutional manner. Following the implementation of [22], we adopt two separable 1D kernels as a replacement of 2D kernel for efficiency. Details of the backbone structure can be found in the supplementary material.

Unfolding Module. In this module, we adopt optical flow to represent the per-pixel position correspondence between a pair of input images. Based on the estimated optical flow, the deformation in the degraded image I_i can be addressed by a warping operation, as shown in Fig. 4. We term this alignment process as unfolding. We adopt a residual variant of U-Net [27] to implement the flow estimation network, where the degraded image I_i and the reference image I_i^p from the interpolation module form the input pair. Different from classical networks for optical flow estimation, our unfolding module can directly estimate optical flow at full image resolution, which eliminates the checkerboard effects caused by subsequent up-sampling. More specifically, the network is composed of four down-sampling and four up-sampling blocks. Each block contains three regular convolutional layers stacking with a skip connection.

Fusion Module. Intuitively, to fill up the corrupted content in the unfolded image I_i^u from the unfolding module, one can directly crop the corresponding part in the interpolated image I_i^p and stitch it with the rest part of I_i^u. However, direct stitching is not optimal since the unfolded image may not be exactly aligned with the interpolated one. On the other hand, in addition to the content deformation and corruption, SFF degradation usually decreases the contrast of image. The above two issues call for an independent module to serve as an advanced solution for image fusion. In our implementation, a simplified version of U-Net [27] is adopted here, which addresses the potential misalignment and the low contrast issues for a better restoration performance (as detailed in Sect. 5.5).

Loss Function. We train the above three modules independently and cascade them in the order of interpolation, unfolding, and fusion. In the proposed restoration framework, we adopt the L_1 distance between the input and the groundtruth as the loss function for each module. Specifically, given the simulated input with SFF artifacts, its artifacts-free version is adopted to supervise the interpolation and fusion modules, and the simulated inverse optical flow is adopted to supervise the unfolding module.

5 Experiments

5.1 Data Preparation

In the training and validation phase, all data we use is from FAFB, the first ssTEM volume of a complete adult drosophila brain imaged at 4×4 nm resolution and sectioned at 40 nm thickness [34]. There are a total of 7062 sections in FAFB, and the original resolution of each section is 286720×155648 which is partitioned into 8192×8192 images, resulting in 40 TB data in storage. The SFF artifacts frequently occur in FAFB, which severely hinder the analysis and visualization on this valuable data. We select a central cube out of the raw volume as our main experimental data (approximately 150 GB) to demonstrate the effectiveness of the proposed method, yet the results could generalize to other portions of the volume. We further partition the selected cube into 512×512 images for easy manipulation. From this cube, we randomly select 4000 artifacts-free samples to generate the training data. Each sample contains three consecutive images, the middle image is used for the simulation of SFF degradation. In addition, we also select 100 samples as validation data for hyper-parameter tuning.

In the test phase, we use the public data from the CREMI challenge in 2016 [3] that aims to facilitate the neuron reconstruction in ssTEM images. The CREMI dataset is also for adult drosophila brain and with the same imaging resolution as FAFB. It consists of three subsets corresponding to different neuron types, each containing 125 images for training and 125 images for testing. Since the training images have manually obtained segmentation labels, we adopt them as the test images in our task to quantitatively evaluate the proposed method in terms of both image restoration quality and neuron segmentation accuracy.

Specifically, we select a few samples out of an image bunch in each subset for SFF simulation, and the restored images are evaluated against the original ones for calculating quantitative metrics. Besides the synthetic test data used above, we also select a few number of real samples with SFF artifacts from FAFB (no overlap with the training and validation sample) for test, where only qualitative results are reported.

5.2   Implementation Details

During the statistical analysis of SFF degradation (in Sect. 3.1), we adopt a block matching method to generate a sparse optical flow between the degraded image and the interpolation result. The block size is set to 71×71 with an overlap of 41 pixels, and the maximum search range is 21 pixels. The estimated sparse optical flow is then upsampled to a dense one at the original resolution. During SFF simulation (in Sect. 3.2), there are three parameters, $e.g.$, the width of dark line w_1, the width of corrupted region w_2 and the decay factor of optical flow α. To increase the diversity of simulated samples, we set the three parameters to be within a dynamic range. The ranges of w_1, w_2 and α are $(5, 30)$, $(w_1, 80)$ and $(-0.1, -0.0001)$, respectively. During each iteration of the training process, we randomly generate a set of parameters to simulate SFF degradation on the training samples.

We adopt the same training setting for the three modules in our restoration framework. The resolution of input image to the network is set as 256, which is obtained by random cropping from the training samples. We train these networks using Adam [14] with $\beta_1 = 0.9$, $\beta_2 = 0.999$, a learning rate of 0.0001, and a batch size of 32 on four *NVIDIA Titan Xp* GPUs. We perform random rotation and flip in the training phase for data augmentation. In addition, we perform random contrast and brightness adjustment to address the contrast and brightness variation around the dark line. We adopt the early termination strategy to obtain an optimal model on the validation data. Each module requires about one day for training. In the inference phase, it takes 0.42 s to process an image with 2048×2048 resolution.

5.3   Restoration on Synthetic Data

To quantitatively evaluate the proposed method, we first conduct experiments on three subsets from CREMI (termed as A, B, and C), where the degraded images (SFF) are synthesized from the artifacts-free ones. For each subset, we select 25 out of 125 samples for simulation. Four representative restoration strategies are adopted for comparison, including: (a) the substitution strategy (Sub.) which directly replaces the degraded image with the previous adjacent image, (b) the inpainting strategy which directly repairs the dark line, without consideration of the surrounding deformation. We adopt the state-of-the-art image inpainting method, $i.e.$, Partial Convolution (PC) [17], as a representative. We fine-tune the PC model pretrained on natural images with our simulated training samples. (c) Inpainting with unfolding (PC-unfold), where the PC model is fine-tuned

Table 1. Quantitative comparisons of restoration results on synthetic data from CREMI dataset, in terms of both image restoration quality and neuron segmentation accuracy. For each subset, 25 out of 125 images are selected to synthesize SFF artifacts.

CREMI	Metric	Method					
		SFF	Sub. [16]	PC [17]	PC-unfold	Interp [22]	Ours
A	PSNR ↑	15.57	19.57	16.35	18.01	22.70	**26.20**
	SSIM ↑	0.5615	0.4353	0.5665	0.7517	0.6595	**0.8261**
	FID ↓	229.71	33.30	36.83	62.44	144.69	**27.80**
	VOI ↓	2.6780	1.2505	2.6440	0.9507	0.8967	**0.7833**
	ARAND ↓	0.4971	0.2881	0.4783	0.1442	0.1518	**0.0968**
B	PSNR ↑	15.11	18.16	16.43	18.17	22.22	**26.81**
	SSIM ↑	0.5842	0.3586	0.6161	0.7532	0.6041	**0.8202**
	FID ↓	260.27	50.54	41.78	53.57	175.26	**38.17**
	VOI ↓	4.0629	3.8864	3.7147	3.4817	3.1898	**3.0957**
	ARAND ↓	0.5806	0.5855	0.3931	0.4355	0.3680	**0.3517**
C	PSNR ↑	14.52	17.74	15.09	16.97	21.96	**25.74**
	SSIM ↑	0.4988	0.3066	0.5037	0.7257	0.5766	**0.7957**
	FID ↓	335.38	42.70	44.60	73.33	168.13	**42.26**
	VOI ↓	4.5572	3.6755	4.5789	3.4882	3.1606	**3.0825**
	ARAND ↓	0.4244	0.4280	0.4308	0.3309	0.2835	**0.2789**
Inference time (s)		*	*	0.1235	0.5130	0.0075	0.4251

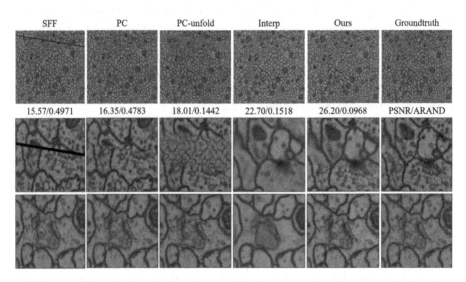

Fig. 5. Visual comparison of restoration results on *synthetic* test data. More visual comparison results on the synthetic test data can be found in the supplementary material. (Color figure online)

with the unfolded images obtained from our proposed unfolding module. (d) The interpolation strategy (Interp) [22] which completely discards the degraded image and generates the result from two adjacent images.

SFF   PC   PC-unfold   Interp   Ours   Subsequent

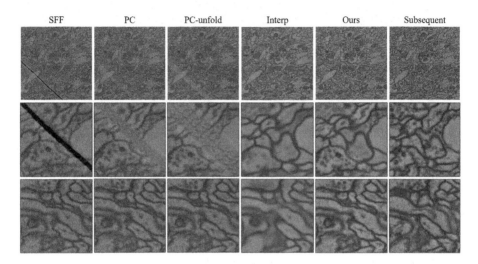

Fig. 6. Visual comparison of restoration results on *real* test data. More visual comparisons can be found in the supplementary material. Note that, due to the lack of groundtruth, we include the subsequent image to the degraded one as a reference here. (Color figure online)

As shown in Table 1, the quantitative results on the three subsets demonstrate that our proposed method outperforms the other restoration strategies by a large margin, in terms of image restoration quality. Besides two widely used fidelity metrics PSNR and SSIM, we also adopt a perceptual index for evaluation, *i.e.*, Fréchet Inception Distance (FID) [12], which is generated by computing the feature distance between the restored and groundtruth images. This perceptual index validates the superiority of the proposed method again.

As can be observed from the visual results in Fig. 5 (red box), the plain inpainting method PC only fills in the dark line but cannot address the surrounding deformation. Although the deformation can be largely corrected by PC-unfold with our unfolding module, it still cannot repair the complex neuron structure in the corrupted region only depending on the degraded image itself. On the other hand, as can be seen from Fig. 5 (green box), although the interpolation result gets rid of misalignment with adjacent images, it loses fine details in the original degraded images away from the dark line. In contrast, these details are well preserved in our result, while both the corrupted and deformed regions are recovered.

To validate the effectiveness of our proposed restoration method in the subsequent analysis tasks, we also conduct segmentation experiments on the above

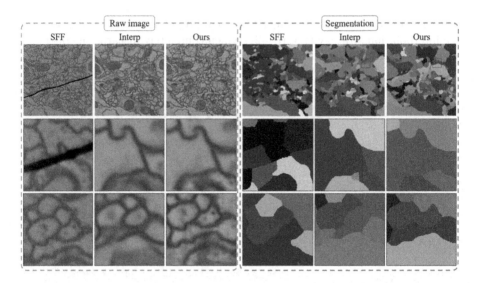

Fig. 7. Exemplar segmentation results on real data with SFF degradation. We adopt a representative segmentation method [8] for evaluation. Each pseudo color represents one neuron. Red box denotes a region near the dark line with severe deformation. Green box denotes a region away from the dark line with fine details. (Color figure online)

restored images. To this end, we utilize a state-of-the-art neuron segmentation method [8], in which a 3D U-Net architecture is used to predict the affinity maps of ssTEM images and the final segmentation results are obtained after the operations of seeded watershed and agglomeration. We use two common metrics for evaluation of segmentation accuracy: adapted Rand error (ARAND) [1] and variation of information (VOI) [23]. As shown in Table 1, quantitative segmentation results demonstrate that our restoration method preserves more details that are useful for subsequent analysis tasks compared with other restoration strategies.

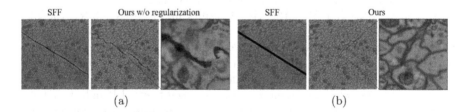

Fig. 8. A visual example to illustrate the effect of regularization of corruption. By covering a straight black line (b) on top of the realistic dark line in the original input (a), a significant improvement of performance is achieved.

Table 2. Ablation results for each module in the proposed restoration framework.

Interpolation	Unfolding	Fusion	PSNR	SSIM	FID	VOI	Rand
✓			22.70	0.6595	144.69	0.8967	0.1518
✓	✓		18.30	0.7393	115.55	0.9416	0.1430
✓		✓	24.68	0.7767	80.54	0.8301	0.1254
	✓	✓	25.94	0.8058	39.96	0.8534	0.1259
✓	✓	✓	**26.20**	**0.8261**	**27.80**	**0.7833**	**0.0968**

5.4 Restoration on Real Data

To evaluate the generalization capability of the proposed method, we further conduct experiments on real degraded images with SFF artifacts from FAFB. As shown in Fig. 6, our proposed method generalizes well to real data. Compared with baseline methods, the superior perceptual quality of recovered images demonstrates the advantage of our proposed method. In addition to perceptual quality, we also conduct segmentation experiments on the recovered images. Due to the lack of groundtruth segmentation labels, we provide qualitative results on one exemplar case to demonstrate the superiority of our proposed method. As shown in Fig. 7 (red box), the interpolation method and our proposed method both obtain good segmentation results near the dark line. Nevertheless, our method preserves more details away from the dark line, which avoids merge errors introduced by interpolation, as shown in Fig. 7 (green box).

5.5 Ablation Study

Function of Module. To verify the function of each module in our restoration framework, we conduct ablation experiments as shown in Table 2. Without the fusion module (Interpolation + Unfolding), the direct stitching cannot ensure the continuity of neuron structure. On the other hand, the issue of low contrast is left unsolved. Therefore, the PSNR score is obviously lower than other settings. Without the unfolding module (Interpolation + Fusion), the useful information in the surrounding area of the dark line cannot be utilized for fusion due to the deformation error. Without the interpolation module (Unfolding + Fusion), the corrupted content cannot be well recovered.

Regularization of Corruption. As described in Sect. 3.2, based on the localization results of Hough transformation, the realistic dark line is automatically covered by a straight black line in the test phase. This regularization process is essential to the restoration performance, which bridges the gap between the synthetic corruption and the realistic one. We demonstrate the effectiveness of this regularization process in Fig. 8.

6   Conclusion

In this paper, we present the first learning-based restoration framework to address content deformation and corruption in ssTEM images, relying on effective modeling and simulation of SFF degradation. Evaluated on both synthetic and real test data, we demonstrate the advantage of our proposed method in terms of both image restoration quality and neuron segmentation accuracy. We believe the proposed method could benefit the future research of neuron morphology and connectomics using ssTEM images.

Acknowledgments. This work was supported in part by Key Area R&D Program of Guangdong Province with grant No. 2018B030338001, Anhui Provincial Natural Science Foundation under grant No. 1908085QF256 and the Fundamental Research Funds for the Central Universities under Grant WK2380000002.

References

1. Arganda-Carreras, I., et al.: Crowdsourcing the creation of image segmentation algorithms for connectomics. Fronti. Neuroanat. **9**, 142 (2015)
2. Bock, D.D., et al.: Network anatomy and in vivo physiology of visual cortical neurons. Nature **471**(7337), 177 (2011)
3. CREMI: Miccal challenge on circuit reconstruction from electron microscopy images. https://cremi.org/ (2016)
4. Denk, W., Horstmann, H.: Serial block-face scanning electron microscopy to reconstruct three-dimensional tissue nanostructure. PLOS Biol. **2**(11), e329 (2004)
5. Duda, R.O., Hart, P.E.: Use of the hough transformation to detect lines and curves in pictures. Commun. ACM **15**(1), 11–15 (1972)
6. Eberle, A., Mikula, S., Schalek, R., Lichtman, J., Tate, M.K., Zeidler, D.: High-resolution, high-throughput imaging with a multibeam scanning electron microscope. J. Microsc. **259**(2), 114–120 (2015)
7. Feist, A., Echternkamp, K.E., Schauss, J., Yalunin, S.V., Schäfer, S., Ropers, C.: Quantum coherent optical phase modulation in an ultrafast transmission electron microscope. Nature **521**(7551), 200 (2015)
8. Funke, J., et al.: Large scale image segmentation with structured loss based deep learning for connectome reconstruction. IEEE Trans. Pattern Anal. Mach. Intell. **41**(7), 1669–1680 (2019)
9. Funke, J.: Automatic neuron reconstruction from anisotropic electron microscopy volumes. Ph.D. thesis, ETH Zurich (2014)
10. Hanslovsky, P., Bogovic, J.A., Saalfeld, S.: Image-based correction of continuous and discontinuous non-planar axial distortion in serial section microscopy. Bioinformatics **33**(9), 1379–1386 (2017)
11. Harris, K.M., Perry, E., Bourne, J., Feinberg, M., Ostroff, L., Hurlburt, J.: Uniform serial sectioning for transmission electron microscopy. J. Neurosci. **26**(47), 12101–12103 (2006)
12. Heusel, M., Ramsauer, H., Unterthiner, T., Nessler, B., Hochreiter, S.: GANs trained by a two time-scale update rule converge to a local nash equilibrium. In: Advances in Neural Information Processing Systems, vol. 30, pp. 6626–6637 (2017)

13. Khalilian-Gourtani, A., Tepper, M., Minden, V., Chklovskii, D.B.: Strip the stripes: artifact detection and removal for scanning electron microscopy imaging. In: ICASSP 2019–2019 IEEE International Conference on Acoustics, Speech and Signal Processing (ICASSP), pp. 1060–1064 (2019)
14. Kingma, D.P., Ba, J.: Adam: a method for stochastic optimization. arXiv preprint arXiv:1412.6980 (2014)
15. Knott, G., Marchman, H., Wall, D., Lich, B.: Serial section scanning electron microscopy of adult brain tissue using focused ion beam milling. J. Neurosci. **28**(12), 2959–2964 (2008)
16. Li, P.H., et al.: Automated reconstruction of a serial-section EM drosophila brain with flood-filling networks and local realignment, p. 605634. bioRxiv (2019)
17. Liu, G., Reda, F.A., Shih, K.J., Wang, T.-C., Tao, A., Catanzaro, B.: Image inpainting for irregular holes using partial convolutions. In: Ferrari, V., Hebert, M., Sminchisescu, C., Weiss, Y. (eds.) ECCV 2018. LNCS, vol. 11215, pp. 89–105. Springer, Cham (2018). https://doi.org/10.1007/978-3-030-01252-6_6
18. Maraghechi, S., Hoefnagels, J., Peerlings, R., Rokoš, O., Geers, M.: Correction of scanning electron microscope imaging artifacts in a novel digital image correlation framework. Exp. Mech. **59**(4), 489–516 (2019)
19. Minh Quan, T., et al.: Removing imaging artifacts in electron microscopy using an asymmetrically cyclic adversarial network without paired training data. In: Proceedings of the IEEE/CVF International Conference on Computer Vision (ICCV) Workshops, October 2019
20. Nazeri, K., Ng, E., Joseph, T., Qureshi, F., Ebrahimi, M.: EdgeConnect: structure guided image inpainting using edge prediction. In: Proceedings of the IEEE/CVF International Conference on Computer Vision (ICCV) Workshops, October 2019
21. Niklaus, S., Mai, L., Liu, F.: Video frame interpolation via adaptive convolution. In: 2017 IEEE Conference on Computer Vision and Pattern Recognition (CVPR), pp. 2270–2279 (2017)
22. Niklaus, S., Mai, L., Liu, F.: Video frame interpolation via adaptive separable convolution. In: 2017 IEEE International Conference on Computer Vision (ICCV), pp. 261–270 (2017)
23. Nunez-Iglesias, J., Ryan Kennedy, T.P., Shi, J., Chklovskii, D.B.: Machine learning of hierarchical clustering to segment 2D and 3D images. PLoS ONE **8**(8), e71715 (2013)
24. Pathak, D., Krähenbühl, P., Donahue, J., Darrell, T., Efros, A.A.: Context encoders: feature learning by inpainting. In: 2016 IEEE Conference on Computer Vision and Pattern Recognition (CVPR), pp. 2536–2544 (2016)
25. Popovych, S., Alexander Bae, J., Seung, H.S.: Caesar: Segment-wise alignment method for solving discontinuous deformations. In: 2020 IEEE 17th International Symposium on Biomedical Imaging (ISBI), pp. 1214–1218 (2020)
26. Roels, J., et al.: An overview of state-of-the-art image restoration in electron microscopy. J. Microsc. **271**(3), 239–254 (2018)
27. Ronneberger, O., Fischer, P., Brox, T.: U-Net: convolutional networks for biomedical image segmentation. In: Navab, N., Hornegger, J., Wells, W.M., Frangi, A.F. (eds.) MICCAI 2015. LNCS, vol. 9351, pp. 234–241. Springer, Cham (2015). https://doi.org/10.1007/978-3-319-24574-4_28
28. Saalfeld, S., Fetter, R., Cardona, A., Tomancak, P.: Elastic volume reconstruction from series of ultra-thin microscopy sections. Nat. Methods **9**(7), 717–720 (2012)
29. Schorb, M., Haberbosch, I., Hagen, W.J., Schwab, Y., Mastronarde, D.N.: Software tools for automated transmission electron microscopy. Nat. Methods **16**(6), 471–477 (2019)

30. Zhu, S., Ma, K.-K.: A new diamond search algorithm for fast block matching motion estimation. In: Proceedings of ICICS, 1997 International Conference on Information, Communications and Signal Processing. Theme: Trends in Information Systems Engineering and Wireless Multimedia Communications, Cat, vol. 1, pp. 292–296 (1997)

31. Tapia, J.C., et al.: High-contrast en bloc staining of neuronal tissue for field emission scanning electron microscopy. Nat. Protoc. **7**(2), 193 (2012)

32. Yang, C., Lu, X., Lin, Z., Shechtman, E., Wang, O., Li, H.: High-resolution image inpainting using multi-scale neural patch synthesis. In: 2017 IEEE Conference on Computer Vision and Pattern Recognition (CVPR), pp. 4076–4084 (2017)

33. Zeng, Y., Fu, J., Chao, H., Guo, B.: Learning pyramid-context encoder network for high-quality image inpainting. In: 2019 IEEE/CVF Conference on Computer Vision and Pattern Recognition (CVPR), pp. 1486–1494 (2019)

34. Zheng, Z., et al.: A complete electron microscopy volume of the brain of adult drosophila melanogaster. Cell **174**(3), 730–743 (2018)

Learning to Segment Microscopy Images with Lazy Labels

Rihuan Ke[1(✉)] , Aurélie Bugeau[2] , Nicolas Papadakis[3] , Peter Schuetz[4],
and Carola-Bibiane Schönlieb[1]

[1] DAMTP, University of Cambridge, Cambridge CB3 0WA, UK
rk621@cam.ac.uk
[2] LaBR, University of Bordeaux, UMR 5800, 33400 Talence, France
[3] CNRS, University of Bordeaux, IMB, UMR 5251, 33400 Talence, France
[4] Unilever R&D Colworth, Bedford MK44 1LQ, UK

Abstract. The need for labour intensive pixel-wise annotation is a major limitation of many fully supervised learning methods for segmenting bioimages that can contain numerous object instances with thin separations. In this paper, we introduce a deep convolutional neural network for microscopy image segmentation. Annotation issues are circumvented by letting the network being trainable on coarse labels combined with only a very small number of images with pixel-wise annotations. We call this new labelling strategy 'lazy' labels. Image segmentation is stratified into three connected tasks: rough inner region detection, object separation and pixel-wise segmentation. These tasks are learned in an end-to-end multi-task learning framework. The method is demonstrated on two microscopy datasets, where we show that the model gives accurate segmentation results even if exact boundary labels are missing for a majority of annotated data. It brings more flexibility and efficiency for training deep neural networks that are data hungry and is applicable to biomedical images with poor contrast at the object boundaries or with diverse textures and repeated patterns.

Keywords: Microscopy images · Multi-task learning · Convolutional neural networks · Image segmentation

1 Introduction

Image segmentation is a crucial step in many microscopy image analysis problems. It has been an active research field in the past decades. Deep learning approaches play an increasingly important role and have become state-of-the-art in various segmentation tasks [12,17,21,27,40]. However, the segmentation of microscopy images is very challenging not only due to the fact that these images are often of low contrast with complex instance structures, but also because of the difficulty in obtaining ground truth pixel-wise annotations [2,15] which hinders the applications of recent powerful but data-hungry deep learning techniques.

© Springer Nature Switzerland AG 2020
A. Bartoli and A. Fusiello (Eds.): ECCV 2020 Workshops, LNCS 12535, pp. 411–428, 2020.
https://doi.org/10.1007/978-3-030-66415-2_27

In this paper, we propose a simple yet effective multi-task learning approach for microscopy image segmentation. We address the problem of finding segmentation with accurate object boundaries from mainly rough labels. The labels are all pixel-wise and contain considerable information about individual objects, but they are created relatively easily. The method is different from pseudo labelling (PL) approaches, which generate fake training segmentation masks from coarse labels and may induce a bias in the masks for microscopy data.

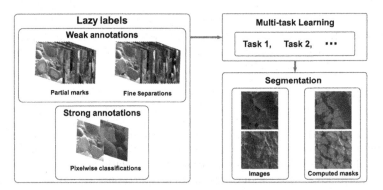

Fig. 1. Multi-task learning for image segmentation with lazy labels. The figure uses Scanning Electron Microscopy (SEM) images of food microstructures as an example and demonstrates a segmentation problem of three classes, namely air bubbles (green), ice crystals (red) and background respectively. Most of the training data are weak annotations containing (i) partial marks of ice crystals and/or air bubbles instances and (ii) fine separation marks of boundaries shared by different instances. Only a *few* strongly annotated images are used. On the bottom right SEM images and their corresponding segmentation outputs from the learned multi-task model are shown. (Color figure online)

To circumvent the need for a massive set of ground truth segmentation masks, we rather develop a segmentation approach that we split into three relevant tasks: detection, separation and segmentation (cf. Fig. 1). Doing so, we obtain a weakly supervised learning approach that is trained with what we call "lazy" labels. These lazy labels contain a lot of coarse annotations of class instances, together with a few accurately annotated images that can be obtained from the coarse labels in a semi-automated way. Contrary to PL approaches, only a very limited number of accurate annotation are considered. In the following, we will refer to weak (resp. strong) annotations for coarse (resp. accurate) labels and denote them as WL (resp. SL).

We reformulate the segmentation problem into several more tractable tasks that are trainable on less expensive annotations, and therefore reduce the overall annotation cost. The *first task* detects and classifies each object by roughly determining its inner region with an under-segmentation mask. Instance counting can be obtained as a by-product of this task. As the main objective is instance detection, exact labels for the whole object or its boundary are not necessary at this stage. We use instead weakly annotated images in which a rough region inside each object is marked, cf. the most top left part of Figure 1. For segmentation problems with a dense population of instances, such as the food components (see e.g., Fig. 1), cells [13,33], glandular tissue, or people in a crowd [42], separating objects sharing a common boundary is a well known challenge. We can optionally perform a *second task* that focuses on the separation of instances that are connected without a clear boundary dividing them. Also for this task we rely on WL to reduce the burden of manual annotations: touching interfaces are specified with rough scribbles, cf. top left part of Fig. 1. Note that this task is suitable for applications with instances that are occasionally connected without clear boundaries. One can alternatively choose to have fewer labelled samples in this task if the annotation cost per sample is higher. The *third task* finally tackles pixel-wise classification of instances. It requires strong annotations that are accurate up to the object boundaries. Thanks to the information brought by weak annotations, we here just need a very small set of accurate segmentation masks, cf. bottom left part of Fig. 1. To that end, we propose to refine some of the coarse labels resulting from task 1 using a semi-automatic segmentation method which requires additional manual intervention.

The three tasks are handled by a single deep neural network and are jointly optimized using a cross entropy loss. In this work we use a network architecture inspired by U-net [33] which is widely used for segmenting objects in microscopy images. While all three tasks share the same contracting path, we introduce a new multi-task block for the expansive path. The network has three outputs and is fed with a combination of WL and SL described above. Obtaining accurate segmentation labels for training is usually a hard and time consuming task. We here demonstrate that having exact labels for a small subset of the whole training set does not degrade training performances. We evaluate the proposed approach on two microscopy image datasets, namely the segmentation of SEM images of food microstructure and stained histology images of glandular tissues.

The contributions of the paper are threefold. (1) We propose a decomposition of the segmentation problems into three tasks and a corresponding user friendly labelling strategy. (2) We develop a simple and effective multi-task learning framework that learns directly from the coarse and strong manual labels and is trained end-to-end. (3) Our approach outperforms the pseudo label approaches on the microscopy image segmentation problems being considered.

2   Related Work

In image segmentation problems, one needs to classify an image at pixel level. It is a vast topic with a diversity of algorithms and applications being considered, including traditional unsupervised methods like k-means clustering [29] that splits the image into homogeneous regions according to image low level features, curve evolution based methods like snakes [7], graph-cut based methods [5,24, 34], just to name a few. Interactive approaches like snakes or Grabcut enable getting involved users' knowledge by means of initializing regions or putting constraints on the segmentation results. For biological imaging, the applications of biological prior knowledge, such as shape statistics [14], semantic information [24] and atlas [10], is effective for automatic segmentation approaches.

2.1 Deep Neural Networks for Segmentation

In the last years, numerous deep convolutional neural network (DCNN) approaches have been developed for segmenting complex images, especially in the semantic setting. In this work, we rely more specifically on fully convolutional networks (FCN) [28], that replace the last few fully connected layers of a conventional classification network by up-sampling layers and convolutional layers, to preserve spatial information. FCNs have many variants for semantic segmentation. The DeepLab [9] uses a technique called atrous convolution to handle spatial information together with a fully connected conditional random field (CRF) [8] for refining the segmentation results. Fully connected CRF can be used as post-processing or can be integrated into the network architecture, allowing for end-to-end training [46].

One type of FCNs commonly used in microscopy and biomedical image segmentation are encoder-decoder networks [1,33]. They have multiple up-sampling layers for better localizing boundary details. One of the most well-known models is the U-net [33]. It is a fully convolutional network made of a contracting path, which brings the input images into very low resolution features with a sequence of down-sampling layers, and an expansive path that has an equal amount of up-sampling layers. At each resolution scale, the features on the contracting path are merged with the corresponding up-sampled layers via long skip connections to recover detailed structural information, e.g., boundaries of cells, after down-sampling.

2.2 Weakly Supervised Learning and Multi-task Learning

Standard supervision for semantic segmentation relies on a set of image and ground truth segmentation pairs. The learning process contains an optimization step that minimizes the distance between the outputs and the ground truths. There has been a growing interest in weakly supervised learning, motivated by the heavy cost of pixel-level annotation needed for fully supervised methods. Weakly supervised learning uses weak annotations such as image-level labels [17,25,30–32,36,47], bounding boxes[21,35], scribbles [26] and points [3].

Many weakly supervised deep learning methods for segmentation are built on top of a classification network. The training of such networks may be realized using segmentation masks explicitly generated from weak annotations [21,25, 40,43,44]. The segmentation masks can be improved recursively, which involves several rounds of training of the segmentation network [11,19,43]. Composite losses from some predefined guiding principles are also proposed as supervision from the weak signals [23,23,38].

Multi-task techniques aim to boost the segmentation performance via learning jointly from several relevant tasks. Tailored to the problems and the individual tasks of interest, deep convolutional networks have been designed, for example, the stacked U-net for extracting roads from satellite imagery [39], the two stage 3D U-net framework for 3D CT and MR data segmentation [41], encoder-decoder networks for depth regression, semantic and instance segmentation [20], or the cascade multi-task network for the segmentation of building footprint [4].

Segmentation of Microscopy and Biomedical Images. Various multi-task deep learning methods have been developed for processing microscopy images and biomedical images. An image level lesion detection task [32] is investigated for the segmentation of retinal red/bright lesions. The work [30] considers to jointly segment and classify brain tumours. A deep learning model is developed in [48] to simultaneously predict the segmentation maps and contour maps for pelvic CT images. In [15], an auxiliary task that predicts centre point vectors for nuclei segmentation in 3D microscopy images is proposed. Denoising tasks, which aims to improve the image quality, can also be integrated for better microscopy image segmentation [6].

In this work, the learning is carried out in a weakly supervised fashion with weak labels from closely related tasks. Nevertheless, the proposed method exploits cheap and coarse pixel-wise labels instead of very sparse image level annotations and is more specialized in distinguishing the different object instances and clarifying their boundaries in microscopy images. The proposed method is completely data-driven and it significantly reduces the annotation cost needed by standard supervision. We aim at obtaining segmentation with accurate object boundaries from mainly coarse pixel-wise labels.

3   Multi-task Learning Framework

The objective of fully supervised learning for segmentation is to approximate the conditional probability distribution of the segmentation mask given the image. Let $s^{(3)}$ be the ground truth segmentation mask and I be the image, then the segmentation task aims to estimate $p(s^{(3)} \mid I)$ based on a set of sample images $\mathcal{I} = \{I_1, I_2, \cdots, I_n\}$ and the corresponding labels $\{s_1^{(3)}, s_2^{(3)}, \cdots, s_n^{(3)}\}$. The set $\mathcal{I}$ is randomly drawn from an unknown distribution. In our setting, having the whole set of segmentation labels $\{s_i^{(3)}\}_{1,\cdots,n}$ is impractical, and we introduce two auxiliary tasks for which the labels can be more easily generated to achieve an overall small cost on labelling.

For a given image $I \in \mathcal{I}$, we denote as $s^{(1)}$ the rough instance detection mask, and $s^{(2)}$ a map containing some interfaces shared by touching objects. All labels $s^{(1)}, s^{(2)}, s^{(3)}$ are represented in one-hot vectors. For the first task, the contours of the objects are not treated carefully, resulting in a coarse label mask $s^{(1)}$ that misses most of the boundary pixels, cf left of Fig. 1. In the second task, the separation mask $s^{(2)}$ only specifies connected objects without clear boundaries rather than their whole contours. Let $\mathcal{I}_k \subset \mathcal{I}$ denote the subset of images labelled for task k $(k = 1, 2, 3)$. As we collect a different amount of annotations for each task, the number of annotated images $|\mathcal{I}_k|$ may not be the same for different k. Typically the number of images with strong annotations satisfies $|\mathcal{I}_3| \ll n$, as the annotation cost per sample is higher.

The set of samples in $\mathcal{I}_3$ for segmentation being small, the computation of an accurate approximation of the true probability distribution $p(s^{(3)} \mid I)$ is a challenging issue. Given that much more samples of $s^{(1)}$ and $s^{(2)}$ are observed, it is simpler to learn the statistics of these *weak* labels. Therefore, in a multi-task learning setting, one also aims at approximating the conditional probabilities $p(s^{(1)} \mid I)$ and $p(s^{(2)} \mid I)$ for the other two tasks, or the joint probability $p(s^{(1)}, s^{(2)}, s^{(3)} \mid I)$. The three tasks can be related to each other as follows. First, by the definition of the detection task, one can see that $p(s^{(3)} = z \mid s^{(1)} = x) = 0$ for x and z satisfying $x_{i,c} = 1$ and $z_{i,c} = 0$ for some pixel i and class c other than the background. Next, the map of interfaces $s^{(2)}$ indicates small gaps between two connected instances, and is therefore a subset of boundary pixels of the mask $s^{(3)}$.

Let us now consider the probabilities given by the models $p(s^{(k)} \mid I; \theta)$ $(k = 1, 2, 3)$ parameterized by θ, that will consist of network parameters in our setting. We do not optimize θ for individual tasks, but instead consider a joint probability $p(s^{(1)}, s^{(2)}, s^{(3)} \mid I; \theta)$, so that the parameter θ is shared among all tasks. Assuming that $s^{(1)}$ (rough under-segmented instance detection) and $s^{(2)}$ (a subset of shared boundaries) are conditionally independent given image I, and if the samples are i.i.d., we define the maximum likelihood (ML) estimator for θ as

$$\theta_{\mathrm{ML}} = \arg\max_\theta \sum_{I \in \mathcal{I}} \left(\log p\left(s^{(3)} \mid s^{(1)}, s^{(2)}, I; \theta\right) + \sum_{k=1}^{2} \log p\left(s^{(k)} \mid I; \theta\right) \right). \tag{1}$$

The set $\mathcal{I}_3$ may not be evenly distributed across $\mathcal{I}$, but we assume that it is generated by a fixed distribution as well. Provided that the term $\{p(s^{(3)} \mid s^{(1)}, s^{(2)}, I)\}_{I \in \mathcal{I}}$ can be approximated correctly by $p(s^{(3)} \mid s^{(1)}, s^{(2)}, I; \theta)$ even if θ is computed without $s^{(3)}$ specified for $\mathcal{I} \backslash \mathcal{I}_3$, then

$$\sum_{I \in \mathcal{I}} \log p\left(s^{(3)} \mid s^{(1)}, s^{(2)}, I; \theta\right) \propto \sum_{I \in \mathcal{I}_3} \log p\left(s^{(3)} \mid s^{(1)}, s^{(2)}, I; \theta\right). \tag{2}$$

Finally assuming that the segmentation mask does not depend on $s^{(1)}$ or $s^{(2)}$ given $I \in \mathcal{I}_3$, and if $|\mathcal{I}_1|, |\mathcal{I}_2|$ are large enough, then from Eqs. (1), and (2), we approximate the ML estimator by

$$\hat{\theta} = \arg\max_\theta \sum_{k=1}^{3} \sum_{I \in \mathcal{I}_k} \alpha_k \log p\left(s^{(k)} \mid I; \theta\right) \tag{3}$$

in which $\alpha_1, \alpha_2, \alpha_3$ are non negative constants.

3.1 Loss Function

Let the outputs of the approximation models be denoted respectively by $h_\theta^{(1)}(I)$, $h_\theta^{(2)}(I)$, and $h_\theta^{(3)}(I)$, with $\left[h_\theta^{(k)}(I)\right]_{i,c}$ the estimated probability of pixel i to be in class c of task k. For each task k, the log likelihood function related to the label $s^{(k)}$ writes

$$\log p\left(s^{(k)} \mid I; \boldsymbol{\theta}\right) = \sum_i \sum_{c \in C_k} s_{i,c}^{(k)} \log \left[h_{\boldsymbol{\theta}}^{(k)}(I)\right]_{i,c}, \quad k = 1, 2, 3, \tag{4}$$

in which $s_{i,c}^{(k)}$ denotes the element of the label $s^{(k)}$ at pixel i for class c and C_k is the set of classes for task k. For example, for SEM images of ice cream (see details in Sect. 4.1), we have three classes including air bubbles, ice crystals and the rest (background or parts of the objects ignored by the weak labels), so $C_1, C_3 = \{1, 2, 3\}$. For the separation task, there are only two classes for pixels (belonging or not to a touching interface) and $C_2 = \{1, 2\}$. According to Eq. (3), the network is trained by minimizing the weighted cross entropy loss:

$$L(\boldsymbol{\theta}) = -\sum_{I \in \mathcal{I}} \sum_{k=1}^{3} \alpha_k \mathbb{1}_{\mathcal{I}_k}(I) \log p\left(s^{(k)} \mid I; \boldsymbol{\theta}\right), \tag{5}$$

Here $\mathbb{1}_{\mathcal{I}_k}(\cdot)$ is an indicator function which is 1 if $I \in \mathcal{I}_k$ and 0 otherwise.

3.2 Multi-task Network

We follow a convolutional encoder-decoder network structure for multi-task learning. The network architecture is illustrated in Fig. 2. As an extension of the U-net structure for multiple tasks, we only have one contracting path that encodes shared features representation for all the tasks. On the expansive branch, we introduce a multi-task block at each resolution to support different learning purposes (blue blocks in Fig. 2). Every multi-task block runs three paths, with three inputs and three corresponding outputs, and it consists of several sub-blocks.

In each multi-task block, the detection task (task 1) and the segmentation task (task 3) have a common path similar to the decoder part of the standard U-net. They share the same weights and use the same concatenation with feature maps from contracting path via the skip connections. However, we insert an additional residual sub-block for the segmentation task. The residual sub-block provides extra network parameters to learn information not known from the detection task, e.g. object boundary localization. The path for the separation task (task 2) is built on the top of detection/segmentation ones. It is also a U-net decoder block structure, but the long skip connections start from the sub-blocks of the detection/segmentation paths instead of the contracting path. The connections extract higher resolution features from the segmentation task and use them in the separation task.

To formulate the multi-task blocks, let x_l and z_l denote respectively the output of the detection path and segmentation path at the multi-task block l, and let c_l be the feature maps received from the contracting path with the skip connections. Then for task 1 and task 3 we have

$$\begin{cases} x_{l+1} = F_{W_l}(x_l, c_l), \\ z_{l+\frac{1}{2}} = F_{W_l}(z_l, c_l), \quad z_{l+1} = z_{l+\frac{1}{2}} + F_{W_{l+\frac{1}{2}}}(z_{l+\frac{1}{2}}), \end{cases} \tag{6}$$

in which $W_l, W_{l+1/2} \in \boldsymbol{\theta}$ are subsets of network parameters and F_{W_l}, $F_{W_{l+\frac{1}{2}}}$ are respectively determined by a sequence of layers of the network (cf. small grey blocks on the right of Fig. 2). For task 2 the output at l^{th} block $\boldsymbol{y}_{l+1}$ is computed as $\boldsymbol{y}_{l+1} = G_{\tilde{W}_l}(\boldsymbol{z}_{l+1}, \boldsymbol{y}_l)$ with additional network parameters $\tilde{W}_l \in \boldsymbol{\theta}$. Finally, after the last multi-task block, softmax layers are added, outputting a probability map for each task.

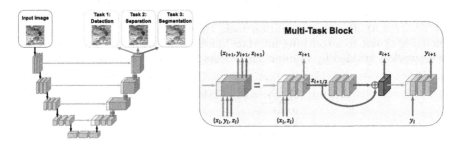

Fig. 2. Architecture of the multi-task U-net. The left part of the network is a contracting path similar to the standard U-net. For multi-task learning, we construct several expansive paths with specific multi-task blocks. At each resolution, task 1 (Detection in yellow) and task 3 (Segmentation in red) run through a common sub-block, but the red path learns an additional residual to better localize object boundaries. Long skip connections with the layers from contracting path are built for yellow/red paths via concatenation. Task 2 (Separation, in green) mainly follows a separated expansive path, with its own up-sampled blocks. A link with the last layer of task 3 is added via a skip connection in order to integrate accurate boundaries in the separation task. (Color figure online)

Implementation Details. We implement a multi-task U-net with 6 levels of spatial resolution and input images of size 256×256. A sequence of down-sampling via max-pooling with pooling size 2×2 is used for the contracting path of the network. Different from the conventional U-net [33], each small grey block (see Fig. 2) consists of a convolution layer and a batch normalization [18], followed by a leaky ReLU activation with a leakiness parameter 0.01. The same setting is also applied to grey sub-blocks of the 4 multi-task blocks. On the expansive path of the network, feature maps are up-sampled (with factor 2×2) by bilinear interpolation from a low resolution multi-task block to the next one.

3.3   Methods for Lazy Labels Generation

We now explain our strategy for generating all the lazy annotations that are used for training. We introduce our method with a data set of ice cream SEM images but any other similar microscopy datasets could be used. Typical images of ice cream samples are shown in the top row of the left part of Fig. 3. The segmentation problem is challenging since the images contain densely distributed small object instances (i.e., air bubble and ice crystals), and poor contrast between the foreground and the background. The sizes of the objects can vary significantly in a single sample. Textures on the surfaces of objects also appear.

As a first step, scribble-based labelling is applied to obtain detection regions of air bubbles and ice crystals for task 1. This can be done in a very fast way as no effort is put on the exact object boundaries. We adopt a lazy strategy by picking out an inner region for each object in the images (see e.g., the second row of the left part of Fig. 3). Though one could get these rough regions as accurate as possible, we delay such refinement to task 3, for better efficiency of the global annotation process. Compared to the commonly used bounding box annotations in computer vision tasks, these labels give more confidence for a particular part of the region of interest.

In the second step, we focus on tailored labels for those instances that are close one to each other (task 2), without a clear boundary separating them. Again, we use scribbles to mark their interface. Examples for such annotations are given in Fig. 3 (top line, right part). The work can be carried out efficiently especially when the target scribbles have a sparse distribution. Lazy manual labelling of tasks 1 and 2 are done independently. It follows the assumption made in Sect. 3 that $s^{(1)}$ and $s^{(2)}$ are conditionally independent given image I.

The precise labels for task 3 are created using interactive segmentation tools. We use Grabcut [16,34] a graph-cut based method. The initial labels obtained from the first step give a good guess of the whole object regions. The Grabcut works well on isolated objects. However, it gives poor results when the objects are close to each other and have boundaries with inhomogeneous colors. As corrections may be needed for each image, only a few images of the whole dataset are processed. A fully segmented example is shown in the last row of Fig. 3.

4   Experiments

In this section, we demonstrate the performance of our approach using two microscopy image datasets. For both, we use strong labels (SL) and weak labels (WL). We prepare the labels and design the network as described in Sect. 3.

4.1   Segmenting SEM Images of Ice Cream

Scanning Electron Microscopy (SEM) constitutes the state-of-the-art for analysing food microstructures as it enables the efficient acquisition of high quality images for food materials, resulting into a huge amount of image data available for analysis. However, to better delineate the microstructures and provide exact statistical information, the partition of the images into different structural components and instances is needed. The structures of food, especially soft solid materials, are usually complex which makes automated segmentation a difficult task. Some SEM images of ice cream in our dataset are shown on the bottom right of Fig. 1. A typical ice cream sample consists of air bubbles, ice crystals and a concentrated unfrozen solution. We treat the solution as the background and aim at detecting and computing a pixel-wise classification for each air bubbles and ice crystals instances.

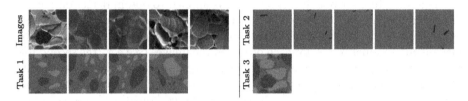

Fig. 3. Example of annotated images. Some of the annotations are not shown because the images are not labelled for the associated tasks. The red color and green color are for air bubbles and ice crystals, respectively. The blue curves in Task 2 are labels for interfaces of touching objects. (Color figure online)

The set of ice-cream SEM dataset consists of 38 wide field-of-view and high resolution images that are split into three sets (53% for training, 16% for validation and 31% testing respectively). Each image contains a rich set of instances with an overall number of instances around 13300 for 2 classes (ice crystals and air bubbles). For comparison, the PASCAL VOC 2012 dataset has 27450 objects in total for 20 classes.

For training the network, data augmentation is applied to prevent overfitting. The size of the raw images is 960×1280. They are rescaled and rotated randomly, and then cropped into an input size of 256×256 for feeding the network. Random flipping is also performed during training. The network is trained using Adam optimizer [22] with a learning rate $r = 2 \times 10^{-4}$ and a batch size of 16.

In the inference phase, the network outputs for each patch a probability map of size 256×256. The patches are then aggregated to obtain a probability map for the whole image. In general, the pixels near the boundaries of each patch are harder to classify. We thus weight the spatial influence of the patches with a Gaussian kernel to emphasize the network prediction at patch center.

We now evaluate the multi-task U-net and compare it to the traditional single task U-net. The performance of each model is tested on 12 wide FoV images, and average results are shown in Table 1. In the table, the dice score for a class c is defined as $d_c = 2 \sum_i x_{i,c} y_{i,c} / (\sum_i x_{i,c} + \sum_i y_{i,c})$ where x is the computed segmentation mask and y the ground truth.

We train a single task U-net (i.e., without the multi-task block) on the weakly labelled set (task 1), with the 15 annotated images. The single task U-net on weak annotations gives an overall dice score at 0.72, the lowest one among the three other methods tested. One reason for the low accuracy of the single task U-net on weak (inaccurate) annotations is that in the training labels, the object boundaries are mostly ignored. Hence the U-net is not trained to recover them, leaving large parts of the object not recognized. Second, we consider strong annotations as training data, without the data of the other tasks, i.e. only 2 images

Table 1. Dice scores of segmentation results on the test images of SEM images of ice cream dataset.

The models	air bubbles	ice crystals	Overall
U-net on WL	0.725	0.706	0.716
U-net on SL	0.837	0.794	0.818
PL approach	0.938	0.909	0.924
Multi-task U-net	**0.953**	**0.931**	**0.944**

Fig. 4. The error bars for the PL and multi-task U-net. The top of each box represent the mean of the scores over 8 different experiments, the minimum and maximum of which are indicated by the whiskers

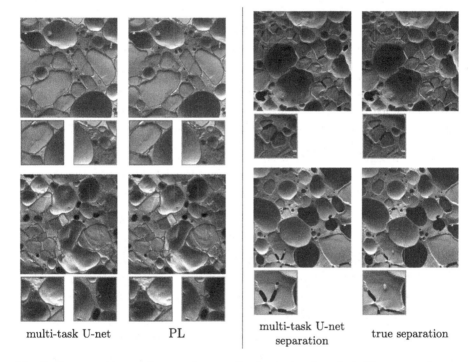

multi-task U-net   PL   multi-task U-net separation   true separation

Fig. 5. Segmentation and separation results (best view in color). First two columns: the computed contours are shown in red for air bubbles and green for ice crystals. While multi-task U-net and PL supervised network both have good performance, PL misclassifies the background near object boundaries. Last two columns: Examples of separation by the multi-task U-net and the ground truth.

with accurate segmentation masks are used. The score of the U-net trained on SL is only 0.82, which is significantly lower than the 0.94 obtained by our multi-task network.

We also compare our multi-task U-net results with one of the major weakly supervised approaches that make use of pseudo labels (PL) (see e.g., [19,21]). In these approaches, the pseudo segmentation masks are created from WLs and are used to feed a segmentation network. Following the work of [21], we use the Grabcut method to create the PLs from the partial masks of task 1. For the small subset of images that are strongly annotated, the full segmentation masks are used instead of PLs. The PLs are created without human correction, and then used for feeding the segmentation network. Here we use the single task U-net for baseline comparisons.

Our multi-task network outperforms the PL approach as shown in Table 1. Figure 4 displays the error bars for the two methods with dice scores collected from 8 different runs. The performance of the PL method relies on the tools used for pseudo segmentation mask generations. If the tools create bias in the pseudo labels, then the learning will be biased as well, which is the case in this example. The images in the left part of Figure 5 show that the predicted label of an object tends to merge with some background pixels when there are edges of another object nearby.

Besides the number of pixels that are correctly classified, the separation of touching instances is also of interest. In addition to the dice scores in Table 1, we study the learning performance of our multi-task network on task 2, which specializes in the separation aspect. The test results on the 12 images give an overall precision of 0.70 of the detected interfaces, while 0.82 of the touching objects are recognized. We show some examples of computed separations and ground truth in the right part of Figure 5.   For the detection task, the network

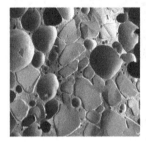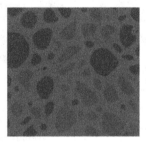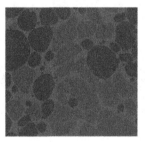

Fig. 6. The image (left), the inaccurate label predicted by the network for the detection task (middle), and the ground truth segmentation mask (right). The red and green colors on the middle and right images stand for air bubbles and ice crystals respectively.

predicts a probability map for the inner regions of the object instances. An output of the network is shown in Figure 6. With partial masks as coarse labels for this task, the network learns to identify the object instances.

4.2 Gland Segmentation on H&E-Stained Images

We apply the approach to the segmentation of tissues in histology images. In this experiment, we use the GlaS challenge dataset [37] that consists of 165 Hematoxylin and eosin (H&E) stained images. The dataset is split into three parts, with 85 images for training, and 60 for offsite test and 20 images for onsite test (we will call the latter two sets Test part A and Test part B respectively in the following).

Apart from the SL available from the dataset, we create a set of a weak labels for the detection task and separation task. These weak labels together with a part of the strong labels are used for training the multi-task U-net.

Table 2. Average dice score for segmentation of gland. Results of two sets of methods, weakly supervised (WS) and strongly supervised (SS) are displayed. Our method uses both SL and WL. The ratio of strong labels (SL) is increased from 2.4% to 100%, and the scores of the methods are reported here for two parts A and B of the test sets, as split in [37].

SL Ratio			2.4%	4.7%	9.4%	100%
Test Part A	WS	Ours	**0.866**	**0.889**	**0.915**	**0.921**
		Single task	0.700	0.749	0.840	0.921
		PL	0.799	0.812	0.820	
	SS	MDUnet				0.920
Test Part B	WS	Ours	**0.751**	**0.872**	**0.904**	**0.910**
		Single task	0.658	0.766	0.824	0.908
		PL	0.773	0.770	0.782	
	SS	MDUnet				0.871

In this experiment, we test the algorithm on different ratios of SL, and compare it with the baseline U-net (single task), PL approach (where PL are generated in the same way as the ones for the SEM dataset), and a fully supervised approach called Multi-scale Densely Connected U-Net (MDUnet) [45]. The results on two sets of test data are reported in Table 2. As the SL ratios increase from 2.4% to 9.4%, an improvement of performance of the multi-task U-net is gained. When it reaches 9.4% SL, the multi-task framework achieves comparable score with the fully supervised version, and outperforms the PL approach by a significant margin. We emphasize that the 9.4% SL and WL can be obtained several times faster than the 100% SL used for fully supervised learning. Example of segmentation results are displayed in Figure 7.

5 Conclusion

In this paper, we develop a multi-task learning framework for microscopy image segmentation, which relaxes the requirement for numerous and accurate annotations to train the network. It is therefore suitable for segmentation problem with

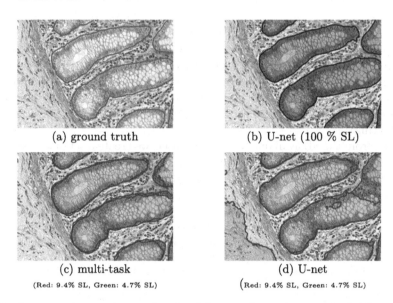

(a) ground truth

(b) U-net (100 % SL)

(c) multi-task
(Red: 9.4% SL, Green: 4.7% SL)

(d) U-net
(Red: 9.4% SL, Green: 4.7% SL)

Fig. 7. Segmentation results on the gland dataset (best view in color). The ground truth and the results. For (c) and (d), Red contour denotes the results from 9.4% strong labels; Green contour denotes results from 4.7% strong labels. (Color figure online)

a dense population of object instances. The model separates the segmentation problem into three smaller tasks. One of them is dedicated to the instance detection and therefore does not need exact boundary information. This gives potential flexibility as one could concentrate on the classification and rough location of the instances during data collection. The second one focuses on the separation of objects sharing a common boundary. The final task aims at extracting pixel-wise boundary information. Thanks to the information shared within the multi-task learning, this accurate segmentation can be obtained using very few annotated data. Our model is end-to-end and requires no reprocessing for the weak labels. For the partial masks that ignore boundary pixels, the annotation can also be done when the boundaries of object are hard to detect. In the future, we could like to extend the proposed approach for solving 3D segmentation problems in biomedical images where labelling a single 3D image needs much more manual work.

Acknowledgments. RK and CBS acknowledge support from the EPSRC grant EP/T003553/1. CBS additionally acknowledges support from the Leverhulme Trust project on 'Breaking the non-convexity barrier', the Philip Leverhulme Prize, the EPSRC grant EP/S026045/1, the EPSRC Centre Nr. EP/N014588/1, the RISE projects CHiPS and NoMADS, the Cantab Capital Institute for the Mathematics of Information and the Alan Turing Institute, Royal Society Wolfson fellowship. AB and NP acknowledge support from the EU Horizon 2020 research and innovation programme NoMADS (Marie Skłodowska-Curie grant agreement No. 777826).

References

1. Badrinarayanan, V., Kendall, A., Cipolla, R.: Segnet: A deep convolutional encoder-decoder architecture for image segmentation. IEEE transactions on pattern analysis and machine intelligence **39**(12), 2481–2495 (2017)
2. Bajcsy, P., Feldman, S., Majurski, M., Snyder, K., Brady, M.: Approaches totraining multiclass semantic image segmentation of damage in concrete.Journal of Microscopy (2020)
3. Bearman, A., Russakovsky, O., Ferrari, V., Fei-Fei, L.: What's the point: Semantic segmentation with point supervision. In: European conference on computer vision. pp. 549–565. Springer (2016)
4. Bischke, B., Helber, P., Folz, J., Borth, D., Dengel, A.: Multi-task learning for segmentation of building footprints with deep neural networks. In: 2019 IEEE International Conference on Image Processing (ICIP). pp. 1480–1484. IEEE (2019)
5. Boykov, Y., Funka-Lea, G.: Graph cuts and efficient nd image segmentation. International journal of computer vision **70**(2), 109–131 (2006)
6. Buchholz, T.O., Prakash, M., Krull, A., Jug, F.: Denoiseg: Joint denoising and segmentation. arXiv preprint arXiv:2005.02987 (2020)
7. Caselles, V., Kimmel, R., Sapiro, G.: Geodesic active contours. International journal of computer vision **22**(1), 61–79 (1997)
8. Chen, L.C., Papandreou, G., Kokkinos, I., Murphy, K., Yuille, A.L.: Semantic image segmentation with deep convolutional nets and fully connected crfs. arXiv preprint arXiv:1412.7062 (2014)
9. Chen, L.C., Papandreou, G., Kokkinos, I., Murphy, K., Yuille, A.L.: Deeplab: Semantic image segmentation with deep convolutional nets, atrous convolution, and fully connected crfs. IEEE transactions on pattern analysis and machine intelligence **40**(4), 834–848 (2018)
10. Ciofolo, C., Barillot, C.: Atlas-based segmentation of 3d cerebral structures with competitive level sets and fuzzy control. Medical image analysis **13**(3), 456–470 (2009)
11. Ezhov, M., Zakirov, A., Gusarev, M.: Coarse-to-fine volumetric segmentation of teeth in cone-beam ct. arXiv preprint arXiv:1810.10293 (2018)
12. Ghosh, A., Ehrlich, M., Shah, S., Davis, L., Chellappa, R.: Stacked u-nets for ground material segmentation in remote sensing imagery. In: Proceedings of the IEEE Conference on Computer Vision and Pattern Recognition Workshops. pp. 257–261 (2018)
13. Guerrero-Pena, F.A., Fernandez, P.D.M., Ren, T.I., Yui, M., Rothenberg, E., Cunha, A.: Multiclass weighted loss for instance segmentation of cluttered cells. In: 2018 25th IEEE International Conference on Image Processing (ICIP). pp. 2451–2455. IEEE (2018)

14. Heimann, T., Meinzer, H.P.: Statistical shape models for 3d medical image segmentation: a review. Medical image analysis **13**(4), 543–563 (2009)
15. Hirsch, P., Kainmueller, D.: An auxiliary task for learning nuclei segmentation in 3d microscopy images. arXiv preprint arXiv:2002.02857 (2020)
16. Hong, S., Noh, H., Han, B.: Decoupled deep neural network for semi-supervised semantic segmentation. In: Advances in neural information processing systems. pp. 1495–1503 (2015)
17. Huang, Z., Wang, X., Wang, J., Liu, W., Wang, J.: Weakly-supervised semantic segmentation network with deep seeded region growing. In: Proceedings of the IEEE Conference on Computer Vision and Pattern Recognition. pp. 7014–7023 (2018)
18. Ioffe, S., Szegedy, C.: Batch normalization: Accelerating deep network training by reducing internal covariate shift. arXiv preprint arXiv:1502.03167 (2015)
19. Jing, L., Chen, Y., Tian, Y.: Coarse-to-fine semantic segmentation from image-level labels. arXiv preprint arXiv:1812.10885 (2018)
20. Kendall, A., Gal, Y., Cipolla, R.: Multi-task learning using uncertainty to weigh losses for scene geometry and semantics. In: Proceedings of the IEEE Conference on Computer Vision and Pattern Recognition. pp. 7482–7491 (2018)
21. Khoreva, A., Benenson, R., Hosang, J., Hein, M., Schiele, B.: Simple does it: Weakly supervised instance and semantic segmentation. In: Proceedings of the IEEE conference on computer vision and pattern recognition. pp. 876–885 (2017)
22. Kingma, D.P., Ba, J.: Adam: A method for stochastic optimization. arXiv preprint arXiv:1412.6980 (2014)
23. Kolesnikov, A., Lampert, C.H.: Seed, expand and constrain: Three principles for weakly-supervised image segmentation. In: European Conference on Computer Vision. pp. 695–711. Springer (2016)
24. Krasowski, N., Beier, T., Knott, G., Köthe, U., Hamprecht, F.A., Kreshuk, A.: Neuron segmentation with high-level biological priors. IEEE transactions on medical imaging **37**(4), 829–839 (2017)
25. Lee, J., Kim, E., Lee, S., Lee, J., Yoon, S.: Ficklenet: Weakly and semi-supervised semantic image segmentation using stochastic inference. arXiv preprint arXiv:1902.10421 (2019)
26. Lin, D., Dai, J., Jia, J., He, K., Sun, J.: Scribblesup: Scribble-supervised convolutional networks for semantic segmentation. In: Proceedings of the IEEE Conference on Computer Vision and Pattern Recognition. pp. 3159–3167 (2016)
27. Litjens, G., Kooi, T., Bejnordi, B.E., Setio, A.A.A., Ciompi, F., Ghafoorian, M., Van Der Laak, J.A., Van Ginneken, B., Sánchez, C.I.: A survey on deep learning in medical image analysis. Medical image analysis **42**, 60–88 (2017)
28. Long, J., Shelhamer, E., Darrell, T.: Fully convolutional networks for semantic segmentation. In: Proceedings of the IEEE conference on computer vision and pattern recognition. pp. 3431–3440 (2015)
29. MacQueen, J., et al.: Some methods for classification and analysis of multivariate observations. In: Proceedings of the fifth Berkeley symposium on mathematical statistics and probability. vol. 1, pp. 281–297. Oakland, CA, USA (1967)
30. Mlynarski, P., Delingette, H., Criminisi, A., Ayache, N.: Deep learning with mixed supervision for brain tumor segmentation. arXiv preprint arXiv:1812.04571 (2018)
31. Papandreou, G., Chen, L.C., Murphy, K.P., Yuille, A.L.: Weakly-and semi-supervised learning of a deep convolutional network for semantic image segmentation. In: Proceedings of the IEEE international conference on computer vision. pp. 1742–1750 (2015)

32. Playout, C., Duval, R., Cheriet, F.: A novel weakly supervised multitaskarchitecture for retinal lesions segmentation on fundus images. IEEEtransactions on medical imaging (2019)
33. Ronneberger, O., Fischer, P., Brox, T.: U-net: Convolutional networks for biomedical image segmentation. In: International Conference on Medical image computing and computer-assisted intervention. pp. 234–241. Springer (2015)
34. Rother, C., Kolmogorov, V., Blake, A.: Grabcut: Interactive foreground extraction using iterated graph cuts. In: ACM transactions on graphics (TOG). vol. 23, pp. 309–314. ACM (2004)
35. Shah, M.P., Merchant, S., Awate, S.P.: Ms-net: Mixed-supervision fully-convolutional networks for full-resolution segmentation. In: International Conference on Medical Image Computing and Computer-Assisted Intervention. pp. 379–387. Springer (2018)
36. Shin, S.Y., Lee, S., Yun, I.D., Kim, S.M., Lee, K.M.: Joint weakly and semi-supervised deep learning for localization and classification of masses in breast ultrasound images. IEEE transactions on medical imaging $38(3)$, 762–774 (2019)
37. Sirinukunwattana, K., Pluim, J.P., Chen, H., Qi, X., Heng, P.A., Guo, Y.B., Wang, L.Y., Matuszewski, B.J., Bruni, E., Sanchez, U., et al.: Gland segmentation in colon histology images: The glas challenge contest. Medical image analysis 35, 489–502 (2017)
38. Sun, F., Li, W.: Saliency guided deep network for weakly-supervised image segmentation. Pattern Recognition Letters 120, 62–68 (2019)
39. Sun, T., Chen, Z., Yang, W., Wang, Y.: Stacked u-nets with multi-output for road extraction. In: 2018 IEEE/CVF Conference on Computer Vision and Pattern Recognition Workshops (CVPRW). pp. 187–1874. IEEE (2018)
40. Tsutsui, S., Kerola, T., Saito, S., Crandall, D.J.: Minimizing supervision for free-space segmentation. In: Proceedings of the IEEE Conference on Computer Vision and Pattern Recognition Workshops. pp. 988–997 (2018)
41. Wang, C., MacGillivray, T., Macnaught, G., Yang, G., Newby, D.: A two-stage 3d unet framework for multi-class segmentation on full resolution image. arXiv preprint arXiv:1804.04341 (2018)
42. Wang, X., Xiao, T., Jiang, Y., Shao, S., Sun, J., Shen, C.: Repulsion loss: Detecting pedestrians in a crowd. In: Proceedings of the IEEE Conference on Computer Vision and Pattern Recognition. pp. 7774–7783 (2018)
43. Wei, Y., Liang, X., Chen, Y., Shen, X., Cheng, M.M., Feng, J., Zhao, Y., Yan, S.: Stc: A simple to complex framework for weakly-supervised semantic segmentation. IEEE transactions on pattern analysis and machine intelligence $39(11)$, 2314–2320 (2017)
44. Wei, Y., Xiao, H., Shi, H., Jie, Z., Feng, J., Huang, T.S.: Revisiting dilated convolution: A simple approach for weakly-and semi-supervised semantic segmentation. In: Proceedings of the IEEE Conference on Computer Vision and Pattern Recognition. pp. 7268–7277 (2018)
45. Zhang, J., Jin, Y., Xu, J., Xu, X., Zhang, Y.: Mdu-net: Multi-scale densely connected u-net for biomedical image segmentation. arXiv preprint arXiv:1812.00352 (2018)
46. Zheng, S., Jayasumana, S., Romera-Paredes, B., Vineet, V., Su, Z., Du, D., Huang, C., Torr, P.H.: Conditional random fields as recurrent neural networks. In: Proceedings of the IEEE international conference on computer vision. pp. 1529–1537 (2015)

47. Zhou, J., Luo, L.Y., Dou, Q., Chen, H., Chen, C., Li, G.J., Jiang, Z.F., Heng,P.A.: Weakly supervised 3d deep learning for breast cancer classification andlocalization of the lesions in mr images. Journal of Magnetic ResonanceImaging (2019)

48. Zhou, S., Nie, D., Adeli, E., Yin, J., Lian, J., Shen, D.: High-resolution encoder-decoder networks for low-contrast medical image segmentation. IEEE Transactions on Image Processing **29**, 461–475 (2019)

Multi-CryoGAN: Reconstruction of Continuous Conformations in Cryo-EM Using Generative Adversarial Networks

Harshit Gupta⬤, Thong H. Phan⬤, Jaejun Yoo$^{(\boxtimes)}$⬤, and Michael Unser⬤

École Polytechnique Fédérale de Lausanne (EPFL), Lausanne, Switzerland
harshit.gupta.cor@gmail.com, {huy.thong,jaejun.yoo}@epfl.ch

Abstract. We propose a deep-learning-based reconstruction method for cryo-electron microscopy (Cryo-EM) that can model multiple conformations of a nonrigid biomolecule in a standalone manner. Cryo-EM produces many noisy projections from separate instances of the same but randomly oriented biomolecule. Current methods rely on pose and conformation estimation which are inefficient for the reconstruction of continuous conformations that carry valuable information. We introduce Multi-CryoGAN, which sidesteps the additional processing by casting the volume reconstruction into the distribution matching problem. By introducing a manifold mapping module, Multi-CryoGAN can learn continuous structural heterogeneity *without pose estimation* nor *clustering*. We also give a theoretical guarantee of recovery of the true conformations. Our method can successfully reconstruct 3D protein complexes on synthetic 2D Cryo-EM datasets for both continuous and discrete structural variability scenarios. Multi-CryoGAN is the first model that can reconstruct continuous conformations of a biomolecule from Cryo-EM images in a fully unsupervised and end-to-end manner.

Keywords: Cryo-EM · Inverse problem · Image reconstruction · Generative adversarial networks · Continuous protein conformations

1 Introduction

The determination of the structure of nonrigid macromolecules is an important aspect of structural biology and is fundamental in our understanding of biological mechanisms and in drug discovery [3]. Among other popular techniques such as X-ray crystallography and nuclear magnetic resonance spectroscopy, Cryo-EM has emerged as a unique method to determine molecular structures at unprecedented high resolutions. It is widely applicable to proteins that are difficult to crystallize or have large structures. Cryo-EM produces a large number (from 10^4

Electronic supplementary material The online version of this chapter (https://doi.org/10.1007/978-3-030-66415-2_28) contains supplementary material, which is available to authorized users.

© Springer Nature Switzerland AG 2020
A. Bartoli and A. Fusiello (Eds.): ECCV 2020 Workshops, LNCS 12535, pp. 429–444, 2020.
https://doi.org/10.1007/978-3-030-66415-2_28

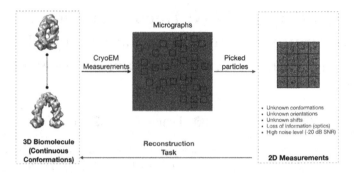

Fig. 1. Reconstruction task of Cryo-EM. Many samples of a biomolecule (which may exhibit continuously varying conformations) are frozen in vitreous ice. These are then imaged/projected using an electron beam to get 2D micrographs. The 2D images containing projection of a single sample are then picked out (black-box). The task then is to reconstruct the conformations of the biomolecule from these measurements.

to 10^7) of tomographic projections of the molecules dispersed in a frozen solution. The reconstruction of 3D molecular structures from these data involves three main challenges: possible structural heterogeneity of the molecule, random locations and orientations of the molecules in the ice, and an extremely poor signal-to-noise ratio (SNR), which can be as low as to -20 dB (Fig. 1). In fact, the reconstruction of continuously varying conformations of a nonrigid molecule is still an open problem in the field [6,17]. A solution would considerably enhance our understanding of the functions and behaviors of many biomolecules.

Most current methods [18,19] find the 3D structure by maximizing the likelihood of the data. They employ an iterative optimization scheme that alternatively estimates the distribution of poses (or orientations) and reconstructs a 3D structure until a criterion is satisfied (Fig. 2(a)). To address the structural variability of protein complexes, these methods typically use discrete clustering approaches. However, the pose estimation and clustering steps are computationally heavy and include heuristics. This makes these methods inefficient when the molecule has continuous variations or a large set of discrete conformations.

Recently, two deep-learning-based reconstruction methods that require no prior-training nor additional training data have been introduced. On one hand, CryoDRGN [26] uses a variational auto-encoder (VAE) to model continuous structural variability, avoiding the heuristic clustering step. It is a likelihood-based method that requires pose estimation using an external routine like a branch-and bound-method [18]. This additional processing step can complicate the reconstruction procedure and limit the flexibility of the model. On the other hand, Gupta *et al.* [10] have recently proposed CryoGAN. It addresses the problem under a generative adversarial framework [8]. CryoGAN learns to reconstruct a 3D structure whose randomly projected 2D Cryo-EM images match the acquired data in a distributional sense (Fig. 2(b)). Due to this likelihood-free characteristic, CryoGAN does not require any additional processing step such

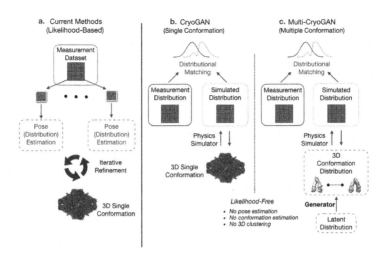

Fig. 2. Schematic overview of the reconstruction methods in Cryo-EM. (a) Current methods; (b) CryoGAN; (c) proposed method (Multi-CryoGAN).

as pose estimation, while it can be directly deployed on the Cryo-EM measurements. This largely helps simplify the reconstruction procedure. However, its application is limited to the reconstruction of a single conformation.

In this paper, we combine the advantages of CryoDRGN and CryoGAN. We propose an unsupervised deep-learning-based method, called Multi-CryoGAN. It can reconstruct continuously varying conformations of a molecule in a truly standalone and likelihood-free manner. Using a convolutional neural network (CNN), it directly learns a mapping from a latent space to the 3D conformation distribution. Unlike current methods, it requires no calculation of pose or conformation estimation for each projection, while it has the capacity to reconstruct low-dimensional but complicated conformation manifolds [11].

Using synthetic Cryo-EM data as our benchmark, we show that our method can reconstruct the conformation manifold for both continuous and discrete conformation distributions. In the discrete case, it also reconstructs the corresponding probabilities. To the best of our knowledge, this is the first standalone method that can construct whole manifold of the biomolecule conformations.

2   Related Work

Traditional Cryo-EM Image Reconstruction. A detailed survey of the classical methods is provided in [21,22]. Most of them fall into the maximum-likelihood (ML) framework and rely on either expectation-maximization (ML-EM) [19] or gradient descent (the first stage of [18]). In the context of heterogeneous conformation reconstruction, a conjugate-gradient descent is used to estimate the volume covariance matrix [1]. The eigendecomposition of this matrix contains information about the conformation distribution which is then input to the ML framework. In [5], a conformation manifold is generated for each group of projections with similar poses. This data-clustering approach assumes orientation rather than structural heterogeneity to be the dominant cause for variations among the projection images, a strong constraint. In addition, the reconstruction of 3D movies from multiple 2D manifolds can be computationally expensive. In another method, Moscovich *et al.* [16] compute the graph Laplacian eigenvectors of the conformations using covariance estimation. In [14], the problem of heterogeneous reconstructions is reformulated as the search for a homogeneous high-dimensional structure that represents all the states, called a hypermolecule, which is characterized by a basis of hypercomponents. This allows for reconstruction of high-dimensional conformation manifolds but requires assumptions on the variations of the conformations as a prior in their Bayesian formulation.

One of the main drawbacks of these methods is that they require marginalization over the space of poses for each projection image, which is computationally demanding and potentially inaccurate. In addition, because they rely on 3D clustering to deal with structural variations of protein complexes, these methods become inefficient for a large set of discrete conformations and struggle to recover a continuum of conformations.

Deep Learning for Cryo-EM Reconstructions. In addition to CryoDRGN and CryoGAN that have already been discussed in the introduction, there is a third described in [15]. It uses a VAE and a framework based on a generative adversarial network (GAN) to learn the latent distribution of the acquired data. This representation is then used to estimate the orientation and other important parameters for each projection image.

Deep Learning to Recover a 3D Object from 2D Projections. The implicit or explicit recovery of 3D shapes from 2D views is an important problem in computer vision. Many deep-learning algorithms have been proposed for this [7,23,24]. Taking inspiration from compressed sensing, Bora *et al.* [4] have recently introduced a GAN framework that can recover an original distribution from the measurements through a forward model. While these approaches would in principle be applicable, they consider a forward model that is too simple for Cryo-EM, where a contrast transfer function (CTF) must be taken into account and where the noise is orders of magnitude stronger (e.g. with a typical SNR of -10 to -20 dB.

3 Background and Preliminaries

3.1 Image-Formation Model

The aim of Cryo-EM is to reconstruct the 3D molecular structure from the measurements $\{y^1_{data}, \ldots, y^Q_{data}\}$, where Q is typically between 10^4 to 10^7. Each measurement $y^q \in \mathbb{R}^{N \times N}$ is given by

$$y^q = \underbrace{C_{c^q} * S_{t^q} P_{\theta^q}}_{H_{\varphi^q}} \{x^q\} + n^q, \tag{1}$$

where

- $\mathbf{x}^q \in \mathbb{R}^{N \times N \times N}$ is a separate instance of the 3D molecular structure;
- $\mathbf{n}^q \in \mathbb{R}^{N \times N}$ is the noise;
- $\mathbf{H}_{\varphi^q}$ is the measurement operator which depends on the imaging parameters $\varphi^q = (\boldsymbol{\theta}^q, \mathbf{t}^q, \mathbf{c}^q) \in \mathbb{R}^8$ and involves three operations.
  - The term $\mathbf{P}_{\boldsymbol{\theta}^q}\{\mathbf{x}^q\}$ is the tomographic projection of $\mathbf{x}^q$ rotated by $\boldsymbol{\theta}^q = (\theta_1^q, \theta_2^q, \theta_3^q)$.
  - The operator $\mathbf{S}_{\mathbf{t}^q}$ shifts the projected image by $\mathbf{t}^q = (t_1^q, t_2^q)$. This shift arises from off-centered particle picking.
  - The Fourier transform of the resulting image is then modulated by the CTF $\hat{\mathbf{C}}_{\mathbf{c}^q}$ with defocus parameters $\mathbf{c}^q = (d_1^q, d_2^q, \alpha_{\mathrm{ast}}^q)$ and thereafter subjected to inverse Fourier Transform.

For more details, please see *Supplementary: Image Formation* in [10]. The challenge of Cryo-EM is that, for each measurement $\mathbf{y}^q$, the structure $\mathbf{x}^q$ and the imaging parameters $(\boldsymbol{\theta}^q, \mathbf{t}^q)$ are unknown, the CTF is a band pass filter with multiple radial zero frequencies that incur irretrievable loss of information, and the energy of the noise is multiple times (~ 10 to 100 times) that of the signal which corresponds to SNRs of -10 to -20 dB. In the homogeneous case (single conformation), all $\mathbf{x}^q$ are identical. But in the heterogeneous case (multiple conformations), each $\mathbf{x}^q$ represents a different conformation of the same biomolecule.

Stochastic Modeling. We denote the probability distribution over the conformation landscape by $p_{\mathrm{conf}}(\mathbf{x})$ from which a conformation $\mathbf{x}^q$ is assumed to be sampled from. We assume that the imaging parameters and the noise are sampled from known distributions $p_{\varphi} = p_{\boldsymbol{\theta}} p_{\mathbf{t}} p_{\mathbf{c}}$ and $p_{\mathbf{n}}$, respectively. For a given conformation distribution $p_{\mathrm{conf}}(\mathbf{x})$, this stochastic forward model induces a distribution over the measurements which we denote by $p(\mathbf{y})$. We denote by $p_{\mathrm{conf}}^{\mathrm{data}}(\mathbf{x})$ the true conformation distribution from which the data distribution $p_{\mathrm{data}}(\mathbf{y})$ is acquired such that $\{\mathbf{y}_{\mathrm{data}}^1, \ldots, \mathbf{y}_{\mathrm{data}}^Q\} \sim p_{\mathrm{data}}(\mathbf{y})$. The distribution $p_{\mathrm{conf}}^{\mathrm{data}}(\mathbf{x})$ is unknown and needs to be recovered.

The classical methods are likelihood-based and rely on the estimation of imaging parameters $(\boldsymbol{\theta}^q, \mathbf{t}^q)$ (or a distribution over them) and of the conformation class for each measurement image $\mathbf{y}^q$. This information is then utilized to reconstruct the multiple discrete conformations. Our method, in contrast, is built upon the insight that, to recover $p_{\mathrm{conf}}^{\mathrm{data}}(\mathbf{x})$ it is sufficient to find a $p_{\mathrm{conf}}^{\mathrm{gen}}(\mathbf{x})$ whose corresponding measurement distribution $p_{\mathrm{gen}}(\mathbf{y})$ is equal to $p_{\mathrm{data}}(\mathbf{y})$ (see Theorem 1). This does away with pose (or distributions over the poses) estimation and conformation clustering for each measurement.

3.2   CryoGAN

Our scheme is extension of the CryoGAN [10] method, which is applicable only for the homogeneous case $p_{\mathrm{conf}}^{\mathrm{data}}(\mathbf{x}) = \delta(\mathbf{x} - \mathbf{x}_{\mathrm{data}})$, where $\mathbf{x}_{\mathrm{data}}$ is the true 3D structure. CryoGAN tackles the challenge by casting the reconstruction problem

Multi-CryoGAN

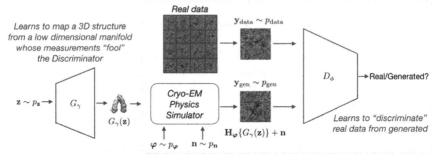

a. Conformation manifold mapper G_γ b. CryoGAN

Fig. 3. Schematic illustration of Multi-CryoGAN and its components. (a) Conformation manifold mapper; (b) CryoGAN.

as a distribution-matching problem (Fig. 2(b)). More specifically, it learns to reconstruct the 3D volume $\mathbf{x}^*$ whose simulated projection set (measurement distribution) is most similar to the real projection data in a distributional sense, such that

$$\mathbf{x}^* = \arg\min_{\mathbf{x}} \mathrm{WD}(p_{\mathrm{data}}(\mathbf{y})\|p_{\mathrm{gen}}(\mathbf{y};\mathbf{x})). \qquad (2)$$

Here, $p_{\mathrm{gen}}(\mathbf{y};\mathbf{x})$ is the distribution generated from the Cryo-EM physics simulator and WD refers to the Wasserstein distance [25]. This goal is achieved by solving the min-max optimization problem:

$$\mathbf{x}^* = \arg\min_{\mathbf{x}} \underbrace{\max_{D_\phi:\|D_\phi\|_L \leq 1} (\mathbb{E}_{\mathbf{y}\sim p_{\mathrm{data}}(\mathbf{y})}[D_\phi(\mathbf{y})] - \mathbb{E}_{\mathbf{y}\sim p_{\mathrm{gen}}(\mathbf{y};\mathbf{x})}[D_\phi(\mathbf{y})])}_{\mathrm{WD}(p_{\mathrm{data}}(\mathbf{y})\|p_{\mathrm{gen}}(\mathbf{y};\mathbf{x}))}, \quad (3)$$

where D_ϕ is a neural network with parameters ϕ that is constrained to have Lipschitz constant $\|D_\phi\|_L \leq 1$ [9] (Fig. 3(b)). Here, D_ϕ learns to differentiate between the real projection $\mathbf{y}$ and the simulated projection $\mathbf{H}_\varphi\{\mathbf{x}\}$ and scores the realness of given samples. As the discriminative power of D_ϕ becomes stronger (`maximization`) and the underlying volume estimate $\mathbf{x}$ is updated accordingly (`minimization`), $p_{\mathrm{data}}(\mathbf{y})$ and $p_{\mathrm{gen}}(\mathbf{y};\mathbf{x})$ become indistinguishable so that the algorithm recovers $\mathbf{x}^* = \mathbf{x}_{\mathrm{data}}$.

4   Method

4.1   Parameterization of the Conformation Manifold

CryoGAN successfully reconstructs the volumetric structure of a protein by finding a single volume $\mathbf{x}$ that explains the entire set of projections, which is adequate when all the imaged particles are identical (homogeneous case). However, in reality, many biomolecules have nonrigid structures, which carry vital information.

Algorithm 1: Samples from the generated distribution $p_{\text{gen}}(\mathbf{y}; G_\gamma)$.

Input: latent distribution $p_{\mathbf{z}}$; angle distribution p_θ; translation distribution $p_{\mathbf{t}}$; CTF parameters distribution $p_{\mathbf{c}}$; noise distribution $p_{\mathbf{n}}$

Output: Simulated projection $\mathbf{y}_{\text{gen}}$

1. Sample $\mathbf{z} \sim p_{\mathbf{z}}$.
2. Feed $\mathbf{z}$ into generator network to get $\mathbf{x} = G_\gamma(\mathbf{z})$.
3. Sample the imaging parameters $\varphi = [\boldsymbol{\theta}, \mathbf{t}, \mathbf{c}]$.
   - Sample the Euler angles $\boldsymbol{\theta} = (\theta_1, \theta_2, \theta_3) \sim p_\theta$.
   - Sample the 2D shifts $\mathbf{t} = (t_1, t_2) \sim p_{\mathbf{t}}$.
   - Sample the CTF parameters $\mathbf{c} = (d_1, d_2, \alpha_{\text{ast}}) \sim p_{\mathbf{c}}$.

4. Sample the noise $\mathbf{n} \sim p_{\mathbf{n}}$.
5. Generate $\mathbf{y}_{\text{gen}} = \mathbf{H}_\varphi \mathbf{x} + \mathbf{n}$ based on (8).

return $\mathbf{y}_{\text{gen}}$

To address this, we introduce a manifold-learning module G_γ that uses a CNN with learnable weights γ (Fig. 3 (a)). Sampling from $p_{\text{conf}}(\mathbf{x})$ is then equivalent to getting $G_\gamma(\mathbf{z})$, where $\mathbf{z}$ is sampled from a prior distribution. Therefore, $\mathbf{y} \sim p_{\text{gen}}(\mathbf{y})$ is obtained by evaluating $\mathbf{H}_\varphi \{G_\gamma(\mathbf{z})\} + \mathbf{n}$, where $(\mathbf{n}, \mathbf{z}, \varphi)$ are sampled from their distributions (see Algorithm 1 and Fig. 3). To explicitly show this dependency to G_γ, we hereafter denote the generated distribution of projection data by $p_{\text{gen}}(\mathbf{y}; G_\gamma)$.

4.2 Optimization Scheme

We now find G_{γ^*} such that the distance between $p_{\text{data}}(\mathbf{y})$ and $p_{\text{gen}}(\mathbf{y}; G_\gamma)$ is minimized, which results in the min-max optimization problem [2]

$$G_{\gamma^*} = \arg\min_{G_\gamma} \text{WD}(p_{\text{data}}(\mathbf{y}) \| p_{\text{gen}}(\mathbf{y}; G_\gamma)) \tag{4}$$

$$= \arg\min_{G_\gamma} \underbrace{\max_{D_\phi : \|D_\phi\|_L \leq 1} \left(\mathbb{E}_{\mathbf{y} \sim p_{\text{data}}}[D_\phi(\mathbf{y})] - \mathbb{E}_{\mathbf{y} \sim p_{\text{gen}}(\mathbf{y}; G_\gamma)}[D_\phi(\mathbf{y})] \right)}_{\text{WD}(p_{\text{data}}(\mathbf{y}) \| p_{\text{gen}}(\mathbf{y}; G_\gamma))} . \tag{5}$$

As will be discussed in Theorem 1, the global minimizer G_{γ^*} of (5) indeed captures the true conformation landscape $p_{\text{conf}}(\mathbf{x})$, which is achieved when D_ϕ is no longer able to differentiate the samples from $p_{\text{data}}(\mathbf{y})$ and $p_{\text{gen}}(\mathbf{y}; G_{\gamma^*})$.

The framework of Multi-CryoGAN is similar to the conventional Wasserstein GANs [2], however, it is crucial to note the difference between the two. In the latter, G directly outputs the samples from $p_{\text{gen}}(\mathbf{y})$, whereas ours outputs the samples $\mathbf{x}$ from the conformation distribution $p_{\text{conf}}(\mathbf{x})$ whose stochastic projections are the samples of $p_{\text{gen}}(\mathbf{y})$. The conventional schemes only helps one to generate samples which are similar to the real data but does not recover the underlying conformation landscape. Our proposed scheme includes the physics of Cryo-EM, which ties $p_{\text{gen}}(\mathbf{y})$ with the conformation landscape $p_{\text{conf}}(\mathbf{x})$ and is thus able to recover it (see Theorem 1 in Sect. 6).

Algorithm 2: Reconstruction of multiple conformations using Multi-CryoGAN.

Input: Dataset $\{\mathbf{y}_{\text{data}}^1, \ldots, \mathbf{y}_{\text{data}}^Q\}$; training parameters: number of steps k to apply to the discriminator and penalty parameter λ

Output: A mapping G from the latent space to the 3D conformation space.

for n_{train} training iterations **do**

    for k steps **do**

        – sample from real data: $\{\mathbf{y}_{\text{data}}^1, \ldots, \mathbf{y}_{\text{data}}^B\}$.

        – sample from generated data: $\{\mathbf{y}_{\text{gen}}^1, \ldots, \mathbf{y}_{\text{gen}}^B\} \sim p_{\text{gen}}(\mathbf{y}; G_\gamma)$ (see Algorithm 1).

        – sample from $\{\kappa_1, \ldots, \kappa_B\} \sim \text{U}[0, 1]$.

        – compute $\mathbf{y}_{\text{int}}^b = \kappa_b \cdot \mathbf{y}_{\text{batch}}^b + (1 - \kappa_b) \cdot \mathbf{y}_{\text{gen}}^b$ for all $b \in \{1, \ldots, B\}$.

        – update the discriminator D_ϕ by gradient ascent on the loss (5) complemented with the gradient penalty term from [9].

    end

    – sample generated data: $\{\mathbf{y}_{\text{gen}}^1, \ldots, \mathbf{y}_{\text{gen}}^B\} \sim p_{\text{gen}}(\mathbf{y}; G_\gamma)$ (see Algorithm 1).

    – update the Generator G_γ by gradient descent on the loss (5).

end

return $\mathbf{G}_{\gamma^*}$

---

5   Experiments and Results

We evaluate the performance of the proposed algorithm on synthetic datasets obtained from a protein with multiple conformations. We synthesize two datasets: one continuum of configurations and one where the particles can only take a discrete number of states. During reconstruction, no assumption is made of their continuous or discrete nature, which suggests that our method is capable of learning different conformation distribution behaviors.

Dataset. For each dataset, we generate 100,000 simulated projections from the *in vivo* conformation variation of the heat-shock protein *Hsp90*. The Coulomb density maps of each conformation are created by the code provided in [20] with slight modifications. The conformation variation of this protein is represented by the bond-angle β, which describes the work cycle of the molecule, where the two subunits continuously vary between fully closed ($\beta = 0°$, protein database entry *2cg9*) and fully opened ($\beta = 20°$). We sample $\beta \sim \text{Uniform}(0°, 20°)$ for the continuous case and $\beta \sim 20° * \text{Bernoulli}(0.75)$ for the discrete case. Here, Uniform(a, b) is the uniform distribution between a and b, and Bernoulli(p) denotes the Bernoulli distribution with parameter p. A conformation is generated with $(32 \times 32 \times 32)$ voxels, where the size of each voxel is 5 Å. A 2D projection with random orientation of this conformation is obtained ((32×32) image, Fig. 4b). The orientation is sampled from a uniform distribution over $SO(3)$. Then, the CTF is applied to this projection image with a defocus uniformly sampled between [1.0 μm, 2.0 μm], assuming that the horizontal and vertical defocus values are the same and there is no astigmatism. Translations/shifts are disabled in these experiments. Finally, Gaussian noise was added to the CTF-modulated images, resulting in an SNR of approximately -10 dB.

Implementation Details. The reconstruction of the conformation is done by solving (5) using Algorithm 2. For both continuous and discrete conformations, we use the same distribution $p_{\mathbf{z}} \sim$ Uniform $(\mathbf{z}_0, \mathbf{z}_1)$, where $\mathbf{z}_0, \mathbf{z}_1 \in \mathbb{R}^{32 \times 32 \times 32}$ are randomly chosen from Uniform$(0, 0.025)$ and fixed throughout the process. Thus, we do not impose any prior knowledge whether the landscape is continuous or discrete. Note that the latent variables implicitly lie on a 1D manifold embedded in a high dimensional space. As we shall see later, this is sufficiently rich to represent the variation of interest in this synthetic dataset. The architecture of D, G, and training details are provided in the supplementary material.

Metric. We deploy two metrics based on Fourier-Shell Correlation (FSC). The FSC between two structures $\mathbf{x}_1$ and $\mathbf{x}_2$ is given by

$$\text{FSC}(\omega, \mathbf{x}_1, \mathbf{x}_2) = \frac{\langle \mathbf{V}^{\omega}_{\hat{\mathbf{x}}_1}, \mathbf{V}^{\omega}_{\hat{\mathbf{x}}_2} \rangle}{\|\mathbf{V}^{\omega}_{\hat{\mathbf{x}}_1}\| \|\mathbf{V}^{\omega}_{\hat{\mathbf{x}}_2}\|} \tag{6}$$

where $\mathbf{V}^{\omega}_{\hat{\mathbf{x}}}$ is the vectorization of the shell of $\hat{\mathbf{x}}$ at radius ω and $\hat{\mathbf{x}}$ is the 3D Fourier transform of $\mathbf{x}$. As first metric, we use the FSC between a reconstructed conformation and the corresponding ground truth conformation. This metric encapsulates the structural quality of an individual reconstructed conformation.

To evaluate the landscape of the conformations, we propose a second metric that we call the matrix $\mathbf{M} \in \mathbb{R}^{L \times L}$ of FSC cross conformations (FSCCC). Its entries are given by

$$\mathbf{M}[m, n] = \text{AreaFSC}(\mathbf{x}_m, \mathbf{x}_n) = \int_{0 < \omega \leq \omega_c} \text{FSC}(\omega, \mathbf{x}_m, \mathbf{x}_n) \mathrm{d}\omega \tag{7}$$

where $\mathbf{x}_m$ and $\mathbf{x}_n$ are samples in the reconstructed conformation manifold and ω_c is the normalized Nyquist frequency. We determine it for the reconstructed landscape by setting $\mathbf{x}_m = G_{\gamma^*}((1 - \alpha_m) * \mathbf{z}_0 + \alpha_m * \mathbf{z}_1)$ where $\alpha_m = (m/L)$ for $m \in \{0, \ldots, L\}$. The matrix $\mathbf{M}$ encapsulates how similar $\mathbf{x}_m$ is compared to other structures $\mathbf{x}_n$ across the manifold ($\mathbf{M}[m, n]$ is proportional to the similarity between $\mathbf{x}_m$ and $\mathbf{x}_n$), hence allowing for a visualization of the manifold.

For the continuous conformation, it is useful to compare the FSCCC of our reconstructions with that of the ground truth. To that end, we also evaluate $\mathbf{M}[m, n]$ when $\mathbf{x}_m$ corresponds to the bond-angle $\beta = 20°(m/L)$, where $m \in \{0, \ldots, L\}$. In our experiments, we used $L = 20$ for all FSCCC calculations.

5.1   Continuous Conformations

We give in Fig. 4 a qualitative comparison between the ground truth conformation variation, as the angle β goes from $0°$ to $20°$, and the reconstructions $G(\gamma^*)(\mathbf{z})$, where $\mathbf{z} = (1 - \alpha)\mathbf{z}_0 + \alpha\mathbf{z}_1$ and α goes from 0 to 1. Our method successfully reconstructs a manifold that exhibits smooth continuous conformation variation (Fig. 4(a)), where the input parameter α has direct control over

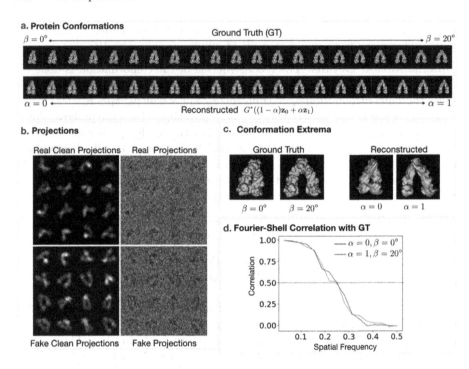

Fig. 4. Continuous conformations experiment. (a) Comparison between the ground truth conformation manifold and the reconstructed conformation manifold $G^*(\mathbf{z})$ where $\mathbf{z} = (1 - \alpha)\mathbf{z}_0 + \alpha\mathbf{z}_1$, $\alpha \in [0, 1]$. (b, left-column) Clean projections of random samples from the ground truth and reconstructed manifold; (b, right-column) their CTF-modulated and noise-corrupted projection. These are the real and generated samples that are fed to D_ϕ. (c) Ground truth with angles $0°$ and $20°$, and the reconstruction corresponding to the endpoints in the latent space. (d) The FSC between them.

the bond-angle for the reconstruction. This shows that not only the true conformation landscape has been captured, but its factor of variation has been meaningfully encoded by the latent variables $\mathbf{z}$. The similarity between simulated projections and the ground truth data in Fig. 4(b) suggests that the algorithm has achieved $p_{\text{data}} = p_{\text{gen}}$. Moreover, their underlying distributions of noiseless projections are also similar, in accordance to the property discussed in Theorem 1.

We also evaluate the structural quality of reconstruction for certain representative individual conformations. In Fig. 4(c), the extreme conformations for the ground truth $\beta = 0°$ and $\beta = 20°$ and the reconstructions $\alpha = 0$ and $\alpha = 1$ are shown. Their FSC plot reach the value 0.5 after the normalized frequency of 0.25, so that at least half of the Nyquist resolution is achieved (Fig. 4(d)). All these results are further confirmed by the very similar FSCCC matrix for ground truth and reconstruction in Fig. 5(a). This implies that the reconstruction manifold successfully approximates the continuous ground truth.

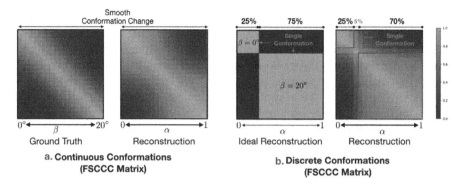

a. **Continuous Conformations**
   (FSCCC Matrix)

b. **Discrete Conformations**
   (FSCCC Matrix)

Fig. 5. The FSC cross-conformation (FSCCC) matrix in (7). (a) Continuous conformation with (left) ground truth and (right) reconstruction. It shows that the reconstructed conformations smoothly vary (without forming clusters) similar to the ground truth case. (b) Discrete conformation case with (left) ideal reconstruction and (right) obtained reconstruction. The ideal reconstruction describes the case where 25% and 75% of latent space would have mapped to the two distinct conformations without any transitions. The obtained reconstruction case can be seen to be very similar to the ideal case with 25% and 70% being the latent space occupied by the two conformations.

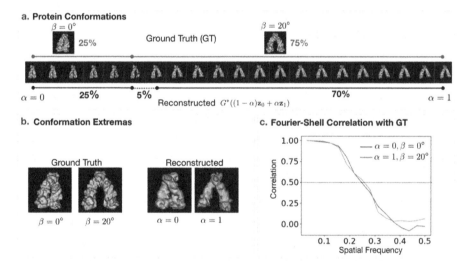

Fig. 6. Discrete conformations experiment. (a) Comparison between the ground truth (GT) taking only two conformations with probability 0.25 and 0.75 and the reconstructed conformations $G_{\gamma^*}(\mathbf{z})$, where $\mathbf{z} = (1-\alpha)\mathbf{z}_0 + \alpha\mathbf{z}_1$, $\alpha \in [0,1]$. (b) The GT with bond angles $0°$ and $20°$ and their reconstruction. (c) FSC of the structures in (b).

5.2   Discrete Conformations

We present in Fig. 6 the reconstruction results for the discrete case, where our proposed method successfully recovers not only the conformations but also their probabilities. About 70% of the reconstructed landscape matches the configuration for $\beta = 20°$, while 25% matches $\beta = 0°$. The remaining 5% of the landscape corresponds to a relatively abrupt transition between them. This suggests that our model distribution $p_{\text{conf}}(\mathbf{x})$ closely follows the ground truth Bernoulli distribution. This is further supported in Fig. 5(b), where the reconstructed FSCCC matrix greatly resembles the ideal reconstruction case. Ideally, one would expect the two conformations to occupy 25% and 75% of the latent space without having any transition conformations. The structural quality of these two recovered configurations with respect to the corresponding ground truth are given in Fig. 6(b). Their FSC show that at least half of the Nyquist resolution is achieved (Fig. 6(c)).

The FSCCC reconstruction matrix (Fig. 5(b)) validates the fact that the reconstructed structures cluster into two main conformations. We use it to determine the probabilities of these cluster/conformations. We determine the probability of a conformation using its first and last row (similarity of the conformations with respect to the extreme conformations). We consider that a conformation $\mathbf{x}_n$ belongs to the first cluster/conformation if $\mathbf{M}[0, n] > 0.5$ and $\mathbf{M}[20, n] < 0.5$. If the case is reversed ($\mathbf{M}[0, n] < 0.5$ and $\mathbf{M}[20, n] > 0.5$), then it belongs to the second cluster/conformation. Otherwise, it is considered as a transitioning conformation. This yields that the first 25% and the last 70% structures cluster together to form the $\beta = 0°$ and $\beta = 20°$ conformations, respectively, and the middle 5% are the transitioning conformations.

6   Theoretical Guarantee of Recovery

Our experiments illustrate that enforcing a match between the distribution of the simulated measurements and that of the data is sufficient to reconstruct the true conformations. We now prove this mathematically. For the homogeneous case, the proof is already discussed in [10, Theorem 1] which we now extend to the heterogeneous case. We switch to a continuous-domain formulation of the Cryo-EM problem while noting that the result is transferable to the discrete-domain as well, albeit with some discretization error.

Notations and Preliminaries. We denote by $\mathcal{L}_2(\mathbb{R}^3)$ the space of 3D structures $f : \mathbb{R}^3 \to \mathbb{R}$ with finite energy $\|f\|_{L_2} < \infty$. The imaging parameters φ are assumed to lie in $\mathcal{B} \subset \mathbb{R}^8$. We denote by $\mathcal{L}_2(\mathbb{R}^2)$ the space of 2D measurements with finite energy. Each individual continuous-domain Cryo-EM measurement $y \in \mathcal{L}_2(\mathbb{R}^2)$ is given by

$$y = H_\varphi\{f\} + n, \tag{8}$$

where $f \in \mathcal{L}_2(\mathbb{R}^3)$ is some conformation of the biomolecule sampled from the probability measure $\tilde{\mathbb{P}}_{\text{conf}}$ on $\mathcal{L}_2(\mathbb{R}^3)$, the imaging parameters φ are sampled from p_φ, and n is sampled from the noise probability measure $\mathbb{P}_n$ on $\mathcal{L}_2(\mathbb{R}^2)$.

We define $[f] := \{r_{\mathbf{A}}\{f\} : r_{\mathbf{A}} \in O\}$ as the set of all the rotated-reflected version of f. There, O is the set of all the rotated-reflected versions over

$\mathcal{L}_2(\mathbb{R}^3)$. We define the space $\sum \mathcal{L}_2(\mathbb{R}^3) = \mathcal{L}_2(\mathbb{R}^3)/O$ as the quotient space of the shapes. For any $\tilde{\mathbb{P}}_{\mathrm{conf}}$ defined over $\mathcal{L}_2(\mathbb{R}^3)$, an equivalent $\mathbb{P}_{\mathrm{conf}}$ exists over $\sum \mathcal{L}_2(\mathbb{R}^3)$. Since we are interested only in the shape of conformations of the biomolecule, we will only focus on recovering $\mathbb{P}_{\mathrm{conf}}$. We denote by Ψ the probability measure on $\mathcal{B} \in \mathbb{R}^8$. The measure Ψ is associated to the density function p_φ. Both of these induce a probability measure $\mathbb{P}_{\mathrm{clean}}$ on the space $\mathcal{L}_2^2 = \{f : \mathbb{R}^3 \to \mathbb{R}^2 \text{ s.t. } \|f\|_{L_2} < \infty\}$ through the forward operator. This is given by $\mathbb{P}_{\mathrm{clean}}[A] = (\mathbb{P}_{\mathrm{conf}} \times \Psi)[([f], \varphi) \in (\sum \mathcal{L}_2 \times \mathcal{B}) : H_\varphi f \in A]$ for any Borel measure set $A \in \mathcal{L}_2(\mathbb{R}^2)$. We denote $\mathbb{P}_{\mathrm{meas}}$ as the probability measure of the noisy measurements.

Theorem 1. *Let $\mathbb{P}_{\mathrm{conf}}^{\mathrm{data}}$ and $\mathbb{P}_{\mathrm{conf}}^{\mathrm{gen}}$ be the true and the reconstructed conformation probability measures on the quotient space of 3D structures $\sum \mathcal{L}_2(\mathbb{R}^3)$, respectively. We assume that they are atomic and that they are supported only on nonnegative-valued shapes. Let $\mathbb{P}_{\mathrm{meas}}^{\mathrm{data}}$ and $\mathbb{P}_{\mathrm{meas}}^{\mathrm{gen}}$ be the probability measures of the noisy Cryo-EM measurements obtained from $\mathbb{P}_{\mathrm{conf}}^{\mathrm{data}}$ and $\mathbb{P}_{\mathrm{conf}}^{\mathrm{gen}}$, respectively.*

Make the following physical assumptions:

1. *the noise probability measure $\mathbb{P}_n$ is such that its characteristic functional vanishes nowhere in its domain and that its sample n is pointwise-defined everywhere;*
2. *the distributions p_θ, $p_\mathbf{t}$, and $p_\mathbf{c}$ are bounded;*
3. *for any two $\mathbf{c}_1, \mathbf{c}_2 \sim p_\mathbf{c}, \mathbf{c}_1 \neq \mathbf{c}_2$, the CTFs $\hat{\mathbf{C}}_{\mathbf{c}_1}$ and $\hat{\mathbf{C}}_{\mathbf{c}_2}$ share no common zero frequencies.*

Then, it holds that

$$\mathbb{P}_{\mathrm{meas}}^{\mathrm{data}} = \mathbb{P}_{\mathrm{meas}}^{\mathrm{gen}} \Rightarrow \mathbb{P}_{\mathrm{conf}}^{\mathrm{data}} = \mathbb{P}_{\mathrm{conf}}^{\mathrm{gen}}. \tag{9}$$

Proof. We first prove that $\mathbb{P}_{\mathrm{meas}}^{\mathrm{data}} = \mathbb{P}_{\mathrm{meas}}^{\mathrm{gen}} \Rightarrow \mathbb{P}_{\mathrm{clean}}^{\mathrm{data}} = \mathbb{P}_{\mathrm{clean}}^{\mathrm{gen}}$. Note that, due to the independence of clean measurements and noise, we have that

$$\hat{\mathbb{P}}_{\mathrm{meas}}^{\mathrm{data}} = \hat{\mathbb{P}}_{\mathrm{clean}}^{\mathrm{data}} \, \hat{\mathbb{P}}_n$$
$$\hat{\mathbb{P}}_{\mathrm{meas}}^{\mathrm{gen}} = \hat{\mathbb{P}}_{\mathrm{clean}}^{\mathrm{gen}} \, \hat{\mathbb{P}}_n. \tag{10}$$

From the assumption that $\hat{\mathbb{P}}_n$ is nonzero everywhere, we deduce that $\hat{\mathbb{P}}_{\mathrm{clean}}^{\mathrm{data}} = \hat{\mathbb{P}}_{\mathrm{clean}}^{\mathrm{gen}}$. This proves the first step.

To prove the next step, we invoke Theorem 4 in [10] which states that any two probability measures $\mathbb{P}_{\mathrm{clean}}^1$ and $\mathbb{P}_{\mathrm{clean}}^2$ that correspond to Dirac probability measures $\mathbb{P}_{\mathrm{conf}}^1$ and $\mathbb{P}_{\mathrm{conf}}^2$ on $\sum \mathcal{L}_2(\mathbb{R}^3)$, respectively, are mutually singular (zero measure of the common support) if and only if the latter are distinct. We denote the relation of mutual singularity by $\perp$.

Since $\mathbb{P}_{\mathrm{conf}}^{\mathrm{data}}$ is an atomic measure (countable weighted sum of distinct Dirac measures), the corresponding $\mathbb{P}_{\mathrm{clean}}^{\mathrm{data}}$ is composed of a countable sum of mutually singular measures. The same is true for $\mathbb{P}_{\mathrm{clean}}^{\mathrm{gen}}$ since it is equal to $\mathbb{P}_{\mathrm{clean}}^{\mathrm{data}}$.

We proceed by contradiction. We denote by $\mathrm{Supp}\{\mathbb{P}\}$ the support of the measure $\mathbb{P}$. Assume that $\mathrm{Supp}\{\mathbb{P}_{\mathrm{conf}}^{\mathrm{data}}\} \neq \mathrm{Supp}\{\mathbb{P}_{\mathrm{conf}}^{\mathrm{gen}}\}$. Let us define $\mathcal{S}_1 = \mathrm{Supp}\{\mathbb{P}_{\mathrm{conf}}^{\mathrm{gen}}\} \cap \mathrm{Supp}\{\mathbb{P}_{\mathrm{conf}}^{\mathrm{data}}\}^C$. For any $[f] \in \mathcal{S}_1$, we denote by $\mathbb{P}_{\mathrm{clean}}^f$ its noiseless probability measure. Since $f \in \mathcal{S}_1$, it is distinct from any constituent Dirac measure in $\mathbb{P}_{\text{conf}}^{\text{data}}$. Therefore, by using [10, Theorem 4], $\mathbb{P}_{\text{clean}}^{f}$ is mutually singular to each of the constituent mutually singular measures of $\mathbb{P}_{\text{clean}}^{\text{data}}$, implying that $\mathbb{P}_{\text{clean}}^{f} \perp \mathbb{P}_{\text{clean}}^{\text{data}}$.

From $\text{Supp}\{\mathbb{P}_{\text{clean}}^{f}\} \subset \text{Supp}\{\mathbb{P}_{\text{clean}}^{\text{gen}}\}$, it follows that $\mathbb{P}_{\text{clean}}^{\text{data}} \neq \mathbb{P}_{\text{clean}}^{\text{gen}}$, which raises a contradiction. Therefore, the set $\mathcal{S}_1$ is empty. The same can be proved for the set $\mathcal{S}_2 = \text{Supp}\{\mathbb{P}_{\text{conf}}^{\text{gen}}\}^{C} \cap \text{Supp}\{\mathbb{P}_{\text{conf}}^{\text{data}}\}$. Therefore, $\text{Supp}\{\mathbb{P}_{\text{conf}}^{\text{gen}}\} = \text{Supp}\{\mathbb{P}_{\text{conf}}^{\text{data}}\}$, which means that the location of their constituent Dirac measures are the same. To maintain $\mathbb{P}_{\text{clean}}^{\text{data}} = \mathbb{P}_{\text{clean}}^{\text{gen}}$, the weight of their constituent Dirac measures have to be the same, too. This concludes the proof. ∎

In essence, Theorem 1 claims that a reconstructed manifold of conformations recovers the true conformations if its measurements match the acquired data in a distributional sense. Though the result assumes the true conformation landscape to be discrete (atomic measure), it holds for an infinite number of discrete conformations which could be arbitrarily close/similar to each other and is thus relevant to continuously varying conformations. We leave the proof of the latter case to future works.

7 Conclusion

We have introduced a novel algorithm named Multi-CryoGAN. It can reconstruct the 3D continuous conformation manifold of a protein from a sufficiently rich set of 2D Cryo-EM data. By matching the simulated Cryo-EM projections with the acquired data distribution, Multi-CryoGAN naturally learns to generate a set of 3D conformations in a likelihood-free way. This allows us to reconstruct both continuous and discrete conformations without any prior assumption on the conformation landscape, data preprocessing steps, nor external algorithms such as pose estimation. Our experiments shows that Multi-CryoGAN successfully recovers the molecular conformation manifold, including the underlying distribution. We believe that, with a better incorporation of state-of-the art GAN architectures [12,13], Multi-CryoGAN could become an efficient and user-friendly method to reconstruct heterogeneous biomolecules in Cryo-EM.

Acknowledgments. This work is funded by H2020-ERC Grant 692726 (Global-BioIm).

References

1. Andén, J., Katsevich, E., Singer, A.: Covariance estimation using conjugate gradient for 3D classification in cryo-EM. In: 12th International Symposium on Biomedical Imaging (ISBI), pp. 200–204. IEEE (2015)
2. Arjovsky, M., Chintala, S., Bottou, L.: Wasserstein generative adversarial networks. In: International Conference on Machine Learning, pp. 214–223 (2017)

3. Bendory, T., Bartesaghi, A., Singer, A.: Single-particle cryo-electron microscopy: mathematical theory, computational challenges, and opportunities. IEEE Signal Process. Mag. **37**(2), 58–76 (2020)

4. Bora, A., Price, E., Dimakis, A.G.: AmbientGAN: generative models from lossy measurements. In: International Conference on Learning Representations, vol. 2, pp. 5–15 (2018)

5. Dashti, A., et al.: Trajectories of the ribosome as a Brownian nanomachine. Proc. Natl. Acad. Sci. **111**(49), 17492–17497 (2014)

6. Frank, J., Ourmazd, A.: Continuous changes in structure mapped by manifold embedding of single-particle data in Cryo-EM. Methods **100**, 61–67 (2016)

7. Gadelha, M., Maji, S., Wang, R.: 3D shape induction from 2D views of multiple objects. In: 2017 International Conference on 3D Vision (3DV), pp. 402–411 (2017)

8. Goodfellow, I., et al.: Generative adversarial nets. In: Advances in Neural Information Processing Systems, pp. 2672–2680 (2014)

9. Gulrajani, I., Ahmed, F., Arjovsky, M., Dumoulin, V., Courville, A.C.: Improved training of Wasserstein GANs. In: Advances in Neural Information Processing Systems, pp. 5767–5777 (2017)

10. Gupta, H., McCann, M.T., Donati, L., Unser, M.: CryoGAN: a new reconstruction paradigm for single-particle Cryo-EM via deep adversarial learning. BioRxiv (2020)

11. Hornik, K., Stinchcombe, M., White, H.: Multilayer feedforward networks are universal approximators. Neural Netw. **2**(5), 359–366 (1989)

12. Karras, T., Aila, T., Laine, S., Lehtinen, J.: Progressive growing of GANs for improved quality, stability, and variation. arXiv:1710.10196 (2017)

13. Karras, T., Laine, S., Aila, T.: A style-based generator architecture for generative adversarial networks. In: Proceedings of the IEEE Conference on Computer Vision and Pattern Recognition, pp. 4401–4410 (2019)

14. Lederman, R.R., Andén, J., Singer, A.: Hyper-molecules: on the representation and recovery of dynamical structures for applications in flexible macro-molecules in cryo-EM. Inverse Prob. **36**(4), 044005 (2020)

15. Miolane, N., Poitevin, F., Li, Y.T., Holmes, S.: Estimation of orientation and camera parameters from cryo-electron microscopy images with variational autoencoders and generative adversarial networks. In: Proceedings of the IEEE/CVF Conference on Computer Vision and Pattern Recognition Workshops, pp. 970–971 (2020)

16. Moscovich, A., Halevi, A., Andén, J., Singer, A.: Cryo-EM reconstruction of continuous heterogeneity by Laplacian spectral volumes. Inverse Prob. **36**(2), 024003 (2020)

17. Ourmazd, A.: Cryo-EM, XFELs and the structure conundrum in structural biology. Nat. Methods **16**(10), 941–944 (2019)

18. Punjani, A., Rubinstein, J.L., Fleet, D.J., Brubaker, M.A.: cryoSPARC: algorithms for rapid unsupervised cryo-EM structure determination. Nat. Methods **14**(3), 290–296 (2017)

19. Scheres, S.H.: RELION: Implementation of a Bayesian approach to cryo-EM structure determination. J. Struct. Biol. **180**(3), 519–530 (2012)

20. Seitz, E., Acosta-Reyes, F., Schwander, P., Frank, J.: Simulation of cryo-EM ensembles from atomic models of molecules exhibiting continuous conformations. BioRxiv p. 864116 (2019)

21. Singer, A., Sigworth, F.J.: Computational methods for single-particle electron cryomicroscopy. Ann. Rev. Biomed. Data Sci. **3** (2020)

22. Sorzano, C.O.S., et al.: Survey of the analysis of continuous conformational variability of biological macromolecules by electron microscopy. Acta Crystallogr. Sect. F Struct. Biol. Commun. **75**(1), 19–32 (2019)

23. Tewari, A., et al.: State of the art on neural rendering. arXiv preprint arXiv:2004.03805 (2020)

24. Tulsiani, S., Efros, A.A., Malik, J.: Multi-view consistency as supervisory signal for learning shape and pose prediction. In: Proceedings of the IEEE Conference on Computer Vision and Pattern Recognition, pp. 2897–2905 (2018)

25. Villani, C.: Optimal Transport: Old and New, vol. 338. Springer, Heidelberg (2008)

26. Zhong, E.D., Bepler, T., Berger, B., Davis, J.H.: CryoDRGN: reconstruction of heterogeneous structures from cryo-electron micrographs using neural networks. bioRxiv (2020)

Probabilistic Deep Learning for Instance Segmentation

Josef Lorenz Rumberger[2,3]([✉]) [iD], Lisa Mais[1,3] [iD], and Dagmar Kainmueller[1,3] [iD]

[1] Berlin Institute of Health, Berlin, Germany
[2] Charité University Hospital, Berlin, Germany
[3] Max Delbrück Center for Molecular Medicine, Berlin, Germany
{joseflorenz.rumberger,lisa.mais,dagmar.kainmueller}@mdc-berlin.de

Abstract. Probabilistic convolutional neural networks, which predict distributions of predictions instead of point estimates, led to recent advances in many areas of computer vision, from image reconstruction to semantic segmentation. Besides state of the art benchmark results, these networks made it possible to quantify local uncertainties in the predictions. These were used in active learning frameworks to target the labeling efforts of specialist annotators or to assess the quality of a prediction in a safety-critical environment. However, for instance segmentation problems these methods are not frequently used so far. We seek to close this gap by proposing a generic method to obtain model-inherent uncertainty estimates within proposal-free instance segmentation models. Furthermore, we analyze the quality of the uncertainty estimates with a metric adapted from semantic segmentation. We evaluate our method on the BBBC010 C. elegans dataset, where it yields competitive performance while also predicting uncertainty estimates that carry information about object-level inaccuracies like false splits and false merges. We perform a simulation to show the potential use of such uncertainty estimates in guided proofreading.

Keywords: Instance segmentation · Probabilistic deep learning · Bayesian inference · Digital microscopy

1 Introduction

Probabilistic deep learning models predict distributions of predictions instead of single point estimates and make it possible to quantify the inherent uncertainty of predictions. They were successfully applied for computer vision tasks such as image classification [2,8–10,16], semantic segmentation [9,14,23] and regression problems like instance counting [27] and depth regression [14]. They achieve state-of-the-art results for complex problems such as predicting distributions for segmentation tasks that carry inherent ambiguities [17] or microscopy image restoration [32]. However, said probabilistic deep learning methods are not directly applicable to proposal-free instance segmentation approaches, which

© Springer Nature Switzerland AG 2020
A. Bartoli and A. Fusiello (Eds.): ECCV 2020 Workshops, LNCS 12535, pp. 445–457, 2020.
https://doi.org/10.1007/978-3-030-66415-2_29

define the current state-of-the-art in some applications from the bio-medical domain [7,13,18]. This is due to the fact that proposal-free instance segmentation methods require some form of inference to yield instance segmentations from network predictions. Thus, to the best of our knowledge, probabilistic deep learning models have not yet been studied in the context of proposal-free instance segmentation methods. To close the gap, this work (1) makes use of a probabilistic CNN to estimate the local uncertainty of a metric-learning based instance segmentation model, and (2) examines these estimates with regard to their informativeness on local inaccuracies. On the challenging BBBC010 C. elegans dataset [31], our model achieves competitive performance in terms of accuracy, while additionally providing estimates of object-level inaccuracies like false splits or false merges.

The concept of sampling hypotheses from a probabilistic model for instance segmentation was already used in [6]. However, their work differs in that the candidates are not obtained from an end-to-end trainable model as well as they do not consider uncertainties. The idea of predicting local uncertainty estimates for instance segmentation tasks has been proposed by [22], who use dropout sampling on a Mask-RCNN [12]. However, their model employs a proposal-based segmentation method, which carries the critical disadvantage that bounding-boxes are required to be sufficient region proposals for objects. For thin, long and curvy structures which often arise in the bio-medical domain, bounding-boxes frequently contain large parts of instances of the same category, which deteriorates segmentation performance [4]. Furthermore, [22] limit their analysis of results to a probabilistic object detection benchmark [11], while we quantitatively assess the quality of the uncertainty estimates for instance segmentation.

Overcoming the limitations of proposal-based methods, state-of-the-art proposal-free models use a CNN to learn a representation of the data that allows instances to be separated. Popular approaches are based on learning and post-processing a watershed energy map [1,34], an affinity-graph [7,13,20,33] or a metric space [3,4] into binary maps for each instance. However, this binary output misses information like confidence scores or uncertainty measures.

In Bayesian machine learning, at least two kinds of uncertainties are distinguished that together make up the predictive uncertainty: Data uncertainty (i.e. aleatoric uncertainty) accounts for uncertainty in the predictions due to noise in the observation and measurement process of data or ambiguities in the annotation process and thus does not decrease with more training data [14]. Model uncertainty (i.e. epistemic uncertainty) captures the uncertainty about the model architecture and parameters. To account for it, parameters are modeled as probability distributions. The more data gets acquired and used for training, the more precise one can estimate their distribution and the smaller the variance becomes. Thus, model uncertainty decreases with more data. This work focuses on the estimation of model uncertainty, since it is high in applications with small annotated datasets and sparse samples [14], which is typical for bio-medical image data.

Model uncertainty is commonly estimated by approximating the unknown parameter distributions by simple variational distributions. A popular choice is the Gaussian distribution, which has proven effective but leads to high computational complexity and memory consumption [2,15]. To overcome these limitations, [8,16] proposed to use Bernoulli distributions instead, which can be implemented via Dropout. For our work, we use the Concrete Dropout model [9], because it has shown high quality uncertainty estimates for semantic segmentation tasks [23] and provides learnable dropout rates due to its variational interpretation. To our knowledge it has not been applied to instance segmentation tasks yet.

In order to obtain draws from the posterior predictive distribution in the Concrete Dropout model [9], a single input image is passed several times through the network, each time with a new realization of model parameters. Our proposed pipeline post-processes each such draw into binary instance maps, and agglomerates the resulting predictions into a single probabilistic prediction for each instance. Our proposed pipeline is constructed to be model-agnostic and applicable to any CNN-based proposal-free instance segmentation method. For showcasing, in this work, we pick a metric learning model, because (1) post-processing is fast and simple, and (2) metric learning models have shown competitive performance on several challenging datasets [3,21].

In summary, the key contributions of this work are:

- To the best of our knowledge, this is the first work that uses a Bayesian approximate CNN in conjunction with proposal-free instance segmentation.
- We adapt a metric for quantitative comparison of different uncertainty estimates [23], originally proposed for semantic segmentation, to the case of instance segmentation.

2   Methodology

This section describes the models employed in our proposed pipeline, as well as the associated loss functions and post-processing steps. Furthermore, we propose an adaptation of an uncertainty evaluation metric for probabilistic semantic segmentation [23] to the case of instance segmentation.

2.1   Models and Losses

Metric Learning with the Discriminative Loss Function: We follow the proposal-free instance segmentation approach of [3], a metric learning method which predicts, for each pixel, a vector in an embedding space, and trains for embedding vectors that belong to the same instance to be close to their mean, while mean embeddings of different instances are trained to be far apart. This is achieved by means of the *discriminative loss function*, which consists of three terms that are jointly optimized: The variance term (1) pulls embeddings $e_{c,i}$ towards their instance center μ_c in the embedding space:

$$L_{var} = \frac{1}{C} \sum_{c=1}^{C} \frac{1}{N_c} \sum_{i=1}^{N_c} ||\mu_c - e_{c,i}||^2 \tag{1}$$

with C the total number of instances and N_c the total number of pixels of instance c. The original formulation in [3] included a hinge that set the loss to zero for embeddings that are sufficiently near the center. However, in our work we use the version proposed by [21], which excludes the hinge in order to have lower intra-cluster variance of embeddings which is desirable to prevent false splits during post-processing. The distance term (2) penalizes cluster centers c_A

$$L_{dist} = \frac{1}{C(C-1)} \sum_{c_A=1}^{C} \sum_{\substack{c_B=1 \\ c_A \neq c_B}}^{C} [2\delta_d - ||\mu_{c_A} - \mu_{c_B}||^2]_+^2 \tag{2}$$

and c_B for lying closer together than $2\delta_d$ and therefore pushes clusters away from each other. Distance hinge parameter $\delta_d = 4$ is used to push clusters sufficiently far apart. Choosing this hyperparameter posed a trade-off: a higher δ_d led to a wider separation of clusters and increased segmentation performance but also made the loss have high jumps between samples, which led to training instability.

The last part is the regularization term (3), which penalizes the absolute sum of the embedding centers and therefore draws them towards the origin.

$$L_{reg} = \frac{1}{C} \sum_{c=1}^{C} ||\mu_c||^2 \tag{3}$$

In order to jointly optimize, the terms are weighted as follows: $L_{disc} = L_{var} + L_{dist} + 0.001 \cdot L_{reg}$. Besides this loss, a three-class cross-entropy loss function is added in order to learn to distinguish background, foreground and pixels that belong to overlapping instances.

Baseline Model: We employ a U-Net [29] as backbone architecture, which is a popular choice for pixel-wise prediction tasks [7,13,21]. The network is trained with weight decay to make it comparable to our Concrete Dropout Model described in the following.

Concrete Dropout Model: We employ the Concrete Dropout model proposed in [9]. The remainder of this Section is our attempt to motivate and describe this model to the unfamiliar reader by means of intuitions (where possible) and technical details (where necessary). It does not intend to serve as comprehensive summary of the respective theory, for which we refer the reader to [9].

It has been shown that Dropout, when combined with weight decay, can be interpreted as a method for approximating the posterior of a deep Gaussian Process [8]. This finding provides a theoretical basis to the practical approach of assessing output uncertainties from multiple predictions with Dropout at test time.

An intuitive detail of the respective theory is that a single weight matrix $M \in \mathbf{R}^{k \times l}$, together with a dropout rate p, implements a distribution over weight matrices of the form $M \cdot \mathrm{diag}([z]_{i=1}^{k})$, with $z_i \sim \mathrm{Bernoulli}(1-p)$. This, however, entails that for fixed dropout rates, output uncertainty scales with weight magnitude. This is not desired, as high magnitude weights may be necessary to explain the data well. This discrepancy, formally due to a lack of *calibration* of Gaussian Processes, motivates the need for dropout rates to be learnt from data, simultaneously with respective weight matrices [9].

A respective training objective has been proposed in [9] (cf. Eq. 1–4 therein). The objective stems, as in [8], from the idea of approximating the posterior of a deep Gaussian Process, with the difference that [9] consider learnable dropout rates. The objective combines a data term, which pulls dropout rates towards zero, and boils down to a standard SSD loss for regression problems, with a regularizer, which keeps the joint distribution over all weight matrices close to the prior of a Gaussian Process. This regularizer effectively pulls dropout rates towards 0.5 (i.e. maximum entropy), and is weighted by one over the number of training samples available, which entails higher dropout rates/higher uncertainty for smaller training set size. More specifically, the regularizer takes the following form (see Eq. 3, 4 in [9]):

$$L_{concrete} = \frac{1}{N} \cdot \left(\sum_{l=1}^{L} \frac{\iota^2(1 - p_l)}{2} ||M_l||^2 - \zeta F_l \mathcal{H}(p_l) \right) \tag{4}$$

$$\text{with } \mathcal{H}(p_l) := -p_l ln(p_l) - (1 - p_l) ln(1 - p_l) \tag{5}$$

The first part of the function is an l_2 regularizer on the pre-dropout weight matrices M_l, with l denoting one of the L layers of the network, p_l the dropout probability of the respective layer, N as the number of training examples, and ι the prior length scale, which is treated as a hyperparameter that controls the strength of the regularizer. The second term serves as a dropout rate regularizer. It captures the entropy $\mathcal{H}(p_l)$ of a Bernoulli random variable with probability p_l, where high entropy is rewarded. Hyperparameter ζ serves as weight for this term. It is furthermore scaled by the number of nodes per layer F_l, thus encouraging higher dropout rates for layers with more nodes. Note that hyperparameter ι gauges a weight decay effect implemented by this loss. Thus we set it equal to the weight decay factor in our baseline model to yield comparable effects.

To form the loss for our Concrete Dropout Metric Learning model, we add $L_{concrete}$ to the discriminative loss from our baseline metric learning objective, which is a weighted sum of terms 1, 2, and 3. Note that term 1 is a sum of squared differences loss and hence constitutes a theoretically sound data term in the Concrete Dropout objective. It is, however, unclear if terms 2 and 3 can be grounded in the same theoretical framework, and a respective analysis is subject to future work.

Our loss has the drawback that it is difficult to optimize w.r.t. dropout rates. (Note that each of its four terms depends on the dropout rates.) The Concrete Dropout Model [9] alleviates this drawback by interpreting each (layer-individual) dropout rate as parameter of a Concrete distribution, which is a continuous relaxation of the respective Bernoulli distribution.

2.2   Post-processing

To yield an instance segmentation from predictions of embeddings, mean-shift clustering [5] is applied to find cluster centers in the embedding space. All embeddings within a given threshold of a cluster center are gathered and their corresponding pixels represent an instance. In [3], the authors propose to use the hinge parameter from the variance term L_{var} as the mean-shift bandwidth and the clustering threshold during post-processing. In the version of the loss used here, Eq. (1) does not have a hinge parameter. Therefore, the mean-shift bandwidth and the clustering threshold are treated as hyperparameters and found through a grid-search with 2-fold cross-validation on the test set. To reduce the number of embeddings considered during clustering, embeddings at pixels classified as background are ignored. To tackle overlapping instances, we propose a straightforward heuristic: For each pixel, background/foreground/overlap probabilites are predicted in addition to embeddings. We add the resulting overlap map, containing all overlaps of a sample, to each individual instance map. In each instance-plus-overlap map, the connected component with the largest Intersection-over-Union (IoU) with the respective sole instance map is selected as the final instance prediction.

2.3   Inference on the Probabilistic Model

Every sample is passed eight times through the Concrete Dropout U-Net to obtain draws from the posterior predictive distribution. Each draw is post-processed individually by mean-shift clustering, such that eight instance segmentations are generated.

We propose a straightforward approach to capture uncertainties on the instance level: Each instance contained in a single draw is first converted into an individual binary segmentation. The draw that contains the highest total number of instances is the base of the following agglomeration. By choosing this draw as the first base map, the agglomeration strategy decreases false negatives in the final prediction. A second draw of instance segmentation maps is taken and a linear assignment problem is constructed by calculating the IoU for every pair of instance segmentations that belong to the different draws. To solve this, the Kuhn-Mukres (i.e. Hungarian) matching algorithm [24] is employed, which finds a mapping from the base set of instances to the other, such that the sum of IoUs is maximized. To aggregate more draws, a union of the two maps is calculated and taken as the base for subsequent steps. When all draws are agglomerated, the summed up instance maps are divided by the number of draws from the posterior to obtain a probability for every pixel to belong to a given instance.

2.4   Uncertainty Evaluation

In order to quantitatively assess the quality of our uncertainty estimates, we adapt the metrics presented by Mukhoti and Gal [23] to the task of instance segmentation. The authors state that good uncertainty estimates should be high where the model is inaccurate and low where it is accurate, and formalize this into the following conditional probabilities:

1. $p(accurate|certain)$: the probability that a models predictions are accurate given it is certain about it.
2. $p(uncertain|inaccurate)$: the probability that a model is uncertain, given its prediction is inaccurate.

The authors use the metrics on 2×2 image patches and calculate the mean accuracy and uncertainty of each patch. Then both mean values are converted into binary variables by thresholding the accuracy with 0.5 and plotting the probabilities as a function of uncertainty thresholds. Here, these metrics are adapted for instance segmentation by focusing on the precision instead of the accuracy. We do so, by excluding all image patches that neither belong to the foreground in the prediction nor the groundtruth and calculate the following for every instance prediction:

$$p(accurate|certain) = \frac{n_{ac}}{(n_{ac} + n_{ic})} \tag{6}$$

$$p(uncertain|inaccurate) = \frac{n_{iu}}{(n_{ic} + n_{iu})} \tag{7}$$

With n_{ac} the number of patches that are accurate and certain, n_{iu} the patches that were inaccurate and uncertain and the two undesired cases (accurate and uncertain, inaccurate and certain). Both probabilities are then combined into a single metric, called Patch Accuracy vs. Patch Uncertainty (PAvPU) [23]:

$$PAvPU = \frac{(n_{ac} + n_{iu})}{(n_{ac} + n_{au} + n_{ic} + n_{iu})} \tag{8}$$

Since instance segmentation pipelines return at least binary maps for every instance, we interpret a single segmentation of an instance as a draw from a Bernoulli random variable and use the entropy of the Bernoulli as stated in Eq. (5) as the uncertainty measure [23]. Instead of 2×2 image patches, we use 4×4 patches for the calculation of the metrics, since we are more interested in larger structural errors and less in small boundary adherence errors.

3   Results

This section presents benchmark results of the proposed methods on the worm-bodies BBBC010 dataset [31] and an analysis of the uncertainty estimates. Both models employ a 5-level U-Net architecture with same-padding, ReLU activation functions in all hidden layers and filter size $(3, 3)$. The number of filters increases from 19 in the first layer with each down-sampling step by factor 2 and decreases vice versa for up-sampling steps. The l_2 weight decay is weighted with factor 1e−6 for the baseline model and the corresponding parameter in the concrete dropout model is set to $\iota^2 = 1e-6$ to get a comparable regularization.

The dropout rate regularizer hyperparameter is set to $\zeta = 1e-3$ which leads to dropout rates up to 50% in deep layers of the U-Net and near zero rates for in- and output layers. Models are trained for 800,000 iterations on random 512×512 pixel sized slices of the data to predict 16 dimensional embedding vectors. Standard data augmentations, including random rotations and elastic deformations, are applied. The dataset is split into a train and a test set with 50 samples each as in [13, 26, 35] and we perform 2-fold cross-validation to determine hyperparameters.

Table 1. Evaluation on the BBBC010 test set and comparison to state of the art. Especially our Concrete Dropout model yields competitive performance while also predicting uncertainty estimates that carry information about object-level inaccuracies. Best results are shown in bold, second best are underlined.

Models	avAP$_{[0.5:0.95]}$	AP$_{.5}$	AP$_{.75}$	Recall$_{.8}$	avAP$_{dsb[0.5:0.95]}$
Semi-conv Ops [26]	0.569	0.885	0.661	–	–
SON [35]	–	–	–	~0.7	–
Discrim. loss (from [18])	0.343	0.624	0.380	–	–
Harmonic embed. [18]	0.724	0.900	0.723	–	–
PatchPerPix [13]	**0.775**	0.939	**0.891**	**0.895**	**0.727**
Baseline model	0.761	<u>0.963</u>	0.879	<u>0.81</u>	0.686
Concrete dropout model	<u>0.770</u>	**0.974**	<u>0.883</u>	<u>0.81</u>	<u>0.703</u>

Table 1 shows benchmark results against state of the art models. Since the benchmark metrics require binary instance segmentation maps, the probabilities that the Concrete U-Net pipeline predicts are binarized with threshold 0.75. The presented average precision (AP, avAP) scores are the widely used MS COCO evaluation metrics [19]. As a complement, avAP$_{dsb[0.5:0.95]}$ follows the Kaggle 2018 data science bowl definition for AP scores ($AP_{dsb} = \frac{TP}{TP+FP+FN}$) that also accounts for false negatives.

The concrete dropout model performs in all metrics slightly better than the baseline model, as expected based on [9, 23]. Both models outperform the competitors semi-convolutional operators [26], singling-out networks (SON) [35] and Harmonic Embeddings [18] by a large margin. Especially interesting is the comparison with the recent Harmonic Embedding model [18]. In their evaluation on BBBC010, [18] furthermore compared their model to a vanilla discriminative loss model, which yielded considerably lower performance than our models. The main differences of our baseline model when compared to the vanilla discriminative loss model employed in [18] are that we perform hyperparameter optimization on the mean-shift bandwidth and clustering threshold, and don't employ a hinge parameter in Eq. 1. That said, both of our models show slightly lower performance than the recent PatchPerPix [13] except for the AP50 score, in which it constitutes the new state of the art.

Figure 1 shows samples from the test set with associated predictions from the Concrete U-Net. Typical error cases of the predictions are false merges and incomplete segmentations, that occur where two or more worms overlap each other. The latter error also happens where worms are blurry due to movement or poor focus. At first glance, the uncertainty estimates shown in Fig. 1 are informative on the locations of segmentation errors. A quantitative analysis of (un)certainty vs. (in)accuracy is presented in Fig. 2. It shows the uncertainty metrics plotted against a raising entropy threshold. The smaller the entropy threshold, the more patches have an entropy exceeding the threshold and are thus denoted as uncertain. For threshold 0.05, the probability that a patch whose entropy is below 0.05 is accurate is 0.962 (red curve). The probability that a given inaccurate patch is classified as uncertain at the 0.05 entropy threshold is 0.559 (gray curve). Therefore not all inaccurate patches can be targeted with this approach, even on this low threshold.

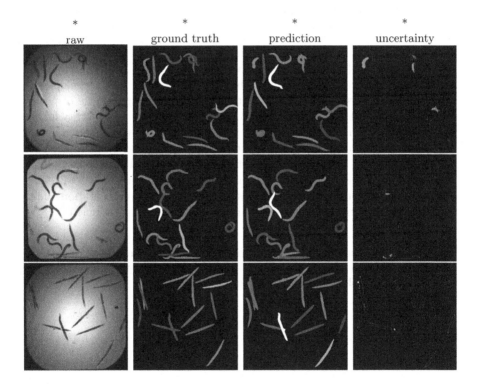

Fig. 1. Test set samples 'C10', 'C04' and 'D21', from top to bottom. SYTOX staining in red in the raw input images, predictions and associated uncertainties from the Concrete Dropout model

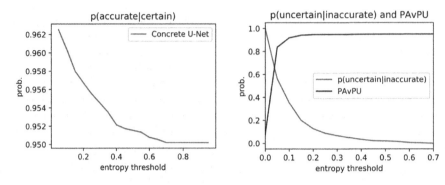

Fig. 2. Uncertainty metrics for test set predictions of the Concrete Dropout model (Color figure online)

3.1 Simulation Experiment

One potential use-case for the uncertainty estimates is proof-reading guidance. In order to assess its feasibility, the following simulation experiment is conducted: For the $\{5, 10, 15, 20\}$ highest uncertainty peak patches among all samples of the dataset, instance predictions that lie within or directly neighbor the peak patches are corrected. In this simulation, correction is done by swapping the respective instance maps with their ground truth counterparts. Benchmark metrics for this simulation are shown in Table 2. With increasing number of simulated corrections, the evaluation metrics raise substantially when considering that corrections are only done on one or two instances at a time. This shows that the uncertainty estimates carry enough information to guide manual proof-readers towards segmentation errors, and that a significant boost in segmentation quality can be obtained by means of a small number of targeted inspections.

Table 2. Simulation results. Number of corrections denotes the number of uncertainty peak patches that were used for error correction guidance. Concrete U-Net probabilities are binarized by thresholding with 0.75 to calculate the metrics

Corrections	$\mathrm{avAP}_{[0.5:0.95]}$	$\mathrm{AP}_{.5}$	$\mathrm{AP}_{.75}$	$\mathrm{Recall}_{.8}$	$\mathrm{avAP}_{dsb[0.5:0.95]}$
0	0.770	0.974	0.883	0.81	0.703
5	0.779	0.977	0.883	0.82	0.714
10	0.782	0.978	0.896	0.83	0.719
15	0.786	0.980	0.897	0.83	0.725
20	0.791	0.982	0.903	0.84	0.735

4   Discussion

The proposed models show high benchmark results when compared to other methods. The difference between our models and the discriminative loss model in [18] suggests that further performance gains could be reached by incorporating estimates for the mean-shift bandwidth and the clustering threshold into the model optimization objective. During hyperparameter search, we also observed that a smaller mean-shift bandwidth could be vital to prevent false-merges in some instance pairs, whereas it led to false-splits in others. This leads to the hypothesis, that an instance specific bandwidth could further boost performance.

Regarding the concrete dropout model, it shows slightly higher performance and its uncertainty estimates are informative on local inaccuracies. Nevertheless, there still exist many regions in the predictions that have both, segmentation errors and low local uncertainty estimates. To get more informative uncertainty estimates, a loss function that incorporates the quantification of data uncertainty [28] into the model could be used instead of the discriminative loss function. Furthermore, Variational Inference techniques like Concrete Dropout have the drawback that their approximate posterior is just a local approximation of the full posterior distribution. Therefore global features of the posterior are neglected [25], which reduces the quality of the uncertainty estimates. One could improve on that by using an ensemble of probabilistic models [30]. The models of the ensemble explore various local optima in the loss landscape and therefore various regions of the posterior distribution. Thus, the ensemble better reflects global features of the posterior, while still approximating the local features reasonably well. The simulation experiment intended to show a possible use case of the proposed method for practitioners. Other use cases are imaginable, like Bayesian active learning [10], which ranks unlabeled samples based on their uncertainty estimates to point the annotation efforts towards more informative samples during data generation.

5   Conclusion

We presented a practical method for uncertainty quantification in the context of proposal-free instance segmentation. We adapted a metric that evaluates uncertainty quality from semantic to instance segmentation. Furthermore, we also adapted a metric learning method to be able to cope with overlapping instances. This work is just a first step towards a probabilistic interpretation of instance segmentation methods and an important future research topic is the formulation of a loss function that incorporates data related uncertainty estimates.

Acknowledgments. J.R. was funded by the German Research Foundation DFG RTG 2424 *CompCancer*. L.M. and D.K. were funded by the Berlin Institute of Health and the Max Delbrueck Center for Molecular Medicine and were supported by the HHMI Janelia Visiting Scientist Program.

References

1. Bai, M., Urtasun, R.: Deep watershed transform for instance segmentation. In: Proceedings of the IEEE Conference on Computer Vision and Pattern Recognition, pp. 5221–5229 (2017)
2. Blundell, C., Cornebise, J., Kavukcuoglu, K., Wierstra, D.: Weight uncertainty in neural networks. arXiv preprint arXiv:1505.05424 (2015)
3. De Brabandere, B., Neven, D., Van Gool, L.: Semantic instance segmentation with a discriminative loss function. arXiv preprint arXiv:1708.02551 (2017)
4. Fathi, A., et al.: Semantic instance segmentation via deep metric learning. arXiv preprint arXiv:1703.10277 (2017)
5. Fukunaga, K., Hostetler, L.: The estimation of the gradient of a density function, with applications in pattern recognition. IEEE Trans. Inf. Theory $21(1)$, 32–40 (1975)
6. Funke, J., et al.: Candidate sampling for neuron reconstruction from anisotropic electron microscopy volumes. In: Golland, P., Hata, N., Barillot, C., Hornegger, J., Howe, R. (eds.) MICCAI 2014. LNCS, vol. 8673, pp. 17–24. Springer, Cham (2014). https://doi.org/10.1007/978-3-319-10404-1_3
7. Funke, J., et al.: Large scale image segmentation with structured loss based deep learning for connectome reconstruction. IEEE Trans. Pattern Anal. Mach. Intell. $41(7)$, 1669–1680 (2018)
8. Gal, Y., Ghahramani, Z.: Dropout as a Bayesian approximation: representing model uncertainty in deep learning. In: Proceedings of the International Conference on Machine Learning, pp. 1050–1059 (2016)
9. Gal, Y., Hron, J., Kendall, A.: Concrete dropout. In: Advances in Neural Information Processing Systems, pp. 3581–3590 (2017)
10. Gal, Y., Islam, R., Ghahramani, Z.: Deep Bayesian active learning with image data. In: Proceedings of the 34th International Conference on Machine Learning-Volume 70, pp. 1183–1192. JMLR. org (2017)
11. Hall, D., et al.: Probabilistic object detection: definition and evaluation. In: Proceedings of the IEEE Winter Conference on Applications of Computer Vision, pp. 1031–1040 (2020)
12. He, K., Gkioxari, G., Dollár, P., Girshick, R.: Mask R-CNN. In: Proceedings of the IEEE International Conference on Computer Vision, pp. 2961–2969 (2017)
13. Hirsch, P., Mais, L., Kainmueller, D.: PatchPerPix for instance segmentation. arXiv preprint arXiv:2001.07626 (2020)
14. Kendall, A., Gal, Y.: What uncertainties do we need in Bayesian deep learning for computer vision? In: Advances in Neural Information Processing Systems, pp. 5574–5584 (2017)
15. Kingma, D.P., Welling, M.: Auto-encoding variational Bayes. arXiv preprint arXiv:1312.6114 (2013)
16. Kingma, D.P., Salimans, T., Welling, M.: Variational dropout and the local reparameterization trick. In: Advances in Neural Information Processing Systems, pp. 2575–2583 (2015)
17. Kohl, S., et al.: A probabilistic U-Net for segmentation of ambiguous images. In: Advances in Neural Information Processing Systems, pp. 6965–6975 (2018)
18. Kulikov, V., Lempitsky, V.: Instance segmentation of biological images using harmonic embeddings. In: Proceedings of the IEEE/CVF Conference on Computer Vision and Pattern Recognition, pp. 3843–3851 (2020)

19. Lin, T.-Y., et al.: Microsoft COCO: common objects in context. In: Fleet, D., Pajdla, T., Schiele, B., Tuytelaars, T. (eds.) ECCV 2014. LNCS, vol. 8693, pp. 740–755. Springer, Cham (2014). https://doi.org/10.1007/978-3-319-10602-1_48

20. Liu, Y., et al.: Affinity derivation and graph merge for instance segmentation. In: Ferrari, V., Hebert, M., Sminchisescu, C., Weiss, Y. (eds.) ECCV 2018. LNCS, vol. 11207, pp. 708–724. Springer, Cham (2018). https://doi.org/10.1007/978-3-030-01219-9_42

21. Luther, K., Seung, H.S.: Learning metric graphs for neuron segmentation in electron microscopy images. In: Proceedings of the IEEE International Symposium on Biomedical Imaging, pp. 244–248. IEEE (2019)

22. Morrison, D., Milan, A., Antonakos, N.: Estimating uncertainty in instance segmentation using dropout sampling. In: CVPR Robotic Vision Probabilistic Object Detection Challenge (2019)

23. Mukhoti, J., Gal, Y.: Evaluating Bayesian deep learning methods for semantic segmentation. arXiv preprint arXiv:1811.12709 (2018)

24. Munkres, J.: Algorithms for the assignment and transportation problems. J. Soc. Ind. Appl. Math. **5**(1), 32–38 (1957)

25. Murphy, K.P.: Machine Learning: A Probabilistic Perspective. MIT Press, Cambridge (2012)

26. Novotny, D., Albanie, S., Larlus, D., Vedaldi, A.: Semi-convolutional operators for instance segmentation. In: Ferrari, V., Hebert, M., Sminchisescu, C., Weiss, Y. (eds.) ECCV 2018. LNCS, vol. 11205, pp. 89–105. Springer, Cham (2018). https://doi.org/10.1007/978-3-030-01246-5_6

27. Oh, M.H., Olsen, P.A., Ramamurthy, K.N.: Crowd counting with decomposed uncertainty. In: AAAI, pp. 11799–11806 (2020)

28. Oh, S.J., Murphy, K., Pan, J., Roth, J., Schroff, F., Gallagher, A.: Modeling uncertainty with hedged instance embedding. arXiv preprint arXiv:1810.00319 (2018)

29. Ronneberger, O., Fischer, P., Brox, T.: U-Net: convolutional networks for biomedical image segmentation. In: Navab, N., Hornegger, J., Wells, W.M., Frangi, A.F. (eds.) MICCAI 2015. LNCS, vol. 9351, pp. 234–241. Springer, Cham (2015). https://doi.org/10.1007/978-3-319-24574-4_28

30. Smith, L., Gal, Y.: Understanding measures of uncertainty for adversarial example detection. arXiv preprint arXiv:1803.08533 (2018)

31. Wählby, C., et al.: An image analysis toolbox for high-throughput c. elegans assays. Nat. Methods **9**(7), 714 (2012)

32. Weigert, M., et al.: Content-aware image restoration: pushing the limits of fluorescence microscopy. Nat. Methods **15**(12), 1090–1097 (2018)

33. Wolf, S., et al.: The mutex watershed: efficient, parameter-free image partitioning. In: Ferrari, V., Hebert, M., Sminchisescu, C., Weiss, Y. (eds.) ECCV 2018. LNCS, vol. 11208, pp. 571–587. Springer, Cham (2018). https://doi.org/10.1007/978-3-030-01225-0_34

34. Wolf, S., Schott, L., Kothe, U., Hamprecht, F.: Learned watershed: end-to-end learning of seeded segmentation. In: Proceedings of the IEEE International Conference on Computer Vision, pp. 2011–2019 (2017)

35. Yurchenko, V., Lempitsky, V.: Parsing images of overlapping organisms with deep singling-out networks. In: Proceedings of the IEEE Conference on Computer Vision and Pattern Recognition, pp. 6280–6288 (2017)

Registration of Multi-modal Volumetric Images by Establishing Cell Correspondence

Manan Lalit[1,2] , Mette Handberg-Thorsager[1,2] , Yu-Wen Hsieh[1,2] ,
Florian Jug[1,2,3(✉)] , and Pavel Tomancak[1,2(✉)]

[1] Max Planck Institute of Molecular Cell Biology and Genetics, Dresden, Germany
{jug,tomancak}@mpi-cbg.de
[2] Center for Systems Biology Dresden (CSBD), Dresden, Germany
[3] Fondazione Human Technopole, Milan, Italy

Abstract. Early development of an animal from an egg involves a rapid increase in cell number and several cell fate specification events accompanied by dynamic morphogenetic changes. In order to correlate the morphological changes with the genetic events, one typically needs to monitor the living system with several imaging modalities offering different spatial and temporal resolution. Live imaging allows monitoring the embryo at a high temporal resolution and observing the morphological changes. On the other hand, confocal images of specimens fixed and stained for the expression of certain genes enable observing the transcription states of an embryo at specific time points during development with high spatial resolution. The two imaging modalities cannot, by definition, be applied to the same specimen and thus, separately obtained images of different specimens need to be registered. Biologically, the most meaningful way to register the images is by identifying cellular correspondences between these two imaging modalities. In this way, one can bring the two sources of information into a single domain and combine dynamic information on morphogenesis with static gene expression data. Here we propose a new computational pipeline for identifying cell-to-cell correspondences between images from multiple modalities and for using these correspondences to register 3D images within and across imaging modalities. We demonstrate this pipeline by combining four-dimensional recording of embryogenesis of Spiralian annelid ragworm *Platynereis dumerilii* with three-dimensional scans of fixed *Platynereis dumerilii* embryos stained for the expression of a variety of important developmental genes. We compare our approach with methods for aligning point clouds and show that we match the accuracy of these state-of-the-art registration pipelines on synthetic data. We show that our approach outperforms these methods on real biological imaging datasets. Importantly, our approach uniquely provides, in addition to the registration, also the non-redundant matching of corresponding, biologically meaningful entities within the registered specimen which is the prerequisite for generating biological insights from the combined datasets. The complete pipeline is available for public use through a Fiji plugin.

Keywords: Image registration · *Platynereis dumerilii* · Shape context

© Springer Nature Switzerland AG 2020
A. Bartoli and A. Fusiello (Eds.): ECCV 2020 Workshops, LNCS 12535, pp. 458–473, 2020.
https://doi.org/10.1007/978-3-030-66415-2_30

1   Introduction

Development of an animal embryo is a highly dynamic process spanning several temporal and spatial scales, and involves a series of dynamic morphogenetic events that are driven by gene regulatory networks encoded by the genome. One of the major challenges in developmental biology is to correlate the morphological changes with the underlying gene activities [18]. Recent advances in fluorescence microscopy, such as light-sheet microscopy [12,23], allows investigating the spatio-temporal dynamics of cells in entire developing organisms and in a time-resolved manner. The three-dimensional time-lapse data produced by light-sheet microscopes contain information about positions, trajectories, and divisions of most cells in the embryo during development. However, such data sets typically lack information about gene activities in the living system.

The molecular information is provided by complementary approaches, such as confocal imaging of fixed specimens, stained for expression of a certain gene (following the molecular protocols of whole-mount *in-situ* hybridization (ISH)). The three-dimensional images of the fixed and stained embryos contain information about the spatial position of all cells or nuclei and in addition some cells are specifically labelled to indicate the expression of a gene of interest. Images of many such stained specimen showing expression of different genes at a particular stage of development can be readily collected. In order to systematically connect the molecular state of a cell to its fate during embryo morphogenesis, one needs to detect the cells in both live and fixed imaging modalities and identify cell-to-cell correspondences. This can be achieved, in principle, by aligning the images. However, the process of chemical fixation during ISH leads to a global and non-linear deformation of the specimen. Additionally, the round embryos are scanned in random orientations, and each specimen is a distinct individual showing stochastic differences in numbers and positions of the cells. This makes the problem of image registration in this context non-trivial.

We reasoned that since the primary objective is to transfer information between the imaging modalities and since cells (or nuclei) are the units of biological interest, it is more important to establish precise correspondences between equivalent cells across specimens and modalities, and that once this is achieved the registration will be obtained implicitly (Fig. 1A). We aimed to solve two matching and registration problems. Firstly, intramodal registration, where different fixed embryos stained for different gene expression patterns are registered to one reference specimen (Fig. 1B). When successful, the intramodal registration will transfer information about expression of multiple genes derived from distinct staining and imaging experiments to a single reference atlas. Secondly, intermodal registration where individual fixed and stained specimens are registered to an appropriate matching time-point of a time-lapse series of the same animal species imaged live (Fig. 1C). When successful, the intermodal registration will transfer gene expression information from fixed data to the live imaged specimen where it can be propagated along the developmental trajectories of the cells. In both cases, the common denominator are the labelled nuclei and the task is to establish the correspondences between them as precisely as possible.

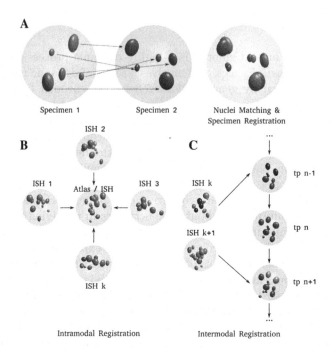

Fig. 1. Establishing cell correspondences enables registration. (A) 2-D schematic illustrating the idea: two distinct specimens (left: source and middle: target) are compared in order to estimate pair-wise cell nuclei correspondences and an optimal transform that registers the source onto the target (right) (B, C) 2-D schematics illustrating the two use cases: (B) images of distinct, independent *in-situ* specimens, acquired through confocal microscopy are registered to each other, which enables formation of an average, virtual atlas. (C) images of *in-situ* specimens, acquired through confocal microscopy are registered to the appropriate frame (tp: time point) in a time-lapse movie acquired through SPIM imaging. Nuclei indicated in darker shades are the ones expressing the gene being investigated. In both cases, the information about gene expression is transferred from the source nucleus to the corresponding target nucleus.

To address these challenges, we developed a new computational pipeline to identify cell-to-cell correspondences between images from the same and multiple imaging modalities and use these correspondences to register the images. We demonstrate the results of the pipeline on fixed ISH images of the embryos of the marine annelid worm *Platynereis dumerillii* at 16 16 h post fertilization (hpf) and the corresponding long term time-lapse acquired with light-sheet microscopy. This worm is particularly suitable for demonstrating our approach because its embryonic development is highly stereotypic, meaning that the number, arrangement and dynamic behaviour of cells is highly similar across individuals.

We compare our algorithm with methods for matching point clouds from computer vision such as Coherent Point Drift [19] and a variant of ICP (which we refer to as PCA-ICP) and show that our method outperforms the accuracy of these state-of-the-art global registration pipelines on real biological data. We also perform a series of controlled experiments on synthetic data in order to demonstrate that our method is robust to initial conditions, and noisy nuclei detections. Importantly, the pipeline is made available to the biology research community through an easy-to-use plugin distributed on the Fiji platform [21], accessible through the project page https://juglab.github.io/PlatyMatch.

2   Related Work

2.1   Registration Approaches Applied to Images of *Platynereis Dumerilii* embryonic and larval development

Platynereis dumerillii has been a playground for image registration approaches in the recent years, due to the efforts to infer gene regulatory networks underlying neuronal development by registering ISH expression patterns. Most of this work has emphasized non-linear registration of an *in-situ* specimen to a virtual atlas. For instance, a new computational protocol was developed to obtain a virtual, high resolution gene expression atlas for the brain sub-regions in embryos at 48 hpf and onwards [22]. The reference signal used in this protocol was the larval axonal scaffold and ciliary band cells stained with an acetylated-tubulin antibody. This signal has a very distinctive 3D shape within the larva and so this approach relied on intensity based registration where linear transformations were initially applied on the source image to obtain a coarse, global registration. This was then followed by applying a non-linear, deformable transformation which employed mutual-information as the image similarity metric [26].

Another approach, more related to the path we took, leveraged the DAPI image channel (which localises the cell nuclei) to obtain registration of high-quality whole-body scans to a virtual atlas for embryos at stages 48 and 72 hpf [2] and for a larva at 144 hpf [25]. Also these approaches relied on voxel intensities of the DAPI channel rather than on the matching of segmented nuclei as in our approach. Most similar to our work is the approach of [27] where the early lineages of developing embryos were linked to gene expression ISH data by identifying corresponding nuclei between embryos imaged in two modalities based on their shape, staining intensity, and relative position.

The embryo specimens targeted in our study are spherical and highly symmetrical, lack distinctive features such as a prominent ciliary band and the nuclei are densely packed. Therefore, intensity-based registration approaches using DAPI or neuronal marker channel either fail or perform poorly on such data. Contrary to these approaches and driven by the objective to, first and foremost, transfer gene expression information with cellular resolution between modalities, we adopt a matching-by-detection workflow, where we first detect nuclei in the source and target DAPI image channels and use the detections to estimate an initial transform. We then refine this transform and estimate optimal pair-wise nuclear correspondences. Therefore, after the nuclei detection step, the problem is cast into the realm of point cloud geometric registration methods that has received substantial attention in both biological and computer vision research communities. We discuss the existing approaches in the following two sections.

2.2   Matching of Cells or Nuclei in Biological Specimens

The work on matching nuclei between biological specimens has focused mainly on *Caenorhabditis elegans* (*C. elegans*) model system that exhibits perfectly stereotypic mode of development, and in fact, every single cell in the animal has its own name. Using this information, a digital atlas was constructed, which labels each nucleus segmentation in a three-dimensional image with an appropriate name. This was initially achieved using a relatively simple RANSAC based matching scheme [16] and was later extended by an active graph matching approach to jointly segment and annotate nuclei of the larva [14]. The *C. elegans* pipelines work well partly due to the highly distinctive overall shape of the larvae and non-homogenous distribution of the nuclei. Another example of matching nuclei between biological specimens uses identification of symmetry plane, to pair cells between multiple, independent time-lapse movies showing ascidian development [17]. These publications however emphasized nuclei detection and matching between images arising from the same modality. We are not aware of any automated strategy that identified nuclear correspondences between images from different modalities, as we attempt to do (see Fig. 1C).

2.3   Approaches to Point Cloud Matching in Computer Vision

In computer vision, a typical workflow for matching point clouds estimates a rigid or affine transform in order to perform an initial global alignment, which is followed by a local refinement of the initial transform through the Iterative Closest Point (ICP) algorithm. Many global alignment methods identify point-to-point matches based on geometric descriptors [9]. Once candidate correspondences are collected, alignment is estimated from a sparse subset of correspondences and then validated on the entire cloud. This iterative process typically employs variants of RANSAC [7].

One example of geometric descriptors is Shape Context, which was introduced by [3] for measuring similarity between two dimensional point clouds and was employed for registering surfaces in biomedical applications [1,24]. This work was further extended for use with three dimensional point clouds by [8] and employed for the recognition of three dimensional objects.

A prominent example of geometric descriptor matching inspired by the computer vision work and applied biological image analysis is the bead-based registration of multiview light sheet (Selective Plane Illumination Microscopy (SPIM)) data [20]. Here, fluorescent beads embedded around a specimen are used as fiduciary markers to achieve registration of 3D scans of the same specimen from multiple imaging angles (referred to as views). This is achieved by building rotation, translation and scale-invariant bead descriptors in local bead neighbourhoods, which enables identification of corresponding beads in multiple views and thus allows image registration and subsequent fusion of the views. The approach was extended to multiview registration using nuclei segmented within the specimen instead of beads [13], however the approach is not robust enough to enable registration across different specimen and/or imaging modalities.

A second body of approaches estimate the optimal transform between the source and target point clouds in a single step. One such example is Coherent Point Drift (CPD) algorithm [19] where the alignment of two point clouds is considered as a probability density estimation problem : gaussian mixture model centroids (representing the first point cloud) are fitted to the data (the second point cloud) by maximizing the likelihood. CPD has also been used to perform non-rigid registration of features extracted from biomedical images [6,11]. In this paper, we use CPD as one of the baselines to benchmark the performance of our approach.

It is important to note, that in computer vision, matching of interest points represented by geometric descriptors is not the goal but rather the means to register underlying objects or shapes in the images and volumes. Therefore, using a subset of descriptors to achieve the registration is perfectly acceptable and in fact many of the schemes rely on pruning correspondence candidates in the descriptor space to a highly reliable subset. By contrast, in biology, the nuclei that form the basis of the descriptors are at the same time the entities of interest and the goal is to match most, if not all of them, accurately.

3   Our Method

The core of our method is to match the nuclei in the various imaged specimens by means of building the shape context descriptors in a coordinate frame of reference that is unique to each nucleus. This makes the problem of matching rotationally invariant (Fig. 2 (II)). The descriptors are then matched in the descriptor space by finding the corresponding closest descriptor in the two specimens and these initial correspondences are pruned by RANSAC to achieve an initial guess of the registration. This alignment is next refined by ICP (Fig. 2 (III)). The performance of this part of the pipeline is compared to two baselines, PCA-ICP and CPD (run in affine mode), which are also able to estimate an optimal alignment. At this point, we diverge from the classical approach and evaluate the correspondences through a maximum bipartite matching to achieve the goal of matching every single nucleus from one specimen to a corresponding nucleus in the other (Fig. 2 (IV)). The pipeline relies on an efficient nucleus detection method. We present one possible approach based on scale-space theory (Fig. 2 (I)) but in principle any detection approach can be used to identify feature points to which the Shape Context descriptors would be attached. Also optionally, after the maximum bipartite matching, the estimated correspondence can be used to non-linearly deform the actual images to achieve a visually more convincing overlap of corresponding nuclei (Fig. 2 (V)). The individual steps of the pipeline are described in detail in the following subsections.

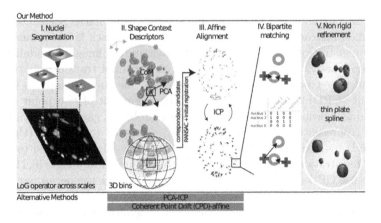

Fig. 2. Overview of the elements of the proposed registration pipeline. Figure illustrating the key elements of our pipeline: (*I*) A two dimensional slice of a volumetric image of the DAPI channel in a fixed *Platynereis* specimen. The operators which provide the strongest local response are shown for three exemplary cell nuclei. (*II*) In order to ensure that the shape context geometric descriptor is rotationally covariant, we modify the original coordinate system (shown in gray, top left) to obtain a unique coordinate system (show in black) for each nucleus detection. The Z-axis is defined by the vector joining the center of mass of the point cloud to the point of interest, the X-axis is defined along the projection of the first principal component of the complete point cloud evaluated orthogonal to the Z-axis. The Y-axis is evaluated as a cross product of the first two vectors. Next, the neighbourhood around each nucleus detection is binned in order to compute the shape context signature for each detection. The resulting shape context descriptors from the two clouds are compared to establish correspondence candidates. These are pruned by RANSAC filtering and subsequently used to estimate a global affine transform which coarsely registers the source point cloud to the target point cloud. (*III*) Next, Iterative Closest Point (ICP) algorithm is used to obtain a tighter fit between the two clouds of nuclei detections. The procedure involves the iterative identification of the nearest neighbours (indicated by black arrows), followed by the estimation of the transform parameters. (*IV*) At this stage, a Maximum Bipartite Matching is performed between the transformed source cloud of cell nuclei detections and the static target cloud of cell nuclei detections, by employing the Hungarian Algorithm for optimization. (*V*) Since the two specimens are distinct individuals, non-linear differences would persist despite the preceding linear registration. We improve the quality of the registration at this stage by employing a non-linear transform (such as thin plate spline and free-form deformation) that uses the correspondences evaluated from the previous step as ground truth control points to estimate the parameters of the transform.

3.1   Detecting Nuclei

Following the scale-space theory [15], we assume that the fluorescent cell nuclei visible in the DAPI image channel inherently possess a range of scales or sizes, and that each distinct cell nucleus achieves an extremal response at a scale σ proportional to the size of that cell nucleus (Fig. 2 (I)). We compute the trace of the scale-normalized Hessian matrix H of the gaussian-smoothened image $L(x, y, z, \sigma)$ which is equivalent to the convolution ($\circledast$) response resulting from the scale-normalized Laplacian of Gaussian kernel and the image $I(x, y, z)$ i.e.

$$\text{trace}\left(H_{\text{norm}}L\left(x, y, z, \sigma\right)\right) = \sigma^2 \left(L_{xx} + L_{yy} + L_{zz}\right), \text{ where}$$

$$L_{kk} = \frac{\partial^2 G_\sigma}{\partial k^2} \circledast I(x, y, z), \; k \in \{x, y, z\} \text{ and} \tag{1}$$

$$G_\sigma(x, y, z) = \frac{1}{(2\pi\sigma^2)^{\frac{3}{2}}} e^{-\frac{x^2+y^2+z^2}{2\sigma^2}}.$$

The cell nuclei centroid locations (and additional scale information) are then estimated as the local minima of the 4D (x, y, z, σ) space. At this stage, some of the detections might overlap especially in dense regions. To address this, first we employ the assumption that the estimated spherical radius $\hat{r}$ of a cell nucleus is related to its estimated scale $\hat{\sigma}$ through the following relation $\hat{r} = \sqrt{3}\hat{\sigma}$. Next we state a relation drawn from algebra that if d is the distance between two spheres with radii r_1 and r_2 (and corresponding volumes V_1 and V_2, respectively), and provided that $d < r_1 + r_2$, the volume of intersection V_i of these two spheres is calculated as in [4], by:

$$V_i = \frac{\pi}{12d}(r_1 + r_2 - d)^2 \left(d^2 + 2d(r_1 + r_2) - 3(r_1 - r_2)\right)^2. \tag{2}$$

Spheres for which $V_i < t \times \min(V_1, V_2)$ are suppressed greedily, by employing a non-maximum suppression step. In our experiments, we use the threshold $t = 0.05$. An optional manual curation of the nuclei detections is made possible through our Fiji plugin.

3.2   Finding Corresponding Nuclei Between Two Point Clouds

Estimating a Global Affine Transform. In this section, we will provide the details of our implementation of the 3D shape context geometric descriptor, which is a signature obtained uniquely for all feature points in the source and target point clouds. This descriptor takes as input a point cloud P (which represents the nuclei detections described in the previous section) and a basis point p, and captures the regional shape of the scene at p using the distribution of points in a support region surrounding p. The support region is discretized into bins, and a histogram is formed by counting the number of point neighbours falling within each bin. As in [3], in order to be more sensitive to nearby points, we use a log-polar coordinate system (Fig. 2 (II), bottom). In our experiments, we build a 3D histogram with 5 equally spaced log-radius bins and 6 and 12 equally spaced elevation (θ) and azimuth (ϕ) bins respectively.

For each basis point p, we define a unique right-handed coordinate system: the Z-axis is defined by the vector joining the center of mass of the point cloud to the point of interest, the X-axis is defined along the projection of the first principal component of all point locations in P, evaluated orthogonal to the Z-axis. The Y-axis is evaluated as a cross product of the first two vectors (Fig. 2

(II), top). Since the sign of the first principal component vector is a 'numerical accident' and thus not repeatable, we use both possibilities and evaluate two shape context descriptors for each feature point in the source cloud. Building such a unique coordinate system for each feature point ensures that the shape context descriptor is rotationally invariant. Additionally since the chemical fixation introduces shrinking of the embryo volume (the intermodal registration use case, see Fig. 1C) and since the embryo volume may considerably differ across a population (intramodal use case, see Fig. 1B), an additional normalization of the shape context descriptor is performed to achieve scale invariance. This is done by normalizing all the radial distances between p and its neighbours by the mean distance between all point pairs arising in the point cloud. Similar to [3], we use the χ^2 metric to identify the cost of matching two points p_i and q_j arising from two different point clouds i.e.

$$
C_{ij} := C\left(p_i, q_j\right) = \frac{1}{2} \sum_{k=1}^{K} \frac{\left(h_i\left(k\right) - h_j\left(k\right)\right)^2}{h_i\left(k\right) + h_j\left(k\right)}, \tag{3}
$$

where $h_i\left(k\right)$ and $h_j\left(k\right)$ denote the K-bin normalized histogram at p_i and q_j respectively. By comparing shape contexts resulting from the two clouds of cell nuclei detections, we obtain an initial temporary set of correspondences. These are filtered to obtain a set of inlier point correspondences using RANSAC [7]. In our experiments, we specified an affine transform model, which requires a sampling of 4 pairs of corresponding points. We executed RANSAC for 20000 trials, used the Moore-Penrose Inverse operation to estimate the affine transform between the two sets of corresponding locations, and allowed an inlier cutoff of 15 pixels L_2 distance between the transformed and the target nucleus locations.

Obtaining a Tighter Fit with ICP. The previous step provides us a good initial alignment. Next, we employ ICP which alternates between establishing correspondences via closest-point lookups (see Fig. 2 (III)) and recomputing the optimal transform based on the current set of correspondences. Typically, one employs Horn's approach [10] to estimate strictly-rigid transform parameters. We see equivalently accurate results with iteratively estimating an affine transform, which we compute by employing the Moore-Penrose Inverse operation between the current set of correspondences.

Estimating the Complete Set of Correspondences. We build a $M \times N$-sized cost matrix C where the entry C_{ij} is the euclidean distance between the ith transformed source cell nucleus detection and the jth target cell nucleus detection. Next, we employ the Hungarian Algorithm to perform a maximum bipartite matching and estimate correspondences $\hat{X}$ (see Fig. 2 (IV)):

$$
\hat{X} = \arg\min_X \sum_{i=1}^{M} \sum_{j=1}^{N} C_{ij} X_{ij}, \text{ where } X_{ij} \in \{0,1\} \text{ s.t. } \sum_{k=1}^{k=M} X_{ik} \le 1, \sum_{k=1}^{k=N} X_{kj} \le 1.
$$

3.3   Estimating a Non-linear Transform

Since the two specimens being registered are distinct individuals, non-linear differences would persist despite the preceding, linear (affine) registration. We improve the quality of the image registration at this stage by implementing an optional non-linear transform (for example the thin-plate spline transform or the free-form deformation). The correspondences evaluated from the previous step are used as ground truth control points to estimate the parameters of the transform function.

4   Materials

To test our method, we are using two sets of real biological specimen. Firstly, representing the fixed biological specimen containing information about gene expression, we collected whole-mount specimens of *Platynereis dumerilii* stained with ISH probes for several different, developmentally regulated transcription factors at the specific developmental stage of 16 hpf. These specimens were scanned in 3D by laser scanning confocal microscopy resulting in three-dimensional images containing the DAPI (nucleus) channel used in our registration as a common reference and the gene expression channel. Secondly, representing the live imaging modality, we obtained access to a recording capturing the embryological development of the *Platynereis dumerilii* at cellular resolution *in toto* [23] using a SimView light sheet microscope. The embryos were injected with a fluorescent nuclear tracer prior to imaging and thus the time-lapse movie visualizes all the nuclei in the embryo throughout development. This movie includes the 16 hpf stage of *Platynereis* development providing an appropriate inter-modal target to register the fixed specimen to on the basis of the common nuclear signal.

5   Results

We evaluate our proposed strategy on real and simulated data and compare against two competitive baselines. The first baseline, which we refer to as *PCA-ICP* is an extension of ICP and includes a robust initialization prior to performing ICP. The center of mass of the source point cloud is translated to the location of the center of mass of the target point cloud. Next, the translated source point cloud is rotated about its new center of mass such that its three principal component vectors align with the three principal component vectors of the target point cloud. In order to ensure that the orthogonal system forming the three principal components is not mirrored along any axis, we consider all 8 possibilities for the obtained principal component vectors of the source point cloud. We initialize ICP from these 8 setups and iteratively estimated a *similar* transform (scale, rotation and translation). Finally, the configuration which provides the least L_2 euclidean distance between the two sets of correspondences obtained through nearest neighbour lookups upon the termination of ICP, is kept and the rest of the configurations are discarded. The second baseline is *Coherent Point*

Drift (CPD) [19]. In our experiments, we executed CPD in the Affine mode with normalization set to 1, maximum iterations equal to 100 and tolerance equal to $1e^{-10}$. We use two metrics in order to quantify the performance of all considered methods: (*i*) *Matching Accuracy* which we define as the ratio of the true positive matches and the total number of inlier matches, and (*ii*) *Average Registration Error* which we define as the average L_2 euclidean distance between a set of ground truth landmarks arising from the two point clouds, evaluated after the completion of the registration pipeline. A higher Matching Accuracy and a lower Average Registration Error are desirable readouts to demonstrate better performance.

5.1   Experiments on Real Data

For the intramodal registration use case (see Figs. 3A & 3C), nuclei detections arising from 11 images of *in-situ* specimens were registered to nuclei detections arising from the image of a typical, target *in-situ* specimen. Since for real data the true correspondences are not known, we asked expert biologists to manually identify 12 corresponding landmark nuclei. This set represents ground truth landmarks against which we evaluated the results of our registration based on the average L_2 euclidean distance of proposed landmark correspondences (Source landmarks are labeled 1, . . . 12 and Target Landmarks are similarly labeled 1', . . . 12' in Fig. 3).

For the intermodal registration use case (see Figs. 3B & 3D), nuclei detections arising from 7 confocal images of *in-situ* specimens are registered to the corresponding frame from the time lapse movie which contains an equivalent number of nuclei. They were similarly evaluated on the average L_2 euclidean distance in the positions of landmarks identified in the movies by the expert annotators. We noticed that instead of directly registering an *in-situ* specimen to its corresponding time point in the developmental trajectory of the live embryo, our pipeline gives better results if we map the *in-situ* specimen to a reference atlas, then apply a pre-computed affine transform to the atlas to transform it to the domain of the live embryo and lastly refine this coarse registration using ICP.

The results show that after applying our proposed pipeline, the average registration error of corresponding landmarks is around 25 and 35 pixels for our intramodal and intermodal registration use cases respectively (Fig. 3E). The accuracy is significantly better compared to the baseline methods. The exemplary intramodal image shows good overlap of the nuclear intensities (Fig. 3C). The displacement of the corresponding landmarks (denoted by the yellow unprimed numbers) is better in the left part of the specimen compared to the right part. This suggests that significant non-linear deformation occurred during the staining process and our current pipeline relying on affine models is unable to undo this deformation. For the intermodal registration, the pipeline clearly compensated for the mismatch in scale between the fixed and live specimen (Fig. 3D). The remaining error is, similar to the intramodal case, likely due to non-linear distortions. In terms of matching accuracy after performing maximum bipartite matching, our method outperforms the baselines. Since the

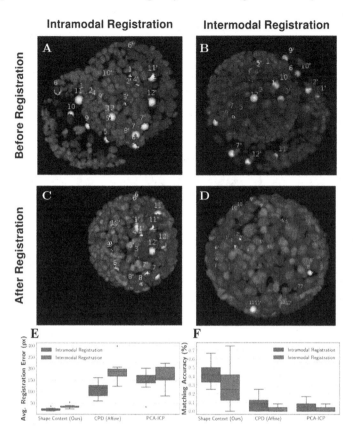

Fig. 3. Experiments on real data. (A) DAPI channels indicating cell nuclei for two distinct *in-situ* specimens before registration (source: green, target: magenta). (B) DAPI channels indicating cell nuclei for an *in-situ* specimen (source: green) before it was registered to the corresponding frame containing equivalent number of cell nuclei, in the time-lapse movie (target: magenta). Landmarks for source image are indicated as yellow spheres and labeled from 1, ... 12. Similarly, landmarks for the target image are labeled from 1', ... 12'. (C) Specimen shown in (A) after intramodal registration using our proposed pipeline. (D) Specimen shown in (B) after intermodal registration using our proposed pipeline. (E) Plot indicating the average Euclidean distance between landmarks after applying different registration pipelines. (F) Plot indicating the percent of correct correspondences between landmarks, evaluated through Maximum Bipartite Matching, after applying different registration pipelines. (Color figure online)

matching accuracy is estimated on only 12 corresponding landmarks, which represents only 3.6 % of the total matched nuclei, it is likely subject to sampling error. This is reflected by the broad spread of accuracy for both inter- and intramodal use cases (Fig. 3F).

Since obtaining a larger set of ground truth correspondences is not practical we turn next to evaluating the approach on synthetic data.

5.2   Experiments on Simulated Data

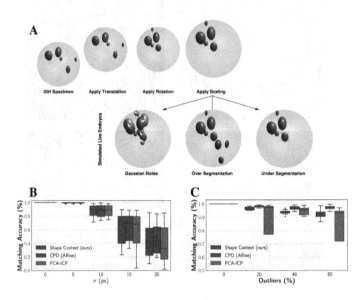

Fig. 4. Experiments on simulated data. Synthetic 'live' embryos are simulated by manipulating cell nuclei detections from real *in-situ* specimens globally and locally. (A) First all cell nuclei detections of an *in-situ* specimen were translationally offset, next the translated point cloud was randomly rotated by an angle $\in \{-\pi/6, \pi/6\}$ about a random axis passing through the center of mass of the translated point cloud, and finally, the translated and rotated point cloud was scaled by a random factor. (B) The above globally transformed nuclei are provided independent gaussian noise. Plot indicating the percent of correct correspondences between all pairs evaluated through Maximum Bipartite Matching, after applying different registration pipelines. (C) The above globally transformed nuclei are corrupted with excess outliers. Plot indicating the percent of correct correspondences between all pairs evaluated through Maximum Bipartite Matching, after applying different registration pipelines.

Starting from the nuclei detection on real fixed embryos, we generated simulated ground truth data by random translation, rotation and scaling operations, followed by (*i*) adding gaussian noise to the location of individual segments (i.e. nuclei) and (*ii*) randomly adding nuclei (Fig. 4A). The simulated embryos are meant to resemble the live-imaged embryos which in real scenarios are also rotated, translated and scaled compared to the fixed specimens and may have extraneous or missing nuclei due to biological variability or segmentation errors.

Robustness to Gaussian Noise. The synthetic 'live embryos' were generated by manipulating nuclei detections from multiple, independent *in-situ* specimens. First, the nuclei detections of each *in-situ* specimen are provided a random translation offset, next the translated point cloud is rotated by a random angle between $-30°$ and $+30°$ about an arbitrary axis passing through the center of mass of the point cloud, and finally, the translated and rotated point cloud is scaled by a random factor (See Fig. 4A). After these global transformations, we add gaussian noise to each individual detection in order to vary their positions independently along the X, Y and Z axes. We evaluate five levels of Gaussian noise with standard deviations [0:5:20]. The results of evaluation of matching accuracy with respect to different levels of Gaussian noise show that all methods provide equivalent performance (Fig. 4B). The matching accuracy starts to break down when the magnitude of gaussian noise is greater than 10 pixels.

Robustness to Outliers. In order to test robustness against over or under-segmentation of nuclei, we add outliers to both the source fixed *in-situ* volumes and the corresponding simulated 'live embryo'. New outlier points are generated by sampling existing points and adding a new point at a standard deviation of 20 pixels from their locations. The results show that the CPD Affine method performs the best in the presence of outliers, while our approach is more stable compared to the PCA-ICP (Fig. 4C).

6   Discussion

Our method showed promising results on real biological data in terms of average registration error and provided equivalent performance when compared to state of the art methods on simulated data. The pipeline offers several entry points for further improvement towards achieving more precise one-to-one matching of cells within and across imaging modalities for separate biological specimens. One area open for future investigations is certainly obtaining more accurate initial segmentations. Another performance boost may come from the definition of the 3D geometric descriptor. Our implementation of shape context as a 3D geometric descriptor draws from [3]. We use a log-polar coordinate system and build 3D histograms by evenly dividing the azimuth and elevation axis. This creates bins with unequal sizes, especially near the poles, and makes the matching of feature points non robust to noisy detections. This drawback could be addressed through two approaches: (i) employing the optimal transport distance [5] between two 3D histograms would provide a more natural way of comparing two histograms as opposed to the current χ^2 squared distance formulation, (ii) opting for a more uniform binning scheme (see for example, [28]) would eliminate the issue of noisy detections jumping arbitrarily between bins near the poles. Finally, the method will benefit from non-linear refinement as the specimens are often deformed in an unpredictable manner during the staining and imaging protocols.

By establishing nuclei correspondences between images of *in-situ* specimens and the time lapse movies, biologists will be able to transfer the gene expression information from the fixed specimens to the dynamic cell lineage tree generated by performing cell tracing on the time-lapse movie. This will enable biologists to study the molecular underpinning of dynamic morphogenetic processes occurring during embryo development.

References

1. Acosta, O., et al.: 3D shape context surface registration for cortical mapping (2010). In: IEEE International Symposium on Biomedical Imaging (ISBI) 2010
2. Asadulina, A., Panzera, A., Veraszto, C., Lieblig, C., Jekely, G.: Whole-body gene expression pattern registration in Platynereis larvae. EvoDevo **3**(27), (2012)
3. Belongie, S., Puzicha, J., Malik, J.: Shape matching and object recognition using shape contexts. IEEE Trans. Pattern Anal. Mach. Intell. **24**(4), 509–522 (2002)
4. Bondi, A.: Van der Waals volumes and radii. J. Phys. Chem. **68**, 441–451 (1964)
5. Cuturi, M.: Sinkhorn distances: lightspeed computation of optimal transportation distances. In: Advances in Neural Information Processing Systems, vol. 26, pp. 2292–2300 (2013)
6. Farnia, P., Ahmadian, A., Khoshnevisan, A., Jaberzadeh, N., Kazerooni, A.: An efficient point based registration of intra-operative ultrasound images with MR images for computation of brain shift; a phantom study. In: Annual International Conference of the IEEE Engineering in Medicine and Biology Society (2011)
7. Fischler, M.A., Bolles, R.C.: Random sample consensus: a paradigm for model fitting with applications to image analysis and automated cartography. Commun. ACM **24**(6), 381–395 (1981)
8. Frome, A., Huber, D., Kolluri, R., Bülow, T., Malik, J.: Recognizing objects in range data using regional point descriptors. In: Pajdla, T., Matas, J. (eds.) ECCV 2004. LNCS, vol. 3023, pp. 224–237. Springer, Heidelberg (2004). https://doi.org/10.1007/978-3-540-24672-5_18
9. Guo, Y., Bennamoun, M., Sohel, F.: A comprehensive performance evaluation of 3D local feature descriptors. Int. J. Comput. Vis. **116**, 66–89 (2016). https://doi.org/10.1007/s11263-015-0824-y
10. Horn, B.: Closed-form solution of absolute orientation using unit quaternions. J. Opt. Soc. Am. **4**, 629–642 (1987)
11. Hu, Y., Rijkhorst, E.-J., Manber, R., Hawkes, D., Barratt, D.: Deformable Vessel-Based Registration Using Landmark-Guided Coherent Point Drift. In: Liao, H., Edwards, P.J.E., Pan, X., Fan, Y., Yang, G.-Z. (eds.) MIAR 2010. LNCS, vol. 6326, pp. 60–69. Springer, Heidelberg (2010). https://doi.org/10.1007/978-3-642-15699-1_7
12. Huisken, J., Swoger, J., Bene, F.D., Wittbrodt, J., Stelzer, E.H.K.: Optical sectioning deep inside live embryos by selective plane illumination microscopy. Science **305**, 1007–1009 (2004)
13. Hörl, D., et al.: BigStitcher: reconstructing high-resolution image datasets of cleared and expanded samples. Nat. Methods **16**, 870–874 (2019)
14. Kainmueller, D., Jug, F., Rother, C., Myers, G.: Active graph matching for automatic joint segmentation and annotation of *C. elegans*. In: Golland, P., Hata, N., Barillot, C., Hornegger, J., Howe, R. (eds.) MICCAI 2014. LNCS, vol. 8673, pp. 81–88. Springer, Cham (2014). https://doi.org/10.1007/978-3-319-10404-1_11
15. Lindeberg, T.: Feature detection with automatic scale selection. Int. J. Comput. Vis. **30**(2), 70–116 (1998). https://doi.org/10.1023/A:1008045108935
16. Long, F., Peng, H., Liu, X., Kim, S.K., Myers, E.: A 3D digital atlas of C. elegans and its application to single-cell analyses. Nat. Methods **6**(4), 667–672 (2009)
17. Michelin, G., Guignard, L., Fiuza, U.M., Lemaire, P., Godin, C., Malandain, G.: Cell pairings for ascidian embryo registration. In: 2015 IEEE International Symposium on Biomedical Imaging (ISBI) (2015)

18. Munro, E., Robin, F., Lemaire, P.: Cellular morphogenesis in ascidians: how to shape a simple tadpole. Curr. Opin. Genet. Dev. **16**(4), 399–405 (2006)
19. Myronenko, A., Song, X.: Point set registration: coherent point drift. IEEE Trans. Pattern Anal. Mach. Intell. **32**(12), 2262–2275 (2010)
20. Preibisch, S., Saalfeld, S., Schindelin, J., Tomancak, P.: Software for bead-based registration of selective plane illumination microscopy data. Nat. Methods **7**, 418–419 (2010)
21. Schindelin, J., et al.: Fiji: an open-source platform for biological-image analysis. Nat. Methods **9**(7), 676–682 (2012)
22. Tomer, R., Denes, A.S., Tessmar-Raible, K., Arendt, D.: Profiling by image registration reveals common origin of annelid mushroom bodies and vertebrate pallium. Cell **142**(5), 800–809 (2010)
23. Tomer, R., Khairy, K., Amat, F., Keller, P.J.: Quantitative high-speed imaging of entire developing embryos with simultaneous multiview light-sheet microscopy. Nat. Methods **9**, 755–763 (2012)
24. Urschler, M., Bischof, H.: Registering 3D lung surfaces using the shape context approach. In: Proceedings of the International Conference on Medical Image Understanding and Analysis, pp. 512–215 (2004)
25. Vergara, H.M., et al.: Whole-organism cellular gene-expression atlas reveals conserved cell types in the ventral nerve cord of Platynereis dumerilii. Proc. Natl. Acad. Sci. **114**(23), 5878–5885 (2017)
26. Viola, P.A., Wells, W.M.: Alignment by maximization of mutual information. Int. J. Comput. Vis. **24**, 137–154 (1995). https://doi.org/10.1023/A:1007958904918
27. Vopalensky, P., Tosches, M., Achim, K., Thorsager, M.H., Arendt, D.: From spiral cleavage to bilateral symmetry: the developmental cell lineage of the annelid brain. BMC Biol. **17**, 81 (2019). https://doi.org/10.1186/s12915-019-0705-x
28. Zhong, Y.: Intrinsic shape signatures: a shape descriptor for 3D object recognition. In: International Conference on Computer Vision ICCV Workshop (2009)

W2S: Microscopy Data with Joint Denoising and Super-Resolution for Widefield to SIM Mapping

Ruofan Zhou, Majed El Helou$^{(\boxtimes)}$, Daniel Sage, Thierry Laroche, Arne Seitz, and Sabine Süsstrunk

École Poletechnique Fédérale de Lausanne (EPFL), Lausanne, Switzerland
{ruofan.zhou,majed.elhelou,sabine.susstrunk}@epfl.ch

Abstract. In fluorescence microscopy live-cell imaging, there is a critical trade-off between the signal-to-noise ratio and spatial resolution on one side, and the integrity of the biological sample on the other side. To obtain clean high-resolution (HR) images, one can either use microscopy techniques, such as structured-illumination microscopy (SIM), or apply denoising and super-resolution (SR) algorithms. However, the former option requires multiple shots that can damage the samples, and although efficient deep learning based algorithms exist for the latter option, no benchmark exists to evaluate these algorithms on the joint denoising and SR (JDSR) tasks.

To study JDSR on microscopy data, we propose such a novel JDSR dataset, **W**idefield2**S**IM (W2S), acquired using a conventional fluorescence widefield and SIM imaging. W2S includes 144,000 real fluorescence microscopy images, resulting in a total of 360 sets of images. A set is comprised of noisy low-resolution (LR) widefield images with different noise levels, a noise-free LR image, and a corresponding high-quality HR SIM image. W2S allows us to benchmark the combinations of 6 denoising methods and 6 SR methods. We show that state-of-the-art SR networks perform very poorly on noisy inputs. Our evaluation also reveals that applying the best denoiser in terms of reconstruction error followed by the best SR method does not necessarily yield the best final result. Both quantitative and qualitative results show that SR networks are sensitive to noise and the sequential application of denoising and SR algorithms is sub-optimal. Lastly, we demonstrate that SR networks retrained end-to-end for JDSR outperform any combination of state-of-the-art deep denoising and SR networks (Code and data available at https://github. com/IVRL/w2s).

Keywords: Image restoration dataset · Denoising · Super-resolution · Microscopy imaging · Joint optimization

R. Zhou and M. El Helou—The first two authors have similar contributions.

Electronic supplementary material The online version of this chapter (https:// doi.org/10.1007/978-3-030-66415-2_31) contains supplementary material, which is available to authorized users.

A. Bartoli and A. Fusiello (Eds.): ECCV 2020 Workshops, LNCS 12535, pp. 474–491, 2020.
https://doi.org/10.1007/978-3-030-66415-2_31

1   Introduction

Fluorescence microscopy allows to visualize sub-cellular structures and protein-protein interaction at the molecular scale. However, due to the weak signals and diffraction limit, fluorescence microscopy images suffer from high noise and limited resolution. One way to obtain high-quality, high-resolution (HR) microscopy images is to leverage super-resolution fluorescence microscopy, such as structure illumination microscopy (SIM) [17]. This technique requires multiple captures with several parameters requiring expert tuning to get high-quality images. Multiple or high-intensity-light acquisitions can cause photo-bleach and even damage the samples. The imaged cells could be affected and, if imaged in sequence for live tracking, possibly killed. This is because a single SIM acquisition already requires a set of captures with varying structured illumination. Hence, a large set of SIM captures would add up to high illumination and an overhead in capture time that is detrimental to imaging and tracking of live cells. Therefore, developing an algorithm to effectively denoise and super-resolve a fluorescence microscopy image is of great importance to biomedical research. However, a high-quality dataset is needed to benchmark and evaluate joint denoising and super-resolution (JDSR) on microscopy data.

Deep-learning-based methods in denoising [2,12,41,49] and SR [45,54,55] today are outperforming classical signal processing approaches. A major limitation in the literature is, however, the fact that these two restoration tasks are addressed separately. This is in great part due to a missing dataset that would allow both to train and to evaluate JDSR. Such a dataset must contain aligned pairs of LR and HR images, with noise and noise-free LR images, to allow retraining retrain prior denoising and SR methods for benchmarking the consecutive application of a denoiser and an SR network as well as candidate one-shot JDSR methods.

In this paper, we present such a dataset, which, to the best of our knowledge, is the first JDSR dataset. This dataset allows us to evaluate the existing denoising and SR algorithms on microscopy data. We leverage widefield microscopy and SIM techniques to acquire data fulfilling the described requirements above. Our noisy LR images are captured using widefield imaging of human cells. We capture a total of 400 replica raw images per field of view. We average several of the LR images to obtain images with different noise levels, and all of the 400 replicas to obtain the noise-free LR image. Using SIM imaging [17], we obtain the corresponding high-quality HR images. Our resulting **W**idefield**2S**IM (W2S) dataset consists of 360 sets of LR and HR image pairs, with different fields of view and acquisition wavelengths. Visual examples of the images in W2S are shown in Fig. 1.

We leverage our JDSR dataset to benchmark different approaches for denoising and SR restoration on microscopy images. We compare the sequential use of different denoisers and SR methods, of directly using an SR method on a noisy LR image, and of using SR methods on the noise-free LR images of our dataset for reference. We additionally evaluate the performance of retraining SR networks on our JDSR dataset. Results show a significant drop in the performance of SR networks when the low-resolution (LR) input is noisy compared to it being noise-free. We also find that the consecutive application of denoising and SR achieves better results. It is, however, not as performing in terms of RMSE and perceptual texture reconstruction as training a single model on the JDSR task, due to the accumulation of error. The best results are thus obtained by training a single network for the joint optimization of denoising and SR.

In summary, we create a microscopy JDSR dataset, W2S, containing noisy images with 5 noise levels, noise-free LR images, and the corresponding high-quality HR images. We analyze our dataset by comparing the noise magnitude and the blur kernel of our images to those of existing denoising and SR datasets. We benchmark state-of-the-art denoising and SR algorithms on W2S, by evaluating different settings and on different noise levels. Results show the networks can benefit from joint optimization.

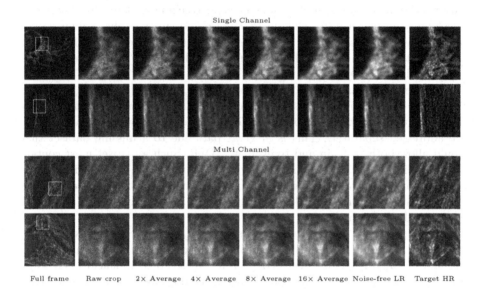

Fig. 1. Example of image sets in the proposed W2S. We obtain LR images with 5 different noise levels by either taking a single raw image or averaging different numbers of raw images of the same field of view. The more images we average, the lower the noise level, as shown in the different columns of the figure. The noise-free LR images are the average of 400 raw images, and the HR images are obtained using structured-illumination microscopy (SIM) [17]. The multi-channel images are formed by mapping the three single-channel images of different wavelengths to RGB. A gamma correction is applied for better visualization. Best viewed on screen.

2   Related Work

2.1   Biomedical Imaging Techniques for Denoising and Super-Resolution

Image averaging of multiple shots is one of the most employed methods to obtain a clean microscopy image. This is due to its reliability and to avoid the potential blurring or over-smoothing effects of denoisers. For microscopy experiments requiring long observation and minimal degradation of specimens, low-light conditions and short exposure times are, however, preferred as multiple shots might damage the samples. To reduce the noise influence and increase the resolution, denoising methods and SR imaging techniques are leveraged.

To recover a clean image from a single shot, different denoising methods have been designed, including PURE-LET [29], EPLL [59], and BM3D [8]. Although these methods provide promising results, recent deep learning methods outperform them by a big margin [53]. To achieve resolution higher than that imposed by the diffraction limit, a variety of SR microscopy techniques exist, which achieve SR either by spatially modulating the fluorescence emission using patterned illumination (*e.g.*, STED [19] and SIM [17]), or by stochastically switching on and off individual molecules using photo-switchable probes (*e.g.*, STORM [37]), or photo-convertible fluorescent proteins (*e.g.*, PALM [40]). However, all of these methods require multiple shots over a period of time, which is not suitable for live cells because of the motion and potential damage to the cell. Thus, in this work, we aim to develop a deep learning method to reconstruct HR images from a single microscopy capture.

2.2   Datasets for Denoising and Super-Resolution

Several datasets have commonly been used in benchmarking SR and denoising, including Set5 [4], Set14 [48], BSD300 [31], Urban100 [20], Manga109 [32], and DIV2K [42]. None of these datasets are optimized for microscopy and they only allow for synthetic evaluation. Specifically, the noisy inputs are generated by adding Gaussian noise for testing denoising algorithms, and the LR images are generated by downsampling the blurred HR images for testing SR methods. These degradation models deviate from the degradations encountered in real image capture [6]. To better take into account realistic imaging characteristics and thus evaluate denoising and SR methods in real scenarios, real-world denoising and SR datasets have recently been proposed. Here we discuss these real datasets and compare them to our proposed W2S.

Real Denoising Dataset. Only a few datasets allow to quantitatively evaluate denoising algorithms on real images, such as DND [35] and SSID [1]. These datasets capture images with different noise levels, for instance by changing the ISO setting at capture. More related to our work, Zhang *et al.* [53] collect a dataset of microscopy images. All three datasets are designed only for denoising, and no HR images are provided that would allow them to be used for SR evaluation. According to our benchmark results, the best denoising algorithm does not necessarily provide the best input for the downstream SR task, and the JDSR learning is the best overall approach. This suggests a dataset on joint denoising and SR can provide a more comprehensive benchmark for image restoration.

Real Super-Resolution Dataset. Recently, capturing LR and HR image pairs by changing camera parameters has been proposed. Chen *et al.* collect 100 pairs of images of printed postcards placed at different distances. SR-RAW [52] consists of 500 real scenes captured with multiple focal lengths. Although this dataset provides real LR-HR pairs, it suffers from misalignment due to the inevitable perspective changes or lens distortion. Cai *et al.* thus introduce an iterative image registration scheme into the registration of another dataset, RealSR [5]. However, to have high-quality images, all these datasets are captured with low ISO setting, and the images thus contain very little noise as shown in our analysis. Qian *et al.* propose a dataset for joint demosaicing, denoising and SR [36], but the noise in their dataset is simulated by adding white Gaussian noise. Contrary to these datasets, our proposed W2S is constructed using SR microscopy techniques [17], all pairs of images are well aligned, and it contains raw LR images with different noise levels and the noise-free LR images, thus enabling the benchmarking of both denoising and SR under real settings.

2.3   Deep Learning Based Image Restoration

Deep learning based methods have shown promising results on various image restoration tasks, including denoising and SR. We briefly present prior work and the existing problems that motivate joint optimization.

Deep Learning for Denoising. Recent deep learning approaches for image denoising achieve state-of-the-art results on recovering the noise-free images from images with additive noise. Whether based on residual learning [49], using memory blocks [41], bottleneck architecture [46], attention mechanisms [2], internally modeling Gaussian noise parameters [12], these deep learning methods all require training data. For real-world raw-image denoising, the training data should include noisy images with a Poisson noise component, and a corresponding aligned noise-free image, which is not easy to acquire. Some recent self-supervised methods can learn without having training targets [3,24,26], however, their performance does not match that of supervised methods. We hence focus on the better-performing supervised methods in our benchmark, since targets are available. All these networks are typically evaluated only on the denoising task, often only on the one they are trained on. They optimize for minimal squared pixel error, leading to potentially smoothed out results that favour reconstruction error at the expense of detail preservation. When a subsequent task such as SR is then applied on the denoised outputs from these networks, the quality of the final results does not, as we see in our benchmark, necessarily correspond to the denoising performance of the different approaches. This highlights the need for a more comprehensive perspective that jointly considers both restoration tasks.

Deep Learning for Super-Resolution. Since the first convolutional neural network for SR [10] outperformed conventional methods on synthetic datasets, many new architectures [22, 28, 39, 43, 45, 54, 55] and loss functions [21, 25, 38, 51, 56] have been proposed to improve the effectiveness and the efficiency of the networks. To enable the SR networks generalize better on the real-world LR images where the degradation is unknown, works have been done on kernel prediction [5, 16] and kernel modeling [50, 58]. However, most of the SR networks assume that the LR images are noise-free or contain additive Gaussian noise with very small variance. Their predictions are easily affected by noise if the distribution of the noise is different from their assumptions [7]. This again motivates a joint approach developed for the denoising and SR tasks.

Joint Optimization in Deep Image Restoration. Although a connection can be drawn between the denoising and super-resolution tasks in the frequency domain [13], their joint optimization was not studied before due to the lack of a real benchmark. Recent studies have shown the performance of joint optimization in image restoration, for example, the joint demosaicing and denoising [15, 23], joint demosaicing and super-resolution [52, 57]. All these methods show that the joint solution outperforms the sequential application of the two stages. More relevant to JDSR, Xie *et al.* [47] present a dictionary learning approach with constraints tailored for depth maps, and Miao *et al.* [33] propose a cascade of two networks for joint denoising and deblurring, evaluated on synthetic data only. Similarly, our results show that a joint solution for denoising and SR also obtains better results than any sequential application. Note that our W2S dataset al.lows us to draw such conclusions on *real* data, rather than degraded data obtained through simulation.

3   Joint Denoising and Super-Resolution Dataset for Widefield to SIM Mapping

In this section, we describe the experimental setup that we use to acquire the sets of LR and HR images and present an analysis of the noise levels and blur kernels of our dataset.

3.1   Structured-Illumination Microscopy

Structured-illumination microscopy (SIM) is a technique used in microscopy imaging that allows samples to be captured with a higher resolution than the one imposed by the physical limits of the imaging system [17]. Its operation is based on the interference principle of the Moiré effect. We present how SIM works in more detail in our supplementary material. We use SIM to extend the resolution of standard widefield microscopy images. This allows us to obtain aligned LR and HR image pairs to create our dataset. The acquisition details are described in the next section.

3.2 Data Acquisition

We capture the LR images of the W2S dataset using widefield microscopy [44]. Images are acquired with a high-quality commercial fluorescence microscope and with real biological samples, namely, human cells.

Widefield Images. A time-lapse widefield of 400 images is acquired using a Nikon SIM setup (Eclipse T1) microscope. The details of the setup are given in the supplementary material. In total, we capture 120 different fields-of-view (FOVs), each FOV with 400 captures in 3 different wavelengths. All images are *raw*, *i.e.*, are linear with respect to focal plane illuminance, and are made up of 512×512 pixels. We generate different noise-level images by averaging 2, 4, 8, and 16 raw images of the same FOV. The larger the number of averaged raw images is, the lower the noise level. The noise-free LR image is estimated as the average of all 400 captures of a single FOV. Examples of images with different noise levels and the corresponding noise-free LR images are presented in Fig. 1.

SIM Imaging. The HR images are captured using SIM imaging. We acquire the SIM images using the same Nikon SIM setup (Eclipse T1) microscope as above. We present the details of the setup in the supplementary material. The HR images have a resolution that is higher by a factor of 2, resulting in 1024×1024 pixel images.

3.3 Data Analysis

W2S includes 120 different FOVs, each FOV is captured in 3 channels, corresponding to the wavelengths 488nm, 561nm and 640nm. As the texture of the cells is different and independent across different channels, the different channels can be considered as different images, thus resulting in 360 views. For each view, 1 HR image and 400 LR images are captured. We obtain LR images with different noise levels by averaging different numbers of images of the same FOV and the same channel. In summary, W2S provides 360 different sets of images, each image set includes LR images with 5 different noise levels (corresponding to 1, 2, 4, 8, and 16 averaged LR images), the corresponding noise-free LR image (averaged over 400 LR images) and the corresponding HR image acquired with SIM. The LR images have dimensions 512×512, and the HR images 1024×1024.

To quantitatively evaluate the difficulty of recovering the HR image from the noisy LR observation in W2S, we analyze the degradation model relating the LR observations to their corresponding HR images. We adopt a commonly used degradation model [6,10,16,58], with an additional noise component,

$$I_{LR}^{noisy} = (I_{HR} \circledast k) \downarrow_m + n, \tag{1}$$

where I_{LR}^{noisy} and I_{HR} correspond, respectively, to the noisy LR observation and the HR image, $\circledast$ is the convolution operation, k is a blur kernel, $\downarrow_m$ is a downsampling operation with a factor of m, and n is the additive noise. Note that n is usually assumed to be zero in most of the SR networks' degradation models, while it is not the case for our dataset. As the downsampling factor m is equal to the targeted super-resolution factor, it is well defined for each dataset. We thus analyze in what follows the two unknown variables of the degradation model for W2S; namely the noise n and the blur kernel k. Comparing to other denoising datasets, W2S contains 400 noisy images for each view, DND [7] contains only 1, SSID [1] contains 150, and FMD [53], which also uses widefield imaging, contains 50. W2S can thus provide a wide range of noise levels by averaging a varying number of images out of the 400. In addition, W2S provides LR and HR image pairs that do not suffer from misalignment problems often encountered in SR datasets.

Noise Estimation. We use the noise modeling method in [14] to estimate the noise magnitude in raw images taken from W2S, from the denoising dataset FMD [53], and from the SR datasets RealSR [5] and City100 [6]. The approach of [14] models the noise as Poisson-Gaussian. The measured noisy pixel intensity is given by $y = x + n_P(x) + n_G$, where x is the noise-free pixel intensity, n_G is zero-mean Gaussian noise, and $x + n_P(x)$ follows a Poisson distribution of mean ax for some $a > 0$. This approach yields an estimate for the parameter a of the Poisson distribution. We evaluate the Poisson parameter of the noisy images from the three noise levels (obtained by averaging 1, 4 and 8 images) of W2S, the raw noisy images of FMD, and the LR images of the SR datasets for comparison. We show the mean of the estimated noise magnitude for the different datasets in Fig. 2(a). We see that the raw noisy images of W2S have a high noise level, comparable to that of FMD. On the other hand, the estimated noise parameters of the SR datasets are almost zero, up to small imprecision, and are thus significantly lower than even the estimated noise magnitude of the LR images from the lowest noise level in W2S. Our evaluation highlights the fact that the additive noise component is not taken into consideration in current state-of-the-art SR datasets. The learning-based SR methods using these datasets are consequently not tailored to deal with noisy inputs that are common in many practical applications, leading to potentially poor performance. In contrast, W2S contains images with high (and low) noise magnitude comparable to the noise magnitude of a recent denoising dataset [53].

Blur Kernel Estimation. We estimate the blur kernel k shown in Eq. (1) as

$$k = \underset{k}{argmin}||I_{LR}^{noise-free} \uparrow^{bic} -k \circledast I_{HR}||_2^2, \tag{2}$$

where $I_{LR}^{noise-free} \uparrow^{bic}$ is the noise-free LR image upscaled using bicubic interpolation. We solve for k directly in the frequency domain using the Fast Fourier Transform [11]. The estimated blur kernel is visualized in Fig. 2(b). For the purpose of comparison, we show the estimated blur kernel from two SR datasets: RealSR [5] and City100 [6]. We also visualize the two other blur kernels: the MATLAB bicubic kernel that is commonly used in the synthetic SR datasets, and the Gaussian blur kernel with a sigma of 2.0, which is the largest kernel used by the state-of-the-art blind SR network [16] for the upscaling factor of 2. From the visualization we clearly see the bicubic kernel and Gaussian blur kernel

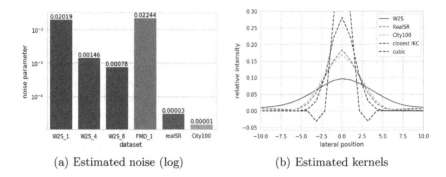

(a) Estimated noise (log) (b) Estimated kernels

Fig. 2. Noise and kernel estimation on images from different datasets. A comparably-high noise level and a wide kernel indicate that the HR images of W2S are challenging to recover from the noisy LR observation.

that are commonly used in synthetic datasets are very different from the blur kernels of real captures. The blur kernel of W2S has a long tail compared to the blur kernels estimated from the other SR datasets, illustrating that more high-frequency information is removed for the LR images in W2S. This is because a wider space-domain filter corresponds to a narrower frequency-domain low pass, and vice versa. Hence, the recovery of HR images from such LR images is significantly more challenging.

Compared to the SR datasets, the LR and HR pairs in W2S are well-aligned during the capture process, and no further registration is needed. Furthermore, to obtain high-quality images, the SR datasets are captured under high ISO and contain almost zero noise, whereas W2S contains LR images with different noise levels. This makes it a more comprehensive benchmark for testing under different imaging conditions. Moreover, as shown in Sect. 3.3, the estimated blur kernel of W2S is wider than that of other datasets, and hence it averages pixels over a larger window, filtering out more frequency components and making W2S a more challenging dataset for SR.

4 Benchmark

We benchmark on the sequential application of state-of-the-art denoising and SR algorithms on W2S using RMSE and SSIM. Note that we do not consider the inverse order, *i.e.*, first applying SR methods on noisy images, as this amplifies the noise and causes a large increase in RMSE as shown in the last row of Table 2. With current methods, it would be extremely hard for a subsequent denoiser to recover the original clean signal.

4.1 Setup

We split W2S into two disjoint training and test sets. The training set consists of 240 LR and HR image sets, and the test set consists of 120 sets of images, with no overlap between the two sets. We retrain the learning-based methods on the training set, and the evaluation of all methods is carried out on the test set.

For denoising, we evaluate different approaches from both classical methods and deep-learning methods. We use a method tailored to address Poisson denoising, PURE-LET [29], and the classical Gaussian denoising methods EPLL [59] and BM3D [8]. The Gaussian denoisers are combined with the Anscombe variance-stabilization transform (VST) [30] to first modify the distribution of the image noise into a Gaussian distribution, denoise, and then invert the result back with the inverse VST. We estimate the noise magnitude using the method in [14], to be used as input for both the denoiser and for the VST when the latter is needed. We also use the state-of-the-art deep-learning methods MemNet [41], DnCNN [49], and RIDNet [2]. For a fair comparison with the traditional non-blind methods that are given a noise estimate, we separately train each of these denoising methods for every noise level, and test with the appropriate model per noise level. The training details are presented in the supplementary material.

We use six state-of-the-art SR networks for the benchmark: four pixel-wise distortion based SR networks, RCAN [54], RDN [55], SAN [9], SRFBN [27], and two perceptually-optimized SR networks, EPSR [43] and ESRGAN [45]. The networks are trained for SR and the inputs are assumed to be noise-free, *i.e.*, they are trained to map from the noise-free LR images to the high-quality HR images. All these networks are trained using the same settings, the details of which are presented in the supplementary material.

Table 1. RMSE/SSIM results on denoising the W2S test images. We benchmark three classical methods and three deep learning based methods. The larger the number of averaged raw images is, the lower the noise level. [†]The learning based methods are trained for each noise level separately. An interesting observation is that the best RMSE results (in red) do not necessarily give the best result after the downstream SR method as show in Table 2. We highlight the results under the highest noise level with gray background for easier comparison with Table 2.

	Method	Number of raw images averaged before denoising				
		1	2	4	8	16
Denoisers	PURE-LET [29]	0.089/0.864	0.076/0.899	0.062/0.928	0.052/0.944	0.044/0.958
	VST+EPLL [59]	0.083/0.887	0.074/0.916	0.061/0.937	0.051/0.951	0.044/0.962
	VST+BM3D [8]	0.080/0.897	0.072/0.921	0.059/0.939	0.050/0.953	0.043/0.962
	MemNet[†] [41]	0.090/0.901	0.072/0.909	0.063/0.925	0.059/0.944	0.059/0.944
	DnCNN[†] [49]	0.078/0.907	0.061/0.926	0.049/0.944	0.041/0.954	0.033/0.964
	RIDNet[†] [2]	0.076/0.910	0.060/0.928	0.049/0.943	0.041/0.955	0.034/0.964

4.2   Results and Discussion

We apply the denoising algorithms on the noisy LR images, and calculate the RMSE and SSIM values between the denoised image and the corresponding noise-free LR image in the test set of W2S. The results of the 6 benchmarked denoising algorithms are shown in Table 1. DnCNN and RIDNet outperform the classical denoising methods for all noise levels. Although MemNet achieves worse results than the classical denoising methods in terms of RMSE and SSIM, the results of MemNet contain fewer artifacts as shown in Fig. 3.

One interesting observation is that a better denoising with a lower RMSE or a higher SSIM, in some cases, results in unwanted smoothing in the form of a local filtering that incurs a loss of detail. Although the RMSE results of DnCNN are not the best (Table 1), when they are used downstream by the SR networks in Table 2, the DnCNN denoised images achieve the best final performance.

Qualitative denoising results are shown in the first row of Fig. 3. We note that the artifacts created by denoising algorithms are amplified when SR methods are applied on the denoised results (e.g., (a) and (b) of Fig. 3). Although the denoised images are close to the clean LR image according to the evaluation metrics, the SR network is unable to recover faithful texture from these denoised images as the denoising algorithms remove part of the high-frequency information.

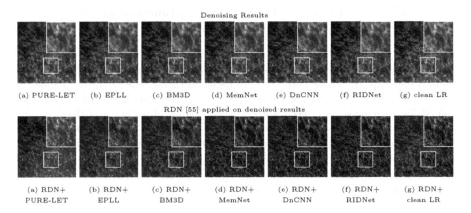

Fig. 3. The first row shows qualitative results of the denoising algorithms on a test LR image with the highest noise level. The second row shows qualitative results of the SR network RDN [55] applied on top of the denoised results. RDN amplifies the artifacts created by PURE-LET and EPLL, and is unable to recover faithful texture when the input image is over-smoothed by denoising algorithms. A gamma correction is applied for better visualization. Best viewed on screen.

The SR networks are applied on the denoised results of the denoising algorithms, and are evaluated using RMSE and SSIM. We also include the results of applying the SR networks on the noise-free LR images. As mentioned above, we notice that there is a significant drop in performance when the SR networks

Table 2. RMSE/SSIM results on the sequential application of denoising and SR methods on the W2S test images with the highest noise level, corresponding to the first column of Table 1. We omit the leading '0' in the results for better readability. For each SR method, we highlight the best RMSE value in red. The SR networks applied on the denoised results are trained to map the noise-free LR images to the high-quality HR images.

		Super-resolution networks					
		RCAN	RDN	SAN	SRFBN	EPSR	ESRGAN
Denoisers	PURE-LET	.432/.697	.458/.695	.452/.693	.444/.694	.658/.594	.508/.646
	VST+EPLL	.425/.716	.434/.711	.438/.707	.442/.710	.503/.682	.485/.703
	VST+BM3D	.399/.753	.398/.748	.418/.745	.387/.746	.476/.698	.405/.716
	MemNet	.374/.755	.392/.749	.387/.746	.377/.752	.411/.713	.392/.719
	DnCNN	.357/.756	.365/.749	.363/.753	.358/.754	.402/.719	.373/.726
	RIDNet	.358/.756	.371/.747	.364/.752	.362/.753	.411/.710	.379/.725
	Noise-free LR	.255/.836	.251/.837	.258/.834	.257/.833	.302/.812	.289/.813
	Noisy LR	.608/.382	.589/.387	.582/.388	.587/.380	.627/.318	.815/.279

are given the denoised LR images instead of the noise-free LR images as shown in Table 1. For example, applying RDN on noise-free LR images results in the SSIM value of 0.836, while the SSIM value of the same network applied to the denoised results of RIDNet on the lowest noise level is 0.756 (shown in the first row, last column in Table 3). This illustrates that the SR networks are strongly affected by noise or over-smoothing in the inputs. We also notice that a better SR network according to the evaluation on a single SR task does not necessarily provide better final results when applied on the denoised images. Although RDN outperforms RCAN in both RMSE and SSIM when applied on noise-free LR images, RCAN is more robust when the input is a denoised image. Among all the distortion-based SR networks, RCAN shows the most robustness as it outperforms all other networks in terms of RMSE and SSIM when applied on denoised LR images. As mentioned above, another interesting observation is that although DnCNN results in lower RMSE and higher SSIM than other networks for denoising at the highest noise level, DnCNN still provides a better input for the SR networks. We note generally that better denoisers according to the denoising benchmark do not necessarily provide better denoised images for the downstream SR task. Although the denoised results from MemNet have larger RMSE than the conventional methods, as shown in Table 1, the SR results on MemNet's denoised images achieve higher quality based on RMSE and SSIM.

Qualitative results are given in Fig. 4, where for each SR network we show the results for the denoising algorithm that achieves the highest RMSE value for the joint task (*i.e.*, using the denoised results of DnCNN). We note that none of networks is able to produce results with detailed texture. As denoising algorithms remove some high-frequency signals along with noise, the SR results from the distortion-based networks are blurry and many texture details are lost. Although the perception-based methods (EPSR and ESRGAN) are able to produce sharp results, they fail to reproduce faithful texture and suffer a drop in SSIM.

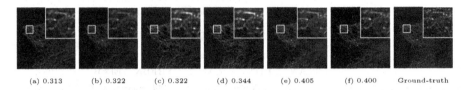

(a) 0.313   (b) 0.322   (c) 0.322   (d) 0.344   (e) 0.405   (f) 0.400   Ground-truth

Fig. 4. Qualitative results with the corresponding RMSE values on the sequential application of denoising and SR algorithms on the W2S test images with the highest noise level. (a) DnCNN+RCAN, (b) DnCNN+RDN, (c) DnCNN+SAN, (d) DnCNN+SRFBN (e) DnCNN+EPSR, (f) DnCNN+ESRGAN. A gamma correction is applied for better visualization. Best viewed on screen.

4.3   Joint Denoising and Super-Resolution (JDSR)

Our benchmark results in Sect. 4 show that the successive application of denoising and SR algorithms does not produce the highest-quality HR outputs. In this section, we demonstrate that it is more effective to train a JDSR model that directly transforms the noisy LR image into an HR image.

4.4   Training Setup

For JDSR, we adopt a 16-layer RRDB network [45]. To enable the network to better recover texture, we replace the GAN loss in the training with a novel texture loss. The GAN loss often results in SR networks producing realistic but fake textures that are different from the ground-truth and may result in a significant drop in SSIM [45]. Instead, we introduce a texture loss that exploits the features' second-order statistics to help the network produce high-quality and real textures. This choice is motivated by the fact that second-order descriptors have proven effective for tasks such as texture recognition [18]. We leverage the difference in second-order statistics of VGG features to measure the similarity of the texture between the reconstructed HR image and the ground-truth HR image. The texture loss is defined as

$$\mathcal{L}_{texture} = ||Cov(\phi(I_{SR})) - Cov(\phi(I_{HR}))||_2^2, \tag{3}$$

where I_{SR} is the estimated result from the network for JDSR and I_{HR} is the ground-truth HR image, $\phi(\cdot)$ is a neural network feature space, and $Cov(\cdot)$ computes the covariance. We follow the implementation of MPN-CONV [34] for the forward and backward feature covariance calculation. To improve visual quality, we further incorporate a perceptual loss to the training objective

$$\mathcal{L}_{perceptual} = ||\phi(I_{SR}) - \phi(I_{HR})||_2^2. \tag{4}$$

Table 3. JDSR RMSE/SSIM results on the W2S test set. †The denoising networks are retrained per noise level. ‡The SR networks are trained to map noise-free LR images to HR images. *The networks trained for JDSR are also retrained per noise level.

Method	Number of raw images averaged before JDSR				#Parameters
	1	2	4	8	
DnCNN†+RCAN‡	0.357/0.756	0.348/0.779	0.332/0.797	0.320/0.813	0.5M+15M
DnCNN†+ESRGAN‡	0.373/0.726	0.364/0.770	0.349/0.787	0.340/0.797	0.5M+18M
JDSR-RCAN*	0.343/0.767	0.330/0.780	0.314/0.799	0.308/0.814	15M
JDSR-ESRGAN*	0.351/0.758	0.339/0.771	0.336/0.788	0.322/0.798	18M
Ours*	0.340/0.760	0.326/0.779	0.318/0.797	0.310/0.801	11M

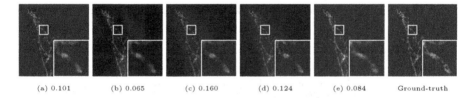

(a) 0.101   (b) 0.065   (c) 0.160   (d) 0.124   (e) 0.084   Ground-truth

Fig. 5. Qualitative results, with RMSE, of denoising and SR on the W2S test images with the highest noise level. (a) DnCNN+RCAN, (b) RCAN, (c) DnCNN+ESRGAN, (d) ESRGAN, (e) a 16-layer RRDB [45] trained with texture loss. A gamma correction is applied for better visualization (best on screen).

Our final loss function is then given by

$$\mathcal{L} = \mathcal{L}_1 + \alpha \cdot \mathcal{L}_{perceptual} + \beta \cdot \mathcal{L}_{texture}, \tag{5}$$

where $\mathcal{L}_1$ represents the $\ell 1$ loss between the estimated image and the ground-truth. We empirically set $\alpha = 0.05$ and $\beta = 0.05$. We follow the same training setup as the experiments in Sect. 4. For comparison, we also train RCAN [55] and ESRGAN [45] on JDSR.

4.5   Results and Discussion

The quantitative results of different methods are reported in Table 3. The results indicate that comparing to the sequential application of denoising and SR, a single network trained on JDSR is more effective even though it has fewer parameters. GAN-based methods generate fake textures and lead to low SSIM scores. Our model, trained with texture loss, is able to effectively recover high-fidelity texture information even when high noise levels are present in the LR inputs. We show the qualitative results of JDSR on the highest noise level (which corresponds to the first column of Table 1) in Fig. 5. We see that other networks have difficulties to recover the shape of the cells in the presence of noise, whereas our method trained with texture loss is able to generate a higher-quality HR image with faithful texture.

5 Conclusion

We propose the first joint denoising and SR microscopy dataset, **Widefield2SIM**. We use image averaging to obtain LR images with different noise levels and the noise-free LR. The HR images are obtained with SIM imaging. With W2S, we benchmark the combination of various denoising and SR methods. Our results indicate that SR networks are very sensitive to noise, and that the consecutive application of two approaches is sub-optimal and suffers from the accumulation of errors from both stages. We also observe form the experimental results that the networks benefit from joint optimization for denoising and SR. W2S is publicly available, and we believe it will be useful in advancing image restoration in medical imaging. Although the data is limited to the domain of microscopy data, it can be a useful dataset for benchmarking deep denoising and SR algorithms.

References

1. Abdelhamed, A., Lin, S., Brown, M.S.: A high-quality denoising dataset for smartphone cameras. In: CVPR (2018)
2. Anwar, S., Barnes, N.: Real image denoising with feature attention. In: ICCV (2019)
3. Batson, J., Royer, L.: Noise2Self: Blind denoising by self-supervision. In: ICML (2019)
4. Bevilacqua, M., Roumy, A., Guillemot, C., Alberi-Morel, M.L.: Low-complexity single-image super-resolution based on nonnegative neighbor embedding (2012)
5. Cai, J., Zeng, H., Yong, H., Cao, Z., Zhang, L.: Toward real-world single image super-resolution: a new benchmark and a new model. In: CVPR (2019)
6. Chen, C., Xiong, Z., Tian, X., Zha, Z.J., Wu, F.: Camera lens super-resolution. In: CVPR (2019)
7. Choi, J.H., Zhang, H., Kim, J.H., Hsieh, C.J., Lee, J.S.: Evaluating robustness of deep image super-resolution against adversarial attacks. In: ICCV (2019)
8. Dabov, K., Foi, A., Katkovnik, V., Egiazarian, K.: Image denoising by sparse 3-D transform-domain collaborative filtering. IEEE Trans. Image Process. **16**, 2080–2095 (2007)
9. Dai, T., Cai, J., Zhang, Y., Xia, S.T., Zhang, L.: Second-order attention network for single image super-resolution. In: CVPR (2019)
10. Dong, C., Loy, C.C., He, K., Tang, X.: Learning a deep convolutional network for image super-resolution. In: Fleet, D., Pajdla, T., Schiele, B., Tuytelaars, T. (eds.) ECCV 2014. LNCS, vol. 8692, pp. 184–199. Springer, Cham (2014). https://doi.org/10.1007/978-3-319-10593-2_13
11. El Helou, M., Dümbgen, F., Achanta, R., Süsstrunk, S.: Fourier-domain optimization for image processing. arXiv preprint arXiv:1809.04187 (2018)
12. El Helou, M., Süsstrunk, S.: Blind universal Bayesian image denoising with Gaussian noise level learning. IEEE Trans. Image Process. **29**, 4885–4897 (2020)
13. El Helou, M., Zhou, R., Süsstrunk, S.: Stochastic frequency masking to improve super-resolution and denoising networks. In: Vedaldi, A., Bischof, H., Brox, T., Frahm, J.-M. (eds.) ECCV 2020. LNCS, vol. 12361, pp. 749–766. Springer, Cham (2020). https://doi.org/10.1007/978-3-030-58517-4_44

14. Foi, A., Trimeche, M., Katkovnik, V., Egiazarian, K.: Practical Poissonian-Gaussian noise modeling and fitting for single-image raw-data. IEEE Trans. Image Process. **17**, 1737–1754 (2008)
15. Gharbi, M., Chaurasia, G., Paris, S., Durand, F.: Deep joint demosaicking and denoising. TOG **35**, 1–12 (2016)
16. Gu, J., Lu, H., Zuo, W., Dong, C.: Blind super-resolution with iterative kernel correction. In: CVPR (2019)
17. Gustafsson, M.G.: Surpassing the lateral resolution limit by a factor of two using structured illumination microscopy. J. Microsc. **198**, 82–87 (2000)
18. Harandi, M., Salzmann, M., Porikli, F.: Bregman divergences for infinite dimensional covariance matrices. In: CVPR (2014)
19. Hein, B., Willig, K.I., Hell, S.W.: Stimulated emission depletion (STED) nanoscopy of a fluorescent protein-labeled organelle inside a living cell. Proc. Natl. Acad. Sci. **105**(38), 14271–14276 (2008)
20. Huang, J.B., Singh, A., Ahuja, N.: Single image super-resolution from transformed self-exemplars. In: CVPR (2015)
21. Johnson, J., Alahi, A., Fei-Fei, L.: Perceptual losses for real-time style transfer and super-resolution. In: Leibe, B., Matas, J., Sebe, N., Welling, M. (eds.) ECCV 2016. LNCS, vol. 9906, pp. 694–711. Springer, Cham (2016). https://doi.org/10.1007/978-3-319-46475-6_43
22. Kim, J., Kwon Lee, J., Mu Lee, K.: Accurate image super-resolution using very deep convolutional networks. In: CVPR (2016)
23. Klatzer, T., Hammernik, K., Knobelreiter, P., Pock, T.: Learning joint demosaicing and denoising based on sequential energy minimization. In: ICCP (2016)
24. Krull, A., Buchholz, T.O., Jug, F.: Noise2Void-learning denoising from single noisy images. In: CVPR (2019)
25. Ledig, C., et al.: Photo-realistic single image super-resolution using a generative adversarial network. In: CVPR (2017)
26. Lehtinen, J., et al.: Noise2Noise: learning image restoration without clean data. In: ICML (2018)
27. Li, Z., Yang, J., Liu, Z., Yang, X., Jeon, G., Wu, W.: Feedback network for image super-resolution. In: CVPR (2019)
28. Lim, B., Son, S., Kim, H., Nah, S., Mu Lee, K.: Enhanced deep residual networks for single image super-resolution. In: CVPR (2017)
29. Luisier, F., Blu, T., Unser, M.: Image denoising in mixed Poisson-Gaussian noise. IEEE Trans. Image Process. (2011)
30. Makitalo, M., Foi, A.: Optimal inversion of the generalized Anscombe transformation for Poisson-Gaussian noise. IEEE Trans. Image Process. **22**(1), 91–103 (2012)
31. Martin, D., Fowlkes, C., Tal, D., Malik, J.: A database of human segmented natural images and its application to evaluating segmentation algorithms and measuring ecological statistics. In: ICCV (2001)
32. Matsui, Y., et al.: Sketch-based manga retrieval using manga109 dataset. Multimedia Tools Appl. **76**(20), 21811–21838 (2016). https://doi.org/10.1007/s11042-016-4020-z
33. Miao, S., Zhu, Y.: Handling noise in image deblurring via joint learning. arXiv preprint (2020)
34. Li, P., Xie, J., Wang, Q., Zuo, W.: Is second-order information helpful for large-scale visual recognition? In: ICCV (2017)
35. Plotz, T., Roth, S.: Benchmarking denoising algorithms with real photographs. In: CVPR (2017)

36. Qian, G., Gu, J., Ren, J.S., Dong, C., Zhao, F., Lin, J.: Trinity of pixel enhancement: a joint solution for demosaicking, denoising and super-resolution. arXiv preprint (2019)
37. Rust, M.J., Bates, M., Zhuang, X.: Sub-diffraction-limit imaging by stochastic optical reconstruction microscopy (STORM). Nat. Methods **3**(10), 793–796 (2006)
38. Sajjadi, M.S., Scholkopf, B., Hirsch, M.: EnhanceNet: single image super-resolution through automated texture synthesis. In: ICCV (2017)
39. Shi, W., et al.: Real-time single image and video super-resolution using an efficient sub-pixel convolutional neural network. In: CVPR (2016)
40. Shroff, H., Galbraith, C.G., Galbraith, J.A., Betzig, E.: Live-cell photoactivated localization microscopy of nanoscale adhesion dynamics. Nat. Methods **5**(5), 417–423 (2008)
41. Tai, Y., Yang, J., Liu, X., Xu, C.: MemNet: a persistent memory network for image restoration. In: ICCV (2017)
42. Timofte, R., Gu, S., Wu, J., Van Gool, L.: NTIRE 2018 challenge on single image super-resolution: methods and results. In: CVPRW (2018)
43. Vasu, S., Thekke Madam, N., Rajagopalan, A.N.: Analyzing perception-distortion tradeoff using enhanced perceptual super-resolution network. In: Leal-Taixé, L., Roth, S. (eds.) ECCV 2018. LNCS, vol. 11133, pp. 114–131. Springer, Cham (2019). https://doi.org/10.1007/978-3-030-11021-5_8
44. Verveer, P.J., Gemkow, M.J., Jovin, T.M.: A comparison of image restoration approaches applied to three-dimensional confocal and wide-field fluorescence microscopy. J. Microsc. **193**(50–61), 6 (1999)
45. Wang, X., et al.: ESRGAN: enhanced super-resolution generative adversarial networks. In: Leal-Taixé, L., Roth, S. (eds.) ECCV 2018. LNCS, vol. 11133, pp. 63–79. Springer, Cham (2019). https://doi.org/10.1007/978-3-030-11021-5_5
46. Weigert, M., et al.: Content-aware image restoration: pushing the limits of fluorescence microscopy. Nat. Methods **15**(12), 1090–1097 (2018)
47. Xie, J., Feris, R.S., Yu, S.S., Sun, M.T.: Joint super resolution and denoising from a single depth image. TMM (2015)
48. Zeyde, R., Elad, M., Protter, M.: On single image scale-up using sparse-representations. In: Boissonnat, J.-D., et al. (eds.) Curves and Surfaces 2010. LNCS, vol. 6920, pp. 711–730. Springer, Heidelberg (2012). https://doi.org/10.1007/978-3-642-27413-8_47
49. Zhang, K., Zuo, W., Chen, Y., Meng, D., Zhang, L.: Beyond a Gaussian denoiser: residual learning of deep CNN for image denoising. IEEE Trans. Image Process. **2**(5), 10 (2017)
50. Zhang, K., Zuo, W., Zhang, L.: Deep plug-and-play super-resolution for arbitrary blur kernels. In: CVPR (2019)
51. Zhang, W., Liu, Y., Dong, C., Qiao, Y.: RankSRGAN: generative adversarial networks with ranker for image super-resolution. In: ICCV (2019)
52. Zhang, X., Chen, Q., Ng, R., Koltun, V.: Zoom to learn, learn to zoom. In: CVPR (2019)
53. Zhang, Y., et al.: A Poisson-Gaussian denoising dataset with real fluorescence microscopy images. In: CVPR (2019)
54. Zhang, Y., Li, K., Li, K., Wang, L., Zhong, B., Fu, Y.: Image super-resolution using very deep residual channel attention networks. In: Ferrari, V., Hebert, M., Sminchisescu, C., Weiss, Y. (eds.) ECCV 2018. LNCS, vol. 11211, pp. 294–310. Springer, Cham (2018). https://doi.org/10.1007/978-3-030-01234-2_18
55. Zhang, Y., Tian, Y., Kong, Y., Zhong, B., Fu, Y.: Residual dense network for image super-resolution. In: TPAMI (2020)

56. Zhang, Z., Wang, Z., Lin, Z., Qi, H.: Image super-resolution by neural texture transfer. In: CVPR (2019)
57. Zhou, R., Achanta, R., Süsstrunk, S.: Deep residual network for joint demosaicing and super-resolution. In: Color and Imaging Conference (2018)
58. Zhou, R., Süsstrunk, S.: Kernel modeling super-resolution on real low-resolution images. In: ICCV (2019)
59. Zoran, D., Weiss, Y.: From learning models of natural image patches to whole image restoration. In: ICCV (2011)

W03 - Egocentric Perception, Interaction, and Computing

W03 - Egocentric Perception, Interaction, and Computing

The 7th International Workshop on Egocentric Perception, Interaction, and Computing (EPIC 2020) held at the European Conference on Computer Vision (ECCV 2020) aimed to bring together the different communities which are relevant to egocentric perception, including Computer Vision, Machine Learning, Multimedia, Augmented and Virtual Reality, Human Computer Interaction, and Visual Sciences. The main goal of the workshop was to provide a discussion forum and facilitate the interaction between researchers coming from different areas of competence. We invited researchers interested in these topics to join the EPIC 2020 workshop, submitting ongoing and recently published ideas, demos, and applications in support of human performance through egocentric sensing.

For EPIC 2020, the call for participation was issued March 2020, inviting researchers to submit both extended abstracts and full papers to the workshop. In total, five full papers were submitted and two extended abstracts were submitted to the workshop.

The organizers selected a team of 10 reviewers for the valid submissions (one full paper was desk rejected) with every full paper receiving at least two reviews. Once submitted, the workshop organizers selected two full papers for publication/presentation and two extended abstracts for presentation at the workshop.

The workshop contained four keynotes from Bernard Ghanem (KAUST, Saudi Arabia), Federica Bogo (Microsoft Research, Switzerland), Richard Newcombe (FRL, USA), and Jitendra Malik (University of California, Berkeley and FAIR, USA), as well as presentations from the two accepted full papers and two accepted extended abstracts, as well as small spotlight talks from accepted papers at the main conference from seven authors.

August 2020

Michael Wray
Dima Damen
Hazel Doughty
Antonino Furnari
Walterio Mayol-Cuevas
Giovanni Maria Farinella
David Crandall
Kristen Grauman

An Investigation of Deep Visual Architectures Based on Preprocess Using the Retinal Transform

Álvaro Mendes Samagaio[(✉)] and Jan Paul Siebert

Computer Vision for Autonomous Systems Group, School of Computing Science, University of Glasgow, Glasgow, UK
alvaromiguelsamagaio@gmail.com

Abstract. This work investigates the utility of a biologically motivated software retina model to pre-process and compress visual information prior to training and classification by means of a deep convolutional neural networks (CNNs) in the context of object recognition in robotics and egocentric perception. We captured a dataset of video clips in a standard office environment by means of a hand-held high-resolution digital camera using uncontrolled illumination. Individual video sequences for each of 20 objects were captured over the observable view hemisphere for each object and several sequences were captured per object to serve training and validation within an object recognition task. A key objective of this project is to investigate appropriate network architectures for processing retina transformed input images and in particular to determine the utility of spatio-temporal CNNs versus simple feed-forward CNNs. A number of different CNN architectures were devised and compared in their classification performance accordingly. The project demonstrated that the image classification task could be conducted with an accuracy exceeding 98% under varying lighting conditions when the object was viewed from distances similar to that when trained.

Keywords: Deep learning · Retina · Visual cortex · CNN · Retinal transform · Image classification

1 Introduction and Motivation

Deep learning based approaches are now well established as the state-of-the-art methodology for solving computer vision problems [10]. Recent advances in computing science and hardware performance have allowed the scientific community to develop higher performance and more robust algorithms. However, as computing power has increased so has the quantity of data that networks are required to process, as has the size of the networks themselves. Standard images now routinely exceed 2 million pixels and this can constitute an issue when real-time performance is required in image based tasks using deep learning approaches. Recent improvements in computing power have underpinned the wide adoption of CNN's (Convolutional Neural Networks) [9]. However, these models remain very limited in terms of the size of images they can process in a single pass [17].

© Springer Nature Switzerland AG 2020
A. Bartoli and A. Fusiello (Eds.): ECCV 2020 Workshops, LNCS 12535, pp. 495–510, 2020.
https://doi.org/10.1007/978-3-030-66415-2_32

In this investigation we validate a computational model of a space-variant visual pathway architecture, based on a *software retina pre-processor,* that we can find in the mammalian vision system [14]. By re-sampling input images by means of our software retina it is possible to reduce the overall size of the output *cortex images* and thereby memory requirements, when using deep learning frameworks, without the need to sacrifice high frequency image information at the centre of the field of view [12]. The software retina is a biologically inspired subsampling method that employs a set of overlapping Gaussian receptive fields spatially configured in such way that approximately mimics a natural human retina tessellation. In this case the sampling density of the receptive fields decreases with increasing visual angle from the optical axis, i.e. visual eccentricity, while at the same time the receptive field size increases (until 15° of eccentricity, beyond which the sampling density continues to decrease but the receptive field size remains approximately constant). The application of retina pre-processing brings a number of other advantages when used in combination with CNN's. The images that the CNN's process are the result retinal sampling and spatial remapping within a model of the mammalian retino-cortical transform [15] which is very similar to the log-polar mapping, and this affords both a very substantial data reduction and also a degree of invaraince to scale and in-plane rotation changes within the input image. The retino-cortical transform leads to the correspondence between a rotation or a scale change in the real image and a shift in the cortical output space and accordingly partial scale and rotation invariance. In addition to processing the input image by means of our retino-cortical transform mode, we also implement a subset of the retinal ganglion cells involved in mediating colour perception. Colour perception in humans involves processing opponent colours (blue vs yellow, red vs green, black vs white). In this case we have employed a model similar to that found in the LAB colour space which utilises a monochromatic intensity channel in addition to the above two colour opponent channels.

The primary contribution of the study reported here is to determine if the retina pre-processor is viable within practical vision systems to support robotic and egocentric visual perception, based on:

– Data (video clips) collected by hand within an unconstrained scenario allowing variation in object scale, rotation, lighting, view sphere sampling etc. for training and classification
– Viable sizes of data set and numbers of object classes typically used in robotics hand-eye applications

Accordingly, we investigate the following issues:

– Appropriate network architectures for processing the cortical images generated by the retina, including single frame, multi-frame and recurrent spatio-temporal processing
– Training & recognition performance for hand-held capture around the observable object view hemisphere
– Degree of scale and rotation invariance achieved in practice

- Level of classification performance & data reduction that can be obtained
- Overall system robustness and viability in the above practical use-case

2   Previous Work

2.1   The Mammalian Vision System

The part of the visual system that contacts light first is the retina, which is organised in layers and is located at the back of the eye. These layers correspond to sheets of neurons that possess different functions [3]. At the back of the retina it is possible to find the photoreceptor cells, that transduce the light signal (photons) into neuronal signal [1]. It is possible to identify two photoreceptor types: the cones and the rods, and both of these photoreceptor types transduce light signals into changes in cell membrane potential voltages. These changes are then transmitted along the cell membrane to the bipolar neurons that make up the inner layer of the retina. While rods are specialised for low light or scotopic vision – night time vision, cones are specialised high light or photopic vision – daytime vision. Rods are not present in the fovea, that is the region of the retina in which the cells are most tightly packed. This area enables the perception of fine details and high frequency content in the visual data [2]. The contrary happens in the periphery of the retina, where the cells are more sparsely distributed. The intermediate neurons that transmit the signal to the brain follow the same retinotopic topology. These topologies associated with the attention mechanisms allow the retina to control the amount of visual information that goes onto the brain [4]. The focus of this investigation is the photopic vision, meaning that the rods pathway will not be considered.

It is crucial to understand the concept of receptive fields when studying the intermediate neuronal cells. The final layer of these cells comprise retinal ganglion cells (RGCs) and these cells have the function of relaying the visual signal to the brain. RGCs receive information from various surrounding photoreceptor cells that can be grouped together forming a cluster - RGCs receptive field. Similarly to the retina, the RGCs receptive fields' size increases eccentrically from the fovea, in which the RGCs only receive information from a single photoreceptor cell [6]. This particular physiological property underpins the perception of fine details.

2.2   Retino-Cortical Mapping

Having captured a retinal image projection by means of the photoreceptive cells, the visual signal is then passed to the primary visual cortex (V1). In this portion of the brain, the signal is split into two equal parts and then projected onto each cortical hemisphere in V1. Here the images take the appearance of a spatial complex logarithmic mapping. This mapping appears to facilitate mammalian visual perception and also potentially contributes to scale and rotation invariance in biological visual systems [14]. The retino-cortical mapping is responsible for the *cortical magnification* of the fovea within the visual cortex. In this case, the fovea appears magnified compared to the remaining peripheral zone of the retina, which in turn becomes progressively more compressed as a function of visual eccentricity, matching the retinal photoreceptor sampling density reduction within the retina. This mapping property confers efficient data handling and reduction capabilities. The log-polar mapping also allows for a reduction in the order of the number of required receptive fields, from $O(r)$ to $O(r^2)$, using O notation to consider computational complexity [16]. This way, the number of receptive fields required to sample a (centrally fixated) contour becomes constant. Therefore, the space-variant sampling observed in the retina provides full resolution in the fovea and the capacity for a human to undertake tasks requiring perception of very fine detail and, at the same time, being aware of the surrounding environment, by simulating a static zoom lens [12]. The log-polar mapping proposed by [15] introduces some properties in the vision system. These properties are referred to as edge invariance. A local rotation of the edge contour of the input, while keeping the same fixation point, will produce a translation in the θ axis of the cortical view. In the same way, maintaining the fixation point and performing a scale change in the input will result in a translation along the ρ axis in the cortical mapping. The cortex also demonstrates projective invariance since objects located in the same ground plane as the observer will retain their local shape appearance. With this in mind, it is evident that this mapping conserves only local angles, and the local shape of the objects, being certainly important in the scale, rotation and projection invariance of features, although deforming their global shape. In order to create a model that follows the biological structure more closely, the pure log-polar mapping was altered to prevent the singularity in the center of the fovea. Schwartz [cite] proposed the introduction of a small constant in the horizontal axis of the mapping. This way, the fovea becomes almost linear while the periphery approaches a pure log-polar mapping. A consequence of the foveated nature of the visual system is the capability of conducting visual search by focusing the retina's fovea in different locations. This allows for high visual acuity and important feature retention in the foveal region. This way, the retina grants an attention mechanism that can be very useful to explore the scene when and where required, providing context and adequate understanding. The work of Ram and Siebert in [13], under a computer vision framework, shows that the implementation of the retina based on SIFT descriptors improved recognition against the implementation of SIFT processing in full resolution images, by smoothing out irrelevant peripheral detail.

2.3   A 50k Node Retina Model

This work is based in the implementation of a 50k node retina developed by Ozimek [11] in which the fovea corresponds to approximately 10% of the retina's area. This implementation achieves roughly 16.7 fold of data reduction. Thus, the need to apply a window approach to CNN architectures that is scanned over an image pyramid no longer applies, allowing for a great improve in the CNN efficiency. The retina implementation used has also been enhanced to prevent aliasing artefacts that might appear due to the retinal sampling and cortical image generation [12]. This retina model is supported by hardware acceleration (NVIDIA Graphics Processor Units) so that it becomes possible to take advantage of real-time applications such as robotics and mobile developments.

3   Approach

Given the goal to validate the retina and cortical image generation process in a deep learning framework for robot vision application, this investigation was guided by the following pipeline: The first step was to create the dataset in which the training and testing will be performed. The dataset collected was the Intern's Objects Dataset, described in Sect. 3.1. Then, the videos taken for the dataset were preprocessed, meaning that frames were extracted and the retino-cortical transformations applied. The main focus of the work was to train and test in the cortical space, since it affords a major reduction in information. Pre-processed cortical images were processed by the different Deep Learning architectures highlighted in Sect. 3.2.

3.1   Dataset Collection - Intern's Objects Dataset

In order to achieve the goals proposed in this investigation, we created a dataset called Intern's Objects which comprises frames taken from video clips capturing 20 everyday life objects - cup, case, bag, mouse, hole-punch, stapler, beer, mug, coffee, tea, tape, umbrella, spray, deodorant, book, tissues, eraser, speaker, sleeping bag and a watch. The dataset was divided in training and test sets and each of these two groups consisted of different video clips. These video clips were captured by directing a hand-held camera in a continuous motion around the visible view-hemisphere each given object. Regarding the training set, video clips were acquired around the object several times, during which, several variables changed, such as the angle between the camera and the object and the distance to the object. In this way we are able to modify the size (proportion) and the position of the object in the field of view. Each of the objects was recorded twice in this manner, the only difference between the two capture runs being the lighting conditions (lights turned on and turned off but with natural light coming from one side). The background view varies slightly in each video clip, due to the camera being hand held. Each object is standing on the centre of the same wooden table. The test set was captured using the same protocol as the training set (round motion) and was structured to challenge the hypothesis that we seek to test, accordingly the test set was split in four different video clips: The first two clips are very similar to the training set, varying only the lighting conditions between them - standard test set. The remaining two clips were captured to test the hypothesis that the retino-cortical mapping grants a degree of rotational and scalar invariance. With this in mind, these clips captured rotated views of the object as well as different distances between the object and the camera, to validate the recognition performance for previously unseen scale and rotation changes. All of the clips were recorded using an action camera, the GoPro

Hero5, in full HD spatial resolution (1080p) and 30fps temporal resolution using a linear field of view. The training set contains around 3000 frames for each of the objects (divided between the two video clips). The test sets contain slightly fewer frames. The dataset is specified in Table 1. By means of the above video clip sequences we can perform a series of experiments using any combination of these sets. Following the data collection process, the final dataset consists of 6 video clips of each object (total - 120) and additional 6 video clips of the background without an object. The frames were then processed to produce the images illustrated in Figs. 1, 2 and 3.

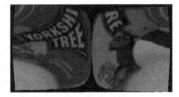

Fig. 1. Tea class example images: (left) Original image; (center) Retinal image; (right) Cortical image

Fig. 2. Tea class rotated example images: (left) Original image; (center) Retinal image; (right) Cortical image

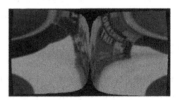

Fig. 3. Tea class scale/distance change example images: (left) Original image; (center) Retinal image; (right) Cortical image

Table 1. Intern's objects dataset information: numbers of images per object class for specific image capture conditions

Labels	Train light	Train dark	Test light	Test dark	Test rotation	Test scale	Train total	Test total
bag	2092	1793	1148	1118	1475	1423	3885	2266
beer	2169	1854	1350	1279	1513	1160	4023	2629
book	2115	1891	1405	1376	1397	1375	4006	2781
case	1962	1747	1266	1180	1294	1338	3709	2446
coffee	2081	1881	1315	1393	1449	1270	3962	2708
cup	2233	2079	1700	1572	1260	1473	4312	3272
deodorant	1934	1937	1166	1246	1336	1443	3871	2412
eraser	2115	1971	1365	1375	1292	1235	4086	2740
hole	2120	2051	1272	1095	1434	1313	4171	2367
mouse	2008	1602	1321	1177	1350	1192	3610	2498
mug	2092	1999	1246	1420	1408	1299	4091	2666
sleep	2087	2102	1535	1377	1333	1278	4189	2912
speaker	2044	2068	1345	1340	1275	1258	4112	2685
spray	2032	1867	1546	1395	1551	1343	3899	2941
stapler	2078	2093	1659	1099	1422	1175	4171	2758
tape	1991	2220	1350	1450	1331	1308	4211	2800
tea	2212	2049	1372	1408	1405	1310	4261	2780
tissues	2005	2104	1358	1384	1266	1546	4109	2742
umbrella	1931	2216	1266	1304	1257	1262	4147	2570
watch	2014	1930	1556	1669	1492	1285	3944	3225
Total	41315	39454	27541	26657	27540	26286	80769	54198

3.2   Neural Network Architectures

In this section we present the different network architectures used to perform classification in the dataset mentioned in Sect. 3.1. We used two different approaches: classifying frame by frame and classifying short video snippets (3–5 frames) along different time intervals to assess the possibility of video classification using the cortical image space. In every architecture presented we use batch normalisation before every input layer so that the training session is faster [7]. We also found that the results were actually better with it. After the first experiments we performed another set of tests with the 1A configuration described in Table 6. In this case we followed three different approaches: in the first we trained with the background class and then performed the normal testing sessions; the second alternative approach was to train on the rotation and scale sets and then perform the test session on the test and training sets and the last approach consisted in training and testing using fixation crop images (similar to the retinal images but without the sampling effect) instead of the transformed cortical images. The main modules of the used network architectures are presented in Tables 3, 4, 5, 6 and 7.

Table 2. Network architectures

CNN + FC	Residual CNN + FC	CNN + LSTM	Residual CNN + LSTM
Convolutional layers-Table 3	Residual layers-Table 5	Convolutional layers-Table 3	Residual layers-Table 5
Flatten			
Fully Connected Layers-Table 4		Recurrent (LSTM) Layers-Table 7	

CNN + Fully Connected Layers. Based on the work of Hristozova [5], the first approach was an altered version of the network presented in her work. The cortical images produced by the retina used in that work were $399 \times 752p$, which differs from the retina implemented in this paper that produces $257 \times 490p$ images. Due to this alteration the final network had only 6 convolutional layers and 3 fully connected layers, as specified in the Table 2, first column. This is the basic approach to image classification. We tried to classify each frame as one of the 20 classes and we used different configurations that are presented in Table 6. This architecture was not expected to produce the best results overall, since it is the simplest.

Residual CNN + Fully Connected Layers. To improve our network, we implemented residual skip layers in the basic CNN architecture described in the section above. Consequently, the final network architecture is defined in Table 2, second column. This way we could make the network deeper (from 6 convolutional layers to 12) which allowed for better feature extraction while keeping it easy and fast to train. The network will combine information from previous layers to create a latent space better mosaic that is then classified. Therefore, it is expected that this approach would produce improved results over those obtained from the simpler non-residual network. In order to compare both networks we used the same configurations highlighted in Table 6.

CNN + Long Short-Term Memory Units. With the goal of testing whether the spatio-temporal information could be extracted and learned from the cortical image space, we implemented a recurrent model capable of retaining memory about the frames that it has already seen. In this case, the strategy was to use the already trained convolutional layers from Sect. 3.2 and replace the fully connected end layers that perform classification with 2 stacked recurrent units. The output from the last CNN layer is then flattened and that corresponds to the input of the first recurrent layer. The recurrent units that were implemented in this investigation were Long Short-Term Memory units (LSTM). These units have the particularity of keeping a memory state and learning what to forget and what to remember from frame to frame. After the last LSTM layer there is one fully connected layer that performs the final classification. This FC layer receives the output from the LSTM layer after the last frame of the sequence as it is possible to confirm in Table 2, third column. In this case, we adopted a different configuration since the inputs should be sequential groups of frames.

Table 3. Convolutional layers

Layer	Type	Filter size
C1	Conv2D	$32 \times 3 \times 3$
M1	MaxPool	2×2
C2	Conv2D	$32 \times 3 \times 3$
M2	MaxPool	2×2
C3	Conv2D	$64 \times 3 \times 3$
M3	MaxPool	2×2
C4	Conv2D	$64 \times 3 \times 3$
M4	MaxPool	2×2
C5	Conv2D	$128 \times 3 \times 3$
M5	MaxPool	2×2
C6	Conv2D	$256 \times 3 \times 3$
M6	MaxPool	2×2

Table 4. Fully connected layers

Layer	Type	Filter size
FC1	Fully Connected Dropout 0.5	132
FC2	Fully Connected Dropout 0.5	132
FC3	Fully Connected	20

Hence, the configuration that we used can be observed in Table 8. In each one of these cases we generated every possible 3–5 frames sequence which further increases the training data, as well as the validation and test data. This way, we guarantee that the network sees all of the changes that can happen in the corresponding span of frames that we introduce.

Residual CNN + Long Short-Term Memory Units. Finally we implemented the also pre-trained Residual CNN layers with the LSTM layers at the end. The architecture is represented in Table 2, fourth column. The configurations used were the same as the ones used in the previous section (Table 8).

4   Results and Discussion

In this section the results achieved using the different approaches described in Sect. 3 are analysed. The training data was divided randomly between the actual training set (80%) and the validation set (20%).

Table 5. Residual convolutional layers

Layer	Type	Filter size
RC1	Conv2D	$32 \times 3 \times 3$
RM1	MaxPool	2×2
RC2	Conv2D	$32 \times 3 \times 3$
RC3	Conv2D	$32 \times 3 \times 3$
SL1	Skip Layer	+RM1
RM2	MaxPool	2×2
RC4	Conv2D	$64 \times 3 \times 3$
RC5	Conv2D	$64 \times 3 \times 3$
RC6	Conv2D	$64 \times 3 \times 3$
SL2	Skip Layer	+RC4
RM3	MaxPool	2×2
RC7	Conv2D	$128 \times 3 \times 3$
RC8	Conv2D	$128 \times 3 \times 3$
RC9	Conv2D	$128 \times 3 \times 3$
SL3	Skip Layer	+RC7
RM4	MaxPool	2×2
RC10	Conv2D	$256 \times 3 \times 3$
RC11	Conv2D	$256 \times 3 \times 3$
RC12	Conv2D	$256 \times 3 \times 3$
SL4	Skip Layer	+RC10
RM5	MaxPool	2×2

Table 6. Configurations for FC ending architecture

Label	Number of frames	Colour space	Input size
1A	1	Greyscale	$1 \times 257 \times 490$
1B	3	Greyscale	$3 \times 257 \times 490$
1C	1	RGB	$3 \times 257 \times 490$
1D	1	Colour Opponency	$3 \times 257 \times 490$

4.1   CNN + FC

As expected, the results produced by this approach (Table 9) when considering the test set were not the best overall. Comparing the different configurations of inputs, we can observe that the results that give the best accuracy for the normal test set utilises colour opponent channels. However we can also note that the performance improvement compared to utilising a single greyscale frame is just 1.7%. Since the colour opponent model utilises the greyscale frame as

Table 7. Recurrent (LSTM) layers

Layer	Type	Filter size
LSTM1	Dropout 0.5 LSTM	150
LSTM2	Dropout 0.5 LSTM	150
FC1	Dropout 0.5 Fully Connected	20

Table 8. Configurations for LSTM ending architecture

Label	Number of frames	Colour space	Input size
2A	3	Greyscale	$3 \times 257 \times 490$
2B	5	Greyscale	$5 \times 257 \times 490$

one of its input channels, we hypothesise that the most important features are contained within this greyscale input, the two colour opponent channels just slightly increase the overall accuracy by complementing with colour and texture information.

4.2   Residual CNN + FC

The accuracy values for all of the configurations were better than the those presented in Sect. 4.1. This means that the residual layers and the increased depth actually help the classification task by combining high level and low level features. Looking at the results in Table 9 it is possible to notice that the pattern of classifications from the *CNN + FC* architecture is the same. The best configuration still utilises colour opponent channels, preceded in performance by the greyscale only configuration. There was also an increase in classification performance on the rotation and scale sets, with the exception of the colour opponency frame. This could be due to excessive overfitting to the training data, as shown in the results of the test set. By examining the confusion matrix and per class accuracy for the colour opponency frame configuration, it is possible to observe that the most successfully recognised class is the "Tissues" class and that might be because it was the only yellow object in the whole dataset, making it easier for the network to distinguish between similarly shaped objects. Once again, the distance dataset presents the lowest classification performance results. The network is does not appear to be able to use the colour information from the RGB frames in order to increase classification accuracy.

4.3   CNN + LSTM

In this subsection we analyse the results obtained by the recurrent networks to draw conclusions about the spatio-temporal information captured by the network. When comparing the results from the 3 frames approach (CNN + FC — 1B) from Sect. 3.2 with the one in Table 9 (CNN + LSTM — 2A), it is evident that there is an increase in the recognition rate in all of the tests. This increase was probably caused by the data sequence memory capability of the recurrent neural networks namely, the LSTM units. The spatio-temporal information from the frames was extracted more effectively when the images were processed sequentially and the network could therefore save a memory state, as opposed to when the frames were stacked together and the network had to process them simultaneously. However, the test accuracy was not higher than the values obtained by the network configurations that used either a single grayscale frame, or a single frame represented by colour opponent channels. Also, when comparing the results between the 5 frame and 3 frame approach (2A and 2B) we can observe that the test results are better for the 2B approach – 95.9% versus 94.8% – meaning that the network is able to perform better with more information. Nonetheless, the results for the rotation and scale datasets are not coherent, which could show signs of overfitting to the training set. Accordingly, by having more information available in each training instance, the network may be adapting itself better to the data, losing some capacity to generalise over new data. Overall, there was not a significant increase in the results for the rotation and scale dataset obtained by this method when compared to the method using only fully connected layers.

4.4   Residual CNN + LSTM

The last approach used to validate the experiment is described in the last subsection of Sect. 3.2 and analysed in this section. Table 9, *Res CNN + LSTM*, allows us to observe that the results obtained by the residual CNN combined with LSTM layers method are better than the results obtained without the residual layers but retaining the LSTM. Once again the residual architecture presents better results than those of the non residual network configurations in all the datasets. When comparing method 2A with method 2B, it is noticeable that the pattern verified using CNN + LSTM architecture repeats itself. The rotation set accuracy increased from 3 frames to 5 frames while the scale set accuracy decreased, once more suggesting an overfitting to the training data. Another issue that could affect and possibly decrease the accuracy of the 5 frame configuration is the reduction in batch size that was necessary due to memory limitations. Looking at 1B approach results for the method using both residual and fully connected layers, it is evident that the classification performance obtained with the LSTM layers improved in every dataset. The data sequence memory of the LSTM units showed once again that it is capable of increasing the recognition rates. It was anticipated that the best overall results would be produced by this architecture however, the actual best test set accuracy - approach 1D with Residual CNN + FC layers - and rotation set accuracy – approach 1D with CNN + FC layers - were obtained using non recurrent layers However, the best scale set accuracy was obtained using recurrent layers and the 3 frames approach. This result supports the hypothesis that the LSTM memory state and the fact that sequence of 3 frames is uninterrupted allows the network to produce a better prediction when presented with variation in spatial scale.

Table 9. Classification % for each model & test set averaged over 3 training runs.

Architecture	Configuration	Training	Validation	Test	Rotate	Scale
CNN + FC	1A	99.2	99.5	95.6	58	44
	1B	98.8	99.1	93.7	50	35
	1C	97.3	99.2	88.0	48	23
	1D	99.3	99.8	97.3	72	48
Res CNN + FC	1A	99.2	99.9	97.7	60	50
	1B	97.4	99.9	96.0	59	44
	1C	99.1	99.9	89.2	50	26
	1D	99.4	99.2	98.5	71	41
CNN + LSTM	2A	99.5	99.9	94.8	57	44
	2B	99.5	99.9	95.8	58	42
Res CNN + LSTM	2A	99.2	99.9	97.4	61	62
	2B	99.9	99.9	97.8	62	57
Res CNN + FC	X	99.6	99.9	97.2	59	52
	Y	99.5	99.9	69.6	–	–
	Z	94.5	99.3	95.4	56	36

4.5 Alternative Approaches

The alternative approaches described in Sect. 3.2 produced the results shown in Table 9. For all of these alternative tests we use the Residual CNN + FC architecture. The approach X adopts the Background class and in this case we can observe that the classification performance values were only higher on the scale dataset, when comparing to approach 1A which does not employ a Background class. This suggests that the background class does not help the network differentiate between the classes, by learning information cues that should be the same for each one of them. The appearance of the frames' background wasn't exactly the same for every video since the camera was handheld. In approach Y we used the rotation and scale sets for training and the original training and test sets for testing. It is evident in approach Y that the performance results are far behind those obtained by not reverse training the model. Hence, this suggests that it is better for the network to learn all the views of the object before trying to deal with scale and rotation variation than the other way around.

In the final experiment Z we employed *fixation crop* images (i.e. raw untransformed input images occupying the same field of view as sampled by the retina) to determine the cortical transform was affording any degree of rotation and scale invariance. However, the fixation crop images had to be resized to the same number of pixels as the cortical images. Analysing the results and comparing them with the approach 1B for the same architecture we can claim that the cortical mapping does indeed confer partial rotation and scale invariance to the model, since the fixation crop accuracy values are lower than those of the retina preprocessed networks.

5   Conclusions

The results obtained from his study indicate that the retino-cortical transform can be applied within a realistic scenario to reduce the data size of high resolution 1080p images to achieve data efficient deep learning in the context of robotic and egocentric perception. Our investigation demonstrated that video clip collection for training and testing networks to classify a set of 20 objects can be accomplished quickly and conveniently using a simple hand-held video camera and natural and standard office illumination. The results achieved were very satisfactory, exceeding 98% accuracy for the principal training and test sets, as these were captured at approximately similar object distances. While the results obtained for the scale and rotation sets do not achieve the same level of classification performance, they are significantly better than those obtained without using the retino-cortical transform and clearly demonstrate that the transform does indeed confer a significant degree of scale invariance and a substantial degree of rotation invariance. It should be borne in mind that when testing the scale and rotation invariance properties of the networks developed here, no prior training or data augmentation with scaled and rotated images has been applied. Accordingly, we would anticipate that standard approaches to data augmentation prior to the retino-cortical transformation would further improve the performance of our networks under these conditions. Use of recurrent networks in this study did not demonstrate significantly improved results over pure feed forward, non-recurrent, networks. Why this should be the case is worthy of further investigation, particularly with regard to sampling the input images non-uniformly over an exponential time window, as opposed to linearly over time as in this study. In this way, it may be possible to capture both slow and fast changes in the input image stream. In addition, we propose to investigate developing loss functions tuned for video sequences that disentangle the static shape and surface texture characteristics of objects from their changing appearance characteristics during pose changes. The implementation of an automatic mechanism for gaze control has the potential to further improve the performance of our networks by directing the high-resolution fovea to the most diagnostic locations to best disambiguate objects to be recognised. Our current work is based on deep reinforcement methods for simultaneously learning object class labels and where to drive the fixation of the retina. Our most recent work [8] has investigated a complete implementation of all the colour processing cell species found in the human retina and also a multi-resolution retina architecture. This first study of the retina applied to a realistic object classification problem within a normal uncontrolled office environment has demonstrated the viability retina-prepossessing approach in this context. Accordingly, current and following work is building on the above results to investigate improved biologically inspired network architectures, increasing the scaling of problems that can be tackled by this approach and applications within robotics and egocentric perception systems.

References

1. Baylor, D.A., Nunn, B.J., Schnapf, J.L., Ylor, D.A.B.: The photocurrent, noise and spectral sensitivity of rods of the Monkey Macaca Fascilcularis. Technical report (1984)
2. Briggs, F.: Mammalian visual system organization subject: sensory systems online publication mammalian visual system organization mammalian visual system. Organization (2017). https://doi.org/10.1093/acrefore/9780190264086.013.66
3. Cajal, S.R.: La rétine des vertébrés (1933). https://books.google.co.uk/books?id=y4i6nQEACAAJ
4. Curcio, C.A., Allen, K.A.: Topography of ganglion cells in human retina. J. Comp. Neurol. **300**(1), 5–25 (1990). https://doi.org/10.1002/cne.903000103
5. Hristozova, N., Ozimek, P., Siebert, J.P.: Efficient egocentric visual perception combining eye-tracking, a software retina and deep learning. Technical report. http://www.dcs.gla.ac.uk
6. Hubel, D.H.: Eye, Brain, and Vision. Scientific American Library Series, no. 22. Scientific American Library/Scientific American Books, New York (1995)
7. Ioffe, S., Szegedy, C.: Batch Normalization: Accelerating Deep Network Training by Reducing Internal Covariate Shift, February 2015. http://arxiv.org/abs/1502.03167
8. Killick, G.: Biomimetic convolutional neural network pipelines for image classification. MSc Dissertation, University of Glasgow (2020)
9. Krizhevsky, A., Sutskever, I., Hinton, G.E.: ImageNet Classification with Deep Convolutional Neural Networks. Technical report. http://code.google.com/p/cuda-convnet/
10. Mahajan, D., et al.: Exploring the limits of weakly supervised pretraining. Technical report (2018). https://arxiv.org/pdf/1805.00932v1.pdf
11. Ozimek, P., Balog, L., Wong, R., Esparon, T., Siebert, J.P.: Egocentric Perception using a biologically inspired software retina integrated with a deep CNN. In: International Conference on Computer Vision 2017, ICCV 2017, Second International Workshop on Egocentric Perception, Interaction and Computing, September 2017. http://eprints.gla.ac.uk/148802/
12. Ozimek, P., Hristozova, N., Balog, L., Siebert, J.P.: A space-variant visual pathway model for data efficient deep learning. Front. Cell. Neurosci. **13**, 36 (2019). https://doi.org/10.3389/fncel.2019.00036. https://www.frontiersin.org/article/10.3389/fncel.2019.00036
13. Ram, I., Siebert, J.P.: Point-based matching applied to images generated by log(z) and log(z+alpha) forms of artificial retina. In: Third World Congress on Nature and Biologically Inspired Computing, pp. 451–458 (2011). https://doi.org/10.1109/NaBIC.2011.6089629

14. Schwartz, E.L.: Spatial mapping in the primate sensory projection: analytic structure and relevance to perception. Biol. Cybern. **25**(4), 181–194 (1977). https://doi.org/10.1007/BF01885636

15. Schwartz, E.L.: Computational anatomy and functional architecture of striate cortex: a spatial mapping approach to perceptual coding. Vis. Res. **20**(8), 645–669 (1980). https://doi.org/10.1016/0042-6989(80)90090-5. https://linkinghub.elsevier.com/retrieve/pii/0042698980900905

16. Wilson, J.C., Hodgson, R.M.: Log-polar mapping applied to pattern representation and recognition, January 1992

17. Wu, S., Mengdan, Z., Chen, G., Chen, K.: A new approach to compute CNNs for extremely large images, pp. 39–48, June 2017. https://doi.org/10.1145/3132847.3132872

Data Augmentation Techniques for the Video Question Answering Task

Alex Falcon[1,2]($\boxtimes$) ⑩, Oswald Lanz[1] ⑩, and Giuseppe Serra[2] ⑩

[1] Fondazione Bruno Kessler, 38123 Trento, Italy
lanz@fbk.eu
[2] University of Udine, 33100 Udine, Italy
falcon.alex@spes.uniud.it, giuseppe.serra@uniud.it

Abstract. Video Question Answering (VideoQA) is a task that requires a model to analyze and understand both the visual content given by the input video and the textual part given by the question, and the interaction between them in order to produce a meaningful answer. In our work we focus on the Egocentric VideoQA task, which exploits first-person videos, because of the importance of such task which can have impact on many different fields, such as those pertaining the social assistance and the industrial training. Recently, an Egocentric VideoQA dataset, called EgoVQA, has been released. Given its small size, models tend to overfit quickly. To alleviate this problem, we propose several augmentation techniques which give us a +5.5% improvement on the final accuracy over the considered baseline.

Keywords: Vision and language · Video Question Answering · Egocentric vision · Data augmentation

1 Introduction

Video Question Answering (VideoQA) is a task that aims at building models capable of providing a meaningful and coherent answer to a visual contents-related question, exploiting both spatial and temporal information given by the video data. VideoQA is receiving attention from both the Computer Vision and the Natural Language Processing communities, due to the availability of both textual and visual data which require to be jointly attended to in order to give the correct answer [7,8,28].

Recent advancements in the VideoQA task have also been achieved thanks to the creation of several public datasets, such as TGIF-QA [12] and MSVD-QA [28], which focus on web scraped video that are often recorded from a third-person perspective. Even more recently, Fan released in [6] EgoVQA, an Egocentric VideoQA dataset which provided the basis to study the importance of such task. In fact, several fields can benefit from advancements in the Egocentric VideoQA task: for example, the industrial training of workers, who may require help in understanding how to perform a certain task given what they

© Springer Nature Switzerland AG 2020
A. Bartoli and A. Fusiello (Eds.): ECCV 2020 Workshops, LNCS 12535, pp. 511–525, 2020.
https://doi.org/10.1007/978-3-030-66415-2_33

see from their own perspective; and the preventive medicine field, where Egocentric VideoQA makes it possible to identify sedentary and nutrition-related behaviours, and help elderly people prevent cognitive and functional decline by letting them review lifelogs [5]. Differently from third-person VideoQA, in the egocentric setting some types of questions can not be posed, such as those pertaining the camera wearer (*e.g.* "what am I wearing?"). Moreover, if the question asks to identify an item the camera wearer is playing with, hands occlusion may partially hide the item, making it hard to recognize.

Data augmentation techniques have proven particularly helpful in several Computer Vision tasks, such as image classification [18]. Not only they can be helpful to avoid overfitting and thus make the model more general, they can also be used to solve class imbalance in classification problems by synthesizing new samples in the smaller classes [24]. With respect to third-person VideoQA datasets, EgoVQA is a small dataset comprising around 600 question-answer pairs and the same number of clips. Since data augmentation is helpful in such contexts but to the best of our knowledge its effectiveness has never been systematically investigated for the Egocentric VideoQA nor for the VideoQA task, in this work we propose several data augmentation techniques which exploit characteristics given by the task itself. In particular, by exploiting the EgoVQA dataset [6] we show their impact on the final performance obtained by the ST-VQA model [12], which is proven to be effective in the study made by Fan.

The main contributions of this paper can be summarized as follows:

- we propose several data augmentation techniques which are purposefully designed for the VideoQA task;
- we show the usefulness of our proposed augmentation techniques on the recently released EgoVQA dataset and try to explain why we observe such improvements;
- we achieve a new state-of-the-art accuracy on the EgoVQA dataset;
- we will release code and pretrained models to support research in this important field.

The rest of the paper is organized as follows: in Sect. 2, we introduce the related work to the topics involved in this study, namely Egocentric VideoQA and data augmentation techniques; in Sect. 3, we detail both our proposed augmentation techniques and the architecture we use; Sect. 4 covers the experiments performed and the discussion of the results that we obtained; finally, Sect. 5 draws the conclusions of this study.

2   Related Work

In this section we will discuss the work related to the two main topics involved in our study, *i.e.* Video Question Answering, and data augmentation techniques.

2.1 VideoQA

Recently, VideoQA has received a lot of attention [6–8,12,28] from researchers both in Computer Vision and NLP fields. Several reasons can be related to this interest, such as the challenges offered by this task and the availability of several datasets, *e.g.* TGIF-QA [12], MSRVTT-QA [28], MSVD-QA [28], ActivityNet-QA [29], and TVQA+ [20], populated by many thousands of examples to learn from.

Modern approaches to this task involve a wide selection of different techniques. Jang *et al.* proposed in [12] to use both temporal attention and spatial attention, in order to learn which frames and which regions in each frame are more important to solve the task. Later on, attention mechanisms have been also used as a cross-modality fusion mechanism [11], and to learn QA-aware representations of both the visual and the textual data [16,20]. Because of the heterogeneous nature of the appearance and motion feature which are usually extracted from the video clips, Fan *et al.* [7] also propose to use memory modules, coupled with attention mechanisms, to compute a joint representation of these two types of features. Moreover, to compute the final answer for the given video and question there are multiple approaches. Simpler ones propose to use fully connected networks coupled with non-linear functions [12], but also more complex solutions have been proposed, *e.g.* based on reasoning techniques which exploit multiple steps LSTM-based neural networks [7] or graphs to better encode the relationships between the visual and textual data [11,13].

Finally, given the multitude of VideoQA datasets, there can also be multiple types of information to exploit. In fact, not only clips, questions, and answers are exploited to solve this task: as an example, TVQA+ [20] also provides subtitles and bounding boxes, by using which it is possible to improve the grounding capabilities of the VideoQA model in both the temporal and the spatial domain.

2.2 Egocentric VideoQA

On the other hand, Egocentric VideoQA was a completely unexplored field until very recently, when Fan released the EgoVQA dataset in [6]. Yet, considering the recent advancements in several fields of the egocentric vision, such as action recognition and action anticipation [3,4], Egocentric VideoQA also plays a primary role in the understanding of the complex interactions of the first-person videos.

Both VideoQA and Egocentric VideoQA usually deal with two main types of tasks: the "open-ended" and the "multiple choice" task [6,12,28,29]. Given a visual contents-related question, the difference between the two is due to how the answer is chosen: in the former, an answer set is generated from the most frequent words (*e.g.* top-1000 [28,29]) in the training set and the model needs to choose the correct answer from it, *i.e.* it is usually treated as a multi-class classification problem; in the latter the model needs to select the correct answer from a small pool of candidate answers (*e.g.* five choices [6,12]), which are usually different for every question. In this work we focus on the multiple choice task.

Together with the release of the EgoVQA dataset, Fan also provided in [6] a baseline made of four models borrowed from the VideoQA literature [7,8,12]. These models use the same backbone, which consists in a frozen, pretrained VGG-16 [25] to extract the frame-level features; a frozen, pretrained C3D [27] to extract the video-level features; and a pretrained GloVe [23] to compute the word embeddings. The four models can be seen as extensions of a basic encoder-decoder architecture (referred to as "ST-VQA without attention" in [6]): "ST-VQA with attention" is based on [12] and uses a temporal attention module to attend to the most important frames in the input clip; "CoMem" is based on [8] and involves the usage of two memory layers to generate attention cues starting from both the motion and appearance features; and finally "HME-VQA" [7] uses two heterogeneous memory layers and a multi-step reasoning module. In [6], Fan shows that these four models achieve similar performance despite the introduction of several cutting-edge modules. Because of this reason and because of its simplicity, in this work we focus on the "ST-VQA with attention" model.

2.3   Data Augmentation Techniques

Several Computer Vision tasks, such as image classification [18] and handwritten digit classification [19], have seen great improvements by exploiting data augmentation techniques, through which the size of the training set can be expanded artificially by several orders of magnitude. This leads to models which are far less susceptible to overfitting and more prone to give better results during the testing phase.

Whereas papers about first- or third-person VideoQA never mention any augmentation technique, in the VisualQA task, which deals with question-answering over images, there are some papers which try to tackle this opportunity by exploiting template-based models or generative approaches. Using a semantic tuple extraction pipeline, Mahendru *et al.* [22] extract from each question a premise, *i.e.* a tuple made of either an object, or an object and an attribute, or two objects and a relation between them, from which new question-answer pairs are constructed using previously built templates. Kafle *et al.* [14] proposes two techniques. The first is a template-based method which exploits the COCO dataset [21] and its segmentation annotations to generate new question-answer pairs of four different types, exploiting several different templates for each type. The second approach is based on sequentially generating new questions (and related answers) by conditioning an LSTM-based network on the image features from the "VQA dataset" [1]. Both these methods focus on creating new question-answer pairs for the same image, either by exploiting purely linguistic aspects or by using visual information to better guide the generation. Yet, as the authors report in the respective papers, these methods although reliable are not error-free [14,22]; on the other hand, our proposed techniques are simple yet effective and they do not raise issues. In particular, in our study we focus on exploiting both the visual and the textual data, although we do not create new questions: two of our techniques create new candidate answers for the same

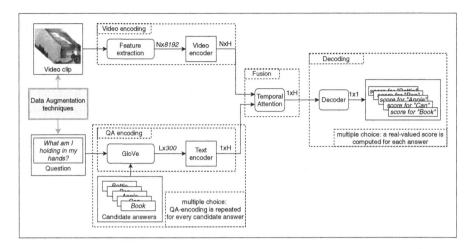

Fig. 1. Model used in our study. The feature extraction module is made of VGG and C3D, whose output consists of N frames and 8192 features. The output of GloVe consists in fixed-length vectors (*i.e.* embeddings) of embedding size $E = 300$. The Video Encoder and the Text Encoder have a similar structure, but whereas the output of the former is a sequence (length N) of hidden states, the output of the latter is a single hidden state. The Decoder outputs a single real-valued score for each candidate answer.

question to strengthen the understanding of the concepts contained in the question and to better distinguish between the correct and the wrong answers, and the other technique creates "new" clips by horizontally flipping the frames and consistently updating both the question and the candidate answers accordingly.

3 Methodology

In this section we will introduce and describe the proposed augmentation techniques. Moreover, we will also discuss and describe the model used in our study, which is called "ST-VQA" and was initially introduced by Jang *et al.* in [12].

3.1 Augmentation Techniques

Following the work made in [6], we are working on the multiple choice setting. For each video and question five candidate answers are provided, of which only one is correct. The wrong answers are randomly sampled from a candidate pool based on the question type, *i.e.* if the question requires to recognize an action, the five candidate answers (both the right one and the four wrong) are actions. By doing so, the model is encouraged to understand the visual contents in order to reply to the question, avoiding the exploitation of pure textual information (*e.g.* exploiting the question type to filter out some of the candidate answers).

We propose to use three simple augmentation techniques designed for the VideoQA task and which exploit the multiple choice setting: resampling, mirroring, and horizontal flip. This is not only helpful when dealing with the overfitting, but can also give the model a better understanding of what the questions is asking for and make the model more robust to variations in the input frames.

Resampling. Given a question Q, in the multiple choice setting a handful (*e.g.* 5 choices [6,12]) of candidate answers are considered. The first technique consists in fixing the correct answer and then *resampling* the wrong ones. By doing so, using the same video and question, we can show the model several more examples of what is *not* the correct answer. This should give the model the ability to better distinguish what the question is and is not asking for. An example is shown in Fig. 2.

Considering that the amount of possible tuples of wrong answers is exponentially big, we are restricting the pool of wrong answers to those pertaining the same question type of Q. Moreover, we are not considering all the possible tuples in the pool.

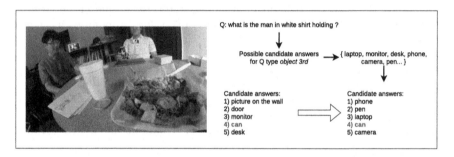

Fig. 2. Example of the "resampling" technique applied to a video clip in the EgoVQA dataset.

Mirroring. Given a question, it may be that the correct answer in the rows of the dataset is often placed in the same position. This can create biases in the model which may tend to prefer an answer simply based on its position (w.r.t. the order of the candidates). To relieve some of this bias we propose the *mirroring* technique, which consists in simply adding a row to the dataset where the order of the candidate answers (and the label value) is mirrored. An example is shown in Fig. 3.

Horizontal Flip. One of the most common image data augmentation techniques consists in horizontally flipping the images, which often improves the model performance thanks to the availability of newly created images which are

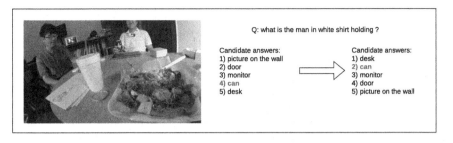

Fig. 3. Example of the "mirroring" technique applied to a video clip in the EgoVQA dataset.

taken both from the left and from the right. This technique may prove useful in a VideoQA setting as well, but it should not be applied lightly because it is a non-label preserving transformation: horizontally flipping the considered frame means that an object which was on the left side of the frame appears on the right side after the transformation, and viceversa, eventually creating wrong labels if not updated correctly. Thus, when flipping the frames in the video clip both the question and the candidate answers likely need to be updated (*e.g.* Fig. 4).

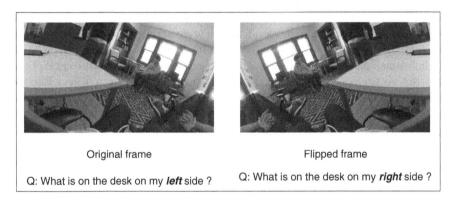

Fig. 4. Example of the "horizontal flip" technique applied to a frame of a video clip in the EgoVQA dataset.

3.2 QA Encoding: Word Embedding and Text Encoder

As shown in Fig. 1, it can be seen as made of four blocks: Question-Answer (QA) Encoding, Video Encoding, Fusion, and Decoding. To compute the word embeddings for the question and the answers, we consider GloVe [23], pretrained on the Common Crawl dataset[1], which outputs a vector of size $E = 300$ for

---

[1] The Common Crawl dataset is available at http://commoncrawl.org.

each word in both the question and the answers. Since GloVe is not contextual, question and answer can be given in input to the model either separately or jointly obtaining the same embedding.

First of all, the question and the candidate answer are tokenized, *i.e.* they are split in sub-word tokens and then each of them receives an identifier, based on the vocabulary used by GloVe. Let $q_1 \ldots q_m$ and $a_1 \ldots a_n$ be the sequence of m words of the question and n words of (one of the candidate) answer, and let $L = m + n$. Thus, let $\phi_q \in \mathbb{R}^{m \times E}$ be the question embedding, and $\phi_a \in \mathbb{R}^{n \times E}$ be the answer embedding. The final question-answer embedding is computed as their concatenation, *i.e.* $\phi_w = [\phi_q, \phi_a] \in \mathbb{R}^{L \times E}$.

Then the Text Encoder, consisting of two stacked LSTM networks, is applied to ϕ_w. By concatenating the *last* hidden state of both the LSTM networks we obtain the encoded textual features $\epsilon_w \in \mathbb{R}^{1 \times H}$, where H is the hidden size.

3.3 Video Encoding

From each input video clip, both motion and appearance features are obtained in the Video Encoding module. In particular, the appearance features are computed as the *fc7* activations ($\phi_a \in \mathbb{R}^{N \times 4,096}$) extracted from a frozen VGG-16 [25], pretrained on ImageNet [18]. We use VGG because we want to keep the positional information extracted by the convolutional layers, which would be otherwise lost in deeper networks, such as ResNet [9], that exploit a global pooling layer before the FC layers. Similarly, the motion features are computed as the *fc7* activations ($\phi_m \in \mathbb{R}^{N \times 4,096}$), extracted from a frozen C3D [27], pretrained on Sports1M [15] and fine-tuned on UCF101 [26]. Finally we concatenate these features and obtain a feature vector $\phi_{a,m} \in \mathbb{R}^{N \times 8,192}$, which is then encoded by a Video Encoder module, consisting of two stacked LSTM networks. The only difference between the Text and Video Encoder module is that the output ϵ_v of the latter consists in the concatenation of the full sequence of hidden states from both the networks, and not only the last hidden state. Thus $\epsilon_v \in \mathbb{R}^{N \times H}$ represents the encoded video features.

3.4 Fusion

Depending on the question (and eventually the candidate answer), a frame may be more or less relevant. To try and exploit this information, the fusion block consists of a temporal attention module that lets the model learn automatically which frames are more important based on both the encoded video features and the textual features. In particular, the temporal attention module is based on the works by Bahdanau *et al.* [2] and by Hori *et al.* [10]. It receives in input the encoded video features ϵ_v and the encoded textual features ϵ_w, and can be described by the following equations:

$$\omega_s = tanh(\epsilon_v W_v + \epsilon_w W_w + b_s)W_s \tag{1}$$

$$\alpha = softmax(\omega_s) \tag{2}$$

$$\omega_a = \mathbb{1}(\alpha \circ \epsilon_v) \tag{3}$$

where $W_v, W_w \in \mathbb{R}^{H \times h}$, $W_s \in \mathbb{R}^h$ are learnable weight matrices, $b_s \in \mathbb{R}^{1 \times h}$ is a learnable bias. $\mathbb{1}$ is a row of ones $(1^{1 \times N})$. $\circ$ represents the element-wise multiplication operator. The output of the Fusion module is a feature vector $\omega_a \in \mathbb{R}^{1 \times H}$.

3.5 Decoding

Finally, the decoding step considers both the attended features computed by the Fusion block and the encoded textual features, as proposed by Fan [6]. In our multiple choice setting, the decoding is performed five times, *i.e.* for each Q-A pair, with different textual features producing five different scores, one per candidate answer. It can be described by the following equations:

$$d_f = tanh(\omega_a W_a + b_a) \tag{4}$$

$$d_r = (d_f \circ \epsilon_w)W_d + b_d \tag{5}$$

where $W_a \in \mathbb{R}^{H \times H}$ and $W_d \in \mathbb{R}^{H \times 1}$ are learnable weight matrices, $b_a \in \mathbb{R}^{1 \times H}$ and $b_d \in \mathbb{R}$ are learnable biases, $d_f \in \mathbb{R}^{1 \times H}$, $d_r \in \mathbb{R}$. d_r can be seen as the score obtained by testing a specific candidate answer (out of the five possible choices related to the given question).

3.6 Loss Function

The model is trained using a pairwise hinge loss, as is done in [6,12]. The loss function can be defined as follows:

$$\mathcal{L}_{c,r} = \begin{cases} 0 & \text{if } c = r \\ max(0, 1 + s_n - s_p) & \text{if } c \neq r \end{cases} \tag{6}$$

where s_n and s_p are respectively the scores d_f computed by the decoder for the choice $c \in \{1, \dots, 5\}$ and the right answer r.

$$\mathcal{L} = \sum_{q \in \mathcal{Q}} \sum_{c=1}^{5} \mathcal{L}_{c,r} \tag{7}$$

Here $\mathcal{Q}$ is the set of the questions, and r is the right answer for the question q.

4 Results

In this section we briefly describe the dataset used to perform the experiments, and we discuss both the overall results and the per question type results.

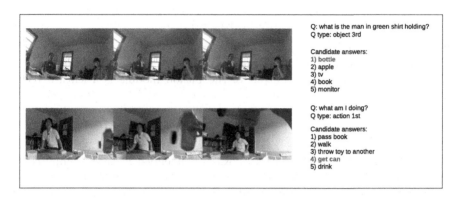

Q: what is the man in green shirt holding?
Q type: object 3rd

Candidate answers:
1) bottle
2) apple
3) tv
4) book
5) monitor

Q: what am I doing?
Q type: action 1st

Candidate answers:
1) pass book
2) walk
3) throw toy to another
4) get can
5) drink

Fig. 5. Samples of video clips, questions, and candidate answers from the EgoVQA dataset.

Table 1. Description of the question types available for testing in the EgoVQA dataset.

Code	Question type	Quantity	Example
Act_{1st}	Action 1st	67	"what am I doing"
Act_{3rd}	Action 3rd	108	"what is the man in red clothes doing"
Obj_{1st}	Object 1st	54	"what am I holding in my hands"
Obj_{3rd}	Object 3rd	86	"what is placed on the desk"
Who_{1st}	Who 1st	13	"who am I talking with"
Who_{3rd}	Who 3rd	63	"who is eating salad"
Cnt	Count	64	"how many people am I talking with"
Col	Color	31	"what is the color of the toy in my hands"

4.1 EgoVQA Dataset

The EgoVQA dataset was recently presented by Fan [6]. It features more than 600 QA pairs and the same number of clips, which are 20–100 s long and are obtained by 16 egocentric videos (5–10 min long) based on 8 different scenarios. An example of these egocentric videos and QA pairs can be seen in Fig. 5. The questions can be grouped in eight major types and they are described in Table 1.

4.2 Implementation Details

In our setting, we fixed $H = 512$ and $h = 256$. To optimize the parameters we used the Adam [17] optimizer with a fixed learning rate of 10^{-3} and a batch size of 8. To implement our solution we used Python 2.7, Numpy 1.16, and PyTorch 1.4. A PyTorch implementation will be made available to further boost the research in this important area at https://github.com/aranciokov/EgoVQA-DataAug.

4.3   Discussion of the Results

Table 2 shows the results obtained for each of the three splits proposed in [6] by applying, with different combinations, our proposed augmentation techniques. Table 3 shows the results based on the question type, whose details (and codes, such as "Act_{1st}" and "Act_{3rd}") are defined in Table 1.

Table 2. Per-question type results obtained by applying the proposed techniques on the EgoVQA dataset.

Augmentation	Accuracy (%) on split			
	0	1	2	Avg
ST-VQA [12]	31.82	37.57	27.27	32.22
+ mirroring	32.58	40.46	23.53	32.19
+ resampling	26.52	28.90	29.41	28.28
+ mirroring	37.88	36.42	**30.48**	34.93
+ horizontal-flip	34.09	41.62	25.13	33.61
+ resampling	37.12	35.26	25.67	32.68
+ mirroring	**40.91**	**43.35**	28.88	**37.71**

Overall it can be seen that, when used in conjunction, the proposed augmentation techniques help improving the performance obtained by the considered model.

Looking at the results per question type, it is possible to notice that:

- the "resampling" technique is particularly helpful when it comes to counting objects ("Cnt") and identifying objects used by actors in front of the camera wearer ("Obj_{3rd}");
- the "mirroring" technique shows sensible improvements during the identification of actors, both when the camera wearer is interacting with them ("Who_{1st}") and when they are performing certain actions in front of the camera wearer itself ("Who_{3rd}").
- the "horizontal flip" technique is especially helpful when the model needs to identify the actions performed by ("Act_{1st}") and the objects over which the action is performed by the camera wearer ("Obj_{1st}"). Moreover, it gives the model a great boost in recognizing colors ("Col").

Both in the questions of type "Cnt" and "Obj_{3rd}" the model is required to recognize an object: in fact, whereas in the latter the model needs to identify an object by distinguishing among the five candidates, the former also requires the model to understand what such object is in order to count how many times it occurs in the scene. It is interesting to notice that in our "resampling" technique we are not augmenting the questions of type "Cnt", because the only five possible candidate answers in the dataset for such question type are the numbers from

Table 3. Per-question type results obtained by applying the proposed techniques on the EgoVQA dataset.

Augmentation	Question type accuracy (%)							
	Act_{1st}	Act_{3rd}	Obj_{1st}	Obj_{3rd}	Who_{1st}	Who_{3rd}	Cnt	Col
ST-VQA [12]	28.36	30.56	31.48	31.40	46.15	34.92	35.94	32.26
+ mirroring	26.87	33.33	35.19	27.91	**53.85**	46.03	26.56	19.35
+ resampling	26.87	26.85	20.37	37.21	15.38	23.81	**43.75**	16.13
+ mirroring	25.37	34.26	25.93	41.86	15.38	**47.62**	42.19	19.35
+ horizontal-flip	**40.30**	36.11	**42.59**	25.58	38.46	28.57	26.56	**41.94**
+ resampling	34.33	37.96	22.22	33.72	15.38	33.33	29.69	29.03
+ mirroring	31.34	**39.81**	22.22	**44.19**	15.38	**47.62**	37.50	38.71

"one" to "five". Thus, since we are able to observe this improvement in both these question types, it likely implies that such data augmentation technique helps the model to better distinguish among the different objects available in the dataset because it provides several more examples where the model needs to understand which object is the right one among several (wrong) candidate answers.

In the case of the question types "Who_{3rd}" and "Act_{3rd}" the improved performance may be due to two aspects: first of all, since they both require to recognize actions performed by actors in front of the camera wearer, the accuracy gain obtained in one type transfers (to some extent) to the other type, and viceversa; and then to the "mirroring" technique, since it is possible to observe that in the training set there is a bias in both question types towards one of the last two labels. In particular, over the three training splits, the last candidate answer is the correct one 60 times over 203 questions (29.55%) of type "Act_{3rd}", whereas "Who_{3rd}" counts 27 instances of the second-to-last candidate answer over a total of 77 questions (35.06%). Using the "mirroring" technique it is thus possible to reduce this bias, making the model more robust. It is interesting to notice that in both these question types, the addition of the "horizontal flip" technique gives a further boost in the accuracy of the model. This is likely related to the fact that several questions in the training data also contain a positional information ("left", "right") of the actor involved: in particular, for the type "Act_{3rd}" there are respectively 16 and 20 questions mentioning "left" or "right" over a total of 203 questions, whereas for "Who_{3rd}" there are respectively 2 and 3 over a total of 77.

In the question type "Who_{1st}" we can observe a sensible improvement with the "mirroring" technique. Although the reason are likely similar (considering that the first candidate answer is the right one 11 times over 20 instances for "Who_{1st}"), we prefer not to make any conclusive claim given that there is only a total of 20 instances in the training set and 13 in the testing set.

The "horizontal flip" technique shines when asked to recognize which object the camera wearer is interacting with ("Obj_{1st}") and to identify which action

($"Act_{1st}"$) is performed by the camera wearer itself. The improvement over the former question type may be explained by the fact that several of its questions in the training data involve a positional information: in particular, there are respectively 18 and 9 questions containing "left" or "right" over a total of 86 questions. On the other hand, the great improvement in the latter question type (whose questions are almost all of the form "what am I doing") is likely justifiable by the greater amount of different visual data available for training.

Finally, among our proposed techniques, only the "horizontal flip" seems to cope well with "Col" questions. This question type is particularly tough because it requires the model to recognize the object which the question is referring to, the action which is performed over the object (35/49 total instances), and sometimes even the colors of the clothes of the actor (*e.g.* "what is the color of the cup held by the man in *black* jacket", 13/49 total instances). First of all, the "mirroring" technique does not help: the training split are slightly biased towards the first and the last labels (respectively, 10 and 16 over 49 instances), meaning that in this case the proposed technique does not resolve the bias towards these two labels. Secondly, our "resampling" technique is not helping because there are only six unique colors in the dataset, thus it is not creating enough new rows. Thirdly, only 2 over a total of 54 questions of this type in the training data contain "left" or "right", likely implying that the improvement obtained by the "horizontal flip" technique is due to having more visual data which forces the model to better understand where to look for the object targeted by the question.

5   Conclusion

Egocentric VideoQA is a task introduced recently in [6] which specializes the VideoQA task in an egocentric setting. It is a challenging task where a model needs to understand both the visual and the textual content of the question, and then needs to jointly attend to both of them in order to produce a coherent answer. In this paper we propose several data augmentation techniques purposefully designed for the VideoQA task. The "mirroring" technique tries to partially remove the ordering bias in the multiple choice setting. The "resampling" technique exploits the training dataset to create new question-answer pairs by substituting the wrong candidate answers with different candidates from the same question type, in order to feed the network with more examples of what is not the target of the question. Finally, the "horizontal-flip" technique exploits both the visual and the textual content of each row in the dataset, and aims at giving the model the ability to differentiate between "left" and "right". To show the effectiveness of these techniques, we test them on the recently released EgoVQA dataset and show that we are able to achieve a sensible improvement (+5.5%) in the accuracy of the model.

As a future work, we are both considering to explore our proposed augmentation techniques with other architectures, such as the HME-VQA model [7], and to replicate these experiments in third-person VideoQA datasets. Moreover, we are considering several different augmentation techniques that deal with the linguistic aspects and the visual information both separately and jointly. In particular, we think that considering them jointly is of most interest because of the inherent characteristics of the problem setting, which requires the model to understand both linguistic and visual clues together. A purely linguistics technique which we plan to explore consists in a variation of the "mirroring" technique which *permutes* the candidate answers instead of simply mirroring them: this should reduce the ordering bias in all the possible situations, even those where the "mirroring" technique is weaker. Then, considering that the egocentric camera may not be aligned at all times due to the camera wearer moving in the scenario, a simple technique which might help consists in applying small rotations to the video clips, similarly to what is done in image tasks. Finally, reversing the video clips and updating both question and the candidate answers accordingly (*e.g.* by "reversing" the name of the actions performed in the video clip) may give the model a more clear understanding of the actions, while better exploiting the sequential nature of the visual data.

References

1. Antol, S., et al.: VQA: visual question answering. In: Proceedings of the IEEE International Conference on Computer Vision, pp. 2425–2433 (2015)
2. Bahdanau, D., Cho, K., Bengio, Y.: Neural machine translation by jointly learning to align and translate. In: ICLR (2015)
3. Damen, D., et al.: Rescaling egocentric vision. CoRR abs/2006.13256 (2020)
4. Damen, D., et al.: The epic-kitchens dataset: Collection, challenges and baselines. IEEE Trans. Pattern Anal. Mach. Intell. (TPAMI) (2020). https://doi.org/10.1109/TPAMI.2020.2991965
5. Doherty, A.R., et al.: Wearable cameras in health: the state of the art and future possibilities. Am. J. Prev. Med. **44**(3), 320–323 (2013)
6. Fan, C.: EgoVQA -an egocentric video question answering benchmark dataset. In: ICCV Workshop (2019)
7. Fan, C., Zhang, X., Zhang, S., Wang, W., Zhang, C., Huang, H.: Heterogeneous memory enhanced multimodal attention model for video question answering. In: CVPR (2019)
8. Gao, J., Ge, R., Chen, K., Nevatia, R.: Motion-appearance co-memory networks for video question answering. In: CVPR (2018)
9. He, K., Zhang, X., Ren, S., Sun, J.: Deep residual learning for image recognition. In: Proceedings of the IEEE Conference on Computer Vision and Pattern Recognition, pp. 770–778 (2016)
10. Hori, C., et al.: Attention-based multimodal fusion for video description. In: ICCV (2017)
11. Huang, D., Chen, P., Zeng, R., Du, Q., Tan, M., Gan, C.: Location-aware graph convolutional networks for video question answering. In: AAAI, pp. 11021–11028 (2020)
12. Jang, Y., Song, Y., Yu, Y., Kim, Y., Kim, G.: TGIF-QA: toward spatio-temporal reasoning in visual question answering. In: CVPR (2017)
13. Jiang, P., Han, Y.: Reasoning with heterogeneous graph alignment for video question answering. In: AAAI, pp. 11109–11116 (2020)

14. Kafle, K., Yousefhussien, M., Kanan, C.: Data augmentation for visual question answering. In: Proceedings of the 10th International Conference on Natural Language Generation, pp. 198–202 (2017)
15. Karpathy, A., Toderici, G., Shetty, S., Leung, T., Sukthankar, R., Fei-Fei, L.: Large-scale video classification with convolutional neural networks. In: CVPR (2014)
16. Kim, H., Tang, Z., Bansal, M.: Dense-caption matching and frame-selection gating for temporal localization in VideoQA. arXiv preprint arXiv:2005.06409 (2020)
17. Kingma, D.P., Ba, J.: Adam: a method for stochastic optimization. In: ICLR (2015)
18. Krizhevsky, A., Sutskever, I., Hinton, G.E.: ImageNet classification with deep convolutional neural networks. In: NeurIPS (2012)
19. LeCun, Y., Bottou, L., Bengio, Y., Haffner, P.: Gradient-based learning applied to document recognition. Proc. IEEE **86**(11), 2278–2324 (1998)
20. Lei, J., Yu, L., Berg, T.L., Bansal, M.: TVQA+: spatio-temporal grounding for video question answering. arXiv preprint arXiv:1904.11574 (2019)
21. Lin, T.-Y., et al.: Microsoft COCO: common objects in context. In: Fleet, D., Pajdla, T., Schiele, B., Tuytelaars, T. (eds.) ECCV 2014. LNCS, vol. 8693, pp. 740–755. Springer, Cham (2014). https://doi.org/10.1007/978-3-319-10602-1_48
22. Mahendru, A., Prabhu, V., Mohapatra, A., Batra, D., Lee, S.: The promise of premise: harnessing question premises in visual question answering. In: Proceedings of the 2017 Conference on Empirical Methods in Natural Language Processing, pp. 926–935 (2017)
23. Pennington, J., Socher, R., Manning, C.D.: GloVe: global vectors for word representation. In: EMNLP (2014)
24. Shorten, C., Khoshgoftaar, T.M.: A survey on image data augmentation for deep learning. J. Big Data **6**(1), 60 (2019)
25. Simonyan, K., Zisserman, A.: Very deep convolutional networks for large-scale image recognition. In: ICLR (2015)
26. Soomro, K., Zamir, A.R., Shah, M.: UCF101: a dataset of 101 human actions classes from videos in the wild. arXiv:1212.0402 (2012)
27. Tran, D., Bourdev, L., Fergus, R., Torresani, L., Paluri, M.: Learning spatiotemporal features with 3D convolutional networks. In: ICCV (2015)
28. Xu, D., et al.: Video question answering via gradually refined attention over appearance and motion. In: ACM Multimedia (2017)
29. Yu, Z., et al.: ActivityNet-QA: a dataset for understanding complex web videos via question answering. In: Proceedings of the AAAI Conference on Artificial Intelligence, vol. 33, pp. 9127–9134 (2019)

W05 - Eye Gaze in VR, AR, and in the Wild

W05 - Eye Gaze in VR, AR, and in the Wild

With the advent of consumer products, AR and VR as a form of immersive technology is gaining mainstream attention. However, immersive technology is still in its infancy, as both users and developers figure out the right recipe for the technology to garner mass appeal. Eye tracking, a technology that measures where an individual is looking and can enable inference of user attention, could be a key driver of mass appeal for the next generation of immersive technologies, provided user awareness and privacy related to eye-tracking features are taken into account.

Open forums of discussion provide opportunities to further improve eye tracking technology, especially in areas like scale and generalization challenges in the next generation of AR and VR systems. For that reason, Facebook organized the first challenge "Eye Tracking for VR and AR (OpenEDS)" at the ICCV 2019 and the independent GAZE committee organized a workshop titled "Gaze Estimation and Prediction in the Wild (GAZE)".

For 2020, the Facebook and GAZE committees partnered to host a joint workshop titled "Eye Gaze in VR, AR, and in the Wild" at the biennial European Conference on Computer Vision (ECCV 2020) conference. The workshop hosted two tracks: the first focused on gaze estimation and prediction methods, with a focus on accuracy and robustness in natural settings (in-the-wild); the second track focused on the scale and generalization problem for eye tracking systems operating on AR and VR platforms.

August 2020

Hyung Jin Chang
Seonwook Park
Xucong Zhang
Otmar Hilliges
Aleš Leonardis
Robert Cavin
Cristina Palmero
Jixu Chen
Alexander Fix
Elias Guestrin
Oleg Komogortsev
Kapil Krishnakumar
Abhishek Sharma
Yiru Shen
Tarek Hefny
Karsten Behrendt
Sachin S. Talathi

Efficiency in Real-Time Webcam Gaze Tracking

Amogh Gudi[1,2(✉)], Xin Li[1,2], and Jan van Gemert[2]

[1] Vicarious Perception Technologies [VicarVision], Amsterdam, The Netherlands
{amogh,xin}@vicarvision.nl
[2] Delft University of Technology [TU Delft], Delft, The Netherlands
j.c.vangemert@tudelft.nl

Abstract. Efficiency and ease of use are essential for practical applications of camera based eye/gaze-tracking. Gaze tracking involves estimating where a person is looking on a screen based on face images from a computer-facing camera. In this paper we investigate two complementary forms of efficiency in gaze tracking: 1. The computational efficiency of the system which is dominated by the inference speed of a CNN predicting gaze-vectors; 2. The usability efficiency which is determined by the tediousness of the mandatory calibration of the gaze-vector to a computer screen. To do so, we evaluate the computational speed/accuracy trade-off for the CNN and the calibration effort/accuracy trade-off for screen calibration. For the CNN, we evaluate the full face, two-eyes, and single eye input. For screen calibration, we measure the number of calibration points needed and evaluate three types of calibration: 1. pure geometry, 2. pure machine learning, and 3. hybrid geometric regression. Results suggest that a single eye input and geometric regression calibration achieve the best trade-off.

1 Introduction

In a typical computer-facing scenario, the task of gaze-tracking involves estimating where a subject's gaze is pointing based on images of the subject captured via the webcam. This is commonly in the form of a gaze vector, which determines the pitch and yaw of the gaze with respect to the camera [30]. A more complete form of gaze tracking further extends this by also computing at which specific point the subject is looking at on a screen in front of the subject [11,28]. This is achieved by estimating the position of the said screen w.r.t. the camera (a.k.a. screen calibration), which is not precisely known beforehand. We present a study of some core choices in the design of gaze estimation methods in combination with screen calibration techniques (see Fig. 1), leaning towards an efficient real-time camera-to-screen gaze-tracking system.

Computational Efficiency: Input Size. Deep networks, and CNNs in particular, improved accuracy in gaze estimation where CNN inference speed is to

A. Bartoli and A. Fusiello (Eds.): ECCV 2020 Workshops, LNCS 12535, pp. 529–543, 2020.
https://doi.org/10.1007/978-3-030-66415-2_34

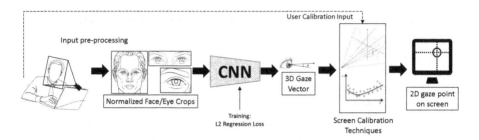

Fig. 1. An overview illustration of our camera-to-screen gaze tracking pipeline under study. (Left to right) Images captured by the webcam are first pre-processed to create a normalized image of face and eyes. These images are used for training a convolutional neural network (using L2 loss) to predict the 3D gaze vector. With the cooperation of the user, the predicted gaze vectors can finally be projected on to the screen where he/she is looking using proposed screen calibration techniques.

a large extent determined by the input image size. The input size for gaze estimation can vary beyond just the image of the eye(s) [16,30], but also include the whole eye region [28], the whole face, and even the full camera image [11]. Yet, the larger the input image, the slower the inference speed. We study the impact of various input types and sizes with varying amounts of facial contextual information to determine their speed/accuracy trade-off.

Usability Efficiency: Manual Effort in Screen Calibration. Most work focus on gaze vectors estimation [2,5,13,15,27]. However, predicting the *gaze-point*, point on a screen in front of the subject where he/she is looking, is a more intuitive and directly useful result for gaze tracking based applications, especially in a computer-facing/human-computer interaction scenario. If the relative locations and pose of the camera w.r.t to the screen were exactly known, projecting the gaze-vector to a point on screen would be straightforward. However, this transformation is typically not known in real-world scenarios, and hence must also be implicitly or explicitly estimated through an additional calibration step. This calibration step needs to be repeated for every setup. Unlike gaze-vector prediction, you cannot have a "pre-trained" screen calibration method. In practice, every-time a new eye-tracking session starts, the first step would be to ask the user to look at and annotate predefined points for calibration. Therefore, obtaining calibration data is a major usability bottleneck since it requires cooperation of the user every time, which in practice varies. Here, we study usability efficiency as a trade-off between the number of calibration points and accuracy.

We consider three types of calibration. Geometry based modelling methods have the advantage that maximum expert/geometrical prior knowledge can be embedded into the system. On the other hand, such mathematical models are rigid and based on strong assumptions, which may not always hold. In contrast, calibration methods based on machine learning require no prior domain knowledge and limited hand-crafted modelling. However, they may be more data-dependent in order to learn the underlying geometry. In this paper, we evaluate the efficiency trade-off of various calibration techniques including a hybrid approach between machine learning regression and geometric modelling.

Contributions. We have the following three contributions:

(i) We evaluate computational system efficiency by studying the balance of gains from context-rich inputs vs their drawbacks. We study their individual impact on the system's accuracy w.r.t. their computational load to determine their efficiency and help practitioners find the right trade-off.
(ii) We demonstrate three practical screen calibration techniques that can be use to convert the predicted gaze-vectors to points-on-screen, thereby performing the task of complete camera-to-screen gaze tracking.
(iii) We evaluate the usability efficiency of these calibration methods to determine how well they utilize expensive user-assisted calibration data. This topic has received little attention in literature, and we present one of the first reports on explicit webcam-based screen calibration methods.

2 Related Work

Existing methods for gaze tracking can be roughly categorized into model-based and appearance-based methods. The former [25,26] generates a geometric model for eye to predict gaze, while the latter [23] makes a direct mapping between the raw pixels and the gaze angles. Appearance driven methods have generally surpassed classical model-based methods for gaze estimation.

Appearance-Based CNN Gaze-Tracking. As deep learning methods have shown their potentials in many areas, some appearance-based CNN networks are shown to work effectively for the task of gaze prediction.

Zhang et al. [29,30] proposed the first deep learning model for appearance-based gaze prediction. Park et al. [16] proposed a combined hourglass [12] and DenseNet [7] network to take advantage of auxiliary supervision based on the gaze-map, which is two 2D projection binary mask of the iris and eyeball. Cheng et al. [3] introduced ARE-Net, which is divided into two smaller modules: one is to find directions from each eye individually, and the other is to estimate the reliability of each eye. Deng and Zhu et al. [4] define two CNNs to generate head and gaze angles respectively, which are aggregated by a geometrically constrained transform layer. Ranjan et al. [17] clustered the head pose into different groups and used a branching structure for different groups. Chen et al. [2] proposed Dilated-Nets to extract high level features by adding dilated convolution. We build upon these foundations where we evaluate the speed vs accuracy trade-off in a real-time setting. The image input size has a huge effect on processing speed, and we control the input image size by varying eye/face context.

The seminal work of Zhang *et al.* [29,30] utilized minimal context by only using the grayscale eye image and head pose as input. Krafka *et al.* [11] presented a more context-dependent multi-model CNN to extract information from two single eye images, face image and face grid (a binary mask of the face area in an image). To investigate how the different face region contributes to the gaze prediction, a full-face appearance-based CNN with spatial weights was proposed [28]. Here, we investigate the contribution of context in the real-time setting by explicitly focusing on the speed/accuracy trade-off.

A GPU based real-time gaze tracking method was presented in [5]. This was implemented in a model ensemble fashion, taking two eye patches and head pose vector as input, and achieved good performance on several datasets [5,22,30] for person-independent gaze estimation. In addition, [2,13] have included some results about the improvements that can be obtained from different inputs. In our work, we perform an ablation study and add the dimension of computation load of each input type. Our insight in the cost vs benefit trade-off may help design efficient gaze tracking software that can run real-time beyond expensive GPUs, on regular CPUs which have wider potential in real world applications.

Screen Calibration: Estimating Point-of-gaze. In a classical geometry-based model, projecting any gaze-vector to a point on a screen requires a fully-calibrated system. This includes knowing the screen position and pose in the camera coordinate system. Using a mirror-based calibration technique [18], the corresponding position of camera and screen can be attained. This method needs to be re-applied for different computer and camera setting, which is non-trivial and time-consuming. During human-computer interactions, information like mouse clicks may also provide useful information for screen calibration [14]. This is, however, strongly based on the assumption that people are always looking at the mouse cursor during the click.

Several machine learning models are free of rigid geometric modelling while showing good performance. Methods like second order polynomial regression [9] and Gaussian process regression [24] have been applied to predict gaze more universally. WebGazer [14] trains regression models to map pupil positions and eye features to 2D screen locations directly without any explicit 3D geometry. As deep learning features have shown robustness in different areas, other inputs can be mixed with CNN-based features for implicit calibration, as done in [11,28]. CNN features from the eyes and face are used as inputs to a support vector regressor to directly estimate gaze-point coordinates without an explicit calibration step. These methods take advantage of being free of rigid modelling and show good performance. On the other hand, training directly on CNN features makes this calibration technique non-modular since it is designed specific to a particular gaze-prediction CNN. In our work, we evaluate data-efficiency for modular screen calibration techniques that convert gaze-vectors to gaze-points based on geometric modelling, machine learning, and a mix of geometry and regression. We explicitly focus on real world efficiency which for calibration is not determined by processing speed, but measured in how many annotations are required to obtain reasonable accuracy.

3   Setup

The pipeline contains three parts, as illustrated in Fig. 1:

1. Input pre-processing by finding and normalizing the facial images;
2. A CNN that takes these facial images as input to predict the gaze vector;
3. Screen calibration and converting gaze-vectors to points on the screen.

3.1   Input Pre-processing

The input to the system is obtained from facial images of subjects. Through a face finding and facial landmark detection algorithm [1], the face and its key parts are localized in the image. Following the procedure described by Sugano et al. [22], the detected 2D landmarks are fitted onto a 3D model of the face. This way, the facial landmarks are roughly localized in the 3D camera coordinate space. By comparing the 3D face model and 2D landmarks, the head rotation matrix $\mathbf{R}$ and translation vector $\mathbf{T}$, and the 3D eye locations $\mathbf{e}$ are obtained in 3D camera coordinate space. A standardized view of the face is now obtained by defining a fixed distance d between the eye centres and the camera centre and using a scale matrix $\mathbf{S} = \text{diag}(1, 1, \frac{d}{||\mathbf{e}||})$. The obtained conversion matrix $\mathbf{M} = \mathbf{S} \cdot \mathbf{R}$ is used to apply perspective warping to obtain a normalized image without roll (in-plane rotation). For training, the corresponding ground truth vector $\mathbf{g}$ is similarly transformed: $\mathbf{M} \cdot \mathbf{g}$.

3.2   CNN Prediction of Gaze Vectors

We use a VGG16 [20] network architecture with BatchNorm [8] to predict the pitch and yaw angles of the gaze vector with respect to the camera from the normalized pre-processed images.

Training. Following the prior work in[30], the network was pre-trained on ImageNet [19]. For all the experiments conducted in this work, we set the following hyperparameters for the training of the network for gaze-vector prediction:

(i) Adam optimizer with default settings [10];
(ii) a validation error based stopping criteria with a patience of 5 epochs;
(iii) learning rate of 10^{-5}, decaying by 0.1 if validation error plateaus;
(iv) simple data augmentation with mirroring and gaussian noise ($\sigma = 0.01$).

Inference. This trained deep neural network can now make prediction of the gaze vector. The predicted gaze-vector (in the form of pitch and yaw angles) are with respect to the 'virtual' camera corresponding to the normalized images. The predicted virtual gaze vectors can be transformed back to the actual gaze vector with respect to the real camera using the transformation parameters obtained during image pre-processing. These vectors can then be projected onto a point on the screen after screen calibration.

3.3 Screen Calibration: Gaze Vectors to Gaze Points

To project the predicted 3D gaze vectors (in the camera coordinate space) to 2D gaze-points on a screen, the position of the screen with respect to the camera must be known which is difficult to obtain in real world settings. The aim of screen calibration is to estimate this geometric relation between the camera and the screen coordinate systems such that the predicted gaze vectors in camera coordinates are calibrated to gaze-points in screen coordinates. Because we focus on the task of eye-tracking in a computer-facing scenario, we can simplify the setup by making some assumptions based on typical webcam-monitor placement (such as for built-in laptop webcams or external webcams mounted on monitors):

  (i) the roll and yaw angles between the camera and the screen are 0°,
 (ii) the intrinsic camera matrix parameters are known, and
(iii) the 3D location of the eye is roughly known w.r.t the camera (estimated by the eye landmarks in the face modelling step in camera coordinate space).

With these assumptions in place, we can design user-aided calibration techniques where the user cooperates by looking at predefined positions on the screen.

4 Screen Calibration Methods

As calibration is tedious and needs to be performed multiple times, we evaluate efficiency in terms of how much manual effort is required for three calibration versions:

1. calibration by geometry;
2. calibration by machine learning;
3. calibration by a hybrid: geometry and regression.

4.1 Geometry-Based Calibration

To perfectly project a gaze-vector w.r.t the camera to a point on a screen, we are essentially required to determine the transformation parameters between the camera coordinate system (CCS) and the screen coordinate system (SCS). With our assumptions about roll and yaw in place, this transformation can be expressed by the rotation matrix $\mathbf{R}$ and the translation vector $\mathbf{T}$ between the camera and the screen:

$$\mathbf{R} = \begin{bmatrix} 1 & 0 & 0 \\ 0 & \cos(\rho) & -\sin(\rho) \\ 0 & \sin(\rho) & \cos(\rho) \end{bmatrix} \quad \& \quad \mathbf{T} = \begin{bmatrix} \Delta x \ \Delta y \ \Delta z \end{bmatrix}^T, \tag{1}$$

where ρ denotes the vertical pitch angle (about the x-axis; along the y-axis) between the camera and the screen norm, and $\Delta x, \Delta y, \Delta z$ represent the translational displacement between the camera and the screen. An illustration of the geometric setup is shown in Fig. 2 (2a and 2b).

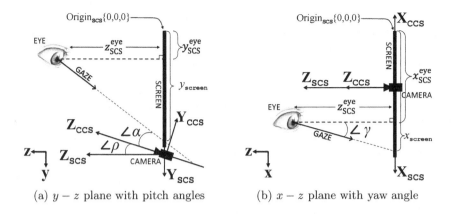

(a) $y - z$ plane with pitch angles   (b) $x - z$ plane with yaw angle

Fig. 2. A illustration of the geometric setup between the eye and the screen in the screen coordinate space. $\{\mathbf{X}, \mathbf{Y}, \mathbf{Z}\}_{\text{ccs}}$ and $\{\mathbf{X}, \mathbf{Y}, \mathbf{Z}\}_{\text{scs}}$ represent the directions of the $\{x, y, z\}$-axes of the camera and screen coordinate systems respectively; Origin$_{\text{scs}}$ represents the origin of the screen coordinate system.

Step 1. The first step is to estimate the location of eye in the screen coordinate system. This can be done with the aid of the user, who is asked to sit at a preset distance z from the screen and look perpendicular at the screen plane (such that the angle between the gaze vector and the screen plane becomes 90°). He is then instructed to mark the point of gaze on the screen, denoted by $\{x, y\}$. In this situation, these marked screen coordinates would directly correspond to the x and y coordinates of the eye in the screen coordinate space. Thus, the eye location can be determined as: $\mathbf{e}_{\text{scs}} = \{x_{\text{scs}}^{\text{eye}}, y_{\text{scs}}^{\text{eye}}, z_{\text{scs}}^{\text{eye}}\} = \{x, y, z\}$.

During this time, the rough 3D location of the eye in the camera coordinate system is also obtained (from the eye landmarks of the face modelling step) and represented as $\mathbf{e}_{\text{ccs}}$. With this pair of corresponding eye locations obtained, the translation vector T can be expressed by:

$$\mathbf{e}_{\text{scs}} = \mathbf{R} \cdot \mathbf{e}_{\text{ccs}} + \mathbf{T} \quad \Longrightarrow \quad \mathbf{T} = \mathbf{e}_{\text{scs}} - \mathbf{R} \cdot \mathbf{e}_{\text{ccs}}. \tag{2}$$

Step 2. Next, without (significantly) changing the head/eye position, the user is asked to look at a different pre-determined point on the screen $\{x_{\text{screen}}, y_{\text{screen}}\}$.

During this time, the gaze estimation system is used to obtain the gaze direction vector in the camera coordinate system:

$$\mathbf{g}_{\text{ccs}} = \begin{bmatrix} x_{\text{ccs}}^{\text{gaze}} & y_{\text{ccs}}^{\text{gaze}} & z_{\text{ccs}}^{\text{gaze}} \end{bmatrix}^T. \tag{3}$$

This is a normalized direction vector whose values denote a point on a unit sphere. Both the pitch α (about the x-axis) and the yaw γ (about the y-axis) angles of the gaze w.r.t the camera can be re-obtained from this gaze direction vector:

$$\alpha = \arctan 2(-y_{\text{ccs}}^{\text{gaze}}, z_{\text{ccs}}^{\text{gaze}}) \quad \& \quad \gamma = \arctan 2(x_{\text{ccs}}^{\text{gaze}}, z_{\text{ccs}}^{\text{gaze}}). \tag{4}$$

Once α is determined, we can calculate the camera pitch angle ρ between the camera and the screen:

$$\rho = \arctan(\frac{-y_{\text{SCS}}^{\text{eye}} + y_{\text{screen}}}{z_{\text{SCS}}^{\text{eye}}}) - \alpha. \tag{5}$$

Using this in Eq. 1, the rotation matrix $\mathbf{R}$ can be fully determined. This known $\mathbf{R}$ can now be plugged into Eq. 2 to also determine the translation vector $\mathbf{T}$. This procedure can be repeated for multiple calibration points in order to obtain a more robust aggregate estimate of the transformation parameters.

Step 3. Once calibration is complete, any new eye location $\hat{\mathbf{e}}_{\text{CCS}}$ can be converted to the screen coordinate space:

$$\hat{\mathbf{e}}_{\text{SCS}} = [\hat{x}_{\text{SCS}}^{\text{eye}}, \hat{y}_{\text{SCS}}^{\text{eye}}, \hat{z}_{\text{SCS}}^{\text{eye}}] = \mathbf{R} \cdot \hat{\mathbf{e}}_{\text{CCS}} + \mathbf{T}. \tag{6}$$

Using the associated new gaze angles $\hat{\alpha}$ and $\hat{\gamma}$, the point of gaze on the screen can be obtained:

$$\hat{x}_{\text{screen}} = \hat{z}_{\text{SCS}}^{\text{eye}} \cdot tan(\hat{\gamma}) + \hat{x}_{\text{SCS}}^{\text{eye}} \quad \& \quad \hat{y}_{\text{screen}} = \hat{z}_{\text{SCS}}^{\text{eye}} \cdot tan(\hat{\alpha} + \rho) + \hat{y}_{\text{SCS}}^{\text{eye}}, \tag{7}$$

4.2 Machine Learning (ML)-Based Calibration

Since the task of gaze vector to gaze point calibration requires learning the mapping between two sets of coordinates, this can be treated as a regression problem. In our implementation, we use a linear ridge regression model for this task for it's ability to avoid overfitting when training samples are scarce. The input to this calibration model includes the predicted gaze-vector angles and the 3D location of the eye, all in the camera coordinate system. The outputs are the 2D coordinates of the gaze-point on the screen in the screen coordinate system.

During calibration, the user is asked to look at a number of predefined points on the screen (such that they span the full region of the screen) while their gaze and eye locations are estimated and recorded for each of these points. These calibration samples are then used to train the model. Given enough training/calibration points, this model is expected to implicitly learn the mapping between the two coordinate systems.

4.3 'Hybrid' Geometric Regression Calibration

To combine the benefits of geometry based prior knowledge with ML based regression, a hybrid geometric regression technique can be derived where machine learning is used to infer the required geometric transformation parameters.

As before, we assume the roll and yaw angles between the camera and the screen are 0°. The only unknown between the pose of the camera w.r.t the screen is the pitch angle ρ. The rotation and translation matrices are the same as given by Eq. 1, and the formulations of gaze pitch and yaw angles α and γ stay the same as defined by Eq. 4.

Again, during calibration the user is asked to look at a number of varied predefined points on the screen while their gaze directions and eye locations are recorded. These data samples are then used to jointly minimize the reprojection errors (squared Euclidean distance) to learn the required transformation parameters $\rho, \Delta x, \Delta y, \Delta z$:

$$\underset{\rho, \Delta x, \Delta y, \Delta z}{\arg\min} \sum_{i=1}^{N} \left((x_{\text{point}}^i - \hat{x}_{\text{screen}}^i)^2 + (y_{\text{point}}^i - \hat{y}_{\text{screen}}^i)^2 \right), \tag{8}$$

where N is the number of training/calibration points; $\{x_{\text{point}}, y_{\text{point}}\}$ denote the ground truth screen points, while $\{\hat{x}_{\text{screen}}, \hat{y}_{\text{screen}}\}$ are the predicted gaze points on screen as estimated using Eq. 7.

We solve this minimization problem by differential evolution [21].

5   Experiments and Results

5.1   Datasets

We perform all our experiments on two publicly available gaze-tracking datasets:

MPIIFaceGaze [28]. This dataset is an extended version of MPIIGaze [30] with available human face region. It contains 37,667 images from 15 different participants. The images have variations in illumination, personal appearance, head pose and camera-screen settings. The ground truth gaze target on the screen is given as a 3D point in the camera coordinate system.

EYEDIAP [6]. This dataset provides 94 video clips recorded in different environments from 16 participants. It has two kinds of gaze targets: screen points and 3D floating targets. It also has two types of head movement conditions: static and moving head poses. For our experiments, we choose the screen point target with both the static and moving head poses, which contains 218,812 images.

5.2   Exp 1: Speed/Accuracy Trade-Off for Varying Input Sizes

Setup. A gaze vector can be predicted based on various input image sizes, as illustrated in Fig. 3:

- Full face image: The largest and most informative input;
- Two eyes: Medium sized and informative;
- Single eye: Minimal information and smallest size.

To assess the performance gains of different input sizes vs their computational loads and accuracy, we setup an experiment where we vary the input training and testing data to the neural network while keeping all other settings fixed. We then measure the accuracy of the system and compute their individual inference-time computational loads.

For this experiment, we individually train our deep network on each of the multiple types and sizes of the pre-processed inputs shown in Fig. 3. In order to obtain a reliable error metric, we perform 5-fold cross-validation training. This experiment is repeated for both the MPIIFaceGaze and EYEDIAP dataset.

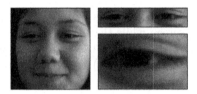

Fig. 3. Examples of three input types used in the experiments: (left) face crop, sized 224×224 and 112×112; (right top) two eyes region crop, sized 180×60 and 90×30; (right bottom) single eye crop, sized 60×32 and 30×18.

Results. The results of this experiment can be seen in Fig. 4. As expected, we observed that the lowest error rates are obtained by the largest size of input data with the maximum amount of context: the full face image. We also observe that using this input type results in the highest amount of computation load.

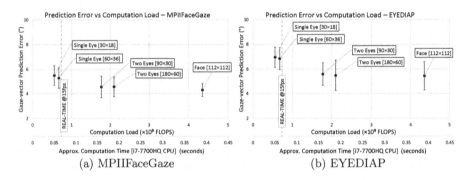

(a) MPIIFaceGaze   (b) EYEDIAP

Fig. 4. Scatter plots of the performance of a VGG16 based gaze tracking network trained on different input types vs their computation load/time in FLOPS/seconds. The error bars represent the standard deviation of the errors (5-fold cross-validation). While the computation cost of these inputs vastly vary, they all perform in roughly the same range of accuracy. The red dashed line represents approximate real-time computation at 15 fps on an *Intel i7* CPU.

As we reduce the input sizes, the accuracy only marginally degrades while the computation load gets cut down severely. In fact, even if we simply use a crop of the eye region or just the crop of a single eye, we obtain accuracies comparable to that from full face input albeit with a fraction of the computation.

5.3   Exp 2: Usability/Accuracy Trade-Off for Screen Calibration

Setup. To evaluate the three screen calibration techniques proposed, we train and test them individually using calibration data samples from MPIIFaceGaze and EYEDIAP. This data for screen calibration consists of pairs of gaze-vectors and their corresponding ground truth screen points. Using this, calibration methods are trained to predict the 2D screen points from the 3D gaze-vectors. We evaluate on noise-free ground truth gaze-vectors and on realistic predicted gaze-vectors (using 30×18 eye crop as input) so as to assess the accuracy of the complete camera-to-screen eye-tracking pipeline. As training data, we obtain calibration data pairs of gaze-vectors and points such that they are spread out evenly over the screen area. This is done by dividing the screen in an evenly-spaced grid and extracting the same number of points from each grid region.

Results. The results of these experiments can be seen in Table 1 for a fixed calibration training set size of 100 samples. For the 'theoretical' task of predicting gaze-points from noise-free ground truth gaze-vectors, we see in Table 1a that the hybrid geometric regression method outperforms others. We see that the gap in performance is smaller when head poses are static, while the hybrid method does better for moving head poses. This suggests that for the simplest evaluation on static head poses with noise-free gaze-vectors, all methods perform well; however, as movement is introduced, the limitations of the purely geometric method and the advantage of hybrid method becomes clearer.

Table 1. Performance of calibration methods (trained with 100 samples) on different datasets and conditions expressed in gaze-point prediction errors (in mm). Hybrid geometric regression technique signifcantly outperforms both purely geometric and purely machine learning (M.L.) based calibration methods in most conditions. Legend: [Static] denotes static head poses, [Moving] denotes moving head poses.

Ground Truth Gaze Vector Dataset	Screen Calibration Method [Prediction Error in mm]		
	Pure Geometric	Pure M.L.	Hybrid Geo. Reg.
MPIIFaceGaze	N/A	9.27	**1.23**
EYEDIAP [Static]	5.98	2.73	**2.35**
EYEDIAP [Moving]	22.45	8.55	**2.39**

Predicted Gaze Vector Dataset	Screen Calibration Method [Prediction Error in mm]		
	Pure Geometric	Pure M.L.	Hybrid Geo. Reg.
MPIIFaceGaze	N/A	50.92	**42.19**
EYEDIAP [Static]	67.72	80.63	**61.6**
EYEDIAP [Moving]	101.53	**82.7**	86.37

(a) Groundtruth gaze-vector calibration   (b) Predicted gaze-vector calibration

When calibration is performed on actual gaze-vectors predicted by the system, overall accuracy deteriorates by one to three orders of magnitude. Comparing the methods, the hybrid geometric regression method also does well compared to others in most conditions, as seen in Table 1b. We observe that the purely geometric method actually copes better than the ML based method when head poses are static. However, it's performance severely degrades with moving head poses. Also, the ML method is able to marginally outperform the hybrid method on moving head poses. This is likely because given sufficient training samples, the ML method is able to learn features from the input that the other—more rigid—methods cannot do. Note that only the EYEDIAP dataset results are reported for the purely geometric technique, since this method can only be trained on static head poses and MPIIFaceGaze does not have any static head poses (the geometric method can still be tested on the moving head poses of EYEDIAP).

To assess the efficiency of these calibration methods, we must ascertain the least amount of calibration samples required with which satisfactory performance can still be attained. This can be assessed by observing the learning curves of the calibration methods, where the prediction errors of the methods are plotted against the number of calibration/training points used. This is shown in Fig. 5.

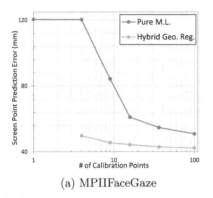
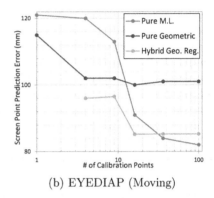

(a) MPIIFaceGaze   (b) EYEDIAP (Moving)

Fig. 5. Learning curves of the calibration techniques on MPIIFaceGaze and EYEDIAP dataset (log scale). The purely geometric method performs better than ML method when calibration data is scarce, but does not improve further when more data is available. The ML method improves greatly when calibration data becomes abundant. The hybrid geometric regression method performs the best over a wide range of calibration data points used.

The hybrid method is able to outperform both the other methods even when a low number of calibration points are available. An interesting observation seen in Fig. 5b is that the purely geometric method actually performs better than the ML method when the number of calibration points is low ($\lesssim 9$). This can be due to the rigid and pre-defined nature of the geometric model which has prior knowledge strongly imparted into it. On the other hand, the ML model requires more data points to learn the underlying geometry from scratch. This is also seen in the results: as the number of points increase, the ML model's performance improves while the geometric model stagnates. Overall, the lower error rate of the hybrid model over a broad range of used calibration points affirms its strengths over the overtly rigid geometric model and the purely data-driven ML approach.

6   Discussion

The experiments related to input types and sizes produce some insightful and promising results. The comparison between them with respect to their performance vs their computational load indicate that the heavier processing of larger inputs with more contextual information is not worth the performance gain they produce. Roughly the same accuracies can be obtained by a system that relies only on eye image crops. In contrast, the gap in the computational load between these two input types is a factor of 20. This supports our idea that for an objective measurement task like gaze-vector prediction, the value of context is limited. These results can help in guiding the design of eye tracking systems meant for real-time applications where efficiency is key.

Outputs in the form of gaze-vectors are not always readily useful in a computer-facing scenario: they need to be projected onto the screen to actually determine where the person is looking. This area has received little attention in literature, and our experiments provide some insight. Our comparison of three calibration techniques show that a hybrid geometric regression method gives the overall best performance over a wide range of available calibration data points. Our results show that purely geometric modelling works better when calibration points are very few, while a purely ML method outperforms it when more points become available. However, a hybrid model offers a robust trade-off between them.

7 Conclusion

In this work, we explored the value of visual context in input for the task of gaze tracking from camera images. Our study gives an overview of the accuracy different types and sizes of inputs can achieve, in relation to the amount of computation their analysis requires. The results strongly showed that the improvement obtained from large input sizes with rich contextual information is limited while their computational load is prohibitively high. Additionally, we explored three screen calibration techniques that project gaze-vectors onto screens without knowing the exact transformations, achieved with the cooperation of the user. We showed that in most cases, a hybrid geometric regression method outperforms a purely geometric or machine learning based calibration while generally requiring less calibration data points and thus being more efficient.

Acknowledgement. The authors are grateful to Messrs. Tom Viering, Nikolaas Steenbergen, Mihail Bazhba, Tim Rietveld, and Hans Tangelder for their most valuable inputs :)

References

1. Baltrušaitis, T., Robinson, P., Morency, L.-P.: Continuous conditional neural fields for structured regression. In: Fleet, D., Pajdla, T., Schiele, B., Tuytelaars, T. (eds.) ECCV 2014. LNCS, vol. 8692, pp. 593–608. Springer, Cham (2014). https://doi.org/10.1007/978-3-319-10593-2_39
2. Chen, Z., Shi, B.E.: Appearance-based gaze estimation using dilated-convolutions. In: Jawahar, C.V., Li, H., Mori, G., Schindler, K. (eds.) ACCV 2018. LNCS, vol. 11366, pp. 309–324. Springer, Cham (2019). https://doi.org/10.1007/978-3-030-20876-9_20

3. Cheng, Y., Lu, F., Zhang, X.: Appearance-based gaze estimation via evaluation-guided asymmetric regression. In: Ferrari, V., Hebert, M., Sminchisescu, C., Weiss, Y. (eds.) Computer Vision – ECCV 2018. LNCS, vol. 11218, pp. 105–121. Springer, Cham (2018). https://doi.org/10.1007/978-3-030-01264-9_7

4. Deng, H., Zhu, W.: Monocular free-head 3D gaze tracking with deep learning and geometry constraints. In: Proceedings of the International Conference on Computer Vision, pp. 3162–3171 (2017)

5. Fischer, T., Chang, H.J., Demiris, Y.: RT-GENE: real-time eye gaze estimation in natural environments. In: Ferrari, V., Hebert, M., Sminchisescu, C., Weiss, Y. (eds.) ECCV 2018. LNCS, vol. 11214, pp. 339–357. Springer, Cham (2018). https://doi.org/10.1007/978-3-030-01249-6_21

6. Funes Mora, K.A., Monay, F., Odobez, J.M.: EYEDIAP: a database for the development and evaluation of gaze estimation algorithms from RGB and RGB-D cameras. In: Proceedings of the ACM Symposium on Eye Tracking Research and Applications (2014)

7. Huang, G., Liu, Z., Van Der Maaten, L., Weinberger, K.Q.: Densely connected convolutional networks. In: Proceedings of the IEEE Conference on Computer Vision and Pattern Recognition, pp. 4700–4708 (2017)

8. Ioffe, S., Szegedy, C.: Batch normalization: accelerating deep network training by reducing internal covariate shift. In: International Conference on Machine Learning, pp. 448–456 (2015)

9. Kasprowski, P., Harezlak, K., Stasch, M.: Guidelines for the eye tracker calibration using points of regard. In: Piętka, E., Kawa, J., Wieclawek, W. (eds.) Information Technologies in Biomedicine, vol. 4, pp. 225–236. Springer, Cham (2014). https://doi.org/10.1007/978-3-319-06596-0_21

10. Kingma, D.P., Ba, J.: Adam: a method for stochastic optimization. In: Proceedings of the International Conference on Learning Representations (2015)

11. Krafka, K., et al.: Eye tracking for everyone. In: IEEE Conference on Computer Vision and Pattern Recognition (CVPR) (2016)

12. Newell, A., Yang, K., Deng, J.: Stacked hourglass networks for human pose estimation. In: Leibe, B., Matas, J., Sebe, N., Welling, M. (eds.) ECCV 2016. LNCS, vol. 9912, pp. 483–499. Springer, Cham (2016). https://doi.org/10.1007/978-3-319-46484-8_29

13. Palmero, C., Selva, J., Bagheri, M.A., Escalera, S.: Recurrent CNN for 3D gaze estimation using appearance and shape cues. In: British Machine Vision Conference (BMVC) (2018)

14. Papoutsaki, A., Sangkloy, P., Laskey, J., Daskalova, N., Huang, J., Hays, J.: Webgazer: scalable webcam eye tracking using user interactions. In: Proceedings of the International Joint Conference on Artificial Intelligence, pp. 3839–3845 (2016)

15. Park, S., Mello, S.D., Molchanov, P., Iqbal, U., Hilliges, O., Kautz, J.: Few-shot adaptive gaze estimation. In: Proceedings of the IEEE International Conference on Computer Vision, pp. 9368–9377 (2019)

16. Park, S., Spurr, A., Hilliges, O.: Deep pictorial gaze estimation. In: European Conference on Computer Vision (2018)

17. Ranjan, R., De Mello, S., Kautz, J.: Light-weight head pose invariant gaze tracking. In: Proceedings of the IEEE Conference on Computer Vision and Pattern Recognition Workshops, pp. 2156–2164 (2018)

18. Rodrigues, R., Barreto, J.P., Nunes, U.: Camera pose estimation using images of planar mirror reflections. In: Daniilidis, K., Maragos, P., Paragios, N. (eds.) ECCV 2010. LNCS, vol. 6314, pp. 382–395. Springer, Heidelberg (2010). https://doi.org/10.1007/978-3-642-15561-1_28

19. Russakovsky, O.: ImageNet large scale visual recognition challenge. Int. J. Comput. Vis. **115**(3), 211–252 (2015). https://doi.org/10.1007/s11263-015-0816-y
20. Simonyan, K., Zisserman, A.: Very deep convolutional networks for large-scale image recognition. In: International Conference on Learning Representations (2015)
21. Storn, R., Price, K.: Differential evolution - a simple and efficient heuristic for global optimization over continuous spaces. J. Global Optim. **11**, 341–359 (1997)
22. Sugano, Y., Matsushita, Y., Sato, Y.: Learning-by-synthesis for appearance-based 3D gaze estimation. In: Proceedings of the IEEE Conference on Computer Vision and Pattern Recognition, pp. 1821–1828 (2014)
23. Tan, K.H., Kriegman, D.J., Ahuja, N.: Appearance-based eye gaze estimation. In: Proceedings of Sixth IEEE Workshop on Applications of Computer Vision, WACV 2002, pp. 191–195. IEEE (2002)
24. Tripathi, S., Guenter, B.: A statistical approach to continuous self-calibrating eye gaze tracking for head-mounted virtual reality systems. In: IEEE Winter Conference on Applications of Computer Vision (WACV), pp. 862–870. IEEE (2017)
25. Wood, E., Baltrušaitis, T., Morency, L.P., Robinson, P., Bulling, A.: A 3D morphable model of the eye region. In: Proceedings of the 37th Annual Conference of the European Association for Computer Graphics: Posters, pp. 35–36 (2016)
26. Wood, E., Bulling, A.: EyeTab: model-based gaze estimation on unmodified tablet computers. In: Proceedings of the Symposium on Eye Tracking Research and Applications, pp. 207–210. ACM (2014)
27. Yu, Y., Liu, G., Odobez, J.M.: Improving few-shot user-specific gaze adaptation via gaze redirection synthesis. In: Proceedings of the IEEE Conference on Computer Vision and Pattern Recognition, pp. 11937–11946 (2019)
28. Zhang, X., Sugano, Y., Fritz, M., Bulling, A.: It's written all over your face: full-face appearance-based gaze estimation. In: IEEE Conference on Computer Vision and Pattern Recognition Workshops (CVPRW), pp. 2299–2308 (2017)
29. Zhang, X., Sugano, Y., Fritz, M., Bulling, A.: Appearance-based gaze estimation in the wild. In: Proceedings of the IEEE Conference on Computer Vision and Pattern Recognition, pp. 4511–4520 (2015)
30. Zhang, X., Sugano, Y., Fritz, M., Bulling, A.: MPIIGaze: real-world dataset and deep appearance-based gaze estimation. IEEE Trans. Pattern Anal. Mach. Intell. **41**(1), 162–175 (2017)

Hierarchical HMM for Eye Movement Classification

Ye Zhu$^{(\boxtimes)}$, Yan Yan, and Oleg Komogortsev

Texas State University, San Marcos, USA
{ye.zhu,tom_yan,ok}@txstate.edu

Abstract. In this work, we tackle the problem of ternary eye movement classification, which aims to separate fixations, saccades and smooth pursuits from the raw eye positional data. The efficient classification of these different types of eye movements helps to better analyze and utilize the eye tracking data. Different from the existing methods that detect eye movement by several pre-defined threshold values, we propose a hierarchical Hidden Markov Model (HMM) statistical algorithm for detecting fixations, saccades and smooth pursuits. The proposed algorithm leverages different features from the recorded raw eye tracking data with a hierarchical classification strategy, separating one type of eye movement each time. Experimental results demonstrate the effectiveness and robustness of the proposed method by achieving competitive or better performance compared to the state-of-the-art methods.

Keywords: Hidden Markov Model · Eye movement · Fixation · Smooth pursuit · Saccade · Classification

1 Introduction

Eye tracking technology, which aims to measure the location where a person is looking at, has been widely applied in various research and application fields including the human-computer interaction [4,16], AR/VR [6,11], behavioral psychology [3,10] and usability studies [8,21] in recent years. With the arising attention and interests from researchers, eye tracking is becoming a potential and promising driver for future immersive technologies.

One fundamental and significant research topic in eye tracking is to identify different types of eye movements. There are several primary types of eye movements: fixations correspond to the situation where the visual gaze is maintained on a single location, saccades are fast movements of the eyes that rapidly change the point of fixation, and smooth pursuits are defined as slower tracking movements of the eyes designed to keep a moving stimulus on the fovea [17]. Ternary eye movement classification [13], which seeks to classify three primary types of eye movement, *i.e.*, fixations, saccades and smooth pursuits, is essential to the above applications.

In this work, we tackle the problem of ternary eye movement classification from a probabilistic perspective by adopting the Hidden Markov Model in a

© Springer Nature Switzerland AG 2020
A. Bartoli and A. Fusiello (Eds.): ECCV 2020 Workshops, LNCS 12535, pp. 544–554, 2020.
https://doi.org/10.1007/978-3-030-66415-2_35

hierarchical way. The hierarchical structure makes it possible to consider several different data features in different stages of classification. The usage of the Viterbi [9] and Baum-Welch algorithms [2] allows us to avoid the inconvenience of selecting thresholds and to improve the robustness of the classification method. Experiments show that our proposed hierarchical HMM is able to achieve competitive performance compared to the state-of-the-art methods.

2   Related Work

One of the most common and intuitive methods to separate fixations from saccades is threshold based algorithms, such as Velocity Threshold Identification (I-VT) [1] and Dispersion Threshold Identification (I-DT) [18]. The former method assumes the velocity of saccades should be larger than the velocity of fixations, while the latter one relies on the difference of duration and positional dispersion between fixations and saccades. However, the above single threshold-based algorithms are unable to accurately separate smooth pursuits from fixations due to a variety of artifacts usually present in the captured eye movement signal.

The existing threshold-based state-of-the-art methods for ternary eye movement classification mainly combine several different single threshold-based algorithms. Velocity Velocity Threshold Identification (I-VVT) [13] is a basic algorithm that adopts two velocity thresholds, the data points with higher velocity than the larger velocity threshold are classified as saccades, the points with a lower velocity than the smaller threshold are classified as fixations, while the remaining points are considered as smooth pursuits. Velocity Dispersion Threshold Identification (I-VDT) [13] is another algorithm that combines I-VT and I-DT [13]. I-VDT firstly filters out saccades by I-VT, I-DT is then further used to separate fixations from smooth pursuits. Velocity Movement Pattern Identification (I-VMP) [12] uses I-VT to identify saccades, and then employs movement direction information to separate fixations from smooth pursuits. All the methods mentioned above rely on empirically selected thresholds to provide a meaningful classification on a targeted dataset.

In addition to the above threshold-based algorithms, video-based methods to detect gazes have also been exploited in [7,14]. Dewhurst *et al.* [5] proposes to use geometric vectors to detect eye movements. Nystrom *et al.* [15] aims to classify fixations, saccades and glissades using an adaptive velocity-based algorithm. Santini *et al.* [20] propose Bayesian method (I-BDT) to identify fixations, saccades and smooth pursuits. Another branch of more recent work adopts machine learning techniques to tackle the problem. Identification using Random Forest machine learning technique (IRF) is used to classify fixations, saccades and post-saccadic oscillations in [24]. Zembly *et al.* [23] further propose a gaze-Net to realize end-to-end eye-movement event detection with deep neural networks. Startsev *et al.* [22] tackles the problem of ternary eye movement classification with a 1D-CNN with BLSTM.

Although the machine learning methods, especially deep learning, have been widely applied in multiple research fields including the eye movement classification, one major drawback of these techniques is that the training process relies

---

Algorithm 1. Hierarchical HMM for Ternary Eye Movement Classification

---

Require: Eye positional data sequence.

1: ε_n number of epochs for n-th HMM, initial start probability vector π, initial transition probability matrix **A** and initial emission probability matrix **B**. Note that initial parameters for HMM will be optimized and updated by Baum-Welch algorithm.

2: **Step 0: Pre-processing**

3: Compute the position, velocity, acceleration feature sequences for input eye tracking data

4: Select appropriate features for classification.

5: **Step 1: Rough classification**

6: Initialize the parameters of the first HMM for selected feature

7: $e_1 \leftarrow 0$

8: **for** $e_1 < \varepsilon_1$ **do**

9:     Viterbi algorithm

10:     Baum-Welch algorithm

11:     Filter saccades

12: **end for**

13: **Step 2: Refined classification**

14: Initialize the parameters of the second HMM for selected feature, define a threshold value $\mathcal{T}$ of the first feature for fine-tuning

15: $e_2 \leftarrow 0$

16: **for** $e_2 < \varepsilon_2$ **do**

17:     Viterbi algorithm

18:     Baum-Welch algorithm

19:     Classify fixations and smooth pursuits

20:     Fine-tune the classification results by $\mathcal{T}$

21: **end for**

22: **Step 3: Merge function**

23: Merge classified points into complete fixations, saccades and smooth pursuits

24: **return** List of classification results

---

on a large amount of data. In addition, the neural networks usually require retraining when applied in a different task or dataset. Our focus in this work is to improve the robustness and performance of the threshold-based state-of-the-art methods, even competing with the recent machine learning based state-of-the-art performance.

3   Methodology

We present the proposed hierarchical HMM method in this section and provide the corresponding pseudo-code.

Hidden Markov Model is a statistical model for time series based on the Markov process with hidden states. The principle of HMM is to determine the hidden state with a maximum probability according to the observable sequence. A traditional HMM takes three sets of parameters as input, which are the start probability vector, transition probability matrix and emission probability matrix.

The ternary eye movement classification can be formulated as a first-order three-state HMM problem, whose hidden states are fixations, saccades and smooth pursuits.

HMM relies on the distinguishable probability distributions of features to correctly define different hidden states, otherwise, the maximum probability of each hidden state may be incorrect if several hidden states have similar probabilities given a certain observation sequence. In the case of eye movement classification, the probability distributions for different movement types are usually represented by continuous Gaussian distributions [19]. The main challenge of ternary eye movement classification using the existing methods lies in the bias of the features for smooth pursuits. While fixations and saccades have very different positional dispersion and velocity features, smooth pursuits are rather ambiguous in terms of dispersion and velocity since the steady state of a smooth pursuit often contains corrective saccades, making the eye tracking data very noisy. To this end, we propose to leverage different features and introduce a hierarchical strategy to tackle the problem. The core idea of our hierarchical HMM is to perform HMM classification for multiple times using different features.

Firstly, we start by analyzing different features of the raw eye positional data as the pre-processing step, e.g., positions, velocities, accelerations, whose objective is to determine the features that can be used to separate three eye movement types in latter steps. After selecting appropriate features from pre-processing, we then perform a first-stage rough classification on the eye tracking data to separate saccades from the fixations and smooth pursuits using the HMM. We refer the first-stage as rough classification due to the reason that the classification results will be fine-tuned in the latter stage. The second-stage classification adopts a different feature to separate fixations and smooth pursuits by another HMM, using the first feature as a fine-tuning criterion at the same time. The final step of our hierarchical HMM is to merge the classified points into complete eye movements with temporal duration criterion.

The pseudo-code of the proposed method is presented in Algorithm 1. Compared with the existing threshold-based state-of-the-art methods, our method has several advantages: 1) the usage of HMM avoids the tedious work to select threshold values as for previous algorithms and improves the robustness of the proposed classification method; 2) a hierarchical strategy with fine-tuning technique further improves the classification performance. In the meanwhile, the proposed method does not rely on a large amount of data for training and leverages several different features in a coherent way (e.g., use the feature from the first stage as a fine-tuning criterion in the second stage). Experimental results prove that our proposed hierarchical HMM method is simple, straightforward yet effective.

4   Experiments

In this section, we present the experimental results obtained with the proposed method. The comparisons with other state-of-the-art methods demonstrate the robustness and effectiveness of the proposed hierarchical HMM. An ablation study is also included to show the contributions of the hierarchical structure.

4.1 Dataset and Evaluation Metrics

We use the eye tracking dataset recorded from the previous research work [13] for experiments. The data are recorded by the EyeLink 100 eyetracker 1000 Hz on a 21-in monitor and contain 11 subjects, containing the human annotations as either clean or noisy data.

Seven different behavior scores are used as quantitative evaluation metrics for our experiments. Saccade quantitative score (SQnS), fixation quantitative score (FQnS) and smooth pursuit quantitative score (PQnS) are used to measure the amount of saccades, fixation and smooth pursuit behavior in response to a stimulus, respectively [12]. Fixation qualitative score (FQlS), smooth pursuit qualitative score for positional accuracy (PQlS_P) and for velocity accuracy (PQlS_V) are to compare the proximity of the detected smooth pursuit signal with the signal presented in the stimuli [13]. Misclassified fixation score (MisFix) of the smooth pursuit is defined as the ration between misclassified smooth pursuit points and the total number of fixation points in the stimuli. The ideal number of MisFix should consider the practical latency situation where smooth pursuit continues when the stimulus changes from smooth pursuit to fixation.

4.2 Implementations

We compare our proposed hierarchical HMM method with six baselines: I-VVT [13], I-VDT [13], I-VMP [12], IRF [24], I-BDT [20] and 1DCNN [22]. Among six baselines, I-VVT, I-VDT, and I-VMP are threshold-based methods, all the threshold values are optimized for different subject recordings in our experiments for comparison. For the IRF, we use the same selected features (*i.e.*, velocity and position) as for our proposed method. For the I-BDT, we follow the parameters chosen in [20]. For the 1DCNN, we take the pre-trained model and fine-tune it on the recording data (except those used for testing) from our targeted dataset.

For our proposed hierarchical HMM method, we use the parameters reported in [19] for the initialization in HMM. We iterate the Viterbi and Baum-Welch algorithms for 3 times to learn and update the parameters for HMM. For the final Merge function, the merge time interval threshold we use is 75 ms, the merge distance is $0.5°$.

4.3 Experimental Results

Results on Behavior Scores. We compare the behavior scores obtained via the proposed hierarchical HMM and other methods on the subjects that are manually evaluated as "Medium", "Good" and "Bad" in the eye tracking data quality [13]. The ideal scores and manually classification results are also reported for comparison, the calculation of the ideal behavior scores considers the actual

Table 1. Comparison of behavior scores obtained by different methods for the subjects that are manually evaluated as "Medium", "Good" and "Bad", respectively. Our proposed method achieves competitive performance close to the human evaluation (*i.e.*, Manual).

Behavior scores	Ideal	Manual [13]	Ours	I-VVT [13]	I-VDT [13]	I-VMP [12]	IRF [24]	I-BDT [20]	1DCNN [22]
SQnS	100%	84%	83%	86%	82%	78%	86%	84%	88%
FQnS	84%	63%	63%	16%	79%	66%	78%	68%	59%
PQnS	52%	47%	48%	38%	55%	61%	43%	51%	40%
MixFix	7.1%	13%	8.9%	42%	6%	22%	12%	15%	20%
FQlS	0°	0.46°	0.4°	0.5°	0.5°	0.5°	0.5°	0.5°	0.4°
PQlS_P	0°	3.07°	3.2°	3.2°	3.2°	3.4°	3.1°	3.2°	3.2°
PQlS_V	0°/s	39°/s	30°/s	16°/s	47°/s	40°/s	38°/s	33°/s	40°/s
Behavior scores	Ideal	Manual [13]	Ours	I-VVT [13]	I-VDT [13]	I-VMP [12]	IRF [24]	I-BDT [20]	1DCNN [22]
SQnS	100%	96%	91%	96%	90%	90%	88%	92%	90%
FQnS	84%	71%	74%	30%	82%	69%	82%	80%	78%
PQnS	52%	39%	44%	39%	30%	56%	58%	42%	38%
MixFix	7.1%	6%	5.5%	50%	4.4%	21%	12%	8%	16%
FQlS	0°	0.44°	0.3°	0.4°	0.4°	0.4°	0.4°	0.4°	0.4°
PQlS_P	0°	3.15°	2.9°	3.6°	3.2°	3.7°	3.7°	3.3°	3.4°
PQlS_V	0°/s	23°/s	25°/s	16°/s	44°/s	40°/s	38°/s	37°/s	32°/s
Behavior scores	Ideal	Manual [13]	Ours	I-VVT [13]	I-VDT [13]	I-VMP [13]	IRF [24]	I-BDT [20]	1DCNN [22]
SQnS	100%	89%	77%	85%	78%	76%	75%	82%	85%
FQnS	84%	42%	49%	17%	61%	51%	43%	45%	50%
PQnS	52%	40%	37%	33%	31%	47%	44%	38%	33%
MixFix	7.1%	33%	11%	33%	22%	36%	18%	12%	20%
FQlS	0°	0.58°	0.5°	0.6°	0.6°	0.6°	0.6°	0.6°	0.6°
PQlS_P	0°	2.58°	2.6°	3.2°	3.5°	3.6°	3.3°	2.4°	3.5°
PQlS_V	0°/s	30°/s	32°/s	16°/s	81°/s	52°/s	46°/s	31°/s	44°/s

physiological reactions of humans when performing different types of eye movements.

The quantitative experimental results are presented in Table 1. We observe that our proposed hierarchical HMM method achieves promising performance compared to the other baseline methods and is much closer to human performance (*i.e.*, manual). The I-VVT algorithm has the worst performance as expected due to the reason that it only considers the velocity and separates the eye movement by two simple threshold values. The I-VDT and I-VMP are able to separate fixations, saccades and smooth pursuits, and I-VDT has better performance compared to I-VMP, achieving a MixFix score closer to the ideal one. Other machine learning based methods IRF, I-BDT, and 1DCNN achieve good performance in general, however, we observe that 1DCNN has relatively lower behavior scores in detecting the smooth pursuit. One possible reason for its under-performance could be the limited amount of smooth pursuit data for fine-tuning. Since the original model was trained in a different dataset and fine-tuned in our experiments, the transfer learning process may be less effective compared

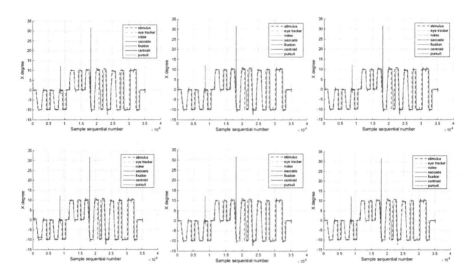

Fig. 1. Qualitative classification results on the "good quality" subject obtained by the proposed hierarchical HMM, I-VDT, I-VMP, IRF, I-BDT and 1D CNN, respectively. Best viewed in color. (Color figure online)

to retraining the entire network model, which also reveals one of the limitations about those "data-driven" methods. Our proposed hierarchical HMM outperforms the other considered methods, especially the threshold-based methods, in most of the behavior scores and it is worth noting that the proposed method has behavior scores close to the manually evaluated results.

We also present the qualitative classification results in Fig. 1. Due to the reason that the I-VVT basically fails in this ternary eye movement classification task, we do not include the classification results of I-VVT in the figure. The qualitative results are consistent with the previous quantitative results. Notably, our proposed hierarchical HMM method succeeds in detecting an unexpected saccade while most of the other methods detect it either as smooth pursuit or fixation.

Feature Analysis. In our experiments, we propose a pre-processing step to analyze different features from raw data. We analyze velocity, position and acceleration features, and select velocity and position as two features used for the hierarchical HMM method. Velocity feature is used in the first HMM to filter saccades and also as the fine-tuning criterion in the second stage HMM, and the position features are used as the main feature to separate fixations and smooth pursuits in the second stage classification.

We present the visualization results of the data features on the subject that is manually evaluated as "good" in [13] using the K-Means cluster algorithm. The main objective of the pre-processing is to select appropriate features for further classifications using HMM. In the meanwhile, it also provides some insights for the reasons why the classic threshold-based methods may fail to achieve good performance in the ternary eye movement classification task. As shown in Fig. 2, the actual eye tracking data usually contain a lot of noises. Take the velocity feature as an example, there exist some noisy data points with extremely large velocities and the difference of velocity among fixations, saccades and smooth pursuits are not very clear. Therefore, the statistical methods with predefined threshold values are more likely to fail in this case.

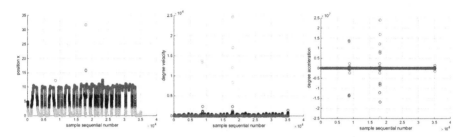

Fig. 2. Visualization of clustering results for position, velocity and acceleration features, respectively, on the data recording from the subject manually evaluated as "good" [13]. Different colors represent different clusters automatically detected by K-means. Best viewed in color. (Color figure online)

Ablation Study Analysis. We conduct an additional ablation experiment to demonstrate the necessity and effectiveness of the proposed hierarchical structure. Since the fixations, saccades and smooth pursuits in the ternary eye movement classification can be considered as three hidden states of an HMM, we use the velocity feature as the criterion and adopt a single-stage HMM with the three-hidden state to do the same task, in other words, we remove the second stage classification from the proposed method.

Table 2. Comparison of behavior scores for the subject manually evaluated as "good" in the ablation study. Our proposed hierarchical strategy especially contributes to the classification of smooth pursuits.

Behavior scores	Ideal	Manual [13]	Ours	3-state HMM
SQnS	100%	96%	91%	91%
FQnS	84%	71%	74%	54%
PQnS	52%	39%	44%	25%
MixFix	7.1%	6%	5.5%	28%
FQlS	0°	0.44°	0.3°	0.4°
PQlS_P	0°	3.15°	2.9°	3.4°
PQlS_V	0°/s	23°/s	25°/s	31°/s

The quantitative and qualitative results are shown in Table 2 and Fig. 3, respectively. Both experimental results prove that the hierarchical structure contributes to better classification results. The qualitative figure shows that the hierarchical structure helps especially with the separation between fixations and smooth pursuits, which further validates the fact that the velocity feature is biased for fixations and smooth pursuits. Therefore, the analysis of different features and the usage of position features in the second stage of the proposed method are necessary and beneficial.

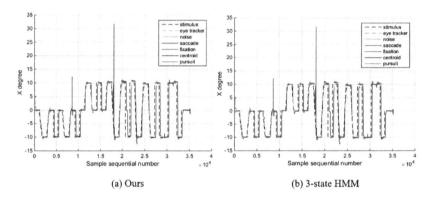

(a) Ours   (b) 3-state HMM

Fig. 3. Qualitative results for the ablation study. Our proposed hierarchical strategy helps with the classification of smooth pursuits, which is consistent with the quantitative results.

5   Conclusions

In this paper, we propose a hierarchical HMM algorithm to tackle the problem of ternary eye movement classification from the raw eye positional data. Different features from the data are considered to realize the multi-stage hierarchical classification. Experiments on multiple data records demonstrate the effectiveness and robustness of the proposed method. Possible future directions of this work could be incorporating the pre-processing step into the proposed method in an automated way, and making efforts to classify more categories of eye movements with a deeper hierarchical structure.

References

1. Bahill, A., Brockenbrough, A., Troost, B.: Variability and development of a normative data base for saccadic eye movements. Invest. Ophthalmol. Vis. Sci. **21**(1), 116–125 (1981)
2. Baum, L.E., Petrie, T., Soules, G., Weiss, N.: A maximization technique occurring in the statistical analysis of probabilistic functions of Markov chains. Ann. Math. Stat. **41**(1), 164–171 (1970)

3. Boraston, Z., Blakemore, S.J.: The application of eye-tracking technology in the study of autism. J. Physiol. **581**(3), 893–898 (2007)
4. Bruneau, D., Sasse, M.A., McCarthy, J.: The eyes never lie: the use of eyetracking data in HCI research. ACM (2002)
5. Dewhurst, R., Nyström, M., Jarodzka, H., Foulsham, T., Johansson, R., Holmqvist, K.: It depends on how you look at it: Scanpath comparison in multiple dimensions with MultiMatch, a vector-based approach. Behav. Res. Methods **44**(4), 1079–1100 (2012). https://doi.org/10.3758/s13428-012-0212-2
6. Duchowski, A.T., Medlin, E., Gramopadhye, A., Melloy, B., Nair, S.: Binocular eye tracking in VR for visual inspection training. In: Proceedings of the ACM Symposium on Virtual Reality Software and Technology, pp. 1–8 (2001)
7. Ebisawa, Y.: Improved video-based eye-gaze detection method. IEEE Trans. Instrum. Meas. **47**(4), 948–955 (1998)
8. Ehmke, C., Wilson, S.: Identifying web usability problems from eye-tracking data. In: Proceedings of the 21st British HCI Group Annual Conference on People and Computers: HCI... But not as We Know it-Volume 1, pp. 119–128. British Computer Society (2007)
9. Forney, G.D.: The viterbi algorithm. Proc. IEEE **61**(3), 268–278 (1973)
10. Granka, L.A., Joachims, T., Gay, G.: Eye-tracking analysis of user behavior in www search. In: Proceedings of the 27th Annual International ACM SIGIR Conference on Research and Development in Information Retrieval, pp. 478–479. ACM (2004)
11. Hickson, S., Dufour, N., Sud, A., Kwatra, V., Essa, I.: Eyemotion: classifying facial expressions in VR using eye-tracking cameras. In: 2019 IEEE Winter Conference on Applications of Computer Vision (WACV), pp. 1626–1635. IEEE (2019)
12. Komogortsev, O.V., Gobert, D.V., Jayarathna, S., Koh, D.H., Gowda, S.M.: Standardization of automated analyses of oculomotor fixation and saccadic behaviors. IEEE Trans. Biomed. Eng. **57**(11), 2635–2645 (2010)
13. Komogortsev, O.V., Karpov, A.: Automated classification and scoring of smooth pursuit eye movements in the presence of fixations and saccades. Behav. Res. Methods **45**(1), 203–215 (2012). https://doi.org/10.3758/s13428-012-0234-9
14. Li, Y., Fathi, A., Rehg, J.M.: Learning to predict gaze in egocentric video. In: Proceedings of the IEEE International Conference on Computer Vision, pp. 3216–3223 (2013)
15. Nyström, M., Holmqvist, K.: An adaptive algorithm for fixation, saccade, and glissade detection in eyetracking data. Behav. Res. Methods **42**(1), 188–204 (2010)
16. Poole, A., Ball, L.J.: Eye tracking in HCI and usability research. In: Encyclopedia of Human Computer Interaction, pp. 211–219. IGI Global (2006)
17. Purves, D.: Neuroscience. Scholarpedia **4**(8), 7204 (2009)
18. Salvucci, D.D., Goldberg, J.H.: Identifying fixations and saccades in eye-tracking protocols. In: Proceedings of the 2000 Symposium on Eye Tracking Research & Applications, pp. 71–78. ACM (2000)
19. Salvucci, D.D., Anderson, J.R.: Tracing eye movement protocols with cognitive process models (1998)
20. Santini, T., Fuhl, W., Kübler, T., Kasneci, E.: Bayesian identification of fixations, saccades, and smooth pursuits. In: Proceedings of the Ninth Biennial ACM Symposium on Eye Tracking Research & Applications, pp. 163–170 (2016)
21. Schiessl, M., Duda, S., Thölke, A., Fischer, R.: Eye tracking and its application in usability and media research. MMI-Interaktiv J. **6**, 41–50 (2003)
22. Startsev, M., Agtzidis, I., Dorr, M.: 1D CNN with BLSTM for automated classification of fixations, saccades, and smooth pursuits. Behav. Res. Methods **51**(2), 556–572 (2019). https://doi.org/10.3758/s13428-018-1144-2

23. Zemblys, R., Niehorster, D.C., Holmqvist, K.: gazeNet: end-to-end eye-movement event detection with deep neural networks. Behav. Res. Methods **51**(2), 840–864 (2018). https://doi.org/10.3758/s13428-018-1133-5
24. Zemblys, R., Niehorster, D.C., Komogortsev, O., Holmqvist, K.: Using machine learning to detect events in eye-tracking data. Behav. Res. Methods **50**(1), 160–181 (2017). https://doi.org/10.3758/s13428-017-0860-3

Domain Adaptation for Eye Segmentation

Yiru Shen[(⊠)], Oleg Komogortsev, and Sachin S. Talathi

Facebook Reality Labs, Redmond, USA
yirus@fb.com, ok1@fb.com, stalathi@fb.com

Abstract. Domain adaptation (DA) has been widely investigated as a framework to alleviate the laborious task of data annotation for image segmentation. Most DA investigations operate under the unsupervised domain adaptation (UDA) setting, where the modeler has access to a large cohort of source domain labeled data and target domain data with no annotations. UDA techniques exhibit poor performance when the domain gap, i.e., the distribution overlap between the data in source and target domain is large. We hypothesize that the DA performance gap can be improved with the availability of a small subset of labeled target domain data. In this paper, we systematically investigate the impact of varying amounts of labeled target domain data on the performance gap for DA. We specifically focus on the problem of segmenting eye-regions from eye images collected using two different head mounted display systems. Source domain is comprised of 12,759 eye images with annotations and target domain is comprised of 4,629 images with varying amounts of annotations. Experiments are performed to compare the impact on DA performance gap under three schemes: unsupervised (UDA), supervised (SDA) and semi-supervised (SSDA) domain adaptation. We evaluate these schemes by measuring the mean intersection-over-union (mIoU) metric. Using only 200 samples of labeled target data under SDA and SSDA schemes, we show an improvement in mIoU of 5.4% and 6.6% respectively, over mIoU of 81.7% under UDA. By using all available labeled target data, models trained under SSDA achieve a competitive mIoU score of 89.8%. Overall, we conclude that availability of a small subset of target domain data with annotations can substantially improve DA performance.

Keywords: Domain adaptation · Eye segmentation

1 Introduction

Semantic segmentation is an important problem in computer vision where the objective is to assign labels to each pixel in an image. In recent years, supervised learning methods using convolutional neural networks (CNNs) have enabled significant improvements in the development of models for semantic segmentation [1,15,24]. However, supervised training of CNNs require a large amount of images with pixel-wise annotations, which, if done manually is time consuming, non-scalable and label inefficient. To ease the problem of pixel-wise image

© Springer Nature Switzerland AG 2020
A. Bartoli and A. Fusiello (Eds.): ECCV 2020 Workshops, LNCS 12535, pp. 555–569, 2020.
https://doi.org/10.1007/978-3-030-66415-2_36

annotation, unsupervised domain adaptation (UDA) methods have been widely studied. UDA methods attempt to align the distribution of features from a different but related, source domain data, which also contains annotations, to target domain data, where no annotations are available [5,31].

UDA predominantly uses adversarial training framework to train models that retain good performance for semantic segmentation on source domain data, while at the same time attempting to reduce the discrepancy in feature distribution in both domains. In doing so the hope is that the trained model can semantically segment the target domain data even in the absence of any labeled data. However, UDA have been shown to fail to learn discriminative class boundaries for target domain data. Furthermore, UDA also fails to train models that can generalize well on the target domain data [25,27,37]. In addition, the setting under which UDA training is appropriate, may not be suitable for real world scenarios where we also have annotations for a subset of target domain data. In this scenario, the following questions are more relevant:

- Given a fixed annotation budget, how much data need to be labeled in order to reach a reasonable performance on the target domain?
- How to optimally utilize the unlabeled data in target domain?

We focus on answering these questions in the eye-region semantic segmentation research area with a long-term goal of increasing the quality of estimated eye gaze data. Obtained eye tracking quality is critical to many applications in VR/AR such as foveated rendering, intent inference, health assessment, and direct gaze interaction. Specifically, in this work, as a first step toward this goal, we conduct experiments to systematically investigate the impact on segmentation performance for varying levels of target data annotations. Three domain adaptation frameworks are investigated: unsupervised domain adaptation (UDA), where large numbers of target images without target annotations are available [5]; supervised domain adaptation (SDA), where a small number of labeled target images are available [27] and semi-supervised domain adaptation (SSDA), where in addition to large numbers of target images without annotations, a small number of target images with annotations are available [28]. We conduct a series of experiments to train semantic segmentation models using all available labeled source domain data while varying the number of labeled target domain data, which are randomly selected as function of DA framework used for training. Specifically, under UDA, we do not use any of the available labeled target domain data and for SDA, we do not use any of the available unlabeled target domain data.

We investigate a single semantic segmentation model architecture which is trained using one of aforementioned DA frameworks. The model consists of two sub-networks: a segmentation network to predict probability map of an input eye image; and a fully convolutional discriminator network that differentiates the probability map in target domain from those in the source domain. We adopt adversarial training based strategy to train the segmentation network so as to fool the discriminator network by producing distribution of probability maps, which are invariant to the change in domain. The intuition for using adversarial training as a mechanism to adapt the probability map for target domain data from the probability map of the source domain data is based on the fact that both the source domain eye images and the target domain eye images exhibit geometric and spatial structural similarities, for example, the ellipse contour of eyes are always maintained [30].

For the training of semantic segmentation model under SSDA, a semi-supervised loss is designed by generating pseudo-labels for eye regions in the unlabeled target images. Specifically, the discriminator network generates confidence score for each pixel of eye image belonging to the source or the target domain. Only the probability map of regions in the image that are identified as close to source domain are treated as pseudo-labels and trained in the segmentation network via a masked cross entropy loss [10]. This semi-supervised loss enables the model to produce probability map of unlabeled target images close to the distribution of source domain images. Note that although both our method and [10] utilize the signals from discriminator to conduct semi-supervised loss, study in [10] only considers leveraging labeled data and unlabeled data in a single domain. Different from [10], the scope of SSDA in our study is to align the data distributions between source and target domain while at the same time further boost the performance for semantic segmentation by conducting semi-supervise learning in the target domain.

Our contributions are summarized as follows:

- A practical framework to train models under DA by optimally utilizing limited images with annotations and the available unlabeled images in the target domain.
- A systematic comparison of UDA, SDA and SSDA performance as function of the number of available images (and labels) in the target domain.

2 Related Work

2.1 Semantic Segmentation of Eye Regions

Segmentation of periocular regions, including pupil, iris, sclera provide comprehensive information on eyes and is critical for gaze estimation [35]. A majority of studies for eye segmentation have focused on segmenting single trait of eye regions, i.e. sclera, iris, etc. Iris segmentation has been investigated for iris recognition that can be used in personal identification and verification [3,26]. In [14], ATT-UNet was proposed to learn discriminative features to separate iris and non-iris pixels. Sclera segmentation is usually considered as a pre-processing step for sclera-based recognition in applications of human identification [16,21,39]. In [34], ScleraSegNet was proposed to achieve competitive performance for sclera segmentation by using channel-wise attention mechanism. However, limited works have focused on semantic segmentation of all eye regions, i.e. pixel-wise classification due to the lack of availability of large-scale datasets with labeled eye images [4,17,18]. Recently, the Open Eye Dataset (OpenEDS) has been released to facilitate development of models for semantic segmentation of all eye regions [7]. Study in [7] trained a modified SegNet model [1] by introducing a multiplicative skip connection between the last layer of the encoder and the decoder network, to classify pupil, iris, sclera and background for each pixel of a given eye image. More recently, few studies have investigated models to segment all eye regions using the OpenEDS datasets, primarily focusing on low computational complexity models [2,11,23].

2.2 Domain Adaptation for Semantic Segmentation

Domain adaptation (DA) has been applied to develop semantic segmentation models to ease the problem of data annotations, by aligning the feature or output pixel-wise class distributions between the source and the target images [5,6,9,25]. Three different schemes of DA have been widely studied: UDA [8,30], SDA [20, 28,33], and SSDA [10,36]. For these schemes, adversarial learning has become a prominent approach in semantic segmentation, where the critical step is to train a domain classifier or discriminator to differentiate whether the representation is from source or target domain [6,9,32]. Due to the assumption of non-availability of supervision from target domain annotations, many recent work has focused on UDA by aligning distributions between source and target domain in feature space [38], or learning discriminative representations in output space [30]. However, UDA can fail to learn discriminative class boundaries on target domains [25]. SDA is applied when limited labeled target images are available. For example, researchers in [28] proposed a hierarchical adaptation method to align feature distributions between abundant labeled synthetic images in source domain and insufficient labeled images in real-world scenario. In SSDA, in addition to a small amount of labelled target images, a larger amount unlabeled target images is available. However, most of the SSDA is applied in applications of classification and object detection [25,29,37], SSDA for semantic segmentation has not been fully explored. In this paper, we revisit this task and compare SSDA with SDA and UDA under varying amount of training images in target domain.

3 Method

3.1 Overview

Figure 1 provides an overview of our proposed model. Our goal is to perform DA from all available labeled images in the source domain, $\mathcal{X}_s$, to produce annotations for unlabeled images in the target domain, $\mathcal{X}_t$, by leveraging the limited number of labeled images available in the target domain. We aim to investigate how the performance on segmentation of unlabeled target domain images is impacted by varying amounts of labeled target domain images.

Our proposed model consists of two sub-networks: a segmentation network SS and a discriminator network D. For segmentation network, we consider the mSegNet architecture for eye segmentation [7], which takes images of dimension $H \times W$ as input and outputs the probability map of dimension $H \times W \times K$,

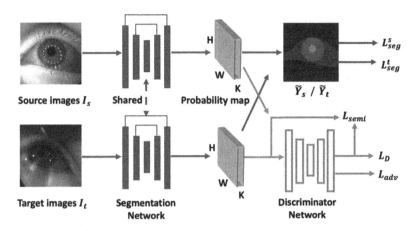

Fig. 1. Architecture overview. Source and target images are passed to segmentation network to produce probability map. Segmentation loss L_{seg} is computed given the probability map. A discriminator loss L_D is calculated to train a discriminator to differentiate whether each pixel is from source or target. An adversarial loss L_{adv} is calculated on target probability map to fool the discriminator. A semi-supervised loss L_{semi} is calculated to boost the training process if additional unlabeled target images are available. Blue arrows indicate forward operation of the segmentation network, while orange arrows indicate that of discriminator. L_{semi} in green is applied when additional unlabeled target images are available, while L_{seg}^t is applied when there are labeled target images. (Color figure online)

where K is the number of semantic categories. For discriminator a.k.a. domain classifier, we design a fully convolutional network, which takes probability maps from segmentation network, and outputs a confidence map of dimension $H \times W$. This map indicates how confident the discriminator about each pixel in the input image belonging to source or target domain.

During training, all labeled images in source domain $\mathcal{X}_s$ are used, and we vary the number of labeled images available for training in the target domain $\mathcal{X}_t$. We denote pairs of images with the per-pixel annotations in $\mathcal{X}_s$ as $\{(I_s, Y_s)\}$; pairs of images with per-pixel annotations and the unlabeled images in $\mathcal{X}_t$ as $\{(I_t, Y_t)\}, \{I_{t,u}\}$, respectively.

Formally, the inputs for three frameworks for domain adaptation are:

- Unsupervised domain adaptation: $\mathcal{X}_s = \{(I_s, Y_s)\}$; $\mathcal{X}_t = \{I_{t,u}\}$
- Supervised domain adaptation: $\mathcal{X}_s = \{(I_s, Y_s)\}$; $\mathcal{X}_t = \{(I_t, Y_t)\}$; $|Y_s| \gg |Y_t|$
- Semi-supervised domain adaptation: $\mathcal{X}_s = \{(I_s, Y_s)\}$; $X_t = \{(I_t, Y_t); I_{t,u}\}$; $|Y_s| \gg |Y_t|, |I_{t,u}| \gg |I_t|$

3.2 Loss Functions

Segmentation Loss. Segmentation loss is the cross-entropy loss that compares probability map with the ground truth annotations.

$$L_{seg}^d = -\sum_{h,w}\sum_{k\in\mathcal{K}} Y_d^{(h,w,k)}\log(\boldsymbol{SS}(I_d)^{(h,w,k)}), \qquad d \in \{\mathcal{X}_s, \mathcal{X}_t\} \tag{1}$$

Adversarial Loss. Adversarial loss is used to train the segmentation network to fool the discriminator, in order to make the distribution of target probability map close to source domain.

$$L_{adv} = -\sum_{h,w}\log(\boldsymbol{D}(\boldsymbol{SS}(I_t))^{(h,w)}) \tag{2}$$

Discriminator Loss. Discriminator is trained to predict the domain label of each pixel in the probability map. We use $Y_t = 0$ for the source domain and $Y_t = 1$ for the target domain.

$$L_D = -\sum_{h,w}(1 - Y_t)\log(1 - \boldsymbol{D}(\boldsymbol{SS}(I_s))^{(h,w)}) + Y_t\log(\boldsymbol{D}(\boldsymbol{SS}(I_t))^{(h,w)}) \tag{3}$$

Semi-supervised Loss. Semi-supervised loss is used to enhance the training process in a self-taught manner using unlabeled images in the target domain [10,13]. The intuition is that the discriminator generates confidence score for each pixel in the images, and those pixels considered close to source domain could be used as additional information to train the segmentation network. Concretely, we highlight regions in target images where the confidence score is above a threshold, T and treat the probability maps from those regions as the pseudo labels. Therefore, the semi-supervised loss is defined as:

$$L_{semi} = -\sum_{h,w}\sum_{k\in\mathcal{K}}[\![\boldsymbol{D}(\boldsymbol{SS}(I_{t,u}))^{(h,w)} > T]\!]\cdot Y_{t,u}^{(h,w,k)}\log(\boldsymbol{SS}(I_{t,u})^{(h,w,k)}) \tag{4}$$

where $[\![\cdot]\!]$ is the indicator function, and $Y_{t,u}^{(h,w,k)} = 1$ if $k = \operatorname{argmax}_k \boldsymbol{SS}(I_{t,u})^{(h,w,k)}$.

3.3 Objective Functions

In this section, we provide the objective functions for three schemes in domain adaptation. Note that the objective function for discriminator is consistent across unsupervised, supervised and semi-supervised domain adaptation. Hence we follow Eq. 3 to train the discriminator and focus more on objective functions of semantic segmentation network $\boldsymbol{SS}$ below.

Unsupervised Domain Adaptation. Only unlabeled images in $\mathcal{X}_t$ are available. The goal is to minimize the per-pixel classifier loss in the source domain. Simultaneously, we want to make the distributions of probability maps in the target domain to closely match the probability map of images in the source domain by training segmentation network and discriminator network in a min-max game. The idea is for the segmentation network to be able to take advantage of labeled source domain images for per-pixel classification of target domain images. Besides, semi-supervised loss in Eq. 4 is used given the unlabeled target images are available. The objective functions to train the **SS** network under UDA can be written as:

$$L_{UDA}^{SS} = L_{seg}^s + \lambda_{adv}L_{adv} + \lambda_{semi}L_{semi} \tag{5}$$

where λ_{semi} is the weight for the pseudo annotations generated from the discriminator.

Supervised Domain Adaptation. Only a small number of labeled images in $\mathcal{X}_t$ are available. Segmentation network is trained to minimized the per-pixel cross-entropy loss in both domains. The objective functions to train the **SS** network under SDA can be written as:

$$L_{SDA}^{SS} = L_{seg}^s + L_{seg}^t + \lambda_{adv}L_{adv} \tag{6}$$

Semi-supervised Domain Adaptation. A majority of unlabeled images are available in $\mathcal{X}_t$, in addition to a small number of labeled images. The goal is to take advantage of the unlabeled images in the target domain to generate pseudo annotations and further improve performance for segmentation of target domain images. As such, the semi-supervised loss is calculated as in Eq. 4. The overall objective function to train the **SS** network under SSDA can be written as:

$$L_{SSDA}^{SS} = L_{seg}^s + L_{seg}^t + \lambda_{semi}L_{semi} + \lambda_{adv}L_{adv} \tag{7}$$

3.4   Network Architecture

Segmentation Network. The mSegNet network is a modified version of SegNet architecture, wherein a multiplicative skip connection between the last layer of the encoder and the decoder network is introduced, to estimate the probability maps for eye regions, including pupil, iris, sclera and background [7]. The network consists a 7-layer encoder module and a 7-layer decoder module, made up of convolutional layers with kernel-size 3×3 and stride of 1. The number of convolution channels in the encoder are 64, 64, 128, 128, 256, 256, 256, while transposed convolutional layer is used in the decoder, which takes the number of channels of 256, 256, 128, 128, 64, 64 and 4.

Discriminator. Discriminator is a fully convolutional network, and we follow the setting in [30] to contain 5 convolution layers, each with kernel 4 × 4 and stride of 2. The number of channels are 64, 128, 256, 512 and 1. Each convolution layer is followed by a leaky-ReLU non-linearity with parameter 0.2 except for the last layer. An up-sampling layer is added to the last convolutional layer to resize the output to the dimension of input.

4   Experiments and Discussion

We begin by introducing the eye data sets in the source and the target domain. We then investigate how segmentation performance as measured by using the mean intersection-over-union (mIoU) metric [15] is affected by varying the amount of training data in the target domain. Next, we run t-SNE analysis [19] to provide visualizations of the learned probability maps to illustrate the improvements for segmentation of eye images in the target domain using the proposed methods.

4.1   Datasets

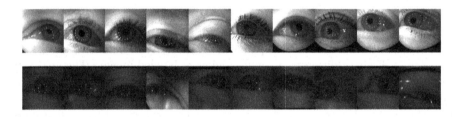

Fig. 2. Top to bottom: examples of images from source and target domain.

Source Domain. OpenEDS is a large scale data set of eye-images captured using a virtual reality (VR) head mounted display mounted with two synchronized eye facing cameras recorded at a frame rate 200 Hz under controlled illumination [7]. This data set was compiled from video capture of the eye-region collected from 152 participants and consists of 12,759 images with pixel-level annotations for key eye-regions: iris, pupil and sclera, at a resolution of 400 × 640 pixels. Following [7], we use the training set of 8,916 images with the corresponding annotations in OpenEDS as the source domain data.

Target Domain. Target domain data was collected using a commercial VR head mounted display that was modified with two eye facing cameras. The data set contains 4,629 eye images at a resolution of 400 × 640 pixels and 4 semantic categories are provided with pixel-level labels (pupil, iris, sclera and background).

For all the experiments we split the data set into images of 1,532, 500 and 2,597 as training, validation and test set. The results of proposed domain adaptation scheme are reported on the test set. Figure 2 shows some examples from the source and target domains respectively.

4.2   Training Details

We use PyTorch to implement the proposed model [22]. ADAM optimizer [12] is used to train both segmentation network and the discriminator network, with initial learning rate of 0.0001, momentum 0.9 and 0.999 respectively, which is decreased using a polynomial decay function with power of 0.9 [30]. We scale the images to a resolution of 184×184 and train the models for 200 epochs with batch size 8. A parameter search is conducted to find the optimal settings of λ_{adv}, λ_{semi} and T. As a result, we set λ_{adv} as 0.001, λ_{semi} as 0.1, T as 0.8. Algorithm 1 demonstrates the process of training models with varying amount of target domain data.

Algorithm 1: Training Protocol

Input Source domain: $\mathcal{X}_s = \{(I_s, Y_s)\}$;
　　　　Target domain (UDA): $\mathcal{X}_t = \{I_{t,u}\}$ or
　　　　Target domain (SDA): $\mathcal{X}_t = \{(I_t, Y_t)\}$ or
　　　　Target domain (SSDA): $\mathcal{X}_t = \{(I_t, Y_t); I_{t,u}\}$;
　　　　N: # unlabeled images selected per iteration;
　　　　M: max iterations;
Model Segmentation network SS; Discriminator D;
for $iteration = 1$ to M **do**
　　Initialize SS with parameters that are pretrained on $\mathcal{X}_s$
　　UDA Training SS and D with $(\mathcal{X}_s, \mathcal{X}_t)$ via Eq. 5
　　SDA Training SS and D with $(\mathcal{X}_s, \mathcal{X}_t)$ via Eq. 6
　　SSDA Training SS and D with $(\mathcal{X}_s, \mathcal{X}_t)$ via Eq. 7
　　Random select N unlabeled images $I_N \in \mathcal{X}_t$ where:
　　UDA $\mathcal{X}_t \leftarrow I_{t,u} \cup I_N$
　　SDA $\mathcal{X}_t \leftarrow (I_t, Y_t) \cup (I_N, Y_N)$, Y_N is corresponding labels of I_N
　　SSDA $\mathcal{X}_t \leftarrow ((I_t, Y_t) \cup (I_N, Y_N); I_{t,u} \setminus I_N)$
end

4.3   Results

Performance with Varying Amount of Training Data. In Fig. 3, we chart the segmentation performance, as measured using mIoU, for the three DA frameworks as function of the number of target domain images available in the training data. We note that the mIoU for models trained under all three frameworks increase and reach a plateau as more images in the target domain are available, the extent of increase is much smaller for model trained under UDA. We note that, with only 200 labeled images in the target domain, mIoU is improved by 5.4% (from 81.7% in UDA to 87.1% in SDA) and 6.6% (from 81.7% UDA to 88.3% SSDA), which affirms our hypothesis that even with limited number of labeled target domain images, domain-specific information can boost the performance for segmentation of images in the target domain.

We note that SSDA consistently outperforms SDA, since SSDA is able to leverage the most available information, including images with annotations as well as all the available unlabeled images. As can be seen from Eqs. 6 and 7, the gain in the performance for SSDA relative to SDA is brought about by the addition of adversarial loss as well as the semi-supervised loss from unlabeled target images. As shown in Algorithm 1, however, as more target images with annotations are provided to train the model, the amount of unlabeled target images are decreased. Therefore, the performance gain brought by the adversarial and semi-supervised loss from unlabeled target images is diminishing. This explains the smaller performance gap between SDA and SSDA as more labeled target images are available during training. We finally note that SSDA is able to achieve a competitive mIoU of 89.5% with only 800 labeled target-domain images (53.2% of total data). It is encouraging to observe that the proposed SSDA can efficiently take advantage of insufficient labels while still producing competitive performance on segmentation task.

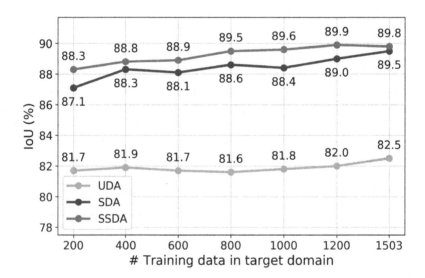

Fig. 3. Performance of three domain adaptation frameworks (unsupervised, supervised and semi-supervised domain adaptation) trained with varying amounts of target images.

Performance with All Training Data. In Table 1, we summarize the performance for all the three frameworks for the case when all training images and the corresponding annotations in target domain are used. Two baseline models are being compared: 1) **Source Only:** we train the segmentation model on source domain data. 2) **Target finetuning:** we train the segmentation model on source domain data and then finetuned the segmentation model using the available labeled target domain data. By comparing UDA with Source only, the adversarial loss directly brings an improvement of 15.2%, from 66.3% of baseline to 81.5% for UDA. With additional labeled target domain data available, by comparing SDA and SSDA with target finetuning, mIoU is increased by 0.2% and 0.5%, respectively. Comparing all the domain adaptation based models, given the context of additional labels, mIoU is increased by 8.3%, from 81.5% for UDA to 89.8% for SSDA. Not only does SSDA consistently performs better than SDA over varying number of labeled target domain training images, SSDA also achieved slightly better performance metrics than SDA (0.3%) and target finetuning (0.5%) when all target domain labeled images are available.

Table 1. Results of models trained with all target images (# images = 1,503).

Models	Overall	Pupil	Iris	Sclera	Background
Source only	66.3	60.7	66.4	48.8	89.5
Target finetuning	89.3	89.6	90.6	79.0	97.5
UDA	82.5	82.3	84.0	64.0	94.1
SDA	89.5	89.8	90.7	78.8	97.8
SSDA	89.8	90.0	90.3	79.3	97.9

Visualization of Segmentation Results. Figure 4 shows segmentation results from models trained under different frameworks. Note that the ground truth labels in the target domain are missing fine-grained boundary, i.e. detailed boundaries of sclera. Domain adaptation is able to transfer the knowledge from source domain with high quality labels to target domain where both quality and quantity are limited, which improves segmenting detailed geometry such as sclera in the target images.

Visualization of Feature Clustering. Figure 5 shows the T-SNE [19] of the embeddings of the per-pixel probability maps of training data in source and target domain. The "perplexity" parameter for T-SNE was set as 20. Qualitative results show that with domain adaptation applied, the distributions of source and target domain probability maps overlap closely.

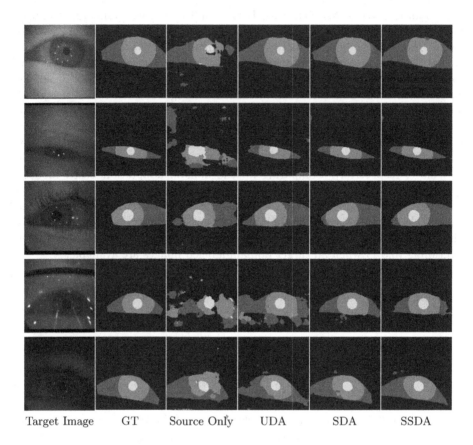

Target Image   GT   Source Only   UDA   SDA   SSDA

Fig. 4. Examples and segmentation results from different models. From left to right: target images, ground truth, source only, predictions of UDA, SDA and SSDA. From top to bottom: regular-opened eyes, half-opened eyes, long eyelashes, eyeglasses, dim light.

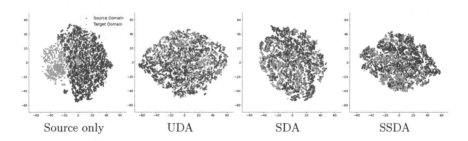

Source only   UDA   SDA   SSDA

Fig. 5. T-SNE visualization. Blue: source domain. Red: target domain. (Color figure online)

5   Conclusions

This paper considers the problem of domain adaptation for eye segmentation, under the setting where a small amount of labeled images and a majority of unlabeled images in the target domain are available. We systematically compare the performance of semantic segmentation models trained under unsupervised (UDA), supervised (SDA) and semi-supervised domain adaptation (SSDA), as a function of the number of labeled target domain training images. Results show that SSDA is able to improve mIoU from 81.7% of UDA to 88.3% when only 200 labeled target domain images are available for training. Furthermore, with 50% of labeled target domain images being used, we are able to achieve a competitive mIoU of 89.5% using SSDA. In conclusion, results demonstrate the benefit to annotate a small number of target domain images to effectively perform domain adaptation. We hope that presented approaches would be useful for eye tracking practitioners that employ segmentation of periocular regions in their eye tracking pipelines to improve data quality of obtained gaze positional data, which is important for providing the broadest possible field of eye tracking-driven applications.

In our future work we plan to explore what aspects of eye tracking data quality (e.g., spatial accuracy, spatial precision, spatial resolution, data loss, etc.) are affected by segmentation of periocular regions and to what degree.

Acknowledgment. We thank our colleagues Jixu Chen and Kapil Krishnakumar for providing target domain data, Robert Cavin, Abhishek Sharma and Elias Guestrin for providing constructive and thoughtful feedback.

References

1. Badrinarayanan, V., Kendall, A., Cipolla, R.: SegNet: a deep convolutional encoder-decoder architecture for image segmentation. IEEE Trans. Pattern Anal. Mach. Intell. **39**(12), 2481–2495 (2017)
2. Boutros, F., Damer, N., Kirchbuchner, F., Kuijper, A.: Eye-MMS: miniature multi-scale segmentation network of key eye-regions in embedded applications. In: Proceedings of the IEEE International Conference on Computer Vision Workshops (2019)
3. Bowyer, K.W., Burge, M.J. (eds.): Handbook of Iris Recognition. ACVPR. Springer, London (2016). https://doi.org/10.1007/978-1-4471-6784-6
4. Das, A., et al.: SSERBC 2017: sclera segmentation and eye recognition benchmarking competition. In: 2017 IEEE International Joint Conference on Biometrics (IJCB), pp. 742–747. IEEE (2017)
5. Ganin, Y., Lempitsky, V.: Unsupervised domain adaptation by backpropagation. In: Bach, F., Blei, D. (eds.) Proceedings of the 32nd International Conference on Machine Learning, Proceedings of Machine Learning Research, Lille, France, vol. 37, pp. 1180–1189. PMLR, 07–09 July 2015
6. Ganin, Y., et al.: Domain-adversarial training of neural networks. J. Mach. Learn. Res. **17**(1), 2030–2096 (2016)
7. Garbin, S.J., et al.: Dataset for eye tracking on a virtual reality platform. In: Symposium on Eye Tracking Research and Applications, pp. 1–10 (2020)

8. Hoffman, J., et al.: CyCADA: cycle-consistent adversarial domain adaptation. In: ICML, vol. 80, pp. 1994–2003. PMLR (2018)
9. Hoffman, J., Wang, D., Yu, F., Darrell, T.: FCNs in the wild: pixel-level adversarial and constraint-based adaptation. arXiv preprint arXiv:1612.02649 (2016)
10. Hung, W.C., Tsai, Y.H., Liou, Y.T., Lin, Y.Y., Yang, M.H.: Adversarial learning for semi-supervised semantic segmentation. In: BMVC, p. 65. BMVA Press (2018)
11. Kim, S.H., Lee, G.S., Yang, H.J., et al.: Eye semantic segmentation with a lightweight model. In: 2019 IEEE/CVF International Conference on Computer Vision Workshop (ICCVW), pp. 3694–3697. IEEE (2019)
12. Kingma, D.P., Ba, J.: Adam: a method for stochastic optimization. In: 3rd International Conference on Learning Representations, ICLR 2015, San Diego, CA, USA, 7–9 May 2015, Conference Track Proceedings (2015). http://arxiv.org/abs/1412.6980
13. Li, Y., Yuan, L., Vasconcelos, N.: Bidirectional learning for domain adaptation of semantic segmentation. In: Proceedings of the IEEE Conference on Computer Vision and Pattern Recognition, pp. 6936–6945 (2019)
14. Lian, S., Luo, Z., Zhong, Z., Lin, X., Su, S., Li, S.: Attention guided U-Net for accurate iris segmentation. J. Vis. Commun. Image Represent. **56**, 296–304 (2018)
15. Long, J., Shelhamer, E., Darrell, T.: Fully convolutional networks for semantic segmentation. In: Proceedings of the IEEE Conference on Computer Vision and Pattern Recognition, pp. 3431–3440 (2015)
16. Lucio, D.R., Laroca, R., Severo, E., Britto, A.S., Menotti, D.: Fully convolutional networks and generative adversarial networks applied to sclera segmentation. In: 2018 IEEE 9th International Conference on Biometrics Theory, Applications and Systems (BTAS), pp. 1–7. IEEE (2018)
17. Luo, B., Shen, J., Cheng, S., Wang, Y., Pantic, M.: Shape constrained network for eye segmentation in the wild. In: The IEEE Winter Conference on Applications of Computer Vision, pp. 1952–1960 (2020)
18. Luo, B., Shen, J., Wang, Y., Pantic, M.: The iBUG eye segmentation dataset. In: 2018 Imperial College Computing Student Workshop (ICCSW 2018). Schloss Dagstuhl-Leibniz-Zentrum fuer Informatik (2019)
19. van der Maaten, L., Hinton, G.: Visualizing data using t-SNE. J. Mach. Learn. Res. **9**, 2579–2605 (2008)
20. Motiian, S., Jones, Q., Iranmanesh, S., Doretto, G.: Few-shot adversarial domain adaptation. In: Advances in Neural Information Processing Systems, pp. 6670–6680 (2017)
21. Naqvi, R.A., Loh, W.K.: Sclera-Net: accurate sclera segmentation in various sensor images based on residual encoder and decoder network. IEEE Access **7**, 98208–98227 (2019)
22. Paszke, A., et al.: Automatic differentiation in PyTorch (2017)
23. Perry, J., Fernandez, A.: MinENet: a dilated CNN for semantic segmentation of eye features. In: Proceedings of the IEEE International Conference on Computer Vision Workshops (2019)
24. Ronneberger, O., Fischer, P., Brox, T.: U-Net: convolutional networks for biomedical image segmentation. In: Navab, N., Hornegger, J., Wells, W.M., Frangi, A.F. (eds.) MICCAI 2015. LNCS, vol. 9351, pp. 234–241. Springer, Cham (2015). https://doi.org/10.1007/978-3-319-24574-4_28
25. Saito, K., Kim, D., Sclaroff, S., Darrell, T., Saenko, K.: Semi-supervised domain adaptation via minimax entropy. In: Proceedings of the IEEE International Conference on Computer Vision, pp. 8050–8058 (2019)

26. Sankowski, W., Grabowski, K., Napieralska, M., Zubert, M., Napieralski, A.: Reliable algorithm for iris segmentation in eye image. Image Vis. Comput. **28**(2), 231–237 (2010)
27. Shu, Y., Cao, Z., Long, M., Wang, J.: Transferable curriculum for weakly-supervised domain adaptation. In: Proceedings of the AAAI Conference on Artificial Intelligence, vol. 33, pp. 4951–4958 (2019)
28. Sun, R., Zhu, X., Wu, C., Huang, C., Shi, J., Ma, L.: Not all areas are equal: transfer learning for semantic segmentation via hierarchical region selection. In: Proceedings of the IEEE Conference on Computer Vision and Pattern Recognition, pp. 4360–4369 (2019)
29. Tang, Y., Wang, J., Gao, B., Dellandréa, E., Gaizauskas, R., Chen, L.: Large scale semi-supervised object detection using visual and semantic knowledge transfer. In: Proceedings of the IEEE Conference on Computer Vision and Pattern Recognition, pp. 2119–2128 (2016)
30. Tsai, Y.H., Hung, W.C., Schulter, S., Sohn, K., Yang, M.H., Chandraker, M.: Learning to adapt structured output space for semantic segmentation. In: Proceedings of the IEEE Conference on Computer Vision and Pattern Recognition, pp. 7472–7481 (2018)
31. Tsai, Y.H., Sohn, K., Schulter, S., Chandraker, M.: Domain adaptation for structured output via discriminative patch representations. In: Proceedings of the IEEE International Conference on Computer Vision, pp. 1456–1465 (2019)
32. Tzeng, E., Hoffman, J., Saenko, K., Darrell, T.: Adversarial discriminative domain adaptation. In: Proceedings of the IEEE Conference on Computer Vision and Pattern Recognition, pp. 7167–7176 (2017)
33. Valindria, V.V., et al.: Domain adaptation for MRI organ segmentation using reverse classification accuracy. arXiv preprint arXiv:1806.00363 (2018)
34. Wang, C., He, Y., Liu, Y., He, Z., He, R., Sun, Z.: ScleraSegNet: an improved U-Net model with attention for accurate sclera segmentation. In: IAPR International Conference on Biometrics, vol. 1 (2019)
35. Wang, J.G., Sung, E., Venkateswarlu, R.: Estimating the eye gaze from one eye. Comput. Vis. Image Underst. **98**(1), 83–103 (2005)
36. Wang, M., Deng, W.: Deep visual domain adaptation: a survey. Neurocomputing **312**, 135–153 (2018)
37. Yao, T., Pan, Y., Ngo, C.W., Li, H., Mei, T.: Semi-supervised domain adaptation with subspace learning for visual recognition. In: Proceedings of the IEEE Conference on Computer Vision and Pattern Recognition, pp. 2142–2150 (2015)
38. Zhang, Y., Qiu, Z., Yao, T., Liu, D., Mei, T.: Fully convolutional adaptation networks for semantic segmentation. In: Proceedings of the IEEE Conference on Computer Vision and Pattern Recognition, pp. 6810–6818 (2018)
39. Zhou, Z., Du, E.Y., Thomas, N.L., Delp, E.J.: A new human identification method: sclera recognition. IEEE Trans. Syst. Man Cybern.-Part A Syst. Hum. **42**(3), 571–583 (2011)

EyeSeg: Fast and Efficient Few-Shot Semantic Segmentation

Jonathan Perry$^{(\boxtimes)}$ ⓘ and Amanda S. Fernandez ⓘ

University of Texas at San Antonio, San Antonio, USA
{jonathan.perry,amanda.fernandez}@utsa.edu
http://www.cs.utsa.edu/~fernandez/vail

Abstract. Semantic segmentation is a key component in eye- and gaze-tracking for virtual reality (VR) and augmented reality (AR) applications. While it is a well-studied computer vision problem, most state-of-the-art models require large amounts of labeled data, which is limited in this specific domain. An additional consideration in eye tracking is the capacity for real-time predictions, necessary for responsive AR/VR interfaces. In this work, we propose EyeSeg, an encoder-decoder architecture designed for accurate pixel-wise few-shot semantic segmentation with limited annotated data. We report results from the OpenEDS2020 Challenge, yielding a 94.5% mean Intersection Over Union (mIOU) score, which is a 10.5% score increase over the baseline approach. The experimental results demonstrate state-of-the-art performance while preserving a low latency framework. Source code is available: http://www.cs.utsa.edu/~fernandez/segmentation.html.

Keywords: Semantic segmentation · Eye tracking · Computer vision · OpenEDS2020

1 Introduction

The concept of foveated rendering has significant potential to improve upon the visual and computational performance of VR/AR applications. A critical underlying component of this technique is eye-tracking, which often relies on semantic segmentation - accurately and efficiently identifying regions of the eye.

Supervised training of neural networks for these segmentation tasks often requires extremely labor-intensive annotations as well as a relatively high volume of samples. In addition, these vision models are intended for embedded systems, such as head-mounted displays (HMD), and therefore we must consider an additional constraint of computational complexity in terms of the number of trainable or learned parameters.

Several efficient approaches to semantic segmentation for eye tracking on HMDs have shown the ability to reduce model complexity and demonstrate accurate performance in terms of mean Intersection Over Union (mIOU) [2,3, 8,11]. However, in this work, we additionally focus on the limited availability of

© Springer Nature Switzerland AG 2020
A. Bartoli and A. Fusiello (Eds.): ECCV 2020 Workshops, LNCS 12535, pp. 570–582, 2020.
https://doi.org/10.1007/978-3-030-66415-2_37

large, fully-labeled datasets for this task. While VR/AR technologies continue to increase in popularity, diversity in implementations reduces the consistency of such available and labeled data for evaluating deep learning models. We therefore explore the efficacy of existing models on datasets with limited amount of labeled data, as defined in the OpenEDS 2020 Challenge for Semantic Segmentation [9], and find a reduction in this measure of performance.

In response, we propose a new encoder-decoder framework, EyeSeg, which is designed for training where there is scarcity of annotated data as well as optimized for embedded systems. Our architecture improves on related state-of-the-art approaches [10,11] in four main ways.

First, it improves upon the constraint of computational complexity by reducing the number of trainable parameters in our framework.

Second, it leverages a customized combined loss function of the standard categorical cross entropy (CCE) and generalised dice loss (GDL) [13].

Third, it applies well-established targeted data manipulation and augmentation techniques which have been demonstrated for their performance optimization [3].

Finally, it will utilize two different training methods to leverage capabilities of semi-supervised learning and identify the performance gain from a standard supervised learning approach.

In evaluation of our proposed approach, we measure the performance of EyeSeg against the Open Eye Dataset [9] for the 2020 Semantic Segmentation Challenge. The performance metric chosen for this challenge is mIOU. Additionally, we compare model complexity, as defined in the previous OpenEDS 2019 challenge [6], in order to thoroughly evaluate and compare with existing approaches. Our method demonstrates a significant improvement over the baseline model, and we additionally compare our proposed method with current state-of-the-art models for eye segmentation.

2   Related Works

As the availability of high-resolution digital media datasets continues to increase, research in segmentation algorithms has kept pace through strategic optimization and deep neural networks. Building on convolutional neural networks (CNNs) and fully convolutional networks (FCNs), segmentation architectures have benefited from techniques in pooling, filtering, and dilation [16]. In this section, related approaches to semantic segmentation are respectively described for accurate pixel-wise classification, complexity reduction methods, and imbalanced class representations.

Encoder-Decoder Frameworks. Convolutional encoder-decoder frameworks have been widely used for robust feature extraction in a range of computer vision applications.

SegNet [1] utilized this framework to improve upon scene understanding with a non-linear upsampling augmentation for FCNs. Chen et al. [5] employed the

DeepLab [4] atrous convolution module, a dilation to further increase the performance of an encoder-decoder framework through exponentially larger receptive fields without an increased computational cost. UNet [12] introduced a patterned encoder-decoder design, containing residual connections in order to maintain spatial information from earlier layers within the encoder. A demonstrable trend in segmentation is to leverage a general encoder-decoder design for increasing performance and decreasing parameterization.

Lightweight Frameworks. The emerging technologies such as AR/VR or autonomous vehicles have shown that model complexity is a key factor in the application of segmentation models in real-world environments. ENet [10] improved upon model complexity towards real-time segmentation for autonomous vehicles. More recently, frameworks have been presented from OpenEDS challenge that improved in both computational complexity and performance capabilities towards AR/VR applications [2,3,8,11]. These frameworks optimize for the number of trainable parameters within a deep neural network.

Semi-supervised and Unsupervised Training. Domain adaptation and self-training have been widely adopted as techniques for a structured method of training with data that has a low amount of labeled samples, and this has been successful in many different domains, including synthetic to real domain adaptation for vehicle video sequences [14]. Recent work on self-training [17] that utilized a student-teacher format demonstrated state-of-the-art performance with a fast training schedule. Both of these works [14,17] used a type of entropy based approximation for determining quality or confidence of inference.

3 EyeSeg Architecture

Our primary aim is to improve the performance and efficiency of semantic segmentation, especially for situations where there is limited availability of labeled data. In this section, we outline our proposed neural network architecture and describe its total loss function.

3.1 Network Architecture

Figure 1 provides a high-level view of the composition of EyeSeg, an encoder-decoder architecture. EyeSeg consists of 4 encoder blocks which store feature maps learned at each step prior to the down sampling portion connecting to the subsequent encoder blocks. The decoder portion of EyeSeg upsamples in a mirrored or patterned fashion with respect to the encoder and utilizes the store feature maps from the encoding as an alternative path to sustain simple high level features.

Recent approaches [3,11] specifically for eye-tracking have shown an increase in accuracy from applying different mechanisms to an encoder-decoder.

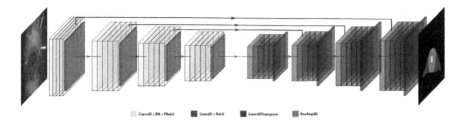

Fig. 1. Visualization of the proposed framework (high-level view). From left: Input image 640 × 640 in gray scale format through 4× Encoder blocks & 4× Decoder blocks to a predicted mask of background, sclera, iris, and pupil.

Similarly, EyeSeg employs two of these components, residual connections and dilated convolutional layers, where these different components combined substantially increase the performance without impacting computational complexity drastically. However, both the encoder and decoder proposed are distinct variants from existing architectures [3,11] in both size and structure. Table 1 outlines the framework of EyeSeg in more detail, breaking down the internal blocks, output sizes, and types within layers.

3.2 Encoder

Each encoder block consists of 4 convolutional layers, that are paired with Parametric rectified Linear Unit (PReLU) [7] and Batch Normalization (BN) layers per convolution. There are 2 variants to the basic structure of an encoder block, which modify one of the convolutional layers. The variant will be a dilated convolutional layer or a pooling layer. Finally, our encoder block utilizes average pooling layers for a more accurate localization than what is provided in max pooling layers. A single encoder block is visualized in Fig. 2.

3.3 Decoder

As illustrated in Fig. 2, the decoder blocks each have 4 primary components that consists of convolutional, activation, upsampling, and normalization layers. Each convolutional layer is paired with a Rectified Linear Unit (ReLU). A convolutional transpose layer is leveraged for the task of upsampling.

With an emphasis in reduction of computational complexity, we forgo the additional BN layers commonly incorporated into the decoder blocks at this stage [11].

In order to sustain spatial information from earlier layers, we implemented residual connections pairing the appropriate encoder blocks to the respective decoder blocks.

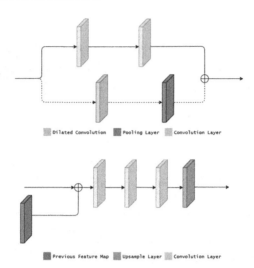

Fig. 2. From top: A flowchart of a single encoder block where the dotted line represents conditionally applicable layers, a single decoder block where the previous feature map shown in red is a residual connection. (Color figure online)

3.4   Loss Function

With motivation from RITNet [3], our implementation utilizes a customized total loss function which can be represented in 2 parts: cross entropy and generalised dice loss. The total loss function for our proposed method is as follows:

$$\mathcal{L}_{loss} = \mathcal{L}_{cce} + \mathcal{L}_{gdl} \tag{1}$$

$$\mathcal{L}_{cce} = -\frac{1}{N}\Sigma_{l=1}\Sigma_{n=1}^{N}r_{ln}\log(p_{ln}) \tag{2}$$

$$\mathcal{L}_{gdl} = 1 - 2\frac{\Sigma_{l=1}w_{l}\Sigma_{n=1}^{N}r_{ln}p_{ln}}{\Sigma_{l=1}w_{l}\Sigma_{n=1}^{N}r_{ln} + p_{ln}} \tag{3}$$

where $\mathcal{L}_{cce}$ is a standard implementation of categorical cross entropy and $\mathcal{L}_{gdl}$ is an implementation of a generalized dice loss function [13] for imbalanced class features. The aim of this combined loss function is to mitigate the over-representation of one or many class features l within each sample n, comparing the ground truth (target) r with the predicted values p.

4   Experiments and Results

4.1   OpenEDS 2020 Challenge Data

In this work, the eye segmentation subset of the Open Eye Dataset 2020 [9] is used for evaluation. This dataset consists of 29,476 images, from 74 different

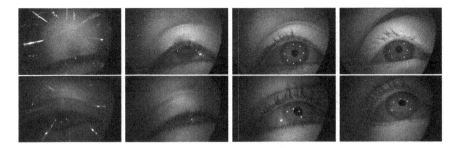

Fig. 3. Sampled images from the OpenEDS2020 [9] dataset. From left column: Image with reflection and participants eye partially open, participants eye occluded by the eye lid, participants eye fully open.

participants, ranging in ethnicity, gender, eye color, age, and accessories (such as make-up and glasses). Shown in Fig. 3, the dataset has variations of images that allow for a range of different real world challenges for applications of eye tracking including participants with accessories, nearly visible eyes, partially occluded eyes, or fully observable eyes. Further, the images were from 200 different sequences of 30 s video recordings from 74 participants. The sequences of recorded sessions contain only a few annotations, where approximately 5% of the entire dataset contained annotations. Labels provided in the form of pixel-level masks denoting eye region, iris, and pupil were manually annotated by two or more individuals. Overall, the labeled portion of this dataset consists of only 2,605 annotated images. For the purposes of challenge, a hidden test set is made unavailable, comprised of five of the annotated images per sequence. This leaves only 1605 total annotated images for our training and validation purposes.

4.2 Data Augmentation

In order to account for the variety of challenging categories within the dataset viewed in Fig. 3, we propose two approaches to reduce the most noticeable undesirable properties of the original images. First, we utilize a technique to amplify the contrast of the image to improve upon low light areas of the original image. Second, we apply image denoising techniques to smooth the prominent reflections caused by participants wearing accessories. Additionally, we performed data manipulation techniques such as horizontal flipped or mirrored samples in order to combat the detriment to training on sparse amount of data. The process of both denoising techniques and contrast amplification are shown in Fig. 4.

Adaptive Histogram Equalization. Contrast Limited Adaptive Histogram Equalization (CLAHE) [15] is an enhancement method for improving the quality of images and video where visibility is less than satisfactory. A key component to CLAHE is the clipped or limited range of its visibility enhancement, whereas the standard AHE algorithm will allow for overly amplified images to occur and

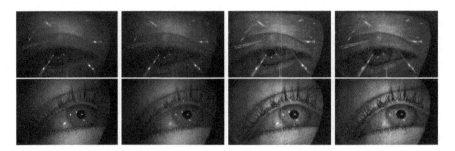

Fig. 4. This figure shows the application of both CLAHE and noise reduction from two different participants where the top row is a participant with reflection from glasses as well as low light and the bottom row is a participant with partial reflection. From Left: Original image, noise reduction applied, CLAHE applied, and fully pre-processed with both CLAHE and noise reduction.

presents the problem of trading off low light samples with over exposed samples. Our proposed EyeSeg utilizes this CLAHE enhancement to employ a contrast amplification to the images in order to achieve more visibility of class features.

Image Noise Reduction. The front facing sensors of AR/VR devices such as HMDs usually are accompanied with both visible and infrared light to illuminate the participants eye for more accurate eye-tracking. The trade-off with deploying these types of emitters will result in glare and reflections from the participants eye itself and additional noise will be caused if the participant is wearing glasses. We use Gaussian filtering as a noise reduction method to provide more clear representations of the images.

4.3   Training

The lack of manually annotated images causes a significant detriment to the capabilities of the traditional methods of training. In this section, we discuss the two experimental training methods applied to EyeSeg. Initially, the 1,605 annotated images are utilized in a supervised learning environment. Second, we describe the semi-supervised training method applied.

Supervised Training. We trained EyeSeg with an ADAM optimizer at a initial learning rate of $1e-3$, and is lowered to $1e-4$ once there is a plateau. The training process is terminated within 200 epochs. The training and inference of this model were performed on the padded image size of 640×640. Our network was tested on the Open Eye Dataset [9] hidden test samples achieving a mIOU score of 0.945 shown in Table 2. Additionally, we trained segmentation models [3,11] from OpenEDS 2019 Challenge [6] using the same training method to encompass a more robust comparison of EyeSeg.

Semi-supervised Training. Utilizing the entire dataset, We trained EyeSeg with pseudo labels generated for the portion of the dataset without annotations. Our proposed semi-supervised method aims to minimize entropy to ensure quality pseudo labels similar to recent works [14,17]. Additionally, we incorporated the 1,605 annotated images from the Open Eye Dataset [9]. The pseudo labels generated were utilized only whenever the entropy of each pixel-wise classification demonstrated high confidence or low entropy. Entropy for a single sample is written as follows:

$$\mathcal{E}_{entropy} = -\Sigma_{i,j=0}^{N}\Sigma_{l=1} p_{ijl} \log(p_{ijl}), \tag{4}$$

where $\mathcal{E}_{entropy}$ will demonstrate the confidence of the pseudo label determined by evaluating each pixel at ij until N^{th} pixel for each label or class l. Entropy will be low if only one class per ij is classified with high confidence from inference. Our network achieved a marginal score increase of 0.06 over the supervised learning method to a total score of 0.951. However, due to a lack of accessibility to the Open Eye Dataset [9] hidden test samples we could not accurately compare this iteration to the previous supervised method.

4.4 Results

The 2020 OpenEDS Semantic Segmentation Challenge includes a leaderboard, ranking submissions by mIOU score. A baseline encoder-decoder network [9] was provided, which was loosely based upon SegNet [1], an encoder-decoder architecture with relatively few parameters and a base score of 0.84. Additional results are provided in Table 2, including mIOU, but also breaking down the performance across the 4 semantic categories and including the number of parameters in the models. Related works in this table include top-performing models from the 2019 OpenEDS challenge, RITNet [3] and MinENet [11]. EyeSeg demonstrates a higher mIOU score, consistently improving across the background, sclera, iris, and pupil semantic categories. While the number of parameters in EyeSeg are streamlined in comparison with related works, the baseline model was significantly smaller than our proposed architecture. This trade-off provides our model with improved performance, but we will discuss further plans to reduce size in the following section.

A visual evaluation of the effectiveness of EyeSeg is shown in Figs. 5 and 6. In Fig. 5, images from the dataset are shown in the first column, followed by the ground truth annotation, and our predicted segmentation. Despite reflections within the eye, partial occlusion by eyelid, and differences in lighting, the EyeSeg predictions are fairly close to the ground truth. In comparison, Fig. 6 looks at challenging edge cases in the dataset - reflections from glasses, severe occlusion by eyelid, and varied lighting. In these instances, the EyeSeg predictions are often close, some degradation exists in the confidence of boundaries, such as in the left side of the final image.

Table 1. Architecture of our proposed method. Output sizes are provided for input size of $640 \times 640 \times 1$.

Name	Type	Output Size
Input		$16 \times 640 \times 640$
Encode Block 1.0		$32 \times 640 \times 640$
Encode Block 1.1		$32 \times 640 \times 640$
Encode Block 1.2		$32 \times 640 \times 640$
Encode Block 1.3	Downsampling	$32 \times 320 \times 320$
Encode Block 2.0		$32 \times 320 \times 320$
Encode Block 2.1	Dilated (2×2)	$32 \times 320 \times 320$
Encode Block 2.2	Dilated (4×4)	$32 \times 320 \times 320$
Encode Block 2.3	Downsampling	$32 \times 160 \times 160$
Encode Block 3.0		$32 \times 160 \times 160$
Encode Block 3.1	Dilated (2×2)	$32 \times 160 \times 160$
Encode Block 3.2	Dilated (4×4)	$32 \times 160 \times 160$
Encode Block 3.3	Downsampling	$32 \times 80 \times 80$
Encode Block 4.0		$32 \times 80 \times 80$
Encode Block 4.1		$32 \times 80 \times 80$
Encode Block 4.2		$32 \times 80 \times 80$
Encode Block 4.3		$32 \times 80 \times 80$
Decode Block 1.0		$32 \times 80 \times 80$
Decode Block 1.1		$32 \times 80 \times 80$
Decode Block 1.2		$32 \times 80 \times 80$
Decode Block 1.3	Upsampling	$32 \times 160 \times 160$
Decode Block 2.0	Residual connection	$64 \times 160 \times 160$
Decode Block 2.1		$32 \times 160 \times 160$
Decode Block 2.2		$32 \times 160 \times 160$
Decode Block 2.3	Upsampling	$32 \times 320 \times 320$
Decode Block 3.0	Residual connection	$64 \times 320 \times 320$
Decode Block 3.1		$32 \times 320 \times 320$
Decode Block 3.2		$32 \times 320 \times 320$
Decode Block 3.3	Upsampling	$32 \times 640 \times 640$
Decode Block 4.0	Residual connection	$64 \times 640 \times 640$
Decode Block 4.1		$32 \times 640 \times 640$
Decode Block 4.2		$32 \times 640 \times 640$
Decode Block 4.3		$32 \times 640 \times 640$
Output		$4 \times 640 \times 640$

Table 2. Comparison of semantic segmentation approaches on the OpenEDS dataset, as of submission. The Baseline is the model provided in the OpenEDS Semantic Segmentation Challenge 2020.

	mIOU	Background	Sclera	Iris	Pupil	#parameters
Baseline [9]	0.84	0.971	0.674	0.835	0.835	40k
MinENet [11]	0.91	0.99	0.83	0.93	0.89	222k
RITNet [3]	0.93	0.99	0.87	0.95	0.915	250k
EyeSeg	**0.945**	0.99	0.89	0.95	0.95	190k

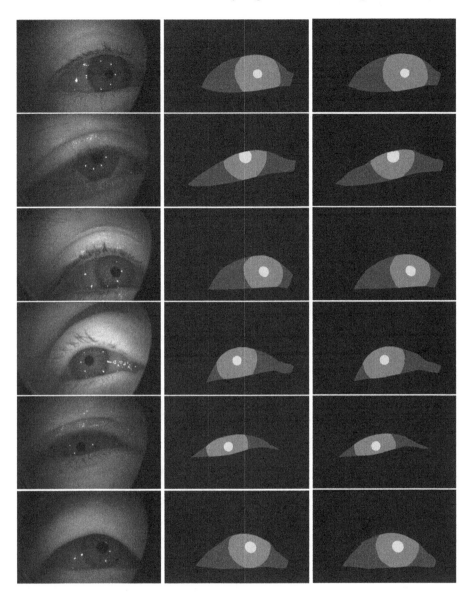

Fig. 5. Results from EyeSeg without any samples that include low visibility or reflections. From Left: Original input image, Ground truth or target value, predictions.

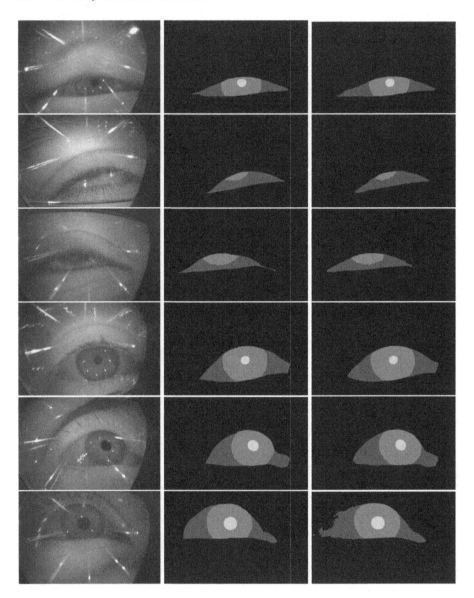

Fig. 6. Results from EyeSeg that posed more of a challenge including low light and reflections. From Left: Original input image, Ground truth or target value, predictions.

5   Conclusion

In this work, we introduce EyeSeg, a generalized method for few-shot segmentation with the use of an efficient encoder-decoder and customized total loss function. We apply EyeSeg to the Open Eye Dataset [9], a challenge for semantic segmentation of eye regions in images taken for VR/AR displays. Our method uses a combined loss function to reduce the impact of imbalanced class features often prevalent in real-world datasets. Additionally, several data augmentation techniques are applied to mitigate the limited amount of labeled data, as well as to accommodate for challenging categories of images within the dataset, such as makeup, glasses, and closed eyelids.

We demonstrate performance of EyeSeg against the baseline implementation on the challenge [9], outperforming by 10.5% mIOU. We also compare with recent related approaches, achieving state-of-the-art performance while maintaining a lightweight design for the capability of real-world use in AR/VR environments.

In future work, we aim to further optimize EyeSeg by reducing the number of parameters, increasing performance per mIOU, and addressing shortcomings on the outlier data, identified in Fig. 6. Since the image data is a sequence of video frames an application of memory units such as an LSTM could have a positive impact on the accuracy of EyeSeg. While our data augmentation methods are beneficial to the performance of EyeSeg in low light or noisy environments, it is not entirely solved and could be addressed from the application of a memory unit or additional pre-processing techniques. We plan apply our approach to further domains and data which contains varying levels of class feature imbalances, and limited labeled data.

References

1. Badrinarayanan, V., Kendall, A., Cipolla, R.: SegNet: a deep convolutional encoder-decoder architecture for image segmentation. IEEE Trans. Pattern Anal. Mach. Intell. **39**(12), 2481–2495 (2017)
2. Boutros, F., Damer, N., Kirchbuchner, F., Kuijper, A.: Eye-MMS: miniature multi-scale segmentation network of key eye-regions in embedded applications. In: The IEEE International Conference on Computer Vision (ICCV) Workshops, October 2019
3. Chaudhary, A.K., et al.: RITnet: real-time semantic segmentation of the eye for gaze tracking. In: 2019 IEEE/CVF International Conference on Computer Vision Workshop (ICCVW), pp. 3698–3702 (2019)
4. Chen, L.C., Papandreou, G., Kokkinos, I., Murphy, K., Yuille, A.: DeepLab: semantic image segmentation with deep convolutional nets, atrous convolution, and fully connected CRFs. IEEE Trans. Pattern Anal. Mach. Intell. (2016). https://doi.org/10.1109/TPAMI.2017.2699184
5. Chen, L.-C., Zhu, Y., Papandreou, G., Schroff, F., Adam, H.: Encoder-decoder with atrous separable convolution for semantic image segmentation. In: Ferrari, V., Hebert, M., Sminchisescu, C., Weiss, Y. (eds.) ECCV 2018. LNCS, vol. 11211, pp. 833–851. Springer, Cham (2018). https://doi.org/10.1007/978-3-030-01234-2_49
6. Garbin, S.J., Shen, Y., Schuetz, I., Cavin, R., Hughes, G., Talathi, S.S.: OpenEDS: open eye dataset. arXiv preprint arXiv:1905.03702 (2019)
7. He, K., Zhang, X., Ren, S., Sun, J.: Delving deep into rectifiers: surpassing human-level performance on ImageNet classification. In: 2015 IEEE International Conference on Computer Vision (ICCV), pp. 1026–1034, December 2015. https://doi.org/10.1109/ICCV.2015.123

8. Kim, S.H., Lee, G.S., Yang, H.J., et al.: Eye semantic segmentation with a lightweight model. In: 2019 IEEE/CVF International Conference on Computer Vision Workshop (ICCVW), pp. 3694–3697. IEEE (2019)

9. Palmero, C., Sharma, A., Behrendt, K., Krishnakumar, K., Komogortsev, O.V., Talathi, S.S.: OpenEDS 2020: Open eyes dataset (2020)

10. Paszke, A., Chaurasia, A., Kim, S., Culurciello, E.: ENet: a deep neural network architecture for real-time semantic segmentation. arXiv preprint arXiv:1606.02147 (2016)

11. Perry, J., Fernandez, A.: MinENet: a dilated CNN for semantic segmentation of eye features. In: The IEEE International Conference on Computer Vision (ICCV) Workshops, October 2019

12. Ronneberger, O., Fischer, P., Brox, T.: U-Net: convolutional networks for biomedical image segmentation. In: Navab, N., Hornegger, J., Wells, W.M., Frangi, A.F. (eds.) MICCAI 2015. LNCS, vol. 9351, pp. 234–241. Springer, Cham (2015). https://doi.org/10.1007/978-3-319-24574-4_28

13. Sudre, C.H., Li, W., Vercauteren, T., Ourselin, S., Jorge Cardoso, M.: Generalised dice overlap as a deep learning loss function for highly unbalanced segmentations. In: Cardoso, M.J., et al. (eds.) DLMIA/ML-CDS -2017. LNCS, vol. 10553, pp. 240–248. Springer, Cham (2017). https://doi.org/10.1007/978-3-319-67558-9_28

14. Vu, T.H., Jain, H., Bucher, M., Cord, M., Pérez, P.: ADVENT: adversarial entropy minimization for domain adaptation in semantic segmentation. In: Proceedings of the IEEE Conference on Computer Vision and Pattern Recognition, pp. 2517–2526 (2019)

15. Yadav, G.: Contrast limited adaptive histogram equalization based enhancement for real time video system, September 2014. https://doi.org/10.1109/ICACCI. 2014.6968381

16. Yu, F., Koltun, V.: Multi-scale context aggregation by dilated convolutions. In: International Conference on Learning Representations (ICLR) (2016)

17. Zou, Y., Yu, Z., Vijaya Kumar, B.V.K., Wang, J.: Unsupervised domain adaptation for semantic segmentation via class-balanced self-training. In: Ferrari, V., Hebert, M., Sminchisescu, C., Weiss, Y. (eds.) ECCV 2018. LNCS, vol. 11207, pp. 297–313. Springer, Cham (2018). https://doi.org/10.1007/978-3-030-01219-9_18

W10 - TASK-CV Workshop and VisDA Challenge

W10 - TASK-CV Workshop and VisDA Challenge

Welcome to the Proceedings of the 7th Workshop on Transferring and Adapting Source Knowledge in Computer Vision (TASK-CV 2020), held in conjunction with the the European Conference on Computer Vision (ECCV 2020).

A key ingredient of the recent successes of computer vision methods is the availability of large sets of annotated data. However, collecting them is prohibitive in many real applications and it is natural to search for an alternative source of knowledge that needs to be transferred or adapted to provide sufficient learning support. The aim of TASK-CV is to bring together researchers in various sub-areas of Transfer Learning and Domain Adaptation (Domain Generalization, Few Show, Multi-Task learning, etc.). In the last years these topics have attracted a lot of attention and by checking the titles and keywords of the papers submitted to the most recent computer vision conferences it is clear that the trend is still growing. In particular, 10% of the ECCV main conference papers are related to the TASK-CV topics and the workshop is an occasion to discuss the new directions and research strategies in this area. The Visual Domain Adaptation Challenge (VisDA) contributed to make the workshop even more relevant in the last four editions: this year it was organized by Kate Saenko, Liang Zheng, Xingchao Peng, and Weijian Deng, and the task of the challenge was domain adaptive instance retrieval for pedestrian re-identification.

We allowed two types of submissions: full papers presenting novel methods and novel applications, and short papers that may contain either already published methods re-tailored for transfer learning and domain adaptation, or ongoing work i.e. new ideas that the authors want to share with the community for which an exhaustive quantitative evaluation is still in process. We are proud that this year's workshop received a total of 23 papers, each of which was sent to two independent reviewers and finally meta-reviewed and selected by the organizers. We accepted 13 contributions (56%): 7 full and 6 short papers. Besides the oral presentation of few selected papers by their authors, the online interactive sessions of the workshop also hosted six invited speakers with interesting talks: Subhransu Maji, Zeynep Akata, Stefano Soatto, Sanja Fidler, DenXin Dai, and Peter Koniusz. We would like to thank all of the speakers for their contribution. An acknowledgement also goes to the reviewers who submitted high-quality assessments in a short period of time, and to the authors for their hard work in submitting high-quality papers. Finally, we thank Naver Labs Europe for sponsoring the Best Paper Award.

August 2020

Tatiana Tommasi
David Vázquez
Antonio M. López
Gabriela Csurka

Class-Imbalanced Domain Adaptation: An Empirical Odyssey

Shuhan Tan[1]([✉]), Xingchao Peng[2], and Kate Saenko[2,3]

[1] Sun Yat-Sen University, Guangzhou, China
tanshh@mail2.sysu.edu.cn
[2] Boston University, Boston, USA
{xpeng,saenko}@bu.edu
[3] MIT-IBM Watson AI Lab, Boston, USA

Abstract. Unsupervised domain adaptation is a promising way to generalize deep models to novel domains. However the current literature assumes that the label distribution is domain-invariant and only aligns the feature distributions or *vice versa*. In this work, we explore the more realistic task of *Class-imbalanced Domain Adaptation*: How to align feature distributions across domains *while* the label distributions of the two domains are also different? Taking a practical step towards this problem, we constructed its first benchmark with 22 cross-domain tasks from 6 real-image datasets. We conducted comprehensive experiments on 10 recent domain adaptation methods and find most of them are very fragile in the face of coexisting feature and label distribution shift. Towards a better solution, we further proposed a feature and label distribution CO-ALignment (COAL) model with a novel combination of existing ideas. COAL is empirically shown to outperform most recent domain adaptation methods on our benchmarks. We believe the provided benchmarks, empirical analysis results, and the COAL baseline could stimulate and facilitate future research towards this important problem.

1 Introduction

The success of deep learning models is highly dependent on the assumption that the training and testing data are *i.i.d* and sampled from the same distribution. In reality, they are typically collected from different but related domains, leading to a phenomenon known as *domain shift* [1]. To bridge the domain gap, Unsupervised Domain Adaptation (UDA) transfers the knowledge learned from a labeled source domain to an unlabeled target domain by statistical distribution alignment [2,3] or adversarial alignment [4–6]. Though recent UDA works

---

S. Tan—Work done while the author was visiting Boston University.

Electronic supplementary material The online version of this chapter (https://doi.org/10.1007/978-3-030-66415-2_38) contains supplementary material, which is available to authorized users.

© Springer Nature Switzerland AG 2020
A. Bartoli and A. Fusiello (Eds.): ECCV 2020 Workshops, LNCS 12535, pp. 585–602, 2020.
https://doi.org/10.1007/978-3-030-66415-2_38

have made great progress, most of them are under the assumption that the prior label distributions of the two domains are identical. Denote the input data as x and output labels as y, and let the source and target domain be characterized by probability distributions p and q, respectively. The majority of UDA methods assume that the conditional label distribution is invariant $(p(y|x) = q(y|x))$, and only the *feature shift* $(p(x) \neq q(x))$ needs to be tackled, neglecting potential *label shift* $(p(y) \neq q(y))$[1]. However, we claim that this assumption makes current UDA methods not applicable in the real world, for the following reasons: **1)** this assumption hardly holds true in real applications, as label shift across domains is commonly seen in the real world. For example, an autonomous driving system should be able to handle constantly changing frequencies of pedestrians and cars when adapting from a rural to a downtown area; or from a rainy to a sunny day. In addition, it is hard to guarantee $p(y) = q(y)$ without any information about $q(y)$ in the real world. **2)** recent theoretical work [9] has demonstrated that if label shift exists, current UDA methods could lead to significant *performance drop*. This is also empirically proved by our experiments. **3)** we cannot check whether label shift exists in real applications. This prevents us from safely applying current UDA methods because we cannot predict the potential risk of performance drop. Therefore, we claim that an applicable UDA method must be able to handle feature shift and label shift at the same time.

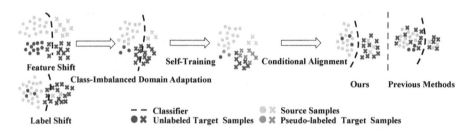

Fig. 1. We propose the Class-imbalanced Domain Adaptation setting, where we consider feature shift and label shift simultaneously. We provide the first empirical evaluation of this setting, showing that existing UDA methods are very fragile in the face of label shift. This is because learning marginal domain-invariant features will incorrectly align samples from different categories, leading to negative transfer. We propose an alternate, more robust approach that combines self-training and conditional feature alignment to tackle feature and label shift.

To formulate the above problem, we propose ***Class-imbalanced Domain Adaptation*** (**CDA**), a more challenging but practical domain adaptation setting where the conditional feature shift and label shift are required to be tackled *simultaneously*. Specifically, in addition to Covariate Shift assumption $(p(x) \neq q(x),\ p(y|x) = q(y|x))$, we further assume $p(x|y) \neq q(x|y)$ and $p(y) \neq q(y)$. The main challenges of CDA are: **1)** label shift hampers the effectiveness of

---

[1] Different from some works [7,8], we do not assume $p(x|y) = q(x|y)$ for label shift.

mainstream domain adaptation methods that only marginally aligns feature distributions, **2)** aligning the conditional feature distributions $(p(x|y), q(x|y))$ is difficult in the presence of label shift, and **3)** when data in one or both of the domains are unequally distributed across different categories, it is difficult to train an unbiased classifier. An overview of CDA is shown in Fig. 1.

Aligned with our idea, several works [10–12] provide theoretical analyses on domain adaptation with both feature and label shift. However, they do not provide sufficient empirical analysis of current UDA methods under this setting. In addition, no practical algorithm that can solve real-world cross-domain problems has been proposed by these works. Therefore, although this problem has been known for years, most recent UDA methods are still not able to handle it. In this paper, we aim raise concerns and interests towards this important problem by taking one practical step. Firstly, we create CDA benchmarks with 22 cross-domain tasks across 6 real-world image classification datasets. We believe this would facilitate future domain adaptation research towards robustly applicable methods. Secondly, we extensively evaluate 10 state-of-the-art domain adaptation methods to analysis how well CDA is solved currently. We find most of these methods cannot handle CDA well and often lead to negative transfer. Thirdly, towards a better solution, we provide a theoretically-motivated novel combination of existing ideas, which works well as a baseline for future research.

In this work, we visited domain adaptation methods in three categories. Mainstream unsupervised domain adaptation aligns the feature distributions of two domains by methods that include minimizing the Maximum Mean Discrepancy [2,3], aligning high-order moments [13,14], or adversarial training [4,5]. However, these models are limited when applied to the CDA task as they only align the feature distribution, ignoring the issue of label shift [9]. Another line of works [7,8] assume that only label shift exists $(p(y) \neq q(y))$ between two domains and the conditional feature distribution is invariant $(p(x|y) = q(x|y))$. These methods have achieved good performance when the data in both domains are sampled from the same feature distribution but under different label distributions. However, these models cannot handle the CDA task as the feature distribution is not well aligned. Recently, several works consider the domain adaptation problem where the categories of the source and target domain are not fully overlapped [15–17]. This setting can be seen as a special case of CDA where for some class i we have either $p(y = i) = 0$ or $q(y = i) = 0$. In our experiments, we showed that 8 out of 10 methods we evaluated on CDA tasks frequently lead to negative transfer (produce worse performance than no-adaptation baseline), while the rest methods only leads to limited improvement over the baseline on average. This limited performance showed that current UDA methods are not robust enough to be practically applied, and motivated us to reconsider the solution to the CDA problem.

We postulate that it is essential to align the conditional feature distributions as well as the label distributions to tackle the CDA task. In this work, we address CDA with feature distribution and label distribution **CO-ALignment** (**COAL**). Specifically, to deal with feature shift and label shift in an unified way, we proposed a simple baseline method that combines the ideas of *prototype-based conditional distribution alignment* [18] and *class-balanced self-training* [19]. First, to tackle feature shift in the context of label shift, it is essential to align the conditional rather than marginal feature distributions, to avoid the negative transfer effects caused by matching marginal feature distributions [9] (illustrated in Fig. 1). To this end, we use a prototype-based method to align the conditional feature distributions of the two domains. The *source* prototypes are computed by learning a similarity-based classifier, which are moved towards the *target* domain with a minimax entropy algorithm [18]. Second, we align the label distributions in the context of feature shift by training the classifier with estimated target label distribution through a class-balanced self-training method [19]. We incorporate the above feature distribution and label distribution alignment into an end-to-end deep learning framework, as illustrated in Fig. 2. Comprehensive experiments on standard cross-domain recognition benchmarks demonstrate that COAL achieves significant improvements over the state-of-the-art methods on the task of CDA.

The main contributions of this paper are highlighted as follows: **1)** to the best of our knowledge, we provide the first set of benchmarks and practical solution for domain adaptation under joint feature and label shift in deep learning, which is important for real-world applications; **2)** we deliver extensive experiments to demonstrate that state-of-the-art methods *fail* to align feature distribution in the presence of label distribution, or vise versa; **3)** we propose a simple yet effective feature and label distribution CO-ALignment (COAL) framework, which could be a useful baseline for future research towards practical domain adaptation. We believe the provided benchmarks, empirical analysis and the baseline model could trigger future research works towards more practical domain adaptation.

2   Related Work

Domain Adaptation for Feature Shift. Domain adaptation aims to transfer the knowledge learned from one or more source domains to a target domain. Recently, many unsupervised domain adaptation methods have been proposed. These methods can be taxonomically divided into three categories [20]. The first category is the discrepancy-based approach, which leverages different measures to align the marginal feature distributions between source and target domains. Commonly used measures include Maximum Mean Discrepancy (MMD) [3,21], $\mathcal{H}$-divergence [22], Kullback-Leibler (KL) divergence [23], and Wasserstein distance [24,25]. The second category is the adversarial-based approach [4,26,27] which uses a domain discriminator to encourage domain confusion via an adversarial objective. The third category is the reconstruction-based approach. Data are reconstructed in the new domain by an encoder-decoder [28,29] or a GAN discriminator, such as dual-GAN [30], cycle-GAN [31], disco-GAN [32], and CyCADA [33]. However, these methods mainly consider aligning the marginal distributions to decrease feature shift, neglecting label shift. To the best of our knowledge, we are the first the propose an end-to-end deep model to tackle both of the two domain shifts between the source and target domains.

Domain Adaptation for Label Shift. Despite its wide applicability, learning under label shift remains under-explored. Existing works tackle this challenge by importance reweighting or target distribution estimation. Specifically, [10] exploit importance reweighting to enhance knowledge transfer under label shift. Recently, [34] introduce a test distribution estimator to detect and correct for label shift. These methods assume that the source and target domains share the same feature distributions and only differ in the marginal label distribution. In this work, we explore transfer learning between domains under label shift and label shift simultaneously. As a special case of label shift, some works consider the domain adaptation problem where the categories in the source domain and target domain are not fully overlapped. [35] propose *open set domain adaptation* where the class set in the source domain is a proper subset of that of the target domain. Conversely, [15] introduce *partial domain adaptation* where the class set of the source domain is a proper superset of that of the target domain. In this direction, [36] introduce a theoretical analysis to show that only learning domain-invariant features is not sufficient to solve domain adaptation task when the label priors are not aligned. In a related work, [12] propose asymmetrically-relaxed distribution alignment to overcome the limitations of standard domain adaptation algorithms which aims to extract domain-invariant representations.

Domain Adaptation with Self-training. In domain adaptation, self-training methods are often utilized to compensate for the lack of categorical information in the target domain. The intuition is to assign pseudo-labels to unlabeled samples based on the predictions of one or more classifiers. [37] leverage an asymmetric tri-training strategy to assign pseudo-labels to the unlabeled target domain. [38] propose to assign pseudo-labels to all target samples and use them to achieve semantic alignment across domains.

Recently, [39] propose to progressively label the target samples and align the prototypes of source domain and target domain to achieve domain alignment. However, to the best of our knowledge, self-training has not been applied for DA with label shift.

3   CO-ALignment of Feature and Label Distribution

In Class-imbalanced Domain Adaptation, we are given a *source* domain $\mathcal{D}_S = \{(x_i^s, y_i^s)_{i=1}^{N_s}\}$ with N_s labeled examples, and a *target* domain $\mathcal{D}_T = \{(x_i^t)_{i=1}^{N_t}\}$ with N_t unlabeled examples. We assume that $p(y|x) = q(y|x)$ but $p(x|y) \neq q(x|y)$, $p(x) \neq q(x)$, and $p(y) \neq q(y)$. We aim to construct an end-to-end deep neural network which is able to transfer the knowledge learned from $\mathcal{D}_S$ to $\mathcal{D}_T$, and train a classifier $y = \theta(x)$ which can minimize task risk in target domain $\epsilon_T(\theta) = \Pr_{(x,y) \sim q}[\theta(x) \neq y]$.

Previous works either focus on aligning the marginal feature distributions [2, 4] or aligning the label distributions [34]. These approaches are not able to fully tackle CDA as they only align one of the two marginal distributions. Motivated by theoretical analysis, in this work we propose to tackle CDA with feature distribution and label distribution CO-ALignment. To this end, we combine the ideas of *prototype-based conditional alignment* [18] and *class-balanced self-training* [19] to tackle feature and label shift respectively. An overview of COAL is shown in Fig. 2.

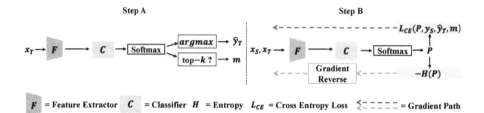

Fig. 2. Overview of the proposed COAL model. Our model is trained iteratively between two steps. In step A, we forward the target samples through our model to generate the pseudo labels and mask. In step B, we train our models by *self-training* with the pseudo-labeled target samples to align the label distributions, and *prototype-based conditional alignment* with the minimax entropy.

3.1   Theoretical Motivations

Conditional Feature Alignment. According to [36], the target error in domain adaptation is bounded by three terms: 1) source error, 2) the discrepancy between the marginal distributions and 3) the distance between the source and target optimal labeling functions. Denote $h \in \mathcal{H}$ as the hypothesis, $\epsilon_S(\cdot)$ and $\epsilon_T(\cdot)$ as the expected error of a labeling function on source and target domain, and f_S and f_T as the optimal labeling functions in the source and target domain. Then, we have:

$$\epsilon_T(h) \leq \epsilon_S(h) + d_{\hat{\mathcal{H}}}(\mathcal{D}_S, \mathcal{D}_T) + \min\{\epsilon_S(f_T), \epsilon_T(f_S)\}, \tag{1}$$

where $d_{\hat{\mathcal{H}}}$ denote the discrepancy of the marginal distributions [36]. The bound demonstrates that the optimal labeling functions f_S and f_T need to generalize well in both domains, such that the term $\min\{\epsilon_S(f_T), \epsilon_T(f_S)\}$ can be bounded. Conventional domain adaptation approaches which only align marginal feature distribution cannot guarantee that $\min\{\epsilon_S(f_T), \epsilon_T(f_S)\}$ is minimized. This motivates us to align the conditional feature distribution, *i.e.* $p(x|y)$ and $q(x|y)$.

Class-Balanced Self-training. Theorem 4.3 in [36] indicates that the target error $\epsilon_T(h)$ can not be minimized if we only align the feature distributions and neglect the shift in label distribution. Denote d_{JS} as the Jensen-Shannon (JS) distance between two distributions, [36] propose:

$$\epsilon_S(h) + \epsilon_T(h) \geq \frac{1}{2}(d_{JS}(p(y), q(y)) - d_{JS}(p(x), q(x)))^2 \tag{2}$$

This theorem demonstrates that when the divergence between label distributions $d_{JS}(p(y), q(y))$ is significant, minimizing the divergence between marginal distributions $d_{JS}(p(x), q(x))$ and the source task error $\epsilon_S(h)$ will enlarge the target task error $\epsilon_T(h)$. Motivated by this, we propose to estimate and align the empirical label distributions with a self-training algorithm.

3.2 Prototype-Based Conditional Alignment for Feature Shift

The mainstream idea in feature-shift oriented methods is to learn domain-invariant features by aligning the marginal feature distributions, which was proved to be inferior in the presence of label shift [9]. Instead, inspired by [18], we align the conditional feature distributions $p(x|y)$ and $q(x|y)$. To this end, we leverage a *similarity-based classifier* to estimate $p(x|y)$, and a minimax entropy algorithm to align it with $q(x|y)$. We achieve conditional feature distribution alignment by aligning the source and target prototypes in an adversarial process.

Similarity-Based Classifier. The architecture of our COAL model contains a feature extractor F and a similarity-based classifier C. Prototype-based classifiers perform well in few-shot learning settings [40], which motivates us to adopt them since in label-shift settings some categories can have low frequencies. Specifically, C is composed of a weight matrix $\mathbf{W} \in \mathbb{R}^{d \times c}$ and a temperature parameter T, where d is the dimension of feature generated by F, and c is the total number of classes. Denote $\mathbf{W}$ as $[\mathbf{w}_1, \mathbf{w}_2, ..., \mathbf{w}_c]$, this matrix can be seen as c d-dimension vectors, one for each category. For each input feature $F(x)$, we compute its similarity with the i_{th} weight vector as $s_i = \frac{F(x)\mathbf{w}_i}{T\|F(x)\|}$. Then, we compute the probability of the sample being labeled as class i by $h_i(x) = \sigma(\frac{F(x)\mathbf{w}_i}{T\|F(x)\|})$, normalizing over all the classes. Finally, we can compute the prototype-based classification loss for $\mathcal{D}_S$ with standard cross-entropy loss:

$$\mathcal{L}_{SC} = \mathbb{E}_{(x,y) \in \mathcal{D}_S} \mathcal{L}_{ce}(h(x), y) \tag{3}$$

The intuition behind this loss is that the higher the confidence of sample x being classified as class i, the closer the embedding of x is to $\mathbf{w}_i$. Hence, when optimizing Eq. 3, we are reducing the intra-class variation by pushing the embedding of each sample x closer to its corresponding weight vector in $\mathbf{W}$. In this way, $\mathbf{w}_i$ can be seen as a representative data point (prototype) for $p(x|y = i)$.

Conditional Alignment by Minimax Entropy. Due to the lack of categorical information in the target domain, it is infeasible to utilize Eq. 3 to obtain target prototypes. Following [18], we tackle this problem by 1) moving each source prototype to be closer to its nearby target samples, and 2) clustering target samples around this moved prototype. We achieve these two objectives jointly by entropy minimax learning. Specifically, for each sample $x^t \in \mathcal{D}_T$ fed into the network, we can compute the mean entropy of the classifier's output by

$$\mathcal{L}_H = \mathbb{E}_{x \in \mathcal{D}_T} H(x) = -\mathbb{E}_{x \in \mathcal{D}_T} \sum_{i=1}^{c} h_i(x) \log h_i(x). \tag{4}$$

Larger $H(x)$ indicates that sample x is similar to all the weight vectors (prototypes) of C. We achieve conditional feature distributions alignment by aligning the source and target prototypes in an adversarial process: (1) we train C to *maximize* $\mathcal{L}_H$, aiming to move the prototypes from the source samples towards the neighboring target samples; (2) we train F to *minimize* $\mathcal{L}_H$, aiming to make the embedding of target samples closer to their nearby prototypes. By training with these two objectives as a min-max game between C and F, we can align source and target prototypes. Specifically, we add a gradient-reverse layer [5] between C and F to flip the sign of gradient.

3.3   Class-Balanced Self-training for Label Shift

As the source label distribution $p(y)$ is different from that of the target $q(y)$, it is not guaranteed that the classifier C which has low risk on $\mathcal{D}_S$ will have low error on $\mathcal{D}_T$. Intuitively, if the classifier is trained with imbalanced source data, the decision boundary will be dominated by the most frequent categories in the training data, leading to a classifier biased towards source label distribution. When the classifier is applied to target domain with a different label distribution, its accuracy will degrade as it is highly biased towards the source domain. To tackle this problem, we use the method in [19] to employ *self-training* to estimate the target label distribution and refine the decision boundary. In addition, we leverage *balanced sampling* of the source data to further facilitate this process.

Self-training. In order to refine the decision boundary, we propose to estimate the target label distribution with self-training. We assign pseudo labels $\hat{y}$ to all the target samples according to the output the classifier C. As we are also aligning the conditional feature distributions $(p(x|y)$ and $q(x|y))$, we assume that the distribution of high-confidence pseudo labels $q(\hat{y})$ can be used as an approximation of the real label distribution $q(y)$ for the target domain. Training C with these pseudo-labeled target samples under approximated target label distribution, we are able to reduce the negative effect of label shift.

   To obtain high-confidence pseudo labels, for each category, we select top $k\%$ of the target samples with the highest confidence scores belonging to that category. We utilize the highest probability in $h(x)$ as the classifier's confidence on sample x. Specifically, for each pseudo-labeled sample $(x, \hat{y})$, we set its selection mask $m = 1$ if $h(x)$ is among the top $k\%$ of all the target samples with the same pseduo-label, otherwise $m = 0$. Denote the pseudo-labeled target set as $\hat{\mathcal{D}}_T = \{(x_i^t, \hat{y}_i^t, m_i)_{i=1}^{N_t}\}$, we leverage the input and pseudo labels from $\hat{\mathcal{D}}_T$ to train the classifier C, aiming to refine the decision boundary with target label distribution. The total loss function for classification is:

$$\mathcal{L}_{ST} = \mathcal{L}_{SC} + \mathbb{E}_{(x,\hat{y},m)\in\hat{\mathcal{D}}_T}\mathcal{L}_{ce}(h(x), \hat{y}) \cdot m \qquad (5)$$

where $\hat{y}$ indicates the pseudo labels and m indicates selection masks. In our approach, we choose the top $k\%$ of the highest confidence target samples *within* each category, instead of universally. This is crucial to estimate the real target label distribution, otherwise, the easy-to-transfer categories will dominate $\hat{\mathcal{D}}_T$, leading to inaccurate estimation of the target label distribution [19]. As training processes, we are able to obtain pseudo labels with higher accuracy. Therefore, we increase k by k_{step} after each epoch until it reaches a threshold k_{max}. Typically, we initialize k with $k_0 = 5$, and set $k_{step} = 5$, $k_{max} = 30$.

Balanced Sampling of Source Data. When coping with label shift, the label distribution of the source domain could be highly imbalanced. A classifier trained on imbalanced categories will make highly-biased predictions for the samples from the target domain [41]. This effect also hinders the self-training process discussed above, as the label distribution estimation will also be biased. To tackle these problems, we apply a balanced mini-batch sampler to generate training data from the source domain and ensure that each source mini-batch contains roughly the same number of samples for each category.

3.4 Training Process

In this section, we combine the above ideas into an end-to-end training pipeline. Denote α as the trade-off between classifier training and feature distribution alignment, we first define the adaptive learning objective as follows:

$$\hat{C} = \arg\min_C \mathcal{L}_{ST} - \alpha\mathcal{L}_H, \qquad \hat{F} = \arg\min_F \mathcal{L}_{ST} + \alpha\mathcal{L}_H. \tag{6}$$

Given input samples from source domain $\mathcal{D}_S$ and target domain $\mathcal{D}_T$, we first pretrain our network F and C with only labeled data $\mathcal{D}_S$. Then, we iterate between **pseudo-label assignment** (step A) and **adaptive learning** (step B). We update the pseudo labels in each epoch as we obtain better feature representations from adaptive learning, which leads to more accurate pseudo labels. On the other hand, better pseudo labels could also facilitate adaptive learning in the next epoch. This process continues until convergence or reaching the maximum number of iterations. An overview of it is shown in Fig. 2.

4   Experiments

In this section, we first construct the CDA benchmarks with 26 cross-domain adaptation tasks based on 4 **Digits** datasets, **Office-Home** [42] and **Domain-Net** [14]. Then we evaluate and analysis 10 representative state-of-the-art domain adaptation methods as well as our COAL baseline. Finally, we provide additional analysis experiments to further explore the CDA problem.

4.1   Class-Imbalanced Domain Adaptation Benchmark

Domain Shift Protocol. Because the images use are already collected from separate feature domains, we only create label shift for each cross-domain task. To create label shift between source and target domains, we sub-sample the current datasets with **R**eversely-unbalanced **S**ource and **U**nbalanced **T**arget (**RS-UT**) protocol. In this setting, both the source and target domains have unbalanced label distribution, while the label distribution of the source domain is a reversed version of that of the target domain. Following [43], the unbalanced label distribution is created by sampling from a Paredo distribution [44]. An illustration of this setting can be found in Fig. 3(b). We refer our reader to supplementary material for detailed data splits and creation process.

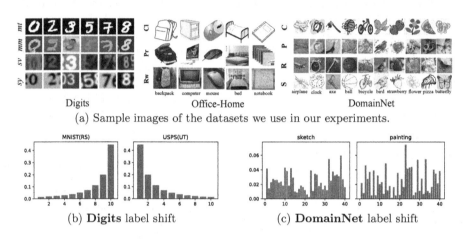

(a) Sample images of the datasets we use in our experiments.

(b) **Digits** label shift   (c) **DomainNet** label shift

Fig. 3. (**a**): Image examples from Digits, Office-Home [42], and DomainNet [14]. (**b**): illustrations of **R**eversely-unbalanced **S**ource (**RS**) and **U**nbalanced **T**arget (**UT**) distribution in MNIST → USPS task. (**c**): Natural label shift of DomainNet.

Digits. We select four digits datasets: MNIST [45], USPS [46], SVHN [47] and Synthetic Digits (SYN) [48] and regard each of them as a separate domain. In this work, we investigate four domain adaptation tasks: **MNIST → USPS**, **USPS → MNIST, SVHN → MNIST**, and **SYN → MNIST**.

Office-Home [42] is a dataset collected in office and home environment with 65 object classes and four domains: Real World (**Rw**), Clipart (**Cl**), Product (**Pr**), Art (**Ar**). Since the "Art" domain is too small to sample an imbalanced subset, we focus on the remaining domains and explore all the six adaptation tasks.

DomainNet [14] is a large-scale testbed for domain adaptation, which contains six domains with about 0.6 million images distributed among 345 classes. Since some domains and classes contains many mislabeled outliers, we select 40 commonly-seen classes from four domains: Real (**R**), Clipart (**C**), Painting (**P**), Sketch (**S**). Different from the two datasets above, the existed label shift in DomainNet is significant enough, as illustrated in Fig. 3(c). Therefore, we use the original label distributions without sub-sampling for this dataset.

Evaluated Methods. To form a comprehensive empirical analysis, we evaluated recent domain adaptation methods from three categories, including **1)** conventional UDA methods that only aligns feature distribution: **DAN** [2],

JAN [21], **DANN** [5], **MCD** [6] and **BSP** [49]; **2)** method that only aligns label distribution: **BBSE**[34]; **3)** methods that align feature distribution while assuming non-overlapping label spaces: **PADA** [15], **ETN** [16] and **UAN** [17]. We also evaluated **FDANN** [12], which relaxes the feature distribution alignment objective in DANN to deal with potential label shift.

Table 1. Per-class mean accuracy on Digits. Our model achieves **84.33%** average accuracy across four tasks, outperforming other evaluated methods.

Methods	USPS → MNIST	MNIST → USPS	SVHN → MNIST	SYN → MNIST	AVG
Source Only	75.31 ± 0.09	87.92 ± 0.74	50.25 ± 0.81	85.74 ± 0.49	74.81
UAN (2019)	55.72 ± 2.06	83.23 ± 0.75	50.20 ± 1.75	71.26 ± 2.82	65.10
ETN (2019)	62.85 ± 2.20	79.27 ± 1.29	52.82 ± 1.50	72.42 ± 6.53	66.84
FDANN (2019)	72.59 ± 1.61	81.62 ± 2.38	45.65 ± 2.93	82.07 ± 1.65	70.48
JAN (2017)	75.75 ± 0.75	78.82 ± 0.93	53.21 ± 3.94	75.64 ± 1.42	70.86
BBSE (2018)	75.01 ± 3.68	78.84 ± 10.73	49.01 ± 2.02	85.69 ± 0.71	72.14
BSP (2019)	71.99 ± 1.52	89.74 ± 0.77	50.61 ± 1.67	77.30 ± 1.20	72.41
PADA (2018)	73.66 ± 0.15	78.59 ± 0.23	54.13 ± 1.61	85.06 ± 0.60	72.86
MCD (2018)	77.18 ± 5.65	85.34 ± 4.07	53.52 ± 4.23	76.37 ± 3.48	73.10
DAN (2015)	79.12 ± 1.34	87.15 ± 1.71	53.63 ± 1.80	80.89 ± 2.00	75.20
DANN (2015)	77.28 ± 2.13	91.88 ± 0.74	57.16 ± 1.83	77.60 ± 1.29	75.98
COAL (Ours)	**88.12 ± 0.37**	**93.04 ± 1.67**	**65.67 ± 1.29**	**90.60 ± 0.44**	**84.33**

Table 2. Per-class mean accuracy on Office-Home dataset. Our model achieve **58.87%** average accuracy across six tasks.

Methods	Rw → Pr	Rw → Cl	Pr → Rw	Pr → Cl	Cl → Rw	Cl → Pr	AVG
Source Only	70.75	35.51	65.65	34.99	51.27	51.11	51.55
BSP (2019)	66.15	23.48	65.42	20.81	34.54	31.04	40.24
PADA (2018)	60.77	32.28	57.09	26.76	40.71	38.34	42.66
BBSE (2018)	61.10	33.27	62.66	31.15	39.70	38.08	44.33
MCD (2018)	66.18	32.32	62.66	28.40	41.41	38.59	44.93
DAN (2015)	67.85	38.17	66.86	34.24	52.95	51.64	45.02
UAN (2019)	70.85	41.15	67.26	36.82	56.24	55.77	48.62
ETN (2019)	71.69	34.03	70.45	**40.74**	**60.48**	55.19	52.14
FDANN (2019)	68.56	40.57	67.32	37.33	55.84	53.67	53.88
JAN (2017)	71.22	43.12	68.20	37.03	57.97	56.80	55.72
DANN (2015)	71.78	**46.08**	67.98	39.45	58.40	57.39	56.85
COAL (Ours)	**73.65**	42.58	**74.46**	40.61	59.22	**62.71**	**58.87**

Implementation Details. We implement all our experiments in Pytorch platform. We used the official implements for all the evaluated methods except for DANN [5], BBSE [34] and FDANN [12], which are reproduced by ourselves. For fair comparison, we use the same backbone networks for all the methods.

Table 3. Per-class mean accuracy on DomainNet dataset with natural label shifts. Our method achieve **75.89%** average accuracy across the 12 experiments. Note that DomainNet contains about 0.6 million images, it is non-trivial to have even one percent performance boost.

Method	R → C	R → P	R → S	C → R	C → P	C → S	P → R	P → C	P → S	S → R	S → C	S → P	AVG
Baseline	58.84	67.89	53.08	76.70	53.55	53.06	84.39	55.55	60.19	74.62	54.60	57.78	62.52
BBSE	55.38	63.62	47.44	64.58	42.18	42.36	81.55	49.04	54.10	68.54	48.19	46.07	55.55
PADA	65.91	67.13	58.43	74.69	53.09	52.86	79.84	59.33	57.87	76.52	66.97	61.08	64.48
MCD	61.97	69.33	56.26	79.78	56.61	53.66	83.38	58.31	60.98	81.74	56.27	66.78	65.42
DAN	64.36	70.65	58.44	79.44	56.78	60.05	84.56	61.62	62.21	79.69	65.01	62.04	67.07
FDANN	66.15	71.80	61.53	81.85	60.06	61.22	84.46	66.81	62.84	81.38	69.62	66.50	69.52
UAN	71.10	68.90	67.10	83.15	63.30	64.66	83.95	65.35	67.06	82.22	70.64	68.09	72.05
JAN	65.57	_73.58_	67.61	85.02	64.96	67.17	_87.06_	67.92	66.10	84.54	72.77	67.51	72.48
ETN	69.22	72.14	63.63	_86.54_	65.33	63.34	85.04	65.69	68.78	84.93	72.17	68.99	73.99
BSP	_67.29_	73.47	69.31	86.50	_67.52_	_70.90_	86.83	70.33	68.75	84.34	72.40	**71.47**	74.09
DANN	63.37	73.56	**72.63**	86.47	65.73	70.58	86.94	_73.19_	_70.15_	_85.73_	**75.16**	70.04	_74.46_
Ours	**73.85**	**75.37**	_70.50_	**89.63**	**69.98**	**71.29**	**89.81**	68.01	**70.49**	**87.97**	_73.21_	_70.53_	**75.89**

Specifically, for the Digits dataset, we adopt the network architecture proposed by [6]. For the other two datasets, we utilize ResNet-50 [50] as our backbone network, and replace the last fully-connected layer with a randomly initialized N-way classifier layer (for N categories). For all the compared methods, we select their hyper-parameters on the validation set of P → C task of DomainNet. We refer our reader to supplementary material for code and parameters of COAL.

Evaluation Metric. When the target domain is highly unbalanced, conventional overall average accuracy that treats every class uniformly is not an appropriate performance metric [51]. Therefore, we follow [52] to use the *per-class* mean accuracy in our main results. Formally, we denote $S_i = \frac{n_{(i,i)}}{n_i}$ as the accuracy for class i, where $n_{(i,j)}$ represents the number of class i samples labeled as class j, and $n_i = \sum_{j=1}^{c} n_{(i,j)}$ represents the number of samples in class i. Then, the per-class mean accuracy is computed as $S = \frac{1}{c} \sum_{i=1}^{c} S_i$.

4.2   Result Analysis

We first show the experimental results on Digits datasets in Table 1. From the results, we can make the following observations: **(1)** Most current domain adaptation methods cannot solve CDA well. On average, 8 of the 10 evaluated domain adaptation methods perform *worse* than the source-only baselines, leading to negative transfer. This result confirmed the theoretical analysis that only aligning marginal feature distribution leads to performance drop under CDA [9]. **(2)** Method that achieve better results on conventional UDA benchmarks does not lead to better results on CDA problem. For example, although MCD is shown to significantly outperform DAN and DANN on several conventional domain adaptation benchmarks [6], its performance is inferior to these older methods in our experiment. We argue that this is because these newer methods achieve better marginal feature distribution alignment, which yet leads to worse performance under label shift. **(3)** Our COAL baseline achieves **84.33%** average accuracy across four experimental setting, outperforming the best-performing method by **8.4%**. This result demonstrate that aligning only the feature distributions or only the label distributions can not fully tackle CDA task. In contrast, our framework co-aligns the conditional feature distributions and label distributions.

Next, we show the experimental results on more challenging real-object datasets, *i.e.*, Office-Home and DomainNet, in Table 2 and Table 3, respectively. In Office-Home experiments, we can also have the above observations. For example, we observe that 7 out of 10 methods lead to negative transfer, which is consistent with the results on Digits dataset. Our COAL framework achieves **58.87%** average accuracy across the six CDA tasks, outperforming other evaluated methods, and has **7.32%** improvement from the source-only result.

In DomainNet experiments, due to smaller degree of label shift, most evaluated methods could outperform the source-only baseline. However, we still observe the negative influence of label shift. First, we observe inferior performance of newer methods to older methods. For example, DANN outperformed MCD by 9.04%, due to the negative effect of stronger marginal alignment in MCD. Moreover, our model get **75.89%** average accuracy across the 12 tasks, outperforming all the compared baselines. This shows the effectiveness of feature and label distribution co-alignment in this dataset. Furthermore, we carefully tuned the hyper-parameters for the evaluated domain adaptation methods to have weaker feature distribution alignment[2]. If we directly apply the parameters set by the authors, many of these models have much worse performance.

Table 4. The performance of five models *w.* or *w/o.* source balanced sampler. We observe a significant performance boost when the source balanced sampler is applied, both for our model and the compared baselines, demonstrating the effectiveness of source balanced sampler to CDA task.

Methods	U → M		S → M		Pr → Cl		Cl → Rw		R → S	
	w/o	with	w/o	with	w/o	with	w/o	with	w/o	with
Source Only	71.35	**75.31**	50.35	50.25	34.99	**34.99**	50.64	**51.11**	50.16	**53.08**
DAN	64.81	**79.12**	22.05	**53.63**	32.93	**34.24**	45.18	**51.64**	64.78	58.44
DANN	42.77	**77.28**	27.60	**57.16**	35.17	**39.45**	47.19	**58.40**	68.92	**72.63**
MCD	20.15	**77.18**	44.83	**53.52**	33.06	28.40	49.57	41.41	58.50	56.26
COAL	87.50	**88.12**	60.12	**65.67**	34.03	**40.61**	57.67	**59.22**	59.23	**70.50**

4.3 Analysis

Effect of Source Balanced Sampler. Source balanced samplers described in Sect. 3.3 can help us tackle the biased-classifier problem caused by the imbalanced data distribution of source domain. A significant performance boost can

---

[2] Please refer to supplementary material for details.

be observed after applying the balanced sampler for our COAL model, as well as the compared baselines. In this section, we specifically show the effect of using source balanced samplers. We show in Table 4 the performance of several methods with and without source balanced samplers on 5 adaptation tasks from multiple datasets. We observe that for 20 of the total 25 tasks (5 models on 5 adaptation tasks), using source balanced samplers will significantly improve the domain adaptation performance. These results show the effectiveness of having a source balanced sampler when tackling CDA task.

Ablation Study. Our COAL method has mainly two objectives: 1) alignment of conditional feature distribution $\mathcal{L}_{ST}$ and 2) alignment of label distribution $\mathcal{L}_H$ To show the importance of these two objectives in CDA, we show the performance of our method without each of these objectives respectively on multiple tasks. The results in Table 5 showed the importance of both objectives. For example, for USPS → MNIST, if we remove the conditional feature distribution alignment objective, the accuracy of our model will drop by 2.6%. Similarly, if we remove the label distribution alignment objective, the accuracy will drop by 2.9%. These results demonstrate that both the alignment of conditional feature distribution and label distribution are important to CDA task.

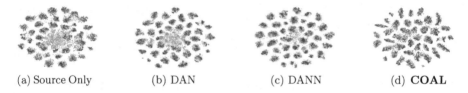

(a) Source Only   (b) DAN   (c) DANN   (d) **COAL**

Fig. 4. t-SNE visualization for features Source Only (baseline), DAN, DANN and COAL on DomainNet task Real → Clipart. Blue and red points represents features from the source domain and target domain, respectively. (Color figure online)

Table 5. Ablation study of different objectives in our method. We randomly select 8 sets of experiments to perform the ablation study.

Methods	U → M	M → U	S → M	Cl → Rw	Pr → Rw	R → C	R → P	P → R	AVG
w/o $\mathcal{L}_{ST}$	85.22	85.94	55.17	58.38	69.39	71.92	74.39	77.45	72.23
w/o $\mathcal{L}_H$	85.57	92.28	63.34	58.17	72.11	71.34	69.92	87.14	74.98
COAL	88.12	93.04	65.67	59.22	74.46	73.85	75.37	89.81	77.44

Feature Visualization. In this section, we plot the learned features with t-SNE [53] in Fig. 4. We investigate the Real to Clipart task in DomainNet experiment with ResNet-50 backbones. From (a)–(d), we observe that our method can better align source and target features in each category, while other methods either leave the feature distributions unaligned, or incorrectly aligned samples in different categories. These results further show the importance of *prototype-based conditional feature alignment* for CDA task.

Different Degrees of Label Shift. We also investigate the effect of different degrees of label shift. Specifically, we create 4 interval degrees of label shift between the **BS-BT** (**B**lanced **S**ource and **B**lanced **T**arget) and RS-UT setting. With these label set settings, we evaluated the performance of different methods. The results show that the performance of previous domain adaptation methods will be significantly affected by label shift, while our method is much more robust. Please refer to supplementary material for detailed setting and results.

5 Conclusion

In this paper, we first propose the Class-imbalanced Domain Adaptation(**CDA**) setting and demonstrate its importance in practical scenarios. Then we provide the first benchmark of this problem, and conduct a comprehensive empirical analysis on recent domain adaptation methods. The result shows that most existing methods are fragile in the face of CDA, which prevents them from being practically applied. Based on theoretical motivations, we propose a feature distribution and label distribution co-alignment framework, which empirically works well as a baseline for future research.

We believe this work takes an important step towards applicable domain adaptation. We hope the provided benchmarks, empirical results and baseline model would stimulate and facilitate future works to design robust algorithms that can handle more realistic problems. An interesting research direction would be better detecting and correcting label shift under feature shift.

References

1. Quionero-Candela, J., Sugiyama, M., Schwaighofer, A., Lawrence, N.D.: Dataset Shift in Machine Learning. The MIT Press, Cambridge (2009)
2. Long, M., Cao, Y., Wang, J., Jordan, M.: Learning transferable features with deep adaptation networks. In: Bach, F., Blei, D. (eds.) Proceedings of the 32nd International Conference on Machine Learning, Proceedings of Machine Learning Research, Lille, France, PMLR, 07–09 July 2015, vol. 37, pp. 97–105 (2015)
3. Tzeng, E., Hoffman, J., Zhang, N., Saenko, K., Darrell, T.: Deep domain confusion: maximizing for domain invariance. arXiv preprint arXiv:1412.3474 (2014)
4. Tzeng, E., Hoffman, J., Saenko, K., Darrell, T.: Adversarial discriminative domain adaptation. In: Computer Vision and Pattern Recognition (CVPR), vol. 1, p. 4 (2017)
5. Ganin, Y., Lempitsky, V.: Unsupervised domain adaptation by backpropagation. In: Bach, F., Blei, D. (eds.) Proceedings of the 32nd International Conference on Machine Learning, Proceedings of Machine Learning Research, Lille, France, PMLR, 07–09 July 2015, vol. 37, pp. 1180–1189 (2015)

6. Saito, K., Watanabe, K., Ushiku, Y., Harada, T.: Maximum classifier discrepancy for unsupervised domain adaptation. In: The IEEE Conference on Computer Vision and Pattern Recognition (CVPR), June 2018
7. Lipton, Z., Wang, Y.X., Smola, A.: Detecting and correcting for label shift with black box predictors. In: Dy, J., Krause, A. (eds.) Proceedings of the 35th International Conference on Machine Learning, Proceedings of Machine Learning Research, Stockholmsmässan, Stockholm Sweden, PMLR, 10–15 July 2018, vol. 80, pp. 3122–3130 (2018)
8. Azizzadenesheli, K., Liu, A., Yang, F., Anandkumar, A.: Regularized learning for domain adaptation under label shifts. In: International Conference on Learning Representations (2019)
9. Zhao, H., Combes, R.T.D., Zhang, K., Gordon, G.: On learning invariant representations for domain adaptation. In: Chaudhuri, K., Salakhutdinov, R. (eds.) Proceedings of the 36th International Conference on Machine Learning, Proceedings of Machine Learning Research., Long Beach, California, USA, PMLR, 09–15 June 2019, vol. 97, pp. 7523–7532 (2019)
10. Zhang, K., Schölkopf, B., Muandet, K., Wang, Z.: Domain adaptation under target and conditional shift. In: International Conference on Machine Learning, pp. 819–827 (2013)
11. Gong, M., Zhang, K., Liu, T., Tao, D., Glymour, C., Schölkopf, B.: Domain adaptation with conditional transferable components. In: International Conference on Machine Learning, pp. 2839–2848 (2016)
12. Wu, Y., Winston, E., Kaushik, D., Lipton, Z.: Domain adaptation with asymmetrically-relaxed distribution alignment. arXiv preprint arXiv:1903.01689 (2019)
13. Zellinger, W., Grubinger, T., Lughofer, E., Natschläger, T., Saminger-Platz, S.: Central moment discrepancy (CMD) for domain-invariant representation learning. CoRR abs/1702.08811 (2017)
14. Peng, X., Bai, Q., Xia, X., Huang, Z., Saenko, K., Wang, B.: Moment matching for multi-source domain adaptation. In: The IEEE International Conference on Computer Vision (ICCV) (2019)
15. Cao, Z., Ma, L., Long, M., Wang, J.: Partial adversarial domain adaptation. In: Proceedings of the European Conference on Computer Vision (ECCV), pp. 135–150 (2018)
16. Cao, Z., You, K., Long, M., Wang, J., Yang, Q.: Learning to transfer examples for partial domain adaptation. In: The IEEE Conference on Computer Vision and Pattern Recognition (CVPR), June 2019
17. You, K., Long, M., Cao, Z., Wang, J., Jordan, M.I.: Universal domain adaptation. In: The IEEE Conference on Computer Vision and Pattern Recognition (CVPR), June 2019
18. Saito, K., Kim, D., Sclaroff, S., Darrell, T., Saenko, K.: Semi-supervised domain adaptation via minimax entropy. In: ICCV (2019)
19. Zou, Y., Yu, Z., Kumar, B.V., Wang, J.: Unsupervised domain adaptation for semantic segmentation via class-balanced self-training. In: Proceedings of the European Conference on Computer Vision (ECCV), pp. 289–305 (2018)
20. Wang, M., Deng, W.: Deep visual domain adaptation: a survey. Neurocomputing 312, 135–153 (2018)
21. Long, M., Zhu, H., Wang, J., Jordan, M.I.: Deep transfer learning with joint adaptation networks. In: Proceedings of the 34th International Conference on Machine Learning, ICML 2017, Sydney, NSW, Australia, 6–11 August 2017, pp. 2208–2217 (2017)

22. Ben-David, S., Blitzer, J., Crammer, K., Kulesza, A., Pereira, F., Vaughan, J.W.: A theory of learning from different domains. Mach. Learn. **79**(1–2), 151–175 (2010)
23. Zhuang, F., Cheng, X., Luo, P., Pan, S.J., He, Q.: Supervised representation learning: transfer learning with deep autoencoders. IJCA **I**, 4119–4125 (2015)
24. Lee, J., Raginsky, M.: Minimax statistical learning with Wasserstein distances. arXiv preprint arXiv:1705.07815 (2017)
25. Shen, J., Qu, Y., Zhang, W., Yu, Y.: Wasserstein distance guided representation learning for domain adaptation. arXiv preprint arXiv:1707.01217 (2017)
26. Liu, M.Y., Tuzel, O.: Coupled generative adversarial networks. In: Advances in Neural Information Processing Systems, pp. 469–477 (2016)
27. Peng, X., Huang, Z., Sun, X., Saenko, K.: Domain agnostic learning with disentangled representations. In: ICML (2019)
28. Bousmalis, K., Trigeorgis, G., Silberman, N., Krishnan, D., Erhan, D.: Domain separation networks. In: Advances in Neural Information Processing Systems, pp. 343–351 (2016)
29. Ghifary, M., Kleijn, W.B., Zhang, M., Balduzzi, D., Li, W.: Deep reconstruction-classification networks for unsupervised domain adaptation. In: Leibe, B., Matas, J., Sebe, N., Welling, M. (eds.) ECCV 2016. LNCS, vol. 9908, pp. 597–613. Springer, Cham (2016). https://doi.org/10.1007/978-3-319-46493-0_36
30. Yi, Z., Zhang, H.R., Tan, P., Gong, M.: DualGAN: unsupervised dual learning for image-to-image translation. In: ICCV, pp. 2868–2876 (2017)
31. Zhu, J.Y., Park, T., Isola, P., Efros, A.A.: Unpaired image-to-image translation using cycle-consistent adversarial networks. In: 2017 IEEE International Conference on Computer Vision (ICCV) (2017)
32. Kim, T., Cha, M., Kim, H., Lee, J.K., Kim, J.: Learning to discover cross-domain relations with generative adversarial networks. In: Precup, D., Teh, Y.W. (eds.) Proceedings of the 34th International Conference on Machine Learning, Proceedings of Machine Learning Research, International Convention Centre, Sydney, Australia, PMLR, 06–11 August 2017, vol. 70, pp. 1857–1865 (2017)
33. Hoffman, J., et al.: CyCADA: cycle-consistent adversarial domain adaptation. In: Dy, J., Krause, A. (eds.) Proceedings of the 35th International Conference on Machine Learning, Proceedings of Machine Learning Research, Stockholmsmässan, Stockholm Sweden, PMLR, 10–15 July 2018, vol. 80, pp. 1989–1998 (2018)
34. Lipton, Z.C., Wang, Y.X., Smola, A.: Detecting and correcting for label shift with black box predictors. arXiv preprint arXiv:1802.03916 (2018)
35. Panareda Busto, P., Gall, J.: Open set domain adaptation. In: Proceedings of the IEEE International Conference on Computer Vision, pp. 754–763 (2017)
36. Zhao, H., Combes, R.T.d., Zhang, K., Gordon, G.J.: On learning invariant representation for domain adaptation. arXiv preprint arXiv:1901.09453 (2019)
37. Saito, K., Ushiku, Y., Harada, T.: Asymmetric tri-training for unsupervised domain adaptation. In: Proceedings of the 34th International Conference on Machine Learning, vol. 70, pp. 2988–2997 (2017). JMLR.org
38. Xie, S., Zheng, Z., Chen, L., Chen, C.: Learning semantic representations for unsupervised domain adaptation. In: International Conference on Machine Learning, pp. 5419–5428 (2018)
39. Chen, C., et al.: Progressive feature alignment for unsupervised domain adaptation. In: Proceedings of the IEEE Conference on Computer Vision and Pattern Recognition, pp. 627–636 (2019)
40. Chen, W.Y., Liu, Y.C., Kira, Z., Wang, Y.C., Huang, J.B.: A closer look at few-shot classification. In: International Conference on Learning Representations (2019)

41. He, H., Garcia, E.A.: Learning from imbalanced data. IEEE Trans. Knowl. Data Eng. **21**(9), 1263–1284 (2009)
42. Venkateswara, H., Eusebio, J., Chakraborty, S., Panchanathan, S.: Deep hashing network for unsupervised domain adaptation. In: IEEE Conference on Computer Vision and Pattern Recognition (CVPR) (2017)
43. Liu, Z., Miao, Z., Zhan, X., Wang, J., Gong, B., Yu, S.X.: Large-scale long-tailed recognition in an open world. In: IEEE Conference on Computer Vision and Pattern Recognition (CVPR) (2019)
44. Reed, W.: The Pareto, Zipf and other power laws. Econ. Lett. **74**, 15–19 (2001)
45. LeCun, Y., Bottou, L., Bengio, Y., Haffner, P.: Gradient-based learning applied to document recognition. Proc. IEEE **86**, 2278–2324 (1998)
46. Hull, J.J.: A database for handwritten text recognition research. IEEE Trans. Pattern Anal. Mach. Intell. **16**(5), 550–554 (1994)
47. Netzer, Y., Wang, T., Coates, A., Bissacco, A., Wu, B., Ng, A.Y.: Reading digits in natural images with unsupervised feature learning (2011)
48. Ganin, Y., Lempitsky, V.S.: Unsupervised domain adaptation by backpropagation. In: ICML (2015)
49. Chen, X., Wang, S., Long, M., Wang, J.: Transferability vs. discriminability: batch spectral penalization for adversarial domain adaptation. In: Chaudhuri, K., Salakhutdinov, R. (eds.) Proceedings of the 36th International Conference on Machine Learning, Proceedings of Machine Learning Research., Long Beach, California, USA, PMLR, 09–15 June 2019, vol. 97, pp. 1081–1090 (2019)
50. He, K., Zhang, X., Ren, S., Sun, J.: Deep residual learning for image recognition. In: 2016 IEEE Conference on Computer Vision and Pattern Recognition (CVPR), pp. 770–778 (2015)
51. He, H., Garcia, E.A.: Learning from imbalanced data. IEEE Trans. Knowl. Data Eng. **21**(9), 1263–1284 (2008)
52. Dong, Q., Gong, S., Zhu, X.: Imbalanced deep learning by minority class incremental rectification. IEEE Trans. Pattern Anal. Mach. Intell. **41**(6), 1367–1381 (2019)
53. van der Maaten, L., Hinton, G.: Visualizing data using t-SNE (2008)

Sequential Learning for Domain Generalization

Da Li[1,2(✉)] , Yongxin Yang[3] , Yi-Zhe Song[3] ,
and Timothy Hospedales[1,2,3]

[1] Samsung AI Center Cambridge, Cambridge, UK
dali.academic@gmail.com
[2] University of Edinburgh, Edinburgh, UK
t.hospedales@ed.ac.uk
[3] SketchX, CVSSP, University of Surrey, Guildford, UK
{yongxin.yang,y.song}@surrey.ac.uk

Abstract. In this paper we propose a sequential learning framework for Domain Generalization (DG), the problem of training a model that is robust to domain shift by design. Various DG approaches have been proposed with different motivating intuitions, but they typically optimize for a single step of domain generalization – training on one set of domains and generalizing to one other. Our sequential learning is inspired by the idea lifelong learning, where accumulated experience means that learning the nth thing becomes easier than the 1st thing. In DG this means encountering a sequence of domains and at each step training to maximise performance on the next domain. The performance at domain n then depends on the previous $n-1$ learning problems. Thus backpropagating through the sequence means optimizing performance not just for the next domain, but all following domains. Training on all such sequences of domains provides dramatically more 'practice' for a base DG learner compared to existing approaches, thus improving performance on a true testing domain. This strategy can be instantiated for different base DG algorithms, but we focus on its application to the recently proposed Meta-Learning Domain generalization (MLDG). We show that for MLDG it leads to a simple to implement and fast algorithm that provides consistent performance improvement on a variety of DG benchmarks.

Keywords: Sequential learning · Meta-learning · Domain generalization

1 Introduction

Contemporary machine learning algorithms provide excellent performance when training and testing data are drawn from the same underlying distribution. How-

Electronic supplementary material The online version of this chapter (https://doi.org/10.1007/978-3-030-66415-2_39) contains supplementary material, which is available to authorized users.

© Springer Nature Switzerland AG 2020
A. Bartoli and A. Fusiello (Eds.): ECCV 2020 Workshops, LNCS 12535, pp. 603–619, 2020.
https://doi.org/10.1007/978-3-030-66415-2_39

ever, it is often impossible to guarantee prior collection of training data that is representative of the environment in which a model will be deployed, and the resulting train-test domain shift leads to significant degradation in performance. The long studied area of *Domain Adaptation* (DA) aims to alleviate this by adapting models to the testing domain [3,4,8,19,20,38]. Meanwhile, the recently topical area of *Domain Generalization* (DG) aims to build or train models that are designed for increased robustness to such domain-shift without requiring adaptation [2,10,15–17,24,35].

A variety of DG methods have been proposed based on different intuitions. To learn a domain-agnostic feature representation, some of these require specific base learner architectures [10,14,24]. Others are model-agnostic modifications to the training procedure of any base learner, for example by via data augmentation [35,40]. Meta-learning (a.k.a learning to learn) has a long history [34,44], with primary application to accelerating learning of new tasks [29,41]. Recently, some researchers proposed meta-learning based methods for DG [2,15]. Different from previous DG methods, these are designed around explicitly mimicking train-test domain-shift during model training, to develop improved robustness to domain-shift at testing. Such meta-learning has an analogy to human learning, where a human's experience of context change provides the opportunity to develop strategies that are more agnostic to context (domain). If a human discovers that their existing problem-solving strategy fails in a new context, they can try to update their strategy to be more context independent, so that next time they arrive in a new context they are more likely to succeed immediately.

While effective, recent meta-DG methods [2,15] provide a 'single-step' of DG meta-learning: training on one set of training domains to optimize performance on a disjoint set of 'validation' domains. However, in human lifelong learning, such learning does not happen once, but sequentially in a continual learning manner. Taking this perspective in algorithm design, one learning update from domain n to $n+1$ should have the opportunity to affect the performance on every subsequent domain encountered, $n+2$ onwards. Such continual learning provides the opportunity for much more feedback to each learning update. In backpropagation, the update at domain $n \rightarrow n+1$ can be informed by its downstream impact on all subsequent updates for all subsequent domains. In this way we can generate more unique episodes for meta-learning, which has improved performance in the more common few-shot applications of meta-learning [21,39,41]. Specifically, in approaches that use a single-pass on all source domains [8,14,23], DG models are trained once for a single objective. Approaches doing one-step meta-learning [2,15] by rotating through meta-train and meta-test (validation) domain splits of N source domains train DG with N distinct domain-shift episodes. Meanwhile within our sequential learning DG framework, by further simulating all possible sequential learning domain sequences, we train with $N!$ distinct domain-shift episodes. This greater diversity of domain-shift training experience enables better generalization to a final true testing domain.

Our proposed framework can be instantiated for multiple base DG algorithms without modifying their underlying design. We focus on its instantiation for a recent architecture-agnostic meta-learning based method MLDG [15], but also show that it can be applied to a traditional architecture based method Undo Bias [14]. In the case of MLDG, we show our sequential-learning generalization S-MLDG, leads to a simple to implement and fast to train meta-learning algorithm that is architecture agnostic and consistently improves performance on a variety of DG benchmarks. This is achieved via a first-order approximation to the full S-MLDG, which leads to a shortest-path descent method analogous to Reptile [26] in few-shot learning.

We summarise our contributions as follows:

- We propose a sequential learning framework for DG that can be applied to different base DG methods. We show that it can be instantiated for at least two different base DG methods, the architecture focused Undo-Bias [14], and the architecture agnostic meta-learning algorithm MLDG [15].
- Our framework improves training by increasing the diversity of unique DG episodes constructed for training the base learner, and enabling future changes in continual-learning performance changes to back-propagate to earlier domain updates.
- We provide an analysis of the proposed S-MLDG, to understand the difference in optimization to the base MLDG algorithm, and to derive a fast first-order approximation FFO-S-MLDG. This algorithm is simple to implement and fast to run, while performing comparably to S-MLDG.
- The resulting S-MLDG and FFO-S-MLDG algorithms provide state of the art performance on three different DG benchmarks.

2 Related Work

Domain Adaptation (DA). Domain adaptation has received great attention from researchers in the past decade [3,4,9,19,20,31,32,38]. Different from domain generalization, domain adaptation assumes that unlabeled target domain data is accessible at training. Various methods have been proposed to tackle domain-shift by reducing discrepancy between source and target domain features. Representative approaches include aligning domains by minimizing distribution shift as measured by MMD [19,38], or performing adversarial training to ensure that in the learned representation space the domains cannot be distinguished [9,32], or learning generative models for cross-domain image synthesis [11,18]. However, data may not be available for the target domain, or it may not be possible to adapt the base model, thus requiring Domain Generalization.

Domain Generalization (DG). A diversity of DG methods have been proposed in recent years [2,10,14–17,23,24,35,40,43]. These are commonly categorized according to their motivating inductive bias, or their architectural assumptions. Common motivating intuitions include feature learning methods [10,17,24] that aim to learn a representation that generates domain invariant features; data augmentation-based methods that aim to improve robustness by synthesizing novel data that better covers the space of domain variability compared to the original source domains [35, 40]; and fusion methods that aim to perform well on test domains by recombining classifiers trained on diverse source domains [22, 43]. Meanwhile in terms of architecture, some methods impose constraints on the specific base classifier architecture to be used [10, 14, 16, 24, 43], while others provide an architecture agnostic DG training strategy [2, 35, 40].

Most of the above methods train a single set of source tasks for a DG objective. Recent meta-learning methods use the set of known source domains to simulate train-test domain-shift and optimize to improve robustness to domain-shift. For example, via gradient alignment [15] or meta-optimizing a robust regularizer for the base model [2]. Our sequential learning framework aims to simulate continual learning over a sequence of domains, and furthermore averages over many such sequences. This provides a greater diversity of distinct domain-shift experiences to learn from, and stronger feedback in the form of the impact of a parameter change not just on the next validation domain, but its subsequent impact on all domains in the continual learning sequence. We show that our framework can be instantiated primarily for the meta-learning method MLDG [15], but also for the classic architecture-specific method Undo-Bias [14], and its recently proposed deep extension [16].

Meta-Learning. Meta-Learning (learning to learn) has a long history [34]. It has recently become widely used in few-shot learning [1, 7, 25, 29] applications. A common meta-optimization strategy is to split training tasks into meta-train and meta-test (validation) task sets, and meta-optimization aims to improve the ability to learn quickly on meta-test tasks given the knowledge in meta-train tasks. This is achieved through various routes, by learning a more general feature embedding [37, 39], learning a more efficient optimizer [1, 29], or even simply learning an effective initial condition for optimization [7, 26]. Several gradient-based meta-learners induce higher-order gradients that increase computational cost, for example MAML [7]. This inspired other studies to develop first order approximations for faster meta-learning; such as Reptile [26] that accelerates MAML. While all these methods meta-optimize for fast adaptation to new tasks, we aim to optimize for domain-generalization: training a model such that it performs well on a novel domain with no opportunity for adaptation. We take inspiration from Reptile [26] to develop a fast implementation of our proposed S-MLDG.

Lifelong Learning. Our sequential learning is inspired by the vision of lifelong learning (LLL) [21, 28, 30, 33]. LLL methods focus on how to accelerate learning of new tasks given a series of sequentially learned previous tasks (and often how to avoid forgetting old tasks). We leverage the idea of optimizing for future performance in a sequence. But different to prior methods: (i) we focus on optimizing for domain invariance, rather than optimizing for speed of learning a new task, and (ii) we back-propagate through the entire sequence of domains so that every update step in the sequence is driven by improving the final domain invariance of the base model. It is important to note that while most lifelong and continual learning studies are oriented around designing a method that is *deployed* in a lifelong learning setting, we address a standard problem setting with a fixed set of source and target (testing) domains. We aim to use sequential training within our given source domains to learn a more robust model that generalizes better to the true testing domain. To this end, since different potential learning sequences affect the outcome in lifelong learning [27], we aim to generate the most unique learning experiences to drive training by simulating all possible sequences through our source domains and optimizing for their expected outcome.

3   Domain Generalization Background

In the domain generalization problem setting, a learner receives N labelled domains (datasets) $\mathcal{D} = [\mathcal{D}_1, \mathcal{D}_2, \cdots, \mathcal{D}_N]$ where $\mathcal{D}_i = (X_i, y_i)$, and aims to produce a model that works for a different *unseen* domain $\mathcal{D}_*$ at testing. We first introduce a simple baseline for DG.

Aggregation Baseline. A simple baseline for DG is to aggregate all domains' data and train a single model on $\mathcal{D}_{\text{agg}} = \mathcal{D}_1 \cup \mathcal{D}_2 \cup \cdots \cup \mathcal{D}_N$. Although not always compared, this obvious baseline often outperforms earlier published DG methods when applied with deep learning [16].

Base Methods. Our sequential learning framework can be applied to generalize MLDG [15] and shallow [14] or deep [16] Undo-Bias. Due to space constraints, we focus on the application to MLDG, and leave application to Undo-Bias to Appendix.

3.1   Meta-Learning Domain Generalization

In contrast to many DG methods [14,16,17,35], which require special designs of model architectures, *Meta-Learning Domain Generalization* (MLDG) [15] proposes an optimization method to achieve DG that is agnostic to base learner architecture. The idea is to mimic (during training) the cross-domain training and testing encountered in the DG setting – by way of meta-training and meta-testing steps.

In each iteration of training it randomly selects one domain $\mathcal{D}_k, k \in [1, N]$ and uses it as the meta-test domain, i.e. $\mathcal{D}_{\text{mtst}} \leftarrow \mathcal{D}_k$ (here $\mathcal{D}_{\text{mtst}}$ can be seen as a kind of *virtual* test domain), and aggregates the remaining to construct the meta-train domain, i.e., $\mathcal{D}_{\text{mtrn}} \leftarrow \bigcup_{i \neq k} \mathcal{D}_i$.

Following the intuition that meta-test will be used to evaluate the effect of the model optimization on meta-train at each iteration, MLDG aims to optimize both the loss on meta-train $\mathcal{L}_1 = \mathcal{L}(\mathcal{D}_{\text{mtrn}}, \theta)$, and loss on meta-test after updating on meta-train $\mathcal{L}_2 = \mathcal{L}(\mathcal{D}_{\text{mtst}}, \theta - \alpha \cdot \nabla_\theta \mathcal{L}_1)$ by one gradient descent step $\alpha \cdot \nabla_\theta \mathcal{L}_1$ with step size α, where $\mathcal{L}(.)$ is the cross-entropy loss here. Overall this leads to optimization of

$$\underset{\theta}{\operatorname{argmin}}\ \mathcal{L}_1(\mathcal{D}_{\text{mtrn}}, \theta) + \beta \mathcal{L}_2(\mathcal{D}_{\text{mtst}}, \theta - \alpha \nabla_\theta \mathcal{L}_1). \tag{1}$$

After training, the base model with parameters θ will be used for true unseen test domain.

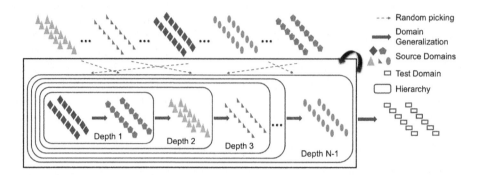

Fig. 1. Schematic illustration of our domain generalization training framework. A base DG method is trained at every step in a sequence of domains. And this is repeated over different random sequences.

4 Sequential Learning Domain Generalization

Domain generalization methods mostly aim to achieve $\min \mathcal{L}(\mathcal{D}_* | \mathcal{D}_{\text{train}})$. I.e., low loss on a testing domain $\mathcal{D}_*$ after training on a set of training domains $\mathcal{D}_{\text{train}}$. Of course this can not be optimized in the conventional way since the target $\mathcal{D}_*$ is not available, so various methods [10,14,24] attempt to achieve this indirectly by various kinds of multi-domain training on the domains in $\mathcal{D}_{\text{train}}$. As outlined in the previous section, meta-learning approaches such as MLDG aim to achieve this by finding a model that performs well over many different meta-train and meta-test splits of the true training domains: $\min \mathbb{E}_{(\mathcal{D}_{\text{mtrn}}, \mathcal{D}_{\text{mtst}}) \sim \mathcal{D}_{\text{train}}} \mathcal{L}(\mathcal{D}_{\text{mtst}} | \mathcal{D}_{\text{mtrn}})$. Inspired by the idea of human lifelong learning-to-learn [36] and the benefit of providing 'more practice' [5,13], we propose to optimize the performance of a sequentially learned DG model at every step of a trajectory p through the domains, averaged over all possible trajectories $\mathcal{P}$. As illustrated in Fig. 1, this corresponds to:

$$\min \mathbb{E}_{p \sim \mathcal{P}} \sum_{d \in p} \mathcal{L}(\mathcal{D}_d | \mathcal{D}_{[:d)}) \tag{2}$$

Here $\mathcal{L}(\mathcal{D}_d | \mathcal{D}_{[:d)})$ denotes the performance on meta-test domain d given a DG model which has been *sequentially* trained on meta-train domains before the arrival of domain d, and p denotes the sequential trajectory. This covers $N!$ distinct DG learning problems (at each incremental step of each possible trajectory p), since the order of the path through any fixed set of source domains matters. The framework is DG-algorithm agnostic in that does not stipulate which DG algorithm should be used at each step. Any base DG algorithm which can be sequentially updated could be used. In this paper we show how to instantiate this idea for both Undo Bias [14] and MLDG [15] DG algorithms.

Algorithm 1: S-MLDG: Sequential Learning MLDG

Input:$\mathcal{D} = [\mathcal{D}_1, \mathcal{D}_2, \ldots, \mathcal{D}_N]$ N source domains.
Initialize: α, β, γ and θ
while *not done training* **do**
 $p = \text{shuffle}([1, 2, \ldots, N])$
 $\tilde{\mathcal{D}} = [\tilde{\mathcal{D}}_1, \tilde{\mathcal{D}}_2, \ldots, \tilde{\mathcal{D}}_N]$ //Sample mini-batches $\tilde{\mathcal{D}}_i$
 $\mathcal{L} = \mathcal{L}(\tilde{\mathcal{D}}_{p[1]}, \theta)$
 for i **in** $[2, 3, \ldots, |p|]$ **do**
  $\mathcal{L} \mathrel{+}= \beta\Big(\mathcal{L}(\tilde{\mathcal{D}}_{p[i]}, \theta - \alpha\nabla_\theta\mathcal{L})\Big)$ //Inner-loop update
 end
 Update $\theta := \theta - \gamma\nabla_\theta\mathcal{L}$ //Meta update
end
Output: θ

4.1 Sequential Learning MLDG (S-MLDG)

Vanilla MLDG already optimizes an expectation over meta-train/meta-test splits over the source domains (Sect. 3.1). At every iteration, it randomly samples one domain as meta-test, and keeps the others as meta-train. But within the meta-train domains, it simply aggregates them. It does not exploit their domain grouping. To instantiate our hierarchical training framework (Eq. 2) for MLDG we imagine *recursively* applying MLDG. For a given meta-test/meta-train split, we apply MLDG again within the meta-train split until there is only a single domain in the meta-train set. This simulates a lifelong DG learning process, where we should succeed at DG between the first and second training domains, and then the result of that should succeed at DG on the third training domain etc. The objective function to optimize for S-MLDG is:

$$
\begin{aligned}
\mathcal{L}_{\text{S-MLDG}} = {} & \mathbb{E}_{p\sim\mathcal{P}} \quad \mathcal{L}_1(\mathcal{D}_{p[1]}, \theta) \\
& + \beta\sum_{i=2}^{N}\mathcal{L}_i(\mathcal{D}_{p[i]}, \theta - \alpha\nabla_\theta\sum_{j=1}^{i-1}\mathcal{L}_j) \\
= {} & \mathbb{E}_{p\sim\mathcal{P}} \quad \mathcal{L}_1(\mathcal{D}_{p[1]}, \theta) \\
& + \beta\mathcal{L}_2(\mathcal{D}_{p[2]}, \theta - \alpha\nabla_\theta\mathcal{L}_1) \\
& + \beta\mathcal{L}_3(\mathcal{D}_{p[3]}, \theta - \alpha\nabla_\theta\sum_{j=1}^{2}\mathcal{L}_j) + \ldots \\
& + \beta\mathcal{L}_N(\mathcal{D}_{p[N]}, \theta - \alpha\nabla_\theta\sum_{j=1}^{N-1}\mathcal{L}_j)
\end{aligned}
\tag{3}
$$

The optimization is carried out over all possible paths p through the training domains. MLDG is model-agnostic and computes a single parameter θ for all domains, so the final θ after optimization is used for inference on unseen domains. The overall algorithm is shown in Algorithm 1.

Algorithm 2: Fast First-Order S-MLDG

Input: $\mathcal{D} = [\mathcal{D}_1, \mathcal{D}_2, \ldots, \mathcal{D}_N]$ N source domains.
Initialize: α, β, γ and θ
while *not done training* **do**
    $\tilde{\theta} = \theta$
    $p = \text{shuffle}([1, 2, \ldots, N])$
    $\tilde{\mathcal{D}} = [\tilde{\mathcal{D}}_1, \tilde{\mathcal{D}}_2, \ldots, \tilde{\mathcal{D}}_N]$ //Sample mini-batches $\tilde{\mathcal{D}}_i$
    for i *in* $[1, N]$ **do**
        $\mathcal{L}_i = \beta\mathcal{L}(\tilde{\mathcal{D}}_{p[i]}, \tilde{\theta})$
        $\tilde{\theta} = \tilde{\theta} - \alpha\nabla_{\tilde{\theta}}\mathcal{L}_i$ //Inner-loop update
    end
    Update $\theta := \theta + \gamma(\tilde{\theta} - \theta)$ //Meta update
end
Output: θ

MLDG. The MLDG mechanism was originally analyzed [15] via a first-order Taylor series. Since MLDG only does one-step DG validation, one domain is sampled as meta-test to split the source domains. Then the objective function is

$$\mathcal{L}_{\text{MLDG}} = \mathcal{L}_1(\mathcal{D}_{\text{mtrn}}, \theta) + \beta\mathcal{L}_2(\mathcal{D}_{\text{mtst}}, \theta - \alpha\nabla_\theta\mathcal{L}_1) \tag{4}$$

After Taylor expansion on the second item, it becomes

$$\mathcal{L}_2(\theta - \alpha\nabla_\theta\mathcal{L}_1) = \mathcal{L}_2(\theta) + \nabla_\theta\mathcal{L}_2 \cdot (-\alpha\nabla_\theta\mathcal{L}_1) \tag{5}$$

and then $\mathcal{L}_{\text{MLDG}}$ becomes

$$\mathcal{L}_{\text{MLDG}} = \mathcal{L}_1(\theta) + \beta\mathcal{L}_2(\theta) - \beta\alpha\nabla_\theta\mathcal{L}_1\nabla_\theta\mathcal{L}_2 \tag{6}$$

This led to MLDG's interpretation as a preference for an optimization path with *aligned* gradients between meta-train and meta-test [15].

S-MLDG. If we use 3 source domains as an example to analyse S-MLDG, the loss function is

$$\begin{aligned}\mathcal{L}_{\text{S-MLDG-3}} = &\mathcal{L}_1(\theta) + \beta\mathcal{L}_2(\theta - \alpha\nabla_\theta\mathcal{L}_1) \\ &+ \beta\mathcal{L}_3(\theta - \alpha\nabla_\theta(\mathcal{L}_1 + \mathcal{L}_2))\end{aligned} \tag{7}$$

The first two items are the same as $\mathcal{L}_{\text{MLDG}}$. Apply Taylor expansion on the third item in $\mathcal{L}_{\text{S-MLDG-3}}$,

$$\mathcal{L}_3(\theta - \alpha\nabla_\theta(\mathcal{L}_1 + \mathcal{L}_2)) = \mathcal{L}_3(\theta) + \nabla_\theta\mathcal{L}_3 \cdot (-\alpha\nabla_\theta(\mathcal{L}_1 + \mathcal{L}_2)) \tag{8}$$

we have,

$$\begin{aligned}\mathcal{L}_{\text{S-MLDG-3}} = &\mathcal{L}_1(\theta) + \beta\mathcal{L}_2(\theta) + \beta\mathcal{L}_3(\theta) \\ &- \beta\alpha\nabla_\theta\mathcal{L}_1\nabla_\theta\mathcal{L}_2 - \beta\alpha\nabla_\theta\mathcal{L}_3\nabla_\theta\mathcal{L}_1 - \beta\alpha\nabla_\theta\mathcal{L}_3\nabla_\theta\mathcal{L}_2\end{aligned} \tag{9}$$

This shows that S-MLDG optimizes all source domains (first three terms), while preferring an optimization path where gradients align across all pairs of domains (second three terms maximising dot products). This is different to MLDG, that only optimizes the inner product of gradients between the current meta-train and meta-test domain splits. In contrast S-MLDG has the chance to optimize for DG on each meta-test domain in the sequential way, thus obtaining more unique experience to 'practice' DG.

A Direct S-MLDG Implementation. A direct implementation of the meta update for S-MLDG in the three domain case would differentiate $\mathcal{L}_{\text{S-MLDG-3}}$ (Eq. 9) w.r.t θ as

$$
\begin{aligned}
\nabla_\theta \mathcal{L}_{\text{S-MLDG-3}} &= \nabla_\theta \mathcal{L}_1(\theta) + \beta \nabla_\theta \mathcal{L}_2(\theta - \alpha \nabla_\theta \mathcal{L}_1) \\
&\quad + \beta \nabla_\theta \mathcal{L}_3(\theta - \alpha \nabla_\theta (\mathcal{L}_1 + \mathcal{L}_2)) \\
&= \frac{\partial \mathcal{L}_1(\theta)}{\partial \theta} + \beta \frac{\partial \mathcal{L}_2(\theta_1)}{\partial \theta_1} \frac{\partial \theta_1}{\partial \theta} + \beta \frac{\partial \mathcal{L}_3(\theta_2)}{\partial \theta_2} \frac{\partial \theta_2}{\partial \theta}
\end{aligned}
\tag{10}
$$

where

$$
\begin{aligned}
\theta_1 &= \theta - \alpha \nabla_\theta \mathcal{L}_1 \\
\theta_2 &= \theta - \alpha \nabla_\theta (\mathcal{L}_1 + \mathcal{L}_2)
\end{aligned}
\tag{11}
$$

However, update steps based on Eq. 10 require high-order gradients when computing $\frac{\partial \theta_1}{\partial \theta}$, $\frac{\partial \theta_2}{\partial \theta}$. These higher-order gradients are expensive to compute.

4.2 First-Order Approximator of S-MLDG

FO-S-MLDG: Similar to [7], we can alleviate the above issue by stopping the gradient of the exposed first derivative items to omit higher-order gradients. I.e., $\nabla_\theta \mathcal{L}_1$ and $\nabla_\theta (\mathcal{L}_1 + \mathcal{L}_2)$ in Eq. 11 are constants when computing $\mathcal{L}_2$ and $\mathcal{L}_3$. Then for FO-S-MLDG, Eq. 10 becomes

$$
\begin{aligned}
\nabla_\theta \mathcal{L}_{\text{S-MLDG-3}} &= \frac{\partial \mathcal{L}_1(\theta)}{\partial \theta} + \beta \frac{\partial \mathcal{L}_2(\theta_1)}{\partial \theta_1} \frac{\partial \theta_1}{\partial \theta} + \beta \frac{\partial \mathcal{L}_3(\theta_2)}{\partial \theta_2} \frac{\partial \theta_2}{\partial \theta} \\
&= \frac{\partial \mathcal{L}_1(\theta)}{\partial \theta} + \beta \frac{\partial \mathcal{L}_2(\theta_1)}{\partial \theta_1} \frac{\partial (\theta - \alpha \nabla_\theta \mathcal{L}_1)}{\partial \theta} \\
&\quad + \beta \frac{\partial \mathcal{L}_3(\theta_2)}{\partial \theta_2} \frac{\partial (\theta - \alpha \nabla_\theta (\mathcal{L}_1 + \mathcal{L}_2))}{\partial \theta} \\
&= \frac{\partial \mathcal{L}_1(\theta)}{\partial \theta} + \beta \frac{\partial \mathcal{L}_2(\theta_1)}{\partial \theta_1} + \beta \frac{\partial \mathcal{L}_3(\theta_2)}{\partial \theta_2}
\end{aligned}
\tag{12}
$$

FO-S-MLDG still follows Algorithm 1, but saves computation by omitting higher-order gradients in back propagation. We use this approximator for S-MLDG by default. However FO-S-MLDG still requires back propagation (as per Eq. 10), to compute gradients of $\mathcal{L}_1$, $\mathcal{L}_2$ and $\mathcal{L}_3$, even though higher-order gradients are ignored.

4.3 Fast First-Order S-MLDG

FFO-S-MLDG: If we look at $\nabla_\theta \mathcal{L}_{\text{S-MLDG-3}}$ in Eq. 12 again, we find

$$
\frac{\partial \mathcal{L}_1(\theta)}{\partial \theta} + \beta \frac{\partial \mathcal{L}_2(\theta_1)}{\partial \theta_1} + \beta \frac{\partial \mathcal{L}_3(\theta_2)}{\partial \theta_2} = \mathcal{L}_1' + \beta \mathcal{L}_2' + \beta \mathcal{L}_3'
\tag{13}
$$

Table 1. Performance on IXMAS action recognition. Leave one camera-view out. Accuracy (%).

Unseen	MMD-AAE [17]	AGG [8]	DANN	CrossGrad [35]	MetaReg [2]	Undo-Bias [14]	S-Undo-Bias	MLDG [15]	FFO-S-MLDG	S-MLDG
4	79.1	80.0	81.6	78.4	79.3	80.7	82.7	79.4	81.1	80.1
3	94.5	94.5	94.5	94.2	94.5	95.3	94.9	95.2	95.1	95.0
2	95.6	99.8	100.0	100.0	99.8	99.9	100.0	99.9	100.0	99.8
1	93.4	93.4	91.7	94.0	92.5	94.8	94.0	95.2	93.1	96.2
0	96.7	93.5	93.5	91.2	92.8	94.2	93.9	90.4	94.3	92.7
Ave.	91.9	92.2	92.3	91.6	91.8	93.0	93.1	91.9	92.7	92.8

Table 2. Performance on VLCS object recognition. Leave one dataset out. Accuracy (%).

Unseen	MMD-AAE [17]	AGG [8]	DANN	CrossGrad [35]	MetaReg [2]	Undo-Bias [14]	S-Undo-Bias	MLDG [15]	FFO-S-MLDG	S-MLDG
V	67.7	65.4	66.4	65.5	65.0	68.1	68.7	67.7	68.1	68.7
L	62.6	60.6	64.0	60.0	60.2	60.3	61.8	61.3	63.1	64.8
C	94.4	93.1	92.6	92.0	92.3	93.7	95.0	94.4	94.8	96.4
S	64.4	65.8	63.6	64.7	64.2	66.0	66.1	65.9	65.2	64.0
Ave.	72.3	71.2	71.7	70.5	70.4	72.0	72.9	72.3	72.8	73.5

This means that one-step meta update of FO-S-MLDG is $\gamma(\mathcal{L}'_1 + \beta\mathcal{L}'_2 + \beta\mathcal{L}'_3)$, where γ is the meta step-size. This indicates that one update of naive first-order S-MLDG is equivalent to updating the parameters towards the result of training on $\mathcal{L}_1(\mathcal{D}_{p[1]})$, $\mathcal{L}_2(\mathcal{D}_{p[2]})$ and $\mathcal{L}_3(\mathcal{D}_{p[3]})$ recursively. In other words, if we regard the initial parameters as θ and the parameters updated recursively on $\mathcal{L}_1(\mathcal{D}_{p[1]})$, $\mathcal{L}_2(\mathcal{D}_{p[2]})$ and $\mathcal{L}_3(\mathcal{D}_{p[3]})$ as $\tilde{\theta}$, then Eq. 13 can be expressed as

$$\mathcal{L}'_1 + \beta\mathcal{L}'_2 + \beta\mathcal{L}'_3 = \theta - \tilde{\theta} \tag{14}$$

This means that we can optimize $\mathcal{L}_1(\mathcal{D}_{p[1]})$, $\mathcal{L}_2(\mathcal{D}_{p[2]})$ and $\mathcal{L}_3(\mathcal{D}_{p[3]})$ in sequence (to obtain $\tilde{\theta}$, and then use the resulting offset vector as the meta-gradient for updating θ). Thus we do not need to backpropagate to explicitly compute the gradients suggested by Eq. 13. The overall flow of FFO-S-MLDG is shown in Algorithm 2.

Link Between Fast First-Order S-MLDG and S-MLDG. We analyze FFO-S-MLDG in Algorithm 2 considering two source domains and derive the expectation of the optimization gradient is

$$\mathbb{E}_{p\sim\mathcal{P}}[g_{p[1]} + g_{p[2]}] = \bar{g}_1 + \bar{g}_2 - \frac{\alpha}{2}\frac{\partial(\bar{g}_1 \cdot \bar{g}_2)}{\partial\tilde{\theta}_1} + \mathcal{O}(\alpha^2) \tag{15}$$

Here $\bar{g}_1$, $\bar{g}_2$ are the gradient updates for the first and second source domains and $\bar{g}_1 \cdot \bar{g}_2$ is the inner product between the two gradients. The gradient $-\frac{\partial(\bar{g}_1 \cdot \bar{g}_2)}{\partial\tilde{\theta}_1}$ is in the direction that maximizes it. This means in expectation of multiple gradient updates FFO-S-MLDG learns to maximize the inner-product between

Table 3. Performance on PACS object recognition across styles (ResNet-18). Accuracy (%).

Unseen	AGG [8]	DANN [8]	CrossGrad [35]	MetaReg [2]	Undo-Bias [14]	S-Undo-Bias	MLDG [15]	FFO-S-MLDG	S-MLDG
A	77.6	81.3	78.7	79.5	78.4	80.6	79.5	80.0	80.5
C	73.9	73.8	73.3	75.4	72.5	76.2	77.3	77.4	77.8
P	94.4	94.0	94.0	94.3	92.8	94.1	94.3	94.6	94.8
S	70.3	74.3	65.1	72.2	73.3	72.2	71.5	73.8	72.8
Ave.	79.1	80.8	77.8	80.4	79.3	80.8	80.7	81.4	81.5

gradients of different domains. Thus it maintains a similar but slightly different objective to S-MLDG, which maximizes the inner-product of gradients in each meta update. More details can be found in Appendix.

5   Experiments

Datasets and Settings. We evaluate our method on three different benchmarks: **IXMAS** [42], where human actions are recognized across different camera views. **VLCS** [6], which requires the domain generalization across different photo datasets. And **PACS** [16] which is a more realistic and challenging cross-domain visual benchmark of images with different style depictions.

Competitors. For comparative evaluation we also evaluate the following competitors:

- **AGG:** A simple but effective baseline of aggregating all source domains' data for training [16].
- **DANN:** Domain adversarial neural networks learns a domain invariant representation such that source domains cannot be distinguished [8].
- **MMD-AAE:** A recent DG method which combines kernel MMD and the adversarial auto encoder [17].
- **CrossGrad:** A recently proposed strategy that learns the manifold of training domains, and uses cross-gradients to generate synthetic data that helps the classifier generalize across the manifold [35].
- **MetaReg:** A latest DG method by meta-learning a regularizer constraining the model parameters to be more domain-generalizable [2].
- **Undo-Bias:** Undo-Bias models [14] each training domain as a linear combination of a domain-agnostic model and domain-specific bias, and then uses the domain-agnostic model for testing. We use the vanilla deep generalization of Undo-Bias explained in [16].
- **MLDG:** A recent DG method that is model-agnostic and meta-learns the domain-generalizable model parameters.

The most related alternatives are **Undo Bias** [14] and **MLDG** [15], which are the models we extend to realize our sequential learning strategy. We re-implement AGG, DANN, CrossGrad, MetaReg, Undo-Bias and MLDG; and report the numbers stated by MMD-AAE.

5.1   Action Recognition Across Camera Views

Setup. The IXMAS dataset contains 11 different human actions recorded by 5 video cameras with different views (referred as 0, ..., 4). The goal is to train an action recognition model on a set of source views (domains), and recognize the action from a novel target view (domain). We follow [17] to keep the first 5 actions and use the same Dense trajectory features as input. For our implementation, we follow [17] to use a one-hidden layer MLP with 2000 hidden neurons as backbone and report the average of 20 runs. In addition, we normalize the hidden embedding by BatchNorm [12] as this gives a good start point for AGG.

Results. From the results in Table 1, we can see that several recent DG methods [2,15,17,35] fail to improve over the strong AGG baseline. Undo-Bias works here, and provides 0.8% improvement. Our extension S-Undo-Bias provides a modest increase of 0.1% on the overall accuracy over vanilla Undo. While the original MLDG [15] fails to improve on the AGG baseline, our S-MLDG provides a 0.9% gain over MLDG and thus improves 0.6% on AGG. Our FFO-S-MLDG runs on par with S-MLDG, demonstrating the efficacy of our approximator. Overall our S-Undo-Bias, FFO-S-MLDG and S-MLDG all provide a gain in performance over the AGG baseline.

5.2   Object Recognition Across Photo Datasets

Setup. VLCS domains share 5 categories: bird, car, chair, dog and person. We use pre-extracted DeCAF6 features and follow [23] to randomly split each domain into train (70%) and test (30%) and do leave-one-out evaluation. We use a 2 fully connected layer architecture with output size of 1024 and 128 with ReLU activation, as per [23] and report the average performance of 20 trials.

Results. In this benchmark, the results in Table 2 show that the simple AGG method works well again. Recent DG methods [2,35] still struggle to beat this baseline. The base DG methods Undo-Bias [14] and MLDG [15] work well here, producing comparable results to the state-of-the-art [17]. Our extensions of these base DG methods, S-Undo-Bias, FFO-S-MLDG and S-MLDG all provide improvements. Overall our S-MLDG performs best, followed closely by S-Undo-Bias and FFO-S-MLDG.

5.3   Object Recognition Across Styles

Setup. The PACS benchmark [16] contains 4 domains: photo, art painting, cartoon and sketch and 7 common categories: 'dog', 'elephant', 'giraffe', 'guitar', 'horse', 'house' and 'person'. [16] showed that this benchmark has much stronger domain shift than others such as Caltech-Office and VLCS. We use a ResNet-18 pre-trained ImageNet as a modern backbone for comparison. We note that MetaReg [2] used a slightly different setup than the official PACS protocol [16], for which their AGG baseline is hard to reproduce. So we stick to the official protocol and rerun MetaReg. To save computational cost, since Undo-Bias and

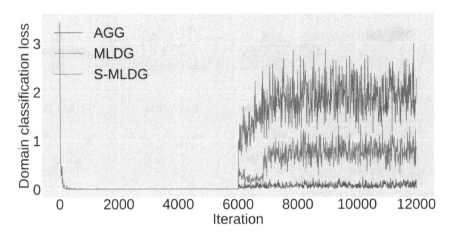

Fig. 2. Domain classification loss analysis on VLCS.

S-Undo-Bias require domain-specific branches that are expensive when applied to ResNet, we only apply these methods to the last ResNet-18 layer – so previous layers are shared as per AGG.

Results. From the results in Table 3, we can see that: (i) Our sequential learning methods S-Undo-Bias and S-MLDG improve on their counterparts Undo-Bias and MLDG, (ii) FFO-S-MLDG performs comparably with S-MLDG, and (iii) S-MLDG and FFO-S-MLDG perform best overall.

5.4 Further Analysis

Analysis for S-MLDG. As shown earlier, MLDG and S-MLDG aim to maximize the inner-product between gradients of different source domains. Intuitively, optimizing this gradient alignment will lead to increase domain invariance [8]. To analyze if this is the case, we use domain-classification loss as a measure of domain invariant feature encoding. We append an additional domain-classifier to the penultimate layer of the original model, creating a domain and category multi-task classifier, where all feature layers are shared. We train the domain classification task for 6000 iterations, then switch to training the category classification task for another 6000 iterations. Using this setup we compare AGG, MLDG and S-MLDG. From Fig. 2, we see that the domain-classification loss decreases rapidly in the first phase: the domain is easy to recognise before DG training. In the second phase we switch on categorization and DG training. S-MLDG and MLDG give higher domain classification loss than AGG – indicating that MLDG and S-MLDG learn features that the domain classifier finds harder to distinguish, and hence are the most domain invariant.

Visualization of Learned Features. We use t-SNE to visualize the feature embedding of a held-out test domain (V) on VLCS, after training models on L, C

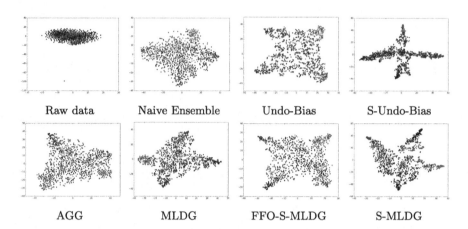

Fig. 3. T-SNE visualization of different models' embeddings of VLCS held-out test data (V) after training on (LCS). Colors represent object categories.

Table 4. Training cost (mins) for PACS with ResNet-18.

AGG	DANN [8]	CrossGrad [35]	MetaReg [2]	MLDG [15]
10.98	11.35	146.51	20.01	49.77
	FFO-S-MLDG	S-MLDG	Undo-Bias [14]	S-Undo-Bias
	11.04	72.64	11.16	11.01

and S. From the results in Fig. 3, we can see that before training the raw test data points are not separable by category. As baselines we also compare AGG and Naive Ensemble (training an ensemble of domain-specific models and averaging their result) for comparison to the models of interest: MLDG, S-MLDG, FFO-S-MLDG, Undo-Bias and S-Undo-Bias. We can see that all these DG methods exhibit better separability than the two baselines, with S-Undo-Bias and S-MLDG providing the sharpest separation.

Computational Cost. A major contribution of this paper is a DG strategy that is not only effective but simple (Algorithm 2) and fast to train. To evaluate this we compare the computational cost of training various methods on PACS with ResNet-18 for $3k$ iterations. We run all the methods on a machine with Intel Xeon(R) CPU (E5-2687W @ 3.10 GHz) and TITAN X (Pascal) GPU. From the results in Table 4, we see that CrossGrad is by far the most expensive with S-MLDG in second place. In contrast, our derived FFO-S-MLDG is not noticeably slower than the baseline and lower-bound, AGG. Undo-Bias and S-Undo-Bias run fast due to only applying them into the last layer. But S-Undo-Bias saves training cost over Undo-Bias as explained in Appendix.

6   Conclusion

We introduced the idea of sequential learning to provide a training regime for a base DG model. This can be seen as generating more unique DG episodes for learning, and as providing more feedback for back-propagation through the chain of domains. Our framework can be applied to different base DG models including Undo-Bias and MLDG. Our final FFO-S-MLDG method provides a simple to implement and fast to train DG method that achieves state of the art results on a variety of benchmarks.

References

1. Andrychowicz, M., et al.: Learning to learn by gradient descent by gradient descent. In: NIPS (2016)
2. Balaji, Y., Sankaranarayanan, S., Chellappa, R.: MetaReg: towards domain generalization using meta-regularization. In: NeurIPS (2018). http://papers.nips.cc/paper/7378-metareg-towards-domain-generalization-using-meta-regularization.pdf
3. Ben-David, S., Blitzer, J., Crammer, K., Pereira, F.: Analysis of representations for domain adaptation. In: NIPS (2006)
4. Bousmalis, K., Trigeorgis, G., Silberman, N., Krishnan, D., Erhan, D.: Domain separation networks. In: NIPS (2016)
5. Doersch, C., Zisserman, A.: Multi-task self-supervised visual learning. In: ICCV (2017)
6. Fang, C., Xu, Y., Rockmore, D.N.: Unbiased metric learning: on the utilization of multiple datasets and web images for softening bias. In: ICCV (2013)
7. Finn, C., Abbeel, P., Levine, S.: Model-agnostic meta-learning for fast adaptation of deep networks. In: ICML (2017)
8. Ganin, Y., et al.: Domain-adversarial training of neural networks. JMLR (2016)
9. Ganin, Y., Lempitsky, V.: Unsupervised domain adaptation by backpropagation. In: ICML (2015)
10. Ghifary, M., Bastiaan Kleijn, W., Zhang, M., Balduzzi, D.: Domain generalization for object recognition with multi-task autoencoders. In: ICCV (2015)
11. Hoffman, J., et al.: CyCADA: cycle-consistent adversarial domain adaptation. In: ICML (2018)
12. Ioffe, S., Szegedy, C.: Batch normalization: accelerating deep network training by reducing internal covariate shift. In: ICML (2015)
13. Jaderberg, M., et al.: Reinforcement learning with unsupervised auxiliary tasks. In: ICLR (2017)
14. Khosla, A., Zhou, T., Malisiewicz, T., Efros, A.A., Torralba, A.: Undoing the damage of dataset bias. In: Fitzgibbon, A., Lazebnik, S., Perona, P., Sato, Y., Schmid, C. (eds.) ECCV 2012. LNCS, vol. 7572, pp. 158–171. Springer, Heidelberg (2012). https://doi.org/10.1007/978-3-642-33718-5_12
15. Li, D., Yang, Y., Song, Y.Z., Hospedales, T.: Learning to generalize: meta-learning for domain generalization. In: AAAI (2018)
16. Li, D., Yang, Y., Song, Y.Z., Hospedales, T.M.: Deeper, broader and artier domain generalization. In: ICCV (2017)
17. Li, H., Jialin Pan, S., Wang, S., Kot, A.C.: Domain generalization with adversarial feature learning. In: CVPR (2018)

18. Liu, M.Y., Tuzel, O.: Coupled generative adversarial networks. In: NIPS (2016)
19. Long, M., Cao, Y., Wang, J., Jordan, M.: Learning transferable features with deep adaptation networks. In: ICML (2015)
20. Long, M., Zhu, H., Wang, J., Jordan, M.I.: Unsupervised domain adaptation with residual transfer networks. In: NIPS (2016)
21. Lopez-Paz, D., Ranzato, M.: Gradient episodic memory for continual learning. In: NIPS (2017)
22. Mancini, M., Bulò, S.R., Caputo, B., Ricci, E.: Best sources forward: domain generalization through source-specific nets. In: ICIP (2018)
23. Motiian, S., Piccirilli, M., Adjeroh, D.A., Doretto, G.: Unified deep supervised domain adaptation and generalization. In: ICCV (2017)
24. Muandet, K., Balduzzi, D., Schölkopf, B.: Domain generalization via invariant feature representation. In: ICML (2013)
25. Munkhdalai, T., Yu, H.: Meta networks. In: ICML (2017)
26. Nichol, A., Achiam, J., Schulman, J.: On first-order meta-learning algorithms. arXiv (2018)
27. Pentina, A., Sharmanska, V., Lampert, C.H.: Curriculum learning of multiple tasks. In: CVPR (2015)
28. Pentina, A., Lampert, C.H.: Lifelong learning with non-i.i.d. tasks. In: NIPS (2015)
29. Ravi, S., Larochelle, H.: Optimization as a model for few-shot learning. In: ICLR (2017)
30. Ruvolo, P., Eaton, E.: ELLA: an efficient lifelong learning algorithm. In: ICML (2013)
31. Saito, K., Ushiku, Y., Harada, T.: Asymmetric tri-training for unsupervised domain adaptation. In: ICML (2017)
32. Saito, K., Watanabe, K., Ushiku, Y., Harada, T.: Maximum classifier discrepancy for unsupervised domain adaptation. In: CVPR (2018)
33. Schmidhuber, J., Zhao, J., Wiering, M.: Shifting inductive bias with success-story algorithm, adaptive Levin search, and incremental self-improvement. Mach. Learn. **28**, 105–130 (1997). https://doi.org/10.1023/A:1007383707642
34. Schmidhuber, J.: On learning how to learn learning strategies. Technical report (1995)
35. Shankar, S., Piratla, V., Chakrabarti, S., Chaudhuri, S., Jyothi, P., Sarawagi, S.: Generalizing across domains via cross-gradient training. In: ICLR (2018)
36. Smith, L.B., Jones, S.S., Landau, B., Gershkoff-Stowe, L., Samuelson, L.: Object name learning provides on-the-job training for attention. Psychol. Sci. **13**(1), 13–19 (2002)
37. Sung, F., Yang, Y., Zhang, L., Xiang, T., Torr, P.H., Hospedales, T.M.: Learning to compare: relation network for few-shot learning. In: CVPR (2018)
38. Tzeng, E., Hoffman, J., Zhang, N., Saenko, K., Darrell, T.: Deep domain confusion: maximizing for domain invariance. arXiv (2014)
39. Vinyals, O., Blundell, C., Lillicrap, T., Kavukcuoglu, K., Wierstra, D.: Matching networks for one shot learning. In: NIPS (2016)
40. Volpi, R., Namkoong, H., Sener, O., Duchi, J., Murino, V., Savarese, S.: Generalizing to unseen domains via adversarial data augmentation. In: NeurIPS (2018)
41. Wei, Y., Zhang, Y., Huang, J., Yang, Q.: Transfer learning via learning to transfer. In: ICML (2018)
42. Weinland, D., Ronfard, R., Boyer, E.: Free viewpoint action recognition using motion history volumes. CVIU **104**(2–3), 249–257 (2006)

43. Xu, Z., Li, W., Niu, L., Xu, D.: Exploiting low-rank structure from latent domains for domain generalization. In: Fleet, D., Pajdla, T., Schiele, B., Tuytelaars, T. (eds.) ECCV 2014. LNCS, vol. 8691, pp. 628–643. Springer, Cham (2014). https://doi.org/10.1007/978-3-319-10578-9_41
44. Yoshua Bengio, S.B., Cloutier, J.: Learning a synaptic learning rule. In: IJCNN (1991)

Generating Visual and Semantic Explanations with Multi-task Network

Wenjia Xu[1,2]([✉]) [iD], Jiuniu Wang[1,2] [iD], Yang Wang[1], Yirong Wu[1,2],
and Zeynep Akata[3] [iD]

[1] Department of Electrical Engineering, University of Chinese Academy of Sciences,
Beijing, China
xuwenjia16@mails.ucas.ac.cn
[2] Aerospace Information Research Institute, Chinese Academy of Sciences,
Beijing, China
[3] Cluster of Excellence Machine Learning, University of Tübingen,
Tübingen, Germany

Abstract. Explaining deep models is desirable especially for improving
the user trust and experience. Much progress has been done recently
towards visually and semantically explaining deep models. However,
establishing the most effective explanation is often human-dependent,
which suffers from the bias of the annotators. To address this issue, we
propose a multitask learning network (MTL-Net) that generates saliency-
based visual explanation as well as attribute-based semantic explanation. Via an integrated evaluation mechanism, our model quantitatively
evaluates the quality of the generated explanations. First, we introduce
attributes to the image classification process and rank the attribute contribution with gradient weighted mapping, then generate semantic explanations with those attributes. Second, we propose a fusion classification
mechanism (FCM) to evaluate three recent saliency-based visual explanation methods by their influence on the classification. Third, we conduct
user studies, quantitative and qualitative evaluations. According to our
results on three benchmark datasets with varying size and granularity,
our attribute-based semantic explanations are not only helpful to the
user but they also improve the classification accuracy of the model, and
our ranking framework detects the best performing visual explanation
method in agreement with the users.

Keywords: Multi-task learning · Explainable AI

1 Introduction

Deep learning has led to remarkable progress in computer vision tasks. However,
despite their superior performance, the black-box nature of deep neural networks harms user trust. In order to build interpretable models that can explain
their behaviour, previous works visually point to the evidence that influences
the network decision [17,24], or provide semantic explanations that justify a

© Springer Nature Switzerland AG 2020
A. Bartoli and A. Fusiello (Eds.): ECCV 2020 Workshops, LNCS 12535, pp. 620–635, 2020.
https://doi.org/10.1007/978-3-030-66415-2_40

category prediction [11,12]. When it comes to generating visual explanations, example methods includes visualizations via gradient flow or filter deconvolution [29,30,39], visualizing class activation maps [28,42] and measuring the effect of perturbations on input images [7,26,27].

However, there is no unified evaluation metric to determine the most effective visualization technique. Although user-studies are widely used to judge the effectiveness of visualization methods, it is unscalable since humans are not on-demand software that can be employed at anytime. Among automatic evaluation methods, RISE [26] proposes to evaluate the influence of different regions to the decision maker by deleting/inserting pixels. However, this requires insertion/removal of many pixel combinations for each image, e.g., several iterations are needed for evaluating one image, which is time consuming.

Natural language explanations [11] is a complementary way to justify neural network decisions. These explanations are usually generated by feeding images and predicted class labels into LSTM [10]. A drawback is that the semantic explanations may lack the class discriminative ability, missing essential details to infer the image label [20].

In this work, our primary aim is to generate visual and semantic explanations that are faithful to the model, then quantitatively and objectively evaluate the justifications of a deep learning based decision maker. To realize this aim, we propose a visual and semantic explanation framework with an integrated quantitative and objective evaluation mechanism without requiring the user in its training or inference steps. We classify and embed attributes to help the category prediction. The semantic explanation is generated based on gradient weighted mapping of the attribute embedding, then evaluated on its image and class relevance. Furthermore, to evaluate the visual explanation methods in this framework, we propose a fusion classification mechanism. The input image is filtered by its visual explanation map and then fed into a classifier. We evaluate the methods based on the classification accuracy.

We argue that an explanation is faithful to the model it is interpreting, if it can help to improve the performance of that black box model. For instance, for the task depicted in Fig. 1, an accurate visual explanation model should attend to the clothing related regions to recognize a "Clothing Store". Hence, if the background and non-relevant pixels are weakened and the clothing related regions are preserved, e.g., the image is filtered through the attention mechanism, the classification result should not be degraded. The same holds for attribute-based justifications. For instance, an accurate semantic explanation of the predicted label "clothing store" should inform the user about the most discriminative attributes such as "enclosed area, cloth, indoor lighting". The effectiveness of this explanation can be verified by feeding these attributes to the network for the classification task.

Our main contributions are summarized as follows. (1) We propose a visual and semantic explanation framework that generates explanations that are fidelity to the model. (2) We design an integrated evaluation mechanism with quantitative and qualitative evaluations as well as user study on the explanations. The quantitative evaluation is automatically performed according to the influence of explanations on classification tasks. (3) We showcase on three datasets that our semantic explanations are faithful to the network they are interpreting. Three representative visual explanations, i.e. Grad-Cam [28], Back-Propagation [29] and RISE [26] are evaluated and the quantitative results agree with the user preference.

2   Related Work

In this section, we summarize the prior work on multitask learning and explainability research related to ours.

Multitask Learning. Multitask learning is a popular method that enables us to train one neural network to do many tasks. Some prior works have shown that learning multiple tasks can improve the generalization of the network and give better performance than doing these tasks separately [2]. For instance, a multitask network for segmentation can improve the performance of object detection while being much faster [4]. In our work, we train a network on image recognition and attributes classification simultaneously, motivated by the fact that sharing lower-level features can benefit these two tasks and result in better performance.

Textual Explanation. Generating semantic explanations has gained interest in the community. Among those, Hendricks et al. [11,12] take the image and its predicted category label as input, and generate explanations with a conditioned LSTM [10]. Although these explanations build a sound basis for enabling user acceptance of the deep models and improving user trust, they cannot guarantee the fidelity to the model. The conditioned LSTM model trained on human annotated captions may generate sentences describing the image content, rather than the real reason for the network decision.

We take advantage of semantic attributes to generate textual explanations. Attributes are human-annotated discriminative visual properties of objects [18, 25]. In zero-shot learning [18,36,37], they are used to build intermediate representations to leverage the lack of labeled images as the model does not have access to any training examples of some classes. [16] and [15] apply attributes as a linguistic explanation for video and image classification. They select the attributes by its interaction information with the input images. Our method differs in that we define the important attributes by how much the they influence the classification task, and aggregate them into a semantic explanation. Image attributes are used to boost the performance for fine-grained classification [6,41], face recognition [13], and image-to-text generation [41]. If annotated on a per-class basis, attributes are both effective and cheap [33].

Visual Explanation. We distinguish between two types of visual explanations: interpretation models and justifying post-hoc reasons. The former visualize the filters and feature maps in CNNs, trying to interpret the knowledge distilled by the model [22,38,39]. For instance, [39] applies a Deconvolutional Network (DeConvNet) to project the feature activations back to the input pixel space. The latter determines which region of the input is responsible for the decision by attaching importance to pixels or image regions. Gradient-based methods back-propagate the loss to the input layer [29,30]. Although these methods generate high-resolution details, they can not localize the image area that the target category focuses on. Visualizing linear combination of the network activations and incorporating them with class-specific weights is another direction [28,42]. Class Activation Mapping (CAM) [42] and its extension Grad-CAM [28] produce class-specific attention maps by performing a weighted combination on forward activation maps. On the other hand, model-agnostic methods propose to explain models by treating them as black boxes. Perturbation-based methods manipulate the input and observe the changes in output [7,8,26,27]. A linear decision model (LIME) [27] feeds super-pixel-masks into the black box and generates attention maps. An extension of LIME, RISE perturbs the input image with random masks and generates weights with the output probabilities, then produce the attention map by the weighted combination of random binary masks. [8] and [7] extends the perturbation to a trainable parameter and generates more smooth masks.

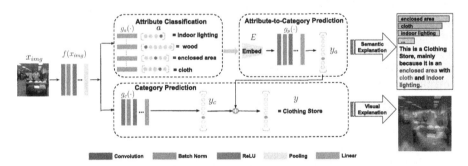

Fig. 1. Overview of our multitask learning network (MTL-Net). $f(x_{img})$ is the feature extraction network in our model. The network contains two pipelines. Category prediction network predicts the label y_c for an input image x_{img}. Attribute classification network predicts the attributes a. Attribute-to-Category prediction network infers the category y_a by attributes embedding. Then the attributes with high contribution to y_a are aggregated into a template-based explanation. Saliency-based visual explanation reflects the attention of y_c on the input image.

Evaluating Explanations. Although visual explanations are an intuitive way of understanding the internal thought process of a neural network, it is not trivial to measure the effectiveness of the visualization method. In recent years, various evaluation methods are performed. The most widely accepted measure of effectiveness is user studies [28,42]. Some explanation methods perform quantitative evaluation methods such as Pointing ame [19,28,40], sanity checks [1], and Deletion-insertion [26].

Since visual explanations reflect the salient area that activates the feature map, improving the visualization would be beneficial for classification. Hence, we propose to fuse the input image with the explanation maps and then measure how the classification accuracy is influenced. The testing procedure is processed only once when evaluating the explanation methods.

3   Visual-Semantically Interpretable Framework

In this section, we introduce how we integrate the attribute prediction with image classification. Then we detail how to generate semantic explanations via attribute contribution. Finally, we present visual explanations generated by various visualization methods and evaluate them with the fusion classification mechanism.

3.1   Multitask Learning Network

Learning multiple complementary tasks would improve the generalization capability of the network, and improve the accuracy of predictions compared to performing these tasks separately. Deep neural networks for image classification uses category level labels as the supervision signal [32,35]. While attributes reveal essential characteristics of objects complementary to image classes [37]. In our multitask learning network (MTL-Net), we combine three modules regarding category and attribute classification within a unified framework as shown in Fig. 1. They are category prediction, attribute classification and attribute-to-category prediction. Given an input image x, our task is to predict the image label y as well as the attribute a with the following steps.

In category prediction, given input image x_{img}, we first extract the image feature v, then we pass these image features into a linear classifier and get the predicted result y_c:

$$v = f(x_{img}), y_c = g_c(v). \tag{1}$$

To predict N_a attributes of one image, we apply a linear classifiers to learn the i_{th} attribute $a_i \in \mathbb{R}^{d_{a_i}}$ from the image feature v:

$$a_i = g_a^i(v). \tag{2}$$

While predicting attributes will help the image encoder $f(\cdot)$ to extract semantic information regarding the attributes and help the image classification process, the predicted attributes are not contributing to the image classification directly. Thus we combine them in the attribute-to-category prediction. We first follow the word embedding method [9] to embed a into a matrix $E \in \mathbb{R}^{N_a \times d}$:

$$E = concat[a_1 \cdot W_1, a_2 \cdot W_2 \cdots, a_{N_a} \cdot W_{N_a}], \tag{3}$$

where W_i is the embedding matrix with dimension $\mathbb{R}^{d_{a_i} \times d}$, and N_a is the number of attributes. Then the embedding E is feed into a linear classifier to get the predicted result y_a:

$$y_a = g_p(E). \tag{4}$$

Thus, the final class prediction is

$$y = y_c + \alpha \cdot y_a, \tag{5}$$

where α is a hyper parameter.

We optimize the MTL-Net with the cross-entropy loss L between predicted class and the ground truth y_{gt}:

$$L = CE(y, y_{gt}). \tag{6}$$

In order to align the predicted attributes with interpretable semantic meaning, we propose to optimize the attribute classification module with human annotated class attributes $\mathcal{A} \in \mathbb{R}^{N_a}$. The attribute classification network would be optimized according to the objective:

$$L_{attri} = \frac{1}{N_a} \sum_{i=1}^{N_a} CE(\mathcal{A}_i, a_i). \tag{7}$$

Our final loss L is the weighted combination of the above three loss: L and L_{attri}:

$$L = L + \beta \cdot L_{attri}. \tag{8}$$

3.2   Interpreting MTL-Net

Here we detail our method for selecting the attributes that make an important impact on the results, and evaluating the image area that the network pays most attention to. We propose a gradient weighted mapping to figure out the attribute contribution in image classification, and generate language explanations using a predefined template. Furthermore, we apply various visualization methods, i.e. Back-propagation (BP) [29], Grad-CAM [28] and RISE [26], to generate saliency maps that provide visual justifications of the network classification process. Then we evaluate them with our fusion classification mechanism.

Generating Attribute-Based Explanations. In attribute-to-category prediction, the predicted image category y_a is inferred by the prediction of the N_a attributes for every image. In order to determine which attributes are most important to the score y_a^k for the predicted class k, we use gradient mapping to generate a saliency map $M^a \in \mathbb{R}^{N_a \times d}$ for attributes embedding E. As is discussed in [30], we only consider the positive impact on increasing the predicted class score, thus the attribute contribution of the i-th attribute a_i is determined as the sum of positive values in $M_i^a \in \mathbb{R}^d$:

$$C_{a_i} = \sum_{j=1}^{d} \mathbb{1}_{M_{ij}^a > 0} \cdot M_{ij}^a, \quad \text{where } M_{ij}^a = \frac{\partial y_a^k}{\partial E_{ij}}. \tag{9}$$

Here M_i^a denotes the gradient of the score y_a^k on the attribute embedding.

The attribute contribution C_{a_i} $(i = 1, \ldots, N_a)$ indicates which attributes have more positive impact on predicting class k. Consequently, we rank the score and pick the attributes with the highest contribution C_{a_i}. And we select the top three attributes to form our semantic explanation.

Generating Visual Explanations. To generate visual explanations and evaluate which one is more fidelity to the network, we utilize three representative methods:

BP. Back-Propagation (BP) [29] computes the saliency map by back propagating the output score y_c into the input layer. Based on the gradient for each pixel (i, j) of the input image, the saliency map is computed as:

$$M_{ij} = max_l \left| \frac{\partial y_c}{\partial x(i, j, l)} \right|, \tag{10}$$

where $max_l |\cdot|$ denotes the maximum magnitude across all color channels l.

Grad-CAM. Grad-CAM [28] uses the gradient flowing back to a specific convolutional layer to calculate the importance weight α_k for every feature map A^k in that layer,

$$\alpha_k = \frac{1}{Z} \sum_i \sum_j \frac{\partial y_c}{\partial A_{ij}^k}. \tag{11}$$

And the final saliency map M is a weighted combination of the feature maps,

$$M = \text{ReLU}(\sum_k \alpha_k A^k), \tag{12}$$

where ReLU means that Grad-CAM only focuses on the features that have a positive influence on network output. And M is resized to the size of input image when we use it.

RISE. For each input image x, RISE [26] generates numerous masks $M^{(i)}$ to cover x. The author assumes that the output score of the whole network for masked image $F(x \odot M^{(i)})$ reflects the importance of that mask, where $\odot$ denotes element-wise multiplication. Thus the final attention map is the weighted sum of these masks,

$$M = \frac{1}{\mathbb{E}[M] \cdot N} \sum_{i=1}^{N} F(x \odot M^{(i)}) \cdot M^{(i)}, \tag{13}$$

where $\mathbb{E}[M]$ denotes the expectation of masks, and N is the number of generated masks.

Evaluating Visual Explanations. The lack of objective evaluation metrics for the performance of these visualization methods may hinder user acceptance. We conjecture that visual justifications would be trustworthy for the user if they improve the performance of the black box neural network on the task that they are visualizing. Hence, we propose image classification as a task to objectively evaluate the visual justification methods without human annotation.

Saliency maps indicate the importance of each pixel and retain the same spatial information as input images. We propose a fusion classification mechanism (FCM), where we overlay the saliency map M onto the raw image x_{img}, and generate the filtered image x_{fuse}. So that visual justifications can be evaluated automatically by training and testing our network on those fused images. We normalized the explanation maps into $[0, 1]$, to make equal compare among every explanation maps. The overlay method is described as,

$$x_{fuse} = (M + \lambda) \odot x_{img}, \tag{14}$$

where λ is a constant parameter that determines how much image content is least preserved, and when $\lambda = 0$ there might be image pixels being removed directly. $\odot$ denotes element-wise multiplication. We then feed the fused image x_{fuse} into the multitask learning network as shown in Fig. 1. Finally, we rank the saliency models based on their performance in classification. The ranking shows us that visual explanation models lead to a higher accuracy can capture important image regions for predicting the right answer.

4   Experiments

In this section, we start by introducing the dataset. We then present our results that validate the proposed multitask learning network (MTL-Net) on three datasets, indicating consistent improvements on single-task networks. Furthermore, our predicted attributes and their aggregated semantic explanations are presented and evaluated. The visual explanations are generated by three well-known visualization methods, and our proposed evaluation technique validates their effectiveness and ranks them based on the class prediction performance.

Datasets. We use three datasets for experimental analysis. CUB-200-2011 (CUB) [34] is a fine-grained dataset for bird classification, with 11,788 images from 200 different types of birds. The dataset consists of 312 binary attributes

Table 1. Ablation study for different settings in MTL-Net. We report the results for baseline models SE-ResNeXt-50 [14], Inception-v4 [31] and PNASNet-5 [21] on y_c. CP represents the accuracy of category prediction y_c trained together with attribute classification, and A2CP represents the accuracy of y when combining category prediction and attribute-to-category prediction. We also report the accuracy for attribute classification in MTL-Net.

Models	SUN	CUB	AwA
SE-ResNeXt-50 [14]	38.28	74.30	94.74
PNASNet-5 [21]	42.53	83.20	95.47
Inception-v4 [31]	35.49	78.90	94.22
CP (ours)	44.70	83.44	**95.71**
A2CP (ours)	**44.90**	**83.77**	95.61
Attribute classification	93.07	88.12	99.03

that describe the color, shape and other characters for 15 body part locations. SUN Attribute Database (SUN) [25] is a fine-grained scene categorization dataset, and consists of 14,340 images from 717 classes (20 images per class). Each image is annotated with 102 attributes that describe the scenes' material and surface properties. Animals with Attributes (AwA) [18,36] is a coarse-grained dataset for animal classification, containing 37,322 images of 50 animal classes. 85 per-class attribute labels [23] are provided in the dataset, describing the appearance and the living habits of the animals.

Implementation Details. The baseline model in MTL-Net is PNASNet-5 [21] pretrained on ImageNet [5] and then finetuned on three datasets separately. The classifier g_i and g_a have the same structure: 2-layer CNN and one linear layer. We train our model with SGD optimizer [3] by setting $momentum = 0.9$, $weight\ decay = 10^{-5}$. We set $\alpha = 1$ and tuned β from 0.1 to 1.5 for different datasets. While evaluating saliency maps, we set λ as a matrix with each element equals to 0.3.

4.1   Evaluating Semantic Explanations

In this section, we quantitatively evaluate our attribute-based semantic explanations in two aspects: the fidelity to the model and the alignment with human annotation. Then we perform human study and qualitative analysis to discuss how well is the semantic explanations when making the network interpretable to users.

Quantitative Analysis. Here, we validate our multitask learning network (MTL-Net) on three benchmark datasets, i.e. CUB, AWA and SUN, for the image classification task.

We report the classification accuracy of category prediction y_c and the final result y, denoted as CP and A2CP respectively, in Table 1. As comparison, we choose the classification accuracy of SE-ResNeXt-50 [14], Inception-v4 [31] and PNASNet-5 [21] as baseline.

Introducing the attribute classification loss to the original image classification task improves the accuracy. As is shown in Table 1, we improve the accuracy of the baseline model on all datasets, achieving 44.70% on SUN, 83.44% on CUB and 95.71% on AWA. These results demonstrate that introducing attributes to image classification not only makes the models more explainable, e.g., predicting attributes such as "white crown, pink legs, white belly" is more informative than only predicting the category "slaty backed gull". After integrating the attribute-to-category prediction and category prediction, the classification accuracy is further improved.

Table 2. The user study for semantic explanations. Image relevance refers to the question: "Does the sentence match the image content?". Class Relevance refers to "Is the explanation reasonable for the prediction?". According to the user, our semantic explanations are image relevant, and reasonable.

Question	Options	Percentage
Image relevance	High	68.4%
	Somewhat	27.4%
	No	4.2%
Class relevance	Yes	68.6%
	No	31.4%

Table 3. The classification accuracy on three datasets (left) and the user study (right) for the fusion classification mechanism (FCM). No-VIS denotes the classification accuracy generated by our MTL-Net. Grad-CAM, BP, RISE refer to the FCM equipped with three visualization methods.

FCM	CUB	AwA	SUN	User study
No-VIS	83.77	95.61	44.90	N/A
Grad-CAM [28]	84.24	96.13	45.27	35.2%
BP [29]	81.87	93.86	41.73	26.2%
RISE [26]	**85.17**	**96.84**	**46.50**	**38.6%**

We also evaluate the predicted attributes by how well they are aligned with their semantic meaning. The predicted attributes in MTL-Net are compared with the class attributes annotated by a human, and we report the attribute classification accuracy on three datasets in Table 1. On average, the predicted attributes are agree with the human annotation, achieving an accuracy of 93.7%, 88.12% and 99.03 for three datasets. Note that the attributes in SUN and AwA dataset are binary, indicating the existence of the attribute. While the attributes in CUB dataset are multi-dimension, for instance, the head color attribute has fifteen options. That can explain why the accuracy of CUB dataset is slightly lower than the other two datasets.

User Study. Semantic explanations are mainly targeted towards the end user and aim to improve the user trust in the machine learning system. To determine if users find our semantic explanations trustworthy, we perform a user study on visual and semantic explanations. CUB being a fine-grained dataset, only bird experts can tell their difference. Hence, we selected 100 images from the scene categorization dataset SUN and the animal classification dataset AwA. Our user group is composed of five university graduates with an average age of 25. In this section, we present our results on semantic explanations for clarity, however, our user study on visual explanations presented in the following section is identical in the number, the demographics, the age and gender of the users as well as the number of images to be evaluated.

In the user study for semantic explanations, our aim is to evaluate two factors: if the semantic explanation is image relevant and how well can they help the user in understanding the black-box model. The annotators are given an image as well as a semantic explanation and are asked to answer two questions, i.e. "Does the sentence match the image?" and "Is the explanation reasonable for the prediction?". We present the results of this study in Table 2. For the question related to the image relevance, 68.4% of the attribute-based semantic explanations are marked as "highly related to the image", while only 4.2% of these results were found irrelevant by the users. For the question relevant to the credibility for the prediction, 68.6% of the sentences are found reasonable for the predicted label. These results complement our prior results and show that our attribute-based semantic explanations are image relevant and reasonable for the label that the model predicts.

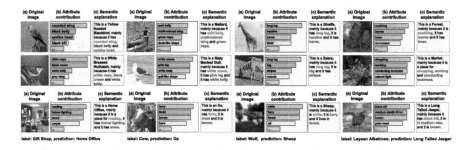

Fig. 2. Our MTL-Net predicts image category and attributes. We select the attributes having the highest contribution on the prediction label (b). The attributes are then aggregated into a template-based semantic explanation (c). The top two rows show our the explanations for right predictions, while the bottom row shows semantic explanations with wrong predictions. The "label" and "prediction" under each image indicates the ground truth label as well as the predicted label.

Qualitative Evaluation. In this section, we evaluate our attribute-based semantic explanations qualitatively, by looking at the sentences generated from the three highest ranked attributes together with the predicted label. Figure 2 shows two rows of example images with their predicted attributes where the label was correctly predicted. In the last row, we present three examples with their highest ranked attributes despite their wrong class predictions.

We observe from both the positive and the negative examples that our explanations correctly reflect the content of the image and the characters of the objects. For instance, in fine-grained bird classification results, our model correctly associates the attributes "green nape, multicolored wing, solid belly" with *Mallard* and "white crown, pink leg and white belly" with *Slaty Backed Gull*.

By looking at the attributes and the predicted label, a user can understand why this prediction was associated with these attributes. For instance, the explanation for zebra points out the most prominent attributes such as "long leg" and "stripe". While for a *Forest* image our model predicts the attributes "soothing, leaves, trees", and for a "Market" it associates "shopping, working and conducting business".

On the other hand, the users might find the reason for a wrong prediction by investigating the semantic explanations. For instance, we observe that due to "reading, indoor lighting, wood", an image for *gift shop* is wrongly predicted as *Home Office*. Arguably, the image looks more like a *home office* than a *gift shop*, i.e. correct class. Similarly, for the *wolf*, due to the unusual color of the animals (i.e., "white") and the tranquillity of the environment (i.e., "forest"), the label is predicted as *sheep*.

Fig. 3. Visual explanations of the correct labels generated by three methods, Grad-CAM, BP and RISE. Images on the left are from CUB, the middle from AWA and the right are from SUN datasets. The bottom row shows the visual explanations for wrong category prediction.

4.2   Evaluating Visual Explanations

To visually justify the classification decision of the model, we use three well-known visual explanation methods, Grad-CAM [28], BP [29], and RISE [26]. We compare them quantitatively in terms of the performance in fusion classification mechanism, through user studies, and qualitatively by visual inspection.

Visual Explanations in Image Classification. We evaluate the fidelity of the visual explanations generated by three models, by using them in the task they are interpreting, i.e. image classification. We first generate the attention maps of the predicted class in our MTL-Net, then fuse the image with the attention map on our fusion classification mechanism (FCM). Then we train and classify the fused image with MTL-Net again, and rank the visual explanations concerning the classification accuracy.

As presented in Table 3, our results indicate that images fused by saliency maps lead to slight improvements in the classification accuracy compared to the case with no fusion. Among the saliency-based explanation methods, RISE [26] consistently achieves better performance than BP and Grad-CAM. BP performs poorly on the fine-grained CUB dataset, and it may because the pixels that BP marks out are spread out all over the bird and they are not distinguishing between similar species (see Fig. 3). On the other hand, we experiment with all these models and indicate that Grad-CAM and BP results are much faster than RISE, since RISE requires multiple times of testing for every image.

User Study. Given three visual explanations, in this section, we aim to determine the visualization that is the most trustworthy for the users. We indicate that "All robots predicted the image as airfield", show visual explanations generated by Grad-CAM, RISE and BP to the annotators, and ask them to answer the question "Which robot do you trust more?". In this way, the annotators first evaluate if the label is correct by looking at certain regions in the image, and then they compare their attention with the visualizations, finally pick the one that matches their mental model. Using 100 images sampled from AWA and SUN datasets, we rank the saliency-based explanation models.

Our results shown in Table 3 (rightmost column) indicate that 38.6% of the annotators trust RISE, 35.2% of them trust Grad-CAM, and 26.2% of them vote for BP. This result is consistent with our fusion classification mechanism (FCM) for evaluating the quality of the saliency-based visualization. As a conclusion, if explanations are helpful for the users, they are expected to perform well in FCM. Although human study is a worthwhile and important evaluation criterion, it can be replaced by our automatic evaluation if time and labor limited.

Qualitative Evaluation. In this section, we evaluate the interpretability of the visual explanations in our MTL-Net. The first two rows in Fig. 3 show the visualizations for the correctly predicted images. In the last row, we present the visualization for images with the wrong prediction.

From both the results with correct and incorrect class prediction, we observe that Grad-CAM and RISE highlight important image regions to explain the network decision, while BP emphasizes a distributed set of pixels that are influential for the classification result. Hence, by looking at the masked image generated by Grad-CAM and RISE, one can easily figure out which part of the image the network focuses on for a particular decision. Generally, Grad-CAM offers a more concentrated focus due to the up-sample operation it takes. While RISE and BP consider more pixels when evaluating the importance.

With the negative predictions presented in the last row, the visual explanations can help the user to understand the causes for wrong predictions.

Indeed, most of the wrong predicted images may be confusing even for a human. For instance, when explaining why the image with the label *Ball Room* is predicted as a *Piano Store*, the visual explanation focuses on the piano and the indoor lighting. The wrong prediction of *Tiger* is an interesting example in that the attention map of Grad-CAM mainly focuses on the stripes that zebra also have. These results indicate that visual explanations can reveal show the weakness of network to the users, e.g., the network typically makes a mistake when it only focuses on wrong details, instead of considering the image as a whole. Moreover, RISE generates more scattered distribution for the wrong predictions. And that might be another clue for identifying wrong classified images.

5 Conclusion

In this work, we propose a visually and semantically interpretable multitask learning network. We introduce attributes into image category prediction and propose a new method to generate attribute-based semantic explanations intuitive for the user. Qualitative evaluations and the user study reveals that our semantic explanations are both class discriminative and image relevant. Moreover, we propose a quantitative evaluation technique to evaluate the effectiveness of visual explanations based on their performance in image classification. Future work includes investigating the network flaws with these explanations and further improve the network.

Acknowledgments. This work has been partially funded by the ERC grant 853489 - DEXIM (Z.A.) and by DFG under Germany's Excellence Strategy EXC number 2064/1 Project number 390727645.

References

1. Adebayo, J., Gilmer, J., Muelly, M., Goodfellow, I., Hardt, M., Kim, B.: Sanity checks for saliency maps. In: Advances in Neural Information Processing Systems, pp. 9505–9515 (2018)
2. Argyriou, A., Evgeniou, T., Pontil, M.: Multi-task feature learning. In: NIPS (2007)
3. Bottou, L.: Large-scale machine learning with stochastic gradient descent. In: Lechevallier, Y., Saporta, G. (eds.) Proceedings of COMPSTAT 2010, pp. 177–186. Springer, Heidelberg (2010). https://doi.org/10.1007/978-3-7908-2604-3_16
4. Dai, J., He, K., Sun, J.: Instance-aware semantic segmentation via multi-task network cascades. In: IEEE CVPR (2016)
5. Deng, J., Dong, W., Socher, R., Li, L.J., Li, K., Fei-Fei, L.: Imagenet: a large-scale hierarchical image database. In: IEEE CVPR (2009)
6. Duan, K., Parikh, D., Crandall, D., Grauman, K.: Discovering localized attributes for fine-grained recognition. In: IEEE CVPR (2012)
7. Fong, R., Patrick, M., Vedaldi, A.: Understanding deep networks via extremal perturbations and smooth masks. In: ICCV (2019)
8. Fong, R.C., Vedaldi, A.: Interpretable explanations of black boxes by meaningful perturbation. In: Proceedings of the IEEE International Conference on Computer Vision, pp. 3429–3437 (2017)

9. Gal, Y., Ghahramani, Z.: A theoretically grounded application of dropout in recurrent neural networks. In: NIPS (2016)

10. Graves, A., Schmidhuber, J.: Framewise phoneme classification with bidirectional LSTM and other neural network architectures. Neural Netw. **18**, 602–610 (2005)

11. Hendricks, L.A., Akata, Z., Rohrbach, M., Donahue, J., Schiele, B., Darrell, T.: Generating visual explanations. In: Leibe, B., Matas, J., Sebe, N., Welling, M. (eds.) ECCV 2016. LNCS, vol. 9908, pp. 3–19. Springer, Cham (2016). https://doi.org/10.1007/978-3-319-46493-0_1

12. Hendricks, L.A., Hu, R., Darrell, T., Akata, Z.: Grounding visual explanations. In: Ferrari, V., Hebert, M., Sminchisescu, C., Weiss, Y. (eds.) ECCV 2018. LNCS, vol. 11206, pp. 269–286. Springer, Cham (2018). https://doi.org/10.1007/978-3-030-01216-8_17

13. Hu, G., et al.: Attribute-enhanced face recognition with neural tensor fusion networks. In: IEEE ICCV (2017)

14. Hu, J., Shen, L., Sun, G.: Squeeze-and-excitation networks. In: IEEE CVPR (2018)

15. Kanehira, A., Harada, T.: Learning to explain with complemental examples. In: IEEE CVPR (2019)

16. Kanehira, A., Takemoto, K., Inayoshi, S., Harada, T.: Multimodal explanations by predicting counterfactuality in videos. In: IEEE CVPR (2019)

17. Kim, J., Rohrbach, A., Darrell, T., Canny, J., Akata, Z.: Textual explanations for self driving vehicles. In: ECCV (2018)

18. Lampert, C.H., Nickisch, H., Harmeling, S.: Learning to detect unseen object classes by between-class attribute transfer. In: IEEE CVPR (2009)

19. Lampert, C.H., Nickisch, H., Harmeling, S.: Attribute-based classification for zero-shot visual object categorization. IEEE TPAMI **36**, 453–465 (2014)

20. Li, Q., Fu, J., Yu, D., Mei, T., Luo, J.: Tell-and-Answer: towards explainable visual question answering using attributes and captions. In: EMNLP (2018)

21. Liu, C., et al.: Progressive neural architecture search. In: ECCV (2018)

22. Olah, C., et al.: The building blocks of interpretability. Distill **3**(3), e10 (2018)

23. Osherson, D.N., Stern, J., Wilkie, O., Stob, M., Smith, E.E.: Default probability. Cogn. Sci. **15**, 251–269 (1991)

24. Park, D.H., et al.: Multimodal explanations: justifying decisions and pointing to the evidence. In: IEEE CVPR (2018)

25. Patterson, G., Xu, C., Su, H., Hays, J.: The sun attribute database: beyond categories for deeper scene understanding. IJCV **108**, 59–81 (2014)

26. Petsiuk, V., Das, A., Saenko, K.: RISE: randomized input sampling for explanation of black-box models. In: BMVC (2018)

27. Ribeiro, M.T., Singh, S., Guestrin, C.: Why should i trust you?: explaining the predictions of any classifier. In: ACM SIGKDD, pp. 1135–1144. ACM (2016)

28. Selvaraju, R.R., Cogswell, M., Das, A., Vedantam, R., Parikh, D., Batra, D.: Gradcam: visual explanations from deep networks via gradient-based localization. In: IEEE ICCV (2017)

29. Simonyan, K., Vedaldi, A., Zisserman, A.: Deep inside convolutional networks: visualising image classification models and saliency maps. CoRR abs/1312.6034 (2013)

30. Springenberg, J.T., Dosovitskiy, A., Brox, T., Riedmiller, M.: Striving for simplicity: the all convolutional net. In: ICLR (2015)

31. Szegedy, C., Ioffe, S., Vanhoucke, V., Alemi, A.A.: Inception-v4, inception-resnet and the impact of residual connections on learning. In: AAAI (2017)

32. Szegedy, C., et al.: Going deeper with convolutions. In: IEEE CVPR (2015)

33. Tokmakov, P., Wang, Y.X., Hebert, M.: Learning compositional representations for few-shot recognition. arXiv preprint arXiv:1812.09213 (2018)
34. Wah, C., Branson, S., Welinder, P., Perona, P., Belongie, S.: The Caltech-UCSD Birds-200-2011 Dataset. Technical Report CNS-TR-2011-001, California Institute of Technology (2011)
35. Wang, Y., Morariu, V.I., Davis, L.S.: Learning a discriminative filter bank within a CNN for fine-grained recognition. In: IEEE CVPR (2018)
36. Xian, Y., Lampert, C.H., Schiele, B., Akata, Z.: Zero-shot learning-a comprehensive evaluation of the good, the bad and the ugly. IEEE TPAMI **41**, 2251–2265 (2018)
37. Xian, Y., Lorenz, T., Schiele, B., Akata, Z.: Feature generating networks for zero-shot learning. In: IEEE CVPR (2018)
38. Xu, K., Park, D.H., Yi, C., Sutton, C.: Interpreting deep classifier by visual distillation of dark knowledge. arXiv preprint arXiv:1803.04042 (2018)
39. Zeiler, M.D., Fergus, R.: Visualizing and understanding convolutional networks. In: Fleet, D., Pajdla, T., Schiele, B., Tuytelaars, T. (eds.) ECCV 2014. LNCS, vol. 8689, pp. 818–833. Springer, Cham (2014). https://doi.org/10.1007/978-3-319-10590-1_53
40. Zhang, J., Bargal, S.A., Lin, Z., Brandt, J., Shen, X., Sclaroff, S.: Top-down neural attention by excitation backprop. IJCV **126**, 1084–1102 (2018)
41. Zhang, N., Paluri, M., Ranzato, M., Darrell, T., Bourdev, L.: Panda: pose aligned networks for deep attribute modeling. In: IEEE CVPR (2014)
42. Zhou, B., Khosla, A., Lapedriza, A., Oliva, A., Torralba, A.: Learning deep features for discriminative localization. In: IEEE CVPR (2016)

SpotPatch: Parameter-Efficient Transfer Learning for Mobile Object Detection

Keren Ye[1(✉)], Adriana Kovashka[1], Mark Sandler[2], Menglong Zhu[2], Andrew Howard[2], and Marco Fornoni[2]

[1] University of Pittsburgh, Pittsburgh, PA 15260, USA
yekeren@cs.pitt.edu
[2] Google Research, 75009 Paris, France

1 Introduction

As mobile hardware technology advances, on-device computation is becoming more and more affordable. On one hand, efficient backbones like MobileNets optimize feature-extraction costs by decomposing convolutions into more efficient operations. On the other hand, one-stage detection approaches like SSD provide mobile-friendly detection heads. As a result, detection models are now massively being moved from server-side to on-device. While this constitutes great progress, it also brings new challenges. Specifically, multiple isolated models are often downloaded to perform related tasks, like detecting faces, products, bar-codes, etc. This rises the questions: Can we represent a diverse set of detection tasks as a small set of "patches" applied to a common model? If the common model also needs to be updated, can we represent the update as a small patch too? To answer the above questions, we studied two experimental scenarios:

1. Adapting a mobile object detector to solve a new task.
2. Updating an existing model, whenever additional training data is available.

To learn the patches, we propose an approach simultaneously optimizing for both accuracy and footprint: (1) for each layer, we compactly represent the patch with scaled 1-bit weight residuals; (2) we employ a gating mechanism to adaptively patch only the most-important layers, reusing the original weights for the remaining ones. We evaluate on object detection tasks, using a setting similar to [9], which we refer to as *"Detection Decathlon"*. We also showcase our method's ability to efficiently update a detector whenever new data becomes available. To the best of our knowledge this is the first systematic study of parameter-efficient transfer learning techniques on object detection tasks.

K. Ye—Work partially done during an internship at Google.

Electronic supplementary material The online version of this chapter (https://doi.org/10.1007/978-3-030-66415-2_41) contains supplementary material, which is available to authorized users.

A. Bartoli and A. Fusiello (Eds.): ECCV 2020 Workshops, LNCS 12535, pp. 636–640, 2020.
https://doi.org/10.1007/978-3-030-66415-2_41

Related Work: [9] proposed a Visual Decathlon benchmark for adapting image classifiers. [1,8] proposed to "patch" MobileNets by fine-tuning only batch-normalization and depthwise-convolution layers. [6,7] used binary masks and simple linear transformations to obtain the target kernels. Their work is similar to quantization approaches [4,5] but using 1-bit representation. [2] built a dynamic routing network to choose for each layer, between fine-tuning it and reusing the pre-trained weights. Our approach differs from the above in that we combine scaled binary mask residuals, with a loss function explicitly minimizing the number of patched layers. This allows our model to adaptively minimize the patch footprint in a task-dependent fashion.

2   Approach

We assume a deep neural network $\mathcal{M}$ of depth N is composed of a set of layers $\boldsymbol{\theta} = \{\mathbf{W}_1, \ldots, \mathbf{W}_N\}$. To adapt $\mathcal{M}$ to solve a new task, we seek a task-specific $\boldsymbol{\theta}' = \{\mathbf{W}_1', \ldots, \mathbf{W}_N'\}$ that optimizes the loss on the target dataset. In addition, since we do not want to fully re-learn $\boldsymbol{\theta}'$, we look for a transformation with minimum cost (in terms of bit-size) to convert the original $\boldsymbol{\theta}$ to $\boldsymbol{\theta}'$. Assume the transformation function can be expressed as $\boldsymbol{\theta}' = f(\boldsymbol{\theta}, \boldsymbol{\gamma})$ where $\boldsymbol{\gamma}$ is an additional set of parameters for the new task, and f is a function that combines the original parameters $\boldsymbol{\theta}$ with the parameter "patch" $\boldsymbol{\gamma}$. Our goal is to reduce the bit-size of $\boldsymbol{\gamma}$. In other words, we want the "footprint" of $\boldsymbol{\gamma}$ to be small.

Task-Specific Weight Transform. Given the weights $\boldsymbol{\theta} = \{\mathbf{W}_1, \ldots, \mathbf{W}_N\}$ shared across tasks, the task-specific weight trasformation adds weight residuals $\omega_i \mathbf{S}_i$ to them: $\mathbf{W}_i' = \mathbf{W}_i + \omega_i \mathbf{S}_i$, where $\mathbf{S}_i$ is a binary mask of the same shape as $\mathbf{W}_i$ with values in $\{-1, +1\}$, and ω_i is a scaling factor. To learn these scaled and zero-centered residuals, we use similar techniques as [6,7], which use real-valued mask variables to achieve differentiable gradients during training. To deploy the model, only the masks $\mathbf{S}_i$ and the per-layer scalar ω_i are used. The incremental footprint of the model is thus $\{\omega_i, \mathbf{S}_i | i \in \{1, \ldots, N\}\}$, or roughly 1-bit per model weight, with a negligible additional cost for the per-layer scalar ω_i.

Spot Patching. Since the difficulty of adapting a detector depends on the target task, we design a gating mechanism to adjust the model patch complexity in a task-dependent fashion: $\mathbf{W}_i' = \mathbf{W}_i + g_i(\omega_i \mathbf{S}_i)$. Simply speaking, we add a *gate* g_i

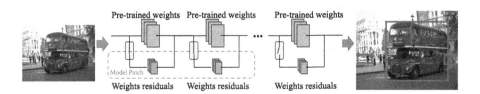

Fig. 1. Our method simultaneously reduces: (1) the bit-size of the weights residuals and (2) the number of patched layers.

for each network layer. The layer uses the original pre-trained weights if the gate value is 0, and weight transform otherwise (Fig. 1). The benefit of gating with g_i is that it allows to search for a task-specific subset of layers to patch, rather than patching all layers. Compared to patching the whole network, it reduces the patch footprint to $\gamma = \{\omega_i, \mathbf{S}_i | i \in \{1, \dots, N\}$ and $g_i = 1\}$. To learn g_i we use similar differentiable binarization techniques as for learning $\mathbf{S}_i$. Furthermore, to force the number of patched layers to be small, we add to the training loss a term minimizing the number of patched layers $\sum_{i=1}^{N} g_i$.

Besides task-specific weight transform and spot patching, we also train task-specific Batch Normalization (BN), which does not significantly increase the footprint. To learn the task-specific patch γ, we minimize the following loss: $L(\gamma) = L_{det}(\gamma) + \lambda_{sps} L_{sps}(\gamma) + \lambda_{adp} L_{adp}(\gamma)$, where: $L_{det}(\gamma)$ is the detection loss optimizing the class confidence scores and box coordinates. $L_{sps}(\gamma) = \sum_{i=1}^{N} g_i$ is the sparsity-inducing loss, pushing the number of patched layers to be small. Finally, $L_{adp}(\gamma) = \sum_{i=1}^{N} \|\omega_i\|_2^2$ is a domain-adaptation loss forcing the scaling factors ω_i to be small, and thus θ' to be similar to θ.

3 Experiments

We consider the two experimental scenarios proposed in the introduction. For the baselines, we compare our method with the following transfer learning methods, which we reproduced in the detection setting:

- FINE-TUNING [3] fine-tunes the whole network.
- TOWER PATCH [3] fine-tunes only the parameters in the detection head.
- BN PATCH [8], DW PATCH [8], and BN+DW PATCH [1,8] learn task-specific BatchNorm, Depthwise, or BatchNorm + Depthwise layers.
- PIGGYBACK [6] and WEIGHTTRANS [7] learn task-specific 1-bit masks and simple linear transformations to obtain the task-related kernels.

Detection Decathlon. We adapt a model trained on OpenImages V4 to nine additional vertical-specific detection tasks: Caltech-UCSD Birds, Stanford Cars, COCO, Stanford Dogs, WiderFace, Kitti, Oxford-IIIT Pet, Retail Product Checkout (RPC), and Pascal VOC. We compare models on the basis of how well they solve all the problems (mAP@0.5), and how small is their relative footprint. Similarly to [9], we use a decathlon-inspired scoring function to evaluate the performance of each method. Scores are normalized so that the FINE-TUNING method reaches 2,500 points. We also report the Score/Footprint ratio [7], which practically measures the performance achieved for every Mb of footprint. Table 1 shows detection decathlon results. Our approach provides the best tradeoff by being parameter-efficient and yet accurate, as measured by the Score/Footprint ratio. This is achieved by learning patches with a task-adaptive footprint, resulting on average in a 24% footprint gain with respect to WEIGHTTRANS.

Model Updating. We pre-trained detection models on 10%, 20%, 40%, and 80% of the COCO data. These models achieved 20.9%, 24.2%, 31.4%, and 35.7%

mAP@0.5 on the COCO17 validation set, respectively. Then, we applied different patching approaches to update these imprecise models, using 100% of the COCO data. We then compared mAP@0.5 of the patched models, as well as the resulting patch footprint. Table 2 shows the results. At 10% training data, we achieve comparable mAP as WeightTrans (32.0% v.s. 32.7%) at a comparable footprint (5.04% v.s. 5.15%). However, when more data is available, the patch footprint generated by our approach is smaller than WeightTrans (2.08% v.s. 5.15%), while accuracy remains comparable (35.8% v.s. 35.9%). Our method can effectively adapt the patch footprint to the amount of new data to be learned by the patch, while maintaining high accuracy.

Table 1. Detection Decathlon. We show footprint, per-dataset mAP, average mAP, score, and score/footprint for each method, best method in **bold**, second-best underline. High score, low footprint, and high score/footprint ratio are good.

Method	Foot-print	Bird	Car	COCO	Dog	Face	Kitti	Pet	RPC	VOC	Avg mAP	Score	Score/Foot-print
Fine-Tuning	9.00	40.8	90.4	39.7	68.5	35.6	71.9	90.9	99.5	68.3	67.3	2500	278
Tower Patch	0.35	10.0	25.5	31.4	21.2	29.1	49.7	66.1	87.6	70.6	43.5	827	2362
Bn Patch	**0.19**	22.6	71.6	30.2	47.8	26.0	50.7	80.6	92.0	**71.1**	54.7	910	<u>4789</u>
Dw Patch	0.34	22.6	69.2	30.7	43.6	26.4	52.1	80.0	92.6	<u>70.8</u>	54.2	898	2642
Bn+Dw Patch	0.50	27.3	80.9	31.0	52.7	28.0	53.1	83.3	95.7	70.6	58.1	1012	2023
Piggyback	<u>0.30</u>	32.2	87.5	32.4	60.8	28.6	57.4	87.7	97.0	66.0	61.1	1353	4509
WeightTrans	0.46	**36.6**	**90.3**	**37.2**	**66.6**	**30.6**	**65.3**	**90.5**	<u>98.7</u>	70.7	**65.2**	**1987**	4319
Ours	0.35	<u>35.8</u>	<u>89.8</u>	<u>36.6</u>	<u>63.3</u>	<u>30.1</u>	<u>64.0</u>	<u>90.3</u>	**98.9**	70.6	<u>64.4</u>	<u>1858</u>	**5310**

Table 2. Model updating. Only one footprint number is shown for the baseline methods since they can only generate a constant-sized model patch.

Method	Footprint (%)				mAP (%)			
	10%	20%	40%	80%	10%	20%	40%	80%
Fine-Tuning			100.0		37.6	37.6	38.2	38.5
Tower Patch			3.85		24.5	26.7	32.4	35.8
Bn Patch			**2.08**		26.2	28.1	33.0	35.8
Dw Patch			3.76		25.9	27.8	33.1	**36.0**
Bn+Dw Patch			5.59		26.8	28.7	33.4	<u>35.9</u>
Piggyback			3.32		26.4	28.3	32.0	35.3
WeightTrans			5.15		**32.7**	**32.4**	**34.8**	<u>35.9</u>
Ours	5.04	4.98	4.21	**2.08**	<u>32.0</u>	<u>31.4</u>	<u>34.2</u>	35.8

References

1. Guo, Y., Li, Y., Feris, R.S., Wang, L., Rosing, T.S.: Depthwise convolution is all you need for learning multiple visual domains. In: AAAI (2019)
2. Guo, Y., Shi, H., Kumar, A., Grauman, K., Rosing, T., Feris, R.: Spottune: Transfer learning through adaptive fine-tuning. In: CVPR (2019)
3. Huang, J., et al.: Speed/accuracy trade-offs for modern convolutional object detectors. In: CVPR (2017)

4. Hubara, I., Courbariaux, M., Soudry, D., El-Yaniv, R., Bengio, Y.: Quantized neural networks: training neural networks with low precision weights and activations. JMLR **18**, 6869–6898 (2017)
5. Jacob, B., et al.: Quantization and training of neural networks for efficient integer-arithmetic-only inference. In: CVPR (2018)
6. Mallya, A., Davis, D., Lazebnik, S.: Piggyback: adapting a single network to multiple tasks by learning to mask weights. In: Ferrari, V., Hebert, M., Sminchisescu, C., Weiss, Y. (eds.) ECCV 2018. LNCS, vol. 11208, pp. 72–88. Springer, Cham (2018). https://doi.org/10.1007/978-3-030-01225-0_5
7. Mancini, M., Ricci, E., Caputo, B., Bulò, S.R.: Adding new tasks to a single network with weight transformations using binary masks. In: Leal-Taixé, L., Roth, S. (eds.) ECCV 2018. LNCS, vol. 11130, pp. 180–189. Springer, Cham (2019). https://doi.org/10.1007/978-3-030-11012-3_14
8. Mudrakarta, P.K., Sandler, M., Zhmoginov, A., Howard, A.: K for the price of 1. Parameter efficient multi-task and transfer learning. In: ICLR (2019)
9. Rebuffi, S.A., Bilen, H., Vedaldi, A.: Learning multiple visual domains with residual adapters. In: NeurIPS (2017)

Using Sentences as Semantic Representations in Large Scale Zero-Shot Learning

Yannick Le Cacheux[1,2(✉)], Hervé Le Borgne[1], and Michel Crupianu[2]

[1] CEA LIST, Gif-sur-Yvette, France
{yannick.lecacheux,herve.le-borgne}@cea.fr
[2] CEDRIC – CNAM, Paris, France
michel.crupianu@cnam.fr

Abstract. Zero-shot learning (ZSL) aims to recognize instances of unseen classes, for which no visual instance is available during training, by learning multimodal relations between samples from seen classes and corresponding class semantic representations. These class representations usually consist of either attributes, which do not scale well to large datasets, or word embeddings, which lead to poorer performance. A good trade-off could be to employ short sentences in natural language as class descriptions. We explore different solutions to use such short descriptions in a ZSL setting and show that while simple methods cannot achieve very good results with sentences alone, a combination of usual word embeddings and sentences can significantly outperform current state-of-the-art.

1 Introduction and Related Work

Zero-shot learning (ZSL) is useful when no visual samples are available for certain classes, provided we have semantic information for these classes [5]. It is then possible to train a model to learn the relations between the visual and semantic features using *seen* classes, for which both modalities are available; these relations can later be employed to classify instances of *unseen* classes, for which no visual sample is available during training, based on their semantic *class prototypes*. The semantic information can consist of vectors of attributes, e.g. binary codes for "has fur", "has stripes", etc. if classes are animal species. In a large-scale setting, it can be impractical to devise and provide attributes for hundreds or even thousands of classes. Word embeddings are then typically used to represent classes. However, a large performance gap still exists between these two types of class representations [6].

   An ideal solution could be to use short natural sentences to describe each class, as this is less time-consuming than providing comprehensive attributes and

Electronic supplementary material The online version of this chapter (https://doi.org/10.1007/978-3-030-66415-2_42) contains supplementary material, which is available to authorized users.

A. Bartoli and A. Fusiello (Eds.): ECCV 2020 Workshops, LNCS 12535, pp. 641–645, 2020.
https://doi.org/10.1007/978-3-030-66415-2_42

can be more visually informative than word embeddings derived from generic text corpora. The use of sentences as class descriptions in ZSL is not well studied. Some works already employed sentences for ZSL, but not in a convenient setting. For instance, [1] relies on 10 sentences *per image* instead of 1 per class. The closest work to ours is probably [4], in which different methods to obtain prototypes from WordNet definitions are evaluated, but reported performance is significantly below that of usual word embeddings.

In this article, we explore several ideas to leverage sentences to build semantic embeddings for ZSL and show that this can significantly outperform previous best reported performance. These proposals are easy to implement and computationally light. We provide the code[1] as well as the corresponding embeddings which can be used out-of-the-box as better quality semantic representations.

2   Sentence-Based Embeddings

A simple baseline consists in averaging the embeddings of the words in the definition, as is usually done for class names consisting of several words: if a sentence s describing a class has N words with respective embeddings $\{\mathbf{w}_1, \ldots, \mathbf{w}_N\}$, then the corresponding semantic representation is $\mathbf{s} = \frac{1}{N} \sum_{n=1}^{N} \mathbf{w}_n$. We call this baseline the *Def average* approach. However, as illustrated in Fig. 1, not all words are equally important in a short sentence description. We therefore explore the use of attention mechanisms: the sentence embedding is a weighted average of the embeddings of its words, so that more important words contribute more to the result. We consider two ways to achieve this.

The first, called the *Def visualness* approach, aims to estimate how "visual" a word is. For a given word w_n, we thus collect the 100 most relevant images from Flickr using the website's search ranking; we obtain a 2048-dimensional representation vector for each image with a pre-trained ResNet, and measure the average distance to their average vector. This constitutes the inverse of the "visualness" v_n of the corresponding word, as we hypothesize that more visually grounded concepts tend to have visual features closer to each other – a qualitative illustration is provided in the supplementary material. We then apply a softmax with temperature τ to weigh each word, τ being a hyper-parameter, so that the resulting sentence embedding is

$$\mathbf{s} = \sum_{n=1}^{N} \frac{\exp(v_n/\tau)}{\sum_{k=1}^{N} \exp(v_k/\tau)} \mathbf{w}_n \tag{1}$$

A second approach called *Def attention* aims to learn to predict the weights v_n from the embeddings $\mathbf{w}_n$ with $v_n = \boldsymbol{\theta}^\top \mathbf{w}_n$. The learned parameters $\boldsymbol{\theta}$ are jointly optimized with the parameters of the ZSL method employed, using the same loss function – a detailed example with a linear model is available in the supplementary material. The sentence embedding can then be obtained using

---

[1] https://github.com/yannick-lc/zsl-sentences.

Table 1. Comparison of approaches on ImageNet with a linear model. Best previous reported result is 14.1 in [3]; the result marked with * corresponds to the same setting (use of *Classname* with Glove embeddings).

	Word2vec	FastText	Glove	Elmo
Classname	12.4	14.8	14.5*	10.9
Classname + Parent	13.4	15.9	15.4	11.4
$Def_{average}$	9.7	10.6	10.0	8.7
$Def_{visualness}$	10.5	10.9	10.5	9.5
$Def_{attention}$	10.5	11.0	10.2	9.5
$Classname + Def_{average}$	14.6	17.2	16.9	12.2
$Classname + Def_{visualness}$	14.8	17.3	16.8	12.1
$Classname + Def_{visualness} + Parent$	**15.4**	**17.8**	**17.3**	**12.5**

Eq. 1 with the estimated v_n and $\tau = 1$ as the temperature is no longer necessary. An illustration of the resulting weights from both approaches is shown in Fig. 1.

The resulting sentence embeddings can then be compared to the standard class prototypes obtained by embedding the class name (***Classname*** approach). Since recent results show that hierarchical and graph relations between classes contain valuable information [3], in the ***Classname+Parent*** approach we combine the *Classname* prototype with the prototype of its parent class.

Finally, we experiment with different combinations of the base approaches: ***Classname+Def*** $_{average}$, ***Classname+Def*** $_{visualness}$ and ***Classname+Def*** $_{visualness}$ ***+Parent***. All these combinations simply consist in finding a value $\mu \in [0, 1]$ such that the combined prototype is $\mathbf{s} = \mu\mathbf{s_1} + (1 - \mu)\mathbf{s_2}$, given $\mathbf{s_1}$ and $\mathbf{s_2}$ the prototypes of two approaches. Here μ is considered to be a hyperparameter.

3 Experimental Protocol and Results

We measure ZSL accuracy on the large-scale ImageNet dataset, with synsets as classnames and corresponding WordNet definitions as sentences. We also use the WordNet hierarchy to determine parent classes. We employ the experimental protocol of [3], with the same ResNet features and, especially, the same train/test splits since [3] evidenced significant structural bias with previous popular splits. Glove embeddings are employed in [3]; we also compare with Word2vec and Fast-Text [2], as well as Elmo contextual embeddings using pre-trained embeddings available on the Internet (details are provided in the supplementary material). The ZSL model is a ridge regression from the semantic to the visual space as it gives good, consistent and easily reproducible results (see [6]). Results for the best setting with other models are provided in the supplementary material.

Results are shown in Table 1. The baseline $Def_{average}$ approach performs poorly compared to the usual *Classname* approach. Attention provides a slight improvement (both $Def_{visualness}$ and $Def_{attention}$ give comparable results), but still does not match *Classname*, even though attention weights seem reasonable (some examples are shown in Fig. 1). The use of parent information improves results compared with *Classname* alone, which is consistent with [3] where the best methods make use of hierarchical relations between classes.

The combination of *Classname* and *Def* approaches brings significantly better scores than either separately. Surprisingly, while any *Def* alone has lower performance than *Classname* alone, the best trade-off obtained by cross-validation is 70% definition and 30% classname in every case, meaning that the definition has a much stronger presence than the class name in the resulting embedding.

The word embedding method has an impact on performance; Fasttext consistently ranks above the other methods. Elmo has surprisingly low performance including with fine-tuned attention, even though one could expect attention to be more effective here as Elmo considers the role of the word in the sentence.

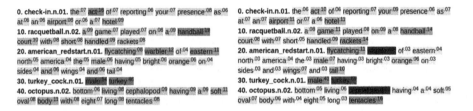

Fig. 1. Illustration of attention scores on some test classes. *Left*: weights from $Def_{visualness}$ after softmax (the temperature is $\tau = 5$ so differences are less pronounced than initially). *Right*: weights learned with $Def_{attention}$, with Fasttext embeddings.

Conclusion. To scale zero-shot learning to very large datasets it is important to solve the problem of providing class prototypes for many classes. The use of class name embeddings scales better than the provision manually-defined attributes but the resulting performance is nowhere near. We suggest that low-cost textual content, consisting in one sentence per class, can bring substantial performance improvements when combined with class name embeddings, when they are processed with the proposed approaches. We make available the improved class prototypes for ImageNet.

References

1. Akata, Z., Reed, S., Walter, D., Lee, H., Schiele, B.: Evaluation of output embeddings for fine-grained image classification. In: Proceedings of the IEEE Conference on Computer Vision and Pattern Recognition, pp. 2927–2936 (2015)
2. Almeida, F., Xexéo, G.: Word embeddings: a survey. arXiv preprint arXiv:1901.09069 (2019)
3. Hascoet, T., Ariki, Y., Takiguchi, T.: On zero-shot recognition of generic objects. In: Computer Vision and Pattern Recognition, pp. 9553–9561 (2019)

4. Hascoet, T., Ariki, Y., Takiguchi, T.: Semantic embeddings of generic objects for zero-shot learning. EURASIP J. Image Video Process. **2019**(1), 1–14 (2019). https://doi.org/10.1186/s13640-018-0371-x
5. Lampert, C.H., Nickisch, H., Harmeling, S.: Attribute-based classification for zero-shot visual object categorization. Pattern Anal. Mach. Intell. **36**(3), 453–465 (2014)
6. Le Cacheux, Y., Popescu, A., Le Borgne, H.: Webly supervised embeddings for large-scale zero-shot learning. arXiv preprint arXiv:2008.02880 (2020)

Adversarial Transfer of Pose Estimation Regression

Boris Chidlovskii$^{(\boxtimes)}$ and Assem Sadek

Naver Labs Europe, chemin Maupertuis 6, 38240 Meylan, France
boris.chidlovskii@naverlabs.com

Abstract. We address the problem of camera pose estimation in visual localization. Current regression-based methods for pose estimation are trained and evaluated scene-wise. They depend on the coordinate frame of the training dataset and show a low generalization across scenes and datasets. We identify the dataset shift an important barrier to generalization and consider transfer learning as an alternative way towards a better reuse of pose estimation models. We revise domain adaptation techniques for classification and extend them to camera pose estimation, which is a multi-regression task. We develop a deep adaptation network for learning scene-invariant image representations and use adversarial learning to generate such representations for model transfer. We enrich the network with self-supervised learning and use the adaptability theory to validate the existence of scene-invariant representation of images in two given scenes. We evaluate our network on two public datasets, Cambridge Landmarks and 7Scene, demonstrate its superiority over several baselines and compare to the state of the art methods.

1 Introduction

Visual localization is the task of accurate camera pose estimation in a known scene. It is a fundamental problem in robotics and computer vision, with multiple applications in autonomous vehicles, structure from motion (SfM), simultaneous localization and mapping (SLAM), Augmented Reality (AR) and Mixed Reality (MR) [30].

Traditional structure-based methods find correspondences between local features extracted from an image by applying image descriptors like SIFT, SURF or ORB [21,26] and 3D geometry of the scene obtained from SfM; obtained 2D-3D matches allow to recover the 6-DoF camera pose. The 3D-based methods are very accurate, their main drawback is that they are scene-specific, i.e., a 3D model needs to be build for each new scene and updated every time the scene changes.

Representing a scene as a simple set of 2D pose-annotated images is more flexible [29,39]. On one side, it represents all information needed to infer the 3D scene geometry; on the other side, it can be easy updated by adding more images. 2D scene representation can be easy adapted to new scenes, comparing to 3D-based methods, and allows to deploy machine learning techniques.

© Springer Nature Switzerland AG 2020
A. Bartoli and A. Fusiello (Eds.): ECCV 2020 Workshops, LNCS 12535, pp. 646–661, 2020.
https://doi.org/10.1007/978-3-030-66415-2_43

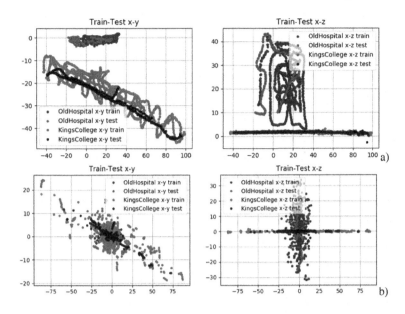

Fig. 1. Joint 2D projections of OldHospital and KingsCollege scenes, in position 3-DoF space. a) absolute poses; b) relative poses. Better seen in colors. (Color figure online)

PoseNet [17] was first to cast camera localization as a regression problem. The trained model learns a mapping from image to absolute pose which is dependent on the coordinate frame of the train set. Models learned on one scene's train set work well on the same scene's test set but fail on other scenes. A typical example is shown in Fig. 1.a) which plots poses of OldHospital and KingsCollege scenes from Cambridge Landmarks dataset in the common coordinate frame. Absolute poses of the two scenes indeed differ in range and spread[1].

The problem of generalization is better addressed by relative pose regression [19]. Visual localization with relative poses commonly relies on image retrieval where a query image is compared against the database of images and its pose is inferred from poses of the retrieved images.

Multiple deep learning methods have been proposed to estimate the relative pose [1,19,24,25,27]. Relative poses indeed reduce the difference in range and spread but solve generalization to some extend only. Figure 1.b) demonstrates the phenomenon by plotting the two scenes' relative poses. 2D projections show that relative poses (blue and green vs red and black) are still sparse. Except the space origin, they still live in different segments of the 6D pose space. Being discriminative, the supervised regression model can not generalize to poses not seen in the train set [29].

State of the art methods, both 3D-based [36] and 2D-based [1,19,39] test with generalization by holding out one scene in Cambridge Landmarks and 7Scenes datasets for evaluation and training a relative pose model on the all remaining

---

[1] We show 3-DoF pose positions only; pose orientations show the same phenomenon.

scenes. Concatenating multiple scenes allow to better populate the relative pose space and to reduce the dataset shift when testing the model on the evaluation scene.

However, this reduction is rather modest. Considering relative poses as 6D points, only 8.8% of relative poses in KingsCollege's test set (see Fig. 1.b) have a 1-top neighbour in OldHospital's train set. This fraction raises to merely 22.6% if the train set is a concatenation of OldHospital's, StMarysChurch's and Shop-Facade's sets. Put all together, relative poses in Cambridge Landmark dataset occupy less than 7% of the 6D sphere with the radius equal to the average distance between two image poses. Taking the curse of dimensionality, many more annotated scenes are needed to densely populate the 6D relative pose space.

We argue that in order to progress in pose generalization one should take into account the *relative pose sparsity* and the *dataset shift* it provokes. One solution is to create a very large dataset (like ImageNet for image classification) by massively annotating scenes with poses. Such annotation is arduous and expensive, so the learner can deploy *multi-task learning* to optimize the performance across m tasks/scenes simultaneously, through some shared knowledge. However, this would not solve the output shift when testing the model on a new, unseen scene.

To cope with the dataset shift in pose estimation, we turn towards *transfer learning* and develop a method to transfer a pose model from one scene to another. Then we extend the method to process multiple scenes as input.

Domain Adaptation. The lack of generalization is a fundamental problem in machine learning. Samples collected in different places and under different conditions results in the dataset bias when a learning method trained on one dataset generalizes poorly across other datasets [32].

Domain adaptation tries to produce good models on a target domain, by training on source labeled images and leveraging unlabeled target images as supplementary information during training. It has demonstrated a significant success in image classification [14], object recognition [23] and semantic segmentation [38]. These successes are due to a common semantic space shared by source and target domains. Common classes implicitly structure the output space [37], where separation between two classes in the source can be transferred to the target. Moreover this knowledge makes possible *unsupervised* domain adaptation [12,14,22].

Unlike classification, domain adaptation for multi-regression tasks is less studied [2,20]. Existing methods still assume the same output space and proceed by adjusting the loss function [11] or by reweighting source instances [9]. In camera pose estimation, one scene does correspond to one domain, but the difference in relative poses breaks the common space assumption. To extend domain adaptation to pose regression, we adopt the principle of domain-invariant representations [22] and use adversarial learning to generate such representations [7,8,12].

Due to the output shift, the source may have only some common poses with the target, so we need supplementary information about target pose space. In this paper we discuss *pose-driven* adaptation, where the transfer task is supported with a small number of target ground truth poses. With such a supplement, we learn scene-invariant representation of source and target images and pivot the source model towards the target.

As an additional contribution, we enrich pose regression with *self-supervised learning* which proved its efficiency in other vision tasks [13,18]. We apply random rotations [13] to an input image and train the network to predict these rotations. The method uses available training data to produce an additional self-supervision signal to the network and allows to learn a more accurate model.

2 Related Work

Traditional structure-based methods rely on SfM to associate 3D points with 2D images represented with their local descriptors. Matches between 2D points in an image and 3D points in the scene are then found by searching through the shared descriptor space. The descriptors can either be hand crafted (e.g., SIFT, SURF or ORB) or learned (e.g., SuperPoint) [21,26]. Given a set of 2D-3D matches, a n-point-pose (PnP) solver estimates candidate poses, and the best pose hypothesis is chosen using RANSAC. The estimated best pose is typically a subject of a further refinement.

PoseNet [17] modified the GoogLeNet architecture, are replaced softmax layers with fully connected (FC) layers to regress the pose. Absolute pose estimation (APE) relies on deep network encodings that are more robust to challenging changes in the scene such as lighting conditions and viewpoint. Comparing to 3D-based methods, a trained model requires less memory and has a constant inference time.

Multiple improvements to the PoseNet include new loss functions [16], adding Long-Short-Term-Memory (LSTM) layers [34], the geometric re-projection error [15], additional data sources and sensor measurements [5] and adversarial learning [6]. Other proposals extend the Posenet-like networks with auxiliary learning [25,33]. They learn additional auxiliary tasks which share representations with absolute pose estimation in order to improve its learning. VLocNet [33] implemented the auxiliary learning approach by jointly learning absolute and relative pose estimations. Its extended version, VLocNet++ [25], added semantic segmentation as a second auxiliary task.

DSAC [3] and DSAC++ [4], combine the pose estimation with structure-based methods and local learning. In such a hybrid paradigm, methods rely on geometrical constraints and utilize a structure-based pipeline where the learning is focused on local computer vision tasks, such as 2D-3D matches.

Instead of working with position and orientation losses separately, [39] learns the relative pose models directly from essential matrices and combines them with geometric models. It attributes the failure of pose regression in competition with geometric methods to the inaccurate feature representation in the last network layer. In this paper we identify the dataset shift as an additional reason of low generalization.

In 3D-based models, generalization efforts count on 3D point clouds and their associations with the scene images. SANet [36] is a scene agnostic neural architecture for camera localization, where model parameters and scenes are independent from each other. The method constructs scene and query feature pyramids, deploys the 3D point clouds at different scales and proceeds with query-scene registration (QSR). It first estimates a scene coordinate map and then computes the camera pose.

Domain Adaptation. Domain adaptation considers the discrepancy between training and testing domains as a fundamental obstacle to generalization [22,35]. The state of the art approaches address the problem by learning domain-invariant feature representations through adversarial deep learning [8,12, 37]. These methods encourage samples from different domains to be non-discriminative with respect to domain labels. Ganin et al. [12] uses a domain classifier to regularize the extracted features to be indiscriminate with respect to the different domains. They assumed the existence of a shared feature space between domains where the distribution divergence is small. The domain-adversarial neural network can be integrated into the standard deep architecture to ensure that the feature distributions over the two domains are made similar.

Initially studied under the assumption of same classes in source and target domains [12], domain adaptation research has recently turned towards more realistic settings. The new extensions address *partial* domain adaptation [7], when the target domain does not include all source classes, *open set* [28] when target include new classes, and *universal* domain adaptation [39] which treat both cases jointly.

3   Camera Pose Regression

Given an RGB image $\mathbf{x} \in \mathbb{R}^{h \times w \times 3}$, our task is to predict the (absolute or relative) camera pose $\mathbf{p} = [\mathbf{t}, \mathbf{q}]$ given by position vector $\mathbf{t} \in \mathbb{R}^3$ and orientation quaternion $\mathbf{q}$, $\mathbf{q} \in \mathbb{R}^4$. The following loss function is used to train a pose regression network [16]

$$L_p(\hat{\mathbf{p}}, \mathbf{p}) = ||\mathbf{t} - \hat{\mathbf{t}}|| e^{-s_t} + s_t + ||\mathbf{q} - \hat{\mathbf{q}}|| e^{-s_q} + s_q, \tag{1}$$

where $\hat{\mathbf{p}} = [\hat{\mathbf{t}}, \hat{\mathbf{q}}]$, and $\hat{\mathbf{t}}$ and $\hat{\mathbf{q}}$ represent the predicted position and orientation, respectively, s_t and s_q are trainable parameters to balance both distances, and $|| \cdot ||$ is the l_1 norm.

All pose regression networks share three main components, namely, encoder, localizer and regressor [29,30]. Given an image $\mathbf{x}$, encoder E is a deep network that extracts visual feature vectors from $\mathbf{x}$, $f = E(\mathbf{x})$. The localizer then uses FC layers to map a visual feature vector to a localization feature vector. Finally, two separate connected layers are used to regress $\hat{\mathbf{t}}$ and $\hat{\mathbf{q}}$, respectively, giving the estimated pose $\hat{\mathbf{p}} = [\hat{\mathbf{t}}, \hat{\mathbf{q}}]$.

Existing variations of this architecture concern the encoder network (GoogleNet [17] and ResNet34 with global average pooling [5]), and the localizer

(1 to 3 FC layers, 1 FC layer extended with 4 LSTMs [34], etc.). We use a configuration that includes the ResNet34 encoder, 1 FC layer localizer and trained with the pose regression loss in Eq. 1.

Absolute pose estimation methods [15,16,34] work with the absolute poses **p**. To get relative poses, we follow AnchorNet [27] and explicitly define a set of anchor points, which correspond to a subset of all training images in the network, we then estimate the pose of a test image with respect to these anchors.

3.1   Self-supervision

We extend our pose estimation network with self-supervised learning [13,18]. This concept proposes to learn image representations by training the network to recognize the geometric transformation applied to an input image. It first defines a set of discrete geometric transformations, then those geometric transformations are applied to each input image. The produced transformed images are fed to the model that is trained to recognize the transformation.

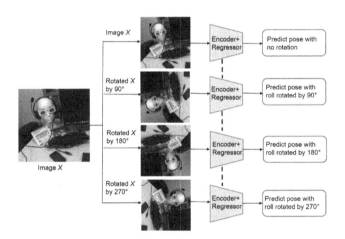

Fig. 2. Self-supervised learning by rotating the input images. The model learns to predict which rotation is applied.

We follow [13] in defining the geometric transformations as the image rotations by 0, 90, 180, and 270 degrees. Unlike [13] where the CNN model is trained on the 4-way image classification task to recognize one of the four image rotations, we train the network to identify the rotation applied to the input image (see Fig. 2). The main argument is that in order a CNN model to be able to recognize the rotation transformation that was applied to an image it will require to understand the concept of the objects present in the image, such as their location and their pose [18].

We crop the input image and rotate it randomly 0, 90, 180 or 270 degrees and expect that the network is able to predict the rotation applied to the image.

Image rotation changes its orientation $\mathbf{q}$ but not position $\mathbf{t}$. We calculate the orientation of the rotated image by first transforming the quaternion $\mathbf{q}$ of the input image in Euler angles $yaw, pitch, roll$. Then, we change $roll$ accordingly to the applied rotation, and transform the new angles back in quaternion $\mathbf{q}'$. Fed with the rotated image, we train the network to predict pose of the rotated image to be $[\mathbf{t}, \mathbf{q}']$.

3.2   Adversarial Pose Adaptation Network

We now consider the task of adapting a pose regression model from one (source) scene to another (target) scene. The task constitutes a source set $D_s = \{(\mathbf{x}_s^i, \mathbf{p}_s^i)\}_{i=1}^{n_s}$ of n_s images with ground truth poses and a target set $D_t = \{\mathbf{x}_t^j\}_{j=1}^{n_t}$ of n_t images. A small number of target images might be labeled with poses, $D_t^a = \{(\mathbf{x}_t^j, \mathbf{p}_t^j)\}_{j=1}^{n_t^a}, n_t^a << n_t$. Unlike the classification and semantic segmentation, where source and target domains share the same classes, we additionally face the output shift, where source and target poses lie in different segments of the coordinate system, i.e. $\{\mathbf{p}_s\} \neq \{\mathbf{p}_t\}$.

Our pose adaptation network enables an end-to-end training of a transferable encoder E and an adaptive pose regressor G_p. Trained on labeled source images and (mostly) unlabeled target images, the network enables an accurate adaptation of the source pose model to the target scene.

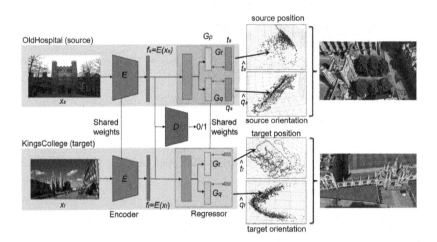

Fig. 3. Adversarial Pose Adaptation Network for the camera pose adaptation, where E is the transferable encoder, G_p is the adaptive pose regressor (including the localizer), D is scene discriminator.

The main problem of domain adaptation is to reduce the discrepancy between the source and target images [22]. Domain adversarial networks [8,12,37] tackle this problem by learning *scene-invariant* image representations in a two-player minimax game. The first player is a scene discriminator G_d trained to distinguish feature representations of source images from target images, and the second player is the encoder E trained simultaneously to deceive the domain discriminator G_d.

Specifically, the scene-invariant image representations $f = E(\mathbf{x})$ are learned in a minimax optimization procedure, where encoder E is trained by maximizing the loss of scene discriminator G_d, while G_d is trained by minimizing its own scene discrimination loss. As the ultimate goal is to learn a source pose regression model and transfer it to target scene, the loss of the source pose regressor G_p should be also minimized.

This leads to the optimization problem over the following terms.

Source Pose Regression. The source regression loss is defined on labeled source images,

$$\mathcal{L}_{pose}^s(E, G_p) = \sum_{\mathbf{x}_i \in D_s} L_p(G_p(E(\mathbf{x}_i)), \mathbf{p}_i), \tag{2}$$

where L_p is the regression loss function defined in Eq. 1.

Scene Discrimination Network. Scene discriminator G_d is trained to distinguish between feature representations of the source and target images, with the adversarial loss

$$\mathcal{L}_{adv}(G_d) = - \sum_{\mathbf{x}_i \in D_s \cup D_t} L_d(G_d(E(\mathbf{x}_i)), \mathbf{d}_i), \tag{3}$$

where L_d is the cross-entropy loss function and $\mathbf{d}_i$ is the scene label (0 for source images and 1 for target images).

Semi-Supervised Adaptation. In classification, domain adaptation can be achieved with the two above terms in an unsupervised way [12,37]. In pose regression, we are supplied with a small number of labeled target images, D_t^a. Then we define a regression term on D_t^a,

$$\mathcal{L}_{pose}^t(E, G_p) = \sum_{\mathbf{x}_j \in D_t^a} L_p(G_p(E(\mathbf{x}_t^j)), \mathbf{p}_t^j), \tag{4}$$

where L_p is the regression loss function defined in Eq. 1. The pose regression loss then includes the source and target terms, $\mathcal{L}_{pose}(E, G_p) = \mathcal{L}_{pose}^s(E, G_p) + \mathcal{L}_{pose}^t(E, G_p)$.

The total loss for training our *adversarial pose adaptation network* (APANet) can be represented as

$$\mathcal{L}_{APANet}(E, G_p, G_d) = \mathcal{L}_{pose}(E, G_p) + \alpha \mathcal{L}_{adv}(E, G_d), \tag{5}$$

where α is a hyper-parameter controlling the importance of the adversarial loss.

The training objective of the minimax game is the following

$$E^*, G_p^* = \arg \min_{E, G_p} \max_{G_d} \mathcal{L}_{APANet}(E, G_d, G_p). \tag{6}$$

Equation 6 is solved by alternating between optimizing E, G_p and G_d until the total loss (5) converges.

The APANet architecture is presented in Fig. 3. The network inputs a batch of source images and a batch of target images. Encoder E generates image representations $f = E(\mathbf{x})$ for both batches. Scene discriminator G_d is trained on image representations f and scene labels $\mathbf{d}$ to distinguish source images from target images. Pose regressor G_p is trained on a full set of source image poses and, when available, a small number of target scene poses. Position regressor G_t and orientation regressor G_q are trained separately. The position and orientation predictions are concatenated to produce the 6-DoF pose estimation, $\hat{\mathbf{p}} = [\hat{\mathbf{t}}, \hat{\mathbf{q}}]$, where $\hat{\mathbf{t}} = G_t(E(\mathbf{x}))$ and $\hat{\mathbf{q}} = G_q(E(\mathbf{x}))$.

3.3 Scene Adaptability

We complete this section by the notion of adaptability which has been introduced [10] to measure the transferability of feature representations from one domain to another. Adaptability is quantified by the error of a joint hypothesis h^* in both domains. The ideal joint hypothesis h^* is found by training on both source and target labeled images. Note that the target labels are only used to reason about the adaptability.

If the joint model shows a low error on source and target test sets, it suggests that an efficient transfer is possible across domains. In the evaluation section, we apply this idea to pose adaptation tasks. We will learn a joint model in order to reason about adaptability of source absolute and relative models to the target scene.

4   Evaluation

Datasets. We test our pose adaptation network on two public datasets, Cambridge Landmarks [17] and 7Scene [31]. *Cambridge Landmarks* [17] is an outdoor dataset collected in four sites around Cambridge University. It is collected using a Google mobile phone while pedestrians walk. The images are captured at the resolution of 1920×1080, the ground truth pose is obtained through VisualSFM software. Each site corresponds to one scene: Old Hospital, King's College, StMary's Church and Shop Facade. We first consider 1-to-1 scenario to test the model transfer from one scene to another. We form twelve pose adaptation tasks, by enumerating all possible *source → target* pairs. Then, we consider n-to-1 scenario and form four adaptation tasks, where one scene is retained as target and three remaining scenes is used as source.

7Scene [31] is an indoor dataset collected with a handheld RGB-D camera. The ground truth pose is generated using the Kinect Fusion approach [31]. The dataset is captured in 7 indoor scenes. In 1-to-1, we form twelve adaptation tasks, with Chess selected as a pivot scene. It constitutes six Chess →X tasks and six $X→$ Chess tasks, where X is one of the six remaining scenes: Fire, Heads, Office, Pumpkin, Red Kitchen or Stairs. In n-to-1 scenario, seven adaptation tasks are formed, one scene is retained as target and six other scenes is used as source.

Table 1. KingsCollege→X adaptation tasks in Cambridge Landmarks dataset. The median position (in meters)/orientation (in degrees) errors are reported.

Source	Case	Method	OldHospital	StMarysChurch	ShopFacade
Kings college	3D	DSAC++ [4]	**0.20/0.30**	**0.13/0.40**	**0.06/0.30**
	APE	No adaptation	38.63/83.04	28.53/110.39	31.21/45.19
		Joint	1.55/5.05	2.19/6.65	1.32/4.58
		SS, $\nu = 0.05$	6.74/14.92	6.76/15.84	6.52/13.10
		APANet, $\nu = 0.05$	3.74/8.27	3.63/10.18	2.61/8.63
		APANetS, $\nu = 0.05$	**3.56/6.71**	**3.58/8.23**	**2.58/7.41**
	RPE	No adaptation	3.63/7.22	3.53/10.39	2.91/8.16
		Joint	0.53/1.59	0.45/1.21	0.42/1.29
		SS, $\nu = 0.05$	1.54/1.82	1.36/3.04	1.18/3.81
		APANet, $\nu = 0.05$	1.03/2.17	0.80/**2.18**	**0.61**/2.62
		APANetS, $\nu = 0.05$	0.98/**1.94**	**0.77**/2.25	0.62/**2.49**
		NC-Essnet* [39]	**0.95**/2.65	1.12/3.64	0.70/3.41

Implementation and Setup. The APANet is implemented in PyTorch. Encoder E is fine-tuned on ResNet-34 network. Pose regressor G_p and scene discriminator G_d are trained from scratch. G_p includes a FC localizer with 1024 nodes and two separate layers G_t and G_q, to regress position and orientation vectors, with 256 nodes each. Scene discriminator G_d is similar to one used in the universal domain adaptation (UDA) network [37]; it includes three FC layers with 1024, 256 and 64 nodes, interleaved with ReLu layers, the drop-out is 0.5.

In the train phase, the network inputs a batch of source images and a batch of target images to fine-tune the encoder E and to train scene discriminator G_d and pose regressor G_p. In the test phase, only encoder E and pose regressor G_p are necessary. A target image $\mathbf{x}_t$ is fed to the network and its pose is estimated as $G_p(E(\mathbf{x}_t))$.

We use the Adam optimizer; we train the network with a learning rate of 10^{-5} and the batch size of 16 images. We initialize parameters s_t and s_q in the pose loss (Eq. 1) with 0 and -1.0 respectively. Hyper-parameter α is set to 1.0. We use the same image pre-processing steps as the state of the art methods [16,29]. For training, we randomly crop the image to 224×224 pixels. For testing, images are cropped to 224×224 pixels at the center of the image. Training images are shuffled before they are fed to the network. In the self-supervised extension (APANetS), the input image is randomly rotated 0, 90, 180, 270 degrees and the network is trained to correctly predict the applied rotation.

Evaluation Options. For all adaptation tasks, we compare five models, including two baselines and two APANet versions. First, we train the *joint model* (see Sect. 3.3) on source and target train sets. This model is an indicator of adaptability of the source model to target scene. The low error on source and target test sets suggests a good generalization and there exists a joint model performing well on both scenes. Second, as the true baseline, we consider a *semi-supervised*

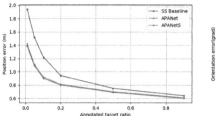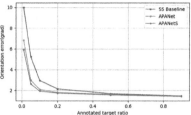

Fig. 4. Position (top) and orientation (bottom) RPE errors of SS, APANet and APANetS for different ν values.

(SS) relaxation of the joint model, where a fraction of target train set is used for training. Denoted ν, where $\nu = \frac{n_t^a}{n_t}$, this fraction varies between 0 (source scene only) and 1 (the joint model).

The third model is the APANet (Sect. 3.2) trained with ν target poses. This model can be considered as the SS baseline extended with adversarial learning of scene-invariant image representations. Finally, we include results of APANet extended with *self-supervision* (Sect. 3.1) and denoted APANetS.

The joint model gives an estimate of the smallest error the optimal joint model could achieve on the target scene. It serves as an indicator for the SS baseline and APANets. All three methods are exposed to the trade-off between ratio ν and accuracy drop with respect to the joint model. In the following, we compare the SS baseline and APANets for the same values of ν, in order to measure the size of target supplement sufficient to ensure an efficient pose mode transfer.

4.1   Evaluation Results

Cambridge Landmarks. Table 1 reports APE and RPE evaluation results for three adaptation tasks KingsCollege $\rightarrow X$. For each task, the table reports the error without adaptation, the joint model error, SS and APANets errors for the selected value $\nu = 0.05$, which corresponds to 5% of target ground truth poses available for training. In all tasks, the joint model error is small and suggests a good adaptability across scenes.

For the ratio $\nu = 0.05$, APANets perform much better that the SS baseline. Averaged over 12 adaptation tasks, the APANet position error is 46.8% (APE) and 52.2% (RPE) lower than the SS position error. Similarly, the APANet orientation error is on average 37.9% and 42.1% lower than the SS orientation error. Self-supervised image rotation in APAnetS further reduces the orientation error but has a negligible impact on the position error.

Impact of Ratio ν. Figure 4 gives an aggregated view by averaging results over twelve adaptation tasks. It compares the SS, APANet ans APANetS errors for ν varying between 0.01 and 0.90.

The figure shows little or no difference when target poses are abundantly available in training. Having 20% or more target poses is sufficient to train a

Table 2. 1-to-1 transfer in 7Scenes dataset. The median position (in meters)/ orientation (in degrees) errors are reported.

Source	Case	Method	Fire	Heads	Office	Pumpkin	Kitchen	Stairs
Chess	3D	DSAC++ [4]	**0.02/0.90**	**0.01/0.80**	**0.03/0.70**	**0.04/1.10**	**0.04/1.10**	**0.09/2.60**
	APE	No Adaptation	2.25/28.32	1.91/31.76	0.30/23.75	1.08/20.82	1.59/23.14	1.83/28.45
		Joint	0.40/9.45	0.28/14.41	0.31/7.20	0.27/4.91	0.34/7.26	0.34/9.52
		SS, $\nu = 0.05$	1.31/19.08	1.04/24.62	0.89/18.36	0.82/11.15	1.01/17.59	1.03/20.12
		APANet, $\nu = 0.05$	0.63/14.67	0.44/17.03	0.42/10.62	0.39/6.87	0.43/10.35	**0.47**/12.28
		APANetS, $\nu = 0.05$	**0.59/12.06**	**0.41**/14.29	**0.41/8.45**	**0.37/5.71**	**0.41/8.66**	0.48/**9.87**
	RPE	No Adaptaption	2.25/28.32	1.91/31.76	1.63/23.75	2.78/20.82	1.89/23.14	1.83/28.45
		Joint	0.11/7.14	0.08/7.14	0.09/4.02	0.09/4.68	0.07/3.86	0.09/8.52
		SS, $\nu = 0.05$	0.41/17.08	0.34/14.62	0.29/14.36	0.42/11.15	0.31/7.75	0.29/19.12
		APANet, $\nu = 0.05$	0.22/10.15	0.17/10.26	**0.14/7.36**	**0.19**/6.46	0.19/5.65	0.17/12.28
		APANetS, $\nu = 0.05$	**0.21**/9.72	0.15/**9.35**	0.15/**6.68**	**0.19**/5.87	**0.16/5.13**	**0.16**/11.77
		NC-EssNet* [39]	0.26/**9.64**	**0.14**/10.66	0.20/**6.68**	0.22/**5.72**	0.22/6.31	0.31/17.88

joint model working well on both scenes. Instead, when the ratio ν is reduced to 1%–5%, the advantage of APANet over the SS baseline becomes multi-fold. We conclude that learning scene-invariant image representations accelerates the adaptation across scenes. Moreover it enables an accurate adaptation with a smaller target supplement.

Note no existing RPE method can perform 1-to-1 scenario without performance drop. Results for 3D-based DSAC++ [4] and 2D-based NC-EssNet [39] are included in Table 1 (and Table 2) for the comparison only. The state of the art NC-EssNet is a supervised learning method trained and tested on the target scene, while APAnet(S) are transfer models trained on the source train set and a fraction of the target train set. Still, all 2D-based methods are less accurate than the best 3D-based models, such as DSAC++ and Active Search.

7Scenes. Table 2 reports evaluation results for 1-to-1 adaptation tasks Chess $\rightarrow X$ in the 7Scene dataset. In all tasks, we again observe a small error of the joint models. For the ratio $\nu = 0.05$, the APANet performs better that the SS baseline and APANetS improves over APANet in the orientation error.

Table 3. RPE transfer in 1-to-1 and n-to-1 scenarios. The median position (in meters)/orientation (in degrees) errors are reported.

Source	Target	Case	Joint	SS,ν=0.05	Apanet,ν=0.05	APANetS,ν=0.05	NC-EssNet*
Cambridge	Cambridge	1-to-1	0.67/1.45	1.90/4.08	0.88/3.07	**0.87/2.86**	
		n-to-1	0.63/1.38	1.51/3.82	**0.81**/2.96	0.82/**2.56**	0.85/2.82
Cambridge	7Scenes	1-to-1	0.36/6.37	1.13/20.10	0.62/13.16	**0.61/12.43**	
		n-to-1	0.31/6.49	1.19/19.01	0.50/13.47	**0.47/12.96**	0.48/32.97
7Scenes	7Scenes	1-to-1	0.13/3.84	0.38/10.22	0.25/7.63	**0.24/7.16**	
		n-to-1	0.13/3.45	0.34/9.30	0.22/7.76	0.22/**7.31**	**0.21**/7.50
7Scenes	Cambridge	1-to-1	1.78/7.42	5.86/27.17	3.95/16.67	**3.91/16.15**	
		n-to-1	1.61/7.55	5.81/26.08	**3.86**/15.86	3.87/**15.56**	7.98/24.35

Cross Dataset Transfer. Table 3 reports evaluation results for 1-to-1 and *n*-to-1 scenarios, on the same or accross the Cambridge and 7Scenes datasets. When we test the APANets in *n*-to-1 scenario on the same dataset, a model is trained with all but one scenes, such a setup is used in the previous methods [19,39]. In evaluations across the two datasets, all scenes from the source dataset are used for training. All the joint models behave well and indicate a good generalization. The self-supervised learning in APANetS reduces the orientation error. Instead, adding more scenes does really help APANet and APANetS.

Again, results of NC-EssNet [39] are included for the comparison only. Being totally supervised, NC-EssNet shows good performance on the same dataset (Cambridge to Cambridge and 7Scenes to 7Scenes) but suffers from the dataset shift in the cross-dataset evaluations, if compared with APANets.

4.2 Discussion

We identify the dataset shift an important barrier to the generalization of pose estimation models. Analysis of the evaluation results in the previous section raises two critical issues. One concerns the target supplement to accompany the adaptation process and to preserve the pose estimation accuracy. We have promoted the pose-driven adaptation, where a small number of target poses guide adaptation with a modest performance drop. This method showed promising results. However, acquiring target ground truth poses is not always possible. We therefore need alternatives to parametrize the model transfer across scenes, where a model-driven regression can alleviate the absence of target poses.

Another issue concerns scene adaptability. Low joint model errors in both datasets have been a strong indicator that the pose regression adaptation is possible and the scene-invariant representations can accelerate the adaptation process. If, instead, the joint model error is higher, it can seriously compromise any chances of an efficient adaptation. In the Cambridge Landmarks dataset, the joint models for the lately added Street scene indeed show a high error. This suggests a low adaptability from and to Street scene. The adaptability theory developed an instance adaptation technique for low adaptability in classification tasks [10]. Therefore it looks important to extend this theory to multi-regression tasks as well.

5   Conclusion

We address the problem of low generalization of the relative pose estimation models. We attribute it to the dataset shift and propose adaptation across scenes as an alternative way towards a better generalization. We extend domain adaptation techniques invented for classification to the multi-regression task and developed a deep network to adapt a pose regression model from one scene to another. The Adversarial Pose Adaptation Network learns scene-invariant image representations and use target scene supplements to guide the transfer of source models to the target scene. We also use the adaptability theory to measure the transferability of feature representations from one scene to another. We validate the superiority of the APANet on Cambridge Landmarks and 7Scene datasets over the baselines and compare them to the state of the art supervised methods.

References

1. Balntas, V., Li, S., Prisacariu, V.: Relocnet: continuous metric learning relocalisation using neural nets. In: European Conference Computer Vision (ECCV), pp. 782–799 (2018)
2. Borchani, H., Varando, G., Bielza, C., Larrañaga, P.: A survey on multi-output regression. Wiley Int. Rev. Data Min. Knowl. Disc. 5(5), 216–233 (2015)
3. Brachmann, E., et al.: DSAC - differentiable RANSAC for camera localization. In: Computer Vision Pattern Recognition (CVPR), pp. 2492–2500 (2017)
4. Brachmann, E., Rother, C.: Learning less is more - 6D camera localization via 3D surface regression. In: Computer Vision Pattern Recognition (CVPR), pp. 4654–4662 (2018)
5. Brahmbhatt, S., Gu, J., Kim, K., Hays, J., Kautz, J.: Geometry-aware learning of maps for camera localization. In: Computer Vision Pattern Recognition (CVPR), pp. 2616–2625 (2018)
6. Bui, M., Baur, C., Navab, N., Ilic, S., Albarqouni, S.: Adversarial networks for camera pose regression and refinement. In: ICCV Workshops, vol. 2019, pp. 3778–3787 (2019)
7. Cao, Z., Ma, L., Long, M., Wang, J.: Partial adversarial domain adaptation. In: European Conference Computer Vision (ECCV), pp. 139–155 (2018)
8. Cao, Z., You, K., Long, M., Wang, J., Yang, Q.: Learning to transfer examples for partial domain adaptation. In: Computer Vision Pattern Recognition (CVPR), pp. 2985–2994 (2019)
9. Chen, X., Monfort, M., Liu, A., Ziebart, B.D.: Robust covariate shift regression. Proc. AISTATS. 51, 1270–1279 (2016)
10. Chen, X., Wang, S., Long, M., Wang, J.: Transferability vs. discriminability: batch spectral penalization for adversarial domain adaptation. In: International Conference on Machine Learning (ICML), vol. 97, pp. 1081–1090 (2019)
11. Cortes, C., Mohri, M.: Domain adaptation in regression. In: Proceedings 22nd International Conference on Algorithmic Learning Theory (2011)
12. Ganin, Y., et al.: Domain-adversarial training of neural networks. J. Mach. Learn. Res. 17, 59:1–59:35 (2016)
13. Gidaris, S., Singh, P., Komodakis, N.: Unsupervised representation learning by predicting image rotations. In: International Conference on Learning Representation (ICLR) (2018)
14. Hoffman, J., Rodner, E., Donahue, J., Darrell, T., Saenko, K.: Efficient learning of domain-invariant image representations. CoRR arXiv:1301.3224 (2013)
15. Kendall, A., Cipolla, R.: Modelling uncertainty in deep learning for camera relocalization. In: IEEE International Conference on Robotics and Automation, ICRA, pp. 4762–4769 (2016)
16. Kendall, A., Cipolla, R.: Geometric loss functions for camera pose regression with deep learning. In: Computer Vision Pattern Recognition (CVPR), pp. 6555–6564 (2017)
17. Kendall, A., Grimes, M., Cipolla, R.: PoseNet: a convolutional network for real-time 6-DOF camera relocalization. In: International Conference on Computer Vision (ICCV), pp. 2938–2946 (2015)

18. Kolesnikov, A., Zhai, X., Beyer, L.: Revisiting self-supervised visual representation learning. In: Computer Vision Pattern Recognition (CVPR), pp. 1920–1929 (2019)
19. Laskar, Z., Melekhov, I., Kalia, S., Kannala, J.: Camera relocalization by computing pairwise relative poses using convolutional neural network. In: IEEE International Conference on Computer Vision Workshops, pp. 929–938 (2017)
20. Lathuilière, S., Mesejo, P., Alameda-Pineda, X., Horaud, R.: A comprehensive analysis of deep regression. CoRR 1803.08450 (2018)
21. Leng, C., Zhang, H., Li, B., Cai, G., Pei, Z., He, L.: Local feature descriptor for image matching: a survey. IEEE Access **7**, 6424–6434 (2019)
22. Long, M., Cao, Y., Wang, J., Jordan, M.I.: Learning transferable features with deep adaptation networks. In: International Conference on Machine Learning (ICML), pp. 97–105 (2015)
23. Long, M., Wang, J., Ding, G., Sun, J., Yu, P.S.: Transfer joint matching for unsupervised domain adaptation. In: Computer Vision Pattern Recognition (CVPR), pp. 1410–1417 (2014)
24. Melekhov, I., Ylioinas, J., Kannala, J., Rahtu, E.: Relative Camera Pose Estimation Using Convolutional Neural Networks. CoRR 1702.01381 (2017)
25. Radwan, N., Valada, A., Burgard, W.: Vlocnet++: deep multitask learning for semantic visual localization and odometry. IEEE Rob. Autom. Lett. **3**(4), 4407–4414 (2018)
26. Rublee, E., Rabaud, V., Konolige, K., Bradski, G.R.: ORB: an efficient alternative to SIFT or SURF. In: International Conference on Computer Vision (ICCV), pp. 2564–2571 (2011)
27. Saha, S., Varma, G., Jawahar, C.V.: Improved visual relocalization by discovering anchor points. In: British Machine Computer Vision (BMVC), p. 164 (2018)
28. Saito, K., Yamamoto, S., Ushiku, Y., Harada, T.: Open set domain adaptation by backpropagation. In: European Conference Computer Vision (ECCV), pp. 153–168 (2018)
29. Sattler, T., Zhou, Q., Pollefeys, M., Leal-Taixe, L.: Understanding the limitations of CNN-based absolute camera pose regression. In: Computer Vision Pattern Recognition (CVPR), pp. 3302–3312 (2019)
30. Shavit, Y., Ferens, R.: Introduction to Camera Pose Estimation with Deep Learning. CoRR 1907.05272 (2019)
31. Shotton, J., Glocker, B., Zach, C., Izadi, S., Criminisi, A., Fitzgibbon, A.W.: Scene coordinate regression forests for camera relocalization in RGB-D images. In: Computer Vision Pattern Recognition (CVPR), pp. 2930–2937 (2013)
32. Torralba, A., Efros, A.A.: Unbiased look at dataset bias. In: Computer Vision Pattern Recognition (CVPR), pp. 1521–1528 (2011)
33. Valada, A., Radwan, N., Burgard, W.: Deep auxiliary learning for visual localization and odometry. In: IEEE International Conference on Robotics and Automation, ICRA, pp. 6939–6946 (2018)
34. Walch, F., Hazirbas, C., Leal-Taixé, L., Sattler, T., Hilsenbeck, S., Cremers, D.: Image-based localization using LSTMs for structured feature correlation. In: International Conference on Computer Vision (ICCV), pp. 627–637 (2017)
35. Wang, M., Deng, W.: Deep visual domain adaptation: a survey. Neurocomputing **312**, 135–153 (2018)
36. Yang, L., Bai, Z., Tang, C., Li, H., Furukawa, Y., Tan, P.: Sanet: scene agnostic network for camera localization. In: European Conference Computer Vision (ECCV), pp. 42–51 (2019)
37. You, K., Long, M., Cao, Z., Wang, J., Jordan, M.I.: Universal domain adaptation. In: Computer Vision Pattern Recognition (CVPR), pp. 2720–2729 (2019)

38. Zhang, Y., David, P., Gong, B.: Curriculum domain adaptation for semantic segmentation of urban scenes. In: International Conference on Computer Vision (ICCV), pp. 2039–2049 (2017)
39. Zhou, Q., Sattler, T., Pollefeys, M., Leal-Taixe, L.: To learn or not to learn: Visual localization from essential matrices. CoRR abs/1908.01293 (2019)

Disentangled Image Generation for Unsupervised Domain Adaptation

Safa Cicek[1(✉)], Ning Xu[2], Zhaowen Wang[2], Hailin Jin[2], and Stefano Soatto[1]

[1] University of California, Los Angeles, USA
{safacicek,soatto}@ucla.edu
[2] Adobe Research, San Jose, USA
{nxu,zhawang,hljin}@adobe.com

Abstract. We explore the use of generative modeling in unsupervised domain adaptation (UDA), where annotated real images are only available in the source domain, and pseudo images are generated in a manner that allows independent control of class (content) and nuisance variability (style). The proposed method differs from existing generative UDA models in that we explicitly disentangle the content and nuisance features at different layers of the generator network. We demonstrate the effectiveness of (pseudo)-conditional generation by showing that it improves upon baseline methods. Moreover, we outperform the previous state-of-the-art with significant margins in recently introduced multi-source domain adaptation (MSDA) tasks, achieving significant error reduction rates of 50.27%, 89.54%, 75.35%, 27.46% and 94.3% in all 5 tasks.

1 Introduction

In unsupervised domain adaptation (UDA), one is given annotated images from a *source* domain (e.g., hand-written digits), and images from a *target* domain (e.g., house numbers) with no annotations. The goal is to leverage annotated samples in the source domain to solve a task in the target domain, for example to classify images into, numbers from 0 to 9.

While the quality of a UDA classifier can be measured by standard classification metrics, these yield no insight on how the classifier manages nuisance variability due to the domain (covariate) shift. Is the classifier agnostic, or invariant, to domain changes? Where is such invariance achieved in the model? We wish to devise a more interpretable model, where we can pinpoint the effects of changes in the domain and class of the input image within the model.

To this end, we develop an architecture that embodies the ability to generate images with controlled class and domain labels. That is, the model can receive as input a domain (e.g. SVHN) and class label (e.g. 5), along with a random vector, and produce as output diverse images of the given class in the given domain for different realizations of the random vector (e.g. different digit fives in SVHN).

Electronic supplementary material The online version of this chapter (https://doi.org/10.1007/978-3-030-66415-2_44) contains supplementary material, which is available to authorized users.

A. Bartoli and A. Fusiello (Eds.): ECCV 2020 Workshops, LNCS 12535, pp. 662–665, 2020.
https://doi.org/10.1007/978-3-030-66415-2_44

2   Proposed Method

To test the hypothesis that a generative model that allows control of a factor, say the class label, independent of another, say the nuisance variability in the images, improves UDA we leverage recent progress in generative model for style transfer [1], where the separation of style and content corresponds to class and nuisances in UDA. If the domain label and the class label can be disentangled, one can easily perform image generation in the unlabeled target domain while respecting the input class label simply by changing the domain label from source to target. As an architectural regularization, we encoded the class label information in the early stages of the decoder ("coarse layers"), while domain variations are fed to the late stages ("fine layers") (see Fig. 1). In the original StyleGAN architecture, the same learned latent vector is fed to all hidden layers of the generator for the task of image generation. As a loss regularization, we proposed the pseudo-conditional GAN loss described next, on top of the standard UDA losses described in Supp. Mat.

To generate images with controlled domain and class labels, we construct a conditional GAN loss where the label space (for conditioning the joint discriminator) is the Cartesian product of the domain label set ($[M] := \{0, \cdots, M-1\}$)

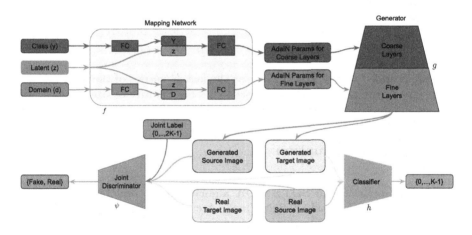

Fig. 1. Detailed overview of the proposed approach. StyleGAN based generator is used to generate images in the UDA setting. The generator (g) is conditioned on both the domain (d) and the class (y) label. The class label (y) is used only to tweak the coarse layer parameters (the red blocks) while the domain label is used only to tweak the fine layer parameters (the purple blocks). The joint discriminator (the green block) tries to distinguish generated images from the reals ones for guiding the generator training. The joint discriminator has $2K$ dimensional output from which the only one is selected conditioned on the joint label. For instance, given that $K = 10$, the source (target) 9 has the joint label of $j = 9(19)$, and jth index of the joint discriminator output is used to tell whether the image is fake or real. For the unlabeled target samples, another classifier (h, the orange block) is used to provide the pseudo-labels for calculating the loss the joint discriminator is minimizing. The classifier (h) is trained on labeled images from both source and target (generated) domains. (Color figure online)

and class label set $([K] := \{0, \cdots, K-1\})$, $[K] \times [M]$. For simplicity, we denote this as a set of scalars, $[MK] = \{0, \cdots, MK-1\}$. We define the "joint label" $j = mK + k \in [MK]$ for the given class label k and domain label m. The joint discriminator (the green block in Fig. 1) tries to distinguish fake and real images with the label $j \in [MK]$. The challenge in the UDA setting is that we do not have access to class labels k (thus joint label j) for the target samples. We will be relying on the pseudo-labels,

$$k(x^t) = \arg \max_k h(x^t)[k] \tag{1}$$

For simplicity, we will assume the number of domains is $M = 2$ where $m = 0$ is for the source and $m = 1$ is for the target domain. Non-saturating GAN loss is used as a conditional GAN loss where the generator g solves the following optimization problem

$$\min_g \mathbb{E}_{P(z,k,m)} \phi(-\psi(g(z,k,m))[mK+k]) \tag{2}$$

where $P(z,k,m) = N(z-0,I)\frac{I(k<K)}{K}\frac{I(m<M)}{M}$, I is indicator function and $\phi(x) = \text{softplus}(x) = \log(\exp(x)+1)$. The joint discriminator ψ competes with the generator g by solving the following optimization,

$$\min_\psi \mathbb{E}_{P(z,k,m)} \phi(\psi(g(z,k,m))[mK+k]) + \mathbb{E}_{(x^s,e_k)\sim P^s} \phi(-\psi(x^s)[k])$$

$$+ \mathbb{E}_{x^t \sim P^t_x} \phi(-\psi(x^t)[K+k(x^t)]) \tag{3}$$

where K is added to $k(x^t)$ from Eq. 1 as the last K entries of the discriminator output are devoted to the target samples. e_k is the identity of size K whose kth element is 1 and 0 elsewhere.

3   Empirical Evaluation

Table 1. Comparison to SOTA UDA algorithms on multi-source domain adaptation (MSDA) task "Digit-Five". Accuracy on the target test data are reported. Algorithms are trained on the 4 labeled source sets and 1 unlabeled target set. NR stands for not reported.

Target dataset	SVHN	SYN-DIGITS	MNIST	USPS	MNIST-M
[2] DCTN	77.5	NR	NR	NR	70.9
[3] M³SDA	81.32	89.58	98.58	96.14	72.82
Ours	**90.71**	**98.91**	**99.65**	**97.20**	**98.45**
Source-only (baseline)	82.50	94.60	99.18	89.48	65.54

The "Digit-Five" dataset proposed by [2,3] combines 5 digit datasets: MNIST, MNIST-M, SVHN, SYN-DIGITS, USPS. For each experiment, one of the five domains is chosen as the target domain in a leave-one-out setting. Following

[2,3], we sample 25000 images for training and 9000 images for testing from each dataset. For USPS, all of the images are used i.e. 7291 for training and 2007 for testing.

The proposed method can be applied to any number of domains with the modest expense of increased output dimension for the joint discriminator. We simply increase d dimension to the number of domains: i.e., we increase the domain label dimension from 2 to 5 for Digit-Five [2,3] experiments where the first $M - 1$ entries of d are used for source domains and the last entry is for the target domain. So, say domain label is $d = [0, 1, 0, 0, 0]$ then the generator produces fake images of the second source domain.

In Table 1, we report the performance of the proposed method along with the source-only baseline. Here the source-only baseline only trained with the labeled source samples using the same network and training procedure as the actual method. We outperform all the previous methods in this task. We outperform the SOA (M^3SDA) in all 5 tasks with error reduction rates of 50.27%, 89.54%, 75.35%, 27.46% and 94.3%. Note that our baseline scores are better than the reported numbers in the earlier MSDA works. The classifier used in the earlier MSDA works has 3 FC layers whose dimensions are too large for digit tasks resulting in over-fitting. Our classifier is a standard, light-weight network used in earlier UDA works.

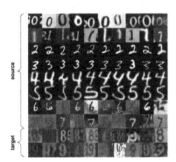

Fig. 2. Conditional image generation results of the proposed method. The first 8 rows and the last 2 rows are generated images for the labeled source domains and **unlabeled** target domain (e.g. SVHN) respectively. Each row shows the generated images for the same domain label and the same class label with different realizations of the latent vector z. The generator can generate a diverse set of samples for each class in the target domain which is utilized by the classifier.

References

1. Karras, T., Laine, S., Aila, T.: A style-based generator architecture for generative adversarial networks. In: Proceedings of the IEEE Conference on Computer Vision and Pattern Recognition, pp. 4401–4410 (2019)
2. Xu, R., Chen, Z., Zuo, W., Yan, J., Lin, L.: Deep cocktail network: multi-source unsupervised domain adaptation with category shift. In: Proceedings of the IEEE Conference on Computer Vision and Pattern Recognition, pp. 3964–3973 (2018)
3. Peng, X., Bai, Q., Xia, X., Huang, Z., Saenko, K., Wang, B.: Moment matching for multi-source domain adaptation. arXiv preprint arXiv:1812.01754 (2018)

Domain Generalization Using Shape Representation

Narges Honarvar Nazari$^{(\boxtimes)}$ and Adriana Kovashka

University of Pittsburgh, Pittsburgh, PA 15260, USA
{nhonarvar,kovashka}@cs.pitt.edu

1 Introduction

CNN-based representations have greatly advanced the state of the art in visual recognition, but the community has primarily focused on the setting where training and test set belong to the same dataset/distribution. However, models trained on one dataset do not generalize well to other datasets [3,5]. *Human* vision, which is robust to data/domain shifts, relies on *shape* in addition to texture/appearance, as shown in prior research. On the other hand, prior work in *computer* vision shows CNN representations are biased towards *texture* [4]. We propose a new shape-based representation which captures the medial axis transform and skeleton of an object. As shown in Fig. 1, shape is more robust to domain shifts than texture. We apply it in the domain generalization (DG) setting: methods are trained on a set of source domains, and are tested on a disjoint domain from which no data is available at training time. Unlike related prior shape work [7,8], which primarily targeted cross-modal retrieval and *scene* classification, our representation is denser than an edge map.

2 Approach

Our hypothesis is that representing a shape through the skeleton of an object will be more robust to domain shifts, compared to (1) representations based on texture, and (2) representations of shape using edgemaps but no specialized shape filters. We capture shape through filters that approximate a Medial Axis Transform, but compute a dense representation of the image, to increase the amount of signal a network can capture. We combine shape and texture branches.

Medial Axis Transform: The object skeleton or medial axis (MA) of an object is the set of points inside the object boundary that represent the overall shape. Every point in the MA is located at the center of the largest circle which can be inscribed within the object boundary. The set of centers and radii of all maximally inscribed (MI) circles represents the Medial Axis Transform (MAT) [1]. Figure 2 (a) shows an example; the red pixels are MA and the green circle is one of the MAT circles (MI disks). Figure 2 (b-d) show situations where a disk is not MI, and the MAT response at the corresponding location should be smaller than MAT response in (a).

© Springer Nature Switzerland AG 2020
A. Bartoli and A. Fusiello (Eds.): ECCV 2020 Workshops, LNCS 12535, pp. 666–670, 2020.
https://doi.org/10.1007/978-3-030-66415-2_45

Fig. 1. Objects of the same category share common shape even if there is large variation in appearance. This figure shows two images from four modalities from the PACS dataset [5], all belonging to the category "dog". Although all images belong to the same category, their textures vary. In the skeletons which do not have color and texture, modalities share more common features; the legs and tails are very similar.

Fig. 2. Different situations of the overlap (or lack of overlap) of filters and the MAT of a horse. Filter responses should be large for (a), medium for (b), and zero for (c, d). (Color figure online)

Filters: We show one of our filters in Fig. 3 (a). To construct filters that match the desired properties in Fig. 2, we first compute the number of circles which can be fitted in a given square. We find the pixels whose coordinates overlap with the equation of each circle. We enforce the constraint that the sum of all values inside the filter should be zero. Since the filters need to mimic the MAT, the outer circles (orange ones in Fig. 3 (a)) need to be positive and inner circles (green ones in Fig. 3 (a)) need to be negative and of the same absolute values. Some pixels overlap with more than one circle (shown in red) and we need to sum the two values for these pixels.

Pipeline: We convert the image to grayscale, and use Canny edge detection to extract the object edges. We then pass the edge image to our shape-based convolutional network with circles filters of different sizes. The MAT filters form one convolutional layer, and are resized to the same largest size (padding with zeros as needed). We use the created filters as kernels weights in a single convolutional layer, followed by an adaptive average pooling and a fully connected (FC) layer. We apply ReLU after the convolution and FC layers, and batch normalization and dropout after the convolutional layers. We combine the shape filter branch with a standard CNN (except its classification layer), using a FC layer fed to a softmax. The full architecture are shown in Fig. 3 (b).

3   Experimental Validation

We show results for the following methods. In Table 1, all methods use AlexNet as the backbone in Fig. 3 (b), and in Table 2 and 3, all methods use ResNet. We pre-train all methods on ImageNet.

Methods Compared: We show results for the following methods.

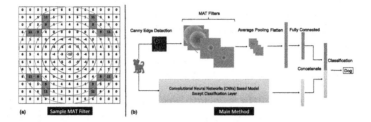

Fig. 3. (a) An example of a circle filter used to compute a MAT. For a line crossing the filter in the middle, the response would be zero as desired. (b) We combine our explicit shape representation (through special filters) with a standard texture-based CNN.

Table 1. PACS results using AlexNet. Best result in **bold**, second-best in *italics*.

Method/Target	Painting	Cartoon	Sketch	Photo	Average
DEEP ALL	0.5934	0.6280	0.4871	*0.8777*	0.6465
EDGE+APP	*0.6124*	0.6373	*0.5489*	**0.8780**	**0.6691**
JIGSAW [2]	0.6064	0.6316	0.5113	0.8774	0.6566
EPISODIC [6]	0.5744	*0.6539*	**0.5970**	0.8158	0.6603
GRADREV [3]	0.5854	0.6314	0.5486	0.8667	0.6580
OURS	**0.6215**	**0.6561**	0.5194	0.8744	*0.6678*

- DEEP ALL: Simply pools data from different source modalities.
- EDGE+APP: We replace the MAT filters in OURS with learnable filters of size 11 × 11 **not constrained** to be circles, in a single convolutional layer.
- JIGSAW [2] solves a self-supervision task (of predicting how pieces of an image were shuffled) in addition to the main classification task.
- EPISODIC training [6] exposes a network to domain shifts through three complementary strategies. It trains multiple domain-specific networks at training time, but only a single network at test time.
- GRADREV [3] negates the gradient from a domain classifier, to encourage similar features in different domains. We perform it over the source domains.

Setup: We use the PACS and Office-Home datasets. **PACS** [5] contains 9,991 images from seven classes and four modalities. **Office-Home** [9] contains 15,500 images from 65 categories and four modalities. The source set contains three domains/modalities used during the training process, and the target modality is only used to test. In all experiments 90% of images in source modalities are used for training, and 10% are used for validation.

Results: For AlexNet on PACS (Table 1), our shape representation achieves competitive results compared to EDGE+APP even though **the number of parameters in Edge+App is almost 2x** OURS. EPISODIC, JIGSAW and GRADREV, are 1–2% stronger than DEEP ALL but 1–2% **weaker than our**

Table 2. PACS results using ResNet-18. Best method in bold, second-best in italics.

Method/Target	Painting	Cartoon	Sketch	Photo	Average
DEEP ALL	0.7278	0.6941	0.5964	0.9497	0.7420
EDGE+APP	0.7402	0.7096	0.6305	*0.9539*	0.7585
JIGSAW [2]	0.7303	0.6925	0.6376	0.9493	0.7524
EPISODIC [6]	**0.7727**	**0.7205**	*0.6475*	0.9032	*0.7609*
OURS	*0.7529*	*0.7201*	**0.6478**	**0.9561**	**0.7692**

Table 3. Office-Home results using ResNet-18.

Method/Target	Art	Clipart	Product	Real-World	Average
DEEP ALL	0.5807	0.4680	*0.7260*	0.7582	0.6332
EDGE+APP	*0.5906*	0.4775	**0.7265**	*0.7612*	*0.6389*
JIGSAW [2]	0.5786	0.4547	0.7182	0.7497	0.6253
EPISODIC [6]	0.5854	*0.4830*	0.7165	0.7449	0.6324
OURS	**0.5929**	**0.4847**	0.7259	**0.7631**	**0.6416**

method on average. For ResNet on PACS and Office-Home, OURS achieves the best average performance. EPISODIC and OURS are similar in Table 2 for two domains, but on average OURS is stronger. EDGE+APP and OURS are similar in Table 3 for three domains, but on average OURS is stronger. **Our method is consistently 1–3% better** than DEEP ALL in all experiments and it achieves very competitive results compared to the DG methods, without the additional losses or specialized training of [2,3], [6]. We drop GRADREV from Table 2 and 3 for space.

Acknowledgement. Our work was funded by UPitt CRDF, Google & Amazon.

References

1. Blum, H.: A transformation for extracting new descriptors of shape. In: Wathen-Dunn, W. (ed.) Models for the Perception of Speech and Visual Form, pp. 362–380. MIT Press, Cambridge (1967)
2. Carlucci, F.M., D'Innocente, A., Bucci, S., Caputo, B., Tommasi, T.: Domain generalization by solving jigsaw puzzles. In: CVPR (2019)
3. Ganin, Y., Lempitsky, V.: Unsupervised domain adaptation by backpropagation. In: ICML (2015)
4. Geirhos, R., Rubisch, P., Michaelis, C., Bethge, M., Wichmann, F.A., Brendel, W.: ImageNet-trained CNNs are biased towards texture. In: ICLR (2019)
5. Li, D., Yang, Y., Song, Y.Z., Hospedales, T.M.: Deeper, broader and artier domain generalization. In: ICCV (2017)
6. Li, D., Zhang, J., Yang, Y., Liu, C., Song, Y.Z., Hospedales, T.M.: Episodic training for domain generalization. In: ICCV (2019)

7. Radenović, F., Tolias, G., Chum, O.: Deep shape matching. In: Ferrari, V., Hebert, M., Sminchisescu, C., Weiss, Y. (eds.) ECCV 2018. LNCS, vol. 11209, pp. 774–791. Springer, Cham (2018). https://doi.org/10.1007/978-3-030-01228-1_46
8. Rezanejad, M., et al.: Scene categorization from contours. In: CVPR (2019)
9. Venkateswara, H., Eusebio, J., Chakraborty, S., Panchanathan, S.: Deep hashing network for unsupervised domain adaptation. In: CVPR (2017)

Bi-Dimensional Feature Alignment
for Cross-Domain Object Detection

Zhen Zhao[1], Yuhong Guo[1,2(✉)], and Jieping Ye[1]

[1] DiDi Chuxing, Beijing, China
{alexzhaozhen,yejieping}@didiglobal.com
[2] Carleton University, Ottawa, Canada
yuhong.guo@carleton.ca

Abstract. Recently the problem of cross-domain object detection has started drawing attention in the computer vision community. In this paper, we propose a novel unsupervised cross-domain detection model that exploits the annotated data in a source domain to train an object detector for a different target domain. The proposed model mitigates the cross-domain representation divergence for object detection by performing cross-domain feature alignment in two dimensions, the depth dimension and the spatial dimension. In the depth dimension of channel layers, it uses inter-channel information to bridge the domain divergence with respect to image style alignment. In the dimension of spatial layers, it deploys spatial attention modules to enhance detection relevant regions and suppress irrelevant regions with respect to cross-domain feature alignment. Experiments are conducted on a number of benchmark cross-domain detection datasets. The empirical results show the proposed method outperforms the state-of-the-art comparison methods.

Keywords: Domain adaptation · Object detection · Style · Attention

1 Introduction

The deployment of supervised deep learning models has led to great advance in many computer vision tasks such as image classification [29], object detection [11,12,22,25], and image segmentation [35]. However, their success relies on the assumptions of standard supervised learning; that is, the deep models need to be trained with a sufficient amount of i.i.d. labeled samples that come from the same distribution as the test data. In practice, due to factors such as the collection means or weather conditions, the operational test dataset can be different from the training dataset, which can significantly degrade the performance of image analysis systems. For example, Fig. 1 presents the direct deployment result of an object detector trained in one domain, Cityscapes, and applied in another domain, Foggy Cityscapes. It shows the detection model trained with images collected in normal weather fails to detect many objects on images collected in foggy weather. Although one can solve this problem by collecting labeled data from the same test dataset, the data annotation/labeling process is typically

© Springer Nature Switzerland AG 2020
A. Bartoli and A. Fusiello (Eds.): ECCV 2020 Workshops, LNCS 12535, pp. 671–686, 2020.
https://doi.org/10.1007/978-3-030-66415-2_46

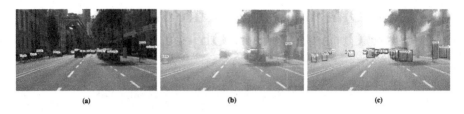

(a)   (b)   (c)

Fig. 1. Example of deployment of supervised object detector with different training and test datasets, Cityscapes and Foggy Cityscapes. (a) Labeled image example from the training set, Cityscapes. (b) The detection result on an image in Foggy Cityscapes using the detector trained on Cityscapes. (c) The ground-truth annotation of the Foggy Cityscapes example. This example shows that weather-induced domain gaps can lead to performance degradation.

time-consuming and expensive. To avoid the expensive needs of repeatedly collecting labeled images, many unsupervised domain adaptation methods have been developed for image segmentation and classification tasks to overcome the cross-domain performance degradation [3,8,23,30,31,34]. However, much less effort has been devoted to the more complex cross-domain object detection task.

As a detector needs to identify both the objects and their precise locations in an image, it is more challenging to design an effective cross-domain detector than a cross-domain classifier. One early work for adaptive object detection [1] adopts a domain adversarial feature alignment strategy at both the global image level and the proposal instance level. However, global image feature alignment is more effective to domain shifts over image appearances and textures, while unsuitable for handling cross-domain spatial distribution divergences. A more recent work [26] improves adaptive detection by deploying strong cross-domain alignment at low-level features such as local textures/colors and weak alignment at high-level global image features. Nevertheless, this work still fails to explore the spatial properties of features which are essential for object detection.

In this paper, we propose a novel end-to-end deep learning model for cross-domain object detection by aligning features from the source and target domains in both the depth and spatial dimensions. Our assumption is that the image representation can be captured from the perspectives of both the semantic contents (e.g.., the objects contained in the image) and the style of the image, and hence cross-domain feature alignment should be addressed from both aspects. Following previous work [9], we represent the style of an image using the inter-channel Gram matrix computed over features in the depth dimension of the feature map, which captures the correlations between the different filter responses along the spatial dimension, and can be adversarially aligned across domains. For cross-domain content feature alignment, we propose to use an attention module along the spatial dimension to enhance features in important regions (e.g.., regions with objects) and suppress features in irrelevant background areas. This attention module not only will guide the domain adaption model to form a region-sensitive domain adversarial feature alignment, but also will be added into the feature representations of the backbone network to facilitate the consequent region proposal and local object classification steps of the detector. Overall the contribution of this work can be summarized as follows: (1) This is the first domain adaptation work that performs cross domain semantic content and style feature alignments separately and simultaneously in the spatial and depth dimensions. (2) We deploy a novel spatial attention module to achieve target region sensitive cross-domain feature alignment. (3) We conduct extensive experiments on benchmark cross-domain detection datasets and the proposed model achieves the state-of-the-art performance.

2   Related Work

Object Detection. The development of convolutional neural networks (CNN) has led to great advance in object detection. Traditional object detection methods use sliding windows and manual feature classification designs [5,7]. In recent years, a two-stage detection strategy based on region of interest (ROI) has gained wide applicability [11,12,25]. The early RCNN model [12] uses selective search to generates a set of region proposals for object detection. Fast-RCNN [11] improves RCNN by identifying region proposals and deploying ROI pooling on the convolutional feature map of CNN. Faster-RCNN [25] combines Region Proposal Network (RPN) and Fast-RCNN to replace the previous selective search and further improve the detection performance. As a landmark detection model, Faster-RCNN provides the basis for many subsequent research studies [14,20,22,24]. This paper and many related unsupervised domain adaptive object detection methods also use Faster-RCNN as the backbone detection model.

Unsupervised Domain Adaptation. Unsupervised domain adaptation, which aims to train a model in a label-rich source domain for using in an unlabeled target domain, has attracted a lot of attention in the computer vision community. Many works have tried to learn cross-domain aligned feature representations [2,6,8,23,28]. The work in [8] used a gradient reversal layer (GRL) to achieve the adversarial feature alignment operation. The authors of [23] proposed a conditional adversarial domain adaptive method by using category predictions as an additional input for the domain discriminator. The work in [2] used an image generation mechanism to achieve cross-domain transformation through image pixel-level alignment. There are also some studies that perform domain adaptation by minimizing various feature distribution distances between different domains, such as the maximum mean discrepancy(MMD) [6] and the Wasserstein distance [28]. However, most of these studies focus on image classification and segmentation tasks.

Domain Adaptation for Object Detection. Although there are many works on cross-domain image classification and segmentation, domain adaption for object detection has just begun to receive attention. One relatively early work [1] proposed to align image-level features and instance-level features with adversarial domain adaptation strategy for adaptive object detection. The work in [16]

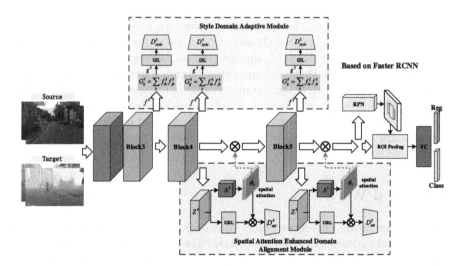

Fig. 2. The model structure of the proposed multi-dimensional domain adaptive object detection method, the Style and Spatial Attention enhanced feature alignment method for Domain Adaptive detection (SSA-DA). The proposed model uses Faster-RCNN as its backbone network and has two major adaptive components: the style domain adaptive module and the spatial attention enhanced domain alignment module.

used image pixel-level transitions and pseudo-labeling to achieve cross-domain weakly supervised target detection. Another work [19] used Cycle-GAN [37] to generate multiple intermediate domain images between the source domain and the target domain to learn domain invariant representations. The work [26] proposed a multi-level adversarial feature alignment strategy, global weak alignment and local strong alignment, to improve cross-domain detection performance. Multi-level alignment is also adopted in [33], which aligns the distributions of local features and global features simultaneously. while another work [38] focused on selective alignment of related areas. The work [36] performs conditional adversarial global feature alignment with dual multi-label prediction. The authors of [15] adopted the idea of layered alignment by adding proportional reduction and weighted gradient inversion layers to achieve domain invariance. Different from these existing methods, our proposed approach induce domain invariant features by enforcing not only multi-level alignments, but also multi-dimensional alignments.

3   Methodology

In this section, we consider the unsupervised cross-domain object detection problem, where the source domain contains fully labeled data and the data in the target domain is entirely unannotated. Let X_s denote the fully annotated source domain data, such that $X_s = \{(x_i^s, \mathbf{b}_i^s, \mathbf{c}_i^s)\}_{i=1}^{n_s}$, where x_i^s denotes the i-th image, $\mathbf{b}_i^s$ and $\mathbf{c}_i^s$ represent the bounding box coordinates and the corresponding labels respectively for the objects contained in the i-th image. Let $X_t = \{x_i^t\}_{i=1}^{n_s}$ represent the unannotated images from the target domain. We aim to develop a good cross-domain detection method that trains an object detector on these available image resources to perform well in the target domain.

In this work we propose a bi-dimensional feature alignment method, a Style and Spatial Attention enhanced feature alignment method for Domain Adaptive detection (SSA-DA). The main idea of SSA-DA is to model the cross-domain style representation divergence and semantic content representation divergence separately by aligning image styles across domains in the depth dimension with style domain adaptive modules and aligning detection effective features in the spatial dimension with spatial attention enhanced domain alignment modules. SSA-DA adopts the widely used Faster-RCNN as the backbone detection network, while the feature alignments can be conducted in multiple layers of the feature extraction subnetwork. The overall structure of the proposed SSA-DA model is illustrated in Fig. 2. As shown in Fig. 2, a style domain adaptive module is added after the 3rd, 4th, and 5th convolution blocks separately, and a spatial attention enhanced domain alignment module is added after the 4th and 5th convolution blocks separately. The details of the two types of adaptive alignment modules and the overall learning problem will be introduced below.

3.1   Depthwise Style Domain Adaptive Module

Image style is an important aspect of the image representation. Variations in image styles can hinder the cross-domain object detection performance. Following a previous work [9], we build the style representation of an image in a feature space that captures texture information, by calculating inter-channel feature correlations, i.e., correlations between different filter responses, in the feature map produced in any layer of the feature extraction network. Let the feature map obtained for a given image x after the l-th convolution block in the backbone of Faster-RCNN be expressed as $Z^l = F^l(x) \in \mathbb{R}^{C^l \times H^l \times W^l}$, where C^l denotes the number of channels (i.e., filters), H^l and W^l denote the spatial dimensions. Each channel contains the responses of the corresponding convolution filter of the current layer. To facilitate calculation, we transform Z^l into a two-dimensional matrix $f^l \in \mathbb{R}^{C^l \times M^l}$ with $M^l = H^l \cdot W^l$, such that the rows and columns of the matrix represent the channel and spatial dimensions respectively. Then the style features can be calculated as a Gram matrix $G^l \in \mathbb{R}^{C \times C}$, such that

$$G_{ij}^l = \sum_k f_{ik}^l f_{jk}^l \tag{1}$$

Each entry G_{ij}^l captures the inter-channel correlations between the i-th and j-th channels. For simplicity, we can further reshape G^l into a vector form $g^l \in \mathbb{R}^{C^{l2} \times 1}$, which contains the style features produced. Such style features capture the texture information but not the location arrangement.

To overcome the cross-domain style variation, we propose to align the style features across domains at a given layer with an adversarial domain adaptation mechanism based on the generative adversarial network (GAN), which can effectively align two distributions [13]. As shown in Fig. 2, a style domain adaptive module is used to induce style-invariant feature representations at multiple convolution blocks. At the l-th block, a domain discriminator D_{style}^l is introduced to predict the domain of an input image style feature vector g^l, with $D_{style}^l(g^l)$ denoting the predicted probability of g^l coming from the source domain. The feature alignment can be achieved through a min-max game between the feature extractor F^l and the domain discriminator D_{style}^l:

$$\min_{F^l} \max_{D_{style}^l} \quad L_{style}^l = \frac{1}{2}(L_{style}^{ls} + L_{style}^{lt}) \tag{2}$$

$$L_{style}^{ls} = \mathbb{E}_{x_s \in X_s}\left[(1 - D_{style}^l(g_s^l))^\gamma \log(D_{style}^l(g_s^l))\right] \tag{3}$$

$$L_{style}^{lt} = \mathbb{E}_{x_t \in X_t}\left[(D_{style}^l(g_t^l))^\gamma \log(1 - D_{style}^l(g_t^l))\right] \tag{4}$$

Here we adopted the focal loss [21,26] in this adversarial training objective to give more weights to examples that are hard to classify by D_{style}^l, and γ is a modulation factor that controls the contribution degree of the hard examples. g_s^l represents the style feature vector generated from the convolutional feature map $F_l(x_s)$ from the source domain and g_t^l represents the style feature vector generated from $F_l(x_t)$ from the target domain. In the adversarial min-max game, the discriminator D_{style}^l tries to maximally discriminate the source domain features from the target domain features, where the feature extractor F^l tries to induces style features such that the discriminator can be maximally confused.

The style representation is a multi-scale representation involving multiple layers of the deep network. The style features obtained from lower level layers reflect more pixel level information, while the style features from higher level layers reflect more image structural information [9,17]. Therefore, we apply adversarial style feature alignment on multiple convolution blocks to obtain stable and multi-scale style domain adaptation of image features. In particular, in the proposed model, one style domain adaptive module is added to each of the convolution block 3, block 4 and block 5 of the backbone network of the Faster-RCNN before the region proposal network (RPN). The overall multi-level style adversarial training loss L_{style} can be summarized as follows:

$$L_{style} = \sum_{l=3}^{5} \min_{F^l} \max_{D_{style}^l} L_{style}^l \tag{5}$$

3.2 Spatial Attention Domain Alignment Module

Image features that reflect the semantic contents of images are essential for object detection and it is important to bridge domain divergence over the image content features. In general, an object is usually localized into some local region in an image and the detector only needs to recognize the relevant regions. Therefore directly aligning the global image features across domains might not be the most suitable strategy as these features can be dominated by irrelevant regions, especially when the objects are small. To address this problem, we propose a spatial attention domain alignment module, as shown in Fig. 2, which learns spatial attention to enhance features from the semantically relevant regions. Specifically, given the feature map produced from the l-th convolution block, $Z^l = F^l(x)$, we follow the method of [32] to generate a spatial attention map $\phi^l \in \mathbb{R}^{1 \times H^l \times W^l}$ using an attention network A^l, such that $\phi^l = A^l(Z^l)$, where A^l is a convolution network with filters of 7×7. Then the spatial attention enhanced feature map can be produced as $Z_\phi^l = \phi^l \otimes Z^l$, where $\otimes$ denotes a replicated element-wise product operator that multiplies ϕ^l to each channel of Z^l. It is expected that the learned spatial attention can enhance features in relevant regions of Z^l and diminish features in irrelevant regions.

Given the spatial attention enhanced feature map Z_ϕ^l, we introduce a domain discriminator D_{att}^l to perform adversarial cross-domain feature alignment. Similar to the adversarial alignment on style features, we use focal loss in the adversarial objective, such that the min-max adversarial optimization problem is formulated as:

$$\min_{F^l} \max_{D_{att}^l} \quad L_{att}^l = \frac{1}{2}(L_{att}^{ls} + L_{att}^{lt}) \tag{6}$$

$$L_{att}^{ls} = \mathbb{E}_{x_s \in X_s}\left[(1 - D_{att}^l(Z_{\phi_s}^l))^\varepsilon \log(D_{att}^l(Z_{\phi_s}^l)) \right] \tag{7}$$

$$L_{att}^{lt} = \mathbb{E}_{x_t \in X_t}\left[(D_{att}^l(Z_{\phi_t}^l))^\varepsilon \log(1 - D_{att}^l(Z_{\phi_t}^l)) \right] \tag{8}$$

where ε is a modulation factor hyperparameter for the focal loss; $Z_{\phi_s}^l$ and $Z_{\phi_t}^l$ are the spatial attention enhanced feature maps from the l-th block for the source domain image x_s and the target domain image x_t respectively. Moreover, as previous work [9,17] has shown effective semantic information that can characterize image content is often available in the deep layers of a convolutional network instead of the shallow layers, we propose to perform multi-level spatial attention enhanced domain alignment on the last two convolution blocks of the Faster-RCNN, i.e., block 4 and block 5 in Fig. 2. The overall spatial attention enhanced adversarial training loss L_{att} can be written as follows:

$$L_{att} = \sum_{l=4}^{5} \min_{F^l} \max_{D_{att}^l} L_{att}^l \tag{9}$$

In this min-max adversarial training, the feature extraction network $\{F^4, F^5\}$ will tries to maximally confuse the domain discriminators $\{D_{att}^4, D_{att}^5\}$ and align the spatial attention enhanced features across domains.

Meanwhile, the spatial attention enhanced feature map Z_ϕ^l at each block will be used as the input for the next block along the backbone of the detection network. The attention enhanced feature map, Z_ϕ^5, at the last convolution block, block 5, will be provided to the region proposal network (RPN) to produce region proposals and perform object classification and bounding box regression.

The object detection loss L_{det} in the source domain can be expressed as:

$$L_{det} = \frac{1}{n_s} \sum_{i=1}^{n_s} L_{cr}(R(Z_{\phi_i}^5), (\mathbf{b}_i^s, \mathbf{c}_i^s)) \tag{10}$$

where R denotes the combined function for the RPN, region classification and regression modules of the Faster-RCNN, and L_{cr} represents all the supervised classification loss and regression loss.

3.3   Overall Adversarial Learning

We combine object detection loss L_{det}, style adversarial training loss L_{style}, and the spatial attention enhanced adversarial training loss L_{att} together to form the following overall adversarial learning objective:

$$L_{all} = L_{det} + \lambda L_{style} + \mu L_{att} \tag{11}$$

where λ and μ are the trade-off parameters to balance different loss terms. This overall learning problem minimizes the detection loss on the labeled source domain data, while bridging the representation gap between the source and target domains from both the style and content perspectives. SGD optimization algorithm is used to perform training, while GRL [8] is adopted to implement the gradient sign flip for the domain discriminator update.

4   Experiments

To evaluate the proposed SSA-DA model, we conducted experiments on benchmark cross-domain detection tasks in three cross-domain variation scenarios: (1) Normal to foggy weather variation. In this scenario, we used the cross-domain detection task of adapting from Cityscapes [4] to Foggy Cityscapes [27]. (2) Virtual to real scene variation. The adaptive detection task from SIM-10K [18] to Cityscapes [4] is used in this scenario. (3) Cross-camera situation. We used data collected with two different cameras to form the cross-domain detection task from KITTI [10] and Cityscapes [4]. We compared our proposed model with the state-of-the-art cross-domain detection methods. In this section, we present our experimental results and discussions.

4.1   Experimental Setup

We followed the same experimental setup as in [1]. The VGG16 [29] model is used as the backbone of the Faster-RCNN detection model and pre-trained on ImageNet. We set the momentum as 0.9, the weight decay as 0.0005, and the total training epoch number as 20. The domain discriminator D_{style}^l has three fully connected layers, while the discriminator D_{att}^l has one convolutional layer and two fully connected layers. For the hyperparameters involved in the proposed method, we set $\lambda = 1$, set γ to 5 on all blocks, and set ε to 5 on block 5, and 4 on block 4. For all the experiments, we used the mean average precision (mAP) with a threshold of 0.5 to evaluate the results.

Table 1. Detection results on the validation set of the Foggy Cityscapes. SD and SA denote the two major components of the proposed SSA-DA: SD denotes style domain adaptive component, and SA denotes spatial attention enhanced feature alignment.

Method	SD	SA	Prson	Rider	Car	Truck	Bus	Train	Mcycle	Bicycle	mAP
Source-only			25.1	32.7	31.0	12.5	23.9	9.1	23.7	29.1	23.4
BDC-Faster [26]			26.4	37.2	42.4	21.2	29.2	12.3	22.6	28.9	27.5
DA-Faster [1]			25.0	31.0	40.5	22.1	35.3	20.2	20.0	27.1	27.6
SC-DA(Type3) [38]			33.5	38.0	**48.5**	26.5	39.0	23.3	28.0	33.6	33.8
MAF [15]			28.2	39.5	43.9	23.8	39.9	33.3	29.2	33.9	34.0
SW-DA [26]			29.9	42.3	43.5	24.5	36.2	32.6	30.0	35.3	34.3
DD-MRL [19]			30.8	40.5	44.3	27.2	38.4	34.5	28.4	32.2	34.6
Dense-DA [33]			33.2	44.2	44.8	28.2	41.8	28.7	30.5	36.5	36.0
SSA-DA	✓		33.3	46.2	44.0	31.1	47.7	36.4	36.1	36.4	38.9
		✓	32.7	47.5	44.9	**36.2**	43.7	23.4	38.0	36.5	37.8
	✓	✓	**33.9**	**48.3**	47.7	35.7	**52.0**	**44.7**	**39.6**	**37.9**	**42.5**

4.2 Normal to Foggy Weather Adaptation

For object detection in real road scenarios, weather condition is a common factor that affects the detection performance. The change of weather conditions can lead to large visual variations in images and videos, which presents an obvious domain shift situation for the deployment of object detectors. To test the proposed adaptive detection model in this scenario, we used the cross-domain detection task from Cityscapes [4] to Foggy Cityscapes [27]. These two datasets have eight object categories: person, rider, car, truck, bus, train, motorcycle and bicycle. Foggy Cityscapes is a fog dataset synthesized from Cityscapes, which can simulate the fog weather condition in real scenes. The Cityscapes are used as the source domain, and the training set of Foggy Cityscapes is used as the target domain. We set the hyperparameter $\mu = 0.5$ in the experiment, and report detection results for all categories on the Foggy Cityscapes validation set.

We compared the proposed SSA-DA method with seven state-of-the-art cross-domain detection methods and one baseline source-only training method. The comparison results are reported in Table 1. As the baseline Source-only is only trained in the source domain without handling the domain shift problem, we can see that all the other domain adaptive detection methods outperform Source-only. By having both style and spatial attention enhanced feature alignments, the proposed SSA-DA greatly improves the cross-domain detection performance, far exceeding all other comparison methods. It outperforms the Source-only baseline by 19.1% in terms of the average mAP. As the domain shift is caused by the fog in this task, there is a significant stylistic difference across domains. We can see that by using the style domain adaptive component (SD) alone in SSA-DA, it has already outperformed the best comparison method by 2.9% in terms of average mAP. Meanwhile by using only the spatial attention enhanced feature alignment component (SA), SSA-DA works very well in the 'truck' category, where all the other comparison methods have poor performance. This suggests

Table 2. Detection results on adaptation from SIM-10k to Cityscapes. SD and SA denote the two major components of the proposed SSA-DA: SD denotes style domain adaptive component, and SA denotes spatial attention enhanced feature alignment.

Method	SD	SA	AP of car
Source-only			34.3
BDC-Faster [26]			31.8
DA-Faster [1]			39.0
MAF [15]			41.1
SW-DA [26]			40.1
SW-DA ($\gamma = 3$) [26]			42.3
SC-DA (Type3) [38]			43.0
Dense-DA (n = 6) [33]			42.8
SSA-DA	✓		42.0
		✓	42.4
	✓	✓	43.8

that the 'truck' object under the condition of fog is very difficult to capture, while the spatial attention mechanism can help mitigate the problem. However, excessive attention alone may have a negative impact on the category of 'train', while the style feature alignment component can help to mitigate this drawback. Overall, by integrating both the SD and SA components, the proposed SSA-DA approach demonstrates great performance.

4.3   Virtual to Real Scene Adaptation

As it is difficult to collect annotated data in many application tasks, it is a good option to use computer-generated labeled virtual image data for training models. However due to the visual difference between the virtual data and real data, the performance of the detection model trained on the virtual data can severely degrade when applying to the real data, hence cross-domain detection techniques are important. In this experiment, we tested the proposed SSA-DA method on the domain adaptive detection task from virtual scenes to real scenes. In particular, we adopted the virtual scene dataset SIM-10K [18] as the source domain, and took the real scene dataset Cityscapes [4] as the target domain, while using the car category detection as the domain adaptive detection task. All training images from both domains were used during training, test evaluation was conducted on the validation set of Cityscapes. Following [1], we set the trade-off hyperparameter $\mu = 0.1$.

Same as above, we compared the proposed SSA-DA with both the Source-only baseline and a number of state-of-the-art methods which were tested on this cross-domain detection task. The comparison results are reported in Table 2. We can see that our proposed SSA-DA improves the performance of the Source-only

Table 3. Detection results of cross camera adaptation from KITTI and Cityscapes. SD and SA denote the two major components of SSA-DA: SD denotes style domain adaptive component, and SA denotes spatial attention enhanced feature alignment.

Method	SD	SA	AP of car
Source-only			30.2
DA-Faster [1]			38.5
SC-DA (Type3) [38]			42.5
MAF [15]			41.0
SSA-DA	✓		42.6
		✓	42.2
	✓	✓	43.3

baseline model by 9.5%, and exceeds the best results of all the other cross-domain detection models. It can also be observed that even with only one of the two components, SD and SA, SSA-DA can still reach a good performance level and outperform some of the latest methods. In addition, we can also observe that the SC-DA(Type3) method produces the best result among the other comparison method and it outperforms the latest Dense-DA method. Meanwhile, SC-DA(Type3) also demonstrates an obvious advantage over other comparison methods in the category 'car' in the experiment of Sect. 4.2. This validates that SC-DA(Type3) is more suitable for small vehicle detection. Nevertheless, our proposed SSA-DA outperforms SC-DA(Type3). These results suggest that the proposed SSA-DA is very effective for cross-domain detection.

4.4   Cross-Camera Adaptation

Due to the variations in camera equipments and collection scenes, real road condition data acquired under similar weather conditions can also have a domain shift problem. In this experiment, we used two real datasets, KITTI [10] and Cityscapes [4], to study cross-domain object detection under cross-camera variations. Following [1], we used the KITTI dataset as the source domain, used the Cityscapes training set as the target domain, and evaluated the performance of adaptive detection models on the validation set of Cityscapes with the category 'car'. The experimental results are reported in Table 3. We can see that the proposed SSA-DA method produced the best result, which is 13.1% higher than the baseline, and outperforms even the more complex SC-DA(Type3) models that are suitable for automotive inspection.

4.5   Qualitative Result Visualization

In addition to the quantitative results reported above, we present an example of the qualitative adaptive detection results in Fig. 3. We can see that the Source-only baseline can only detect objects within close range and missed most objects

(a) Cityscapes   (b) Source-only   (c) DA-Faster   (d) SSA-DA

Fig. 3. Qualitative results of adaptation from Cityscapes to Foggy Cityscapes. (a) Annotated image in Cityscapes. (b) Detection results of Source-only in the target domain. (c) Detection results of DA-Faster. (d) Detection results of SSA-DA. The blue boxes show ground-truth and the green boxes show the detection results. (Color figure online)

far away. DA-Faster was able to detect cars that were a little further away, but it mistakenly classified the motorcycle, and missed the rider, person, as well as many other objects. The proposed SSA-DA model correctly detected motorcycle and rider, and detected more person and car objects in the dense fog.

4.6   Analysis of the SSA-DA Model

The proposed SSA-DA model has two major components, the style domain adaptive module (SD) and the spatial attention enhanced feature alignment module (SA). In this section, we will analyze the impact of these two components at multi-levels on the performance of the SSA-DA model using the adaptive detection task from Cityscapes to Foggy Cityscapes, as well as investigate the impact of the hyperparameters on the components and the full model.

Impact of the Component Modules. We use SD-DA to denote the variant of SSA-DA that drops the SA modules and keeps only the SD modules. Similarly, we use SA-DA to denote the variant of SSA-DA that drops the SD modules and keeps only the SA modules. Both modules are applied on multiple blocks (block 3, 4 and 5) of the backbone detection network, we hence further conducted ablation study to investigate their impacts on lower levels of convolution layers and upper levels of convolution layers. As the style features obtained at the lower levels of the network reflect more detailed pixel level information and the style features at the higher level reflect smooth structural information, we performed ablation study on SD-DA by adding the SD module from the lower level blocks to the higher level blocks. Semantic content information however is more prevalent at higher levels of the extraction network. Hence we performed ablation study on SA-DA by adding the SA module from the higher level blocks to the lower level blocks. The experimental results are reported in Table 4. We can see that the overall performance of the SD-DA gradually increase by adding the SD module into higher level blocks. This proves that simultaneous style feature alignments at multiple blocks from the low level to the high level of the network can benefit the domain adaptation performance. Meanwhile, for SA-DA, adding the SA module to the low level block 3 actually degrades its

Table 4. Detection results of SD-DA and SA-DA with deployment on different blocks.

Method	block5	block4	block3	Person	Rider	Car	Truck	Bus	Train	Mcycle	Bicycle	mAP
SD-DA			✓	32.3	44.4	43.7	28.8	42.6	22.7	33.3	38.2	35.8
		✓	✓	32.5	45.7	43.8	26.8	49.6	30.0	33.5	37.5	37.4
	✓	✓	✓	33.3	46.2	44.0	31.1	47.7	36.4	36.1	36.4	**38.9**
SA-DA	✓			30.6	40.0	40.5	26.4	37.4	27.3	30.0	33.4	33.2
	✓	✓		32.7	47.5	44.9	36.2	43.7	23.4	38.0	36.5	**37.8**
	✓	✓	✓	33.9	47.2	44.9	35.5	41.2	11.9	40.0	35.6	36.3

overall performance. It has a significant negative impact on the 'train' category. This suggests that with spatial attention, it is more suitable to conduct semantic content feature alignment at higher levels of the network.

Parameter Sensitivity on γ and ε. These two parameters are modulation factors for the focal loss used in the adversarial alignment of style features (SD module) and spatial content features respectively (SA module). As they are separately involved in the two modules, we conducted sensitivity experiments for γ and ε using SD-DA and SA-DA respectively.

As the style feature alignment has been shown to be useful at both low and high level blocks, we set γ to the same value for the SD modules added to block 3, 4 and 5. We tested the SD-DA variant by varying the γ within the range of values $[3, 4, 5, 6]$, and the results are reported in Fig. 4(a). We can see SD-DA outperforms Dense-DA for different γ values, especially when $3 \leq \gamma \leq 5$, while the best result is achieved with $\gamma = 5$.

From previous experiments, we can see that the SA module is more suitable for higher level blocks. Hence here we separately investigated the best ε value for the SA module used in block 4 and block 5. First we use a variant, SA-DA(5), which only adds SA module to block 5, to conduct sensitivity experiments by varying ε within the range of values $[3, 4, 5, 6]$. The results are reported in Fig. 4(b). We can see adding SA module only to block 5 leads to degraded results comparing to the full SA-DA. The performance varies with different ε values while the best result is gained with $\varepsilon = 5$. Then we fixed $\varepsilon = 5$ for the SA module in block 5, and tested the full SA-DA method by varying ε in block 4 within the range of values $[3, 4, 5, 6]$. The results are reported in Fig. 4(c). We can see that when $\varepsilon = \{2, 3, 4\}$ on block4, the performance remains at a high level, but will drop when ε continues to increase. Overall, these results suggest γ and ε should not be set to big values as they may overfit the hard examples. A value no larger than 5 would be more suitable.

Parameter Sensitivity on λ and μ. In addition, we conducted experiments on the complete model SSA-DA by adjusting the trade-off parameters λ and μ in Eq. (11). These two parameters control the weight of style domain adaption (SD) and spatial attention enhanced feature alignment (SA) in the detection model. To vary the λ value, we fixed $\mu = 0.5$; to vary the μ value, we fixed $\lambda = 1$. The sensitivity results are reported in Fig. 5. We can see that although the performance varies with different λ and μ values, the performance in general

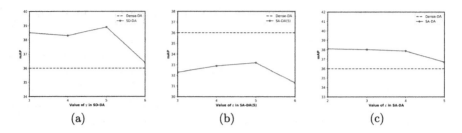

(a)   (b)   (c)

Fig. 4. Parameter sensitivity on γ and ε with adaptation from Cityscapes to Foggy Cityscapes.(a): Sensitivity results over γ with SD-DA. (b): Sensitivity results over ε with SA-DA(5). (c): Sensitivity results over ε with SA-DA.

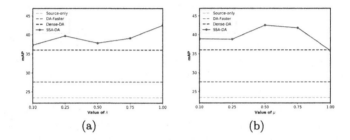

(a)   (b)

Fig. 5. Parameter sensitivity analysis on λ and μ with the adaptation from Cityscapes to Foggy Cityscapes. (a): Sensitivity results over λ. (b): Sensitivity results over μ.

is superior to Dense-DA and DA-Faster for most of range of values, while the best results are obtained when $\lambda = 1$ and $\mu = 0.5$.

5   Conclusion

In this paper, we proposed a novel style and spatial attention enhanced bi-dimensional feature alignment method for domain adaptive detection (SSA-DA). The proposed method deploys two important modules, the style domain adaptive module and the spatial attention enhanced domain alignment module, at multi-levels to align features in both the depth and spatial dimensions across domains. With both the style and spatial attention enhanced content feature alignments, the detector trained in the source domain can be more adaptive to the target domain. We conducted experiments on benchmark datasets in three different cross-domain variation scenarios. The experimental results demonstrated the proposed model achieved the state-of-the-art adaptive detection performance.

References

1. Chen, Y., Li, W., Sakaridis, C., Dai, D., Van Gool, L.: Domain adaptive faster R-CNN for object detection in the wild. In: CVPR (2018)

2. Choi, J., Kim, T., Kim, C.: Self-ensembling with gan-based data augmentation for domain adaptation in semantic segmentation. In: ICCV (2019)
3. Cicek, S., Soatto, S.: Unsupervised domain adaptation via regularized conditional alignment. arXiv preprint arXiv:1905.10885 (2019)
4. Cordts, M., et al.: The cityscapes dataset for semantic urban scene understanding. In: CVPR (2016)
5. Dalal, N., Triggs, B.: Histograms of oriented gradients for human detection. In: CVPR (2005)
6. Dziugaite, G.K., Roy, D.M., Ghahramani, Z.: Training generative neural networks via maximum mean discrepancy optimization. In: UAI (2015)
7. Felzenszwalb, P.F., Girshick, R.B., Mcallester, D., Ramanan, D.: Object detection with discriminatively trained part-based models. PAMI **32**(9), 1627–1645 (2010)
8. Ganin, Y., et al.: Domain-adversarial training of neural networks. JMLR **17**, 2030–2096 (2016)
9. Gatys, L.A., Ecker, A.S., Bethge, M.: Image style transfer using convolutional neural networks. In: CVPR (2016)
10. Geiger, A., Lenz, P., Urtasun, R.: Are we ready for autonomous driving? The KITTI vision benchmark suite. In: CVPR (2012)
11. Girshick, R.: Fast R-CNN. In: ICCV (2015)
12. Girshick, R., Donahue, J., Darrell, T., Malik, J.: Rich feature hierarchies for accurate object detection and semantic segmentation. In: CVPR (2014)
13. Goodfellow, I., et al.: Generative adversarial nets. In: NIPS (2014)
14. He, K., Gkioxari, G., Dollár, P., Girshick, R.: Mask R-CNN. In: ICCV (2017)
15. He, Z., Zhang, L.: Multi-adversarial faster-RCNN for unrestricted object detection. In: ICCV (2019)
16. Inoue, N., Furuta, R., Yamasaki, T., Aizawa, K.: Cross-domain weakly-supervised object detection through progressive domain adaptation. In: CVPR (2018)
17. Johnson, J., Alahi, A., Fei-Fei, L.: Perceptual losses for real-time style transfer and super-resolution. In: Leibe, B., Matas, J., Sebe, N., Welling, M. (eds.) ECCV 2016, Part II. LNCS, vol. 9906, pp. 694–711. Springer, Cham (2016). https://doi.org/10.1007/978-3-319-46475-6_43
18. Johnson-Roberson, M., Barto, C., Mehta, R., Sridhar, S.N., Rosaen, K., Vasudevan, R.: Driving in the matrix: Can virtual worlds replace human-generated annotations for real world tasks? ICRA (2017)
19. Kim, T., Jeong, M., Kim, S., Choi, S., Kim, C.: Diversify and match: a domain adaptive representation learning paradigm for object detection. In: CVPR (2019)
20. Lin, T.Y., Dollár, P., Girshick, R., He, K., Hariharan, B., Belongie, S.: Feature pyramid networks for object detection. In: CVPR (2017)
21. Lin, T.Y., Goyal, P., Girshick, R., He, K., Dollár, P.: Focal loss for dense object detection. In: ICCV (2017)
22. Liu, W., et al.: SSD: single shot multibox detector. In: Leibe, B., Matas, J., Sebe, N., Welling, M. (eds.) ECCV 2016, Part I. LNCS, vol. 9905, pp. 21–37. Springer, Cham (2016). https://doi.org/10.1007/978-3-319-46448-0_2
23. Long, M., Cao, Z., Wang, J., Jordan, M.I.: Conditional adversarial domain adaptation. In: NIPS (2018)
24. Redmon, J., Divvala, S., Girshick, R., Farhadi, A.: You only look once: unified, real-time object detection. In: CVPR (2016)
25. Ren, S., He, K., Girshick, R., Sun, J.: Faster R-CNN: towards real-time object detection with region proposal networks. In: NIPS (2015)
26. Saito, K., Ushiku, Y., Harada, T., Saenko, K.: Strong-weak distribution alignment for adaptive object detection. In: CVPR (2019)

27. Sakaridis, C., Dai, D., Van Gool, L.: Semantic foggy scene understanding with synthetic data. Int. J. Comput. Vis. **126**(9), 973–992 (2018). https://doi.org/10.1007/s11263-018-1072-8

28. Shen, J., Qu, Y., Zhang, W., Yu, Y.: Wasserstein distance guided representation learning for domain adaptation. In: AAAI (2018)

29. Simonyan, K., Zisserman, A.: Very deep convolutional networks for large-scale image recognition. arXiv preprint arXiv:1409.1556 (2014)

30. Tsai, Y.H., Hung, W.C., Schulter, S., Sohn, K., Yang, M.H., Chandraker, M.: Learning to adapt structured output space for semantic segmentation. In: CVPR (2018)

31. Tsai, Y.H., Sohn, K., Schulter, S., Chandraker, M.: Domain adaptation for structured output via discriminative representations. In: ICCV (2019)

32. Woo, S., Park, J., Lee, J.-Y., Kweon, I.S.: CBAM: convolutional block attention module. In: Ferrari, V., Hebert, M., Sminchisescu, C., Weiss, Y. (eds.) ECCV 2018, Part VII. LNCS, vol. 11211, pp. 3–19. Springer, Cham (2018). https://doi.org/10.1007/978-3-030-01234-2_1

33. Xie, R., Yu, F., Wang, J., Wang, Y., Zhang, L.: Multi-level domain adaptive learning for cross-domain detection. In: ICCV (2019)

34. Zhang, Y., David, P., Gong, B.: Curriculum domain adaptation for semantic segmentation of urban scenes. In: ICCV (2017)

35. Zhao, H., Shi, J., Qi, X., Wang, X., Jia, J.: Pyramid scene parsing network. In: CVPR (2017)

36. Zhao, Z., Guo, Y., Shen, H., Ye, J.: Adaptive object detection with dual multi-label prediction. In: Vedaldi, A., Bischof, H., Brox, T., Frahm, J.-M. (eds.) ECCV 2020, Part XXVIII. LNCS, vol. 12373, pp. 54–69. Springer, Cham (2020). https://doi.org/10.1007/978-3-030-58604-1_4

37. Zhu, J.Y., Park, T., Isola, P., Efros, A.A.: Unpaired image-to-image translation using cycle-consistent adversarial networks. In: ICCV (2017)

38. Zhu, X., Pang, J., Yang, C., Shi, J., Lin, D.: Adapting object detectors via selective cross-domain alignment. In: CVPR (2019)

Bayesian Zero-Shot Learning

Sarkhan Badirli[1(✉)], Zeynep Akata[2], and Murat Dundar[3]

[1] Computer Science Department, Purdue University, West Lafayette, USA
s.badirli@gmail.com
[2] Cluster of Excellence Machine Learning, University of Tübingen,
Tübingen, Germany
[3] Computer and Information Science Department, IUPUI, Indianapolis, USA

Abstract. Object classes that surround us have a natural tendency to emerge at varying levels of abstraction. We propose a Bayesian approach to zero-shot learning (ZSL) that introduces the notion of meta-classes and implements a Bayesian hierarchy around these classes to effectively blend data likelihood with local and global priors. Local priors driven by data from seen classes, i.e., classes available at training time, become instrumental in recovering unseen classes, i.e., classes that are missing at training time, in a generalized ZSL (GZSL) setting. Hyperparameters of the Bayesian model offer a convenient way to optimize the trade-off between seen and unseen class accuracy. We conduct experiments on seven benchmark datasets, including a large scale ImageNet and show that our model produces promising results in the challenging GZSL setting.

Keywords: Generalized ZSL · Bayesian hierarchical models

1 Introduction

Natural images exhibit power-law property; hence, in a randomly sampled training set, no training examples are expected to be available for most of the object categories [11,23,31]. This restriction becomes more evident in a fine-grained object recognition task. Zero-shot learning (ZSL), which considers training and test classes, i.e. seen and unseen classes, as two disjoint sets, was introduced to mitigate this limitation [12,20]. The two groups of classes are linked through a shared set of attributes that characterize high level semantic descriptions of all classes. During the training phase, a mapping between examples of seen classes and their corresponding class-based attributes is learned. This mapping is later used to identify examples of unseen classes during the test phase.

The standard ZSL setting restricts test time search space to only unseen classes. This somewhat unrealistic stipulation was later relaxed in the generalized

Electronic supplementary material The online version of this chapter (https://doi.org/10.1007/978-3-030-66415-2_47) contains supplementary material, which is available to authorized users.

A. Bartoli and A. Fusiello (Eds.): ECCV 2020 Workshops, LNCS 12535, pp. 687–703, 2020.
https://doi.org/10.1007/978-3-030-66415-2_47

ZSL (GZSL) setting to include all classes during the test phase [26]. In GZSL, side information, i.e., attributes, are as important as the perceptual representation of images. Attribute vectors are either manually annotated [5,12] or derived from free-form text using word embedding [6,16,29]. Early line of work in ZSL [12] assumes attribute independence and uses probabilistic classifiers to assign images to test classes.

In this paper, we tackle ZSL by introducing a two layer Bayesian hierarchy manifesting over both seen and unseen classes. Our approach is designed to leverage the implicit hierarchy present among classes, especially evident in fine grained data sets [18,21,33]. Unlike earlier approaches, which seek to optimize an embedding between image and semantic spaces, the proposed method assumes that there are latent classes that define the class hierarchy in image space and uses semantic information to build the Bayesian hierarchy around these meta-classes.

Our model uses two types of Bayesian priors: global and local. As the name suggests, global priors are shared across all classes, whereas local priors are only shared among semantically similar classes, which are identified based on the distances between attribute vectors in the Euclidean space. Unlike standard Bayesian models where the posterior predictive distribution establishes a compromise between prior and likelihood, our approach utilizes posterior predictive distributions to reconcile information about local and global priors as well as the likelihood to more effectively accommodate the class hierarchy. In this framework, unseen classes are represented by their corresponding meta classes (see Fig. 1), and test samples are classified based on posterior predictive likelihoods computed for both seen and unseen classes. Our approach achieves significant

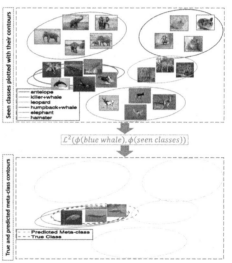

Fig. 1. Meta-classes illustrated in 2D PCA space (reduced from 2048). Only a small subset of seen classes are shown. Contours are derived from class covariance matrices and placed at two standard deviations away from class means. Meta class for *blue whale* (unseen) predicted based on *killer* and *humpback* whales (seen). (Color figure online)

improvements on both seen and unseen class accuracies to achieve the best results on a variety of benchmark datasets among the currently published state of the art methods.

Our contributions are as follows. (1) We propose a hierarchical Bayesian model based on the intuition that actual classes originate from their corresponding local priors, each defined by a meta-class of its own. (2) We derive the posterior predictive distribution (PPD) for a two-layer Gaussian mixture model to effectively blend local and global priors with data likelihood. These PPDs are used to implement a maximum-likelihood classifier, which represents seen classes by their own PPDs and unseen classes by meta-class PPDs. (3) Across seven datasets with varying granularity and sizes, in particular on the large-scale ImageNet dataset, we show that the proposed model is highly competitive against existing inductive techniques in the GZSL setting.

2 Related Work

In this section, we discuss the prior work on zero-shot learning and hierarchical generative models related to ours.

Zero-Shot Learning. There has been an increasing interest in classifying fine grained and large-scale image datasets [18,21,23,33]. This, in turn, led to a surge of interest in ZSL as labeling them is extremely costly. In their seminal paper [12], authors tackle ZSL by implementing a probabilistic classifier for each attribute and then classifying test cases by aggregating attribute probabilities for each class. This approach treats attributes as independent, which is a fairly strong assumption for most real-world data sets. This work was followed by a large body of work that seeks to optimize a mapping from image space, i.e., feature vectors, onto semantic space, i.e., attribute vectors. This line of work can be categorized into two according to whether the mapping is bi-linear [1,2,6,10,22] or non-linear [29,34]. Related to ours, [19,25,39] first maps image and semantic space into an intermediate space and represents unseen classes as a mixture of seen classes. Besides these mainline ZSL studies, a recent study evaluates an extended version of a few-shot learning algorithm for ZSL [30]. This approach learns a deep metric to query images with few shot samples. Extension to ZSL is achieved by replacing few-shot samples with one-shot class attribute vectors.

Generative Models for ZSL. Although most of the early work focused on discriminative models, there are a few studies that use generative models to tackle ZSL [17,32]. The study in [17] uses Normal distributions to model both image features and semantic vectors and learns a multimodal mapping between two spaces. This mapping is optimized by minimizing a similarity based cross domain loss function. In a similar fashion the study in [32] utilizes a regression model to optimize a mapping between class attributes and parameters of class conditional distributions. A comprehensive review of these techniques and their performance on several benchmark data sets can be found in [35].

There are also quite a few techniques that tackle ZSL in a transductive setting. Experiments in [28,37] demonstrate that unlabeled data from unseen classes as well as training data augmented by generative adversarial nets/ variational autoencoders can notably boost the classification accuracy. We believe that this line of work should be treated under a different category as a direct comparison with current ZSL techniques is not possible since similar data augmentation techniques could have most certainly benefited these techniques.

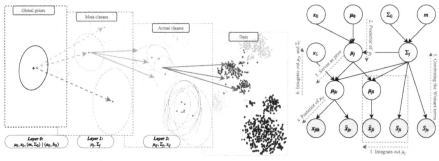

(a) Our Bayesian Zero-Shot Learning Model (BZSL)   (b) Graphical model

Fig. 2. Two-layer Generative Model for Bayesian zero-shot learning (BZSL). (a) Latent (Meta) classes shown by dashed lines in layer 1 are generated from Normal distribution ($N(\mu_j, \Sigma_j)$). Hyperparameters in layer 0 are the priors on the mean and covariance of these Gaussians. The sufficient statistics of actual classes in layer 2 are obtained from layer 1. Finally data samples are drawn from Gaussian distribution with the mean, μ_{ji} and covariance, Σ_j. (b) Conditional hierarchical Gaussian data generation (Likelihood) model and derivation of marginal likelihood.

Bayesian Models. In this paper, we offer a hierarchical Bayesian perspective on ZSL as a promising alternative to earlier approaches. Although hierarchical Bayesian mixture models have been previously explored for a variety of clustering problems [3,4,9,38], their extension to ZSL comes with two distinct features that could help the proposed model prevail over the large body of early work in ZSL. First, as a Bayesian model, ours offers a systematic approach to sharing information between seen and unseen classes as well as within each group through the utilization of local and global priors. Global priors are defined by hyperparameters, whereas local priors are determined by the parameters of the meta classes, which are estimated from corresponding seen classes. Second, as a hierarchical model, it can better accommodate data sets with different levels of class abstractions, i.e., fine-grained vs. coarse-grained data sets, which is particularly appealing for large-scale classification. A hierarchical Bayesian model was previously studied in a one-shot learning setting [24]. Our proposed approach differs from this model in two essential aspects. First, unlike our proposed approach, no semantic information was used when establishing the Bayesian hierarchy in [24], and class discovery was performed in a fully unsupervised fashion. Second, our approach introduces the notion of local prior, which becomes highly instrumental in defining meta-classes and modeling dispersion of classes.

Our Work. Unlike the vast majority of early work, which seeks to optimize a mapping between image features and attribute vectors, our approach readily models class distributions in the feature space by exploiting both local and global priors defined over the parameters of these distributions. Local priors are defined by meta-classes. In the proposed approach, attribute vectors only come into play when determining meta-class memberships of actual classes. Classes with similar attribute vectors are pooled together to derive local priors.

3   Bayesian Zero Shot Learning

Bayesian classification places a shared prior over the parameters of class distributions, which are assumed to be generated independently conditioned on the prior. Imposing the same prior across all classes creates dependencies among them, enabling information propagation and regularization at the same time during model inference. However, in real-world applications the classes are often not generated independently, indeed for a large number of classes different levels of abstraction is expected. On the other hand, availability of semantic side information suggests that there is a deeper level of hierarchy among existing classes than a single global Bayesian prior can explain.

Images from semantically similar classes are embedded close to each other due to their shared latent parameters. When such similarities are not accounted for in the classification model, sample estimates of class parameters derived based on independence assumption among classes become nullified. In other words, knowing the parameters of the global prior may not be sufficient for achieving independence as latent parameters define deeper level hierarchical relationships among classes. Our model resolves this problem by introducing a layer of meta-classes between global prior and actual classes, paving the way for independence and enabling information sharing and propagation across classes.

3.1   Generative Model

Our approach to ZSL employs class similarities by a two layer generative model. As shown in Fig. 2, our model identifies meta-classes that determine groupings among classes. These meta-classes play a key role by acting as a local prior for individual classes, i.e. both seen and unseen classes that belong to the same meta-class inheriting the same local prior. In our framework, the data points with the same local prior that do not belong to any of the seen classes can be considered from unseen classes. If the class groupings can be arranged such that there is only one unseen class associated with each local prior then unseen classes can be uniquely identified. Associating each unseen class with a different local prior forms the basis of our approach. Our generative model is designed as follows:

$$x_{jik} \sim N(\mu_{ji}, \Sigma_j), \quad \mu_{ji} \sim N(\mu_j, \Sigma_j \kappa_1^{-1}), \quad \mu_j \sim N(\mu_0, \Sigma_j \kappa_0^{-1}), \quad \Sigma_j \sim W^{-1}(\Sigma_0, m) \tag{1}$$

with the meta-class index j, the actual class index i, the image index k. We assume that images x_{jik} come from a Gaussian with mean μ_{ji} and covariance matrix Σ_j. They are generated independently conditioned not only on the global prior but also on their corresponding meta-class.

Each meta-class is characterized by the parameters μ_j and Σ_j. μ_0 is the mean of the Gaussian prior defined over the mean vectors of meta-classes, κ_0 is a scaling constant that adjusts the dispersion of the centers of meta classes around μ_0. A smaller value for κ_0 suggests that class centers are expected to be farther apart from each other whereas a larger value suggests they are expected to be closer to each other. On the other hand, Σ_0 and m dictate the expected shape of the class distributions, as under the inverse Wishart distribution assumption the expected covariance is $E(\Sigma|\Sigma_0, m) = \frac{\Sigma_0}{m-D-1}$, where D is the dimension of image feature space. The minimum feasible value of m is equal to $D+2$, and the larger the m is the less individual covariance matrices will deviate from the expected shape.

On the other hand, κ_1 is a scaling constant that adjusts the dispersion of the actual class means around their corresponding meta-class means. A larger κ_1 leads to smaller variations in class means compared to the mean of their corresponding meta classes, suggesting a fine-grained relationship among classes sharing the same meta-class. On the other hand, a smaller κ_1 dictates coarse-grained relationships among classes sharing the same meta-class. In this model, classes with the same meta-class also share the same covariance matrix Σ_j to preserve conjugacy of the model.

To classify test examples, we need the posterior predictive distributions (PPD) of seen and unseen classes which we will explain next. More details about the derivation are provided in the supplementary.

3.2   Posterior Predictive Distribution

In our model, the posterior predictive distribution (PPD) incorporates three sources of information: the data likelihood that arises from the current class, the local prior that results from other classes sharing the same meta class as the current class, and global prior defined in terms of hyperparameters. The derivation in six steps are outlined in Fig. 2(b) and Algorithm[1] 1 describes a pseudo code on deriving PPD for both seen and unseen classes. Class sufficient statistics are summarized by $\bar{x}_{ji}, S_{ji}$ and n_{ji} which represent sample mean, scatter matrix and size of class i of meta-class j, respectively. The notations ω_{jc} and ω_j used in the Algorithm 1 represents the current seen class and unseen class, whose PPD is being derived.

In step 1, we establish the link between class sample mean $\bar{x}_{ji}$ and its corresponding meta-class mean μ_j by marginalizing out the intermediate class mean μ_{ji}. As all of these are Gaussians, this marginalization yields a Gaussian:

$$P(\bar{x}_{ji}|\mu_j, \Sigma_j, \kappa_1) = N(\bar{x}_{ji}|\mu_j, \Sigma_j(\frac{1}{n_{ji}} + \frac{1}{\kappa_1})) \tag{2}$$

In step 2, we use Bayes rule to obtain the posterior distribution of the meta-class mean vector μ_j:

$$P(\mu_j|\mu_0, \Sigma_j, \kappa_0, \kappa_1, \{\bar{x}_{ji}\}_{t_i=j}) = N(\mu_j|\bar{\mu}_j, \bar{\kappa}_j^{-1}\Sigma_j)$$

$$\bar{\mu}_j = \frac{\sum_{i:t_i=j} \frac{n_{ji}\kappa_1}{(n_{ji}+\kappa_1)}\bar{x}_{ji} + \kappa_0\mu_0}{\sum_{i:t_i=j} \frac{n_{ji}\kappa_1}{(n_{ji}+\kappa_1)} + \kappa_0}, \quad \bar{\kappa}_j = (\sum_{i:t_i=j} \frac{n_{ji}\kappa_1}{(n_{ji}+\kappa_1)} + \kappa_0) \tag{3}$$

---

[1] The code is publicly available at GitHub.

Algorithm 1. Modeling seen and unseen classes in BZSL

Input: Training data, $\phi(seen)$, $\phi(unseen)$
Output: PPD parameters for each seen class $(\bar{\mu}_{jc}, \bar{v}_{jc}, \bar{\Sigma}_{jc})$ and unseen class $(\bar{\mu}_j, \bar{v}_j, \bar{\Sigma}_j)$

1: Set hyper-parameters: $\kappa_0, \kappa_1, m, s, K$
2: Compute μ_0 (mean of class means) and Σ_0 (mean of class covariances scaled by s)
3: **for** each seen class ω_{jc} **do**   ▷ Images available
4:     Calculate current class params: $\bar{x}_{jc}, n_{jc}, S_{jc}$
5:     Find K most similar seen classes:
6:     $\mathcal{L}^2(\phi(\omega_{jc}), \phi(seen))$
7:     **for** each selected seen class ω_{ji} **do**
8:         Calculate class params: $\bar{x}_{ji}, n_{ji}, S_{ji}$
9:     **end for**
10:     Calculate intermediate terms: $\tilde{\kappa}_j, \bar{\mu}_j, S_\mu$ (Eq 5,3,6)
11:     Calculate PPD parameters by combining *local prior*
12:     and *data driven likelihood*: $\bar{\mu}_{jc}, \bar{v}_{jc}, \bar{\Sigma}_{jc}$ (Eq 7)
13: **end for**
14: **for** each unseen class ω_j **do**   ▷ No image available
15:     Find K most similar seen classes:
16:     $\mathcal{L}^2(\phi(\omega_j), \phi(seen))$
17:     **for** each selected seen class ω_{ji} **do**
18:         Calculate class params: $\bar{x}_{ji}, n_{ji}, S_{ji}$
19:     **end for**
20:     Calculate intermediate terms: $\tilde{\kappa}_j, S_\mu$ (Eq 5, 6)
21:     Calculate PPD parameters using only *local*
22:     *prior*: $\bar{\mu}_j, \bar{v}_j, \bar{\Sigma}_j$ (Eq 3, 7)
23: **end for**

---

where t_i is the meta-class indicator for class i. Note that the mean $\bar{\mu}_j$ is the weighted average of the prior mean and class means share the same meta-class.

In step 3, we obtain the local prior for class mean vector μ_{jc} by propagating the information from other classes sharing the same meta-class as the current class c. This is achieved by integrating out the meta-class mean vector μ_j.

$$P(\mu_{jc}|\mu_0, \Sigma_j, \kappa_0, \kappa_1, \{\bar{x}_{ji}\}_{t_i=j}) = N(\mu_{jc}|\bar{\mu}_j, \Sigma_j(\bar{\kappa}_j^{-1} + \kappa_1^{-1})) \qquad (4)$$

In step 4, we derive the posterior of the current class mean vector μ_{jc} by combining current class sample mean $\bar{x}_{jc}$ from step 1 and the local prior from step 3.

$$P(\mu_{jc}|\mu_0, \Sigma_j, \kappa_0, \kappa_1, \{\bar{x}_{ji}\}_{t_i=j}, \bar{x}_{jc}) = N(\mu_{jc}|\frac{n_{jc}\bar{x}_{jc}+\tilde{\kappa}_j\bar{\mu}_j}{n_{jc}+\tilde{\kappa}_j}, \Sigma_j(\tilde{\kappa}_j^{-1} + n_{jc}^{-1}))$$

$$\tilde{\kappa}_j = \frac{(\sum_{i:t_i=j}\frac{n_{ji}\kappa_1}{(n_{ji}+\kappa_1)}+\kappa_0)\kappa_1}{\sum_{i:t_i=j}\frac{n_{ji}\kappa_1}{(n_{ji}+\kappa_1)}+\kappa_0+\kappa_1} \qquad (5)$$

In step 5, we derive the posterior distribution of the covariance matrix Σ_j by combining the local prior of the covariance matrix $P(\Sigma_j|\Sigma_0, m)$ with the distribution of the scatter matrices of the classes associated with meta-class j S_{ji} and current class S_{jc}:

$$P(\Sigma_j|\{S_{ji}\}_{t_i=j}, S_{jc}) = IW(\Sigma_j|\bar{S}_c, m + \sum_{i:t_i=j}(n_{ji}-1) + n_{jc})$$

$$\bar{S}_c = \Sigma_0 + \sum_{i:t_i=j} S_{ji} + S_{jc} + S_\mu, \quad S_\mu = \frac{n_{jc}\tilde{\kappa}_j}{\tilde{\kappa}_j+n_{jc}}(\bar{x}_{jc} - \bar{\mu}_j)(\bar{x}_{jc} - \bar{\mu}_j)^T \quad (6)$$

In step 6, we derive the posterior predictive distribution by integrating out meta-class mean vector μ_j and covariance Σ_j in the form of a Student-t distribution as follows.

$$P(x|\{\bar{x}_{ji}, S_{ji}\}_{t_i=j}, \bar{x}_{jc}, S_{jc}, \mu_0, \kappa_0, \kappa_1) = T(x|\bar{\mu}_{jc}, \bar{\Sigma}_{jc}, \bar{v}_{jc})$$

$$\bar{\mu}_{jc} = \frac{n_{jc}\bar{x}_{jc} + \tilde{\kappa}_j\bar{\mu}_j}{n_{jc} + \tilde{\kappa}_j}, \quad \bar{v}_{jc} = n_{jc} + \sum_{i:t_i=j}(n_{ji}-1) + m - D + 1$$

$$\bar{\Sigma}_{jc} = \frac{\Sigma_0 + \sum_{i:t_i=j} S_{ji} + S_{jc} + S_\mu}{\frac{(n_{jc}+\tilde{\kappa}_j)\bar{v}_{jc}}{n_{jc}+\tilde{\kappa}_j+1}} \quad (7)$$

where, $\bar{\mu}_j$, $\tilde{\kappa}_j$ and S_μ are defined as in Eq. (3), (5) and (6) respectively. The index c in Eq. (7) represents the current seen class, whose PPD is being derived. Top K most similar seen classes are identified as the ones with the smallest Euclidean distance to the current class in the attribute space. If the current class is a seen class, PPD takes the form in Eq. (7). When it is an unseen class with no images available in training, the sample statistics of the current class in (7) drops and PPD becomes:

$$P(x|\{\bar{x}_{ji}, S_{ji}\}_{t_i=j}, \mu_0, \kappa_0, \kappa_1) = T(x|\bar{\mu}_j, \bar{\Sigma}_j, \bar{v}_j)$$

$$\bar{v}_j = \sum_{i:t_i=j}(n_{ji}-1) + m - D + 1, \quad \bar{\Sigma}_j = \frac{(\Sigma_0+\sum_{i:t_i=j} S_{ji})(\tilde{\kappa}_j+1)}{\tilde{\kappa}_j\bar{v}_j} \quad (8)$$

where $\bar{\mu}_j$ and $\tilde{\kappa}_j$ are defined as in Eq. (3) and (5), respectively. In this setting, a new image is labeled by evaluating PPDs for seen and unseen classes and assigning the image to the class that generates the maximum likelihood.

3.3   Meta-class Formation

Meta-class for each unseen class is formed by finding K most similar seen classes to the current unseen class using $\mathcal{L}^2$ distance between the attribute vectors (ϕ) of that unseen class and of seen classes. In the case of tie, the least similar class among selected seen classes (K^{th}) is replaced by the next one until tie is broken. These define a local prior in the PPD of the unseen class. Meta-class formation for a seen class follows the same procedure. We use the $\mathcal{L}^2$ distance between the current seen class attribute and other seen class attributes to find K most similar classes. As we have access to seen class samples, the PPD of the seen class (Eq. 7) uses class samples in addition to local and global priors from its meta-class. An illustration for the formation of the meta-class associated with an unseen class *blue whale*, from AWA dataset, is shown in Fig. 1. $\phi(blue\ whale)$

is compared against $\phi(seen)$ in the semantic space, *humpback* and *killer whale* are identified as the two closest matches. Using *humpback* and *killer whale* class samples, the meta-class for *blue whale* is formed as a local prior in the PPD for *blue whale*.

Table 1. $|Y^{all}|$, $|Y^s|$, and $|Y^u|$ denote the number of classes in all, seen and unseen classes, respectively. To clarify the numbers in last 3 columns, we give an illustration on FLO dataset: FLO has total of 102 classes of which 62 are training, 20 are validation (both seen during training) and 20 are test classes (unseen during training).

Dataset	#imgs	Type	#att	$|Y^{all}|$	$|Y^s|$	$|Y^u|$
FLO	8,189	Fine	102	102	62 + 20	20
SUN	14,340	Fine	102	717	580 + 65	72
CUB	11,788	Fine	312	200	100 + 50	50
AWA1	30,475	Coarse	85	50	27 + 13	10
AWA2	37,322	Coarse	85	50	27 + 13	10
aPY	15,339	Coarse	64	32	15 + 5	12
ImageNet	14M	Large	500	21K	1K	20K

4 Experiments

We evaluate the performance of the proposed approach on several benchmark data sets and compare the results with the current state of the art in ZSL.

Datasets and Specifications. Experiments are evaluated on ZSL datasets widely used for benchmarking. Among those, CUB [33], FLO [18] and SUN [21] are medium scale, fine-grained datasets. AWA1 [13] and AWA2 [36] and aPY [5], on the other hand, are coarse-grained datasets. Finally we evaluate our model on ImageNet [23] with more than 14 million images and 21K classes. SUN, AWA1, AWA2, aPY and CUB datasets come with visual attributes whereas FLO uses sentences and ImageNet uses word embeddings as class vectors. We use the publicly available image embeddings of [36], i.e. 2048-dimensional top-layer pooling units of the 101-layered ResNet [7] as feature vectors. Additional information about each dataset including the number of images, number of attributes, and sizes of train, validation, and test class splits are present in Table 1.

For ImageNet following the benchmark in [36] we use all of the images from 1K classes, i.e. seen classes, for training so that we do not violate the zero-shot assumption as ResNet-101 [7] is trained on the same 1K classes from ImageNet. We evaluate the proposed technique in nine different configurations as proposed in [36], all of which differs according to how test class subsets are chosen.

Evaluation Criteria. We use the same evaluation procedure employed in [36] as described below. The standard practice in ZSL literature is to evaluate classification performance by Top-1 accuracy. To avoid large classes dominating the overall accuracy, Top-1 accuracy is separately calculated for each class and the mean of individual class accuracies is used for evaluation. GZSL setting includes both seen and unseen classes in the test phase, hence the search space includes all the classes, i.e. $|Y^{all}|$. Hence, first seen and unseen class accuracies are separately computed and then their harmonic mean is used as the final score for evaluation. For ImageNet, the final score is the average Top-1 accuracy over the images of unseen classes (although the search space is still $|Y^{all}|$) as no images from seen classes are available during testing phase.

Implementation Details. We implement two versions of our model: *unconstrained* (UBZSL) and *constrained* (CBZSL) Bayesian ZSL. For large data sets, e.g. ImageNet, our model in Eq. 1 suffers from the large memory requirement due to the unconstrained structure of the class covariance matrices. To alleviate this problem we developed a scalable version of our model where the covariance matrices are constrained to have diagonal forms. The only difference between these two models is that constrained version uses an Inverse Gamma prior on the diagonal entries of the covariance matrix as opposed to an Inverse Wishart in the unconstrained version. With this revision the generative model in Eq. 1 is updated as follows.

$$x_{jik}^d \sim N(\mu_{ji}^d, \Sigma_j^d), \quad \mu_{ji}^d \sim N(\mu_j^d, \Sigma_j^d \kappa_1^{-1}), \quad \mu_j^d \sim N(\mu_0^d, \Sigma_j^d \kappa_0^{-1}), \quad \Sigma_j^d \sim IG(a_0, b_0)$$

where the superscript d is added to refer to the dth component of each parameter. The Inverse Wishart parameters m and Σ_0 are replaced with the scale (a_0) and shape (b_0) parameters of the Inverse Gamma distribution. The derivation of PPD for the constrained model is in the supplementary.

The hyperparameters of the model are coarsely tuned to maximize the harmonic mean score on the validation set for all datasets but ImageNet. The training, test and validation set splits for these datasets are done according to [36] to maintain a fair comparison. As hyperparameter tuning for ImageNet can be computationally unmanageable and to demonstrate the robustness of the model we used the hyperparameters of the SUN dataset for ImageNet. For CBZSL we utilize all 2048 ResNet features whereas for UBZSL we applied PCA to reduce the dimensionality to 500.

Both UBZSL and CBZSL have four hyperparameters: $\kappa_0, \kappa_1, m, s, K$. Here, K is the selected number of classes most similar to the current class in the attribute space. To simplify the parameter tuning process, we set prior mean, μ_0, to the average of class means. We set Σ_0 to the average of class scatter matrices scaled by a constant s.

4.1 Model Evaluation

In this section, we evaluate our model through an ablation study and investigate the tradeoff between seen and unseen class accuracies.

Model Ablation. Our model formulates zero-shot learning in the framework of hierarchical Bayes. Towards this end, we validate the necessity of each component in our model by eliminating one component at a time and investigating the performance of the model with remaining components on several benchmark datasets.

Table 2. Ablation study (in harmonic mean) on 6 datasets. In the UBZSL (V1) we discard Bayesian aspect and in UBZSL (V2) we impose similar dispersion for meta and actual classes

Method	SUN	CUB	AWA1	AWA2	aPY	FLO
UBZSL (V1)	32.5	24.9	21.1	29.0	10.0	20.5
UBZSL (V2)	3.0	18.3	38.0	40.3	9.5	34.1
UBZSL	32.8	37.5	49.6	49.7	35.4	40.4

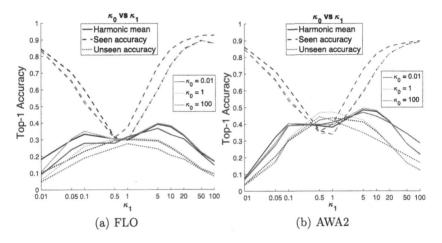

(a) FLO   (b) AWA2

Fig. 3. Variations in seen and unseen class accuracies and their harmonic means with respect to changes in κ_0 and κ_1. Seen and unseen class accuracies are highly sensitive to changes in κ_1 whereas minimal changes are observed wrt changes in κ_0.

Our observations from Table 2 are as follows. (1) If we break the hierarchy by removing the meta-class layer, then actual classes are directly linked to the global prior and same PPD is assigned to all unseen classes. Thus, unseen classes can no longer be distinguished during test time. (2) If we discard the Bayesian aspect by eliminating the global and local priors, each seen class is fit a single Gaussian and each unseen class is fit a GMM with K components. We observe drastic drop in harmonic mean, almost cut in half, in all datasets but SUN (1st row in Table 2: V1). In general, GMM works better on fine grained datasets than coarse grained ones as the distribution produced by a mixture of very similar classes can be better fit by GMM compared to a distribution produced by a mixture of relatively less similar classes. (3) Finally, if we impose similar dispersion for actual and meta classes (by improperly adjusting κ_0 and κ_1) with respect to the center of the data, harmonic mean again suffers significantly (2nd row in Table 2: V2). In particular, results of the SUN dataset suffers the most. The impact of improper tuning of κ_1 is explained in the next section. Unlike V1, UBZSL V2 works better on coarse grained datasets (AWA1, AWA2) as class centers in these datasets are more separated than fine grained ones. As a result the adverse effects of setting $\kappa_1 \ll 1$ in experiments performed with these datasets seem to be less significant[2].

Effect of κ_0 and κ_1. In both of our models (*constrained* and *unconstrained*), different hyperparameter settings can be used to modify the operating point of the classifier to favor seen class accuracy over unseen one or vice versa. In this experiment we investigate the effect of κ_0 and κ_1 on seen and unseen class accuracies. Recall that κ_0 adjusts the dispersion of meta-class centers with respect to the center of the overall data and κ_1 adjusts the dispersion of actual class centers with respect to their corresponding meta class centers. The smaller these parameters are the higher the dispersion will be.

Table 3. GZSL results achieved by the proposed approach (CBZSL and UBZSL) along with results of several other techniques from the literature on SUN, CUB, FLO, AWA1, AWA2, aPY datasets. We measure per-class averages top-1 accuracy on seen classes (**tr**), unseen classes (**ts**) and their harmonic mean (**H**).

Method	SUN			CUB			AWA1			AWA2			aPY			FLO		
	ts	tr	H	ts	tr	H	ts	tr	H	ts	tr	H	ts	tr	H	ts	tr	H
LATEM [34]	14.7	28.8	19.5	15.2	57.3	24.0	7.3	71.7	13.3	11.5	77.3	20.0	0.1	73.0	0.2	6.6	47.6	11.5
ALE [1]	21.8	33.1	26.3	23.7	62.8	34.4	16.8	76.1	27.5	14.0	81.8	23.9	4.6	73.7	8.7	13.3	61.6	21.9
DEVISE [6]	16.9	27.4	20.9	23.8	53.0	32.8	13.4	68.7	22.4	17.1	74.7	27.8	4.9	76.9	9.2	9.9	44.2	16.2
SJE [2]	14.7	30.5	19.8	23.5	59.2	33.6	11.3	74.6	19.6	8.0	73.9	14.4	3.7	55.7	6.9	13.9	47.6	21.5
ESZSL [22]	11.0	27.9	15.8	12.6	63.8	21.0	6.6	75.6	12.1	5.9	77.8	11.0	2.4	70.1	4.6	11.4	56.8	19.0
SYNC [25]	7.9	43.3	13.4	11.5	70.9	19.8	8.9	87.3	16.2	10.0	90.5	18.0	7.4	66.3	13.3	–	–	–
SAE [10]	8.8	18.0	11.8	7.8	54.0	13.6	1.8	77.1	3.5	1.1	82.2	2.2	0.4	80.9	0.9	–	–	–
GFZSL [32]	0.0	39.6	0.0	0.0	45.7	0.0	1.8	80.3	3.5	2.5	80.1	4.8	0.0	83.3	0.0	–	–	–
TCN [8]	31.2	37.3	**34.0**	52.6	52.0	**52.3**	49.4	76.5	**60.0**	61.2	65.8	**63.4**	24.1	64.0	35.1	–	–	–
DCN [14]	25.5	37.0	30.2	28.4	60.7	38.7	25.5	84.2	39.1	–	–	–	14.2	75.0	23.9	–	–	–
REL. NET [30]	11.1	20.0	14.3	14.0	35.7	20.1	22.9	76.9	35.3	18.6	87.3	30.6	11.5	60.9	19.4	13.8	73.8	23.2
CBZSL	29.0	32.7	30.7	21.1	43.5	28.5	38.9	67.2	49.3	34.1	72.5	46.4	18.8	70.8	29.6	31.3	28.5	29.8
UBZSL	31.7	34.0	32.8	31.5	46.3	37.5	38.7	69.3	49.6	37.1	75.1	49.7	24.0	67.4	**35.4**	27.2	78.2	**40.4**

Figure 3 illustrates on FLO and AWA2 that unseen class accuracy is highest when κ_1 is close to 1 and drops significantly lower in both directions, i.e., for $\kappa_1 \ll 1$ and $\kappa_1 \gg 1$. As expected the opposite of this pattern is observed for seen class accuracy. Although both seen and unseen class accuracies are highly sensitive to the selection of κ_1, the changes are marginal with respect to κ_0. Moving κ_1 towards zero encodes a local prior that imposes unrealistically large dispersion for centers of actual-classes sharing the same meta-class, which

---

[2] As [30] uses different set of attributes in their experiments, we rerun their algorithm with the attributes from [35] to maintain a fair comparison.

violates the main assumption of our model that classes sharing the same meta class are semantically similar classes. On the other hand moving κ_1 towards infinity encodes a local prior that imposes limited to no deviation among centers of actual classes which is another extreme that is not true for real-world datasets, i.e. classes are supposed to be statistically identifiable.

In both extremes unrealistic prior assumptions that cannot be reconciled with the characteristics of real-world data sets impede knowledge transfer between seen and unseen classes and lead to poor classification performance on unseen classes. On the other hand, the same extreme assumptions happen to help with seen class accuracies because likelihood and data-driven local priors (both of which lacks for unseen classes) outweigh the effect of unrealistic global prior in posterior predictive distributions.

4.2   Comparison with State of the Art

Results obtained by the proposed CBZSL and UBZSL models on SUN, CUB, FLO, AWA1, AWA2, aPY datasets are presented in Table 3. In addition to all SotA techniques reported in [35] we also included results of more recently published techniques [8,14,30] in this comparison. These results suggest that the proposed unconstrained model (UBZSL) demonstrates better performance than all other techniques but TCN. The constrained version of our model, i.e., CBZSL, also renders comparable results with the unconstrained version of the model despite its simplicity.

Results in Table 3 further show that in all of the experiments, unseen class accuracies achieved by our models are substantially higher than those achieved by all other techniques, but the TCN [8] model. This is achieved while maintaining a comparable performance on seen class accuracies in most of the experiments. Intuitively speaking, the two-level Bayesian hierarchy defined by meta-classes is expected to better manage the open space risk [27] by assigning an image of an unseen class to its meta class as opposed to misclassifying it into one of the seen classes.

4.3 Large-Scale Experiments on ImageNet

ImageNet is currently the most challenging dataset for ZSL. Arguably it constitutes the most natural setup to evaluate ZSL learning performance as it contains 22K classes (1K of which are used to train state of the art deep neural networks) and most of these classes are sparsely populated.

Table 4 summarizes ImageNet results under nine different test set configurations. Our unconstrained model (UBZSL) improves over the state of the art in 2/3 Hop and highly populated test classes. Of particular importance is the highly competitive performance by the constrained model (CBZSL) that improves the current state of the art in all test con-

Table 4. ImageNet results in nine different test phase configurations. Lp and Mp refer to least and most populated classes, respectively. 2/3 Hop represents the classes that are 2/3-hops away from 1K training classes according to the ImageNet label hierarchy. Finally All appears for all 21K ImageNet classes. The results are in top-K accuracy.

Split	UBZSL			CBZSL			SoA from [35]		
	1	5	10	1	5	10	1	5	10
2Hop	2.6	13.1	20.3	3.9	15.0	22.8	2.2	10.3	19.3
3Hop	0.8	4.1	6.9	1.0	4.1	6.9	0.8	3.7	**7.2**
Lp500	1.8	5.1	8.6	2.5	10.2	14.3	1.9	6.1	10.4
Lp1K	1.2	4.6	7.3	2.3	7.3	10.7	1.4	4.8	8.5
Lp5K	0.5	2.0	3.5	0.6	2.4	4.0	0.4	2.2	3.9
Mp500	3.4	17.4	26.5	7.5	25.2	35.0	2.9	14.9	26.6
Mp1K	2.4	13.0	20.2	4.8	17.3	25.5	2.3	11.8	20.7
Mp5K	1.1	6.1	9.9	1.5	6.6	10.5	1.1	6.2	10.0
All	0.3	1.8	3.0	0.4	1.8	2.9	0.3	**2.0**	**3.4**

figurations with respect to Top-1 accuracy (3.9% vs 2.18% on 2Hop, 7.51% vs 2.9% on Mp500, 4.78% vs 2.34% on Mp1000). Our model achieves the best results in eight of the nine test configurations for Top-5 and seven of the nine for Top-10 accuracies. Especially in the least populated (Lp500) classes the accuracy improvement is four percentage points in Top-5 and Top-10 accuracies. In most populated classes (Mp500) the accuracy gets almost doubled, i.e. 25.20% vs 14.86% on Top-5.

These results show that as the number of classes and the average number of samples per class (1300 in ImageNet vs 700 in benchmark datasets) increase, the explicit hierarchy across classes becomes more evident leading to more informative local priors. ImageNet contains both coarse- and fine-grained classes. The results suggest that our technique can be equally effective on datasets with hybrid granularity.

5   Conclusions

Summary of Our Contributions: In this study, we proposed a Bayesian approach to ZSL that relies on the consideration that classes in real-world datasets emerge at different levels of abstraction, and there are meta-classes that inherently organize the class hierarchy in the semantic space. We introduced concepts of local and global priors and showed that knowledge transfer from seen classes to unseen ones could be effectively carried out in the image space by a two-layer GMM. The proposed two-layer GMM offers extreme flexibility in modeling datasets with different characteristics by tuning its hyperparameters, each of which models a different aspect of the data. We performed extensive experiments with benchmark datasets (fine-grained, coarse-grained, and large-scale) to demonstrate the utility of the proposed Bayesian approach for ZSL, which favors the proposed approach over other state-of-the-art inductive ZSL techniques.

Future Research Directions: Recently proposed transductive methods [15, 28, 37] have proved that generating features for unseen classes and treating ZSL as a closed-set classification can produce much better results than running ZSL in an inductive setting. Although features generated by these techniques do not seem to preserve correlation among features and are far from recovering unseen class distributions, they do preserve the relative distances among unseen classes, which in turn helps improve the performance of a softmax classifier in the closed-set setting. Thus, using prototypical feature vectors for unseen classes and integrating these into PPDs can offer significant boost for the performance of the proposed hierarchical Bayesian model. In our future work we aim to demonstrate that these prototypical feature vectors can be easily obtained by solving a simple compressed sensing problem and PPDs updated with these prototypical vectors can be used to generate new features in a probabilistic way. Such an approach can potentially preserve both the correlation among features and the relative distance between classes to generate more realistic features. Although not discussed in current work the proposed framework can be easily and effectively extended for any-shot learning problems, which will be a research direction we will pursue in parallel to probabilistic feature generation.

Acknowledgements. This work has been partially funded by the NSF grant 1252648 - ISS (M. D.), the ERC grant 853489 - DEXIM (Z. A.) and by DFG grant under Germany's Excellence Strategy – EXC number 2064/1 – Project number 390727645.

References

1. Akata, Z., Perronnin, F., Harchaoui, Z., Schmid, C.: Label-embedding for image classification. TPAMI **38**, 1425–1438 (2016)
2. Akata, Z., Reed, S., Walter, D., Lee, H., Schiele, B.: Evaluation of output embeddings for fine-grained image classification. In: CVPR (2015)
3. Dundar, M., Akova, F., Yerebakan, H.Z., Rajwa, B.: A non-parametric Bayesian model for joint cell clustering and cluster matching: identification of anomalous sample phenotypes with random effects. BMC Bioinform. **15**(1), 314 (2014)
4. Dundar, M., Yerebakan, H.Z., Rajwa, B.: Batch discovery of recurring rare classes toward identifying anomalous samples. In: SIGKDD. ACM (2014)
5. Farhadi, A., Endres, I., Hoiem, D., Forsyth, D.: Describing objects by their attributes. In: CVPR (2009)
6. Frome, A., et al.: Devise: a deep visual-semantic embedding model. In: NIPS (2013)
7. He, K., Zhang, X., Ren, S., Sun, J.: Deep residual learning for image recognition. In: CVPR (2016)
8. Jiang, H., Wang, R., Shan, S., Chen, X.: Transferable contrastive network for generalized zero-shot learning. In: ICCV (2019)
9. Kim, S., Smyth, P.: Hierarchical Dirichlet processes with random effects. In: NIPS (2007)
10. Kodirov, E., Xiang, T., Gong, S.: Semantic autoencoder for zero-shot learning. In: CVPR (2017)

11. Krizhevsky, A., Sutskever, I., Hinton, G.E.: Imagenet classification with deep convolutional neural networks. In: NIPS (2012)
12. Lampert, C., Nickisch, H., Harmeling, S.: Learning to detect unseen object classes by between-class attribute transfer. In: CVPR (2009)
13. Lampert, C., Nickisch, H., Harmeling, S.: Attribute-based classification for zero-shot visual object categorization. TPAMI **36**(3), 453–465 (2013)
14. Liu, S., Long, M., Wang, J., Jordan, M.I.: Generalized zero-shot learning with deep calibration network. In: NIPS (2018)
15. Liu, W., et al.: Towards visually explaining variational autoencoders. In: CVPR (2020)
16. Mikolov, T., Chen, K., Corrado, G., Dean, J.: Efficient estimation of word representations in vector space. In: ICLR (2013)
17. Mukherjee, T., Hospedales, T.: Gaussian visual-linguistic embedding for zero-shot recognition. In: EMNLP (2016)
18. Nilsback, M., Zisserman, A.: Automated flower classification over a large number of classes. In: ICCVGI (2008)
19. Norouzi, M., et al.: Zero-shot learning by convex combination of semantic embeddings. In: ICLR (2014)
20. Palatucci, M., Pomerleau, D., Hinton, G.E., Mitchell, T.M.: Zero-shot learning with semantic output codes. In: NIPS (2009)
21. Patterson, G., Hay, J.: Sun attribute database: Discovering, annotating, and recognizing scene attributes. In: CVPR (2012)
22. Romera-Paredes, B., Torr, P.H.: An embarrassingly simple approach to zero-shot learning. In: ICML (2015)
23. Russakovsky, O., et al.: Imagenet large scale visual recognition challenge. IJCV **115**(3), 211–252 (2015). https://doi.org/10.1007/s11263-015-0816-y
24. Salakhutdinov, R., Tenenbaum, J., Torralba, A.: One-shot learning with a hierarchical nonparametric Bayesian model. In: JMLR workshop (2012)
25. Changpinyo, S., Chao, W.-L., Gong, B., Sha, F.: Synthesized classifiers for zero-shot learning. In: CVPR (2016)
26. Scheirer, W.J., Rocha, A., Sapkota, A., Boult, T.E.: Towards open set recognition. TPAMI **36**, 1757–1772 (2013)
27. Scheirer, W.J., Boult, T.E.: Statistical methods for open set recognition. In: CVPR Tutorial (2016)
28. Schonfeld, E., Ebrahimi, S., Sinha, S., Darrel, T., Akata, Z.: Generalized zero- and few-shot learning via aligned variational autoencoders. In: CVPR (2019)
29. Socher, R., Ganjoo, M., Manning, C.D., Ng, A.: Zero-shot learning through cross-modal transfer. In: NIPS (2013)
30. Sung, F., Yang, Y., Zhang, L., Xiang, T., Torr, P.H., Hospedales, T.M.: Learning to compare: relation network for few-shot learning. In: CVPR (2018)
31. Szegedy, C., et al.: Going deeper with convolutions. In: CVPR (2015)
32. Verma, V.K., Rai, P.: A simple exponential family framework for zero-shot learning. In: Ceci, M., Hollmén, J., Todorovski, L., Vens, C., Džeroski, S. (eds.) ECML PKDD 2017, Part II. LNCS (LNAI), vol. 10535, pp. 792–808. Springer, Cham (2017). https://doi.org/10.1007/978-3-319-71246-8_48
33. Welinder, P., et al.: Caltech-UCSD birds 200 (2010). Caltech, Technical report, CNS-TR-2010–001
34. Xian, Y., Akata, Z., Sharma, G., Nguyen, Q., Hein, M., Schiele, B.: Latent embeddings for zero-shot classification. In: CVPR (2016)
35. Xian, Y., Lampert, C., Schiele, B., Akata, Z.: Zero-shot learning - a comprehensive evaluation of the good, the bad and the ugly. TPAMI **41**, 2251–2265 (2018)

36. Xian, Y., Lampert, C.H., Schiele, B., Akata, Z.: Zero shot learning a comprehensive evaluation of the good, the bad and the ugly. In: CVPR (2017)
37. Xian, Y., Lorenz, T., Schiele, B., Akata, Z.: Feature generating networks for zero-shot learning. In: CVPR (2018)
38. Yerebakan, H.Z., Rajwa, B., Dundar, M.: The infinite mixture of infinite gaussian mixtures. In: NIPS (2014)
39. Zhang, Z., Saligrama, V.: Zero-shot learning via semantic similarity embedding. In: ICCV (2015)

Self-Supervision for 3D Real-World Challenges

Antonio Alliegro[1]([✉]), Davide Boscaini[2], and Tatiana Tommasi[1]

[1] Politecnico di Torino, Turin, Italy
{antonio.alliegro,tatiana.tommasi}@polito.it
[2] Fondazione Bruno Kessler, Trento, Italy
dboscaini@fbk.eu

Abstract. We consider several possible scenarios involving synthetic and real-world point clouds where supervised learning fails due to data scarcity and large domain gaps. We propose to enrich standard feature representations by leveraging self-supervision through a multi-task model that can solve a 3D puzzle while learning the main task of shape classification or part segmentation.

1 Introduction

Point clouds are the standard representation for 3D data, but they come with three main drawbacks: they are un-structured, un-ordered and eager for precise manual annotation due to many possible sources of noise. The first two issues make typical convolutional neural networks (CNN) unsuitable, while the third has initially guided research towards very well lab-controlled and synthetic CAD object datasets where labeling is simpler. The most recent results on those testbed are witnessing a trend of performance saturation raising the question of how to move forward. Self-supervised learning is helpful in this respect: a simple task like solving a 3D puzzle leverages on the spatial co-location of shape parts and exploits reliable knowledge on relative point positions at global and local level.

2 Method

We propose a *new multi-task end-to-end deep learning model for point clouds that combines supervised and self-supervised learning* (see Fig. 1). Specifically, we build on top of PointNet [5] and PointNet++ [6] backbones a deep architecture that solves 3D puzzles while jointly training a main supervised task. We show how these two tasks complement each other making the obtained model (a) more

---

Electronic supplementary material The online version of this chapter (https:// doi.org/10.1007/978-3-030-66415-2_48) contains supplementary material, which is available to authorized users.

A. Bartoli and A. Fusiello (Eds.): ECCV 2020 Workshops, LNCS 12535, pp. 704–708, 2020.
https://doi.org/10.1007/978-3-030-66415-2_48

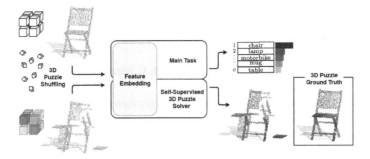

Fig. 1. Overview of the proposed multi-task approach

robust in case of scarce labeled data, (b) easier to transfer for adaptation and (c) more reliable for out of domain generalization. By extensive experiments across three different point clouds datasets we show that our multi-task method defines the new state-of-the-art for both shape classification and part segmentation in the most challenging real world settings.

More formally Our multi-task model can be described as the combination of two parametric non-linear functions: $\mathbf{\Phi}_{\theta_f,\theta_m}$ and $\mathbf{\Psi}_{\theta_f,\theta_p}$, where the subscripts of the parameters θ refer respectively to the feature extraction (f), main task (m), and puzzle solution (p) modules of our deep network. The feature encoder is shared between the two functions. For each sample $\mathbf{x}$ that enters the network, $\mathbf{\Phi}_{\theta_f,\theta_m}(\mathbf{x})$ is the output of the feature extractor and final fully connected part of the network. The loss function $\mathcal{L}_m(\mathbf{\Phi}_{\theta_f,\theta_m}(\mathbf{x}),\mathbf{y})$ measures the prediction error on the main task. The auxiliary function $\mathbf{\Psi}$ deals with a *puzzled* variant $\tilde{\mathbf{x}} = \mathcal{P}(\mathbf{x})$ of the original input point cloud. To get it, we start from $\mathbf{x}$, scale it to unit cube and split each axis into $l = 3$ equal lengths intervals forming l^3 voxels which are labeled according to their original position. Each vertex contained inside a voxel inherits its label. Finally, all the voxels are randomly swapped, producing a new shuffled point cloud. We indicate with $\tilde{S} = \{(\tilde{\mathbf{x}}_i,\tilde{\mathbf{y}}_i)\}_{i=1}^N$ the obtained puzzled samples where the voxel position label for each point is $y_{ik} \in \{1,\ldots,l^3\}$. Once these new displaced data are encoded in the feature latent space, a second network head focuses on solving the 3D puzzle problem by minimizing the auxiliary loss that measures the reordering error $\mathcal{L}_p(\mathbf{\Psi}_{\theta_f,\theta_p}(\tilde{\mathbf{x}}),\tilde{\mathbf{y}})$ in terms of difference between the assigned voxel label and the correct one per point. The training objective is:

$$\arg\min_{\theta_f,\theta_m,\theta_p} \sum_{i=1}^N \mathcal{L}_m(\mathbf{\Phi}_{\theta_f,\theta_m}(\mathbf{x}_i),\mathbf{y}_i) + \alpha\,\mathcal{L}_p(\mathbf{\Psi}_{\theta_f,\theta_p}(\tilde{\mathbf{x}}_i),\tilde{\mathbf{y}}_i), \qquad (1)$$

where both $\mathcal{L}_m$ and $\mathcal{L}_p$ are cross-entropy losses. Note that, while the first loss deals only with original samples, the second involve both original and puzzled samples, given the random nature of the voxel shuffling procedure.

The described learning problem has one main hyper-parameter α, which weights the self-supervised loss, we set $\alpha = 0.6$ for all our analysis.

3 Experiments on Cross-Domain Classification

We evaluate the cross-domain classification performance of our multitask on synthetic and Real-World data respectively from ModelNet40 [10] and ScanObjectNN [8]. The latter offering several splits of the same data with increasing difficulty (OBJ_ONLY to PB_T50_RS_BG) in terms of background, noise, shape perturbation.

Table 1. Shape classification accuracy (%) when training and testing is done on different domains (DG). If the unlabeled target data is provided at training time (DA), our multi-task is able to adapt and reduce the domain gap

Classification - Domain generalization and adaptation						
Method	ModelNet40 →					PB_T50_RS_BG →
		OBJ_ONLY	OBJ_BG	PB_T50_RS	PB_T50_RS_BG	AVG	ModelNet40
PointDAN [7]		56.42	44.84	48.99	34.39	46.16	54.66
PN	Baseline	54.74	43.58	44.96	34.25	44.38	47.43
	Our DG	54.53	49.68	45.22	36.28	46.43	39.30
	Our DA	58.53	47.58	46.70	35.85	47.16	51.54
PN++	Baseline	52.49	44.00	44.83	34.29	43.90	47.66
	Our DG	57.47	52.42	52.84	38.65	50.34	52.88
	Our DA	60.4	53.89	54.66	39.63	**52.14**	**56.07**
3DmFV [1]		30.90	24.00	24.90	16.40	24.05	51.50
PointCNN [4]		32.20	29.50	24.60	19.20	26.37	49.20
DGCNN [9]		49.30	46.70	36.80	27.20	40.00	54.70

Baselines. We use as reference the standard supervised baseline. It is a naïve variant of our method obtained by turning off the puzzle solver ($\alpha = 0$ in Eq. 1).

Domain Generalization. When training and test data are drawn from two very different distributions the model learned on the former one usually fails to generalize to the latter. We consider the DG setting when training on ModelNet40 and testing on ScanObjectNN and report results in Table 1. Our multi-task approach fully trained on only synthetic data shows a significant improvement with respect to the baseline with gains up to 6 and 8 pp in the OBJ_BG and with a still relevant gain of 2 and 4 pp in the most challenging PB_T50_RS_BG, respectively with PN and PN++ encoders. We also consider the inverse generalization direction from PB_T50_RS_BG to ModelNet40 with compelling results.

Unsupervised Domain Adaptation. We also investigated whether our multi-task approach could close the domain gap when unlabeled target data are available at training time, given its unsupervised nature these data are fed to our puzzle solver. DA results in Table 1 provide a positive answer showing a further increase in performance over the DG results. The recent PointDAN method [7] proposed to solve point cloud domain shifts by combining local nodes alignment and global features alignment. Table 1 shows that our multi-task approach largely outperforms this solution. Finally, an overall look at the performance of several recent point cloud networks is provided in the bottom part of Table 1: our multi-task approach establishes the new state-of-the-art.

4   Experiments on Part Segmentation

We focus on the case of scarce labeled data availability when dealing with part segmentation. The quality of the predicted part segmentation is evaluated in terms of the mean Intersection-over-Union (mIoU) metric.

Few-Shot and Semi-supervised. By following [2] we randomly sample 1% and 5% of the ShapeNetPart train set to evaluate the point features in a semi-supervised setting. The results in Table 2 indicate that our multi-task approach, although not improving over the baseline in the few-shot setting, in the semi-supervised setting outperforms the current state of the art in the case of only 1% of supervised data while practically matches it in the 5% case. We plot some visualizations out of our 1% part segmentation experiment in Fig. 2 for chairs and lamps. Regarding chairs, our multi-task approach seems to allow a better recognition of the armrests. Indeed the position of these relative small parts of the chair may be better learned thanks to the auxiliary puzzle solution task. A similar consideration may be done for the lamp basis.

Table 2. Accuracy (mIoU) for part segmentation on ShapeNetPart with limited annotations

Method	1%	5%
SO-Net [3]	64.00	69.00
PointCapsNet [11]	67.00	70.00
CCD [2]	68.20	**77.70**
Baseline	64.52	75.75
Our FS	64.49	75.07
Our SS	**71.95**	77.42

Fig. 2. Part segmentation of chairs and lamps when 1% of training data are available. Baseline prediction (top left) and our approach (bottom right). Black points denotes predictions whose maximum value was not a chair or lamp part

References

1. Ben-Shabat, Y., Lindenbaum, M., Fischer, A.: 3DmFV: three-dimensional point cloud classification in real-time using convolutional neural networks. IEEE RA-L (2018)
2. Hassani, K., Haley, M.: Unsupervised multi-task feature learning on point clouds. In: ICCV (2019)
3. Li, J., Chen, B.M., Hee Lee, G.: SO-Net: self-organizing network for point cloud analysis. In: CVPR (2018)
4. Li, Y., Bu, R., Sun, M., Wu, W., Di, X., Chen, B.: PointCNN: convolution on x-transformed points. In: NIPS (2018)
5. Qi, C.R., Su, H., Mo, K., Guibas, L.J.: PointNet: deep learning on point sets for 3D classification and segmentation. In: CVPR (2017)
6. Qi, C.R., Yi, L., Su, H., Guibas, L.J.: PointNet++: deep hierarchical feature learning on point sets in a metric space. In: NIPS (2017)
7. Qin, C., You, H., Wang, L., Kuo, C.C.J., Fu, Y.: PointDAN: a multi-scale 3D domain adaption network for point cloud representation. In: NIPS (2019)
8. Uy, M.A., Pham, Q.H., Hua, B.S., Nguyen, D.T., Yeung, S.K.: Revisiting point cloud classification: a new benchmark dataset and classification model on real-world data. In: ICCV (2019)
9. Wang, Y., Sun, Y., Liu, Z., Sarma, S.E., Bronstein, M.M., Solomon, J.M.: Dynamic graph CNN for learning on point clouds. TOG **38**, 1–12 (2019)
10. Wu, Z., et al.: 3D ShapeNets: a deep representation for volumetric shapes. In: CVPR (2015)
11. Zhao, Y., Birdal, T., Deng, H., Tombari, F.: 3D point capsule networks. In: CVPR (2019)

Diversified Mutual Learning for Deep Metric Learning

Wonpyo Park[1][✉], Wonjae Kim[1], Kihyun You[2,3], and Minsu Cho[2]

[1] Kakao Corporation, Seongam, Korea
wppark.pio@gmail.com
[2] POSTECH, Pohang, Korea
[3] Kakao Enterprise, Seongam, Korea

Abstract. Mutual learning is an ensemble training strategy to improve generalization by transferring individual knowledge to each other while simultaneously training multiple models. In this work, we propose an effective mutual learning method for deep metric learning, called *Diversified Mutual Metric Learning*, which enhances embedding models with diversified mutual learning. We transfer relational knowledge for deep metric learning by leveraging three kinds of diversities in mutual learning: (1) *model diversity* from different initializations of models (2) *temporal diversity* from different frequencies of parameter update, and (3) *view diversity* from different augmentations of inputs. Our method is particularly adequate for *inductive transfer learning* at the lack of large-scale data, where the embedding model is initialized with a pretrained model and then fine-tuned on a target dataset. Extensive experiments show that our method significantly improves individual models as well as their ensemble. Finally, the proposed method with a conventional triplet loss achieves the state-of-the-art performance of Recall@1 on standard datasets: 69.9 on CUB-200–2011 and 89.1 on CARS-196.

Keywords: Model diversification · Mutual learning · Deep metric learning · Inductive transfer learning

1 Introduction

Mutual learning [39] is an effective ensemble training strategy to improve generalization ability of learners. In mutual learning, multiple models, *i.e.*, *cohort*, learn and teach each other simultaneously during the training time. For example, in the work of [39], two classifier models with different initializations are trained to predict similar output distributions given the same input data. Despite its

---

W. Park and W. Kim—Equal contribution.

Electronic supplementary material The online version of this chapter (https://doi.org/10.1007/978-3-030-66415-2_49) contains supplementary material, which is available to authorized users.

A. Bartoli and A. Fusiello (Eds.): ECCV 2020 Workshops, LNCS 12535, pp. 709–725, 2020.
https://doi.org/10.1007/978-3-030-66415-2_49

effectiveness, previous methods of mutual learning [1,39] are limited to the task of classification. In this work, we propose a mutual learning method for deep metric learning, which aims at learning deep embedding models that maps similar instances to nearby points on a manifold in the embedding space and dissimilar instances apart from each other.

In metric learning, due to small scales of training datasets, the embedding models are commonly initialized using an ImageNet-pretrained network such as Inception-BN [9] or ResNet-50 [6], and then fine-tuned on a target dataset. The common use of inductive transfer learning in metric learning limits the diversity of embedding models in mutual learning. Unlike the previous settings [1,39], where models in the cohort are trained from scratch with diverse initializations, these models with the same backbone network prevents mutual learning from distilling knowledge from each other during training.

To tackle the issue, we propose an effective mutual learning method for deep metric learning, called **D**iversified **M**utual **M**etric **L**earning (DM^2), which enhances the generalization ability of embedding models with diversified mutual learning. We transfer the relational knowledge with pairwise distance between instances while diversifying the parameter update paths of models of the cohort with our three diversities: (1) **model diversity** to leverage different initializations of models (2) **temporal diversity** to diversify the frequency of parameter update, and (3) **view diversity** to exploits diverse inputs with different augmentations. Note that the model diversity is naturally induced by training multiple models as in the work of [39], thus being the core of mutual learning per se. In this work, we introduce two additional diversities to mutual learning and also propose a mutual learning method using relation matrices for metric learning.

Extensive experiments on standard image retrieval datasets [13,17,29] for deep metric learning show that the proposed method, DM^2, significantly improves performance of individual models as well as their ensemble, compared to conventional deep metric learning. Moreover, DM^2 is also an effective regularizer that prevents severe overfitting on small training datasets in the latter part of the training. The benefit of DM^2 monotonically grows as we increase the number of models in the cohort. The proposed method combined with a conventional triplet loss achieves the state-of-the-art performance of Recall@1 on the standard datasets: 69.9 on CUB-200–2011 and 89.1 on CARS-196.

2   Related Work

Our work encompasses the studies of deep metric learning, knowledge transfer, and mutual learning. In this section, we summarize each and explain their relations to our work.

2.1   Deep Metric Learning

Loss Function. Devising a better loss function is one of the main challenges of the recent deep metric learning studies. One family of the loss functions is pairwise distance-based loss which samples positive or negative pairs within a given mini-batch. The objective of the losses is to let the distance of the positive pair to be small, while that of negative pair to be large. Contrastive loss [5] samples pairs of any two examples, while the Triplet loss [26] samples triplets of anchor, positive and negative examples. In Triplet loss, the distance between the anchor and positive is trained to be smaller than the distance between anchor and negative by a certain margin. Extended from the two losses, several methods [17,28,32] are proposed to fully explore pairwise relation within a mini-batch. Other than pairwise ones, losses using proxy [16,21,38] are also proposed. Those methods assign single or multiple proxy vectors for each class, and distances are calculated with the proxies rather than individual embeddings. For our experiments, we adopt a simple pairwise distance-based loss: Triplet loss.

Pair Sampling. Sampling pairs within a mini-batch is crucial to ensure a stable convergence and a better performance for the pairwise distance-based losses. In [26], semi-hard negative examples are sampled rather than the hardest ones, which improves the performance of Triplet loss for very large face datasets. Wu *et al.* [31] propose the distance-weighted sampling considering the distribution of random points in a unit-hyper sphere. Recently, Wang *et al.* [30] cast a sampling problem to a pair weighting problem based on a gradient analysis. Other than the pair sampling within a mini-batch, Suh *et al.* [27] devised a batch sampling method which considers an inter-class relationship. For our experiments, we used the distance-weighted sampling as the pair sampling method.

Multiple Heads. Similarly to the ensemble of totally separate models, multiple heads that share the backbone network for better embeddings is studied intensively in recent work. The main focus is to diversify the output of each head for robust embedding. For example, the importance of data examples is re-weighted differently for each head based on the gradients of other heads [18] or difficulty of the examples [35]. In [25], a dataset is divided into multiple subsets and each head is trained for different subsets. Kim *et al.* [11] applied gating attention mechanism to heads, and the gates are trained to minimize their spatial overlapping regions. In [10], each head receives different high-order moments of the feature. Although multiple heads approach can be adopted to further improve the performance of the deep metric learning model, it is beyond the scope of the proposed method; the models we used in the experiments are single-headed.

2.2 Knowledge Transfer and Mutual Learning

Knowledge Transfer. Knowledge distillation [7,8,14,22,36] aims to transfer the knowledge from a teacher to a student model. The typical setup is to train a student model to mimic the outputs of a teacher model. Traditional knowledge distillation [7] uses class probability as the knowledge to transfer. However, the encoding models of embedding learning do not produce a certain class probability. Therefore in deep metric learning, DarkRank [2] transfers similarity ranks

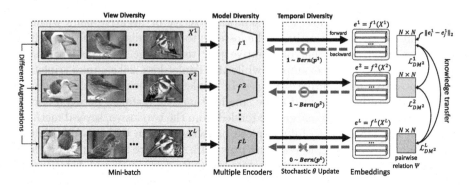

Fig. 1. Given a mini-batch, DM^2 applies view diversity that exploits different augmentations for each encoder which are simultaneously trained. To differentiate the frequency of parameter update among the cohort, DM^2 applies temporal diversity that stochastically updates the parameter of models according to their update probability. The relational knowledge of each model is mutually shared between the models by transferring their pairwise distance of embedding vectors.

between embeddings, and concurrently, relational knowledge distillation [19] and compact networks [34] proposed direct transfer of pairwise distance between embeddings. Multi-type knowledge [14] transfers graph structure lies in instance features and applied the method to the classification task. We also set a pairwise relation matrix of embeddings within a mini-batch, as the knowledge to transfer (Fig. 1).

Mutual Learning. The learning procedure of knowledge distillation requires a fully-trained teacher model. Therefore, practitioners should endure tedious and time-consuming two-stage training pipeline: train a teacher model, and then distill the teacher to a student. To overcome this shortcoming, deep mutual learning [39] proposes a single stage training pipeline for knowledge distillation. The method minimizes the KL divergence between the class probability of multiple models on-the-fly in the training phase, and [1] scaled the method up to hundreds of GPUs by adopting distributed setup. In this online setting of knowledge distillation, there is no clear distinction between the teacher and student as in the classical setting of knowledge distillation, but the *diversity* among the cohort should be guaranteed to make the knowledge be transferred.

Although the effectiveness of mutual learning is proven for classification tasks, all these schematics have not been applied in other tasks. In this paper, DM^2 aims to provide adequate techniques to diversify the models of the cohort. The diversification techniques are particularly important in deep metric learning where the effect of original *model diversity* is not guaranteed due to the same initial weights of the pretrained models.

3 Diversified Mutual Metric Learning

In the framework of the **D**iversified **M**utual **M**etric Learning (DM^2), multiple models, *i.e.*, *cohort*, are involved in the training process and transfer their knowledge to each other. By diversifying the cohort, we have observed more diversity brings better generalization power for every single model and their ensemble. In this section, we will define our objective function and deliver detailed explanations of our three diversities.

Notation. Our cohort is comprised of multiple diverse models f^i, where i is an index that goes from one to the size of the cohort L. The model, or the encoder, f^i maps an image x into an embedding vector $e^i \in \mathbb{R}^d$, *i.e.*, $e^i = f^i(x)$; for deep metric learning, the typical choice of d is 128 or 512 [30,31]. Then the distance between two images x_1 and x_2 can be measured via Euclidean distance of their respective embeddings: $\left\| f^i(x_1) - f^i(x_2) \right\|_2 = \left\| e_1^i - e_2^i \right\|_2$. Deep metric learning loss optimizes the encoders such that the distance between semantically similar images to be close and otherwise far apart. Note that the architecture of the models can either be homogeneous or heterogeneous and its parameters can be diverse or identical, which we will treat as *model diversity* in later Sect. 3.2.

3.1 Objective Function

We set the knowledge to be shared among the cohort to pairwise relation matrix of embeddings that each model produces for a given same mini-batch. More concretely, for embeddings $\{e_k^i\}_{k=1}^N$ of model f^i, where N is the size of mini-batch; we define relation matrix of f^i as $\Psi_{k,l}^i = \left\| e_k^i - e_l^i \right\|_2$. The knowledge sharing from f_k to f_l is defined as a difference between the relation matrix Ψ as follows:

$$\mathcal{L}_{\mathrm{DM}^2}^{l \leftarrow k} = \frac{1}{N^2} \sum_i^N \sum_j^N \left\| \Psi_{i,j}^l - \Psi_{i,j}^k \right\|_2^2. \tag{1}$$

For each model f^l, the knowledge is transferred to all other models in the cohort $\{f^k\}_{k \neq l}$, defining the $\mathcal{L}_{\mathrm{DM}^2}^l$ as follows:

$$\mathcal{L}_{\mathrm{DM}^2}^l = \frac{1}{L-1} \sum_{k \neq l}^L \mathcal{L}_{\mathrm{DM}^2}^{l \leftarrow k}. \tag{2}$$

The final objective $\mathcal{L}^l$ for the model f^l in our diversified mutual metric learning is the sum of the deep metric learning loss $\mathcal{L}_{\mathrm{DML}}^l$ and $\mathcal{L}_{\mathrm{DM}^2}^l$ using Lagrangian multiplier λ_{DM^2}:

$$\mathcal{L}^l = \mathcal{L}_{\mathrm{DML}}^l + \lambda_{\mathrm{DM}^2} \cdot \mathcal{L}_{\mathrm{DM}^2}^l. \tag{3}$$

Algorithm 1. Training procedure of DM^2

Require: L models $\{f^l\}_{l=1}^L$ with parameters $\{\theta^l\}_{l=1}^L$ ▶ *Model Diversity* (MD)
1: **for** the number of iterations **do**
2:      Sample a mini-batch $\{x_k\}_{k=1}^N$ of size N
3:      **for** $l = 1{:}L$ **do**   ▶ Done in parallel
4:          $\{A_k\}_{k=1}^N \sim \mathcal{A}$ ▶ Sample the augmentations
5:          $\{x_k'\}_{k=1}^N = \{A_k(x_k)\}_{k=1}^N$ ▶ *View Diversity* (VD)
6:          $\{e_k^l\}_{k=1}^N = \{f^l(x_k')\}_{k=1}^N$ ▶ Infer embeddings
7:      **end for**
8:      (Barrier) Wait until the inference of entire cohort finished
9:      **for** $l = 1{:}L$ **do**   ▶ Done in parallel
10:          Compute relation matrix Ψ^l from $\{e_k^l\}_{k=1}^N$ for f^l
11:          Compute $\mathcal{L}_{\text{DML}}^l$ using $\{e_k^l\}_{k=1}^N$ for f^l
12:          Compute $\mathcal{L}_{\text{DM}^2}^l$ using Ψ^l and $\{\Psi^k\}_{k \neq l}$ for f^l
13:          $\mathcal{L}^l = \mathcal{L}_{\text{DML}}^l + \lambda_{\text{DM}^2} \cdot \mathcal{L}_{\text{DM}^2}^l$
14:          $p \sim Bern(p^l)$ ▶ Sample p from a Bernoulli distribution
15:          **if** p is 1 **then**   ▶ *Temporal Diversity* (TD)
16:              Update θ^l w.r.t $\mathcal{L}^l$
17:          **end if**
18:      **end for**
19: **end for**

Throughout the experiments, we mostly used Triplet loss [26] for all $\{\mathcal{L}_{\text{DML}}^l\}_{l=1}^L$. We applied a linear warm-up strategy for λ_{DM^2}, which makes each encoder to focus more on the metric learning loss during the early phase of the training.

3.2 Diversities

We now deliver detailed explanations of three diversities that could make the cohort to share more rich knowledge for mutual learning. The combination of the three diversities aims to diversify the update paths of multiple models in the parameter space, and the diversified paths lead to different local optima for each model. Due to the diversities, DM^2 can form a cohort of diverse models. We concretely depict the general framework of DM^2 in Algorithm 1.

Model Diversity. The first is a *model diversity* (MD), which diversifies the cohort by leveraging different initializations of models to encourage the individual models to explore a different region and settle at different local optima after training. MD is naturally induced by the learning procedure of mutual learning [39] where multiple models mutually teach each other. However, this diversity is hard to be achieved in the setting of deep metric learning because models are initialized with the same pretrained parameters, which is essential in inductive transfer learning, do not allow much diverse models, and even forcing them to be identical.

Temporal Diversity. The second is a *temporal diversity* (TD), which diversifies the cohort by discriminating their frequency of parameter update. To discriminate the frequency of parameter update, we let the models stochastically update their parameters according to their update probabilities. We assign diverse update probabilities $\{p^l = 2^{-(l-1)}\}_{l=1}^L$ to each model according to its index. The first model has the highest update probability which is one, and the last model has the lowest update probability. When the update probability is high, the model moves fast along the update path and it results in a more fitted model on the training dataset. On the contrary, when the update probability is low, the model results in less fitted on the dataset. Combining the models with different temporal steps in update helps to form a more diversified cohort.

View Diversity. The third is a *view diversity* (VD), which diversifies the cohort by exploiting diverse inputs with different augmentations for each model. Since the gradients of each model are computed using input data and activations, different augmentations to the mini-batch of each model yield diversified gradients. In addition, VD helps to learn a more robust embedding model, as the models learn to construct the relational structure Ψ that is invariant to noise-injected data.

We defer the implementation details regarding the parameter initialization of the cohort ($\{\theta^l\}_{l=1}^L$) and the augmentation family $\mathcal{A}$ to Sect. 4.2.

3.3 Distributed Parallel Learning

Computing $\mathcal{L}_{DM^2}^l$ requires only the relation matrices from the other encoders $\{\Psi^k\}_{k \neq l}$. Thus the computation including generating mini-batches with different views, inferring the encoder, and calculating the gradients can be done in parallel by allocating encoders to separate computational resources, *e.g.*, GPU. As a result, the training time of DM^2 almost stayed the same compared to that of a single model at the cost of computational resources.

4 Experiment

In this section, we report the efficacy of the DM^2, and the implementation details used in our experiments. We conduct experiments on the following standard datasets for metric learning, and follow the evaluation protocol, and train/test splits as proposed in [17].

4.1 Dataset

We run experiments on four datasets: CUB-200–2011 [29], Cars-196 [13], Stanford Online Products (SOP) [17] and In-Shop Clothes Retrieval (In-Shop) [15].

CUB-200–2011 consists of images of 200 bird species. Half of 200 classes (5,864 images) are used for the training and the remaining 100 classes (5,924 images) are used for the testing. Cars-196 consists of images of 196 car models. As in the CUB-200–2011, half of 196 classes (8,052 images) are used for the training and the remaining 98 classes (8,131 images) are used for the testing.

SOP, which consists of images containing 22,634 product classes, is a much larger dataset compared to the former two datasets. For the training, 11,318 classes (59,551 images) are used and the remaining 11,316 classes (60,499 images) are used for the testing. In-Shop consists of images of 11,735 clothes products. For the training, 3,997 classes (25,882 images) are used, and the remaining 7970 classes (26,830 images) are split into two subsets (query set and gallery set) for the testing.

4.2   Implementation Detail

Following the standard practice of deep metric learning studies, we use the inception network with the batch normalization (BN-Inception) [9] and the ResNet-50 [6] pretrained on ILSVRC 2012-CLS (ImageNet) [24] as our backbone networks. We note again that these pretrained backbones make the effect of model diversity weaker. However at the same time, the pretraining is strongly required for deep metric learning since the size of the datasets is very small to learn a visual feature extraction in an end-to-end manner, which makes the use of ImageNet pretraining essential.

We add a linear projection layer after the last pooling layer of a backbone network and apply l_2 normalization on the output embedding e of the linear projection layer, e.g., $\frac{e}{\|e\|_2}$. We follow the image pre-processing and the augmentation family of a recent state-of-the-art method, HORDE [10], which uses 256×256 random crop from randomly resized images and apply a random horizontal flip.

We train the parameters of batch normalization layers [9] in backbone networks for SOP and In-Shop datasets and freeze them for CUB-200–2011 and Cars-196 datasets following the procedure done in [21,30].

Regardless of the model architecture and dataset choice, the training was done with Adam optimizer [12] with the initial learning rate of $3 \cdot 10^{-5}$ and weight decay of $5 \cdot 10^{-4}$; and the mini-batch size was 120. All experiments were conducted using pytorch 1.3.1 [20] with CUDA 10.2 on Nvidia V100 GPU.

To make enough positive pairs within a mini-batch, we follow the batch construction of FaceNet [26]. For every mini-batch, we randomly sample a certain number of classes and sample 5 images per class for small datasets (CUB-200–2011, Cars-196) and 2 images per class for large datasets (SOP, In-Shop). We apply a warm-up strategy for λ_{DM^2}: linearly increasing it from 0 to 20 for the first three epochs of the training.

Unless specified, we adopt Triplet loss [26] as our deep metric learning loss $\mathcal{L}_{DML}$ and the used distance weighted sampling [31] as a pair sampling method. Note that we carefully tuned our hyper-parameters to build strong baseline models. The source codes will be released in public upon acceptance.

Table 1. Recall@1 performance on CUB-200–2011 and Cars-196 datasets. We used cohort size of 4 for DM^2, and report how the performance increases with the added diversities: model diversity (MD), temporal diversity (TD), and view diversity (VD). We highlight the best performing single model and ensembled model.

Methods	CUB-200–2011					Cars-196				
	Net A	Net B	Net C	Net D	Ens	Net A	Net B	Net C	Net D	Ens
Independent	66.07	66.18	67.21	65.97	70.71	85.24	84.78	85.22	85.26	89.87
MD	67.93	67.12	67.18	67.32	69.60	85.79	86.07	86.31	86.29	87.86
MD+TD	69.45	69.01	68.10	67.59	71.25	89.00	88.49	87.54	86.21	90.49
MD+VD	68.26	68.72	68.42	68.45	71.69	87.91	87.22	87.39	87.49	90.10
MD+TD+VD	**69.90**	69.50	68.86	69.11	**72.89**	**89.10**	88.29	87.65	86.33	**91.48**

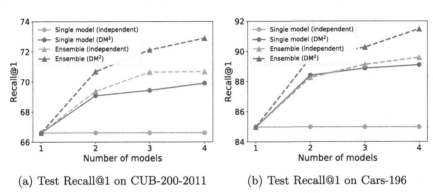

(a) Test Recall@1 on CUB-200-2011   (b) Test Recall@1 on Cars-196

Fig. 2. Plots showing Recall@1 on the evaluation sets of CUB-200–2011 and Cars-196 with various cohort size (1, 2, 3, 4). For the single model results of DM^2, we report the result of the first model ($l=1$) among the multiple models of the cohort.

4.3 Efficacy of Diversities

The result of Table 1 shows the effect of the diversities in the application of mutual learning for deep metric learning. We used a cohort of size 4 for these experiments, the size of the embedding vector is set to 512, and the networks are all BN-Inception architecture initialized with the same parameter pretrained on ImageNet. The only differences in the parameter are induced by the random initialization of the linear projection layer.

The first row of Table 1 (independent) indicates the results of four independently trained networks (*i.e.*, $\lambda_{DM^2} = 0$) and their concatenation-based ensemble performance. The MD only results are identical to the naive adoption of mutual learning. As expected, the naive adoption brings worse result than the individual learning without construct cohort. On the contrary, one can see that the addition of our proposed novel diversities TD, VD brings superior results; and applying all diversities brings the state-of-the-art single model performance (see Table 2).

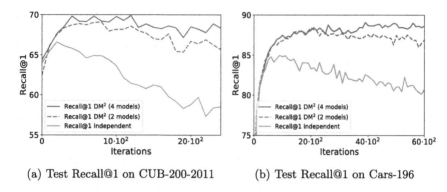

(a) Test Recall@1 on CUB-200-2011   (b) Test Recall@1 on Cars-196

Fig. 3. Plots showing the changes of Recall@1 on the evaluation sets of CUB-200–2011 and Cars-196 during the training. For DM^2, the Recall@1 results are from the first model ($l=1$) among the multiple models of the cohort.

4.4   Scalability of DM^2

Figure 2 delivers how the performance increases for larger cohort size. For both datasets, CUB-200–2011 and Cars-196, the mutual learning with all three diversities (*i.e.*, DM^2) results in a monotonic increment of Recall@1 along with the cohort size. On the contrary, the ensemble performance of individually trained models converges faster at the lower Recall@1 than that of DM^2.

Figure 3 depicts how mutual learning effects as a regularizer, which makes the model more robust. The plot shows individually trained models suffer a problem of severe overfitting on the latter part of the training. However, the model trained with DM^2 keeps outperforming the baseline without severe overfitting. Note that its robustness keeps increase for a larger cohort size.

4.5   Comparison with Other Deep Metric Learning Studies

Since the single model performance improved significantly with DM^2, we compared the recall performance of DM^2 with the current state-of-the-art deep metric learning methods. For fair comparison, the number of embedding dimension and the backbone network were set according to those of compared methods. The cohort size of DM^2 is fixed to 4.

The result of Table 2 shows the recalls measured on CUB-200–2011, Cars-196, and SOP. For ResNet-50, the model of 512 embedding dimension trained with DM^2 and Triplet loss achieves the best Recall@1 for all three datasets: CUB-200–2011, Cars-196, and SOP. For BN-Inception, it achieved the best Recall@1 for CUB-200–2011, Cars-196, and the second best Recall@1 for SOP. For the results on In-shop dataset shown in Table 3, our method achieves comparable performance with the best, less than 1% point difference.

Table 2. Recall@K comparison with state-of-the-art metric learning methods. We divide methods with same backbone network. "Dim." refers to the embedding dimension of the model. Boldface refers to the highest performance among same backbone. The numbers under the datasets refer to recall at K. The results of DM^2 are from a *single* model for fair comparison, which is the first model (l=1) among multiple models of cohort.

B	Methods	Dim.	CUB-200–2011				Cars-196				SOP			
			1	2	4	8	1	2	4	8	1	10	10^2	10^3
ResNet-50	Margin [31]	128	63.6	74.4	83.1	90.0	79.6	86.5	91.9	95.1	72.7	86.2	93.8	98.0
	DC [25]	128	65.9	76.6	84.4	**90.6**	84.6	90.7	94.1	96.5	75.9	88.4	94.9	98.1
	MIC [23]	128	66.1	76.8	**85.6**	-	82.6	89.1	93.2	-	77.2	89.4	95.6	-
	TML [33]	512	62.5	73.9	83.0	89.4	86.3	92.3	**95.4**	**97.3**	78.0	91.2	96.7	99.0
	Triplet + DM^2	128	64.1	75.0	83.2	89.4	84.7	89.9	93.5	95.8	79.9	91.3	96.6	99.0
	Triplet + DM^2	512	**66.7**	**77.1**	85.2	90.5	**86.9**	**92.0**	94.5	96.8	**80.4**	**91.8**	**96.8**	**99.1**
BN-Inception	HTL [4]	512	57.1	68.8	78.7	86.5	81.4	88.0	92.7	95.7	74.8	88.3	94.8	98.4
	NSM [37]	512	59.6	72.0	81.2	88.4	81.7	88.9	93.4	96.0	73.8	88.1	95.0	-
	MS [30]	512	65.7	77.0	86.3	91.2	84.1	90.4	94.0	96.5	78.2	90.5	96.0	98.7
	SoftTriplet [21]	512	65.4	76.5	84.5	90.4	84.5	90.7	94.5	96.9	78.3	90.3	95.9	-
	HORDE [10]	512	66.3	76.7	84.7	90.6	83.9	90.3	94.1	96.3	**80.1**	**91.3**	96.2	98.7
	Triplet + DM^2	512	**69.9**	**79.7**	**86.5**	**91.4**	**89.1**	**93.3**	**95.8**	**97.6**	78.8	90.9	**96.3**	**98.9**

Table 3. Recall@K comparison with state-of-the-art metric learning methods on In-Shop. "Dim." refers to the embedding dimension of the model. Bold-face refers the highest performance among all models. The results of DM^2 are from a *single* model for fair comparison, which is the first model (l=1) among multiple models of cohort.

Methods	Dim.	In-Shop			
		1	10	20	30
FashionNet [15]	4096	53.0	73.0	76.0	77.0
NSM [37]	512	88.6	97.5	98.4	**98.8**
A-BIER [18]	512	83.1	95.1	96.9	97.5
ABE-8 [11]	512	87.3	96.7	97.9	98.2
MS [30]	512	89.7	**97.9**	98.5	**98.8**
HORDE [10]	512	**90.4**	97.8	98.4	98.7
Triplet + DM^2	512	89.6	97.8	**98.6**	**98.8**

4.6 Combining DM^2 with Knowledge Distillation

We combine the ensembled model of DM^2 with *relational knowledge distillation* [19]. For a detailed comparison, we also report the results of other recently proposed knowledge distillation methods for deep metric learning [2,34]. To build a strong teacher model for the distillation methods, we build an ensemble of 4 independently trained models. For DM^2, the cohort of size 4 is mutually trained in parallel. Each model is based on BN-Inception with the embedding dimension of 512, and we transfer the knowledge of the ensemble to a single model (student).

Table 4. Recall@1 comparison between knowledge distillation methods for metric learning and DM^2. **P** indicates whether the teacher and student are trained in a parallel manner. For knowledge distillation methods, **T** and **S** refer to a teacher and a student respectively. For DM^2, **T** and **S** refer to an ensemble and a single model (the first model where l=1) respectively. '→' means that two methods were applied in a sequence.

Methods	P	CUB-200–2011				Cars-196			
		2×models		4×models		2×models		4×models	
		T	S	T	S	T	S	T	S
Baseline	X	-	66.61	-	66.61	-	84.98	-	84.98
RKD-DA [19]	X	69.36	70.13	70.68	70.64	88.26	88.82	89.59	89.95
DarkRank [2]	X	69.36	68.28	70.68	68.09	88.26	87.92	89.59	87.99
Compact [34]	X	69.36	68.70	70.68	67.98	88.26	**89.50**	89.59	89.48
DM^2	O	**70.68**	69.07	**72.89**	69.90	**89.57**	88.41	**91.48**	89.10
DM^2 →RKD-DA	X	**70.68**	**70.54**	**72.89**	**71.22**	**89.57**	89.14	**91.48**	**90.21**

We apply the methods on the final representations, *i.e.*, embeddings. We describe each method briefly as follows:

- **Relational Knowledge Distillation** [19] transfers relative distance and angle formed by points in the embedding space of the model. Between the three combinations of loss they proposed, we used distance and angle combination (RKD-DA). Following the original paper, we did not apply $\mathcal{L}_{\mathrm{DML}}$ while distillation and set the hyper-parameters $\lambda_{\mathrm{distance}}$ and λ_{angle} to 20 and 40 respectively.
- **DarkRank** [2] transfers similarity ranks between examples. Between the two losses they proposed, we use HardRank loss. We carefully tune the hyper-parameters for DarkRnak using a grid search. The hyper-parameters α, β, and $\lambda_{\mathrm{DarkRank}}$ are set to 3, 3, and 1, respectively.
- **Compact Networks** [34] transfers absolute distance among data examples. Following the original paper, the hyper-parameter $\lambda_{\mathrm{compact}}$ is set to 10.

As shown in Table 4, the result of pure DM^2 is on par with other two-staged distilled performances. By combining DM^2 with the RKD-DA, which showed the best distillation performance, the final single model performance reaches the highest Recall@1 71.22 and 90.21 for CUB-200–2011 and Cars-196, respectively. These results show that the ensemble model of DM^2 indeed holds better knowledge to transfer to the student (single model).

4.7   Heterogenous Cohort

For this experiment, we conducted experiments on heterogeneous cohort, which comprises models with diverse architecture. For that, we examined various construction for a cohort of size 4 using BN-Inception and ResNet-50, and the embedding dimension was 512.

Table 5. Recall@1 comparison varying the combinations of heterogenous cohorts. BNI refers to BN-Inception and R50 refers to ResNet-50. × indicates the number of cohorts with the architecture. The results of the Single are from the first model (l=1) among multiple models of cohort. Baseline is a single model trained independently without mutual learning with cohorts.

Combinations	CUB-200–2011			Cars-196		
	BNI (Single)	R50 (Single)	Ens	BNI (Single)	R50 (Single)	Ens
Baseline	66.61	64.01	-	84.98	82.70	-
4×BNI	**69.90**	-	72.89	**89.10**	-	91.48
3×BNI 1×R50	68.69	66.63	**74.17**	87.13	**88.76**	**91.76**
2×BNI 2×R50	68.80	**66.90**	73.90	86.55	87.62	91.46
1×BNI 3×R50	68.08	66.44	71.59	86.41	86.20	90.21
4×R50	-	66.71	68.87	-	86.93	88.62

Table 5 shows the result of the experiment, and the row 4×BNI is identical to our previous experimental setting. We set the temporal diversity individually to the network type, for example, 3×BNI 1×R50 setting has the update rate of $(1, 0.5, 0.25, 1)$ for three BNI and one R50, respectively. The results show a feeble possibility of architecture diversity, and it is noteworthy that the ensemble performances of CUB-200–2011 and Cars-196 were found to be better for the heterogeneous cohort than the homogenous cohort. Also, one can see that the single model performance of R50 grows with the addition of BNI to its cohort.

4.8   DM^2 with Other $\mathcal{L}_{DML}$

DM^2 can be effective with any deep metric learning objectives. To validate that, we conduct an experiment of combining DM^2 with widely adopted deep metric learning losses, *e.g.*, Triplet, Contrastive, Proxy-NCA, Binomial-Deviance, and ArcFace. The size of the cohort is set to 4, the models of BN-Inception with 512 embedding dimension are adopted. We describe the hyper-parameters of each method.

- **Triplet** [26]: we follow the hyper parameter specified in Sect. 4.2.
- **Contrastive** [5]: we set the margin to 1.0 and λ_{DM^2} to 20.
- **Proxy-NCA** [16]: we set the number of the proxies to 1 and λ_{DM^2} to 10^2.
- **Binomial-Deviance** [32]: we set the alpha and beta to 2 and 0.5 respectively following [32], and λ_{DM^2} to 20.
- **ArcFace** [3]: we set the scale factor and margin to 64 and 0.5 respectively following [3], and λ_{DM^2} to $2 \cdot 10^3$.

The result in Table 6 delivers that DM^2 improves the performance of a *single model* with any deep metric learning objectives.

Table 6. Applicability of DM^2 on different deep metric learning objectives. 'Independent' refers to the case when a model is trained independently with only $\mathcal{L}_{\text{DML}}$. For DM^2, the results of Single Model are from the first model ($l=1$) among multiple models of cohort.

(a) Recall@1 on CUB-200-2011.

$\mathcal{L}_{\text{DML}}$	Independent		DM^2	
	Single Model	Ensemble	Single Model	Ensemble
Contrastive [5]	61.97	63.94	66.40	67.45
Triplet [26]	66.07	70.71	69.90	72.89
Binomial-Deviance [32]	64.83	69.01	67.92	69.90
Proxy-NCA [16]	60.11	67.42	64.95	68.67
ArcFace [3]	62.09	69.05	67.74	70.32

(b) Recall@1 on Cars-196.

$\mathcal{L}_{\text{DML}}$	Independent		DM^2	
	Single Model	Ensemble	Single Model	Ensemble
Contrastive [5]	78.93	82.97	83.10	86.72
Triplet [26]	85.24	89.87	89.10	91.48
Binomial-Deviance [32]	83.94	89.16	88.33	89.20
Proxy-NCA [16]	83.70	90.94	86.74	89.01
ArcFace [3]	85.11	91.13	88.45	90.16

5   Conclusion

In this paper, we propose **D**iversified **M**utual **M**etric **L**earning (DM^2) to effectively apply mutual learning on the task of deep metric learning. In DM^2, three diversities are proposed to enrich the knowledge shared by the cohort of mutual learning: model diversity, temporal diversity, and view diversity. We proved that the temporal and view diversities were especially essential to the application of mutual learning since deep metric learning requires pretrained models due to its small size dataset. By combining all three diversities carefully, DM^2 results the state-of-the-art performance on widely adopted deep metric learning datasets: CUB-200–2011 and Cars-196, and SOP.

Acknowledgement. This work was partly supported by Institute for Information & communications Technology Promotion (IITP) grant funded by the Korea government (MSIP) (No. 2019-0-01906, Artificial Intelligence Graduate School Program (POSTECH)) and Basic Science Research Program (NRF-2017R1E1A1A01077999) through the National Research Foundation of Korea (NRF) funded by the Ministry of Science, ICT.

References

1. Anil, R., Pereyra, G., Passos, A., Ormandi, R., Dahl, G.E., Hinton, G.E.: Large scale distributed neural network training through online distillation. arXiv preprint arXiv:1804.03235 (2018)
2. Chen, Y., Wang, N., Zhang, Z.: Darkrank: accelerating deep metric learning via cross sample similarities transfer. In: Thirty-Second AAAI Conference on Artificial Intelligence (2018)
3. Deng, J., Guo, J., Xue, N., Zafeiriou, S.: Arcface: additive angular margin loss for deep face recognition. In: Proceedings of the IEEE Conference on Computer Vision and Pattern Recognition, pp. 4690–4699 (2019)
4. Ge, W.: Deep metric learning with hierarchical triplet loss. In: Proceedings of the European Conference on Computer Vision (ECCV), pp. 269–285 (2018)
5. Hadsell, R., Chopra, S., LeCun, Y.: Dimensionality reduction by learning an invariant mapping. In: 2006 IEEE Computer Society Conference on Computer Vision and Pattern Recognition (CVPR'06), vol. 2, pp. 1735–1742. IEEE (2006)
6. He, K., Zhang, X., Ren, S., Sun, J.: Deep residual learning for image recognition. In: The IEEE Conference on Computer Vision and Pattern Recognition (CVPR) (2016)
7. Hinton, G., Vinyals, O., Dean, J.: Distilling the knowledge in a neural network. arXiv preprint arXiv:1503.02531 (2015)
8. Huang, Z., Wang, N.: Like what you like: knowledge distill via neuron selectivity transfer. arXiv preprint arXiv:1707.01219 (2017)
9. Ioffe, S., Szegedy, C.: Batch normalization: accelerating deep network training by reducing internal covariate shift. arXiv preprint arXiv:1502.03167 (2015)
10. Jacob, P., Picard, D., Histace, A., Klein, E.: Metric learning with horde: high-order regularizer for deep embeddings. In: The IEEE International Conference on Computer Vision (ICCV), Oct 2019
11. Kim, W., Goyal, B., Chawla, K., Lee, J., Kwon, K.: Attention-based ensemble for deep metric learning. In: The European Conference on Computer Vision (ECCV) (2018)
12. Kingma, D.P., Ba, J.: Adam: a method for stochastic optimization. arXiv preprint arXiv:1412.6980 (2014)
13. Krause, J., Stark, M., Deng, J., Fei-Fei, L.: 3d object representations for fine-grained categorization. In: 4th International IEEE Workshop on 3D Representation and Recognition (3dRR-13) (2013)
14. Liu, Y., et al.: Knowledge distillation via instance relationship graph. In: Proceedings of the IEEE Conference on Computer Vision and Pattern Recognition, pp. 7096–7104 (2019)
15. Liu, Z., Luo, P., Qiu, S., Wang, X., Tang, X.: Deepfashion: powering robust clothes recognition and retrieval with rich annotations. In: Proceedings of the IEEE Conference on Computer Vision and Pattern Recognition, pp. 1096–1104 (2016)
16. Movshovitz-Attias, Y., Toshev, A., Leung, T.K., Ioffe, S., Singh, S.: No fuss distance metric learning using proxies. CoRR abs/1703.07464 (2017). http://arxiv.org/abs/1703.07464
17. Oh Song, H., Xiang, Y., Jegelka, S., Savarese, S.: Deep metric learning via lifted structured feature embedding. In: The IEEE Conference on Computer Vision and Pattern Recognition (CVPR) (2016)
18. Opitz, M., Waltner, G., Possegger, H., Bischof, H.: Deep metric learning with bier: boosting independent embeddings robustly. IEEE Trans. Pattern Anal. Mach. Intell. (2018). https://doi.org/10.1109/TPAMI.2018.2848925

19. Park, W., Kim, D., Lu, Y., Cho, M.: Relational knowledge distillation. In: Proceedings of the IEEE Conference on Computer Vision and Pattern Recognition, pp. 3967–3976 (2019)
20. Paszke, A., et al: Automatic differentiation in pytorch (2017)
21. Qian, Q., Shang, L., Sun, B., Hu, J., Li, H., Jin, R.: Softtriple loss: deep metric learning without triplet sampling. arXiv preprint arXiv:1909.05235 (2019)
22. Romero, A., Ballas, N., Kahou, S.E., Chassang, A., Gatta, C., Bengio, Y.: Fitnets: hints for thin deep nets. International Conference on Learning Representations (2015)
23. Roth, K., Brattoli, B., Ommer, B.: Mic: mining interclass characteristics for improved metric learning. In: The IEEE International Conference on Computer Vision (ICCV), October 2019
24. Russakovsky, O., et al.: ImageNet large scale visual recognition challenge. Int. J. Comput. Vis. 115(3), 211–252 (2015). https://doi.org/10.1007/s11263-015-0816-y
25. Sanakoyeu, A., Tschernezki, V., Buchler, U., Ommer, B.: Divide and conquer the embedding space for metric learning. In: Proceedings of the IEEE Conference on Computer Vision and Pattern Recognition, pp. 471–480 (2019)
26. Schroff, F., Kalenichenko, D., Philbin, J.: Facenet: a unified embedding for face recognition and clustering. In: The IEEE Conference on Computer Vision and Pattern Recognition (CVPR) (2015)
27. Suh, Y., Han, B., Kim, W., Lee, K.M.: Stochastic class-based hard example mining for deep metric learning. In: Proceedings of the IEEE Conference on Computer Vision and Pattern Recognition, pp. 7251–7259 (2019)
28. Ustinova, E., Lempitsky, V.: Learning deep embeddings with histogram loss. In: Advances in Neural Information Processing Systems, pp. 4170–4178 (2016)
29. Wah, C., Branson, S., Welinder, P., Perona, P., Belongie, S.: The Caltech-UCSD Birds-200-2011 Dataset. Technical report CNS-TR-2011-001, California Institute of Technology (2011)
30. Wang, X., Han, X., Huang, W., Dong, D., Scott, M.R.: Multi-similarity loss with general pair weighting for deep metric learning. In: Proceedings of the IEEE Conference on Computer Vision and Pattern Recognition, pp. 5022–5030 (2019)
31. Wu, C.Y., Manmatha, R., Smola, A.J., Krahenbuhl, P.: Sampling matters in deep embedding learning. In: The IEEE International Conference on Computer Vision (ICCV) (2017)
32. Yi, D., Lei, Z., Li, S.: Deep metric learning for practical person re-identification. ArXiv e-prints (2014)
33. Yu, B., Tao, D.: Deep metric learning with tuplet margin loss. In: The IEEE International Conference on Computer Vision (ICCV), October 2019
34. Yu, L., Yazici, V.O., Liu, X., Weijer, J.V.D., Cheng, Y., Ramisa, A.: Learning metrics from teachers: compact networks for image embedding. In: Proceedings of the IEEE Conference on Computer Vision and Pattern Recognition, pp. 2907–2916 (2019)
35. Yuan, Y., Yang, K., Zhang, C.: Hard-aware deeply cascaded embedding. In: Proceedings of the IEEE International Conference on Computer Vision, pp. 814–823 (2017)
36. Zagoruyko, S., Komodakis, N.: Paying more attention to attention: improving the performance of convolutional neural networks via attention transfer. International Conference on Learning Representations (2017)
37. Zhai, A., Wu, H.Y.: Classification is a strong baseline for deep metric learning. arXiv preprint arXiv:1811.12649 (2018)

38. Zhai, A., Wu, H.Y., San Francisco, U.: Classification is a strong baseline for deep metric learning (2019)
39. Zhang, Y., Xiang, T., Hospedales, T.M., Lu, H.: Deep mutual learning. In: Proceedings of the IEEE Conference on Computer Vision and Pattern Recognition, pp. 4320–4328 (2018)

Domain Generalization vs Data Augmentation: An Unbiased Perspective

Francesco Cappio Borlino[1]([✉]), Antonio D'Innocente[2,3],
and Tatiana Tommasi[1,3]

[1] Politecnico di Torino, Turin, Italy
francesco.cappio@polito.it, tatiana.tommasi@polito.it
[2] University of Rome Sapienza, Rome, Italy
dinnocente@diag.uniroma1.it
[3] Italian Institute of Technology, Turin, Italy

Abstract. In domain generalization the target domain is not known at training time. We show that a style transfer based data augmentation strategy can be implemented easily and outperforms the current state of the art domain generalization methods. Moreover, we observe that those methods, even if combined with the described data augmentation, do not take advantage of it, indicating the need of new generalization solutions.

Keywords: Domain generalization · Data augmentation · Style transfer

1 Introduction

Domain Generalization (DG) research develops algorithms that are robust to domain shifts with the objective of obtaining good performance on a target domain that is not known at training time. Most of the existing DG strategies try to incorporate the observed data invariances, capturing them at feature [6] or model (meta-learning [5] and self-supervision [11]) level, in the hypothesis that analogous invariances hold for future test domains. An alternative solution consists in extending the source domains by synthesizing new images and including a larger variability in the training set. Some methods do this through generative models which are often difficult to train, but give quite effective results [14]. Still we noticed that newly introduced feature and model-based DG approaches avoid benchmarks against data augmentation strategies [3,10], probably considering them unfair competitors due to the extended training set.

We believe that the field needs some clarification and we dedicate our work on this topic. Specifically our main contributions are: **(1) The proposal of a simple and effective style transfer data augmentation approach for**

Electronic supplementary material The online version of this chapter (https://doi.org/10.1007/978-3-030-66415-2_50) contains supplementary material, which is available to authorized users.

A. Bartoli and A. Fusiello (Eds.): ECCV 2020 Workshops, LNCS 12535, pp. 726–730, 2020.
https://doi.org/10.1007/978-3-030-66415-2_50

domain generalization based on AdaIN [2]. **(2) The design of tailored strategies to integrate style transfer data augmentation with the current state of the art methods.** We show that the original advantage of those techniques almost always disappears when compared with the data augmented baseline. This suggests the need of rethinking domain generalization baselines. On one side simple data augmentation strategies should be envisaged to increase source data variability compatible with orthogonal feature and model generalization approaches. On the other, new cross-source adaptive strategies should be designed to build over images generated by style transfer.

2 Source Augmentation by Style Transfer

We focus on multi-source DG. Our strategy consists in the following two steps.

Training the Style Transfer Model. We use *AdaIN* [2] that allows style transfer in real time, by taking the style from a *style image* and applying it over a *content image*. We train AdaIN on source data: all the train splits of the source domains are used together both as *content dataset* and *style dataset*.

Training the Classification Model on the Augmented Source Data. A standard classification model (AlexNet or ResNet18) is trained on the source domains, exploiting the style transfer model to apply data augmentation. For each image of a training minibatch, we decide with probability p if we want to apply the style transfer. In this case we randomly choose another image of the minibatch and borrow its style. The content image is then substituted with its augmented version. Considering that each batch contains equal parts coming from the different source domains, we obtain a high variability in image styles.

3 Experiments

Datasets. We consider three standard benchmark datasets which differ in number of classes and covered domains. **PACS** [4] contains images of 7 object classes spanning 4 visual domains: Photo, Art Painting, Cartoon, Sketch. **OfficeHome** [9] is similar to PACS, it covers 4 domains (Art, Clipart, Product and Real-World) but shows a much larger set of 65 object classes. **VLCS** [8] is built upon 4 different datasets: PASCAL VOC 2007, Labelme, Caltech and SUN and contains 5 object categories. All its domains are composed of real world photos with the shift mainly due to camera type, illumination conditions, point of view, *etc*. For all our experiments we used the same experimental protocols described in [1], train splits for model training and validation splits for model selection. All our results are average performance over 3 runs.

Reference Methods. We consider as main *Baseline* a classification model learned on all the source data and naïvely applied on the target. We indicate with *Original* the standard data augmentation with horizontal flippling and random cropping, while we use *Stylized* to specify the cases where we add style transfer data augmentation. The behavior of four among the most recent DG methods is evaluated under both these augmentation settings. We integrate the

Table 1. PACS classification accuracy (%). We used AdaIN with $p = 0.75$ for AlexNet-based experiments and $p = 0.90$ for those based on ResNet18.

AlexNet		Painting	Cartoon	Sketch	Photo	Average
Original	Baseline	66.83	70.85	59.75	89.78	71.80
	Rotation	65.66	71.89	62.15	89.88	72.39
	DG-MMLD	69.27	72.83	66.44	88.98	74.38
	Epi-FCR	64.70	72.30	65.00	86.10	72.03
	DDAIG*	62.77	67.06	58.90	86.82	68.89
Stylized	Baseline	71.96	72.47	76.47	88.34	77.31
	Rotation	71.74	73.39	75.98	89.22	77.59
	DG-MMLD	70.50	70.84	75.39	88.43	76.29
	Epi-FCR	65.19	69.54	71.97	83.43	72.53
	DDAIG	69.35	71.10	70.99	87.70	74.79
Mixup	Pixel-level	66.03	68.00	51.18	88.90	68.53
	Feature-level	67.04	69.10	55.40	88.88	70.11
ResNet18						
Original	Baseline	77.28	73.89	67.01	95.83	78.50
	Rotation	78.16	76.64	72.20	95.57	80.64
	DG-MMLD	81.28	77.16	72.29	96.06	81.83
	Epi-FCR	82.10	77.00	73.00	93.90	81.50
	DDAIG*	79.41	74.81	69.29	95.22	79.68
Stylized	Baseline	82.73	77.97	81.61	94.95	84.32
	Rotation	79.51	77.93	82.01	93.55	83.75
	DG-MMLD	80.85	77.10	77.69	95.11	82.69
	Epi-FCR	80.68	78.87	76.57	92.50	82.15
	DDAIG	81.02	78.75	79.67	95.07	83.63
Mixup	Pixel-level	78.09	71.08	66.58	93.85	77.40
	Feature-level	81.20	76.41	69.67	96.31	80.90

Table 2. OfficeHome classification accuracy (%). We used AdaIN with $p = 0.1$.

ResNet18		Art	Clipart	Product	Real World	Average
Original	Baseline	57.14	46.96	73.50	75.72	63.33
	Rotation	55.94	47.26	72.38	74.84	62.61
	DG-MMLD*	58.08	49.32	72.91	74.69	63.75
	Epi-FCR*	53.34	49.66	68.56	70.14	60.43
	DDAIG*	57.79	48.32	73.28	74.99	63.59
Stylized	Baseline	58.71	52.33	72.95	75.00	64.75
	Rotation	57.24	52.15	72.33	73.66	63.85
	DG-MMLD	59.24	49.30	73.56	75.85	64.49
	Epi-FCR	52.97	50.14	67.03	70.66	60.20
	DDAIG	58.21	50.26	73.81	74.99	64.32
Mixup	Feature-level	58.33	39.76	70.96	72.07	60.28

Table 3. VLCS classication accuracy (%). We used AdaIN with p = 0:75.

AlexNet		CALTECH	LABELME	PASCAL	SUN	Average
Original	Baseline	94.89	59.14	71.31	64.64	72.49
	Rotation	94.50	61.27	68.94	63.28	72.00
	DG-MMLD*	96.94	59.10	68.48	62.06	71.64
	Epi-FCR*	91.43	61.36	63.44	60.07	69.07
	DDAIG*	95.75	60.18	65.48	60.78	70.55
Stylized	Baseline	96.86	60.77	68.18	63.42	72.31
	Rotation	96.86	60.77	68.18	63.42	72.31
	DG-MMLD	97.49	61.02	64.23	62.37	71.28
	Epi-FCR	92.69	58.18	62.59	57.87	67.83
	DDAIG	97.48	60.48	65.19	62.57	71.43
Mixup	Feature-level	94.73	62.15	69.82	62.98	72.42

style transfer in each of the considered approaches without undermining their nature: we carefully avoid to mix domains when methods require to access them separately. **DG-MMLD** [7] does not need the source domain labels, thus the style transfer data augmentation is applied exactly as done for the Baseline. **Epi-FCR** [5] is a meta-learning method which splits the network in two modules, each one is trained by pairing it with a partner that is badly tuned for the domain considered in the current learning episode. Since the network is also trained on all source data to build the classification ability, the style transfer data augmentation is applied here and not in the previous step. **DDAIG** [14] is a data augmentation strategy that uses a generator to produce augmented samples. In this method the label classifier is trained on all the source data, both original and synthetic: we further extended this set with style transfer augmented data. **Rotation** [11] exploits self-supervised learning: rotation recognition is combined with classification in a multi-task model. Once again the style transfer data augmentation application is trivial because no domain labels are used. We also experiment with *Mixup* [13] as an alternative to AdaIN for interpolation of source data. We tested data mixing both at pixel and at feature level [12].

We implemented the *Baseline*, *Rotation* and *Mixup*, while we used for the others the code provided by the authors. We report the previously published results whenever possible. We will indicate with a star (*) the results we obtained by running the authors' code.

Results Analysis. Table 1 shows results on PACS. We get two main outcomes. (1) There is an evident improvement in the Baseline performance when using the stylized augmented source data with respect to the original case. (2) All the considered state of the art DG methods benefit from the source augmentation. Indeed in absolute terms their performance grows, but at the same time they lose in effectiveness as they cannot outperform the Baseline any more. Table 2 shows results on OfficeHome. Even if in this case the improvement produced by the source augmentation by style transfer is more limited, the results confirm what already observed for PACS. Table 3 reports results on VLCS. This dataset is particularly challenging and shows a fundamental limit of tackling DG through style transfer data augmentation. Since the domain shift is not originally due to style differences, source augmentation by style transfer does not support generalization. Finally, the results of Mixup show that it is not able to generalize across domains and it might perform even worse than the Original Baseline. Only the feature variant shows some advantage on PACS, so we focused on it in the other tests. Still, its results remain lower than those obtained by the DG methods both with and without style based data augmentation.

4 Conclusions

Among current DG methods some are based on data augmentation and use complex generative approaches, while other propose source feature adaptation and meta-learning strategies. Despite being orthogonal among each other, no previous work tried to integrate them. We investigated here a simple and effective style transfer data augmentation strategy for DG, showing how it overcomes its competitors. However when combined with the most relevant existing DG approaches they lose their original effectiveness, not producing any improvement over the new data augmented baseline. Our work suggests the need of a shading new light on DG problems and calls for novel strategies able to take advantage of the data variability introduced by cross-domain style transfer.

Acknowledgements. Computational resources provided by hpc@polito: (http://hpc. polito.it).

References

1. Carlucci, F.M., D'Innocente, A., Bucci, S., Caputo, B., Tommasi, T.: Domain generalization by solving jigsaw puzzles. In: CVPR (2019)
2. Huang, X., Belongie, S.: Arbitrary style transfer in real-time with adaptive instance normalization. In: ICCV (2017)
3. Huang, Z., Wang, H., Xing, E.P., Huang, D.: Self-challenging improves cross-domain generalization. In: Vedaldi, A., Bischof, H., Brox, T., Frahm, J.-M. (eds.) ECCV 2020. LNCS, vol. 12347, pp. 124–140. Springer, Cham (2020). https://doi. org/10.1007/978-3-030-58536-5_8
4. Li, D., Yang, Y., Song, Y.Z., Hospedales, T.M.: Deeper, broader and artier domain generalization. In: ICCV (2017)

5. Li, D., Zhang, J., Yang, Y., Liu, C., Song, Y.Z., Hospedales, T.M.: Episodic training for domain generalization. In: ICCV (2019)
6. Li, H., Jialin Pan, S., Wang, S., Kot, A.C.: Domain generalization with adversarial feature learning. In: CVPR (2018)
7. Matsuura, T., Harada, T.: Domain generalization using a mixture of multiple latent domains. In: AAAI (2020)
8. Torralba, A., Efros, A.A.: Unbiased look at dataset bias. In: CVPR (2011)
9. Venkateswara, H., Eusebio, J., Chakraborty, S., Panchanathan, S.: Deep hashing network for unsupervised domain adaptation. In: CVPR (2017)
10. Wang, H., Ge, S., Lipton, Z., Xing, E.P.: Learning robust global representations by penalizing local predictive power. In: NeurIPS (2019)
11. Xu, J., Xiao, L., López, A.M.: Self-supervised domain adaptation for computer vision tasks. IEEE Access 7, 156694–156706 (2019)
12. Xu, M., et al.: Adversarial domain adaptation with domain mixup. In: AAAI (2020)
13. Zhang, H., Cisse, M., Dauphin, Y.N., Lopez-Paz, D.: mixup: beyond empirical risk minimization. In: ICLR (2018)
14. Zhou, K., Yang, Y., Hospedales, T., Xiang, T.: Deep domain-adversarial image generation for domain generalisation. In: AAAI (2020)

W11 - Bodily Expressed Emotion Understanding

W11 - Bodily Expressed Emotion Understanding

Humans are arguably innately prepared to possess the ability to comprehend others' emotional expressions from subtle body movements. Many robotic applications become possible if robots or computers can be empowered with this capability. Recognizing human bodily expressions automatically in unconstrained situations, however, is daunting due to the lack of a full understanding of the relationship between body movements and emotional expressions.

The International Workshop on Bodily Expressed Emotion Understanding (BEEU) series aims to serve as a multidisciplinary research and development forum focusing on computer vision and machine learning methods for understanding human bodily expressed emotion. The workshop this year, in conjunction with the European Conference on Computer Vision (ECCV), was the first of this workshop series. Due to the COVID-19 pandemic, the workshop was held virtually. The workshop brought together researchers with different backgrounds, including computer vision, robotics, psychology/psychiatry, graphics, data mining, machine learning, and movement analysis, to discuss approaches to this highly challenging research problem, and present the latest results.

Because the topic was relatively new and many computer vision and machine learning researchers did not have access to a carefully collected dataset, the workshop organizers provided the Body Language Dataset (BoLD) (Luo et al., *International Journal of Computer Vision*, 128(1):1-25, 2020) to the research community. Those receiving the dataset were encouraged to participate in the data modeling challenge. Over 30 teams had downloaded the data. While five teams had completed the challenge, only one team had obtained solid results.

The workshop had outstanding keynote speakers and invited speakers. They are Jitendra Malik (University of California, Berkeley and Facebook, USA), Norman Badler (University of Pennsylvania, USA), both keynote, and Nadia Berthouze (University College London, UK), Kerri L. Johnson (University of California, Los Angeles, USA), Agata Lapedriza (MIT Media Lab, USA and Universitat Oberta de Catalunya, Spain), Xin Lu (Adobe Inc., USA), and Nikolaus Troje (University of York, UK).

We would like to acknowledge our Organizing and Technical Program Committee members. We thank Adam Fineberg for the discussions and Amazon for their funding in support of expanding the BoLD dataset to its current scale. National Science Foundation's CCRI program supported the workshop.

The success of the workshop would not have been possible without the excellent papers and presentations included in the program. We thank all authors and contributors.

August 2020

James Z. Wang
Reginald B. Adams, Jr.
Yelin Kim

Panel: Bodily Expressed Emotion Understanding Research: A Multidisciplinary Perspective

James Z. Wang[1](✉) ⓘ, Norman Badler[2] ⓘ, Nadia Berthouze[3] ⓘ,
Rick O. Gilmore[1] ⓘ, Kerri L. Johnson[4], Agata Lapedriza[5,6] ⓘ, Xin Lu[7],
and Nikolaus Troje[8] ⓘ

[1] The Pennsylvania State University, University Park, PA, USA
jwang@ist.psu.edu
[2] University of Pennsylvania, Philadelphia, PA, USA
[3] University College London, London, UK
[4] University of California, Los Angeles, CA, USA
[5] Massachusetts Institute of Technology, Cambridge, MA, USA
[6] Universitat Oberta de Catalunya, Barcelona, Spain
[7] Adobe Inc, San Jose, CA, USA
[8] York University, Toronto, ON, Canada

Abstract. Developing computational methods for bodily expressed emotion understanding can benefit from knowledge and approaches of multiple fields, including computer vision, robotics, psychology/psychiatry, graphics, data mining, machine learning, and movement analysis. The panel, consisting of active researchers in some closely-related fields, attempts to open a discussion on the future of this new and exciting research area. This paper documents the opinions expressed by the individual panelists.

1 Introduction

Bodily expressed emotion understanding is a complex and highly challenging research problem that requires researchers to use knowledge and approaches from some quite distinct fields. For instance, computer vision, robotics, psychology/psychiatry, graphics, data mining, machine learning, and movement analysis can all play a critical role in solving this problem. Researchers from individual fields have developed theories, techniques, and systems to address this problem. As larger and larger datasets are being collected and shared with the broad research community (*e.g.*, the BoLD dataset made available at the First International Workshop on Bodily Expressed Emotion Understanding (BEEU) [6]), it is becoming important for researchers from different fields to join forces and tackle the problem together.

The goal of this multidisciplinary panel is to open a discussion on how various fields can work together in the current data-driven research environment. Due to the COVID-19 pandemic, the panel was organized as a written panel.

© Springer Nature Switzerland AG 2020
A. Bartoli and A. Fusiello (Eds.): ECCV 2020 Workshops, LNCS 12535, pp. 733–746, 2020.
https://doi.org/10.1007/978-3-030-66415-2_51

Five questions were sent to the panelists a few weeks in advance. Each panelist independently came up with a written response. Recorded in the remaining sections are the opinions of the individual panelists. When the written comments were collected, no length limitation was set. In the interest of keeping a record, the panel chair's personal opinions are also incorporated into this article.

2 Datasets and Benchmarks

What new kind of datasets and/or benchmarks can be helpful in bodily expressed emotion understanding research?

Norman Badler: As we discovered in our work with the effect of personality on motion performance, people can make personality type assessments from observing behaviors, especially when they can compare the same motion between differing personality types. Accordingly, it may be important to understand a subject's baseline personality characteristics (say from the 5-factor OCEAN model) as a tag on acquired motion datasets. Human motions are varied enough that factoring out personality influences may give more accurate base motions from which short-duration behaviors (such as emotional expression) may be more reliably extracted.

Similar observations apply to facial affect. Our work on FacEMOTE led us to believe that some facial action behavior variation could be described globally (per subject) [1]. Combined with the personality work, this implies that facial animation and affect understanding could be aided by an understanding of the subject's personality type. Such personality type annotations are not normally collected in motion datasets.

Nadia Berthouze:

- Naturalistic datasets: understanding emotions expressed through the body as people engaged in their everyday activity
- Situated multimodal datasets to understand affective body in depth and in the context of other modalities and situated interactions
- We need benchmark datasets that help to work across sensor types. Two worlds: computer vision and wearable-based research
- We need datasets across a variety of applications to ensure approaches can be tested for their ability to generalize to different contexts
- It's important that used ethical and GDPR (or related rules) procedure are clearly available with the datasets and consider these issues broadly to ensure that the researchers around the world can download the datasets and use them. At the moment in the UK we are not allow to download benchmark datasets if not compliant with GDPR, or need special permission
- We are missing work on affective hand gestures or touch as an extension of gesture

Rick O. Gilmore: Researchers should commit to and support the creation of an open database of images and videos that can be used to share training and test sets and labelled data in common, interoperable data formats. The database should support API queries so that reproducible computational workflows can be achieved. Researchers who draw upon the database should commit to uploading their results so that the community can accumulate sets of model fits to the same data which can then be compared and contrasted.

Databrary.org, the world's only restricted access data library specialized for storing, streaming, and sharing video could be adapted or extended for this purpose. Databrary has developed a policy framework for sharing identifiable data with participant permission that could also be extended. Indeed, concerns about personal privacy will continue to grow, so I think that collaboration with data repositories that have some of these problems in hand will become increasingly important.

Kerri L. Johnson: One of the challenges for this research is that for many researchers, particularly those who are not at R1 universities[1], the necessary equipment is cost and space prohibitive. Without sizable funding through either equipment grants or start-up funds, many simply cannot conduct research in this area. This makes the publication of open databases of body movements, both in raw and visual formats, all the more important. Unfortunately, many of the existing stimulus archives are insufficient for answering anything but very narrow questions. Expanding those archives to include more variety of movements and more diversity of individuals will expand the research questions that can be probed.

Agata Lapedriza: It depends on the specific problem that one wants to study. For example, to create systems for detecting gestures or body postures that correlate with the intended emotion expression, movies can be an interesting data source. Actors in movies behave in a way that makes it easy for us to understand what they are communicating. That means that, from movies, we should be able of learning some common gestures or body poses that people make for communicating emotions. However, if the goal is to learn a personalized model for recognizing how a specific individual expresses emotion, then it will be necessary to collect data of the specific individual, ideally with self-report emotion expression labels.

Xin Lu: In general, new datasets/benchmarks can be helpful by considering the following aspects: (1) Datasets should include a rich collection of continuous body movements together with contextual information, such as short/long videos with event tags (*e.g.*, cooking, working, delivering a speech or watching movies) or textural descriptions. The camera position (or a relative position of the camera to

---

[1] R1: Doctoral Universities – Very high research activity, as classified in the Carnegie Classification of Institutions of Higher Education.

the person) should be included as metadata of the collected videos. (2) Datasets should include body expressions collected from diverse groups of individuals. Demographic information is helpful for later analyses of body expressions in different age group, gender, and geographic region. (3) To make a dataset useful practically, having instance segmentation masks and information of each instance is crucial.

Creating new datasets by constraining the camera location and the event is another direction to study bodily expressed emotion in context. For instance, video data can be collected by installing a couple of cameras in fixed positions in a few conference rooms, classrooms, gyms, or landmarks.

Besides, body expression closely impact applications in health care and physical therapy. For instance, body expression may indicate a patient's emotional status, such as nervousness, anxiety, and excitement, which may help doctors diagnose and proceed with proper treatments. Also, body expression can help understand the sitting or sleeping posture in daily life, and bad posture could potentially be identified. Changes of body expressions of an individual in a day or a week may also indicate one's emotional status, such as tiredness or elation.

Nikolaus Troje: Depends very much on what we want to learn:

Ideally, we want data bases where we have detailed body kinematics of the individual, on the one hand, and information about the "true" emotional state of that individual, on the other hand. In order to subject such data to end-to-end machine learning, we need large amounts of such data. In order to ensure ecological significance, they should come from naturalistic settings. That way we could learn generative models that can serve both analysis and synthesis.

That would be the ideal scenario, but it is hardly practical. However, there are a number of strategies and workarounds that could help to get close to that point.

On the stimulus side, we may be able to replace the "ground truth" we could obtain with accurate motion capture (MoCap) technology, with information obtained from natural, markerless video. However, that requires better models to be used for markerless pose estimation. Learning these, requires yet other kind of training data and therefore different databases. Here, we need calibrated video, on the one hand, and "ground truth"-delivering motion capture data, on the other hand. However these can come from existing databases and do not necessarily require annotations in terms of emotion. bmlMoVi [3] is a database that has the potential to help learing such models. The large amount of motion capture data merged into AMASS [7] and that fact that it is fully compatible with SMLP [5], a generative model that can reconstruct fully skinned individual body shapes from motion capture data, could also be used to generate large training datasets that connect motion capture data with computer-graphically simulated video data.

On the annotation side, we also may have to make compromises. The "true" emotion of a person is hard to assess. Emotion induction is possible but also tedious and error-prone (*e.g.* [8]). For that reason, the annotation of true emotion of an individual is often replaced with ratings on how this person is perceived. Here, observers watch a video or other visual material and then rate the assumed emotional state of the presented person. Such data can be obtained with crowd sourcing methods. While they do not necessarily speak to the true emotion of the person, they can provide accurate measures as to how a person is being perceived. Wherever a distinction between "true" and perceived emotion is critical, the relation between the two needs to be evaluated in separate experiments which then would require their own databases.

James Z. Wang: Although we have developed and released the BoLD dataset, which is based on videos slips from movies [6], we believe there are at least several ways to substantially expand the data collection effort for bodily expressed emotion understanding.

First, besides collecting emotion annotation data based on the normal or typical population, it would be important to collect a human behavior and interaction dataset based on the *atypical population* (*e.g.*, patients visiting a psychological clinic).

Second, to help bridge the enormous gap between computable human pose and movement information and bodily expressed emotions, it would be helpful to build a middle layer of data with movement annotations. The Laban Movement Analysis can be a valuable way to annotate human movements.

Third, a comprehensive dataset to study emotion should have a significant component that integrates both multi-camera video, depth information, and MoCap data so that more accurate human pose modeling can be possible.

Fourth, a large-scale dataset for machine learning and statistical modeling needs to be relatively balanced in terms of subjects' demographics (*e.g.*, gender, age, race, and ethnic groups) to reduce the potential biases in the learning process.

Finally, a successful dataset should provide user services, such as access and search API and benchmarking, to facilitate multidisciplinary research and stimulate innovation.

3 Challenge Problems

What challenge problems can potentially be important to academia and/or the industry?

Norman Badler: Human performance is multi-layered. An emotional display within an individual may vary due to mood, context, personality, or culture. For example, angry gestures may be subdued in more publicly reverent environments or accentuated in large scale competitive situations. To me, the challenge is managing the layering through explicit representations rather than trying an all-in-one leap via machine learning, perhaps, from observed behavior to hypothesized internal causes or motivation. Such variations are known to exist across cultures, making it crucial that industrial-strength machine vision systems that are being used to watch people make behavioral judgements in the context of who is being observed and what the baselines are for nominal individuals of the community. On the other hand, real stereotypes exist and need to be carefully and deliberately removed from automated assessments that trigger moral or legal actions.

Nadia Berthouze:

- Real-world applications
- Ethical issues
- Working across sensors for ubiquitous leverage of body expressions

Rick O. Gilmore: I think the problem of accurately detecting emotions "in the wild" – from videos collected in every day settings and from possibly mobile cameras – is hard and interesting.

Kerri L. Johnson: To my thinking, a solid appreciation of how dynamic interactions impact social perception lags woefully behind our understanding of face perception. Part of that is undoubtedly due to the costs to entry in this area, relative to face perception research. Increasing accessibility to raw data will help in this regard.

Agata Lapedriza: Automatic emotion recognition in the wild is still immature. There are a lot of problems and challenges that are interesting for academic research, like the ones that I mention below. For the industry, and for society more generally, one of the main challenges is to make sure that emotion recognition software is used correctly and fairly, and just for those applications that benefit humans.

Xin Lu: While we tend to think about patterns across a large group of populations in terms of bodily expressed emotion, a potentially more interesting problem is to analyze bodily expressed emotion of an individual.

People are likely to be more relaxed and share more bodily expressed emotion unconsciously in a private space. Meanwhile, body expression is likely to be richer comparing with publicly shared body expressions. However, data out of those scenarios is usually private. How can we analyze these data assuming we cannot see most of these data due to privacy concerns? If we can, how can individuals benefit from these data? How can we analyze similarities and differences of these data across individuals? How would those patterns benefit the society? I believe this is an interesting direction to explore, which involves sufficient technical challenges and application potentials.

Nikolaus Troje: The distinction between veridical and faked emotion is an interesting topic in itself. Bodily expressions of emotions can be faked, but only to a certain degree. Fake smile has been characterized relatively well and computer vision algorithms have been designed to discriminate fake from veridical smile. However, no database or benchmark framework exists that could be used to systematically test and compare existing algorithms. Extending research to other emotions (in addition to smile) may help to isolate invariants that discriminate between veridical and fake emotional expression in more general terms.

James Z. Wang: Comprehensive understanding of human's bodily expressed emotions has wide-ranging implications. Among others, there are clear applications in healthcare, commerce, robotics, and public safety. I believe computer vision and machine learning researchers need to partner with psychologists and key industrial players to define challenge questions that are both valid from a psychological perspective and meaningful or potentially impactful in the real world. And because real-world applications in this area will have impact to people's lives, it is critical to develop explainable, interpretable, and fair machine learning approaches.

4   Needed Breakthroughs

What theoretical, methodological, or technological breakthroughs are needed?

Norman Badler: Here I am old-fashioned: I like to know what my representation does and doesn't do, how the representation processes and interacts with real data, and what the output states or parameters mean. This goes along with my belief in layering representations to understand interactions, rather than lumping all decisions into a single complex machine learning (ML) system. I think there is a role for ML to develop such smaller steps. Overall the result may be a series of interacting ML systems, but we can look in-between them and gain some useful understanding of what is going on (or dare I say what they are "thinking").

Nadia Berthouze:

- HAR and Affective computing fields should get together. Continuous detection of affect across activities
- Studying affective body expressions as situated interactions rather than based on acontextual models
- Going beyond what we see: looking at physiological and neural mechanisms that drive movements
- Network architectures designed for capturing movement

Rick O. Gilmore: Most computer vision models ignore time, and yet human behaviors and especially bodily expressed emotions are temporally bound phenomena. We need theories that incorporate dynamics and methodological and technological breakthroughs that make this practical.

We need more research that compares model performance to human-labelled ground truth but also models against one another. We also need research with human coders in the loop to determine where specific models fail and why they fail. It may also be useful to compare weighted "mixtures" of models, especially if those are trained on different datasets. This may be an inexpensive way to work toward models that generalize.

Finally, research in this area needs to embrace "multi-modality." By this I mean that humans and non-human animals express emotions through facial expressions, body postures, actions, and vocalizations. Researchers who figure out how to combine these sources will likely create systems that perform better than those that do not embrace a multimodal approach.

Kerri L. Johnson: At present, research in this area remains highly segmented, with vision, computational, and social scientists largely pursuing their work isolated from one another. Consequently, the insights and discoveries can be unnecessarily narrow. While some, including the members of this panel, have sought to bridge the vision, computational, and social approaches, doing so remains rare. It's at this nexus where exciting and groundbreaking discoveries will be made.

Agata Lapedriza: Here are two examples:

- Understanding the context and integrating the analysis of context to the recognition pipeline. A facial expression or a gesture can mean or communicate different messages, depending on the context.
- Dealing with subjectivity: emotion perception is subjective. Annotations on emotions provided by annotators are subjective. This means that the same input can have different labels that are correct, since given a specific input, different people might perceive different emotions and all the perceptions are correct. This is something less common in object or scene categorization, for instance. If an object is a car it is not a bicycle. However, a person can look sad to someone and angry to someone else. And both perceptions are correct, because they are just perceptions. Modelling subjectivity requires learning methods that can correctly deal with different degrees of agreement, different opinions, and different degrees of uncertainty.

Xin Lu: In the past, when we lacked data, we developed methodologies and technologies to overcome the data scarceness. In recent years, when we were equipped with a sufficient amount of data together with computational power, we identified useful patterns and used learned knowledge (usually represented by models) to automate repetitive workflows and alleviate cognitive load for individuals. In the coming future, as the amount of data is soaring, one model may not effectively encode all the information in a complex problem. I believe breakthroughs are needed in the below two aspects: 1. To be able to identify patterns selectively according to context. 2. To be able to incrementally enrich the model as data is accumulating without forgetting existing info.

Nikolaus Troje: Reliable characterization of emotion from non-facial features has interesting applications in situations where the assessment of facial features is difficult because users wear head gear such as shutter glasses or HMDs. Avatars of the players of computer games or users of telecommunication and teleconferencing systems that make use of virtual reality, suffer from rigid facial expression which is hard to assess from the face of a person using gear that covers large parts of the face. Emotions reliably obtained from the kinematics of other parts of the body could be added procedurally to provide avatars not just with bodily expression but also with facial expression.

Motion style, including the body kinematics that express emotion, is often transmitted by minute changes in pose and dynamics. Hands are likely to play an important role in that context. To study hand motion on the stylistic level, we need better models that describe individual shape, pose and motion of hands.

James Z. Wang: As we have discussed in our recent ARBEE work [6], breakthroughs are needed in both computational/technical fields and psychology.

First, vision-based human pose estimation methods are noisy in terms of jitter errors. Emotion understanding demands substantially higher precision of body landmark locations, compared with typical vision applications.

Second, vision-based methods usually address whole-body poses, which have no missing landmarks, and only produce relative coordinates of the landmarks from the pose instead of the actual coordinates in the physical environment. Real-world emotion understanding requires the modeling of the person and the environment together.

Third, while deep learning methods generate better recognition results, they offer minimal explainability and interpretability. In many emotion understanding applications, these characteristics are more important than accuracy. Breakthroughs in core machine learning and statistical modeling are still needed.

Finally, it can be a technical breakthrough to develop an effective data-driven model that can incorporate the person's personality, gender, age, race, ethnic groups, cultural background, and personal characteristics, as well as the context.

5 Vision of the Future

From the perspectives of your field, how do you view the future of this area of research?

Norman Badler: From the perspective of computer graphics motion generation (animation), parameterization is currently the key to automated human (character) behaviors, but the state-of-the-art is not yet sufficient to replace the intuition, skill, and subtlety of computer animators or direct motion capture in the process. Again, I would argue that this is because the transformation from desire to motion expression lacks important input structures. I do not know of any extant computer animation system that allows one to input a character's personality, mood, social context, response history, personal gestural vocabulary, relationship to any interlocutor, and emotional state and then outputs the "right" behavior. I do think this is eventually possible, but it will require deeper understanding of all these influences on behavior which is going to be empirically challenging without some decent annotational bases for learning the complexities of such mappings and their interactions. People learn such associations from years of personal experience, so it must be possible, but even people aren't flawless in their judgements.

Nadia Berthouze:

- Real-world applications to support people in the wild
- Ubiquitous sensors sensing the body in its details

Rick O. Gilmore: I am both optimistic and pessimistic. My optimism stems from the considerable energy, creativity, and inventiveness of researchers in AI. My pessimism stems from the fact that most of the computer vision models I've used are not especially valuable to researchers in the behavioral sciences, and there is not enough collaborative work across these communities to share expertise.

I am of both minds also when I think about the future of data. We need bigger, better, more carefully annotated, and more widely shared data about real-world behavior. The Play & Learning Across a Year (PLAY) project (play-project.org) is one effort to collect data analogous to a "human behaviorome." But COVID-19 has put the effort on hold for now.

Kerri L. Johnson: This is truly an exciting time to be conducting research in this area. The very integration of vision, computational, and social science approaches to understanding dynamic movement is happening, albeit slowly. This has the potential to provide insights that have heretofore been impossible due to computational processing limits and data sets of scale.

Agata Lapedriza: Collecting annotated data is complicated and expensive. I think we need to think about self-supervised approaches or cross-modal approaches for emotion recognition.

Xin Lu: I spent most of my recent years developing machine learning algorithms, systems, and a real-world mobile camera application (the Photoshop Camera). I believe bodily expressed emotion is going to be more fascinating in the mobile and Internet of Things (IoT) era, which has already arrived. Everyone spends more and more time on mobile and IoT devices every day, and people will be interested in knowing more of themselves by way of these devices. Mobile or IoT devices can privately capture personal data and identify useful and interesting patterns without sharing these data to the cloud. The wide availability of these personal devices as well as their powerful computational power make up an unprecedented opportunity to study sensitive and private information such as bodily expressed emotions of individuals.

Nikolaus Troje: I have always been fascinated by the divide between "bodily expressed emotion" and cognitive-rationale responses. For instance, my current work involves immersing experimental participants in virtual reality where we simulate fearful situations. Participants are perfectly aware of the simulated nature of the situation and the fact that they are safe. Nevertheless do they show somatic responses that express true fear: For instance, when navigating a deep abyss on a narrow plank they are sweating, their hard beat increases and their behaviour changes to a point where they are no longer able to function normally [2].

Emotions are to a large degree somatic responses and they express themselves in ways that often are not accessible to cognitive control. Asking someone how she feels is not providing the same information as the one we get when assessing bodily responses. Understanding the latter opens avenues toward treatment of emotional disorders (*e.g.* PTSD, specific phobias). It is also important to assess the affect of fearful situations in the real world, but also in the context of gaming and entertainment, particularly in virtual reality.

A very interesting future area of application is the design of autonomous avatars that also convey emotion in convincing ways.

James Z. Wang: While a comprehensive understanding of human bodily expressed emotion is daunting, I believe we will start to see exciting industrial applications with some limited recognition capabilities. For example, if we limit the emotion categories to a few common emotions, the problem becomes a lot more tractable. As larger and more extensive datasets and more efficient and effective computational modeling techniques become available, the accuracy level and comprehensiveness of the computer-based understanding will continue to improve. When we were developing machine learning and statistical modeling based image annotation in the early 2000s [4], many researchers did not think it was possible to achieve a usable level of accuracy. Today, less than just two decades later, we have already seen wide industrial adoption of machine annotation of images. While emotion understanding is arguably much more subtle and hence more difficult to model than recognizing ordinary objects, I'm optimistic that over time, with increased multidisciplinary collaboration, we will be able to see a number of exciting applications that can improve people's lives.

6 Advice on Innovation

What advice would you offer young researchers trying to work on this topic? How can they be innovative?

Norman Badler: Set yourself a long term goal and evaluate how to proceed toward that goal starting from known techniques combined with integrated insights. For example, don't avoid reading psychology or kinesiology literature. At lot of it won't be useful, but you may get insights from reading outside your academic box. Most of my (ultimately) important computational insights came from works in Natural Language Processing, dance, ergonomics, and communicative and manipulative behaviors. Rather than shun complexity in favor of small piecemeal steps, see if you can manage complexity through problem decomposition and combination. CG is a poster child for combining techniques to achieve stunning results by mixing image processing, ML, algorithmic techniques, human perceptual constraints, and model representations. Animation is moving slowly toward complexities but is mostly still a technique-based discipline. Computer vision is also likely to appreciate that interconnecting and integrating smaller pieces of a technology puzzle into a psychologically justifiable larger framework may be more fruitful than trying to build a single whiz-bang n-layer network with miracle outputs. That's going to take a longer view but I think it's worth trying.

Nadia Berthouze: By thinking to the applications and purpose of use. Understanding the purpose and the context of use may lead to think differently to how we approach the problem. This happened to us in the context of physical rehabilitation. It led to understand the needs for it, its multiple uses, contexts of use and barriers to its use and to challenge current approaches to the problem.

Rick O. Gilmore: There is a huge but largely untapped pool of knowledge in the community of behavioral scientists that could and should inform research in this area. So, my advice would be to make friends with your colleagues in psychology, ethology, communication sciences, and education research.

Kerri L. Johnson: My advice is to follow your interests, not necessarily what seems trendy at the time. That's where you will be your most innovative. Ask the questions that pique your curiosity, and do whatever is necessary to conduct the most rigorous research that will answer them. Be meticulous and careful in your design, analysis, and reporting. And balance your research portfolio with projects that are "reach" projects (long term, higher risk) and those that are safer and faster.

Agata Lapedriza: I would recommend students that want to work in emotion recognition to read a lot of psychology papers and books. We are trying to create machines that do something that is very challenging even for us: we are not that good in inferring other's emotions, and sometimes we are not even able of categorizing how we feel. Being aware of the findings and theories of emotions from a psychology perspective is very important.

Xin Lu: First, think deeply and read broadly. Emotion recognition/analysis is still at an early stage, and an ideal formulation of this problem doesn't exist to a certain extend. Therefore, it is essential to think deeply on fundamentals. Meanwhile, reading broadly to understand theories and methods by way of well-defined problems builds up tools for young researchers to better solve complex and abstract problems.

Second, learn programming tools. As the volume of data is soaring, having a strong expertise in programming is essential for young researchers to be innovative in the digital and mobile world.

Third, scope the problem. An essential skill that young researchers in this area need to have is scoping the problem to one that is solvable. While the sky is big and tool sets are rich, we cannot throw a tool randomly to the sky and expect that a bird is caught. Rather, we need to navigate the tool that pivots towards the scoped problem.

Nikolaus Troje: I think the most interesting aspect of emotion research is the dynamics of emotion emerging in social situations. If two people are communicating, they may enter the encounter with certain emotions that reflect their history until that point, but the really interesting processes are the ones that happen during that interaction. In order to study such processes we need to understand the sophisticated body language between the two interlocutors, we need to understand how it expresses the emotional state of the acting person, and how it affects the emotional state of the perceiving person.

Emotion is nothing static, it is reaction to the (social) environment, on the one hand, but also affects that environment, on the other hand. In order to study the emotional progression of a dyadic encounter we need to be able to monitor both interlocutors simultaneously, and we need to come up with dynamic models that can describe that progression. Of course, things would become even more interesting, if we were able to study emotional progression in larger groups.

In the previous section, I talk about communication with autonomously behaving avatars. Again, for such tools to become truly convincing we have to find ways to not just model emotion in itself, but we have to understand the interactive emotional dynamics emerging during social interactions.

James Z. Wang: First, I'd encourage young researchers interested in this topic to collaborate with those with complementary expertise and take a multidisciplinary approach. The BEEU workshop aims to bring active researchers from different fields together. I believe we are only seeing the beginning of a lot of fruitful collaborations between computer science and behavior sciences.

Second, because this problem is not a typical computer vision or machine learning problem, and it is complex and not well-defined, young researchers need to have the courage to face significant challenges and even many failures. It is likely not going to be a project that can quickly generate publications or citations. It needs patience and perseverance to accomplish novel and significant results that have lasting impacts.

Acknowledgment. The opinions expressed in this article are panelists' personal views. J. Z. Wang is supported by the National Science Foundation under Grant No. 1921783 and the Amazon Research Awards Program. Reginald B. Adams, Jr. and Yelin Kim contributed to the discussions related to the theme of this panel.

References

1. Byun, M., Badler, N.I.: FacEMOTE: qualitative parametric modifiers for facial animations. In: Proceedings of the 2002 ACM SIGGRAPH/Eurographics symposium on Computer animation, pp. 65–71 (2002)
2. Eftekharifar, S., Thaler, A., Troje, N.F.: Contribution of motion parallax and stereopsis to the sense of presence in virtual reality. J. Percept. Imaging, **3**(2), 20502-1–20502-10 (2020)
3. Ghorbani, S., et al.: MoVi: a large multipurpose motion and video dataset. arXiv preprint arXiv:2003.01888 (2020)
4. Li, J., Wang, J.Z.: Automatic linguistic indexing of pictures by a statistical modeling approach. IEEE Trans. Pattern Anal. Mach. Intell. **25**(9), 1075–1088 (2003)
5. Loper, M., Mahmood, N., Romero, J., Pons-Moll, G., Black, M.J.: SMPL: a skinned multi-person linear model. ACM Trans. Graph. **34**(6), 1–16 (2015)
6. Luo, Y., Ye, J., Adams, R.B., Li, J., Newman, M.G., Wang, J.Z.: ARBEE: towards automated recognition of bodily expression of emotion in the wild. Int. J. Comput. Vis. **128**(1), 1–25 (2020)
7. Mahmood, N., Ghorbani, N., Troje, N.F., Pons-Moll, G., Black, M.J.: AMASS: archive of motion capture as surface shapes. In: Proceedings of the IEEE International Conference on Computer Vision, pp. 5442–5451 (2019)
8. Zentner, M., Grandjean, D., Scherer, K.R.: Emotions evoked by the sound of music: characterization, classification, and measurement. Emotion **8**(4), 494 (2008)

Emotion Understanding in Videos Through Body, Context, and Visual-Semantic Embedding Loss

Panagiotis Paraskevas Filntisis[1]([⊠]), Niki Efthymiou[1], Gerasimos Potamianos[2], and Petros Maragos[1]

[1] School of E.C.E., NTUA, Athens, Greece
{filby,nefthymiou}@central.ntua.gr, maragos@cs.ntua.gr
[2] E.C.E. Department, UTH, Volos, Greece
gpotam@ieee.org

Abstract. We present our winning submission to the First International Workshop on Bodily Expressed Emotion Understanding (BEEU) challenge. Based on recent literature on the effect of context/environment on emotion, as well as visual representations with semantic meaning using word embeddings, we extend the framework of Temporal Segment Network to accommodate these. Our method is verified on the validation set of the Body Language Dataset (BoLD) and achieves 0.26235 Emotion Recognition Score on the test set, surpassing the previous best result of 0.2530.

Keywords: Emotion · Body · Context · Visual-semantic · BEEU challenge

1 Introduction

Automatic human affect recognition from visual cues is an important area of computer vision that has attracted increased interest over the last two decades, due to its many applications. Indeed, social robotics [2], psychiatric care [13], and edutainment [10] are all areas that can benefit from automatic recognition of emotion.

Most past approaches to the problem have focused on facial expressions in order to determine the emotional state of the person of interest [7,18,22]. This is reasonable due to the fact that facial expressions have been studied extensively in the psychology and emotion literature [8]. For example, the Facial Action Coding System (FACS) [9] identifies the units of facial movements, based on facial muscle groups. Combinations of the so-called action units (AUs) have also been linked with emotional states with extensions of the basic FACS such as EMFACS (Emotion FACS) [11]. On the other hand, there is no similar established coding system for body expressions, although some have been proposed [4].

© Springer Nature Switzerland AG 2020
A. Bartoli and A. Fusiello (Eds.): ECCV 2020 Workshops, LNCS 12535, pp. 747–755, 2020.
https://doi.org/10.1007/978-3-030-66415-2_52

Compared to facial expression based approaches, recent works have sought alternative modalities and streams of information to detect emotion; one is bodily expressions since many have highlighted the fact that the emotional state is conveyed through bodily expressions as well, and in certain emotions it is the main modality [5, 15, 26], or can be used to correctly disambiguate the corresponding facial expression [1]. Simultaneously, it is important to note that in cases and applications where the emotion needs to be identified, the human body is more frequently available than the face since the face can be occluded, hidden, or far in the distance. Another auxiliary stream of information besides the face and the body that can help in identifying emotions is the context and the surrounding environment of the person [16, 21]. It is apparent that both the place, as well as objects and other humans can influence a person's emotions.

We should also note that inherently emotion recognition is a multi-label problem - the subject might be feeling two or more emotions. This is true, especially when considering an extended set of emotions, as in [19]. The emotions in extended sets do not have the same "semantic" distance between them. For example, anger is more close to annoyance than to happiness. Considering that previous works have showed the superiority of methods that attempt to learn a joint embedding space that contains both word embeddings and visual representations [6, 12, 24], we believe that trying to attach a semantic meaning to the extracted visual feature is a natural way forward.

In this paper, based on the above, we describe the method of our team in the First International Workshop on Bodily Expressed Emotion Understanding (BEEU) challenge. Our method combines Temporal Segment Networks (TSNs) [27] focusing on the body, using the context in each video as an additional stream, and also uses an extra visual-semantic embedding loss, based on GloVE (Global Vectors) [23] word embedding representations. Our experiments in the validation set verify the better performance of our method compared to the traditional TSNs, while our emotion recognition score on the test set was 0.26235.

2   Related Work

While most past approaches in visual detection of affect have been focused on facial expressions [5], recent approaches have started taking into account the body language [15] of the person in question, as well as its surrounding context/environment.

In [14], Gunes and Piccardi introduced a bimodal architecture that takes into account both upper body and facial expressions, in order to detect affect in videos. In [3], Dael et al. analyzed and classified body emotional expressions using a body action and posture coding system which was proposed in [4]. The 3D pose of children was also utilized in [20] by Marinoui et al. to detect emotions in continuous dimensions, while in [10], 2D pose was used and fused with facial expressions for child emotion recognition. Luo et al. [19] introduced a large scale video dataset (BoLD) annotated with categorical and continuous emotions, which is the one used in the BEEU challenge.

Regarding the context modality, Kosti et al. [16] introduced a large scale dataset for emotion recognition (EMOTIC) in different contexts (e.g., other people, places, or objects) and a convolutional neural network (CNN) based two-stream architecture that focused on the body and context of the subjects. The CAER video dataset for context-based emotion recognition was presented in [17], along with a two-stream architecture which employed adaptive-fusion to merge the two steams. In [21], Mittal et al. designed a deep architecture with several branches, focusing on different interpretations of the surrounding context (e.g., environment and interaction context) to significantly increase resulting predictions in the EMOTIC dataset.

Finally, some recent works have also focused on extracting visual representations from images that present the semantic relations found in embeddings built from words. The DeViSE embedding model [12] extracted semantically-meaningful visual representations by introducing a similarity loss between the feature vector extracted from a CNN and the word embedding from a skip-gram text model. Using a similar method, Wei et al. [28] built joint text and visual embeddings as emotion representation from web images, and in [29], Ye and Li built semantic embeddings for a multi-label classification problem.

3 Dataset

The dataset used in the challenge is the BoLD (Body Language Dataset) corpus [19] consisting of 9,876 video clips of humans expressing emotion, primarily through body movements. Each clip can contain multiple characters, yielding a total of 13,239 annotations, split into a training, validation, and test set. The dataset has been annotated by crowdsourcing employing two widely accepted categorizations of emotion. The first one is the categorical annotation with a total of 26 labels first used in [16], by collecting and processing an extensive affective vocabulary. The second annotation regards the continuous emotional dimensions of the VAD (Valence - Arousal - Dominance) Emotional State Model [25]. The methods in the challenge are evaluated using the following Emotion Recognition Score (ERS):

$$\text{ERS} = \frac{1}{2}\left(\text{m}R^2 + \frac{1}{2}(\text{mAP} + \text{mRA})\right) \tag{1}$$

where $\text{m}R^2$ is the mean coefficient of determination (R^2) score for the three dimensional emotions (VAD), and mAP and mRA is the mean Average Precision and the mean area under receiver operating characteristic curve (ROC AUC) of the multilabel categorical predictions.

4 Model Architecture

Our model is based on the TSN architecture [27], which has been widely used in action recognition and can be seen in Fig. 1. During training, K different

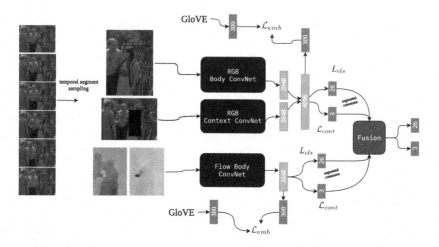

Fig. 1. TSN with two RGB spatial streams (body and context) and one optical flow stream. The final results are obtained using average score fusion.

segments are selected from the input video, and then N consecutive frames are selected from each segment. This is done to deal with the fact that consecutive frames have usually redundant information. Traditionally, two different modalities are used, one is the spatial (RGB) modality and the second one is the optical flow. TSNs have already been shown to achieve good results for the BoLD dataset in its introductory paper [19].

In our approach, we modify the original version of TSNs mainly in two directions:

Context: We introduce one additional stream based on the context-environment surrounding the annotated human. For the RGB modality, we input the context in the network in the same way as in [21], by masking out the instance body (we set all pixels to 0). We call this stream RGB-c, and the body streams RGB-b and Flow-b. During training, the RGB-b and RGB-c streams are combined at the feature level (RGB-bc) and are trained jointly while the Flow-b TSN is trained independently.

Embedding Loss: Our second extension is the introduction of an embedding loss on the feature vector extracted by the Convolutional Neural Network (ConvNet). This is done to exploit the fact that some emotions are closer semantically to others. This is also revealed by examining the correlation matrix of the dataset labels in [19], where some labels occur more frequently in combination with others (e.g. Happiness and Pleasure, Annoyance and Anger, etc.). Due to this result, we try to attach a semantic meaning to the feature vector extracted by the backbone image network.

To implement this, we first obtain for each one of the 26 categorical labels of BoLD their 300-dimensional GloVE word embedding [23]. A PCA-projection

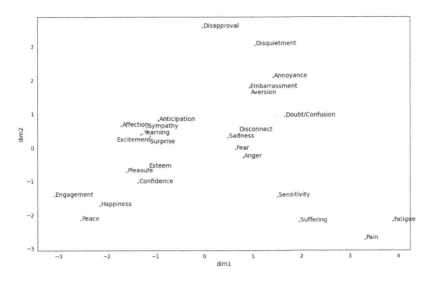

Fig. 2. PCA projection of the categorical emotions GloVE word embeddings.

of the 26 embeddings is shown in Fig. 2, where it is apparent that the distances between embeddings are indicative of their "semantic" distance. We then use a fully connected layer to map the feature extracted from the image to a 300-dimensional space and introduce the following mean-squared based loss:

$$\mathcal{L}_{emb} = ||\boldsymbol{W} f_v(\boldsymbol{x}) - \frac{1}{|K|} \sum_{y \in K} f_w(\boldsymbol{y})||_2 \qquad (2)$$

where $f_v(\boldsymbol{x})$ is the feature vector extracted by applying the convNet on the image $\boldsymbol{x}$, $\boldsymbol{W}$ is a linear transformation from the space of the feature vector to the word embedding space, $f_w(\boldsymbol{y})$ is the word embedding of the label y, and K is the set of all positive labels for the image $\boldsymbol{x}$. That is, we try to reduce the Euclidean distance between the projected image feature and the arithmetic mean of the GloVE embeddings of the positive labels for image/video.

Predictions: Finally, after extracting for each sampled image its feature vector, we use two fully connected layers, one to classify to the 26 different categorical labels, and one to regress over the 3 different categorical emotions. The two TSNs are trained using the following loss:

$$\mathcal{L} = \mathcal{L}_{cls_1} + \mathcal{L}_{cls_2} + \mathcal{L}_{cont} + \mathcal{L}_{emb} \qquad (3)$$

Specifically, since the dataset does not provide explicitly the multilabel targets, but the crowdsourced scores between 0 and 1, we include two different losses for the classification part: $\mathcal{L}_{cls_1}$ that is the binary cross-entropy between the predicted scores and the multilabel target (obtained after thresholding the multilabel scores at 0.5) and $\mathcal{L}_{cls_2}$ that is the mean squared error between the

Table 1. Ablation experiment by training with and without $\mathcal{L}_{emb}$.

	Model	mAP	mRA	mR^2	ERS
Without $\mathcal{L}_{emb}$	RGB-b	0.1567	0.6140	0.0538	0.21955
	Flow-b	0.1444	0.5914	0.0507	0.2093
	RGB-b + Flow-b	0.1623	0.6307	0.078	0.2375
With $\mathcal{L}_{emb}$	RGB-b	0.1564	0.6143	0.0546	0.21997
	Flow-b	0.1465	0.5947	0.0579	0.2142
	RGB-b + Flow-b	**0.1637**	**0.6327**	**0.0874**	**0.2428**

predicted scores and the multilabel scores. We empirically found that the inclusion of $\mathcal{L}_{cls_2}$ slightly boosted performance. For the regression part, $\mathcal{L}_{cont}$ is the mean-squared error between the regressed values and the continuous emotions. Finally $\mathcal{L}_{emb}$ is as in (2).

5   Experimental Results

We train each TSN for 50 epochs using Stochastic Gradient Descent (SGD), with initial learning rate 10^{-3} which drops by a factor of 10 at 20 epochs.[1] The backbone networks used is a residual network (ResNet) with 101 layers for the body convNets and a ResNet with 50 layers for the context convNet. We use the default hyperparameters of TSNs: 3 segments, 1 frame from each segment for the RGB streams, and 5 frames from each segment for the optical flow stream. The consensus used for segment fusion is averaging. For each network, we select the epoch with the best validation ERS. We have also found experimentally that the partialBN (Batch Normalization) technique used in [27] gives a nontrivial boost to the performance of the network.

First, in Table 1 we present two ablation experiments regarding the addition of $\mathcal{L}_{emb}$. We can see that adding the embedding loss increases slightly the performance in the RGB-b stream, and gives a boost to the performance of the Flow-b stream.

Then, in Table 2 we present our experimental results on the validation set of BoLD including the RGB context stream. From the results we can see that including the context along with the body in the RGB modality boosts the validation ERS of the architecture. We also experimented with including the context in the Flow network, but this resulted in worse performance. Our final submission for the test set was the model with the best validation score (0.2439 employing RGB-bc + Flow-b), using 25 segments instead of 3. The results of the different metrics on the test set can also be seen in Table 2, while the final ERS is 0.26235, improving upon the previous best result of 0.2530 [19].

---

[1] PyTorch code available at https://github.com/filby89/NTUA-BEEU-eccv2020.

Table 2. Results on the validation and test set of BoLD including the RGB context stream and $\mathcal{L}_{emb}$.

Set	Model	mAP	mRA	mR^2	ERS
Valid	RGB-c	0.1395	0.5760	0.0365	0.1971
	RGB-bc	0.1566	0.6055	0.0675	0.2243
	RGB-bc + Flow-b	0.1656	0.6266	0.0917	**0.2439**
Test	RGB-bc + Flow-b	0.1796	0.6416	0.1141	**0.26235**

6   Conclusions

In this paper we presented our method submitted at the BEEU challenge, winning first place. Our method extended the TSN framework to include a visual-semantic embedding loss, by utilizing GloVE word embeddings, and also included an additional context stream for the RGB modality. We verified the superiority of our extensions compared to the baseline on the validation set of the challenge, and submitted the best system which achieved 0.26235 Emotion Recognition Score on the BoLD test set, surpassing the previous best result of 0.2530.

Acknowledgments. This research is carried out/funded in the context of the project "Intelligent Child-Robot Interaction System for designing and implementing edutainment scenarios with emphasis on visual information" (MIS 5049533) under the call for proposals "Researchers' support with an emphasis on young researchers- 2nd Cycle". The project is co-financed by Greece and the European Union (European Social Fund-ESF) by the Operational Programme Human Resources Development, Education and Lifelong Learning 2014–2020.

References

1. Aviezer, H., Trope, Y., Todorov, A.: Body cues, not facial expressions, discriminate between intense positive and negative emotions. Science **338**(6111), 1225–1229 (2012)
2. Cavallo, F., Semeraro, F., Fiorini, L., Magyar, G., Sinčák, P., Dario, P.: Emotion modelling for social robotics applications: a review. J. Bionic Eng. **15**(2), 185–203 (2018)
3. Dael, N., Mortillaro, M., Scherer, K.R.: Emotion expression in body action and posture. Emotion **12**(5), 1085 (2012)
4. Dael, N., Mortillaro, M., Scherer, K.R.: The body action and posture coding system (BAP): development and reliability. J. Nonverbal Behav. **36**(2), 97–121 (2012)
5. De Gelder, B.: lWhy bodies? Twelve reasons for including bodily expressions in affective neuroscience. Philos. Trans. R. Soc. Lond. B Biol. Sci. **364**(1535), 3475–3484 (2009)
6. Dong, J., Li, X., Snoek, C.G.: Word2VisualVec: image and video to sentence matching by visual feature prediction. arXiv preprint arXiv:1604.06838 (2016)
7. Du, S., Tao, Y., Martinez, A.M.: Compound facial expressions of emotion. Proc. Natl. Acad. Sci. **111**(15), E1454–E1462 (2014)

8. Ekman, P., Keltner, D.: Universal facial expressions of emotion. In: Segerstrale, U., Molnar, P. (eds.) Nonverbal Communication: Where Nature Meets Culture, pp. 27–46 (1997)

9. Ekman, R.: What the Face Reveals: Basic and Applied Studies of Spontaneous Expression Using the Facial Action Coding System (FACS). Oxford University Press, Oxford (1997)

10. Filntisis, P.P., Efthymiou, N., Koutras, P., Potamianos, G., Maragos, P.: Fusing body posture with facial expressions for joint recognition of affect in child-robot interaction. IEEE Rob. Autom. Lett. **4**(4), 4011–4018 (2019)

11. Friesen, W.V., Ekman, P., et al.: EMFACS-7: emotional facial action coding system. Unpublished manuscript, University of California at San Francisco, vol. 2(36), pp. 1 (1983)

12. Frome, A., et al.: DeViSE: a deep visual-semantic embedding model. In: Advances in Neural Information Processing Systems, pp. 2121–2129 (2013)

13. Gaudelus, B., et al.: Improving facial emotion recognition in schizophrenia: a controlled study comparing specific and attentional focused cognitive remediation. Front. Psychiatry **7**, 105 (2016)

14. Gunes, H., Piccardi, M.: A bimodal face and body gesture database for automatic analysis of human nonverbal affective behavior. In: Proceedings of ICPR, vol. 1, pp. 1148–1153 (2006)

15. Kleinsmith, A., Bianchi-Berthouze, N.: Affective body expression perception and recognition: a survey. IEEE Trans. Affect. Comput. **4**(1), 15–33 (2013)

16. Kosti, R., Alvarez, J.M., Recasens, A., Lapedriza, A.: Emotion recognition in context. In: Proceedings of IEEE Conference on Computer Vision and Pattern Recognition (CVPR), pp. 1960–1968 (2017)

17. Lee, J., Kim, S., Kim, S., Park, J., Sohn, K.: Context-aware emotion recognition networks. In: Proceedings of IEEE International Conference on Computer Vision, pp. 10143–10152 (2019)

18. Lucey, P., Cohn, J.F., Kanade, T., Saragih, J., Ambadar, Z., Matthews, I.: The extended Cohn-Kanade dataset (CK+): a complete dataset for action unit and emotion-specified expression. In: Proceedings of IEEE Computer Society Conference on Computer Vision and Pattern Recognition-Workshops, pp. 94–101 (2010)

19. Luo, Y., Ye, J., Adams, R.B., Li, J., Newman, M.G., Wang, J.Z.: ARBEE: towards automated recognition of bodily expression of emotion in the wild. Int. J. Comput. Vis. **128**(1), 1–25 (2020)

20. Marinoiu, E., Zanfir, M., Olaru, V., Sminchisescu, C.: 3D human sensing, action and emotion recognition in robot assisted therapy of children with autism. In: Proceedings of CVPR,pp. 2158–2167 (2018)

21. Mittal, T., Guhan, P., Bhattacharya, U., Chandra, R., Bera, A., Manocha, D.: EmotiCon: context-aware multimodal emotion recognition using Frege's principle. In: Proceedings of IEEE/CVF Conference on Computer Vision and Pattern Recognition, pp. 14234–14243 (2020)

22. Mollahosseini, A., Hasani, B., Mahoor, M.H.: AffectNet: a database for facial expression, valence, and arousal computing in the wild. IEEE Trans. Affect. Comput. **10**(1), 18–31 (2017)

23. Pennington, J., Socher, R., Manning, C.D.: GloVE: global vectors for word representation. In: Proceedings of Conference on Empirical Methods in Natural Language Processing (EMNLP), pp. 1532–1543 (2014)

24. Ren, Z., Jin, H., Lin, Z., Fang, C., Yuille, A.L.: Multiple instance visual-semantic embedding. In: Proceedings of BMVC (2017)

25. Russell, J.A., Mehrabian, A.: Evidence for a three-factor theory of emotions. J. Res. Pers. **11**(3), 273–294 (1977)
26. Tracy, J.L., Robins, R.W.: Show your pride: Evidence for a discrete emotion expression. Psychol. Sci. **15**(3), 194–197 (2004)
27. Wang, L., et al.: Temporal segment networks: towards good practices for deep action recognition. In: Leibe, B., Matas, J., Sebe, N., Welling, M. (eds.) ECCV 2016. LNCS, vol. 9912, pp. 20–36. Springer, Cham (2016). https://doi.org/10.1007/978-3-319-46484-8_2
28. Wei, Z., et al.: Learning visual emotion representations from web data. In: Proceedings of IEEE/CVF Conference on Computer Vision and Pattern Recognition, pp. 13106–13115 (2020)
29. Yeh, M.C., Li, Y.N.: Multilabel deep visual-semantic embedding. IEEE Trans. Pattern Anal. Mach. Intell. **42**(6), 1530–1536 (2020)

Noisy Student Training Using Body Language Dataset Improves Facial Expression Recognition

Vikas Kumar$^{(\boxtimes)}$ (iD), Shivansh Rao (iD), and Li Yu (iD)

The Pennsylvania State University, University Park, State College, USA
{vuk160,shivanshrao,luy133}@psu.edu

Abstract. Facial expression recognition from videos in the wild is a challenging task due to the lack of abundant labelled training data. Large DNN (deep neural network) architectures and ensemble methods have resulted in better performance, but soon reach saturation at some point due to data inadequacy. In this paper, we use a self-training method that utilizes a combination of a labelled dataset and an unlabelled dataset (Body Language Dataset - BoLD). Experimental analysis shows that training a noisy student network iteratively helps in achieving significantly better results. Additionally, our model isolates different regions of the face and processes them independently using a multi-level attention mechanism which further boosts the performance. Our results show that the proposed method achieves state-of-the-art performance on benchmark datasets CK+ and AFEW 8.0 when compared to single models.

Keywords: Facial expression recognition · Student-teacher network · Semi-supervised learning · Multi-level attention

1 Introduction

Automatic facial expression recognition from images/videos has many applications such as human-computer interaction (HCI), bodily expressed emotions, human behaviour understanding, and has thus gained a lot of attention in academia and industry. Although there has been extensive research on this subject, facial expression recognition in the wild remains a challenging problem because of several factors such as occlusion, illumination, motion blur, subject-specific facial variations, along with the lack of extensive labelled training datasets. Following a similar line of research, our task aims to classify a given video in the wild to one of the seven broad categorical emotions. We propose an efficient model that addresses the challenges posed by videos in the wild while tackling the issue of labelled data inadequacy. The input data used for facial expression recognition can be multi-modal, i.e. it may have visual information

---

V. Kumar and S. Rao—Equal contribution.

© Springer Nature Switzerland AG 2020
A. Bartoli and A. Fusiello (Eds.): ECCV 2020 Workshops, LNCS 12535, pp. 756–773, 2020.
https://doi.org/10.1007/978-3-030-66415-2_53

as well as audio information. However, the scope of this paper is limited to emotion classification using only visual information.

Most of the recent research on the publicly-available AFEW 8.0 (Acted Facial Expressions in the Wild) [1] dataset has focused on improving accuracy without regard to computational complexity, architectural complexity, energy & policy considerations, generality, and training efficiency. Several state-of-the-art methods [2–4] on this dataset have originated from the EmotiW [5] challenge with no clear computational-cost analysis. Fan et al. [2] achieved the highest validation accuracy based on visual cues, but they used a fusion of five different architectures with more than 300 million parameters. In contrast, our proposed method uses a single model with approximately 25 million parameters and comparable performance.

While previous work focused on improving performance by increasing model capacity, our method focuses on better pre-processing, feature selection, and adequate training. Prior research [6–9] uses simple aggregation or averaging operation on features from multiple frames to form a fixed-dimensional feature vector. However, such methods do not account for the fact that a few principal frames in a video can be used to identify the target emotion, while the rest of the frames have a negligible contribution. Frame-attention has been used [10] for selectively processing frames in a video, but it can further be coupled with spatial-attention which could identify the most discriminative regions in a particular frame. We use a three-level attention mechanism in our model: a) spatial-attention block that helps to selectively process feature maps of a frame, b) channel-attention block that focuses on the face regions at a local and a global level, i.e. eyes region (upper face), mouth region (lower face) and whole face, and c) frame-attention block that helps to identify the most important frames in a video.

AFEW 8.0 [1] has several limitations (Sect. 2) that restricts the generalization capabilities of deep learning models. To overcome these limitations, we use an unlabelled subset of the BoLD dataset [11] for semi-supervised learning. Inspired by Xie et al. [12], we use a teacher-student learning method where the training process is iterated by using the same student again as the teacher. During the training of the student, noise is injected into the student model to force it to generalize better than the teacher. Results show that the student performs better with each iteration, hence improving the overall accuracy on the validation set.

The rest of the paper is organized as follows. Section 3 explains the datasets (AFEW 8.0 [1], CK+ [13] and BoLD [11]) that are used for training our model along with the pre-processing pipeline used for face detection, alignment and illumination correction. Section 4.1 explains the backbone network and covers the three types of attention and its importance in detail. Section 4.2 covers the use of the BoLD dataset for iterative training and the experimental results of semi-supervised learning. Section 5.3 compares the results of our methods to other state-of-the-art methods on the AFEW 8.0 dataset. Additionally, we use another benchmark dataset CK+ [13] (posed conditions) as well as perform ablation studies (Sect. 5.4) to prove the validity of our model and training procedure.

2 Related Work

Facial Expression Recognition: A number of methods have been proposed on the AFEW 8.0 dataset [1] since the first EmotiW [5] challenge in 2013. Earlier approaches include non-deep learning methods such as multiple kernel learning [14], least-square regression on grassmanian manifold [15], and feature fusion with kernel learning [16], whereas recent approaches include deep-learning methods such as frame-attention networks [10], multiple spatial-temporal learning [3], and deeply supervised emotion recognition [2]. Although several methods [2–4,17] have achieved impressive results on the AFEW 8.0 dataset, many have used ensemble (fusion) based methods and considered multiple modalities without commenting on the resources and time required to train such models. Spatial-temporal methods [4,18] aim to model motion information or temporal coherency in the videos using 3D Convolution [19] or LSTM (Long short-term memory) [20]. However, owing to computational efficiency and the ability to treat sequential information with a global context, several studies [10,21] related to facial expression recognition have successfully implemented attention-based methods by assigning a weight to each timestep in the video. Similarly, spatial self-attention has been used [21–23] as a means to guide the process of feature extraction and find the importance of each local image feature. Our model builds upon the spatial self-attention mechanism and additionally uses a channel-attention mechanism to exploit the differential effects of facial feedback signals from the upper-face and lower-face regions [24,25].

Training Datasets: Despite being a long-established dataset, AFEW 8.0 [1] has several shortcomings. Firstly, the dataset contains significantly fewer training examples for fear, surprise and disgust categories which makes the dataset imbalanced. Secondly, the videos are extracted from mainstream cinema, and scenes depicting fear are often shot in the dark, which again makes the model biased towards other categories [3,26]. Such limitations warrant the use of additional datasets for better generalization. However, not many in-the-wild labelled video datasets are publicly available for facial expression recognition. Several related datasets [13,27,28] are captured in posed conditions and are restricted to a certain country or community. Aff-Wild2 [29] is another popular dataset, but it contains per-frame annotations, and thus cannot be used in our work which performs video-level classification based on facial expressions. We use an unlabelled portion of the BoLD dataset [11] since the videos are of the desired length and are captured from movies similar to our labelled dataset.

Semi-supervised Learning: The semi-supervised approach is effective in classification problems when the labelled training data is not sufficient. We use noisy student training [12] for semi-supervised learning, in which the trick involves the student to be deliberately noised when it trains on the combined labelled and unlabelled dataset. Input noise is added to the student model in the form of data augmentations, which ensures that different alterations of the same video should

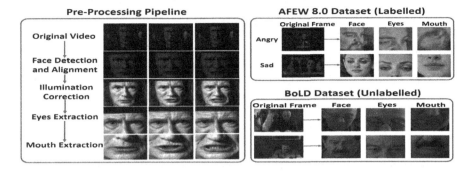

Fig. 1. The pre-processing steps mainly include face detection and alignment (MTCNN [35]), illumination correction (Enlighten-GAN [36]) and landmark-based cropping. Examples from labelled dataset (AFEW 8.0) and unlabelled dataset (BoLD dataset) are shown. As seen in the figure, only videos with a close shot of the face are selected from the BoLD dataset.

have the same emotion, hence making the student model more robust. Additionally, model noise is added in the form of dropout, which forces the student (single model) to match the performance of an ensemble model. Other techniques for semi-supervised learning include self-training [30,31], data-distillation [32] and consistency training [33,34]. Self-training is similar to noisy student training, but it does not use or justify the role of noise in training a powerful student. Data-distillation uses the approach of strengthening the teacher using ensembles instead of weakening the student; however, a smaller student makes it difficult to mimic the teacher. Consistency training adds regularization parameters to the teacher model during training to induce invariance to input and model noise, resulting in confident pseudo-labels. However, such constraints lead to lower accuracy and a less powerful teacher [12].

3   Dataset

In this section, we first describe the datasets that we use in our experiments, followed by the pre-processing pipeline.

Labelled Sets: AFEW 8.0 (Acted Facial Expression in the Wild) [1] contains videos with seven emotion labels, i.e. anger (197 samples), neutral (207 samples), sad (179 samples), fear (127 samples), surprise (120 samples), happiness (212 samples), and disgust (114 samples) from different movies. The train set consists of 773 video samples (46,080 frames), and the validation set consists of 383 video samples (21,157 frames). The results are reported on the validation set since the test set labels are only available to EmotiW challenge [5] participants. Some of the example frames are shown in Fig. 1. CK+ (Cohn Kanade Extended) [13] contains 327 video sequences (5878 frames) divided into seven categories, i.e anger (45 samples), disgust (59 samples), fear (25 samples), happy (69 samples), sad (28 samples), surprise (83 samples), and contempt (18 samples). The motivation behind testing our method on a posed dataset is to establish the robustness of our model and semi-supervised learning method irrespective of the data source. Since CK+ does not have a testing set, we report the average accuracy obtained using 10-fold cross-validation as seen in other studies [10, 37–40].

Unlabelled Set: BoLD (Body Language Dataset) [11] contains videos selected from the publicly available AVA dataset [41], which contains a list of YouTube movie IDs. While the gathered videos are annotated based on body language, the videos having a close shot of the face instead of the whole or partially-occluded body are unlabelled. To create an AFEW-like subset from the BoLD dataset, we impose two conditions to automatically validate a video. Firstly, a video should have f ($\geq$ 30) such consecutive frames where only one actor's face is detected by MTCNN (Multi-task Cascaded Convolutional Networks) [35]. Secondly, the bounding box of the face detected using MTCNN should exceed an occupied area threshold for the majority of those f frames. If the video satisfies the above two conditions, a smaller video with those f frames is added to the unlabelled dataset. Using this procedure, we create a subset of 3450 videos (224,258 frames) from the original BoLD dataset. Some of the examples gathered are shown in Fig. 1.

Pre-processing: Previous work [3, 10] have used CNN-based detector provided by dlib [42] for face alignment. However, the alignment of faces is highly dependent on accurate detection of facial landmarks and CNN-based detector provided by dlib is not reliable for 'in-the-wild' conditions (especially non-frontal faces). We use MTCNN [35] for face detection and alignment. If MTCNN detects multiple faces in a frame, the face with the largest bounding box is selected. After obtaining the facial landmarks, its alignment is corrected using the angle between the line connecting the landmark points of the eyes and the horizontal line. After detection and alignment, the cropped face is resized to 224*224 pixels, which is the input size of our model.

We use the landmarks given by MTCNN to isolate the mouth (lower face) and eyes (upper face) region. The upper face is isolated using the eyes landmarks with the desired left eye normalized co-ordinates being (0.2, 0.6) and right-eye co-ordinates being (0.8, 0.6) in the new frame, which is enough to occlude the lower-half of the face in almost all frames (Fig. 1). A similar procedure is used for occluding the upper-half of the face and isolating the mouth region using left-mouth and right-mouth landmarks. All landmark-based crops are again resized to 224*224 pixels.

As addressed earlier, some of the categories of emotions are often captured in the dark in movies, which requires an illumination correction step. Several methods have been suggested for illumination normalization such as gamma correction [43,44], Difference of Gaussians (DoG) [45] and histogram equalization [46,47] which are effective for facial expression recognition. However, these methods tend to amplify noise, tone distortion, and other artefacts. Hence, we use a state-of-the-art pre-trained deep learning model, i.e. Enlighten-GAN [36] (U-Net [48] as generator) which provides appropriate results (Fig. 1) with uniform illumination and suppressed noise.

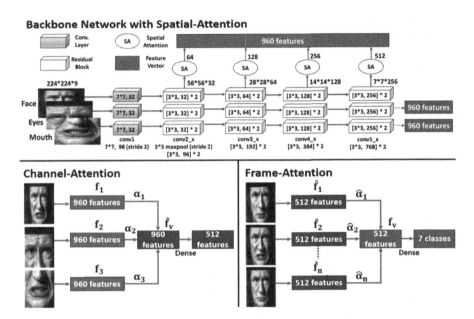

Fig. 2. Figure shows the backbone network (ResNet-18) and the three-level attention mechanism. Inputs are first processed via Spatial-Attention, followed by Channel-Attention and finally by Frame-Attention.

4 Methodology

Our proposed methodology is divided into two phases, i.e. a) architecture implementation that defines the backbone network with the three-level attention mechanism, and b) semi-supervised learning.

4.1 Architecture

Backbone Network: We use ResNet-18 [49] architecture as our backbone network, with minor modifications to increase its computational efficiency. Features from each residual block are combined to form the final feature vector (see Fig. 2). Hence, the final vector has a "multi-level knowledge" from all the residual blocks, ensuring more diverse and robust features. The model is first pre-trained

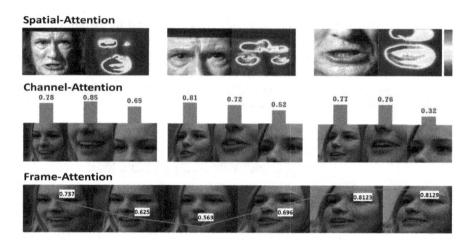

Fig. 3. This figure shows how multi-level attention works in the proposed method. Spatial-attention (from last residual block) chooses the dominant feature maps from each region. Channel-attention picks the most important region that most clearly shows the target emotion. Frame-attention assigns the salient frames a higher weight.

on the FERPlus dataset [50]. Our input at frame-level is an image with nine channels (RGB channels from the face, eyes, and mouth region). To process them independently, the model uses group convolution [51] (groups = 3), i.e. it uses a different set of filters for each of the three regions to get the final output feature maps. Group convolution results in a lower computational cost since each kernel filter does not have to convolve on all the feature maps of the previous layer. Simultaneously, it allows data parallelism where each filter group is learning a unique representation and forms a global (face region) or local (eyes and mouth region) context vector from each frame of a video. To allow more filters per group, we increase the number of filters in each residual block, as shown in Fig. 2.

Spatial-Attention: A common approach in previous methods is a simple aggregation or average pooling of feature maps to form a fixed dimensional feature vector. However, we use spatial-attention [23] that concatenates the feature maps based on the attention weight it has been assigned. Let us assume the output from a residual block is of shape $C = H \times W \times D$ where H and W are the output height and width, and D is the number of output filters. This 3D tensor C is reshaped to a 2D matrix L of shape $R \times D$ where $R = H * W$. The spatial-attention mechanism takes the input matrix L and outputs a weight matrix M of shape $h \times R$ ($h = 2$, h is for multiple hops of attention). Each row of the output matrix represents a different hop of attention, and each column has normalized weights due to softmax (see Eq. 1). The objective is to find the weighted average of R frame descriptors to obtain a vector v of length D (or $h * D$ with multiple hops).

$$M = softmax(W_{s2}tanh(W_{s1}L^T)) \qquad (1)$$
$$v = flatten(M * L) \qquad (2)$$

Equation 1 represents multi-head spatial-attention where W_{s1} is of shape $U \times D$ and W_{s2} is of shape $h \times U$ (U can be set arbitrarily). From this, we obtain flattened vector v using Eq. 2. The spatial-attention module is applied on each residual block (see Fig. 2) and the output vectors are aggregated to obtain a final vector of length $l = 960$ each for face (f_1), eyes (f_2) and mouth (f_3) regions. The advantages of spatial attention can be seen in Fig. 3. While the feature vector from the face is encoded with a global context, the feature maps from the eyes and mouth region have additional information regarding the minute expressions such as furrowed brow or flared nostrils.

Channel-Attention: Let f_1, f_2, and f_3 be the feature vectors obtained from the face, the eyes, and the mouth region respectively. We model the cross-channel interactions using a lightweight attention module. We use two fully-connected layers to obtain a weight α (Eq. 3) for each channel group using which we obtain a weighted average $\hat{f}_v$ (Eq. 4) of the three feature vectors. ReLU (Rectified Linear Unit) activation is used after the first layer to capture non-linear interactions among the channels.

$$\alpha_i = \sigma(w^T(ReLU(W^T f_i))) \qquad (3)$$
$$\hat{f}_v = \frac{\sum_{i=1}^3 \alpha_i * f_i}{\sum_{i=1}^3 \alpha_i} \qquad (4)$$

where σ is the sigmoid activation function, w is a vector of length r (set arbitrarily), and W is a matrix of shape $l \times r$. In Fig. 3, we see that the model assigns more weight to the mouth region instead of the eyes region for an expression depicting happiness which is consistent with our findings that mouth region is more prominent for the happy category (Fig. 5).

Frame-Attention: For a video having n frames, we obtain vector $\hat{f}_i$ of length $\hat{l}$ from each frame after the channel-attention module. Finally, we use frame-attention to assign the most discriminative frames a higher weight. Following a similar intuition as in channel-attention, we use two fully-connected layers to obtain a weight $\hat{\alpha}$ (Eq. 5) for each frame using which we find a weighted average f_v (Eq. 6) of the frame features.

$$\hat{\alpha}_i = \sigma(\hat{w}^T(ReLU(\hat{W}^T \hat{f}_i))) \qquad (5)$$
$$f_v = \frac{\sum_{i=1}^n \hat{\alpha}_i * \hat{f}_i}{\sum_{i=1}^n \hat{\alpha}_i} \qquad (6)$$

where $\hat{w}$ is a vector of length $\hat{r}$ (set arbitrarily), and $\hat{W}$ is a matrix of shape $\hat{l} \times \hat{r}$. Figure 3 shows how the model assigns a higher weight to the frames which distinctively contains expression depicting happiness. The feature vector f_v is passed through a fully-connected layer to obtain the final 7-dimensional output.

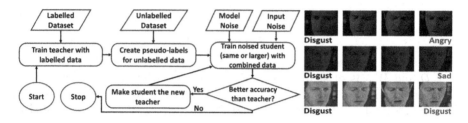

Fig. 4. Semi-supervised algorithm is presented in the flow-chart. We also show an example video from AFEW 8.0 dataset where the frames underwent different augmentations. Predictions without iterative training are shown in red and predictions after iterative training are shown in black. (Color figure online)

Implementation Details: We use weighted cross-entropy as our loss function where class weights are assigned based on number of training samples to alleviate the problem of unbalanced data. Additionally, M (Eq. 1) is regularized by adding the frobenius norm of matrix $MM^T - I$ to the loss function which enforces multi spatial-attention to focus on different regions [23]. We use Adam optimizer with an initial learning rate of 1e−5 (reduced by 40% after every 30 epochs) and the model is trained for 100 epochs. The training takes around 8 min for 1 epoch for AFEW 8.0 training dataset with two NVIDIA Tesla K80 cards.

4.2 Noisy Student Training [12]

Once the model is trained on the labelled set and the best possible model is obtained, we use it as a teacher model to create pseudo-labels on the subset of BoLD dataset that we collected. After generating the pseudo-labels, a student model (same size or larger than teacher) is trained on the combination of labelled and unlabelled dataset. While training the student model, we deliberately add noise in the form of random data augmentations and dropout (with 0.5 probability at the final hidden layer). Random data augmentations (using RandAugment [52]) include transformations such as brightness change, contrast change, translation, sharpness change and flips. RandAugment automatically applies $n \in [2, 4]$ random operations with a random magnitude $m \in [0, 9]$. After the noisy student is trained on the combined data, the trained student becomes the new teacher that generates new pseudo-labels for the unlabelled dataset. The iterative training continues until we observe a saturation in performance. From Fig. 4, we see how noisy training helps the student become more robust with the addition of noise. While the teacher may give different predictions for different alterations of the same video, the student is more accurate and stable with its predictions.

5   Results

In this section, we show the results obtained with and without iterative self-training, followed by comparison with state-of-the-art methods and ablation studies.

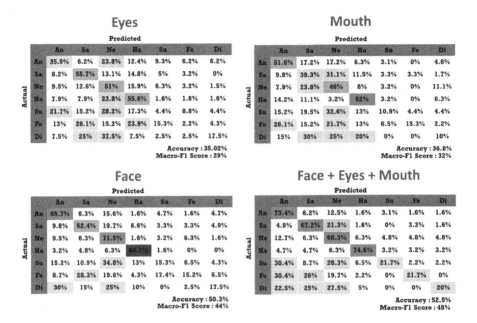

Fig. 5. This figure shows the confusion matrices, the accuracies, and the macro f1 scores achieved on the AFEW 8.0 dataset using different regions of the face. The proposed model (Face + Eyes + Mouth) achieves the highest accuracy. An=Angry, Sa=Sad, Ne=Neutral, Ha=Happy, Su=Surprise, Fe=Fear, Di=Disgust.

5.1 Without Student Training

Figure 5 shows the results of processing individual regions (without group convolution and channel attention) on the AFEW 8.0 dataset, along with the proposed methodology. Our objective is to explore a) if upper face region and lower face regions have different feedback signals that dominate different categories of emotions, and b) if isolating the regions and processing them independently leads to an increase of accuracy. As seen in the confusion matrix (Fig. 5), the eyes region is better than the mouth region in the prediction of sadness and disgust categories. Intuitively, the squinted eyes expression in disgust and the droopy eyelids or furrowed eyebrows expression in sadness makes the eyes region pronounced. On the other hand, the mouth region is comparatively better with categories that require lip movements like happiness, anger, and surprise. Overall, 52.50% accuracy is achieved using the proposed model, which is slightly better than the model that only uses faces. Furthermore, we see a significant increase in the macro f1 score when we include the eyes and mouth region along with faces indicating that the predictions are comparatively more unbiased for the seven categories (an advantage for noisy student training). The proposed model is still biased against fear, surprise, and disgust categories, but performs better than several existing methods [3,26,53] where the reported accuracies for these categories are close to 0%.

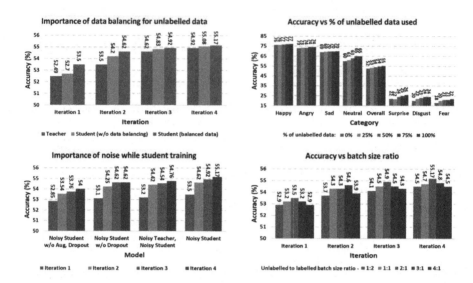

Fig. 6. This figure shows the experimental results of noisy student training for four iterations using AFEW 8.0 and BoLD datset.

5.2 With Iterative Training

Using noisy student training, we report our experimental results for four iterations on the AFEW 8.0 dataset and two iterations on the CK+ dataset.

Data Balancing: Since the model is biased, the number of pseudo-labels in the unlabelled dataset for some categories is smaller than in other categories. We try to match the distribution of the training set by duplicating images of fear, disgust, and surprise categories. Additionally, images of angry, happy, and neutral classes are filtered out based on confidence scores. Figure 6 shows that balancing the pseudo-labels leads to better accuracy in each iteration compared to the student model without data balancing. The same trend is not observed for the CK+ dataset since the pseudo-labels roughly have the same distribution as the training set.

Unlabelled Dataset Size: As stated in the original paper [12], using a large amount of unlabelled data leads to better accuracy. After data balancing, we use a fraction of the BoLD dataset and report the accuracy after several iterations of training until the performance saturates (see Fig. 6). For both CK+ and AFEW 8.0 dataset, we observe that using the whole unlabelled training set is better as opposed to using just a fraction of the dataset. Figure 6 shows a steady increase in all categories and overall accuracy with an increase in data size after four iterations of training on the AFEW 8.0 dataset.

Importance of Noise: Noise helps the student to be more robust than the teacher, as addressed in Sect. 2. The accuracy only reaches 54% on the AFEW 8.0 dataset without noise in student training, and no improvement is seen on the CK+ dataset. However, we achieve an accuracy of 55.17% after noisy training, which shows that input and model perturbations are vital while training the student. Additionally, Fig. 6 shows that it is better when the pseudo-labels are generated without noise, i.e. the teacher remains as powerful as possible.

Table 1. We compare our results to the top-performing *single* models evaluated on the AFEW 8.0 dataset and state-of-the-art models evaluated on the CK+ dataset.

AFEW 8.0		CK+	
Models	Acc.	Models	Acc.
CNN-RNN (2016) [18]	45.43%	Lomo (2016) [40]	92.00%
DSN-HoloNet (2017) [54]	46.47%	CNN + Island Loss (2018) [39]	94.35%
DSN-VGGFace (2018) [2]	48.04%	FAN (2019) (Fusion) [10]	94.80%
VGG-Face + LSTM (2017) [4]	48.60%	Hierarchial DNN (2019) [55]	96.46%
VGG-Face (2019) [21]	49.00%	DTAGN (2015) [38]	97.25%
ResNet-18 (2018) [56]	49.70%	MDSTFN (2019) [57]	98.38%
FAN (2019) [10]	51.18%	Compact CNN (2018) [58]	98.47%
DenseNet-161 (2018) [17]	51.44%	ST Network (2017) [37]	98.47%
Our Model (w/o iter. training)	52.49%	Our Model (w/o iter. learning)	98.77%
VGG-Face + BLSTM (2018) [3]	53.91%	**FAN (2019)** [10]	**99.69%**
Our Model (iter. training)	**55.17%**	**Our Model (iter. learning)**	**99.69%**

Batch Size Ratio: When training on combined data, a batch of labelled images and a batch of unlabelled images are concatenated for each training step. If the batch sizes of labelled and unlabelled sets are equal, the model will complete several epochs of training on labelled data before completing one epoch of training on the BoLD dataset due to its larger size. To balance the number of epochs of training on both datasets, the batch size of the unlabelled set is kept higher than the labelled set. Figure 4 shows that a batch size ratio of 2:1 or 3:1 is ideal for training when AFEW 8.0 is used as the labelled training set. Similarly, a batch size ratio of 5:1 is ideal for the CK+ dataset.

5.3   Comparison with Other Methods

We evaluate our model on the labelled datasets and show a comparison with the existing state-of-the art-methods (Table 1). On the AFEW 8.0 dataset, we achieve an accuracy of 52.5% without iterative training and 55.17% with iterative training. When comparing to existing best single models, our proposed method improves upon the current baseline [3] by 1.6%. Compared to static-based CNN methods that aim to combine frame scores for video-level recognition, we achieve a significant improvement of 3.73% over the previous baseline [17]. We conduct a comparison of performance and speed of the existing state-of-the-art models including fusion methods (only visual modality) with our proposed model. Several methods that show higher validation accuracy have significantly higher computational demand which may be impractical for real-time world applications. For instance, [56] uses an ensemble of 50 models with the same architecture and yet attains a 52.2% validation accuracy. Similarly, [3,17] use a combination of multiple deep learning models where each model has a higher computational cost than ours. We measure the computational complexity of state-of-the-art methods using FLOPS (Floating point operations) and results show that our method is the most optimal based on performance and speed (Fig. 7).

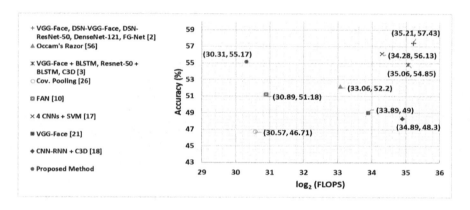

Fig. 7. Comparison of performance (in accuracy) vs computational cost (in FLOPS - Floating point operations) of state-of-the-art models evaluated on AFEW 8.0 dataset. FLOPS for the models are estimated values based on the backbone network unless explicitly specified by the authors. Most optimal models will be closer to the top-left corner.

On the CK+ dataset, our method achieves an on par 10-fold cross-validation accuracy when compared to other state-of-the-art methods. While our model achieves an accuracy of only 98.77% without iterative learning, the accuracy improves by 0.92% when training data of each fold is combined with the unlabelled dataset for two iterations. This confirms our premise that self-training using noisy student is a robust procedure and can be used to increase the performance of a model on several other labelled data sources. Additionally, our results show that one can achieve better performance on a posed dataset when trained with an unlabelled in-the-wild dataset in a semi-supervised manner, which can be an effective alternative to labour-intensive tasks like gathering additional posed samples or labelling data.

5.4   Ablation Studies

Our baseline model is ResNet-18 where the video-level feature vector is an unweighted average of all the frame-level feature vectors. Without sophisticated pre-processing, the baseline achieves an accuracy of 47.5%. To better understand the significance of each component, we record our results after every change to the baseline model (Table 2). Significant improvements are observed when features are concatenated from multiple residual blocks using spatial-attention, and when frame features are combined from multiple regions using group convolution and channel-attention.

Table 2. This table shows the ablation studies conducted with AFEW 8.0 dataset. *Component Importance* shows the increase in accuracy with the addition of each component separately. *Noisy Student Training* shows the increase in accuracy with each loop of iterative learning and the effect of using a larger student.

Component importance		Noisy student training		
Component	Acc.	Iteration	Student	Acc.
ResNet-18 (Baseline)	47.5%	0	–	52.5%
+ MTCNN, Enlighten-GAN (Sect. 3)	48.3%	1	ResNet-18	53.5%
+ Features from all blocks (Sect. 4.1)	49.3%		ResNet-34	53.5%
+ Spatial-Attention (Sect. 4.1)	50.3%	2	ResNet-18	54.6%
+ Multiple Regions (Sect. 4.1)	51.2%		ResNet-34	54.5%
+ Channel-Attention (Sect. 4.1)	51.7%	3	ResNet-18	54.9%
+ Frame-Attention (Sect. 4.1)	52.5%		ResNet-34	54.8%
+ Iteration 1 - Self-training (Sect. 4.2)	53.5%	4	ResNet-18	55.2%
+ Iteration 2 - Self-training (Sect. 4.2)	54.6%		ResNet-34	55.2%
+ Iteration 3 - Self-training (Sect. 4.2)	54.9%	5	ResNet-18	55.2%
+ Iteration 4 - Self-training (Sect. 4.2)	55.2%		ResNet-34	55.2%

Additionally, Table 2 shows the increase in validation accuracy with each loop of iterative learning. As suggested by [12], noisy student learning may perform better if the student is larger in size than the teacher. Since ResNet-34 [49] has a comparatively larger capacity, we report its results besides ResNet-18 as the student model for each iteration. As seen in Table 2, our results do not show improvement when ResNet-18 in our student model is replaced with a larger backbone. A possible explanation is that the unlabelled dataset used by [12] is a hundred times larger than the labelled dataset and using a student with higher capacity may have resulted in better performance. On the contrary, our unlabelled dataset is only four times larger than the labelled dataset. Gathering additional unlabelled samples and using a larger student may result in a further increase in accuracy on the AFEW 8.0 dataset.

6   Conclusion

We propose a multi-level attention model for video-based facial expression recognition. Our contribution is a cost-effective single model that achieves on par performance with state-of-the-art models using two strategies. Firstly, we use attention with multiple sources of information to capture spatially and temporally important features, which is a computationally economical alternative to the fusion of multiple learning models. Secondly, we use self-training to overcome the lack of labelled video datasets for facial expression recognition.

Acknowledgements. The authors acknowledge the contribution of Dr. James Wang for providing the opportunity to work on this project during his course on Artificial Emotion Intelligence taught at the Pennsylvania State University.

References

1. Dhall, A., Goecke, R., Lucey, S., Gedeon, T.: Collecting large, richly annotated facial-expression databases from movies. IEEE Multimedia (3), 34–41 (2012)
2. Fan, Y., Lam, J.C., Li, V.O.: Video-based emotion recognition using deeply-supervised neural networks. In: Proceedings of the 20th ACM International Conference on Multimodal Interaction, pp. 584–588(2018)
3. Lu, C., et al.: Multiple spatio-temporal feature learning for video-based emotion recognition in the wild. In: Proceedings of the 20th ACM International Conference on Multimodal Interaction, pp. 646–652 (2018)
4. Vielzeuf, V., Pateux, S., Jurie, F.: Temporal multimodal fusion for video emotion classification in the wild. In: Proceedings of the 19th ACM International Conference on Multimodal Interaction, pp. 569–576 (2017)
5. Dhall, A.: Emotiw 2019: Automatic emotion, engagement and cohesion prediction tasks. In: 2019 International Conference on Multimodal Interaction, pp. 546–550 (2019)
6. Littlewort, G., Bartlett, M.S., Fasel, I., Susskind, J., Movellan, J.: Dynamics of facial expression extracted automatically from video. In: 2004 Conference on Computer Vision and Pattern Recognition Workshop, p. 80. IEEE (2004)
7. Shan, C., Gong, S., McOwan, P.W.: Facial expression recognition based on local binary patterns: a comprehensive study. Image Vis. Comput. **27**(6), 803–816 (2009)
8. Knyazev, B., Shvetsov, R., Efremova, N., Kuharenko, A.: Convolutional neural networks pretrained on large face recognition datasets for emotion classification from video. arXiv preprint arXiv:1711.04598 (2017)
9. Tang, Y.: Deep learning using linear support vector machines. arXiv preprint arXiv:1306.0239 (2013)
10. Meng, D., Peng, X., Wang, K., Qiao, Y.: frame attention networks for facial expression recognition in videos. In: 2019 IEEE International Conference on Image Processing (ICIP), pp. 3866–3870. IEEE (2019)
11. Luo, Y., Ye, J., Adams, R.B., Li, J., Newman, M.G., Wang, J.Z.: Arbee: towards automated recognition of bodily expression of emotion in the wild. Int. J. Comput. Vis. **128**(1), 1–25 (2020)
12. Xie, Q., Hovy, E., Luong, M.T., Le, Q.V.: Self-training with noisy student improves imagenet classification. arXiv preprint arXiv:1911.04252 (2019)

13. Lucey, P., Cohn, J.F., Kanade, T., Saragih, J., Ambadar, Z., Matthews, I.: The extended cohn-kanade dataset (ck+): a complete dataset for action unit and emotion-specified expression. In: IEEE Computer Society Conference on Computer Vision and Pattern Recognition-Workshops, vol. 2010, pp. 94–101. IEEE (2010)
14. Sikka, K., Dykstra, K., Sathyanarayana, S., Littlewort, G., Bartlett, M.: Multiple kernel learning for emotion recognition in the wild. In: Proceedings of the 15th ACM on International conference on multimodal interaction, pp. 517–524 (2013)
15. Liu, M., Wang, R., Huang, Z., Shan, S., Chen, X.: Partial least squares regression on grassmannian manifold for emotion recognition. In: Proceedings of the 15th ACM on International conference on multimodal interaction, pp. 525–530 (2013)
16. Chen, J., Chen, Z., Chi, Z., Fu, H.: Emotion recognition in the wild with feature fusion and multiple kernel learning. In: Proceedings of the 16th International Conference on Multimodal Interaction, pp. 508–513 (2014)
17. Liu, C., Tang, T., Lv, K., Wang, M.: Multi-feature based emotion recognition for video clips. In: Proceedings of the 20th ACM International Conference on Multimodal Interaction, pp. 630–634 (2018)
18. Fan, Y., Lu, X., Li, D., Liu, Y.: Video-based emotion recognition using cnn-rnn and c3d hybrid networks. In: Proceedings of the 18th ACM International Conference on Multimodal Interaction, pp. 445–450 (2016)
19. Tran, D., Bourdev, L., Fergus, R., Torresani, L., Paluri, M.: Learning spatiotemporal features with 3D convolutional networks. In: Proceedings of the IEEE International Conference on Computer Vision, pp. 4489–4497 (2015)
20. Hochreiter, S., Schmidhuber, J.: Long short-term memory. Neural Comput. 9(8), 1735–1780 (1997)
21. Aminbeidokhti, M., Pedersoli, M., Cardinal, P., Granger, E.: Emotion recognition with spatial attention and temporal softmax pooling. In: Karray, F., Campilho, A., Yu, A. (eds.) ICIAR 2019. LNCS, vol. 11662, pp. 323–331. Springer, Cham (2019). https://doi.org/10.1007/978-3-030-27202-9_29
22. Fang, Y., Gao, J., Huang, C., Peng, H., Wu, R.: Self multi-head attention-based convolutional neural networks for fake news detection. PloS one 14(9), e0222713 (2019)
23. Lin, Z., et al.: A structured self-attentive sentence embedding. arXiv preprint arXiv:1703.03130 (2017)
24. Wang, K., Peng, X., Yang, J., Meng, D., Qiao, Y.: Region attention networks for pose and occlusion robust facial expression recognition. IEEE Trans. Image Process. 29, 4057–4069 (2020)
25. Zeng, X., Wu, Q., Zhang, S., Liu, Z., Zhou, Q., Zhang, M.: A false trail to follow: differential effects of the facial feedback signals from the upper and lower face on the recognition of micro-expressions. Front. Psychol. 9, 2015 (2018)
26. Acharya, D., Huang, Z., Pani Paudel, D., Van Gool, L.: Covariance pooling for facial expression recognition. In: Proceedings of the IEEE Conference on Computer Vision and Pattern Recognition Workshops, pp. 367–374 (2018)
27. Valstar, M., Pantic, M.: Induced disgust, happiness and surprise: an addition to the mmi facial expression database. In: Proceedings of 3rd International Workshop on EMOTION (satellite of LREC): Corpora for Research on Emotion and Affect, Paris, France, p. 65 (2010)
28. Lyons, M.J., Akamatsu, S., Kamachi, M., Gyoba, J., Budynek, J.: The Japanese female facial expression (jaffe) database. In: Proceedings of Third International Conference on Automatic Face and Gesture Recognition, pp. 14–16 (1998)
29. Kollias, D., Zafeiriou, S.: Aff-wild2: extending the aff-wild database for affect recognition. arXiv preprint arXiv:1811.07770 (2018)

30. Yarowsky, D.: Unsupervised word sense disambiguation rivaling supervised methods. In: 33rd Annual Meeting of the Association for Computational Linguistics, pp. 189–196 (1995)

31. Riloff, E.: Automatically generating extraction patterns from untagged text. In: Proceedings of the National Conference on Artificial Intelligence, pp. 1044–1049 (1996)

32. Radosavovic, I., Dollár, P., Girshick, R., Gkioxari, G., He, K.: Data distillation: Towards omni-supervised learning. In: Proceedings of the IEEE Conference on Computer Vision and Pattern Recognition, pp. 4119–4128 (2018)

33. Bachman, P., Alsharif, O., Precup, D.: Learning with pseudo-ensembles. In: Advances in Neural Information Processing Systems, pp. 3365–3373 (2014)

34. Rasmus, A., Berglund, M., Honkala, M., Valpola, H., Raiko, T.: Semi-supervised learning with ladder networks. In: Advances in Neural Information Processing Systems, pp. 3546–3554 (2015)

35. Zhang, K., Zhang, Z., Li, Z., Qiao, Y.: Joint face detection and alignment using multitask cascaded convolutional networks. IEEE Signal Process. Lett. **23**(10), 1499–1503 (2016)

36. Jiang, Y., et al.: Enlightengan: Deep light enhancement without paired supervision. arXiv preprint arXiv:1906.06972 (2019)

37. Zhang, K., Huang, Y., Du, Y., Wang, L.: Facial expression recognition based on deep evolutional spatial-temporal networks. IEEE Trans. Image Process. **26**(9), 4193–4203 (2017)

38. Jung, H., Lee, S., Yim, J., Park, S., Kim, J.: Joint fine-tuning in deep neural networks for facial expression recognition. In: Proceedings of the IEEE International Conference on Computer Vision, pp. 2983–2991 (2015)

39. Cai, J., Meng, Z., Khan, A.S., Li, Z., O'Reilly, J., Tong, Y.: Island loss for learning discriminative features in facial expression recognition. In: 2018 13th IEEE International Conference on Automatic Face & Gesture Recognition (FG 2018), pp. 302–309. IEEE (2018)

40. Sikka, K., Sharma, G., Bartlett, M.: Lomo: latent ordinal model for facial analysis in videos. In: Proceedings of the IEEE Conference on Computer Vision and Pattern Recognition, pp. 5580–5589 (2016)

41. Gu, C., et al.: Ava: a video dataset of spatio-temporally localized atomic visual actions. In: Proceedings of the IEEE Conference on Computer Vision and Pattern Recognition, pp. 6047–6056 (2018)

42. King, D.E.: Dlib-ml: a machine learning toolkit. J. Mach. Learn. Res. **10**(Jul), 1755–1758 (2009)

43. Anila, S., Devarajan, N.: Preprocessing technique for face recognition applications under varying illumination conditions. Glob. J. Comput. Sci. Technol. (2012)

44. Liu, Y., Li, Y., Ma, X., Song, R.: Facial expression recognition with fusion features extracted from salient facial areas. Sensors **17**(4), 712 (2017)

45. Wang, S., Li, W., Wang, Y., Jiang, Y., Jiang, S., Zhao, R.: An improved difference of gaussian filter in face recognition. J. Multimedia **7**(6), 429–433 (2012)

46. Bendjillali, R.I., Beladgham, M., Merit, K., Taleb-Ahmed, A.: Improved facial expression recognition based on dwt feature for deep CNN. Electronics **8**(3), 324 (2019)

47. Karthigayan, M., et al.: Development of a personified face emotion recognition technique using fitness function. Artif. Life Rob. **11**(2), 197–203 (2007)

48. Ronneberger, O., Fischer, P., Brox, T.: U-Net: convolutional networks for biomedical image segmentation. In: Navab, N., Hornegger, J., Wells, W.M., Frangi, A.F. (eds.) MICCAI 2015. LNCS, vol. 9351, pp. 234–241. Springer, Cham (2015). https://doi.org/10.1007/978-3-319-24574-4_28

49. He, K., Zhang, X., Ren, S., Sun, J.: Deep residual learning for image recognition. In: Proceedings of the IEEE Conference on Computer Vision and Pattern Recognition, pp. 770–778 (2016)

50. Barsoum, E., Zhang, C., Ferrer, C.C., Zhang, Z.: Training deep networks for facial expression recognition with crowd-sourced label distribution. In: Proceedings of the 18th ACM International Conference on Multimodal Interaction, pp. 279–283 (2016)

51. Krizhevsky, A., Sutskever, I., Hinton, G.E.: Imagenet classification with deep convolutional neural networks. In: Advances in Neural Information Processing Systems, pp. 1097–1105 (2012)

52. Cubuk, E.D., Zoph, B., Shlens, J., Le, Q.V.: Randaugment: Practical data augmentation with no separate search. arXiv preprint arXiv:1909.13719 (2019)

53. Yan, J., Zheng, W., Cui, Z., Tang, C., Zhang, T., Zong, Y.: Multi-cue fusion for emotion recognition in the wild. Neurocomputing **309**, 27–35 (2018)

54. Hu, P., Cai, D., Wang, S., Yao, A., Chen, Y.: Learning supervised scoring ensemble for emotion recognition in the wild. In: Proceedings of the 19th ACM International Conference on Multimodal Interaction, pp. 553–560 (2017)

55. Kim, J.H., Kim, B.G., Roy, P.P., Jeong, D.M.: Efficient facial expression recognition algorithm based on hierarchical deep neural network structure. IEEE Access **7**, 41273–41285 (2019)

56. Vielzeuf, V., Kervadec, C., Pateux, S., Lechervy, A., Jurie, F.: An occam's razor view on learning audiovisual emotion recognition with small training sets. In: Proceedings of the 20th ACM International Conference on Multimodal Interaction, pp. 589–593 (2018)

57. Sun, N., Li, Q., Huan, R., Liu, J., Han, G.: Deep spatial-temporal feature fusion for facial expression recognition in static images. Pattern Recogn. Lett. **119**, 49–61 (2019)

58. Kuo, C.M., Lai, S.H., Sarkis, M.: A compact deep learning model for robust facial expression recognition. In: Proceedings of the IEEE Conference on Computer Vision and Pattern Recognition Workshops, pp. 2121–2129 (2018)

Emotion Embedded Pose Generation

Amogh Subbakrishna Adishesha$^{(\boxtimes)}$ and Tianxiang Zhao

Pennsylvania State University, University Park, State College, PA 16802, USA
{aus79,tkz5084}@psu.edu

Abstract. Body poses are a rich source of information in the field of sentiment analysis and they complement existing facial emotion recognition tasks by adding significant value especially when faces are not easily available. CCTV recordings and other non-human interfaces are applications where capturing facial expression is challenging and these interfaces can benefit from body pose emotion recognition to conduct context analysis. Another roadblock in this direction is the limited availability of collected and curated datasets of body poses with emotion labels. Addressing these issues, we propose two end-to-end pipelines to generate emotion conditioned human poses corresponding to specific emotion labels. An auxiliary conditional GAN network is presented for pose images and pose skeleton pipelines. The generated images improved emotion classification accuracy by an average of 5.40% across 4 different networks compared to images that were traditionally augmented. Additionally, through image and skeletal augmentation, we achieve state-of-the-art emotion classification results for the BEAST dataset.

Keywords: Emotion recognition · Body pose · Conditional GAN

1 Introduction

Humans express emotions in a multitude of ways including facial, body language, voice tone, words, gestures and even gait. Body language or pose based expressions supplement facial expression and are the most important emotion cues in the absence of facial emotions. This makes it vital in social interactions and understanding subtle expressions. Service animals have been trained to pick up body language cues in order to provide emotional support in times of need. Multiple police agencies are also trained for body posture analysis which is rarely doctored unlike facial expressions. In a high risk settings like airports and banks, closed circuit cameras try covering large areas and are placed at high vantage points. This leaves facial expressions inaccessible due to the distance from the camera. However, body poses and gait are still very viable options for extracting the emotional state of a person in order to asses threat.

In the field of computer vision, affect analysis from facial expression has been widely studied in recent years [29]. However, little attention has been placed on affective body posture and gesture analysis. This has resulted in scarce novel developments related to static emotional human pose generation.

© Springer Nature Switzerland AG 2020
A. Bartoli and A. Fusiello (Eds.): ECCV 2020 Workshops, LNCS 12535, pp. 774–787, 2020.
https://doi.org/10.1007/978-3-030-66415-2_54

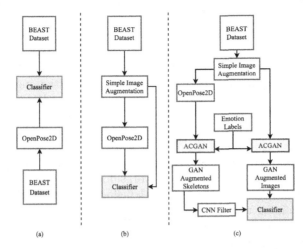

Fig. 1. Overview of our augmentation framework; (a) Classification of image and skeletal poses without augmentation; (b) Classification of image and skeletal poses after simple affine and flip transformation augmentation; (c) Classification of image and skeletal poses with proposed ACGAN augmentation and additional filtering of skeletal poses

Datasets related to static emotional poses are hard to come by as they would require curated stimuli and multiple human actors in lab spaces. Additionally, participants would have to be recruited to assess the poses and label them. Augmentation of data is one of several fundamental procedures used to increase the sampling of input data in learning tasks. This can be as simple as rotation of images or horizontal and vertical flipping. The fundamental issue with such augmentation is that the region of sampling is still constricted around the original data and this can lead to over-fitting of the model. Human poses cannot be flipped vertically as they would lose structural information. Generative adversarial networks (GANs) originally proposed in [14], has shown great success in data generation tasks. This is due to the ability of the network to understand the distribution of the original data and create new samples in the entire distribution space rather than around each sample. Advancements like Conditional-GAN (C-GAN) [22], have helped researchers generate a specific class or category of images and data samples by conditioning the latent space from which it is sampled. Auxiliary Conditional GAN (AC-GAN) is an improvement over C-GAN and uses an auxiliary classifier to help the generator synthesize samples better for each class in the dataset. In our work, we utilize this ability of AC-GAN to synthesize emotion specific images and skeletons of static human poses and validate them using a CNN classifier. This framework is illustrated in Fig. 1(c).

2   Related Work

Emotion recognition is gaining traction in the field of vision as any human computer interface would benefit from knowing the emotion of its user. In this direction, several works have attempted to recognize emotions with a variety of modalities [9]. With the increase of voice assistants, the literature has seen an increase in speech based emotion recognition as done by [11] and [28]. Human body pose emotion analysis is still in its nascent stage and has had some significant works like [3] where an end-to-end emotion recognition model is proposed for CCTV videos. [10] also present a neural network to detect pose emotion. [23] argued that body poses along with audio could perform accurate emotion recognition. However, these works fail to address disparities caused due to the subjectivity in the labelling of the dataset. Most of these works create a custom dataset with very few subjects and repeated stimuli for the subjects. [25] surveys multiple body pose emotion recognition works done on both images and skeletal poses. They point out that subjectivity in labeling of the dataset is an important factor in determining the success of each of mentioned works. They quantitatively argue that neural networks perform far better than classical techniques like SVM, Decision Trees, for body pose emotion classification.

2.1 Image Generation

GANs have exploded the task of image generation with several works like [2,15,30] performing image generation with conditioning of the latent space. More recently, domain translation has gained focus and image based translation has had great success [34]. The usage of GAN for dataset augmentation is not unheard of and has been extensively performed with varying degrees of success like [5,13]. The presence of a constrained feature space due to limited possibilities in spatial arrangement and a hierarchical structure of the body helps the GAN generate realistic body pose images. This has been discussed in [19].

2.2 Pose Skeleton Extraction

Skeleton extraction is a basic problem in the computer vision domain. Multiple works similar to [21] have attempted to identify the location of human joints and construct a connected map of them in 2D or 3D. A well-known method for pose extraction is locating the head using a CNN and then applying a top-down approach to fit other joints to pose in the image. [1,6,7,24] all generate skeletal poses from images. [6] present an easy to use 2D pose extraction model that uses an encoder-decoder architecture for skeletal pose extraction. We use their pre-trained model for extracting skeletal poses from our images.

2.3 Pose Generation

Pose generation has attracted attention in recent years. However, most of the works focus on the action generation task, in combination with skeletal sequence prediction similar to [31]. [4,33] too perform pose skeleton generation in 2D and 3D but do not attempt to translate to the original image or video of the pose. Additionally, although related, very few of them considered the emotions embedded in the pose. Among them, [31,32] are most related to ours. They also exploit using pose skeletons as guiding information for the generation of human pose images however, their objective was focused on generating realistic action sequences and not emotion rich images.

3   Methodology

Our primary objective is to generate new images of human poses corresponding to emotion based labels like *Angry, Happy, Sad* or *Fearful.* It can be mathematically proven that with infinite data samples of a particular distribution, the model can theoretically learn everything from it. However, no dataset can have infinite samples. Augmenting datasets with random images negatively impacts the learning ability of the model as it would be hard to find the true distribution of the data. For this purpose, we need to augment the dataset with class specific images or skeletons in order to maintain the distribution and yet increase the sampling space to help the model learn better. Aiding us in this pursuit is AC-GAN, [26]. The overall architecture of AC-GAN is shown in Fig. 2. The network, similar to C-GAN has a generator synthesizing samples corresponding to a class C and a modified multi-purpose discriminator to pick between real and fake as well as classify the generated sample into the target classes. The objective function of the AC-GAN network is given by Eq. 1 and Eq. 2.

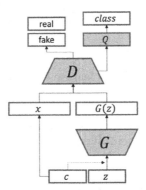

Fig. 2. AC-GAN architecture for image and skeleton generation; Generator (G), Discriminator (D) and Classifier (Q) are shown

$$L_S = \mathbb{E}(\log P(S = real|X_{real})$$
$$+ \, \mathbb{E}(\log P(S = fake|X_{fake}) \tag{1}$$

$$L_C = \mathbb{E}(\log P(C = c|X_{real})$$
$$+ \, \mathbb{E}(\log P(C = c|X_{fake}) \tag{2}$$

As described earlier, the discriminator has two log likelihood functions one for the correct source, L_S and one for the correct class L_C. We use this to augment both skeletons and images to find the best augmentation model for poses with emotions. In Eq. 1, the first term is minimized by the discriminator while the second term is minimized by the generator. S is the source of the image or skeleton. Similarly, in Eq. 2, C is the class of a particular image/skeleton sample and the generator needs to maximize the log likelihood of accurate classification of a synthesized image or skeleton.

3.1 Pose Image Pipeline

Overview. Pose images like any other image set can be augmented by using a generator. This straightforward method can lead to poorly generated images as we are solely relying on visual similarity within image classes and disparity between them. However, this is computationally efficient and consumes the least resources compared to the pose skeleton pipeline.

Model Framework. For the generator, we first concatenate the number of emotion classes to our latent noise vector and make it a 100-dimension sampling space (z). From this, we use a fully connected network to up-sample and feed a convolution block with a BatchNorm layer and a ReLu layer. We pass it through 3 other additional convolution blocks and use Tanh activation in the last block. For the discriminator, we use similar convolution blocks but with Dropout layers. The last layer of our discriminator is a Softmax function to calculate the probabilities of each class by the auxiliary classifier

3.2 Pose Skeleton Pipeline

Overview. Emotion is a high-level feature lying behind human poses, and pose images contain many other features which have little to do with emotions, like gender, clothes, and appearance. Therefore, expecting the GAN to directly generate a full image containing the expected emotion type would be difficult. The generator has too much freedom during the generation process, and the emotion classifier might over-fit to some other features and fail to provide correct supervision. As a result, the obtained images could suffer in quality. Previous works [31,32] have validated that using pose skeletons to guide the generation of pose image could ease this problem. Pose skeletons are a more abstract representation of human poses and it contains all the emotion-related information which makes it suitable for our task.

So, in this branch, we attempt to generate skeletons with emotions, in the same way as the last part. As shown in Fig. 2, we adopt an ACGAN framework and take the emotion embedding along with a random noise vector as input. The generator is required to generate a skeleton map. The discriminator is trained to give two training signals, whether the input skeleton is genuine, and what emotion category is contained in that skeleton.

Pose Extraction. Pose skeleton extraction has long been a popular research direction in the computer vision domain, and there are already several open-source works [6,24] now. In this experiment, we adopt the widely-accepted OpenPose [6] as the pose extractor, and load the parameters from their model pretrained on COCO dataset. This model returns a heatmap of 18 channels, each channel representing a joint in the pose skeleton. The tensor is of shape $[H, W, 18]$, and $[h, w, i]$-th element corresponds to the probability of the i-th joint appearing at position $[h, w]$.

Model Framework. In training the pose generator and discriminator, we keep the same representation form of the skeleton. To illustrate the design of our model, here, we show the transition process of the input vector to a skeleton map during the generation in Fig. 3. It can be seen that the generator takes a 4-layer network, which takes as input a 100-dimension vector (4 dimensions for emotion types, and 96 dimensions for random noise). The output is of the same shape as that extracted by OpenPose.

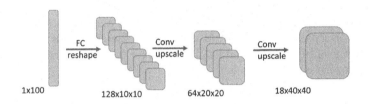

Fig. 3. Overview of the encoding process for pose skeletons

When providing the generated skeleton to guide the generation of pose images, we change the skeleton map into a skeleton image. We find the positions where each joint has the highest probability to appear, and draw a node there. Then, we connect the generated nodes based on predefined human structure hierarchy, as the distance of links between human joints are fixed. Examples of the generated skeleton images can be found in the experiment section.

3.3 Combined Pipeline

Having explored both image based and skeletal pose augmentation techniques, we found empirically that skeletal poses definitely had the advantage of being more robust and accurate. This motivated us to generate skeletons and then convert them images using an established domain translation network, Pix-2-Pix, proposed by [34]. We train this network to take skeletal pose images as input and generate realistic human pose images that resemble the original actors who contributed to the dataset. This has been illustrated in Fig. 4. We used a U-Net 128 architecture for the generator and a patch discriminator for this network. This was trained over 100 epochs.

4   Experiments

4.1   Dataset

Data Overview. Images of human poses that are curated with accurate and definite emotion labels are difficult to collect. Human emotions are non-discrete and are context-related, therefore it is difficult to find one moment and define the pose at this moment as sad or happy. Therefore, data available for this is quite limited. As a result, for this work, we perform the experiments based only on one publicly available dataset, namely the Bodily Expressive Action Stimulus Test (BEAST)[1][8], which contains 254 whole body pose-expression images captured using 46 actors. In the collection of the dataset, the actors are asked to pose in order to emote one specific expression among happy, angry, fearful or sad. These four emotions are chosen deliberately, because they are thought to be most discriminative based purely on the body poses. The images are taken in a professional photo studio, with controlled lighting conditions. The background is set to white, and the camera is also fixed at a static position. After obtaining the images, they are evaluated by 19 volunteers. Each image is shown randomly and the volunteers are asked to classify it into different emotion categories. They report a mean labelling accuracy of 92.5%, and the agreement score, measured using Fleiss Karpa algorithm [12], reaches 0.839. Therefore, it is safe to say that this dataset is of high quality, and can be used for our experiments. An example of each category is shown Fig. 5.

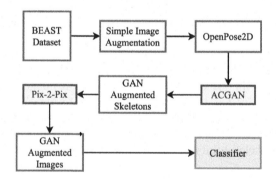

Fig. 4. Overall framework of combined method

---

[1] http://www.beatricedegelder.com/beast.html.

Angry   Fearful   Happy   Sad

Fig. 5. Example images from BEAST dataset

Data Preprocessing. Although the quality of this dataset is high, its size is still small. Only 254 images with about 64 images in each category is not sufficient to train a deep generative model. Addressing that, we apply data augmentation methods to extend the dataset. For each image, we apply two types of perturbations: flip, and affine transformation. Each image is horizontally flipped, which doubles the dataset size, and then affine transformations with 10 different translations, five on the positive X-axis and five on the negative, effectively enlarging the dataset by 11 times. During training, we choose 40 images from each emotion category, and after applying the pre-processing steps, we get a dataset of size 40 (samples)×4 (categories)×2 (flipping)×11 (affine transformation), which is 3520. For testing, we use the remainder of each category, about 24, and pass them through the same augmentation steps to obtain a total of 2068 images. All images are of the resolution 256×256 however they are later resized to fit classifier models.

4.2   Pose Image Results

For training the AC-GAN image generator, we use 600 epochs with learning rate $\alpha = 0.01$ and an ADAM optimizer [18]. The network once trained is queried with emotion labels and 400 new images are generated for each category.

Qualitative Evaluation. The generated images though not visually appealing, contain the relevant image based features to aid in augmentation for classification purposes. We can see in Fig. 6 that the generated images match the emotion class in overall structure. For example, the arms are extended in the happy category (top left), one leg ahead in the angry category (bottom left) and so on.   It is vital to validate the features in the generated images to make sure that they are not random or an amalgamation of all categories. For this purpose, we use t-SNE plots [20], to visualize the learnt parameters in feature space. Figure 7 shows the t-SNE plot for raw data on the left and the generated data on the right. It can be observed that the generator has learnt to separate the 4 classes in feature space and generate images in them.

Fig. 6. Examples of generated pose images along with reference; (a) Happy; (b) Angry; (c) Fearful; (d) Sad

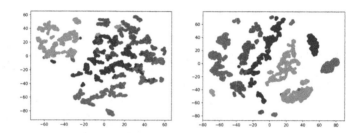

Fig. 7. Obtained t-SNE figures for original and generated image poses. Red: Sad, Green: Fearful, Blue: Angry, Turquoise: Happy. (Color figure online)

Quantitative Evaluation. We also validate our generated images by using them to augment our original dataset and train emotion classification networks. We compare the classification accuracy at different levels of augmentation. There are no generated images in the testset. This helps maintain unbiased evaluation of the images. We used a 3 layer CNN classifier to set our baseline and compared it against ResNet101 [16], Inception-v3 [27] and DenseNet-201 [17]. All reported accuracy percentages are derived by averaging validation accuracy over 200 epochs.

Table 1. Comparison of classification accuracy with different approaches for data augmentation of pose images

Method	# Training samples	Accuracy(3-Layer CNN)
Original	40×4	61.20%
Augmented	880×4	65.70%
GAN Augmented	1220×4	74.12%

Table 1, shows the improvement in classification accuracy of the dataset post GAN augmentation. This only possible if the AC-GAN has generated class-specific images of poses with appropriate features corresponding to the emotion

Table 2. Classification accuracy across deep classification models. ResNet101 performed the best for GAN-Augmented Images

Model	Original	Augmented	GAN Augmented
ResNet-101	68.33%	73.20%	**81.76%**
Inception-V3	**69.21%**	**77.72%**	80.14%
DenseNet-201	66.40%	74.12%	76.34%

categories. Similarly, Table 2 shows the improvement in classification accuracy across all the three deep classification models confirming improvement is not model biased.

4.3 Pose Skeleton Results

In this section, we evaluate the results of the skeleton generation pipeline. The generator is trained using Adaboost optimizer for 5000 epochs, with learning rate initialized to 0.1.

Qualitative Evaluation. In Fig. 8, we provide examples of generated results, for happy, sad, fearful and angry respectively. It can be seen that the model manages to capture the gist of each emotion. For example, for happy, the pose is open and relaxed, as for sad, the head is hanging low with drooping shoulders.

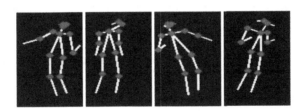

Fig. 8. Examples of generated pose skeletons.

We also visualize the t-SNE clustering of each skeleton's embedding in Fig. 9. We separately train a skeleton classifier, and use its intermediate feature map as the representation of each skeleton sample. On the left t-SNE plot each node represents a sample in the original dataset, while in the right plot, we add the samples representing generated skeletons, with slightly different colors. It can be seen that the generated samples and genuine samples follow very similar distributions. For example, genuine sad (pink) and generated sad (red), have very closely overlapping distributions.

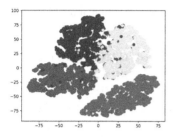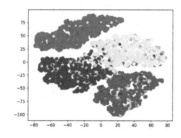

Fig. 9. t-SNE plots for original and generated skeleton poses. Pink: Sad, Blue: Fearful, Yellow: Angry, Green: Happy. (Color figure online)

Quantitative Evaluation. Besides qualitative evaluations, we also perform a simple quantitative evaluation, to show that the generated skeleton can be helpful for augmenting the dataset. Concretely, we split the real dataset into two parts: training set and testing set, and train a classifier in different settings before testing it on the testing set.

The result is summarized in Table 3. It can be seen that the generated skeleton can help improving classifier's performance, which further validates the quality of the generated samples. We observed a large number of unrealistic skeletal poses in our generated set and this motivated us to use a CNN-filter, trained on recognizing between realistic skeletal poses and unrealistic ones. The filter was used in series after the AC-GAN and the resulting clean pose skeletons were also validated with the classifier.

Table 3. Comparison of classification accuracy with different approaches for data augmentation of pose skeletons

Method	# Training samples	Accuracy (3-Layer CNN)
Original	40×4	83.3%
Augmented	880×4	94.6%
GAN Augmented	1200×4	93.0%
Filtered GAN Augmented	1200×4	95.7%

4.4   Combined Model Results

Though the skeletal pose augmentation showed great success in terms of classification accuracy, the generated skeletons cannot be used as images in futures datasets. For this purpose, we convert the generated skeletons back to images using a Pix-2-Pix model and train the classifier again to verify that the effectiveness of the conversion. The qualitative results are shown in Fig. 10.

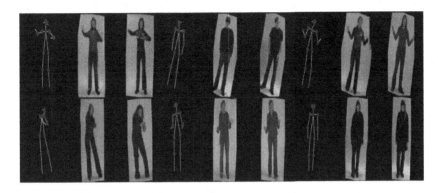

Fig. 10. Results from combined method. Each row has 3 sets and every set consists of an input skeleton, predicted pose image and expected pose image

It can be seen that when the generation of images is constrained on emotion embedded skeletons, the generated images are much more visually appealing and very close to the ground truth. Additionally, even with complicated poses like the bottom left set in Fig. 10, the model generated a pose image extremely similar to the real image captured in lab setting. An interesting outcome from this experiment was the ability of the network to generate pose images with correct genders even without having explicitly given that information. In other words, the network was able to learn what the expected image should look like by only using the skeletal pose as input. The images from this method were also used to train a ResNet-101 classifier and compare against other techniques. We achieved a classification accuracy of 87.56% across the 4 classes when trained with 320 generated images and 880 augmented images. Compared with other results in Table 1, it can be seen to outperform all others by a large margin.

5   Conclusion

In this work, we exploit two approaches for generating body pose images conditioned on emotion labels. In the direction of emotion rich content creation, this work serves as an important tool especially when human poses are involved. Through utilizing skeletons as guiding information, our model successfully generated high-quality pose images that are similar to the original dataset in terms of both visual appearance and statistical distribution. We also prove that the generated data improves the performance of body pose emotion classifiers. In the direction similar to this work, we wish to explore generation of scenes with multiple human subjects and corresponding emotion rich labels.

Acknowledgements. We would like to thank Dr. Sharon Huang and the College of Information Sciences and Technology, Penn State for providing hardware support for this project. We also thank Dr. James Wang and the reviewers for their comments.

References

1. Andriluka, M., Pishchulin, L., Gehler, P., Schiele, B.: 2D human pose estimation: new benchmark and state of the art analysis. In: Proceedings of the IEEE Conference on Computer Vision and Pattern Recognition, pp. 3686–3693 (2014)
2. Arjovsky, M., Chintala, S., Bottou, L.: Wasserstein gan. arXiv preprint arXiv:1701.07875 (2017)
3. Arunnehru, J., Kalaiselvi Geetha, M.: Automatic human emotion recognition in surveillance video. In: Dey, N., Santhi, V. (eds.) Intelligent Techniques in Signal Processing for Multimedia Security. SCI, vol. 660, pp. 321–342. Springer, Cham (2017). https://doi.org/10.1007/978-3-319-44790-2_15
4. Barsoum, E., Kender, J., Liu, Z.: Hp-gan: probabilistic 3D human motion prediction via gan. In: Proceedings of the IEEE Conference on Computer Vision and Pattern Recognition Workshops, pp. 1418–1427 (2018)
5. Bowles, C., et al.: Gan augmentation: Augmenting training data using generative adversarial networks. arXiv preprint arXiv:1810.10863 (2018)
6. Cao, Z., Hidalgo, G., Simon, T., Wei, S.E., Sheikh, Y.: Openpose: real-time multi-person 2D pose estimation using part affinity fields. arXiv preprint arXiv:1812.08008 (2018)
7. Cao, Z., Simon, T., Wei, S.E., Sheikh, Y.: Realtime multi-person 2D pose estimation using part affinity fields. In: Proceedings of the IEEE Conference on Computer Vision and Pattern Recognition, pp. 7291–7299 (2017)
8. De Gelder, B., Van den Stock, J.: The bodily expressive action stimulus test (beast): construction and validation of a stimulus basis for measuring perception of whole body expression of emotions. Front. Psychol. **2**, 181 (2011)
9. El Ayadi, M., Kamel, M.S., Karray, F.: Survey on speech emotion recognition: features, classification schemes, and databases. Pattern Recogn. **44**(3), 572–587 (2011)
10. Elfaramawy, N., Barros, P., Parisi, G.I., Wermter, S.: Emotion recognition from body expressions with a neural network architecture. In: Proceedings of the 5th International Conference on Human Agent Interaction, pp. 143–149 (2017)
11. Fayek, H.M., Lech, M., Cavedon, L.: Evaluating deep learning architectures for speech emotion recognition. Neural Netw. **92**, 60–68 (2017)
12. Fleiss, J.L., Cohen, J.: The equivalence of weighted kappa and the intraclass correlation coefficient as measures of reliability. Educ. Psychol. Measur. **33**(3), 613–619 (1973)
13. Frid-Adar, M., Klang, E., Amitai, M., Goldberger, J., Greenspan, H.: Synthetic data augmentation using gan for improved liver lesion classification. In: 2018 IEEE 15th international symposium on biomedical imaging (ISBI 2018), pp. 289–293. IEEE (2018)
14. Goodfellow, I., et al.: Generative adversarial nets. In: Advances in Neural Information Processing Systems, pp. 2672–2680 (2014)
15. Han, C., et al.: Gan-based synthetic brain mr image generation. In: 2018 IEEE 15th International Symposium on Biomedical Imaging (ISBI 2018), pp. 734–738. IEEE (2018)
16. He, K., Zhang, X., Ren, S., Sun, J.: Deep residual learning for image recognition. In: Proceedings of the IEEE Conference on Computer Vision and Pattern Recognition, pp. 770–778 (2016)
17. Huang, G., Liu, Z., Van Der Maaten, L., Weinberger, K.Q.: Densely connected convolutional networks. In: Proceedings of the IEEE Conference on Computer Vision and Pattern Recognition, pp. 4700–4708 (2017)

18. Kingma, D.P., Ba, J.: Adam: A method for stochastic optimization. arXiv preprint arXiv:1412.6980 (2014)
19. Ma, L., Jia, X., Sun, Q., Schiele, B., Tuytelaars, T., Van Gool, L.: Pose guided person image generation. In: Advances in Neural Information Processing Systems, pp. 406–416 (2017)
20. Maaten, L.V.D., Hinton, G.: Visualizing data using t-SNE. J. Mach. Learn. Res. **9**(Nov), 2579–2605 (2008)
21. Martinez, J., Hossain, R., Romero, J., Little, J.J.: A simple yet effective baseline for 3d human pose estimation. In: Proceedings of the IEEE International Conference on Computer Vision, pp. 2640–2649 (2017)
22. Mirza, M., Osindero, S.: Conditional generative adversarial nets. arXiv preprint arXiv:1411.1784 (2014)
23. Müller, P.M., Amin, S., Verma, P., Andriluka, M., Bulling, A.: Emotion recognition from embedded bodily expressions and speech during dyadic interactions. In: 2015 International Conference on Affective Computing and Intelligent Interaction (ACII), pp. 663–669. IEEE (2015)
24. Newell, A., Yang, K., Deng, J.: Stacked hourglass networks for human pose estimation. In: Leibe, B., Matas, J., Sebe, N., Welling, M. (eds.) ECCV 2016. LNCS, vol. 9912, pp. 483–499. Springer, Cham (2016). https://doi.org/10.1007/978-3-319-46484-8_29
25. Noroozi, F., Kaminska, D., Corneanu, C., Sapinski, T., Escalera, S., Anbarjafari, G.: Survey on emotional body gesture recognition. IEEE Trans. Affect. Comput. (2018)
26. Odena, A., Olah, C., Shlens, J.: Conditional image synthesis with auxiliary classifier gans. In: Proceedings of the 34th International Conference on Machine Learning, vol. 70, pp. 2642–2651. JMLR. org (2017)
27. Szegedy, C., Vanhoucke, V., Ioffe, S., Shlens, J., Wojna, Z.: Rethinking the inception architecture for computer vision. In: Proceedings of the IEEE Conference on Computer Vision and Pattern Recognition, pp. 2818–2826 (2016)
28. Trigeorgis, G., et al.: Adieu features? end-to-end speech emotion recognition using a deep convolutional recurrent network. In: 2016 IEEE International Conference on Acoustics, Speech and Signal Processing (ICASSP), pp. 5200–5204. IEEE (2016)
29. Wu, C.H., Lin, J.C., Wei, W.L.: Survey on audiovisual emotion recognition: databases, features, and data fusion strategies. APSIPA Trans. Signal Inf. Process. **3**, E12 (2014)
30. Xu, T., et al.: Attngan: fine-grained text to image generation with attentional generative adversarial networks. In: Proceedings of the IEEE Conference on Computer Vision and Pattern Recognition, pp. 1316–1324 (2018)
31. Yan, S., Li, Z., Xiong, Y., Yan, H., Lin, D.: Convolutional sequence generation for skeleton-based action synthesis. In: Proceedings of the IEEE International Conference on Computer Vision, pp. 4394–4402 (2019)
32. Yang, C., Wang, Z., Zhu, X., Huang, C., Shi, J., Lin, D.: Pose guided human video generation. In: Proceedings of the European Conference on Computer Vision (ECCV), pp. 201–216 (2018)
33. Zhou, L., Li, W., Zhang, Y., Ogunbona, P., Nguyen, D.T., Zhang, H.: Discriminative key pose extraction using extended lc-ksvd for action recognition. In: 2014 International Conference on Digital Image Computing: Techniques and Applications (DICTA), pp. 1–8. IEEE (2014)
34. Zhu, J.Y., Park, T., Isola, P., Efros, A.A.: Unpaired image-to-image translation using cycle-consistent adversarial networks. In: Proceedings of the IEEE International Conference on Computer Vision, pp. 2223–2232 (2017)

Understanding Political Communication Styles in Televised Debates via Body Movements

Zhiqi Kang[1], Christina Indudhara[2], Kaushik Mahorker[2], Erik P. Bucy[3], and Jungseock Joo[2(✉)]

[1] Université de Technologie de Compiègne, Compiègne, France
[2] UCLA, Los Angeles, USA
jjoo@comm.ucla.edu
[3] Texas Tech University, Lubbock, USA

Abstract. Televised political debates have received much attention by scholars in political communication and social psychology who study nonverbal cues in interpersonal communication and their impact on candidate evaluations. An abundance of political multimedia and new platforms have required leaders to develop an effective and unique communication "style" which may rely on nonverbal devices such as face and body. Emotions conveyed by expressive gestures of candidates during debates have been shown to elicit stronger reactions from the public than rhetorical statements alone. Candidates, for example, may exploit assertive and aggressive gestures to communicate their confidence and attract supporters. Existing studies, however, are based largely on manual coding of human gestures, which may not be scalable or reproducible. The main objectives of our paper are to investigate the role of body movements of candidates using a systematic and automated approach as well as understand the context and effects of gestures. For this analysis, we collected a dataset of political debate videos from the 2020 Democratic presidential primaries and analyzed facial expressions and gestures of candidates. Our preliminary analysis demonstrates that candidates employ gestures to varying extents, and the amount of body movement is correlated with emotions conveyed in the candidates' facial expressions. We discuss our dataset, preliminary results, and future directions in the following sections.

1 Introduction

Televised political debates have received much attention by scholars in political communication and social psychology who study nonverbal cues in interpersonal communication and their impact on candidate evaluations [6–8]. An abundance of political multimedia and new platforms have required leaders to develop an

---

Z. Kang and C. Indudhara—Equal contribution.

A. Bartoli and A. Fusiello (Eds.): ECCV 2020 Workshops, LNCS 12535, pp. 788–793, 2020.
https://doi.org/10.1007/978-3-030-66415-2_55

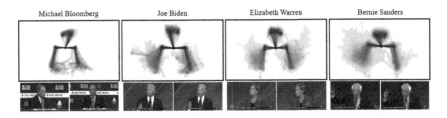

Fig. 1. A visualization of the body movements of candidates with keypoints.

effective and unique communication "style" which may rely on nonverbal devices such as face and body [5,9]. Emotions conveyed by expressive gestures of candidates during debates have been shown to elicit stronger reactions from the public than rhetorical statements alone [1]. Candidates, for example, may exploit assertive and aggressive gestures to communicate their confidence and attract supporters. Existing studies, however, are based largely on manual coding of human gestures, which may not be scalable or reproducible.

The main objectives of our paper are to investigate the role of body movements of candidates using a systematic and automated approach as well as understand the context and effects of gestures. For this analysis, we collected a dataset of political debate videos from the 2020 Democratic presidential primaries and analyzed facial expressions and gestures of candidates. Our preliminary analysis demonstrates that candidates employ gestures to varying extents, and the amount of body movement is correlated with emotions conveyed in the candidates' facial expressions. We discuss our dataset, preliminary results, and future directions in the following sections.

2   Data: Democratic Debate Videos

Democratic Debate Videos. The videos of all 11 debates and 23 candidates were collected via various televised sources (news networks or TV stations such as NBC, ABC, and CBS). The total duration of all videos analyzed is 34 h, 25 min, and 58 s. We also collected timestamped transcripts of each debate with speaker information. We only consider the gestures of candidates while they are speaking.

Facial Expression Recognition Model. In order to understand the relationship between body gestures and facial expressions, we train a CNN to classify facial expression using Expression in-the-Wild (ExpW) Dataset [10] for seven emotion categories: angry, disgust fear, happy, sad, surprise, and neutral. The model assigns emotion scores for each detected face on the screen. We utilize a Resnet-34 [4] pretrained on Imagenet and fine-tuned to the ExpW set.

3   Measuring Body Movements

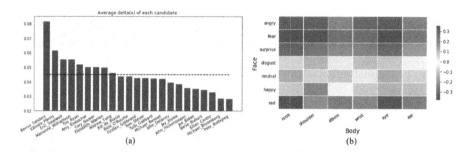

Fig. 2. (a) average $\delta(x)$ of each candidate. The dashed line represents the average of all candidates. (b) correlation coefficients for emotion scores and keypoint movement. Symmetric keypoints from left and right are merged together by showing their average.

To extract the movement from raw videos, we calculate the absolute displacement of the pose keypoints between consecutive frames. This feature is defined as $\delta(x)$, where x represents each 2D keypoint. Using $\delta(x)$ introduces several advantages: (i) compared to the raw keypoints coordinates, $\delta(x)$ is translation invariant, (ii) the displacement of keypoints over a fixed time step is indeed the velocity of the keypoint.

The first step is to extract 2D coordinates of the candidates' body parts in each frame of the debate videos. We utilize OpenPose [2], the state-of-the-art human pose detection system, to extract keypoints from the videos at 15 frames per second. We consider 12 keypoints from the upper part of the body including nose, neck, shoulders, elbows, wrists, eyes and ears. The facial keypoints also serve as a bridge for connecting pose information with the estimated emotion from candidate faces. All sets of keypoints are centered to the neck and the displacement is normalized by a candidate-specific robust nose-to-neck reference distance. On average, more than 97% of the candidates on screen are correctly assigned with information about their poses and expressed emotion.

Is $\delta(x)$ a significant measure? To answer this question, we evaluate it with respect to other properties such as the candidates' habitual behavior. A correlation analysis is then performed to discover a generalized relationship between emotions and body movements.

We calculate the average $\delta(x)$ of each candidate with respect to each emotion across all debate videos, Fig. 2(a). Consistent with our empirical knowledge, candidates such as Bernie Sanders, whose gestures often seem "uncontrolled," have high overall $\delta(x)$, whereas those with slower and tighter gestures such as Michael Bloomberg have a relatively low $\delta(x)$. Figure 1 visualizes this characteristic of candidates by showing the variance of the body, using a similar method in [3]. The deeper the color, the more stable the keypoints. However, the body movement of candidates is related not only to the candidates' habitual behavior, but also to the confrontation during the debates. Hence, we suspect that the real-time emotion would significantly affect their gestures, leading us to the following analysis.

As shown in Fig. 2(b), the seven emotions have different influences on body and face movement. We observe a homogeneous effect of a single emotion on all keypoints. That is, all emotions except neutral, a natural middle ground, have either all positive or all negative correlations with body movement, represented by each of the keypoints. This suggests that the keypoints are positively correlated with each other, which is confirmed by additional tests. Second, the relatively high positive correlation between being angry and having larger body movement depicts a scenario in which the candidate is in firm disagreement with other candidates. Hence, they tend to utilize larger gestures to emphasize their points and convince others. In contrast, a sad emotion suggests a defeated status or even lack of expression in which the candidate tends to reduce their body and head movement. Consequently, this correlation study confirms the validity of recognizing emotions via body and face movement.

Fig. 3. Examples of gestures from different clusters, validated across candidates and debates. Left: *precision grip* gesture. Mid: *waving both hands* gesture. Right: *angrily emphasizing* gesture.

4   Gesture Discovery by Clustering

While the magnitude of gestures provides a helpful cue to understand the communication style of leaders, it will be more useful to differentiate the specific types of gestures the candidates used. Human gestures are so diverse and versatile that it is challenging to predefine a list of recognizable gestures. Therefore, we seek to discover common gesture types used by the candidates by clustering.

Specifically, we crop the original videos and obtain short sequences of body keypoint vectors. Each sample has 15 frames (1 s) of 12 keypoints, yielding a vector of size 180 by concatenating keypoints. This vector is not invariant to temporal shift and we don't know exactly when a gesture begins and ends. Therefore we use overlapping sliding windows to ensure that the complete gesture can be captured. To discard clips with little body movement, we discard 30% of the samples having the smallest sum of $\delta(x)$, leaving 170K valid samples.

Then we apply a K-means clustering with a deliberately defined large value $K = 100$ to separate meaningful clusters with redundant ones from noisy data. Consequently, the algorithm distinguishes several patterns of body gestures as shown in Fig. 3. We reviewed the clusters and identified some having significant emotional associations. For instance, the right cluster in Fig. 3 shows that the vertical displacement of right hands is associated with the angry emotion.

5   Discussion and Future Directions

Our preliminary analysis illustrates robust correlations between facial expressions and body movements. The candidates tended to display more expressive gestures when they were also exhibiting facial displays of anger through their faces. Joe Biden, the presumptive nominee as of July 2020, was among the candidates who showed more calm and reserved body movements, compared to more expressive candidates such as Bernie Sanders or Elizabeth Warren, who needed to draw more attention in the competitive primary process.

We will extend our analysis in several directions in future work. First, we will define a list of meaningful communicative gestures found in debate videos and obtain manual annotations based on the categorization. We will then train a classifier using these annotations and examine the context in which a candidate uses a particular gesture as well as the overall behavioral effect – whether using certain gestures more can help them gain popularity or not. Second, we will further investigate the relation between facial expressions and body gestures. Our current analysis shows correlations between two factors, and we will show the effects of combining them together (e.g. a happy face with an aggressive body gesture) using a multimodal model. Third, we will investigate the generalizability of our model and dataset by cross-training and validation with other benchmark datasets on emotion recognition from human bodies.

Acknowledgement. This work was supported by NSF SBE-SMA #1831848.

References

1. Bucy, E.P., et al.: Performing populism: Trump's transgressive debate style and the dynamics of twitter response. New Media Soc. **22**(4), 634–658 (2020)
2. Cao, Z., Martinez, G.H., Simon, T., Wei, S., Sheikh, Y.A.: Openpose: realtime multi-person 2D pose estimation using part affinity fields. TPAMI **43**, 172–186 (2019)
3. Ginosar, S., Bar, A., Kohavi, G., Chan, C., Owens, A., Malik, J.: Learning individual styles of conversational gesture. In: CVPR, pp. 3497–3506 (2019)
4. He, K., Zhang, X., Ren, S., Sun, J.: Deep residual learning for image recognition. In: Proceedings of the IEEE Conference on Computer Vision and Pattern Recognition, pp. 770–778 (2016)
5. Joo, J., Bucy, E.P., Seidel, C.: Automated coding of televised leader displays: detecting nonverbal political behavior with computer vision and deep learning. Int. J. Commun. (19328036) **13**, 4044–4066 (2019)

6. Patterson, M.L., Churchill, M.E., Burger, G.K., Powell, J.L.: Verbal and nonverbal modality effects on impressions of political candidates: analysis from the 1984 presidential debates. Commun. Monogr. **59**(3), 231–242 (1992)
7. Shah, D.V., Hanna, A., Bucy, E.P., Wells, C., Quevedo, V.: The power of television images in a social media age: linking biobehavioral and computational approaches via the second screen. ANN. Am. Acad. Polit. Soc. Sci. **659**(1), 225–245 (2015)
8. Streeck, J.: Gesture in political communication: a case study of the democratic presidential candidates during the 2004 primary campaign. Res. Lang. Soc. Interact. **41**(2), 154–186 (2008)
9. Xi, N., Ma, D., Liou, M., Steinert-Threlkeld, Z.C., Anastasopoulos, J., Joo, J.: Understanding the political ideology of legislators from social media images. In: Proceedings of the International AAAI Conference on Web and Social Media, vol. 14, pp. 726–737 (2020)
10. Zhang, Z., Luo, P., Loy, C.C., Tang, X.: Learning social relation traits from face images. In: ICCV, pp. 3631–3639 (2015)

Author Index